Advances in Intelligent Systems and Computing

Volume 749

The series "Advances in Intelligent Systems and Computing" contains publications on theory, applications, and design methods of Intelligent Systems and Intelligent Computing. Virtually all disciplines such as engineering, natural sciences, computer and information science, ICT, economics, business, e-commerce, environment, healthcare, life science are covered. The list of topics spans all the areas of modern intelligent systems and computing such as: computational intelligence, soft computing including neural networks, fuzzy systems, evolutionary computing and the fusion of these paradigms, social intelligence, ambient intelligence, computational neuroscience, artificial life, virtual worlds and society, cognitive science and systems, Perception and Vision, DNA and immune based systems, self-organizing and adaptive systems, e-Learning and teaching, human-centered and human-centric computing, recommender systems, intelligent control, robotics and mechatronics including human-machine teaming, knowledge-based paradigms, learning paradigms, machine ethics, intelligent data analysis, knowledge management, intelligent agents, intelligent decision making and support, intelligent network security, trust management, interactive entertainment, Web intelligence and multimedia.

The publications within "Advances in Intelligent Systems and Computing" are primarily proceedings of important conferences, symposia and congresses. They cover significant recent developments in the field, both of a foundational and applicable character. An important characteristic feature of the series is the short publication time and world-wide distribution. This permits a rapid and broad dissemination of research results.

More information about this series at http://www.springer.com/series/11156

Surajit Chattopadhyay · Tamal Roy
Samarjit Sengupta · Christian Berger-Vachon
Editors

Modelling and Simulation in Science, Technology and Engineering Mathematics

Proceedings of the International Conference on Modelling and Simulation (MS-17)

Springer

Editors
Surajit Chattopadhyay
Department of Electrical Engineering
Ghani Khan Choudhury Institute of Engineering and Technology
Malda, West Bengal, India

Tamal Roy
Department of Electrical Engineering
MCKV Institute of Engineering
Howrah, West Bengal, India

Samarjit Sengupta
University of Calcutta
Kolkata, West Bengal, India

Christian Berger-Vachon
University of Lyon
Lyon, France

ISSN 2194-5357 ISSN 2194-5365 (electronic)
Advances in Intelligent Systems and Computing
ISBN 978-3-030-09080-7 ISBN 978-3-319-74808-5 (eBook)
https://doi.org/10.1007/978-3-319-74808-5

This Springer imprint is published by the registered company Springer Nature Switzerland AG
The registered company address is: Gewerbestrasse 11, 6330 Cham, Switzerland

Preface

Application of modelling and simulation in science and technology has undergone a change during the last few decades. During this period, newer ideas have been prescribed, and phenomenal changes have taken place in different directions in R&D activities. In view of this, it becomes important to discuss on the issues of Modelling and Simulation in Science, Technology and Engineering Mathematics. With this motivation, the book entitled *Modelling and Simulation in Science, Technology and Engineering Mathematics* has been edited.

This book contains the research papers presented in International Conference on Modelling and Simulation (MS-17) organized by Association for the Advancement of Modelling and Simulation Techniques in Enterprises (AMSE) in collaboration with The Institution of Engineering and Technology (IET-UK), Kolkata Local Network, on 4–5 November 2017, in Kolkata.

Papers have been divided into following tracks:

- Fuzzy, Optical and Opto Electronic Control of Oscillations
- Power System
- Energy
- Control Techniques
- Neuro Fuzzy, Control System and Optimization
- Computation Technique
- Modelling and Simulation in General Application

Editors of this book would like to acknowledge the support received from authors of the chapters and reviewers for their valuable contribution.

We express our sincere thanks to the members of Springer for their support in publishing the book.

We are sure that this book will give ample scope to the readers to gather knowledge and information on the above subject matters.

Malda, India Surajit Chattopadhyay
Howrah, India Tamal Roy
Kolkata, India Samarjit Sengupta
Lyon, France Christian Berger-Vachon

Contents

About the Editors

Surajit Chattopadhyay has obtained B.Sc. in physics honours from Ramakrishna Mission Vidyamandir (Belur Math), University of Calcutta, in 1998, and then B.Tech., M.Tech. and Ph.D. (Technology) in electrical engineering from the Department of Applied Physics of University of Calcutta in 2001, 2003 and 2010, respectively. He has obtained CEng from Engineering Council, UK, in 2013. He has authored/coauthored around 100 papers published in international and national journals and conferences and 4 books (also with Springer). Seven papers have been selected as "Best Paper" in international level. He has visited many countries for technical interaction like in Lyon (France), Kuala Lumpur (Malaysia), Dhaka (Bangladesh), London and Stevenage (UK) and Negombo (Sri Lanka) and presented his work in different international forums. Presently, he is Dean (Research and Consultancy) and Associate Professor of Electrical Engineering in Ghani Khan Choudhury Institute of Engineering and Technology (under Ministry of HRD, Government of India). He served as Honourable Secretary of the Institution of Engineering and Technology (UK), Kolkata Local Network, from 2013 to 2016 and now Executive Committee Member of the Network. His fields of interest include electric power quality, fault diagnosis, power system protection, signal analysis, robotics application and UAV.

Tamal Roy received his Bachelor of Technology in electrical engineering and Master of Technology in mechatronics engineering from West Bengal University of Technology, Kolkata, in 2005 and 2008, respectively. In 2008, he joined the Department of Electrical Engineering at Hooghly Engineering and Technology College as a Lecturer. Since 2011, he has been working as an Assistant Professor in the Department of Electrical Engineering, MCKV Institute of Engineering. He was awarded his Ph.D. in 2016 in robust control-oriented LFT modelling of nonlinear MIMO system from Jadavpur University. His current research interests include modelling and the robust control of the nonlinear systems, model reference adaptive control.

Samarjit Sengupta holds a B.Sc., B.Tech., M.Tech. and Ph.D. from the University of Calcutta, Kolkata, India. He is currently a Professor of electrical engineering in the Department of Applied Physics at the University of Calcutta. He has published 130 journal papers and eight books on various topics of electrical engineering. His main research interests include power quality instrumentation, power system stability, and security and power system protection. He is a Fellow of IET and IETE, as well as a Senior Member of IEEE. He is former Chairman of IET (UK) Kolkata Network.

Christian Berger-Vachon was born in Lyon, France, in 1944. He received the B.E. in electrical engineering from the University of Lyon, France, in 1965; an engineering degree in Lyon (INSA 1967); a Ph.D. in sciences from the University of Lyon, in 1975; and a MD in 1980.

In 1974, he joined the Department of Electrical Engineering, University of Lyon, as a Lecturer, and in 1989, he was named Professor. He was given the title of Emeritus Professor in 2013 and since then allowed to continue his research activity in the University.

His current research interests include signal processing in electrical machines and in hearing aid devices, and also the construction of models in close connection with the patients' behaviour. He is also involved in sports mechanics and in sports medicine.

He is the General Secretary of AMSE, an international association concerned with the edition of scientific journals and with the organization of scientific conferences throughout the world. He is the Vice-Chairman of IFRATH, a French Association concerned with the use of assistive devices for handicapped people. He was the recipient of the "Academic Palms" awarded by the French Government for his academic career in 2014.

Part I
Fuzzy, Optical and Opto Electronics Control of Oscillations

Studies of Optical Properties of RF Magnetron Sputtered Deposited Zinc Oxide Films

S. K. Nandi

1 Introduction

Zinc oxide (ZnO) is of great interest as a suitable material for high temperature, high power electronic devices either as the active material or as a suitable substrate for epitaxial growth of group III-nitride compounds. With its large, direct band gap $\approx 3 \cdot 4$ eV, a low-power threshold (~160 μJ cm^{-2}) for optical pumping at room temperature and wurtzite crystal structure, ZnO is similar to GaN. Due to its relatively close match in lattice constants, it may be used as a substrate for GaN and AlN epitaxy. As a consequence, there is renewed interest in the properties of ZnO relevant for microelectronic device applications. ZnO thin films have been prepared by a wide variety of techniques, including sputtering, spray-pyrolysis, and electro deposition [1] etc. In particular, the r.f. sputter method has advantages over other processes because of its simplicity [2]. We investigate the optical properties of r.f. magnetron Sputter ZnO/Si films by photoluminescence (PL) measurements, Structure and composition of the ZnO/Si films have been investigated by X-ray diffraction (XRD), atomic force microscopy (AFM), scanning electron microscopy (SEM) and X-ray photoelectron spectroscopy (XPS) for chemical composition.

2 Experiment and Results

The undoped ZnO (100 nm) thin film deposited on Si (100) at 450 °C using 13.56 MHz r.f. magnetron sputtering system with a base pressure of 1.0×10^{-6} Torr., working pressure of 1.0×10^{-2} Torr., used gas of Argon, substrate temperature of 450 °C

S. K. Nandi (✉)
Department of Physics, Rishi Bankim Chandra College, 24-Parganas (North), Naihati 743165, West Bengal, India
e-mail: susantanandi@gmail.com

S. Chattopadhyay et al. (eds.), *Modelling and Simulation in Science, Technology and Engineering Mathematics*, Advances in Intelligent Systems and Computing 749,
https://doi.org/10.1007/978-3-319-74808-5_1

with RF power of 100 W. In the XRD pattern (Fig. 1) a major peak of preferential orientation along (103) and minor one related to (002) of the undoped ZnO films were observed. It indicates that ZnO films are polycrystalline structures.

Figure 2 shows the atomic force micrograph of ZnO film. The scan was taken on a 5 μm × 5 μm area. The statistical information of the topography of the ZnO films as observed from the height histogram of the AFM image are: Rms surface roughness (Zrms) and average roughness (Zav) were found to be 50.9 and 30.4 Å, respectively.

A scanning electron microscopy (SEM) image of the cross-sectional view of ZnO/Si film (Fig. 3) shows columnar growth which indicates an orientation parallel c-axis (002) with thickness 100 nm.

Figure 4 shows core levels of Zn 2p of the ZnO films measured by X-ray photoelectron spectroscopy (PHI-5800). The as-grown ZnO thin film of the peaks of Zn 2p are found to be at 1044·8 eV and 1021·7 eV for Zn $2p_{1/2}$ and Zn $2p_{3/2}$, respectively, with a separation of 23·1 eV between the two peaks which is due to the Zn 2p state.

To investigate the optical properties of the films, photoluminescence (PL) measurements were performed Under the 325 nm excitation, the emission PL spectra of a ZnO film at different temperatures are shown in Fig. 5. From the emission spectra,

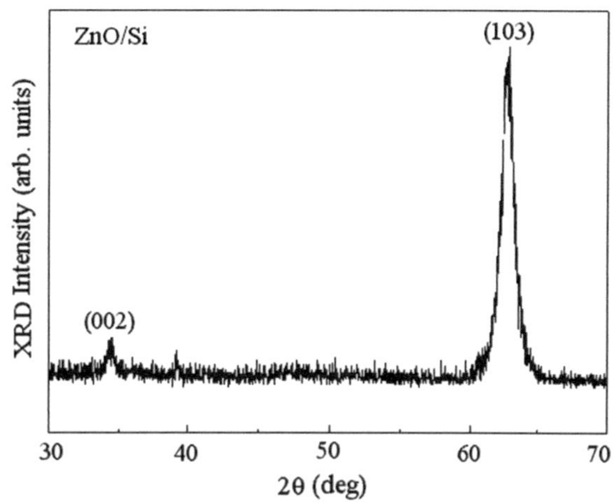

Fig. 1 X-ray diffraction pattern of the as-grown ZnO thin film at 450 °C and r.f. power 100 W

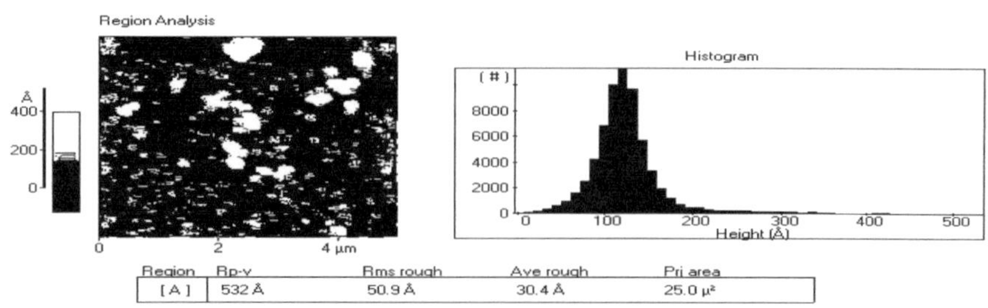

Fig. 2 Two-dimensional AFM image of ZnO Film with scan area of 5 μm × 5 μm

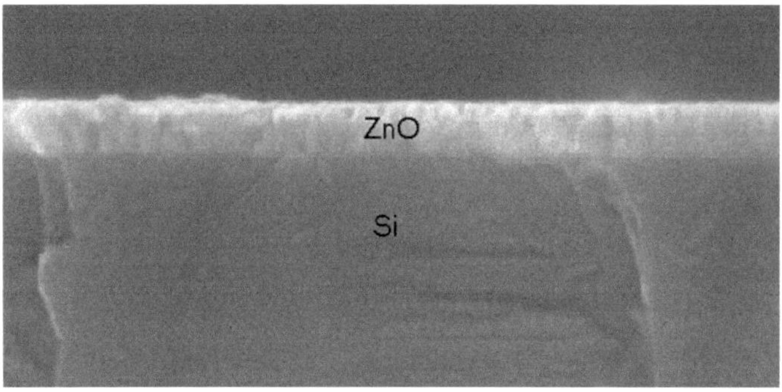

Fig. 3 SEM view of the rf sputtered ZnO film deposited on Si

Fig. 4 XPS core-level of O 1 s and Zn 2p of the as-grown ZnO thin film

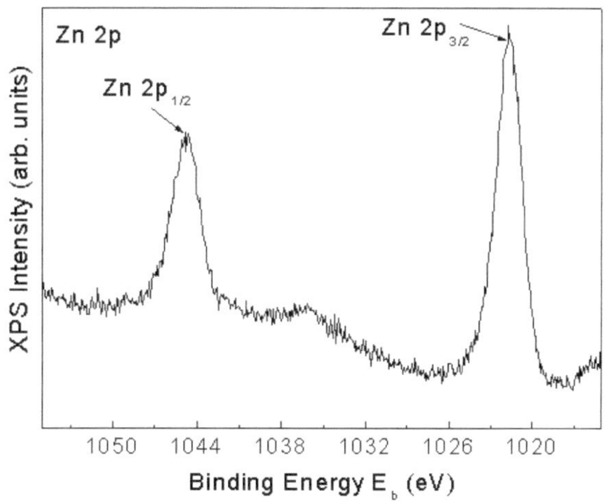

it is clearly found that there is two emission bands peaked at 380 nm (UV band) and 502 nm (green band) for all. The origin of the 380 and 502 nm bands has been ascribed to the band edge radiative recombination and intrinsic defects (mostly O vacancy) of ZnO, respectively in many reports [3]. From Fig. 6, it can be found that the intensity of 380 and 502 nm emission decreases when the sample temperature is increased. When the temperatures are higher than 100 K, the 502 nm emission disappears [4]. Meanwhile, the intensity of 380 nm increases as the sample temperature increases, until it reaches 200 K.

Afterwards, the intensity of 380 nm decreases when the sample temperature continues increasing. The integrated intensities of 380 and 502 nm emission peaks at different temperatures are shown in Fig. 6, which were calculated from the area under the curves of related emission peaks in Fig. 5.

Fig. 5 Emission PL spectra of ZnO films at different temperatures

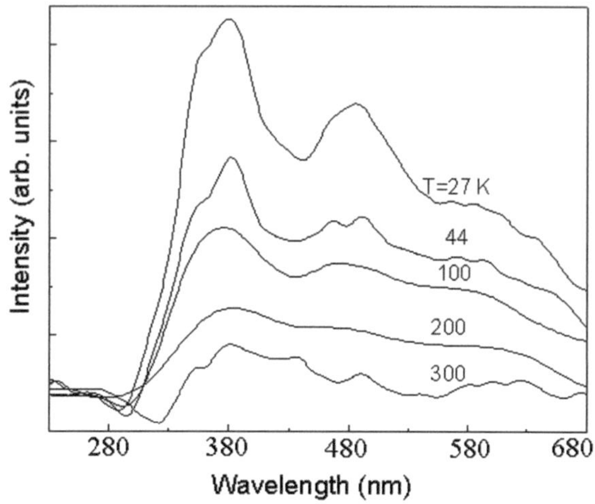

Fig. 6 Integrated intensities of 380 and 502 nm emission peaks at different temperatures

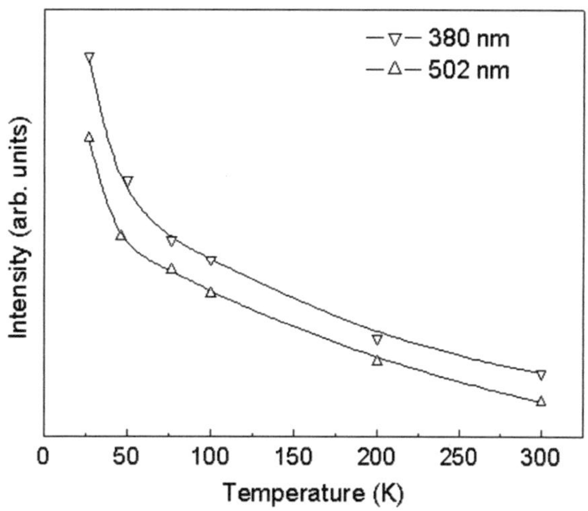

3 Conclusion

The r.f. magnetron sputter ZnO/Si films has been studied. Physical and chemical characterizations of the films were investigated using AFM, SEM, XRD and XPS. Due to its attractive properties ZnO films may have attracted much interest of potential commercial application in Photo voltaic Solar cell and optoelectronic devices, such as light-emitting diodes, laser diodes and UV photo detectors.

References

1. A. Moustaghfir, E. Tomasella, S. Ben Amor, M. Jacquet, J. Cellier, T. Sauvage, Structural and optical studies of ZnO thin films deposited by r.f. magnetron sputtering: influence of annealing. Surf. Coatings Technol. **174–175**, 193–196 (2003)
2. A.E. Rakhshani, Characterization and device applications of p-type ZnO films prepared by thermal oxidation of sputter-deposited zinc oxynitride. J. Alloy. Compd. **695**, 124–132 (2017)
3. W. Water, S.Y. Chu, Physical and structural properties of ZnO sputtered films. Mater. Lett. **55**, 67–72 (2002)
4. S. Tanaka, K. Takahashi, T. Sekiguchi, K. Sumino, J. Tanaka, Cathodoluminescence from fractured surfaces of ZnO varistors. J. Appl. Phys. **77**, 4021–4023 (1995)

Effect of Transmission Delay in a Modified Hybrid Long Loop Phase Lock Loop

Arindum Mukherjee, Shuvajit Roy and B. N. Biswas

1 Introduction

In an ordinary phase lock loop (PLL), it is a common observation that any attempt made to improve the noise squelching property of the loop inevitably leads to lower the capture capability [3, 4]. The earlier works of Biswas et al. [1, 2] overcomes this restriction by using a hybrid long loop (HLL) whereby the limitations imposed on circuit on the capture capability of a conventional phase lock-loop is overcome, and the system response linearity also increases. This paper presents the effect of transmission delay arising due to an IF filter in the loop and hence an attempt has been made here to mitigate this deleterious effect with the help of injection synchronization. Moreover an additional control by incorporating a phase modulator in the loop has been included which will increase the lock range of the loop beyond 90°, thereby reducing the probability of cycle slipping phenomenon. The loop will be now referred to as the modified hybrid long loop PLL (MHLL) because of the presence of this extra phase modulator in the circuit.

Consider the proposed circuit shown in Fig. 1, it consists of an analog mixer, a sinusoidal phase detector (PD), two voltage controlled oscillators (VCO1 and VCO2), two low-pass filters ($F_1(p)$ and $F_2(p)$), a phase modulator (PH. MOD) and an amplifier $\left(K_{inj}\right)$ to control the gain of the injection synchronized path to the VCO1. In addition to these components, a broadband IF filter is inserted which controls the

A. Mukherjee (✉)
Central Institute of Technology, Assam, India
e-mail: a.mukherjee@cit.ac.in

S. Roy
Institute of Radio Physics and Electronics, Kolkata, India
e-mail: roy.shuvajit007@gmail.com

B. N. Biswas
Chairman Education Division, SKFGI, Mankundu, West Bengal, India
e-mail: baidyanathbiswas@gmail.com

S. Chattopadhyay et al. (eds.), *Modelling and Simulation in Science, Technology and Engineering Mathematics*, Advances in Intelligent Systems and Computing 749,
https://doi.org/10.1007/978-3-319-74808-5_2

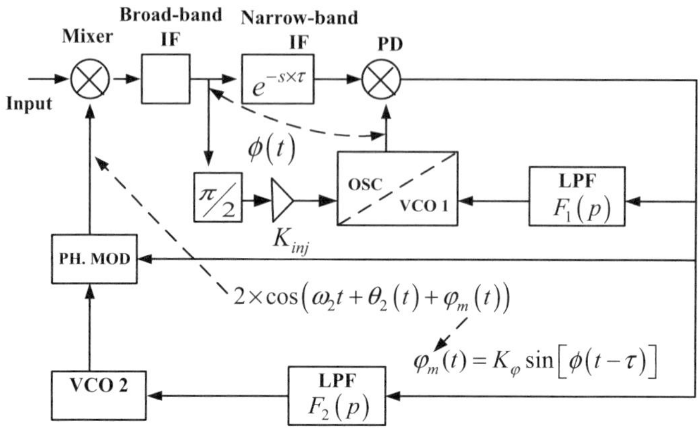

Fig. 1 Modified hybrid long loop PLL

noise bandwidth for the system. The detailed analysis of the noise bandwidth will be reported in a future communication. Finally, the presence of a narrowband tuned circuit introduces transmission delay (τ) in the loop. Output from the mixer feeds simultaneously the PD and the $\pi/2$-phase shifter. For harmonic synchronization of VCO1, this output is injected into the oscillator circuit of the VCO1. The PD output is then delivered through the low-pass filters to the reactance modulators of the two VCO's in order to control the instantaneous frequencies of the VCO's. Thus the error signal from the PD output and the RF signal from the output of the mixer control the instantaneous frequency of VCO1. It is worthwhile to mention here that the low-pass filters are employed to attenuate out the disturbances accompanying the reference signal. With the introduction of the low-pass filters, the capture region of the PLL will be reduced significantly relative to the zone of synchronism (maximum frequency error in the steady state). Again, it is known that for direct synchronization, the capture region and the synchronization region are same. Moreover, the linearity of the proposed system increases as the effective phase error becomes small at the input to the PD, which may be further observed by noting that the heterodyne output of the mixer feeds the PD.

2 Theoretical Analysis

Let us assume the input to be of the form $A\ \sin[\omega_i t + \theta_i(t)]$, and the two VCO outputs are chosen as, VCO1: $2\ \cos[\omega_1 t + \theta_1(t)]$ and VCO2: $2\ \cos[\omega_2 t + \theta_2(t)]$. The phase-detector output is given by

$$\begin{aligned} &A \times \sin[(\omega_i - \omega_1 - \omega_2)t + \theta_i(t) - \theta_1(t) - \theta_2(t) - \varphi_m(t)] \\ &= A \times \sin[\Omega \times t + \theta_i(t) - \theta_1(t) - \theta_2(t) - \varphi_m(t)], \end{aligned} \tag{1}$$

where $\Omega = (\omega_i - \omega_1 - \omega_2)$ is the open loop phase error. Again, since the filter transfer functions are '$F_1(p)$' and '$F_2(p)$' respectively, 'p' being the Heaviside operator, and recognising that the output of the LPF modulates the instantaneous phase of the VCO, one gets

$$\frac{d\theta_2}{dt} = A \times \beta_2 \times F_2(p) \times \sin[\phi(t-\tau)] \tag{2}$$

and

$$\frac{d\theta_1}{dt} = \left(\frac{\omega_0}{2Q}\right) \times K_{inj} \times A\ \sin[\phi(t)] + \{\beta_1 \times F_1(p) \times A\ \sin[\phi(t-\tau)]\} \tag{3}$$

the instantaneous phase equation is given by

$$\begin{aligned}\frac{d\phi}{dt} &= \Omega - \left(\frac{\omega_0}{2Q}\right) \times K_{inj} \times A\ \sin[\phi(t)] \\ &- [\beta_1 \times F_1(p) + \beta_2 \times F_2(p)] \times A\ \sin[\phi(t-\tau)] \\ &- K_\varphi \times \cos[\phi(t-\tau)] \times \frac{d\phi(t-\tau)}{dt} + \frac{d\theta_i}{dt}\end{aligned} \tag{4}$$

This equation is solved numerically to study the variation of phase detector output voltage and its spectrum in absence and presence of delay. It is to be noted that 'ω_0' is the centre frequency of the narrowband IF filter, 'β' is the VCO sensitivity and the low-pass filters are chosen as first order with time constant 'T'.

3 Results and Discussions

It will be reported in a later communication, that injection synchronization reduces the effect of transmission delay $\tau = \left(\frac{2 \times Q}{\omega_0}\right)$ introduced by the narrowband IF filter. It will be also shown that the additional phase control arrangement has more pronounced effect in reducing the third harmonic distortion as compared to the injection synchronized component. The phase detector output in absence and presence of delay are shown in Figs. 2 and 4 respectively. The presences of third harmonic distortion are shown in Figs. 3 and 5 respectively. A numerical experiment has been performed to study the effect of third harmonic distortion (THD) with transmission delay and is shown in Fig. 6 and in Table 1. With the increase in delay, the 3rd harmonic distortion decreases.

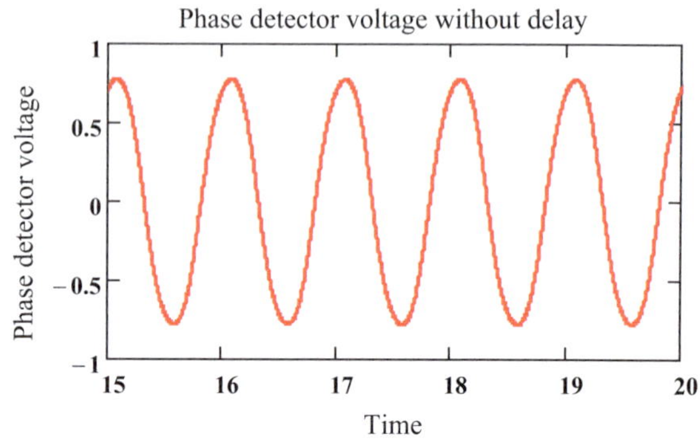

Fig. 2 Phase detector voltage in absence of delay

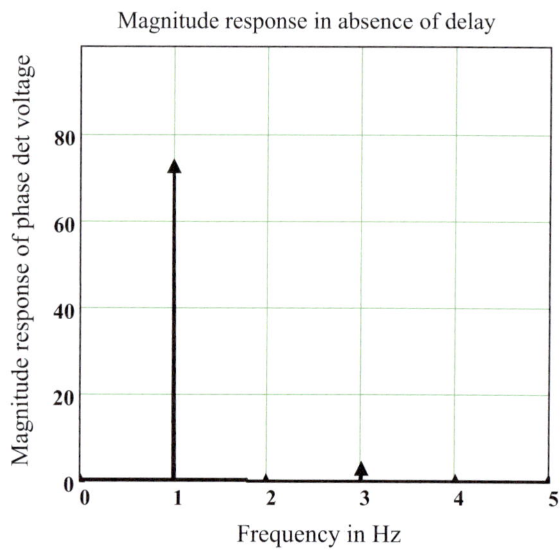

Fig. 3 Spectrum of phase detector output in absence of delay

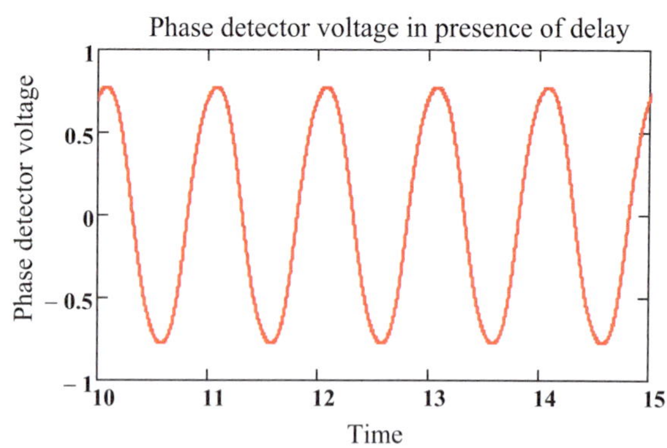

Fig. 4 Phase detector voltage in presence of delay

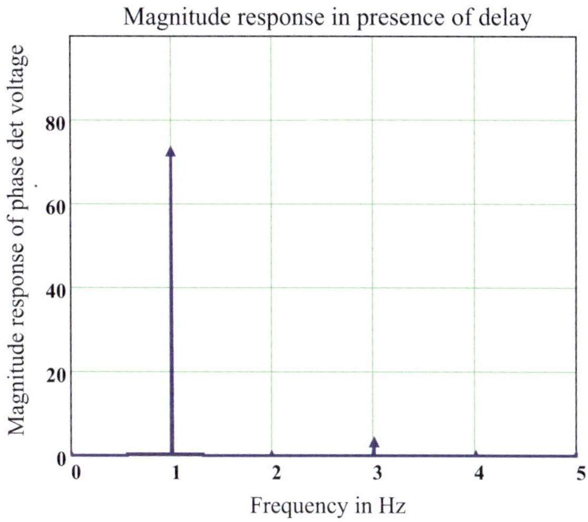

Fig. 5 Spectrum of phase detector output in presence of delay

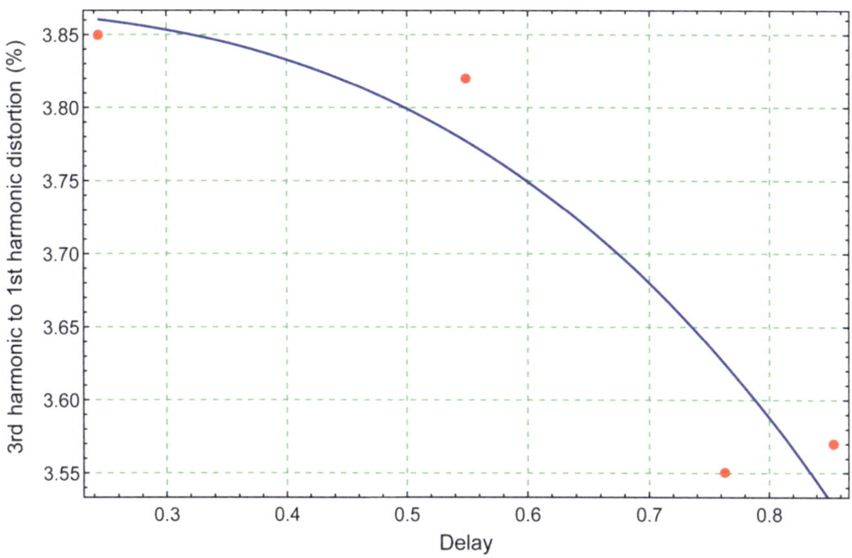

Fig. 6 Third harmonic distortion with delay; red dots are experimental data and blue curve is curve fitting

Table 1 Ratio of 3rd harmonic to fundamental frequency

Without delay	THD %	Delay (s)	THD (%)
	2.5997/69.178 = 4.24	0.031	2.5677/69.226 = 3.71
		0.244	2.6618/69.121 = 3.85
		0.549	2.6131/68.434 = 3.82
		0.763	2.4359/68.552 = 3.55
		0.854	2.4625/68.796 = 3.57

Injection strength = 1.5 V; VCO1 sensitivity = 6 Hz/V; VCO2 sensitivity = 2 Hz/V; Filter time constants = 0.1 s; Modulating signal frequency = 1 Hz; IF filter centre frequency = 10 Hz; Initial detuning = 0

4 Conclusion

A modified HLL is analysed with particular emphasis on the effect of transmission delay in the loop. The presence of third harmonic distortion is reported and its variation with delay is studied. Phase detector output and the corresponding spectrum are also studied in absence and presence of delay. In a future communication, it will be reported that injection synchronization reduces the effect of transmission delay and the phase modulator reduces the loop noise bandwidth.

Acknowledgements The authors are thankful to BoG, Central Institute of Technology and Mr B Guha Mallick, the Chairman of the Supreme Knowledge Group of Institutions, for successfully carrying out this work.

References

1. B.N. Biswas, Combination injection locking with indirect synchronization technique. IEEE Trans. Commun. Technol. (Corres.) **21**, 73 (1967)
2. B.N. Biswas, P. Banerjee, Range extension of a phase-locked loop. *IEEE Trans. Commn.* **COM-21**, 293–296 (1972)
3. B.N. Biswas, *Phase Lock Theories And Application* (Oxford & IBH, New Delhi, 1988)
4. F.M. Gardner, in *Phase Lock Techniques*, 3rd edn. (Wiley, Hoboken, 2005)

Comparative Study of Single Loop OEO Using Static and Dynamic Band Pass Filter

Shantanu Mandal, Kousik Bishayee, C. K. Sarkar, Arindum Mukherjee and B. N. Biswas

1 Introduction

Traditional method of microwave or mm-wave signal generation was realized by means of oscillators based on diodes like Gunn, IMPATT, TRAPATT etc. or using transistors. To achieve the desired frequency range, several stages of frequency multiplication using electronic circuitry needs to be done. Microwave or mm-wave signal can also be generated by beating of two laser beams or by optical injection locking method [1, 2]. These approaches were good and useful for most of traditional applications. However, for many emerging applications of recent days such as in radar, wireless communications, GPS, software defined radio etc., this traditional method fails to produce satisfactory results. Those systems were not only complicated and costly, but also lack of spectral purity, low phase noise and frequency-tunability, which are essential in these modern applications.

Optoelectronic Oscillator (OEO) is the most advanced method for extracting high purity and extremely low phase noise microwave signal proposed by Nakazawa et al.

S. Mandal (✉) · K. Bishayee
Department of ECE, University Institute of Technology,
The University of Burdwan, Burdwan, WB, India
e-mail: mshantanu2253@gmail.com

C. K. Sarkar
Department of Electronics and Telecommunication Engineering,
Jadavpur University, Kolkata, WB, India

A. Mukherjee
Department of ECE, Central Institute of Technology, Kokrajhar, Assam, India

B. N. Biswas
Sir J. C. Bose School of Engineering, SKFGI, Mankundu, WB, India

S. Chattopadhyay et al. (eds.), *Modelling and Simulation in Science, Technology and Engineering Mathematics*, Advances in Intelligent Systems and Computing 749,
https://doi.org/10.1007/978-3-319-74808-5_3

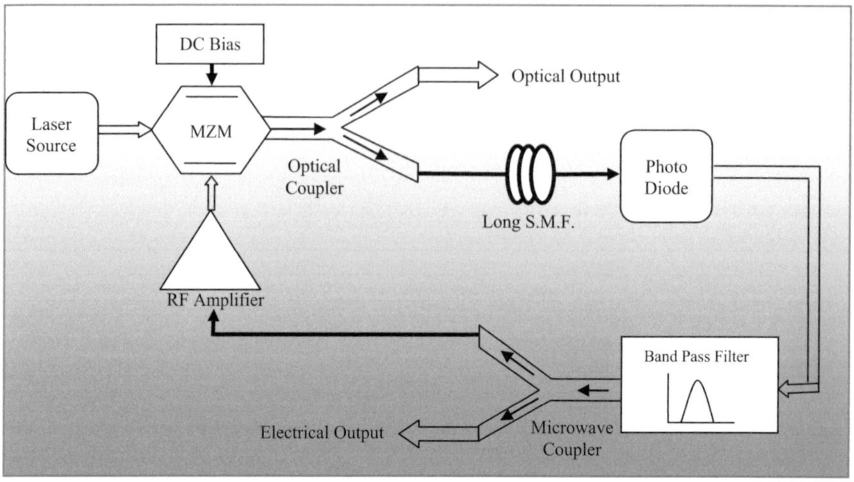

Fig. 1 Conventional single loop optoelectronics oscillator (OEO)

[3] in 1984. The construction, operation and benefits of OEO have been reported elaborately by Yeo and Meleki [4–6].

In single loop OEO (Fig. 1), a highly coherent laser beam is fed to an electro-optic modulator (MZM), the output of which is passed through a long optical fiber and detected with a photo detector. The optical fiber used here as a microwave resonator, which provides low loss and an extremely high Q factor. The output of the photo detector is amplified by an RF amplifier and filtered by a sharp cut off band pass filter and fed back to the electric port of the electro-optic modulator. This PLL configuration supports self-sustained oscillations, at a frequency determined by the fiber delay length, the bias setting of the modulator, and the band pass characteristics of the filter [4–7].

Therefore, in this paper the behavioral pattern of a single loop OEO with a Static RF filter and Dynamic Band Pass filter has been compared. Dynamic Band Pass filter used here is a tuned circuit whose center frequency can be changed with external dc control voltage [8–10]. Steady state amplitude and frequency of the OEO with different type of filter has been derived theoretically. Then the amplitude and frequency variation of the OEO with fiber delay for two different type of filter has been measured. Finally the tracking capability of the OEO with two type of filter is compared for external injection signal of different amplitudes and frequency has been derived theoretically and measured practically. Both theoretical analysis and practical findings are found in good agreement.

2 System Equation of OEO

Let us assume the RF input to the modulating grid of the MZM to be $V_{in}(t) = V(t)e^{j[\omega_1 t+\theta(t)]}$ where, $V(t)$ is the oscillation amplitude with a frequency ω_1 and initial phase θ. The optical power from the electro-optic modulator output port can be obtained

$$P = \frac{1}{2}\alpha P_0\left[1 - \eta \times \sin\pi\left(\frac{V_{in}(t) + V_B}{V_\pi}\right)\right] \tag{1}$$

where α is the fraction of insertion loss of the modulator, V_π is the half-wave voltage, V_B is the bias voltage, P_0 is the input optical power, $V_{in}(t)$ is the input RF voltage to the MZ modulator and η determines the extinction ratio of the modulator. If ρ is the sensitivity of the photo detector and R is the output impedance of the photo detector then the output of the MZ modulator can be written as

$$\begin{aligned} V_0(t) &= -2\eta V_{ph}\cos\left(\frac{\pi V_B}{V_\pi}\right)J_1\left(\frac{\pi V(t-\tau)}{V_\pi}\right)\sin[\omega(t-\tau)] \\ &= \frac{N[V(t-\tau)]}{V(t)}\exp(-s\tau)V_{in}(t) \end{aligned} \tag{2}$$

where, τ is the time delay of the long optical fiber in the loop, $V_{ph} = \frac{R\alpha\rho P_0}{2}$ and

$$N(V(t-\tau)) = -2\eta V_{ph}\cos\left(\frac{\pi V_B}{V_\pi}\right)J_1\left(\frac{\pi V(t-\tau)}{V_\pi}\right) \tag{3}$$

Without loss of any generality, we can assume that $V_B = V_\pi; \eta = 1$ and $\pi V_{ph} = V_\pi$ and $V_\pi = \pi$ then,

$$N(V(t-\tau)) = 2J_1[V(t-\tau)] \tag{4}$$

If the gain of the second order tuned circuit is G_1, gain of the RF amplifier is G_2, frequency detuning is $\Delta\omega$ then, the transfer function can be written as

$$G(s) = \frac{G_1}{1 + Q\left[\frac{s}{\omega_0} + \frac{\omega_0\left(1+\frac{\Delta\omega}{\omega_0}\right)}{s}\right]} \tag{5}$$

where, 'Q' is the quality factor of the tank circuit.

The closed loop equation can be written as

$$V_0(t) = \frac{V_{in}}{G_2 G(s)} = \frac{V_{in}}{G_2 G_1}\left[1 + Q\left\{\frac{s}{\omega_0} + \frac{\omega_0\left(1+\frac{\Delta\omega}{\omega_0}\right)^2}{s}\right\}\right] \tag{6}$$

where $V_{in}(t) = V(t)\exp[j(\omega_0 t + \theta(t)]$ and $G = G_1 G_2$.

The Eq. (6) can be written as with the help of complex frequency [6]

$$\left[\frac{2J_1[V(t-\tau)]}{V(t)}e^{-s\tau}\right] = \frac{1}{G}\left[\begin{array}{l} 1 + \frac{Q}{\omega_0}\left\{\frac{1}{V(t)}\frac{dV}{dt} + j\left(\omega_1 + \frac{d\theta}{dt}\right)\right\} \\ +Q\omega_0\left(1 + \frac{\Delta\omega}{\omega_0}\right)^2\left\{\frac{1}{j\omega_1} + \frac{1}{\omega_1^2}\left(\frac{1}{V}\frac{dV}{dt} + j\frac{d\theta}{dt}\right)\right\} \end{array}\right] \tag{7}$$

Equating the real and imaginary part of Eq. (7) and considering $\omega_1 \approx \omega_0$ we can get

$$\frac{dV}{dt} = \frac{\omega_0}{Q\left(1 + \frac{\Delta\omega}{\omega_0}\right)^2}[2GJ_1[V(t-\tau)]\cos(\omega_0\tau) - V(t)] \tag{8}$$

$$\frac{d\theta}{dt} = \frac{\omega_0\left[\left(1 + \frac{\Delta\omega}{\omega_0}\right)^2 - 1\right]}{\left(1 + \frac{\Delta\omega}{\omega_0}\right)^2 + 1} - \frac{2\omega_0 G}{Q\left[\left(1 + \frac{\Delta\omega}{\omega_0}\right)^2 + 1\right]}\frac{J_1[V(t-\tau)]\sin(\omega_0\tau)}{V(t)} \tag{9}$$

Now in steady state, considering $\frac{dV}{dt} = 0$, the free running amplitude of the oscillation is obtained exactly same as using a static band pass filter [11, 12] given by

$$V(t) = 2\sqrt{2}\sqrt{\left(1 - \frac{1}{G\,\cos(\omega_0\tau)}\right)} \tag{10}$$

The normalized free running frequency is obtained as

$$\frac{\omega_f}{\omega_0} = 1 + \frac{1}{\left[1 + \left(1 + \frac{\Delta\omega}{\omega_0}\right)^2\right]}\left[\left(1 + \frac{\Delta\omega}{\omega_0}\right)^2 - 1 - \frac{\tan(\omega_0\tau)}{Q}\right] \tag{11}$$

Whereas the normalized free running frequency using a static band pass filter obtained earlier [11] is given by

$$\frac{\omega_f}{\omega_0} = 1 - \frac{\tan(\omega_0\tau)}{2Q} \tag{12}$$

Growth of amplitude of oscillation of OEO using static and dynamic BPF is obtained by numerically solving the coupled non-linear delay differential Eqs. (8) and (9) using Wolfram Mathematica 11.1® as shown in Fig. 2. This variation in response is due to the detuning effect of the dynamic filter which has the noticeable difference in the damping factor of the oscillators during the initiation of oscillation.

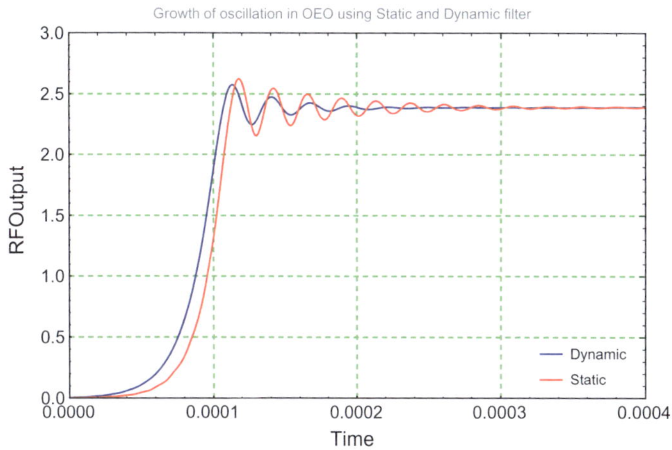

Fig. 2 Growth of amplitude of oscillation of OEO using static and dynamic BPF

Further, when an external synchronization signal is injected into the OEO, it is most important to see the zone over which the sync signal is in lock with the OEO free running signal. In steady state, the non-linear term $\frac{N[V(t-\tau)]}{V(t)} = \frac{2J_1[V(t-\tau)]}{V(t)} = \frac{2J_1[V]}{V}$ can be considered as constant ('C'). Therefore, the normalized lock range of OEO can be written as [11–13].

$$\Omega = \Delta\omega_1 - GC\ \sin(\omega_0\tau) - \sqrt{\left(\frac{GE}{V}\right)^2 - (GC\ \cos\omega_0\tau - 1)^2} \tag{13}$$

where, $\Delta\omega_1 = \left(\frac{2Q}{\omega_0}\right) \times \Delta\omega$

Here the variation of lock range with different fiber delay as well as with different detuning values using static and dynamic filter has been compared experimentally.

3 Experimental Results and Discussion

The experimental validations of the theoretical expressions are obtained using MATLAB® Simulink® environment. The designed experimental setup of the OEO is shown in Fig. 3. Here, the conventional static BPF of earlier work is replaced with the proposed dynamic BPF as shown in the Fig. 3.

The frequency response of the static and dynamic band pass filter is shown in Fig. 4 which shows exactly same type of response in both cases. But in dynamic filter, the center frequency is varying nearly the same amount as frequency detuning (Δw), which is ideal for a dynamic BPF and is the prime goal of using dynamic filter here.

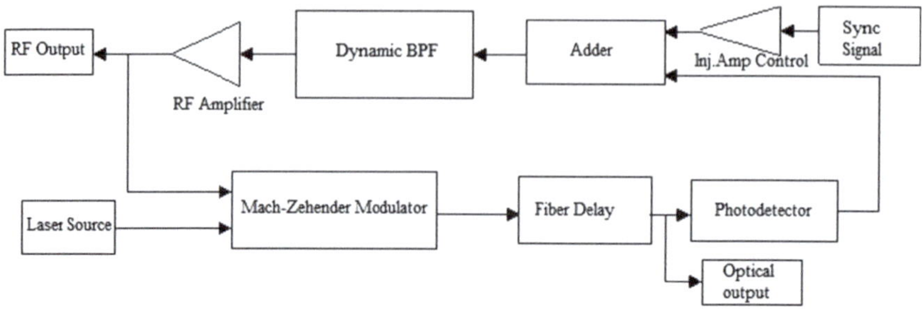

Fig. 3 Simulation block diagram of OEO using dynamic filter

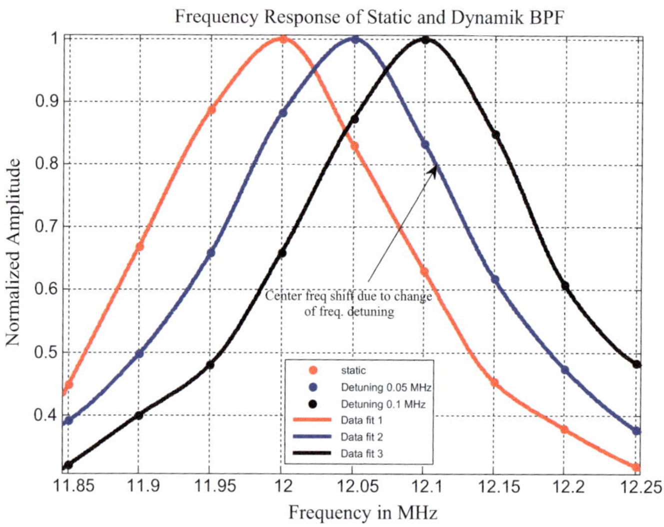

Fig. 4 Frequency response of single loop OEO using static and dynamic filter

The previous observation becomes more prominent when the center frequency variation along with different detuning (Δw) has been measured. Plot of center frequency variation with detuning, considering RF amplifier gain of 3.0 and quality factor (Q) value of 76.9 (Fig. 5), shows completely linear variation.

Next, the experimental variation of normalized amplitude and frequency of the single loop OEO with optical fiber delay is compared for static and dynamic BPF as shown in Figs. 6 and 7. The corresponding experimental data is given in the Tables 1 and 2 respectively where, different frequency detuning of the dynamic filter has been considered. Visualizing the nature of the curves it can be concluded that the use of dynamic filter in OEO reduces its amplitude as well as the frequency variation with delay considerably.

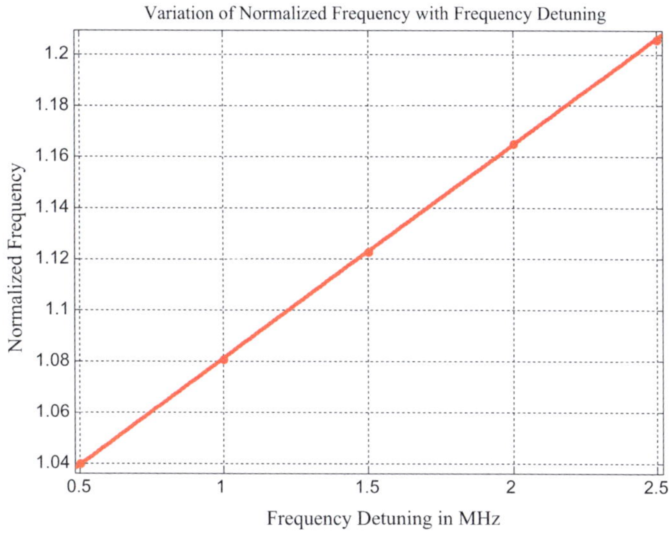

Fig. 5 Frequency Detuning with center frequency shift of dynamic filter

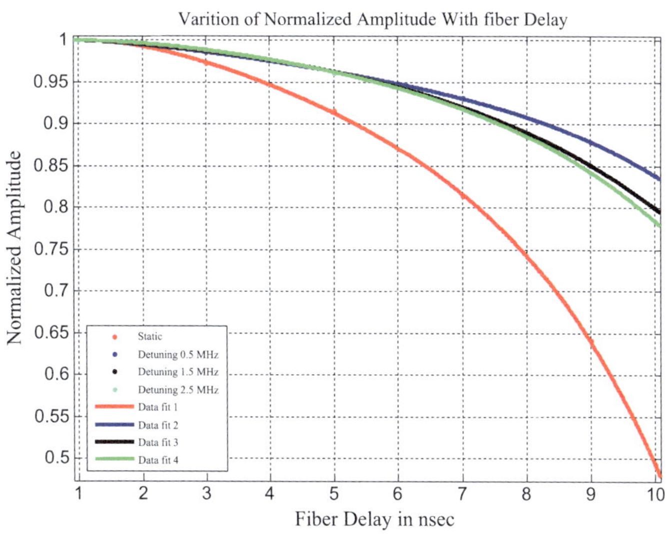

Fig. 6 Experimental variation of normalized amplitude with fiber delay

Finally, the modified single loop OEO is synchronized with the externally injected signal. The dynamic filter adjusts its center frequency in such a way that the center frequency becomes equal to the instantaneous frequency of the injected signal. Once

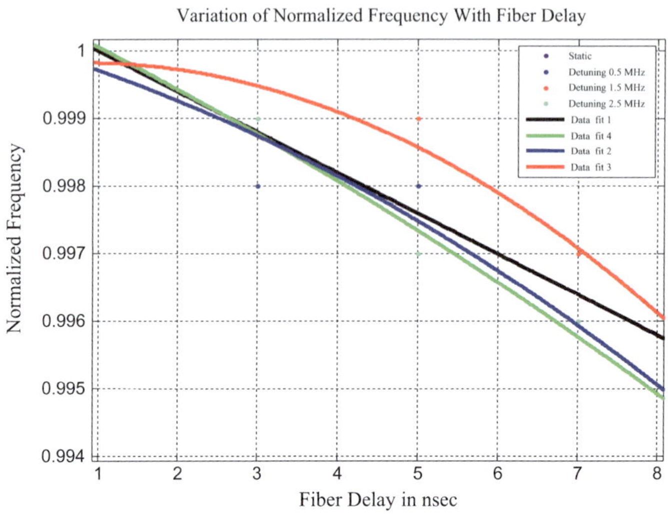

Fig. 7 Experimental variation of normalized frequency with fiber delay

Table 1 Experimental data for normalized amplitude with fiber dela

Fiber delay in nsec	Normalized amplitude of static BPF	Normalized amplitude of dynamic BPF with $\Delta W =$ 2.5 MHz	Normalized amplitude of dynamic BPF with $\Delta W =$ 1.5 MHz	Normalized amplitude of dynamic BPF with $\Delta W =$ 0.5 MHz
1	1	1	1	1
3	0.972	0.989	0.987	0.986
5	0.916	0.961	0.963	0.962
7	0.813	0.919	0.92	0.931
9	0.641	0.841	0.851	0.878
10	0.493	0.785	0.8	0.839

synchronized, the variation of normalized lock range of OEO with fiber delay as well as injection signal amplitude is shown in Figs. 8 and 9 respectively. The nature of the graph shows the exact resemblance with of static and dynamic filter. That means with the use of such a dynamic filter, variable frequency OEO of similar nature to static one, can be obtained.

Table 2 Experimental data for normalized frequency with fiber delay

Fiber delay in nsec	Normalized frequency of static BPF	Normalized frequency of dynamic BPF with ΔW = 2.5 MHz	Normalized frequency of dynamic BPF with ΔW = 1.5 MHz	Normalized frequency of dynamic BPF with ΔW = 0.5 MHz
1	1	1	1	1
3	0.999	0.999	0.999	0.998
5	0.997	0.997	0.999	0.998
7	0.997	0.996	0.997	0.996
9	0.995	0.994	0.995	0.994

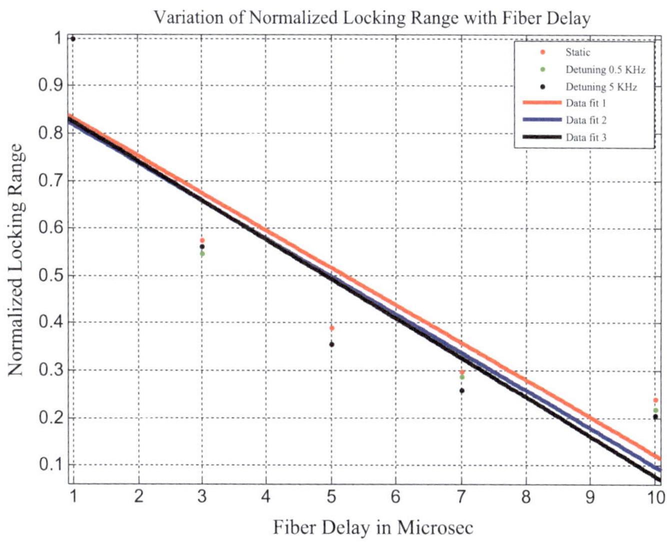

Fig. 8 Experimental variation of lock range with fiber delay

4 Conclusion

This paper considers the effect of synchronizing capability when a conventional RF static band-pass filter is replaced by a dynamic filter in a single loop OEO. System governing equations are derived when the synchronization signal is in the form of angle modulation and an expression for locking range of the OEO has been calculated. In a future communication, the dynamic tracking capability of this filter will be reported.

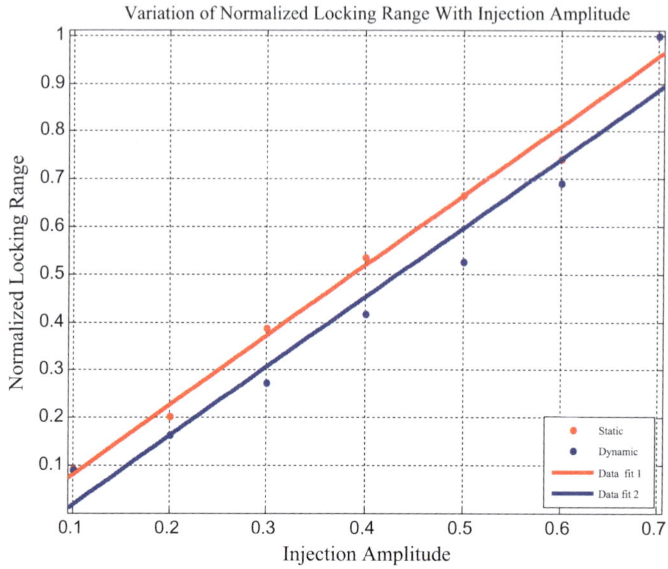

Fig. 9 Experimental variation of lock range with injection amplitude

Acknowledgements The authors are thankful to the management of University Institute of Technology, The University of Burdwan, Burdwan, West Bengal and Central Institute of Technology, Assam for giving an opportunity to carry out this work and also the management of Sir J C Bose School of Engineering for carrying out the work on Sir J C Bose Creativity Centre of Supreme Knowledge Foundation Group of Institution, Mankundu, Hooghly.

References

1. U. Gliese, T.N. Nielsen, S. Nørskov, K.E. Stubkjaer, Multifunctional fiber-optic microwave links based on remote heterodyne detection. IEEE Trans. Microwave Theory Tech. **46**(5), 458–468 (1998)
2. L. Goldberg, H.F. Taylor, J.F. Weller, D.M. Bloom, Microwave signal generation with injection locked laser diodes. Electron. Lett. **19**(13), 491–493 (1983)
3. M. Nakazawa, T. Nakashima, M. Tokuda, An Optoelectronic self-oscillatory circuit with an optical fibre delayed feedback and its injection locking technique. J. Lightwave Technol. LT **2**(5), 719–730 (1984)
4. X.S. Yao, L. Maleki, Optoelectronic microwave oscillator. J. Opt. Soc. Am. B **13**(8), 1725–1735 (1996)
5. X.S. Yao, L. Maleki, L. Devis, Coupled Optoelectronic Oscillators for Generating Both RF Signal and Optical Pulses. J. Lightwave Technol. **18**(1), 73–78 (2000)
6. B.N. Biswas, S. Chatterjee, S Pal, Laser induced microwave oscillator. Int. J. Electron. Commun. Eng. Technol. (IJACET), ISSN 0976–6464(Print), ISSN 0976–6472(Online), **3**(1), (2012), © IAEME
7. I. Gonorovsky, *Radio Circuit and Signals* (Mir Publisher, Moscow, 1974)

8. A. Mukherjee, D. Ghosh, N.R. Das, B.N. Biswas, On a single loop optoelectronic oscillator using variable centre frequency dynamic filter. IJECT **6**(1), Spl-1, Jan-March-2015, ISSN: 2230–7109 (Online) ISSN: 2230-9543(Print)
9. A. Mukherjee, D. Ghosh, B.N. Biswas, On the effect of combining an external synchronizing signal feeding the Mach-Zehnder modulator in an optoelectronic oscillator. Optik Int. J. Light Electron. Optics **127**, 3576–3581 (2015)
10. A. Mukherjee, D. Ghosh, N.R. Das, B.N. Biswas, Harmonic distortion and power relations in a single loop optoelectronic oscillator. Optik Int. J. Light Electron. Optics **127**, 973–980 (2015)
11. K. Bishayee, S. Mandal, A. Mukherjee, S. Pal, B.N. Biswas, Locking phenomenon in a single loop OEO. IJECT **5**(2), Spl-1, Jan–March-2014, ISSN : 2230–7109 (Online) | ISSN: 2230–9543(Print)
12. A. Mukherjee, B.N. Biswas, N.R. Das, A study on the effect of synchronization by an angle modulated signal in a single loop optoelectronic oscillator. Optik Int. J. Light Electron. Optics **126**(19), 1815–1820 (2015)
13. A. Mukherjee, S. Chatterjee, N.R. Das, B.N. Biswas, Laser induced microwave oscillator under the influence of interference. Int. J. Microw. Wirel. Technol. **6**(6), 581–590 (2014)

A Study on the Effect of an External Periodic Signal in a Chaotic Optoelectronic Oscillator

Dia Ghosh, Arindum Mukherjee, Nikhil Ranjan Das and Baidya Nath Biswas

1 Introduction

Over the last few years OEO has seen wide spread application in the field of RADAR, fiber optic communication system, long distance digital communication system, in view of the fact that it has the ability to produce high frequency signal with ultra high spectral purity. This oscillator was first introduced by Neyer and Voges [1]. Posterior to their pioneering work, Yao and Maleki introduced this oscillator as a high quality microwave oscillator [2]. The OEO contains a continuous wave laser source. The optical signal generated from the laser is fed to a Mach-Zehnder modulator (MZM), which is acting as an intensity modulator. The intensity modulated optical signal is passed through an optical fiber delay line and applied to the photo detector. The detected RF signal is then filtered by a band pass filter (BPF). The output of the BPF is fed to the electrical port of the MZM. Generation of high spectrally pure signal is

D. Ghosh (✉)
Department of Electronics and Communication Engineering,
Siliguri Institute of Technology, Siliguri 734009, West Bengal, India
e-mail: dia.slg42@gmail.com

A. Mukherjee
Department of Electronics and Communication Engineering,
Central Institute of Technology, Kokrajhar 783370, Assam, India
e-mail: arindum78@gmail.com

N. R. Das
Institute of Radio Physics and Electronics,
Calcutta University, 92 A.P.C. Road, Kolkata 700009, West Bengal, India
e-mail: nrd@ieee.org

B. N. Biswas
Education Division, SKF Group of Institutions,
Mankundu, Hooghly 712139, West Bengal, India
e-mail: baidyanathbiswas@gmail.com

S. Chattopadhyay et al. (eds.), *Modelling and Simulation in Science, Technology and Engineering Mathematics*, Advances in Intelligent Systems and Computing 749,
https://doi.org/10.1007/978-3-319-74808-5_4

possible due to the long low loss optical fiber delay line in its feedback loop. The long delay line results in a high quality factor and spectral purity. The presence of optical fiber delay line facilitates OEO as a candidate of electro-optical system with delayed feedback. Therefore the study on the complex dynamics of OEO is an important aspect, both from academic and engineering application point of view. Considering the feedback gain as a control parameter Chembo et al. described the generation of chaotic breathers in an OEO [3]. Other schemes for chaotic signal generation and stability analysis in an OEO was also being contemplated [4–9]. By controlling both feedback delay and loop gain the complex dynamics and synchronization property of an OEO was reported [10, 11] but the OEO in this report was implemented using discrete time DSP technology. The oscillator was designed with a laser, electro-optic modulator and a photo-detector but for delay and filtering purpose the DSP board was used. In [12], it has been reported by the present authors that with the variation of loop delay the system loses its stability and following a period doubling route it produces chaotic oscillation.

In the present work we report a study on the complex dynamics of an OEO under the influence of a synchronizing signal. In OEO in order to obtain high spectral purity of the signal a long feedback loop delay is required. But long feedback loop delay produces additional cavity modes. These adjacent cavity modes can even produce unwanted chaotic oscillation. It has been shown that by controlling amplitude of the external signal the chaotic oscillation at the output of the free running oscillator can be destroyed and period -1 oscillation can be produced. Although the method of chaos quenching is not new [13, 14], as far as the knowledge of the authors is concerned, the effect of sync signal to control the chaotic dynamics of the OEO is addressed nowhere.

The paper is organized in the following way: Sect. 2 describes the basic configuration of the oscillator and derivation of the system equation. In Sect. 3 the numerical study is presented. The simulation study is described in Sect. 4. Finally the paper concludes.

2 Derivation of System Equation

Figure 1 shows the basic configuration of an SLOEO. It consists of a continuous wave laser source which is fed into a Mach-Zehnder modulator (MZM), the MZM acts as an intensity modulator of the optical signal. The optical output of the modulator is detected by a photo detector after passing through a long optical delay line. This signal is then passed through an electrical band pass filter (BPF). The output from the BPF is fed back to the electrical port of the MZM. The BPF implemented here using a single tuned circuit.

Let us consider the RF input to the MZM is $V_{in}(t) = V(t)e^{j(\omega_0 t+\theta(t))}$ where $V(t)$ is the amplitude of the signal with free-running frequency ω_0 and the initial phase of $\theta(t)$. The output power of the MZM can be expressed as [15].

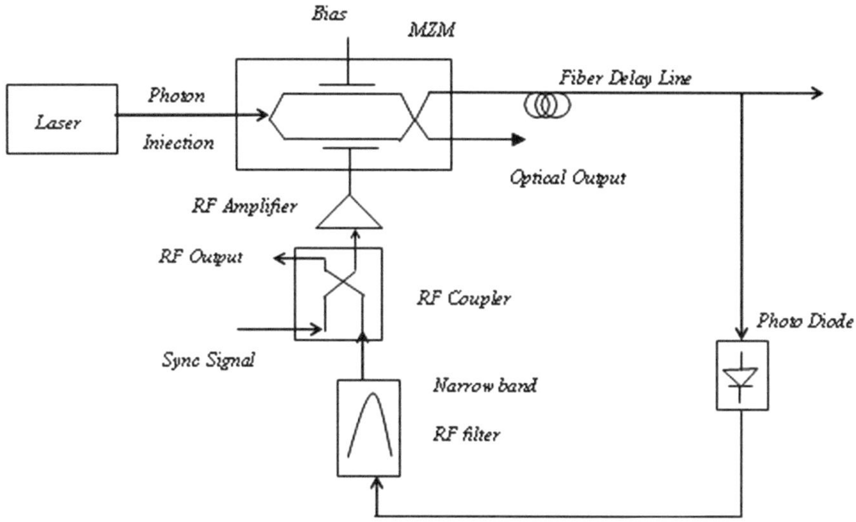

Fig. 1 Basic configuration of a single loop optoelectronic oscillator

$$P(t) = \frac{1}{2}\alpha P_0\left[1 - \eta\,\mathrm{Sin}\,\pi\left(\frac{V_{in}(t) + V_B}{V_\pi}\right)\right] \tag{1}$$

where P_0 is the input optical power, α is the fraction of insertion loss of the modulator, η is the extinction ratio of the modulator, V_B is the bias voltage of the modulator, and V_π is the half wave voltage of the modulator. Therefore the photo-detector output can be expressed as $V_0(t) = \rho R P(t - \tau)$, where ρ is the sensitivity and R is the output impedance of the photo-detector and τ is the time delay resulting from the physical length of the optical fiber in the feed-back loop. Considering all these arguments it can be shown that [15, 16].

$$V_0(t) = V_{ph}\begin{bmatrix} 1 - \eta\,\sin\left(\frac{\pi V_B}{V_\pi}\right)\left\{J_0\left(\frac{\pi V(t-\tau)}{V_\pi}\right) + 2\sum\limits_{m=1}^{\infty} J_{2m}\left(\frac{\pi V(t-\tau)}{V_\pi}\right)\cos[2m\,\omega(t-\tau)]\right\} \\ -2\eta\,\cos\left(\frac{\pi V_B}{V_\pi}\right) \times 2\sum\limits_{m=0}^{\infty} J_{2m+1}\left(\frac{\pi V(t-\tau)}{V_\pi}\right)\sin[(2m+1)\omega(t-\tau)] \end{bmatrix}$$

It is to be noted that with the growth of oscillation amplitude the effective Q value of the tuned circuit becomes narrow and the smaller components of the spectrum are rejected. The highest component of the spectrum only sustains at the output of the oscillator. Thus the output of the MZM is seen to be

$$V_0(t) = -2\eta V_{ph}\cos\left(\frac{\pi V_B}{V_\pi}\right)J_1\left(\frac{\pi V(t-\tau)}{V_\pi}\right)\sin[\omega(t-\tau)] = \frac{N(V(t-\tau))}{V(t)}\exp(-s\tau)V_{in}(t)$$

where $N(V(t-\tau)) = -2\eta V_{ph}\cos\left(\frac{\pi V_B}{V_\pi}\right)J_1\left(\frac{\pi V(t-\tau)}{V_\pi}\right)$ and $V_{ph} = \frac{\alpha R\rho P_0}{2}$, now for simplicity let us consider $\eta = 1; V_B = V_\pi; \pi V_{ph} = V_\pi; V_\pi = \pi$ and $N(V(t-\tau)) = 2J_1(V(t-\tau))$. Here J_1 is the Bessel function of the first kind for order zero.

When the input signal $V_{in}(t)$ passes through the SLOEO the output voltage can be expressed as

$$V_0(t) = \beta(s) \cdot V_{in}(t) \tag{2}$$

$$\beta(s) = \left[\frac{N(V(t-\tau))}{V} G(s) e^{-s\tau} \right] \tag{3}$$

here $G(s)$ is the transfer function of the single tuned circuit and can be written as $G(s) = g_m Z(s)$, g_m is the gain of the tuned circuit.

Using (2) and (3) it can be shown that

$$\frac{1}{Z(s)} = 2g_m \left[\frac{J_1(V(t-\tau))}{V} e^{-s\tau} \right] \tag{4}$$

$$\frac{V}{Z(s)} \cong \left[2J_1(V(t-\tau)) e^{-s\tau} \right] g_m \tag{5}$$

$$\left[\frac{V}{R} + C\frac{dV}{dt} + \frac{1}{L} \int V dt \right] = 2J_1\left[V(t-\tau) e^{-s\tau} \right] g_m \tag{6}$$

To realize the transient behavior, we consider the operation of the system near resonance

$$\left(\frac{1}{R} + j\omega C + \frac{1}{j\omega L} \right) \cong G + 2C(j\omega - j\omega_0) \tag{7}$$

and

$$j\omega = \frac{1}{V(t)} \cdot \frac{dV}{dt} + j\omega_0 + j\frac{d\theta}{dt} \tag{8}$$

using (7), (8) in (6) and equating the real and imaginary part the time varying amplitude and phase of an SLOEO can be written as.

$$\frac{dV}{dt} = \frac{\omega_0}{2Q} [G_1 2J_1[V(t-\tau)] \mathrm{Cos}\,(\omega_0 \tau) - V(t)]$$

$$\frac{d\theta}{dt} = -\frac{\omega_0}{2Q} \frac{G_1 2J_1[V(t-\tau)]}{V} \mathrm{Sin}\,(\omega_0 \tau)] \tag{9}$$

where $G_1 = g_m R$ is gain at resonance.

Considering the following normalized quantities (9) can be rewritten as

$$\begin{aligned}
t' &= \frac{\omega_0 t}{2Q}, \quad \tau' = \frac{\tau}{RC}, \quad b = 2G_1, \\
v &= \frac{V(t)}{V\text{max}}, \quad v(t' - \tau') = \frac{V(t-\tau)}{V\text{max}} \\
\frac{dv}{dt'} &= -v + bJ_1\big[v(t' - \tau')\big]\text{Cos}\,(\omega_0 \tau') \\
\frac{d\theta}{dt'} &= -\frac{b\,J_1\big[v(t' - \tau')\big]}{v}\text{Sin}\,(\omega_0 \tau')
\end{aligned} \tag{10}$$

Now let us consider a synchronizing signal having a form of $S(t) = Ee^{j(\omega_1 t + \psi(t))}$ is injected in the free running oscillator, here E is the amplitude and $\psi(t)$ is the phase of the injected signal. The phase difference between the free running signal and the injected signal is $\phi(t) = \psi(t) - \theta(t)$. Thus it is not difficult to show that the closed loop amplitude and phase equation of the synchronized oscillator will take the following form.

$$\begin{aligned}
\frac{dv}{dt'} &= -v + bJ_1\big[v\big(t' - \tau'\big)\big]\text{Cos}(\omega_0 \tau') + \text{G}_1 e\,\text{Cos}(\phi(t')) \\
\frac{d\phi}{dt'} &= \frac{2Q}{\omega_0}\Omega + \frac{b\,J_1\big[v\big(t' - \tau'\big)\big]}{v}\text{Sin}\,(\omega_0 \tau') - \frac{\text{G}_1 e}{v}\text{Sin}(\phi(t'))
\end{aligned} \tag{11}$$

here $\Omega = \omega_1 - \omega_0$ and e is the normalized amplitude of the sync signal.

3 Numerical Analysis

Equation (10) is the free running system equation of the oscillator. This equation is solved numerically using Mathematica version 10 considering $G_1 = 3.55$, $b = 2G_1 = 7.1$. In our previous work [12] it has been shown that with the variation of feedback loop delay τ' the system produces chaotic oscillation following a period doubling sequence. Figure 2 depicts phase plane plot of the oscillator. In this figure the hyper chaotic oscillation for $\tau\prime = 3.3$ is shown.

The chaotic dynamics is quantified using Lyapunov exponent spectrum, following the technique proposed by Farmer [17]. The spectrum of Lyapunov exponent also ensures the existence of chaotic oscillation beyond $\tau' = 2.3$(Fig. 3). Now at $\tau' = 3.3$ keeping all other parameter values unchanged the external sync signal is injected into the oscillator. The injected signal frequency is same as the free running oscillation frequency. It has been observed using (11) that with suitable control of the sync signal amplitude the chaotic state of the free running oscillator disappears and period -1 oscillation is produced. Figure 4 shows the phase plane plot of the driven oscillator for $e = 2.26$ and $e = 2.31$.

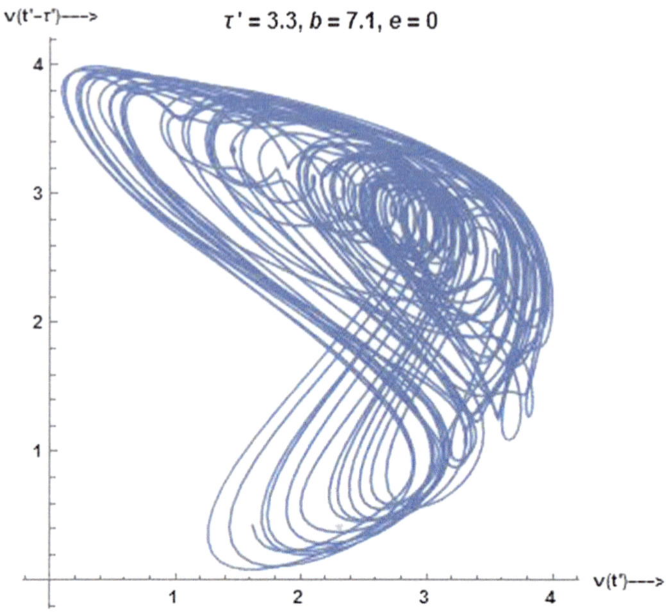

Fig. 2 Numerically obtained phase plane plot of free running oscillator (v, $v(t' - \tau')$ space)

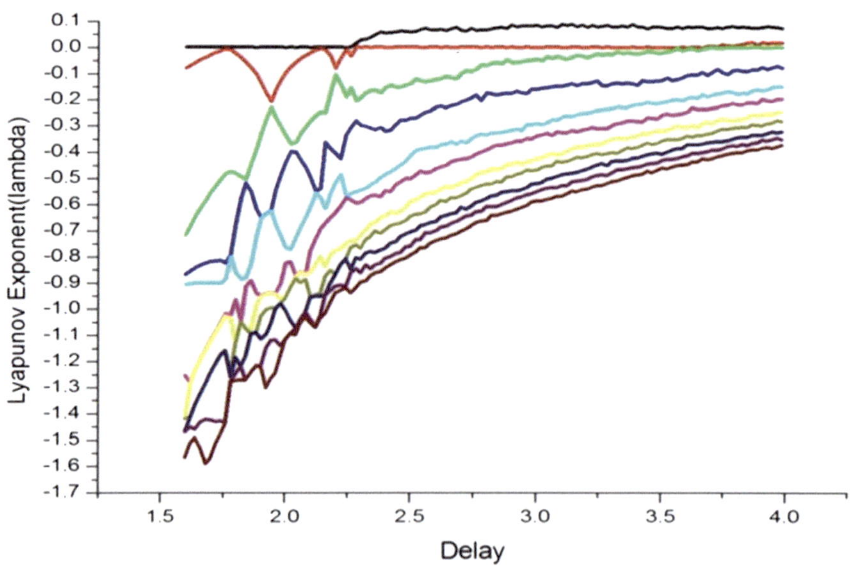

Fig. 3 Lyapunov exponent (λ) with feedback delay

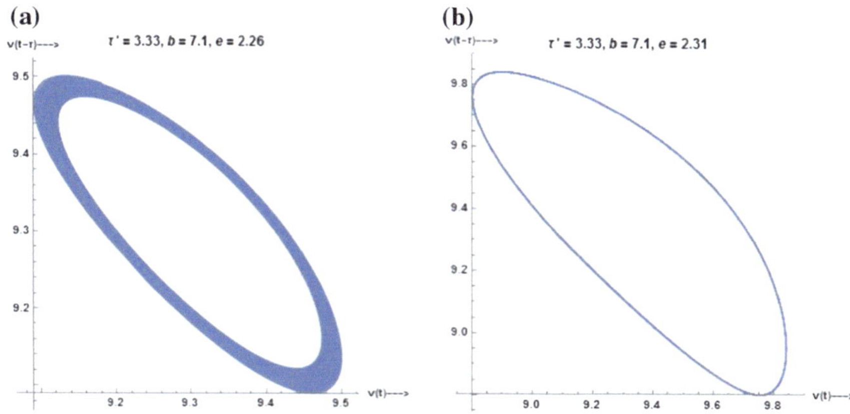

Fig. 4 Numerically obtained phase plane plot of the driven oscillator (v, $v(t' - \tau')$ space)

4 Simulation Study Using MATLAB Simulink Software

The oscillator under study is realized using MATLAB™ 9.0 Simulink software. Figure 5 represents block diagram of the simulation set-up. In general OEO can generate high frequency signal in microwave and mm wave range. However it is difficult to carry out the simulation study in such a high frequency range. To overcome this difficulty the frequency of laser source is chosen as 500 M rad/s and the output signal amplitude of the laser is set at 1.4 V. To design the BPF we have taken C = 1nF, L = 0.2 μH, R = 3 kΩ, with these parameters the operating frequency becomes $f =$ 11.22 MHz, RF gain G_1 is set to 3.55. It can be shown that the fiber delay $\tau = 10$ μs ($\tau' = \frac{\tau}{RC} = 3.33$) produces the chaotic oscillation at the output of the oscillator (Fig. 6). The amplitude of the chaotic oscillation is 5 V. Now we have applied an external RF signal into the oscillator. The operating frequency of the sync signal is kept fixed at $f_s = 11.22$ MHz and the amplitude E is varied. The output spectrum of the driven oscillator is shown in Fig. 7. It can be seen from the figure that at $E = 2.15\ V$ the chaotic oscillation completely disappears but some other adjacent oscillating modes are present. These additional cavity modes are produced due the large feedback loop delay. Now as E is increased further the effect of the side modes are reduced and at $E = 2.55$ V the effect of all side modes are disappeared and single frequency oscillation at 11.22 MHz is achieved.

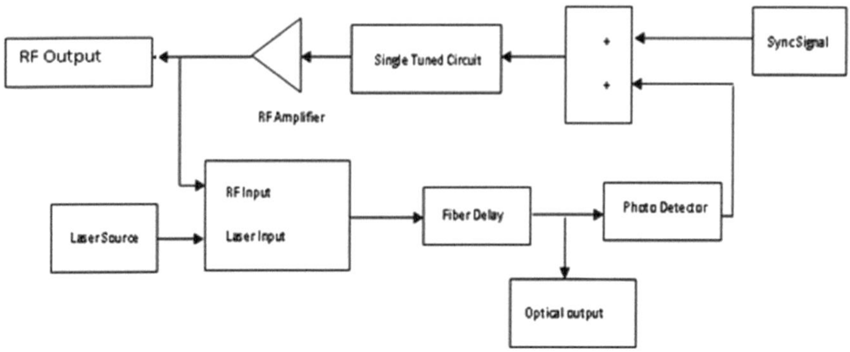

Fig. 5 Schematic representation of simulation set-up

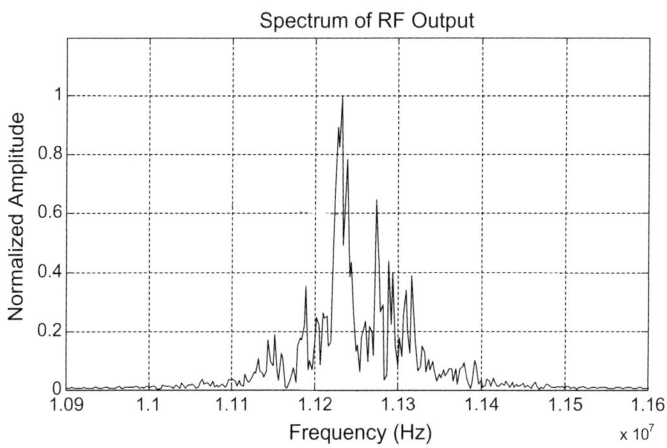

Fig. 6 Chaotic oscillation with C = 1 nF, L = 0.2 μH, R = 3 kΩ, τ = 10 μs

5 Conclusion

In this literature we have studied the dynamics of an SLOEO under the influence of an external sync signal. It has been demonstrated through the numerical and simulation study that the application of the injected signal destroys the chaotic oscillation and suitable control of the injected signal amplitude can produce period −1 oscillation. Optoelectronic oscillator can efficiently produce high frequency signal with high spectral purity. Generation of high spectrally pure signal is possible due to the long low loss optical fiber delay line in its feedback loop. However long feedback loop delay may produce unwanted chaotic oscillation. The proposed technique can be efficiently used to remove chaotic oscillation and produce single frequency oscillation in an OEO.

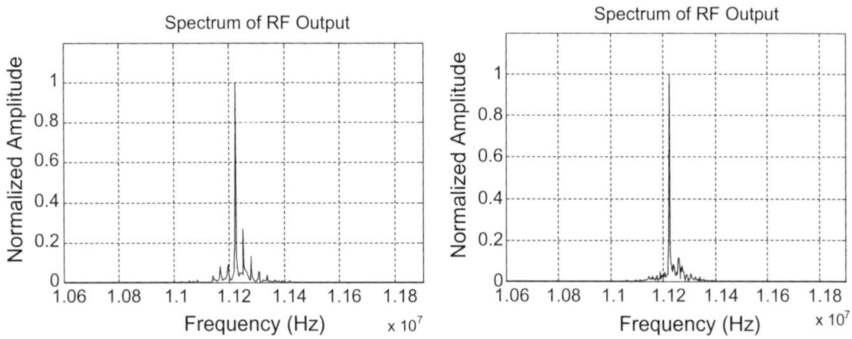

(a) $E = 2.15\ Volt$ (b) $E = 2.55\ Volt$

Fig. 7 RF spectrum of the oscillator obtained from the simulation study with $\tau = 10$ μs and with different values of E, keeping all other parameters unchanged

Acknowledgements The authors are thankful to the management of Siliguri Institute of technology, Siliguri, West Bengal, India, Central Institute of Technology, Assam, India for giving an opportunity to carry out this work. The authors also acknowledge the support from the management of Sir J C Bose School of Engineering to conduct the work at Sir J C Bose Creativity Centre of Supreme Knowledge Foundation Group of Institution, Mankundu, Hooghly, West Bengal, India.

References

1. A. Neyer, E. Voges, High-frequency electro-optic oscillator using an integrated interferometer. Appl. Phys. Lett. **40**, 6–8 (1982)
2. X.S. Yao, L. Maleki, Optoelectronic microwave oscillator. J. Opt. Soc. Am. B. **13**, 1725–1735 (1996)
3. Y. Chembo Koumou, P. Colet, L. Larger, N. Gastaud, Chaotic breathers in delayed electro-optical systems. Phys. Rev. Lett. **95**, 2005
4. Y.K. Chembo, L. Larger, P. Colet, Nonlinear dynamics and spectral stability of optoelectronic microwave oscillator. IEE J. Quantum Electron. **44**, 858–868 (2008)
5. Y.K. Chembo, L. Larger, H. Tavernier, R. Bendoula, E. Rubiola, P. Colet, Dynamic stabilities of microwaves generated with optoelectronic oscillators. Opt. Lett. **32**, 2571–2573 (2007)
6. M. Peil, M. Jacquot, Y.C. Kouomou, L. Larger, T. Erneux, Routes to chaos and multiple time scale dynamics in broadband bandpass nonlinear delay electro-optic oscillators. Phys. Rev. E. **79** (2009)
7. L. Weicker, T. Erneux, M. Jacquot, Y. Chembo, L. Larger, Crenelated fast oscillatory outputs of a two—delay electro-optic oscillator. Phys. Rev. E. **85** (2012)
8. L. Larger, P.A. Lacourt, S. Poinsot, M. Hanna, From flow to map in an experimental high dimensional electro-optic nonlinear delay oscillator. Phys. Rev. Lett. **95** (2005)
9. K.E. Callan, L. Illing, D.J. Gauthier, E. Scholl, Broad band chaos generated by an optoelectronic oscillator, Phys. Rev. Lett. **104** (2010)
10. T.E. Murphy et al., Complex dynamics and synchronization of delayed feedback nonlinear oscillators. Proc. R. Soc. Lond. A **368**, 343–366 (2010)
11. A.B. Cohen, B. Ravoori, T.E. Murphy, R. Roy, Using synchronization for prediction of high dimensional chaotic dynamics. Phys. Rev. Lett. **101** (2008)

12. D. Ghosh, A. Mukherjee, N.R. Das, B.N. Biswas, Study on the Complex Dynamics of a Single Loop Optoelectronic oscillator. URSI Asia-Pacific Radio Science Conference, URSI—APRASC'2016, Seoul, Korea, August, 2016
13. K. Pyragas, Continuous control of Chaos by self controlling feedback. Phys. Lett. A (1992)
14. A.N. Piserchik, B.K. Goswami, Annihilation of one of the coexisting attractors in a bistable system. Phys. Rev. Lett. (2000)
15. B.N. Biswas, S. Chatterjee, S. Pal, Laser induced microwave oscillator. IJECET **3**, 211–219 (2012)
16. A. Mukherjee, B.N. Biswas, N.R. Das, A study on the effect of synchronization by an angle modulated signal in a single loop optoelectronic oscillator. Optik Int. J. Light Electron. Opt. **126** (2015)
17. J.D. Farmer, Chaotic attractor of infinite dimensional dynamical system. Phys. D (1982)

Computation of Current Density in Double Well Resonant Tunneling Diode Using Self-consistency Technique

Biswarup Karmakar, Rupali Lodh, Pradipta Biswas, Subhro Ghosal and Arpan Deyasi

1 Introduction

Resonant tunneling devices are found the interest of both theoretical [1] and experimental researchers [2] for the post decade owing to its novel electronic properties [3], its less complex mechanism supported by the controlled microelectronic growth techniques with various combination of semiconducting materials [4]. Electrical and optical properties of these heterostructure devices can be computed from the knowledge of quantum transport processes, and precise estimation of transmission coefficient is essential for the device with incorporation of physical parameters [5, 6]. Easki and Tsu first proposed a semiconductor symmetric double barrier structure [7] where electronic transport proceeds via resonant tunneling mechanism. This pioneering work makes the road for future research using quantum-confined devices. They showed that a series of energy levels and associated subbands are produced due to the confinement of carriers along one direction of otherwise bulk structures.

Computation for transmission coefficient carried out [8, 9] and later Scandella [10] was without effect of material parameters, which was later realized [11, 12]; who computed resonant tunneling probability in semiconductor double barrier structure for different material parameters. They showed that computation of thermal probability is essential to calculate current from quantum devices. Thermal probability was also computed [13] for thin barrier considering the GaAs/$Al_xGa_{1-x}As$ material composition. Influence of the electron interference effects on the inhomogeneous spatial distribution of the probability current density for the electron waves in semiconductor 2D nanostructures was theoretically investigated [14]. Researchers also

B. Karmakar · R. Lodh · P. Biswas · S. Ghosal
Department of Electronic Science, A.P.C College, Kolkata 700131, India

A. Deyasi (✉)
Department of Electronics and Communication Engineering, RCC Institute of Information Technology, Kolkata 700015, India
e-mail: deyasi_arpan@yahoo.co.in

S. Chattopadhyay et al. (eds.), *Modelling and Simulation in Science, Technology and Engineering Mathematics*, Advances in Intelligent Systems and Computing 749,
https://doi.org/10.1007/978-3-319-74808-5_5

proposed [15] a transition layer model used to calculate resonant tunneling in a double-barrier quantum well system. Modified TBDQW structures are used [16] to design long wavelength semiconductor lasers with low threshold current and small beam divergence.

Recently, photoluminescence spectroscopy of the RTD based THz devices is experimentally measured [17] along with free electron concentration at contact layers. Very high peak current is recently achieved by using novel material as contact [18] in order to improve its candidature to fit into THz range. Triple barrier structure is used for high PVCR [19]; and low cost device is also proposed for high frequency applications [20]. This structure is also able to exhibit microwave generation and detection [21]. In this paper, current density and corresponding peak are calculated for double well resonant tunneling device for different structural parameters, and self-consistency technique is adopted for accurate estimation. Material composition so also modified within type-I limit, and two different sets of dimension are considered for comparative performance estimation related with peak value. Results are significant for application of the device at low bias ranges.

2 Mathematical Modeling

Considering envelope function approximation, electron motion can be written by using time-independent Schrödinger equation

$$-\frac{\hbar^2}{2m^*}\frac{\partial}{\partial z}\left[\frac{1}{m^*(z)}\frac{\partial}{\partial z}\psi(z)\right] + V(z)\psi(z) - q\xi(z)z = E(z)\psi(z) \tag{1}$$

where V(z) is the Hartee-Fock potential represents electrostatic interaction in the quantum device, ζ(z) is the applied field along the direction of wave propagation.

This potential function can be obtained by solving Poisson's equation

$$\frac{d^2V(z)}{dz^2} = \frac{q^2}{\varepsilon_r\varepsilon_0}[N_D(z) - n(z)] \tag{2}$$

where n(z) is confined electron concentration, $N_D(z)$ is the total density of ionized donors.

Thermal equilibrium probability is calculated assuming the physical probable range of wave vector as

$$P = \frac{dk}{2\pi\hbar^2 \ln[1 + \exp(E_F - \hbar^2(n_k - 1)dk + k_{\min})^2]} \tag{3}$$

where dk denotes the range of 'k' values, k is the minimum value of wavevector, E_F is the Fermi energy. Tunneling current density is theoretically defined as the

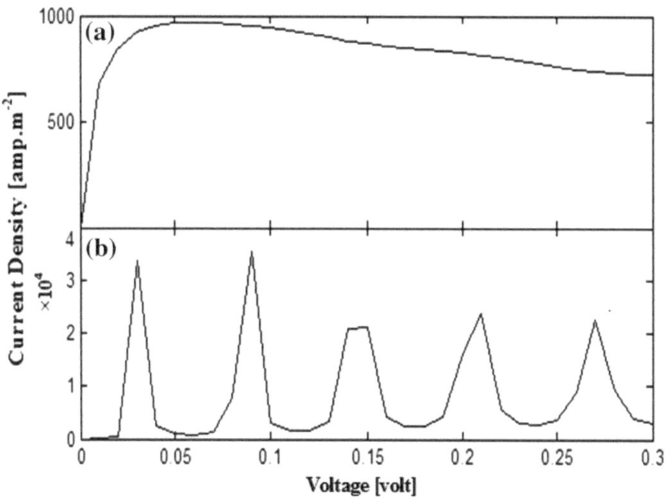

Fig. 1 Current density variation with applied voltage using **a** without self consistency technique; **b** with self consistency technique

probability of finding the electron in a region of space due to the flow of wavevector, either form left or right of the structure. This is defined as

$$J_z = \frac{\hbar}{2m^*}\left[\psi'\frac{\partial\psi}{\partial z} - \psi\frac{\partial\psi'}{\partial z}\right] \tag{4}$$

In practice, it is calculated from the knowledge of Fermi function as

$$J_z = \frac{2q}{h}\int_{U_L}^{\infty}[f(E,\mu_L) - f(E,\mu_R)]T(E)dE \tag{5}$$

3 Results and Discussions

Using Eq. (3), current density is computed for double well RTD using self-consistency method. Figure 1 shows the peak current density variation with applied voltage using (A) without using self-consistency technique and (B) using self-consistency technique. From the plot, it is observed that in case of using without self-consistency technique only one maximum peak is obtained and with further increase in applied voltage peak current density decreases and peak is broadened.

Effect of different material compositions on peak tunneling current density is observed and analyzed from Fig. 2 which shows the current density profile as a function of applied voltage of DBQW structure. From the plot, it is observed that

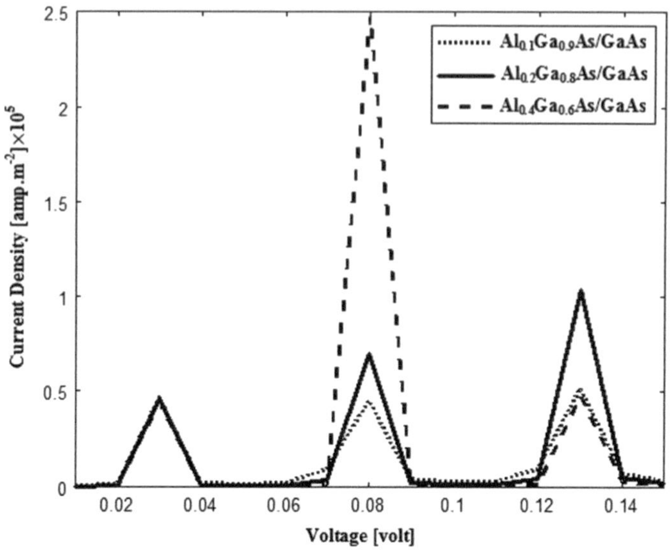

Fig. 2 Current density variation with applied voltage for different material compositions of barrier widths with self consistency technique (for y = 0.1)

maximum peaks are found for Al mole fraction 0.2 and 0.4 at 0.08 V; and for x = 0.1, it is 0.13 V. So, among three set of Al mole fraction maximum tunneling current density is achieved for 0.2 which is 4.416×10^5 amp.m^{-2} at 0.08 V and the current densities are comparatively low for mole fraction 0.4 and 0.1. This is so because when any two Eigenenergy state of DBQW structure matched with each other, the maximum transmission probability occurs i.e. the quantum tunneling phenomenon is happened and maximum value of peak current density is achieved. Then further variation of voltage multiple peaks are achieved but their peak current densities are lower than maximum peak .because the carrier concentration is lowered for such energy state.

Figure 3 shows the current density profile with applied voltage for three different well widths. The peak current density becomes maximum 10.1×10^5 amp.m^{-2} at 0.03 V for well width equals to 8 nm, otherwise it remains less than 10^5 amp.m^{-2} for well widths 4 and 12 nm respectively. This behavior can be well explained following the reason mentioned in the first paragraph of this section. So, among the three well width dimensions, for a particular well width, the transmission probability is maximum; and henceforth, maximum peak current density is achieved. But for other well dimensions, transmission probability is lowered and current density is consequently reduced.

Effect of different middle barrier width on peak tunneling current density is observed and analyzed from Fig. 4. From the result, it is observed that the maximum value of current density is 14.14×10^4 amp.m^{-2} at 0.03 V for barrier width 100 nm and hence peak current density is comparatively lower for 70 nm which is 4.753×10^4

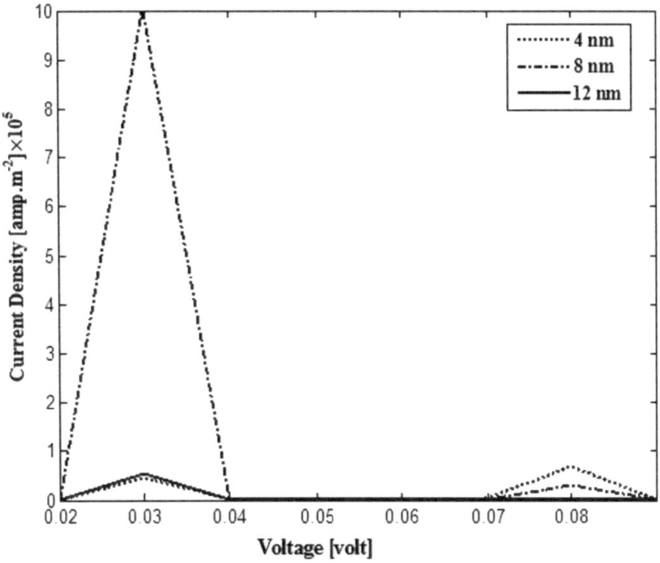

Fig. 3 Current density variation with applied voltage for different well widths with self-consistency technique

amp.m^{-2} and for 40 nm which is 4.091×10^4 amp.m^{-2} both at 0.07 V. For a particular barrier width dimension, the eigenenergy states of DBQW are matched perfectly to each other and maximum transmission probability achieved and the peak current density becomes maximum. But changing the barrier width, transmission probability decreases and corresponding peak current density also decrease.

Effect of different temperature on peak tunneling current density is represented from Fig. 5. It is seen from the figure that at 0.13 V, current density becomes maximum which is 2.414×10^5 amp.m^{-2} for temperature 700 K. Peak current densities for 500 and 300 K are 1.725×10^5 and 1.035×10^5 amp.m^{-2} respectively both at 0.13 V which is comparatively lower than the peak current density achieved at temperature 700 K. Hence peak current density increases with higher temperature.

Figure 6 shows the variation of peak tunneling current density as a function of material composition for two different set of barrier width. For L.B.W = 30 nm, R.B.W = 40 nm peak current density increases with increasing the Al mole fraction up to the limit 0.15 and at 0.17 the peak current density attains maximum value which is 9.77×10^5 amp.m^{-2}. Then further increase of mole fraction peak current density decreases. This is because for this barrier dimensions the quantum encirclement decreases for very high and very low value of Al mole fraction x and hence peak current density also decrease.

For L.B.W = 70 nm, R.B.W = 80 nm, the peak current density remain constant from mole fraction 0.5 to 0.1. After 0.1 the peak current density slowly increase up to 0.3, then increases rapidly. For higher barrier dimension set quantum encirclement

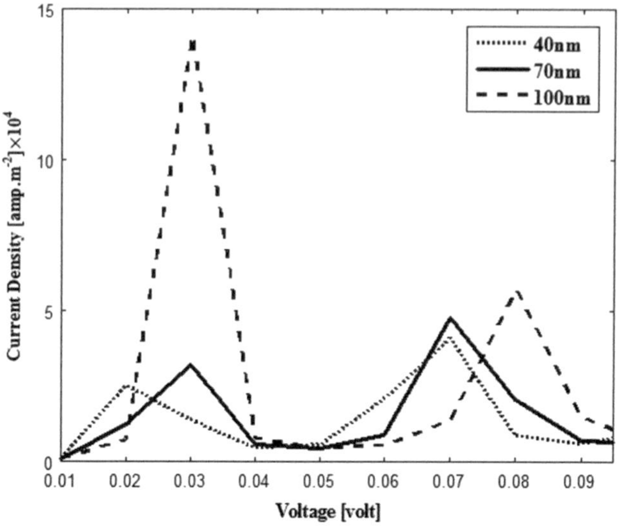

Fig. 4 Current density variation with applied voltage for different middle barrier widths with self-consistency technique

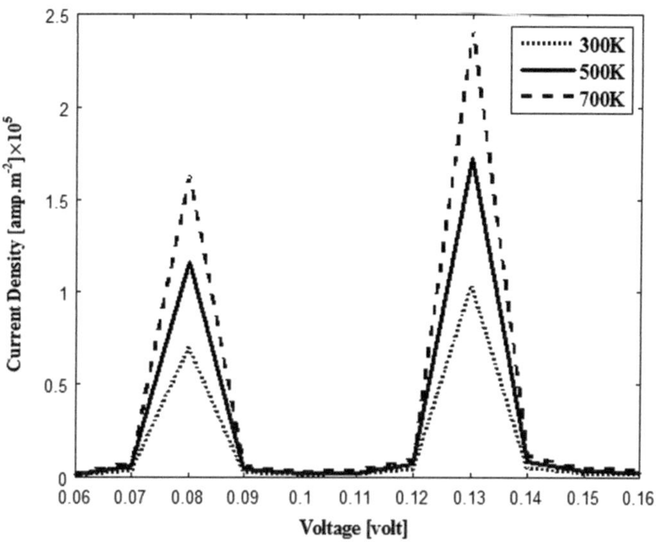

Fig. 5 Current density variation with applied voltage for different temperature with self-consistency technique

increases with increase the Al mole fraction from lower value to higher and the current density also increases.

For L.B.W as 30 nm and R.B.W as 40 nm, peak current density remain constant from well width range 1–9 nm except the range from 4 to 6 nm. In this range peak

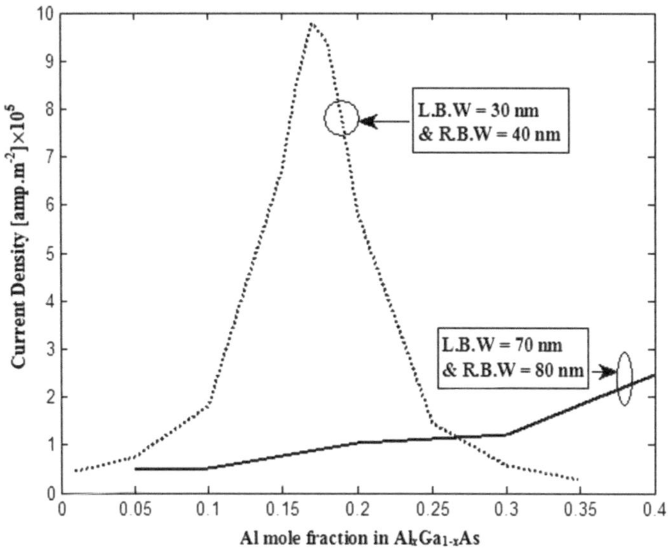

Fig. 6 Variation of peak tunneling current density as a function of material composition for two different set of barrier width

current density first decrease from 4 to 5 nm then it increase from 5 to 6 nm. After 9 nm, peak current density increases rapidly. For this set of barrier width the peak current density is low up to certain limit and nears about show a constant peak current density except the region 4–6 nm and after 6 nm its again show a constant peak current density. In region 4–6 nm there is a dip showing in Fig. 7. In this region first tunneling probability reduced such that the peak current density decreases from 5.5×10^4 amp.m^{-2} to less than 2×10^4 amp.m^{-2} from 4 to 5 nm and minimum 1.714×10^4 amp.m^{-2} at 5 nm then the tunneling probability increases slightly and the peak current density reach 4.681×10^4 amp.m^{-2} at 6 nm. After 9 nm the quantum confinement increases which increase the tunneling probability and the current density increases rapidly.

4 Conclusion

Double well resonant tunneling diode is analytically simulated for different constituent layer widths, and also for different operating temperatures. Peak current densities are obtained at particular bias values, which speak for eigenstates alignment between adjacent quantum wells. Self-consistency technique is incorporated for simulation purpose which provides accurate result regarding the position of the peaks, optimum structural parameters in order to obtain that magnitude, and the junction temperature to obtain measurable current at the applied bias range. It may also

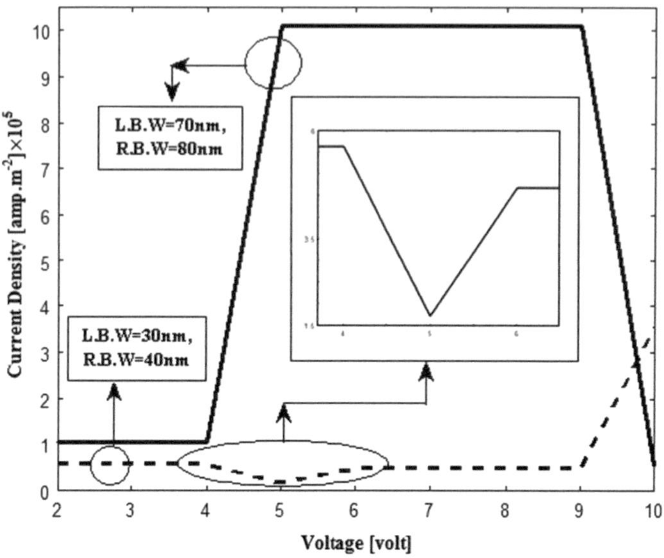

Fig. 7 Variation of peak tunneling current density as a function of well width for the dipping section for L.B.W = 30 nm, R.B.W = 40 nm

be noted that current increases with increase in temperature. Two different dimension set are used for simulation in order to reveal the external influence on electrical properties of the device. Different dimensions of contact regions also help to analyze fluctuations in peak current profile. Thus the device can be operated at those biasing points, where peaks are appeared.

References

1. G. Goldhaber-Gordon, M.S. Montemerlo, J.C. Love, G.J. Opiteck, J.C. Ellenbogen, Overview of nanoelectronic devices. Proc. IEEE **85**, 521–540 (1997)
2. S. Sen, F. Capasso, A.C. Gossard, R.A. Spah, A.I. Hutchinson, S.N.G. Chu, Observation of resonant tunneling through a compositionally graded parabolic quantum well. Appl. Phys. Lett. **51**, 1428–1430 (1987)
3. L. Esaki, L.L. Chang, New transport phenomenon in semiconductor superlattice. Phys. Rev. Lett. **33**, 495–498 (1974)
4. K. Talele, D.S. Patil, Analysis of wavefunction, energy and transmission coefficients in GaN/AlGaN superlattice nanostructures. Prog. Electromagn. Res. **81**, 237–252 (2008)
5. C.E. Simion, C.I. Ciucu, Triple-barrier resonant tunneling: a transfer matrix approach. Rom. Rep. Phys. **59**, 805–817 (2007)
6. A.K. Ghatak, K. Thyagarajan, M.R. Shenoy, A novel numerical technique for solving the one-dimensional Schrödinger equation using matrix approach—application to quantum well structures. IEEE J. Quantum Electron. **24**, 1524–1531 (1988)
7. L. Esaki, R. Tsu, Superlattice and negative differential conductivity in semiconductors. IBM J. Res. Div. **14**, 61–65 (1988)

8. L.A. Chanda, L.F. Eastman, Quantum mechanical reflection at triangular planar-doped' potential barriers for transistors. J. Appl. Phys. **53**, 9165–9169 (1982)
9. D.N. Christodoulides, A.G. Andreou, R.I. Joseph, C.R. Westgate, Analytical calculation of the quantum mechanical transmission coefficient for a triangular, planar-doped potential barrier. Solid State Electron. **28**, 821–822 (1985)
10. L. Scandella, H.J. Güntherodt, Field emission resonances studied with dI/ds(V) and dI/dV(V) curves. Ultramicroscopy **42**, 546–552 (1992)
11. L.L. Chang, L. Esaki, R. Tsu, Resonant tunneling in semiconductor double barriers. Appl. Phys. Lett. **24**, 593–595 (1974)
12. M.A. Reed, R.J. Koestner, M.W. Goodwin, Resonant tunneling through a HeTe/$Hg_{1-x}Cd_xTe$ double barrier, single quantum well structure. J. Vac. Sci. Technol. A **5**, 3147–3149 (1986)
13. R. Wessel, M. Alterelli, Quasi stationary energy level calculation for thin double barrier GaAs-$Ga_{1-x}Al_xAs$ Heterostructures. Phys. Rev. B **39**, 10246–10250 (1989)
14. V.A. Petrov, A.V. Nikitin, in *Penetration of quantum mechanical current density under semi-infinite rectangular potential barrier as the consequence of the interference of the electron waves in semiconductor 2D nanostructures*. Proceedings of SPIE, vol. 7521 (2009)
15. Y. Song, A transition layer model and its application to resonant tunneling in heterostructures. Phys. Lett. A **216**, 183–186 (1996)
16. A. Al-Muhanna, A. Alharbi, A. Salhi, Waveguide design optimization for long wavelength semiconductor lasers with low threshold current and small beam divergence. J. Mod. Phys. **2**, 225–230 (2011)
17. K.J.P. Jacobs, B.J. Stevens, R.A. Hogg, Photoluminescence characterization of high current density resonant tunneling diodes for terahertz applications. IEICE Trans. Electron. **E99.C**(2), 181–188 (2016)
18. Z. Li, H. Tang, H. Liu, Y. Liang, Q. Li, N. An, J. Zeng, W. Wang, Y.Z. Xiong, Improving the peak current density of resonant tunneling diode based on InP substrate. J. Semicond. **38**(6), 064005 (2017)
19. C.C. Yang, Y.K. Su, T.C. Chang, Optimum current-voltage characteristics of GaAs/AlAs intra-band microwave devices. IET Micro. Nano. Lett. **10**(9), 472–475 (2015)
20. K.J.P. Jacobs, B.J. Stevens, O. Wada, T. Mukai, D. Ohnishi, R.A. Hogg, A dual-pass high current density resonant tunneling diode for terahertz wave applications. IEEE Electron. Device Lett. **36**(12), 1295–1298 (2015)
21. G. Keller, A. Tchegho, B. Munstermann, W. Prost, F.J. Tegude, M. Suhara, in *Triple barrier resonant tunneling diodes for microwave signal generation and detection*. European Microwave Integrated Circuits Conference (EuMIC), 2013

Computation of Electrical Parameters for Single-Gate High-K Nanoscale MOSFET with Cylindrical Geometry

Suporna Bhowmick, Debarati Chakraborty and Arpan Deyasi

1 Introduction

Research on nanoscale MOSFET has been initiated a decade ago due to the shrinking gate size [1] with the increasing demand of incorporating more no. of transistors inside the reduced floor area; and this technological improvement is associated with the additional generated complexity in terms of short channel effect [2]. As gate length goes beyond 100 nm, quantum wire is formed in the otherwise bulk channel; and owing to different geometries of the channel, solution of Schrödinger's equation with various boundary conditions becomes more difficult to solve for calculating electrical parameters. This computational problem is solved by adopting Green's function formalism [3], and dissipative effects is considered at both source and drain ends under ballistic limit [4] for near accurate performance estimation. This technique helps to cop up with the ITRS roadmap [1] from theoretical stand-point as predicted in 2007.

Measurement of tunneling current in nano-dimensional MOSFET is the subject of interest [5, 6] as it governs the performance of the device when applied bias is significantly low. The reduction of subthreshold current [7] in short-channel MOSFET is one of the major tasks as depicted in the last decade, and thus gate control plays a major part in this context. This leads to a series of novel proposals as double-gate MOSFET [8, 9], triple-gate MOSFET [10], GAA MOSFET [11] etc. Also high-K dielectrics provide another much-needed breakthrough in context of reduction of subthreshold current [12]. But the interesting fact is that most of the theoretical results available in literatures related with reduction of short-channel effect deals with rectangular structure, which is ideal, and very difficult to reproduce experimentally. In the present paper, electrical parameters in the single-gate nano-dimensional

---

S. Bhowmick · D. Chakraborty · A. Deyasi (✉)
Department of Electronics and Communication Engineering, RCC Institute of Information Technology, Kolkata 700015, India
e-mail: deyasi_arpan@yahoo.co.in

S. Chattopadhyay et al. (eds.), *Modelling and Simulation in Science, Technology and Engineering Mathematics*, Advances in Intelligent Systems and Computing 749,
https://doi.org/10.1007/978-3-319-74808-5_6

MOSFET is calculated using green's function technique; where cylindrical channel geometry is considered. Though a few reports are already available for Cartesian coordinate system, but rarely manuscripts are written considering cylindrical system. Results are calculated considering high-K dielectric, and are compared with that obtained for low-k material.

2 Mathematical Modeling

For computation of drain current, first self-consistent solution of Schrödinger and Poisson equations is to be calculated. Considering the geometry of the structure, Schrödinger equation is given in the following form

$$\begin{bmatrix} -\frac{\hbar^2}{2m_r^*(\theta,z)}\left[\frac{\partial^2}{\partial r^2}+\frac{1}{r}\frac{\partial}{\partial r}\right] \\ -\frac{\hbar^2}{2}\left(\frac{1}{r^2}\frac{\partial}{\partial\theta}\left[\frac{1}{m_\theta^*(r,z)}\right]\frac{\partial}{\partial\theta}\right)- \\ \frac{\hbar^2}{2}\frac{\partial}{\partial z}\left(\frac{1}{m_z^*(r,\theta)}\frac{\partial}{\partial z}\right)+V(x,y,z) \end{bmatrix}\psi(x,y,z)=E\psi(x,y,z) \quad (1)$$

Potential can be computed from Poisson's equation if electron density in the channel is known

$$\nabla^2\psi(r,\theta,z) = -n_{3D,m}/\varepsilon \quad (2)$$

where $\psi(z)$ is the potential that should be determined using self-consistency technique. Self energies at source and drain ends are given by [3].

$$\sum_{S/D} E = \sum_{S/D}[p,q] = -\left(\frac{\hbar^2}{2a^2}\right)a_{mn}(r)|_{r=0}\exp(jk_{m,1}a)\delta_{p,(m-1)R+1}\delta_{q,(m-1)R+1} \quad (3)$$

Retarded Green's function is given by [3]

$$G(E) = [EI - H - \sum_S(E) - \sum_D(E)]^{-1} \quad (4)$$

Finally, drain current is obtained in the form

$$I_{DS} = G_0\frac{k_BT}{q}\sum_i g_i \, \mathrm{In}\frac{1+\exp[(\mu_S - E_{i0})/k_BT]}{1+\exp[(\mu_D - E_{i0})/k_BT]} \quad (5)$$

where E_{i0} represents the minimum energy of i_{th} subband, g_i is the spin degeneracy. After calculating drain current, quantum capacitance and subthreshold swing are obtained as

$$C_Q = q\left(\frac{d|Q|}{dV_G}\right)/\left[q - \left(\frac{d(\mu - \mu_0)}{dV_G}\right)\right] \tag{6}$$

$$S = \left[\frac{1}{k_B T \,\mathrm{In}\,(10)} \frac{d(\mu_s - \mu_0)}{dV_G}\right]^{-1} \tag{7}$$

DIBL is measured by

$$D = -\frac{V_{th}^{DD} - V_{th}^{low}}{V^{DD} - V_D^{low}} \tag{8}$$

3 Results and Discussions

Using Eq. (5), drain current is first calculated as function of both drain voltage and gate voltage. The result is obtained for high-K dielectric, and simultaneously the performance enhancement is measured by comparing with that obtained for conventional low-K dielectric material. Figure 1 shows the drain current variation as a function of drain voltage for two different V_{GS}. The lowest magnitude of V_{GS} is considered as 0.5 V due to the fact that less than the value, the difference of result due to various dielectrics becomes insignificant. It may be observed that replacing SiO_2 by HfO_2, drain current is significantly increased. The saturation value for low-K dielectric is very close to subthreshold region, and thus lowering gate bias may cause a serious problem when applied for digital circuits. The present result shows how significantly the performance is improved for high-K dielectric. Similar distinguishable difference is also observed when transfer characteristics are plotted, as depicted in Fig. 2.

Figure 3 shows the variation of channel transconductance for different channel diameter. It is seen form the plot that with increasing channel dimension, transconductance decreases. With higher dimension, the variation becomes almost linear. Figure 4 shows the variation of quantum capacitance. It is observed that capacitance increases rapidly at lower gate bias, but becomes almost saturated when gate voltage reaches close to 1 V.

Sub threshold swing and DIBL are plotted in Figs. 5 and 6 respectively. For circuit application, it is always desirable to reduce the sub threshold swing, which is achieved by increasing the dielectric constant of the insulating material surrounded the channel. In Fig. 5, it is seen that at larger channel thickness, SS remains very low when HfO_2 is used instead of SiO_2, where the rate of increment with channel dimension is very large. This measurement is very difficult for rectangular channel, as simultaneous tuning of two different confinements leads various results, which is one major disadvantage form fabrication stand-point. The similar nature is also observed in DIBL plot.

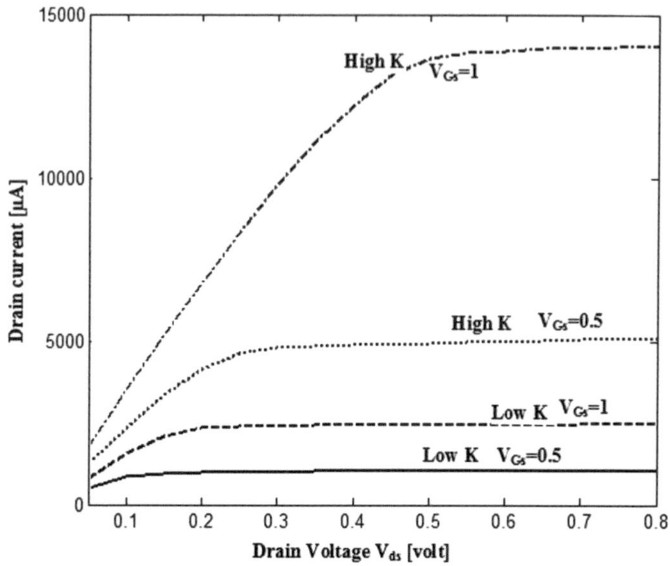

Fig. 1 Static characteristics for two different gate biases with both high-K and low-K dielectrics

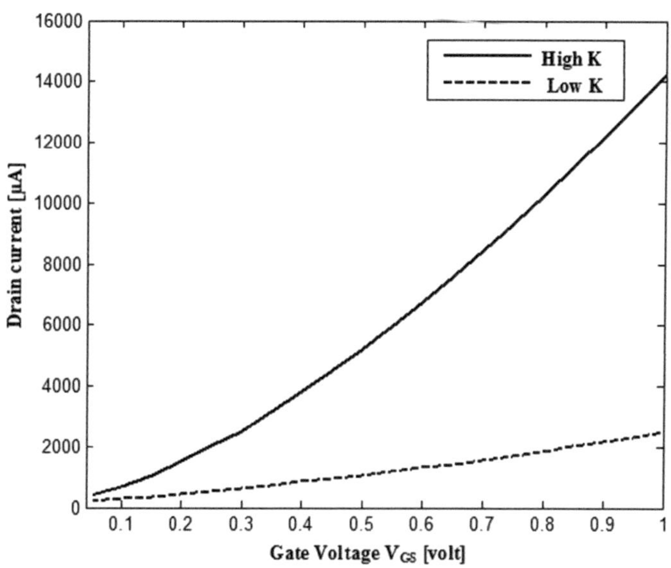

Fig. 2 Transfer characteristics with both high-K and low-K dielectrics

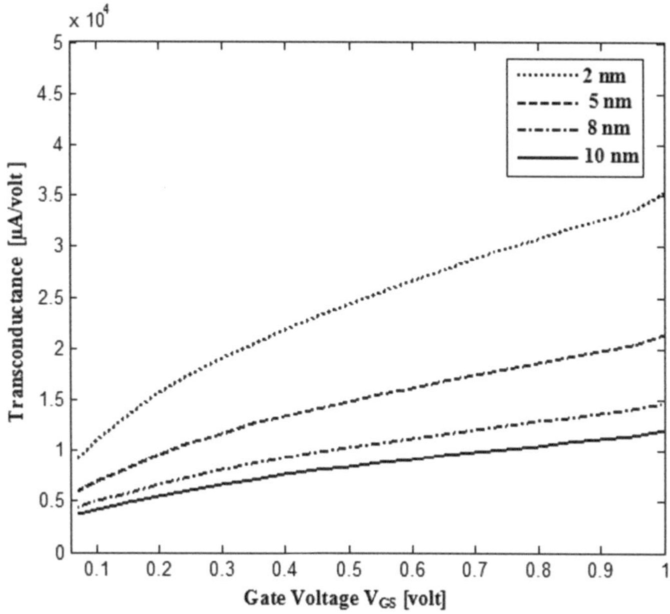

Fig. 3 Transconductance with gate voltage for different channel diameter

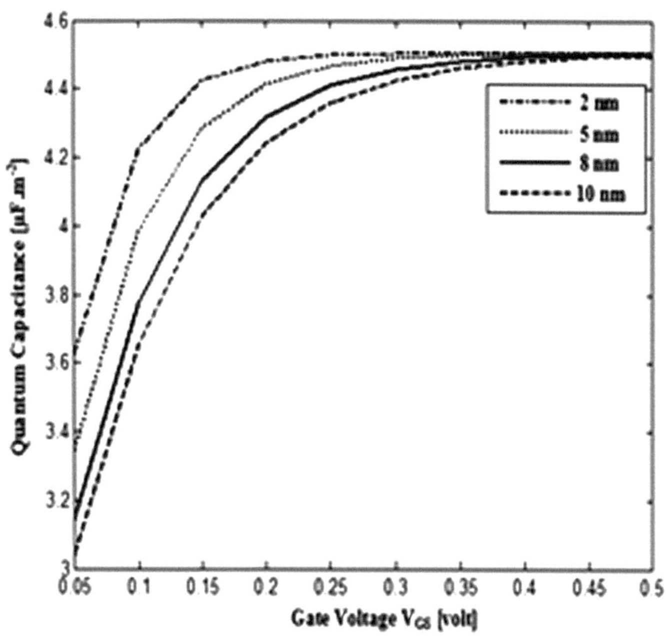

Fig. 4 Quantum capacitance with gate voltage for different channel diameter

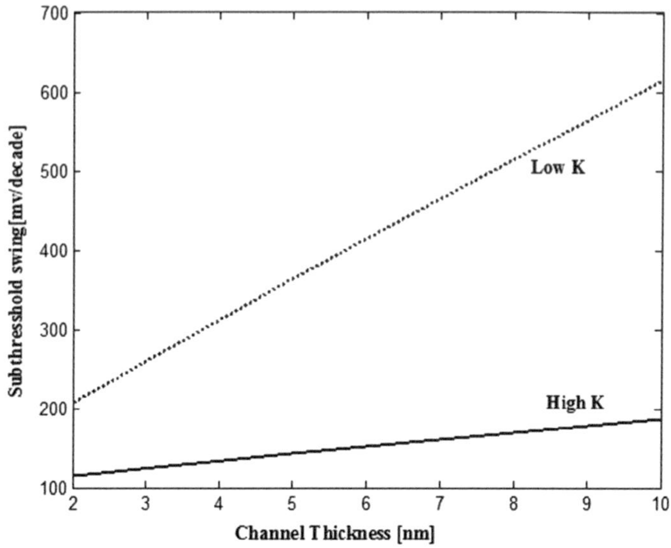

Fig. 5 Sub threshold swing with channel thickness for different dielectrics

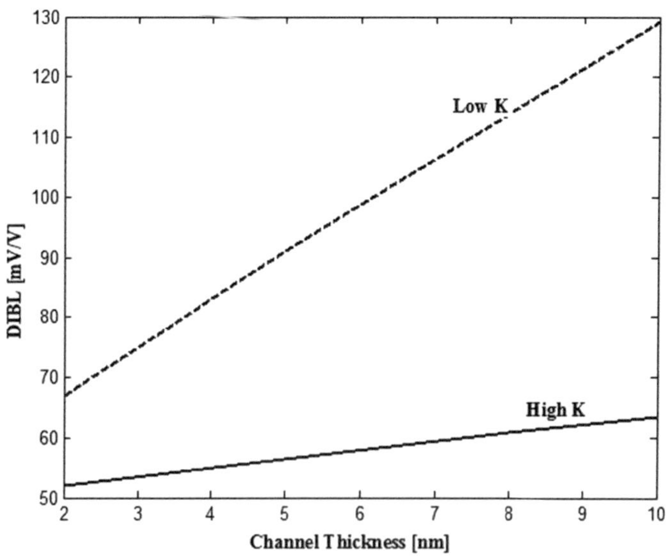

Fig. 6 DIBL with channel thickness for different dielectrics

4 Conclusion

Electrical performance parameters of cylindrical single-gate MOSFET is analytically calculated in presence of low and high-K dielectrics. Result speaks in favor of higher dielectric constant of the insulating material which substantially reduces subthreshold swing and DIBL. Drain current is also considerably increased compared to subthreshold leakage current level even at very low gate bias. Appropriate tailoring of channel diameter effectively controls the device transconductance and quantum capacitance, which are essentially useful for practical implementation of the device in analog and digital circuits respectively.

References

1. http://www.itrs2.net/
2. Y. Taur, T.H. Ning, *Fundamentals of modern VLSI devices* (Cambridge University Press, Cambridge, U.K., 1998)
3. A. Rahman, J. Guo, S. Datta, M. Lundstrom, Theory of ballistic nanotransistors. IEEE Trans. Electron Devices **50**, 1853–1864 (2003)
4. R. Venugopal, Z. Ren, S. Datta, M. Lundstrom, Simulating quantum transport in nanoscale transistors: real versus mode-space approaches. J. Appl. Phys. **92**, 3730–3739 (2002)
5. M. Bella, S. Latreche, Analyze of DGMOS tunneling current through nanoscale gate oxide. Nanosci. Nanotechnol. **6**(1A), 117–121 (2016)
6. M. Chanda, S. De, C.K. Sarkar, Modeling of parameters for nano-scale surrounding-gate MOSFET considering quantum mechanical effect. Int. J. Numer. Model. Electron. Netw. Dev. Fields Spec. Issue Model. High Freq. Silicon Transistors **27**, 883–895 (2014)
7. Y. Swami, S. Rai, Modeling and analysis of sub-surface leakage current in nano-MOSFET under cutoff regime. Superlattices Microstruct. **102**, 259–272 (2017)
8. F. Djeffal, Z. Dibi, M.L. Hafiane, D. Arar, Design and simulation of a nanoelectronic DG MOSFET current source using artificial neural networks. Mater. Sci. Eng., C **27**(5–8), 1111–1116 (2007)
9. B. Baral, A.K. Das, D. De, A. Sarkar, An analytical model of triple-material double-gate metal-oxide-semiconductor field-effect transistor to suppress short-channel effects. Int. J. Numer. Model. Electron. Netw. Dev. Fields **29**(1), 47–62 (2016)
10. P.R. Kumar, S. Mahapatra, Analytical modeling of quantum threshold voltage for triple gate MOSFET. Solid State Electron. **54**(12), 1586–1591 (2010)
11. B. Jena, B.S. Ramkrishna, S. Dash, G.P. Mishra, Conical surrounding gate MOSFET: a possibility in gate-all-around family. Adv. Nat. Sci. Nanosci. Nanotechnol. **7**, 015009 (2016)
12. N.B. Atan, I.B. Ahmad, B.B.Y. Majlis, in *Effects of high-K dielectrics with metal gate for electrical characteristics of 18 nm NMOS device*, IEEE International Conference on Semiconductor Electronics (2014)

Part II
Power System

Fault Diagnosis in Isolated Renewable Energy Conversion System Using Skewness and Kurtosis Assessment

Debopoma Kar Ray, Surajit Chattopadhyay and Samarjit Sengupta

1 Introduction

Renewable energy technology is of great concern in recent days due to the ever-increasing use of fossil fuels and the risk persisting with the rapid depletion of the conventional resources [1] . However, the present trend of developments of non-conventional sources indicates that these will serve as supplements for conventional sources for the coming days. Due to this, it has been of great concern for identifying the various non-linearity in the renewable energy systems and for the condition monitoring of the grid connected and standalone renewable energy networks. Throughout the world, wind energy has become a principle energy source in the world's energy market in more than 70 countries across the universe. A current-source inverter-based standalone WECS (Wind Energy Conversion System) [2] nullifies dump load to avoid surplus power generation. A wind farm associated hybrid energy storage system (HESS) smooth out ripples for reducing impact on the grids [3], wherein a cutoff frequency method optimizes the system, to find the rated power and capacity of HESS. A condition monitoring technique has been seen for an early fault detection to prevent sudden breakdown [4]. Grid-interconnected wind energy system installed in Jordan [5] analysis depicts more percentage error in the estimation of the cost as well as energy extracted per year. A Discrete Wavelet Transform (DWT) based algorithm

D. K. Ray (✉)
EE Department, Faculty, MCKV Institute of Engineering, Howrah, India
e-mail: debopoma86@gmail.com

S. Chattopadhyay
EE Department, Faculty, Ghani Khan Choudhury Institute of Engineering and Technology, Malda, India
e-mail: surajitchattopadhyay@gmail.com

S. Sengupta
Applied Physics Department, Ex-Faculty, University of Calcutta, Kolkata, India
e-mail: samarsgp@rediffmail.com

S. Chattopadhyay et al. (eds.), *Modelling and Simulation in Science, Technology and Engineering Mathematics*, Advances in Intelligent Systems and Computing 749,
https://doi.org/10.1007/978-3-319-74808-5_7

minimizes wind power forecasting errors in WECS [6]. A two-stage optimal power flow iterative algorithm is able to calculate the minimum storage size of a plant during congestion [7]. Baseline principal component analysis (PCA) model can be used for online fault detection in wind turbine [8]. A Fault detection estimator can be used for fault detection at specified location [9]. Various condition monitoring techniques has been seen for increasing the accuracy of wind turbine operation [10]. A SCADA based clustering algorithm and principal components analysis is effective for wind turbine gearbox failure [11]. Various 3 phase induction motor fault diagnosis can be done using Skewness and Kurtosis analysis of the current signatures at different faults occurring in the system [12]. None of studies reviewed so far, deals with the unsymmetrical fault identification in load and source sides of either stand alone or grid interconnected wind energy conversion system, monitoring the Skewness and Kurtosis of the current signatures of the network at normal and in presence of double line (LL), single line to ground (LG) and double line to ground (LLG) faults in the system.

In this paper an attempt has been made for determining the various unsymmetrical faults occurring in a stand-alone wind energy conversion system at generator and load buses of the network, monitoring the Skewness and Kurtosis of the discrete wavelet transform decomposition levels of the bus current signature at normal and in presence of LL, LG, LLG faults in the network. The generator and load bus currents have been acquired at normal and in presence of LL, LG, LLG faults in source and load sides of the network, considered one at a time. These currents have been assessed using Multi-Resolution Analysis of Discrete Wavelet Transform (MRA of DWT). The wavelet decomposition levels obtained from this analysis were analyzed using statistical Skewness and Kurtosis value monitoring technique. Monitoring the Skewness and Kurtosis value of DWT level coefficients, changes obtained at fault from normal has been recorded and corresponding features have been extracted for exact identification of the various faults occurring in the system.

2 Wind Energy Conversion System Under Analysis

A stand alone wind energy conversion system (WECS) has been modeled and used for the analysis purpose. Figure 1 depicts the block diagram of the system and Table 1 shows the network ratings.

In the above Table 1 p.u. voltage = 440 V and 1 p.u. power = 300 kVA.

3 Theoretical Backgrounds

In this analysis MRA of DWT statistical monitoring has been done. A discrete wavelet transform is given by the expression [13]:

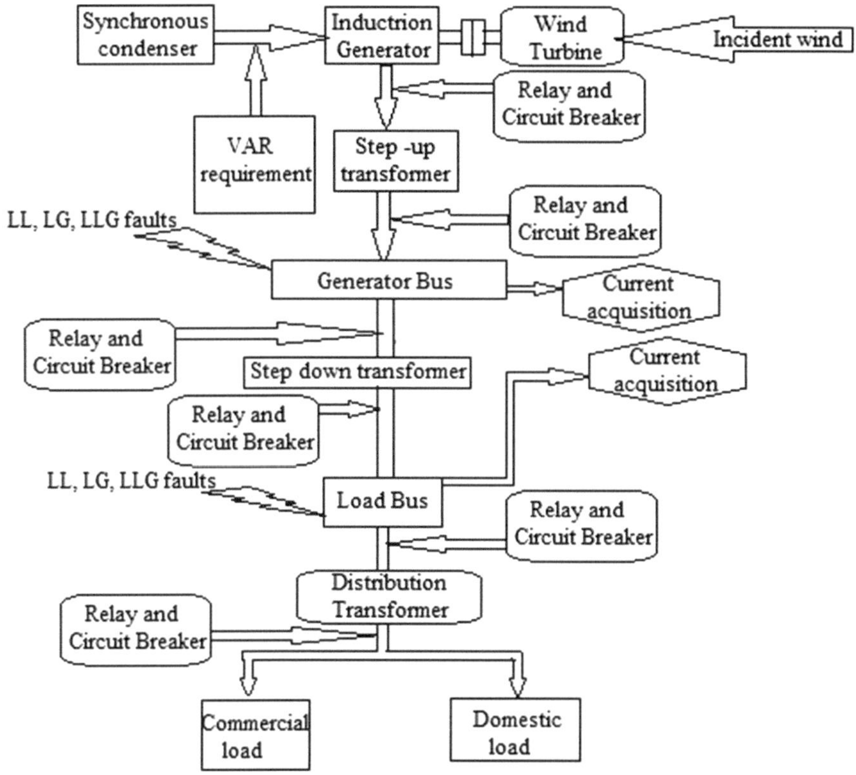

Fig. 1 Block diagram of stand-alone WECS

$$DWT(m,n) = \int_{-\infty}^{\infty} \frac{1}{\sqrt{a_0^m}} f(t)\, g(a_0^{-m} t - n b_0)\, d(t) \quad (1)$$

The wavelet representation is discrete in DWT is discrete and represent the correlation between the original signal and wavelets for different combinations m and n. The digital signal to be analyzed is then decomposed into successive scales. After decomposing the signal into successive levels, the approximate and detailed coefficients obtained has been analyzed and Skewness and Kurtosis values have been calculated.

Skewness [13] defines, how much a distribution is symmetrical/asymmetrical over a sample mean and positive skewness refers to the spreading of data to the right of the mean.

Kurtosis [13] defines how much a distribution is having extension towards right or left of a sample mean.

Table 1 System specifications

Equipments	Specifications
Wind turbine	4 blades, shaft speed-10 m/s
Asynchronous generator	300 kVA, 440 V, 50 Hz, stator resistance and inductance (p.u.): 0.016, 0.06, rotor resistance and inductance (p.u.): 0.015, 0.06, mutual inductance (p.u.): 3.5
Synchronous condenser	300 kVA, 440 V, 50 Hz, stator resistance: 0.017 Ω, reactances (p.u.): Xd = 3.23, Xd′-0.21, Xd″ = 0.15, Xd‴ = 2.79, Xq″ = 0.37, Xl = 0.09
Var compensator	440 V, 50 Hz, 75 KVar
Step up transformer	440 V/33 kV
Step down transformer	33 kV/11 kV
Distribution transformer	11 kV/440 V
Relay	Operating time = 10 ms, % voltage sag = 63% of supply voltage
Circuit breaker	Breaker resistance = 0.001 Ω, transition time = 0.2 s
Load	25, 10 kW

4 Determination of MRA of DWT Coefficients at Normal and Fault

The current signatures of the generator and load buses of the network at normal and at fault have been acquired and assessed using MRA of DWT. The wavelet decomposition and the approximate and detailed coefficients for each decomposition level have been presented in Figs. 2, 3, 4, 5, 6, 7, 8, 9, 10, 11, 12, 13, 14 and 15. In this analysis Daubechies 20 mother wavelet has been used.

The Figs. 2, 3, 4, 5, 6, 7, 8, 9, 10, 11, 12, 13, 14 and 15 have been observed and it has been inferred that, the features for each case study are distinctively different. But specific identification cannot be done from this monitoring technique. Thus for more specific analysis, the various wavelet decomposition levels have been assessed calculating the statistical Skewness and Kurtosis value, which has been provided in succeeding section.

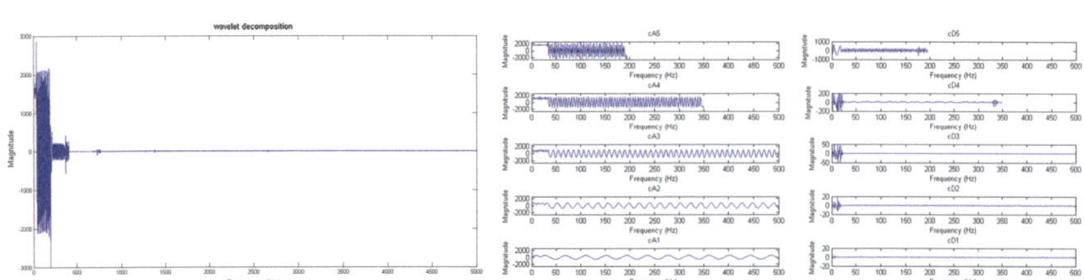

Fig. 2 Source current wavelet decomposition and approximate and detailed coefficients at normal condition

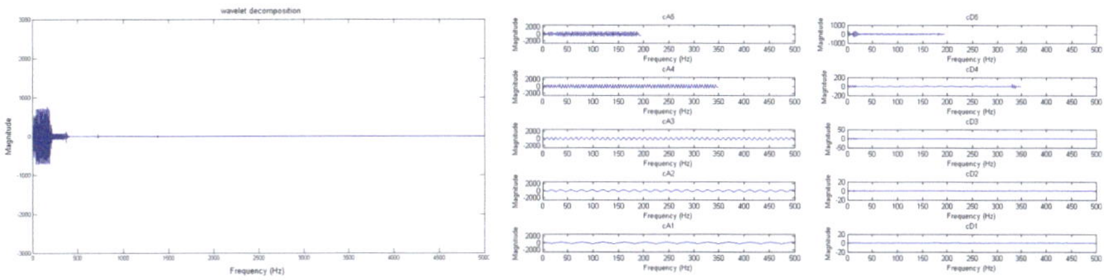

Fig. 3 Load current wavelet decomposition and approximate and detailed coefficients at normal condition

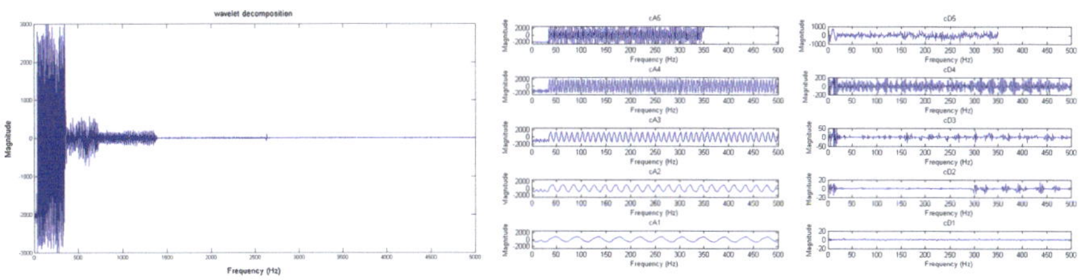

Fig. 4 Source current wavelet decomposition and approximate and detailed coefficients at LG fault in generator bus

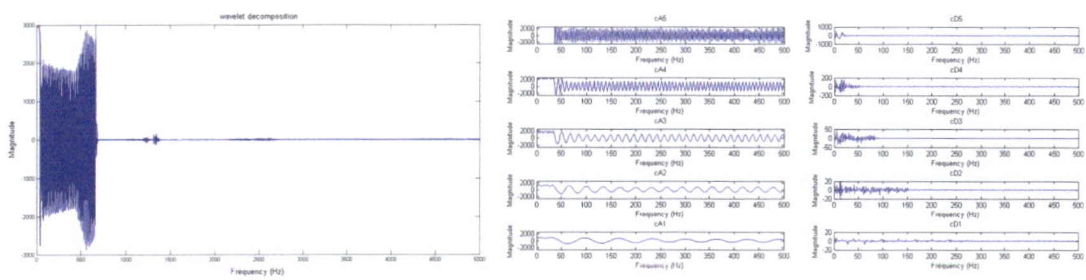

Fig. 5 Source current wavelet decomposition and approximate and detailed coefficients at LL fault in generator bus

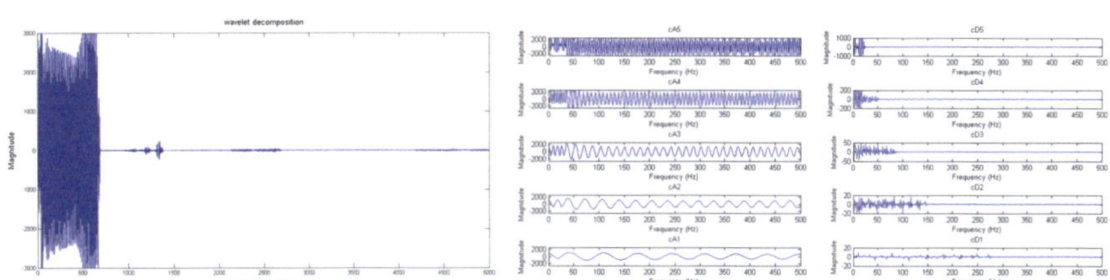

Fig. 6 Source current wavelet decomposition and approximate and detailed coefficients at LLG fault in generator bus

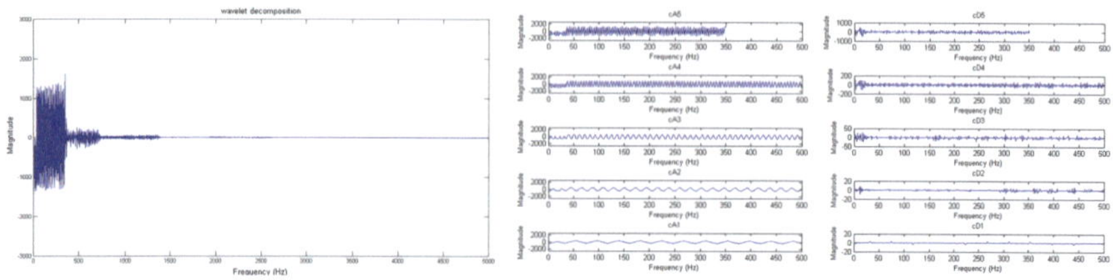

Fig. 7 Load current wavelet decomposition and approximate and detailed coefficients at LG fault in generator bus

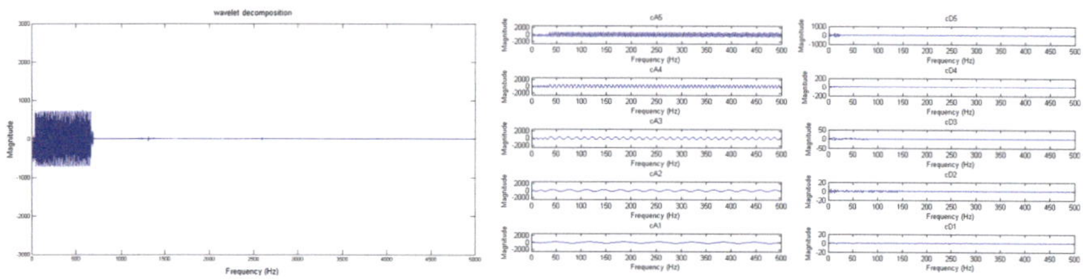

Fig. 8 Load current wavelet decomposition and approximate and detailed coefficients at LL fault in generator bus

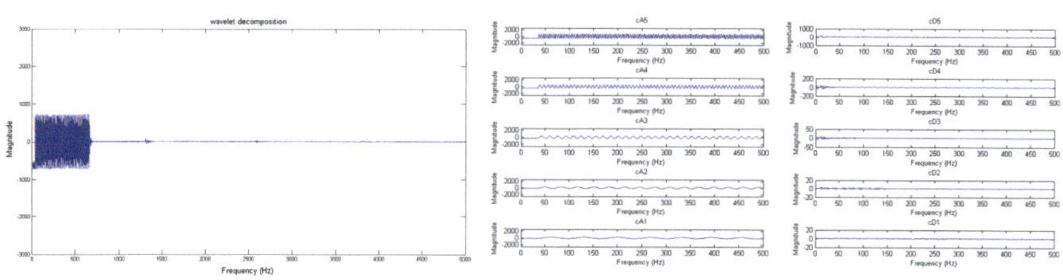

Fig. 9 Load current wavelet decomposition and approximate and detailed coefficients at LLG fault in generator bus

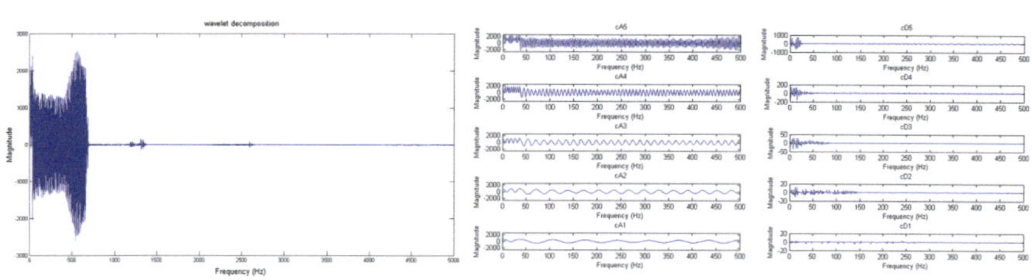

Fig. 10 Source current wavelet decomposition and approximate and detailed coefficients at LG fault in load bus

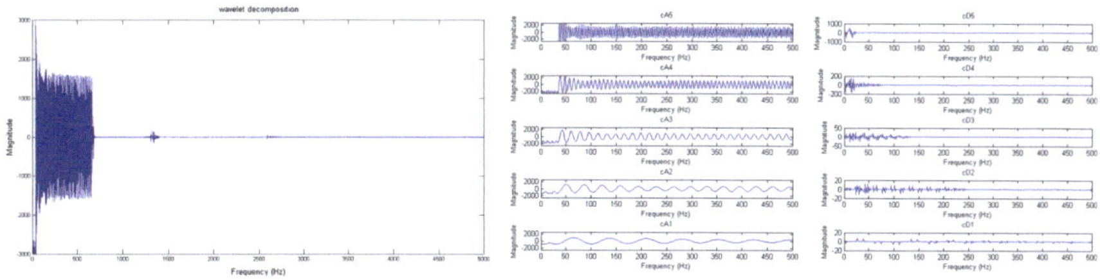

Fig. 11 Source current wavelet decomposition and approximate and detailed coefficients at LL fault in load bus

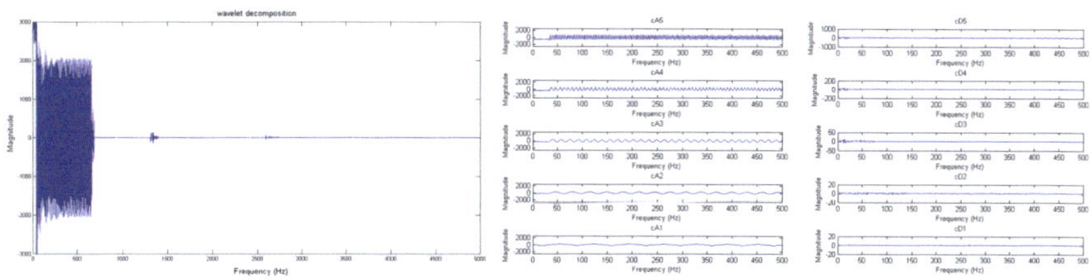

Fig. 12 Source current wavelet decomposition and approximate and detailed coefficients at LLG fault in load bus

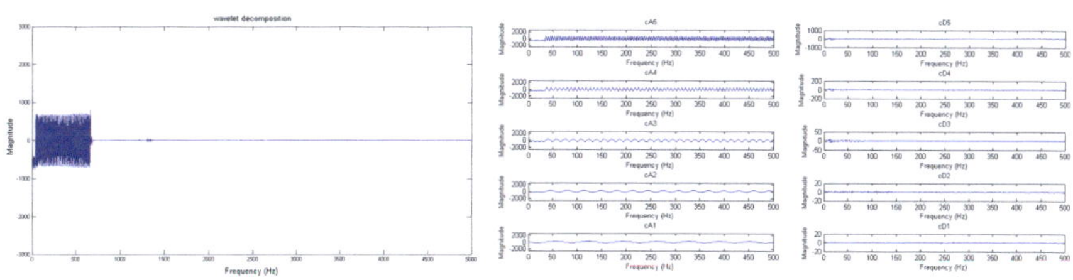

Fig. 13 Load current wavelet decomposition and approximate and detailed coefficients at LG fault in load bus

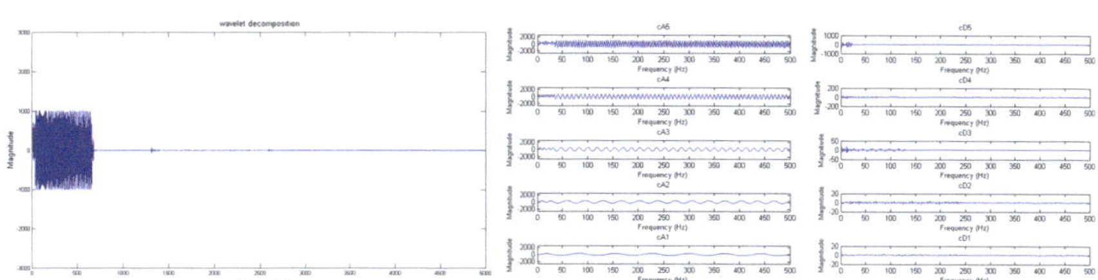

Fig. 14 Load current wavelet decomposition and approximate and detailed coefficients at LL fault in load bus

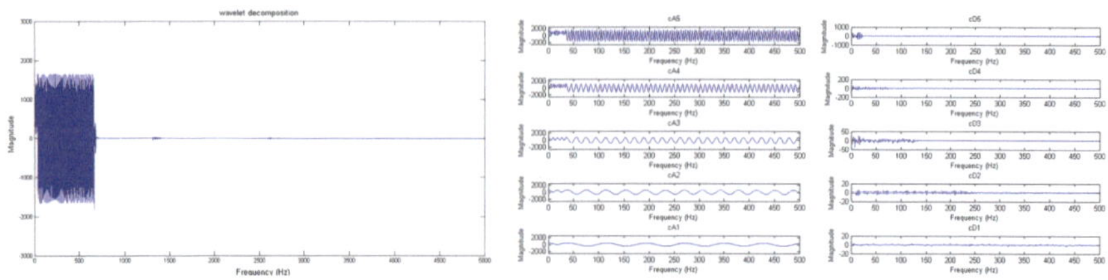

Fig. 15 Load current wavelet decomposition and approximate and detailed coefficients at LLG fault in load bus

5 Determination of Skewness and Kurtosis Values at Normal and Fault

The calculated Skewness and Kurtosis values for normal and at fault in generator and load buses of the network were presented in matrix form in Tables 2, 3, 4 and 5. In each of the matrices, the Skewness and Kurtosis coefficients can be demonstrated as:

$$(SK)_{MxN} = \begin{pmatrix} Sa1 & Sd1 & Ka1 & Kd1 \\ Sa2 & Sd2 & Ka2 & Kd2 \\ Sa3 & Sd3 & Ka3 & Kd3 \\ Sa4 & Sd4 & Ka4 & Kd4 \\ Sa5 & Sd5 & Ka5 & Kd5 \end{pmatrix}$$

$(SK)_{MxN}$ denote the Skewness and Kurtosis matrix for M rows and N columns

Table 2 Skewness and Kurtosis matrix for source side and load side current DWT decomposition levels at healthy condition

Case study	$(SK)_{MxN}$
For source side current	$\begin{pmatrix} -0.01339 & -0.17514 & 1.495922 & 99.79288 \\ -0.03295 & -1.32368 & 1.492209 & 56.12955 \\ -0.06995 & 0.242319 & 1.489988 & 35.51113 \\ -0.14407 & 0.171094 & 1.486702 & 24.42911 \\ -0.31513 & -0.10621 & 1.580086 & 3.469751 \end{pmatrix}$
For load side current	$\begin{pmatrix} -0.00145 & -0.6324 & 1.502547 & 103.3608 \\ 0.001528 & 2.518314 & 1.504211 & 135.469 \\ 0.007446 & 0.716615 & 1.506871 & 99.88604 \\ 0.020698 & 1.285346 & 1.50683 & 35.7106 \\ 0.04426 & -0.07544 & 1.543202 & 6.345082 \end{pmatrix}$

Table 3 Skewness and Kurtosis matrix for source side and load side current DWT decomposition levels at LG fault in generator and load buses, considered one at a time

Case study	$(SK)_{MxN}$ *generator bus fault*	$(SK)_{MxN}$ *load bus fault*
For source side current	$\begin{pmatrix} 0.004051 & -0.31478 & 1.498862 & 15.29745 \\ 0.013732 & 0.887511 & 1.502256 & 35.58154 \\ 0.032777 & 0.445972 & 1.546337 & 20.73361 \\ 0.072961 & -0.02592 & 1.54543 & 2.373368 \\ 0.119334 & -0.00456 & 1.529321 & 3.051837 \end{pmatrix}$	$\begin{pmatrix} -0.00221 & 0.226096 & 1.973001 & 16.09459 \\ -0.005 & 0.276028 & 1.972289 & 11.87682 \\ -0.01059 & 0.170794 & 1.9694 & 16.54555 \\ -0.01734 & 0.60522 & 1.960292 & 40.44648 \\ -0.05017 & -0.07076 & 1.972842 & 44.31766 \end{pmatrix}$
For load side current	$\begin{pmatrix} 0.019569 & -0.05502 & 1.587121 & 9.0657 \\ 0.025333 & -0.0356 & 1.588903 & 6.804406 \\ 0.036437 & 0.104321 & 1.602708 & 5.568114 \\ 0.044845 & -0.03039 & 1.50174 & 2.908291 \\ 0.112922 & -0.01862 & 1.589668 & 5.139542 \end{pmatrix}$	$\begin{pmatrix} 0.002851 & -0.01353 & 1.496464 & 15.34837 \\ 0.00752 & -0.07094 & 1.495122 & 6.461887 \\ 0.016793 & -0.09026 & 1.49299 & 9.549587 \\ 0.035813 & -0.05297 & 1.496044 & 30.60578 \\ 0.074567 & -0.29069 & 1.494198 & 43.0236 \end{pmatrix}$

Table 4 Skewness and Kurtosis matrix for source side and load side current DWT decomposition levels at LL fault in generator and load buses, considered one at a time

Case study	$(SK)_{MxN}$ *generator bus fault*	$(SK)_{MxN}$ *load bus fault*
For source side current	$\begin{pmatrix} 0.00891 & -0.17975 & 1.731736 & 17.57612 \\ 0.019178 & 0.016043 & 1.754151 & 10.52569 \\ 0.03673 & 0.170514 & 1.791537 & 8.758571 \\ 0.061719 & 0.625364 & 1.840528 & 42.78958 \\ 0.083803 & -0.55682 & 1.862598 & 54.79014 \end{pmatrix}$	$\begin{pmatrix} 0.00891 & -0.17975 & 1.731736 & 17.57612 \\ 0.019178 & 0.016043 & 1.754151 & 10.52569 \\ 0.03673 & 0.170514 & 1.791537 & 8.758571 \\ 0.061719 & 0.625364 & 1.840528 & 42.78958 \\ 0.083803 & -0.55682 & 1.862598 & 54.79014 \end{pmatrix}$
For load side current	$\begin{pmatrix} -0.03514 & -0.13522 & 1.847466 & 225.1158 \\ -0.07652 & -0.73658 & 1.955504 & 121.8549 \\ -0.14812 & -0.42996 & 2.131499 & 76.75545 \\ -0.25532 & -0.51682 & 2.369899 & 72.38661 \\ -0.35902 & 0.278571 & 2.496105 & 70.70538 \end{pmatrix}$	$\begin{pmatrix} 0.001318 & -0.28492 & 1.501157 & 119.2922 \\ 0.000043 & -2.78988 & 1.502868 & 199.5778 \\ -0.00244 & 0.243855 & 1.506193 & 95.43031 \\ -0.0078 & -1.20755 & 1.512086 & 46.56893 \\ -0.02874 & 0.141903 & 1.540432 & 40.20908 \end{pmatrix}$

Table 5 Skewness and Kurtosis matrix for source side and load side current DWT decomposition levels at LLG fault in generator and load buses, considered one at a time

Case study	$(SK)_{MxN}$ *generator bus fault*	$(SK)_{MxN}$ *load bus fault*
For source side current	$\begin{pmatrix} 0.002516 & 0.038965 & 1.606521 & 15.6858 \\ 0.002713 & -0.13023 & 1.60732 & 7.428806 \\ 0.00319 & 0.264675 & 1.608419 & 9.356144 \\ 0.005379 & 0.086461 & 1.607178 & 53.28107 \\ -0.00538 & -0.18838 & 1.61077 & 42.75261 \end{pmatrix}$	$\begin{pmatrix} 0.016009 & 4.371866 & 1.660812 & 287.9464 \\ 0.034012 & -1.76333 & 1.708085 & 118.3351 \\ 0.06608 & 0.003883 & 1.789696 & 80.40759 \\ 0.117219 & 0.093933 & 1.922053 & 60.74021 \\ 0.163501 & 0.125944 & 1.992691 & 55.86645 \end{pmatrix}$
For load side current	$\begin{pmatrix} 0.003112 & 0.010786 & 1.499355 & 15.34421 \\ 0.008699 & 0.375878 & 1.498318 & 17.91369 \\ 0.01975 & 0.24549 & 1.496731 & 27.4928 \\ 0.041983 & 0.063178 & 1.493616 & 41.66792 \\ 0.084355 & 0.360648 & 1.489268 & 46.23959 \end{pmatrix}$	$\begin{pmatrix} -0.00497 & 2.537022 & 1.500994 & 181.2674 \\ -0.00856 & -0.4281 & 1.501842 & 51.58309 \\ -0.01564 & -1.19686 & 1.503619 & 108.3703 \\ -0.02965 & 0.890177 & 1.508668 & 43.55096 \\ -0.06337 & -0.21793 & 1.520679 & 45.34609 \end{pmatrix}$

$Sa1----Sa5$	denote the Skewness values for approximation coefficients for 5 decomposition levels
$Sd1----Sd5$	denote the Skewness values for detailed coefficients for 5 decomposition levels
$Ka1----Ka5$	denote the Kurtosis values for approximation coefficients for 5 decomposition levels
$Kd1----Kd5$	denote the Kurtosis values for detailed coefficients for 5 decomposition levels.

Monitoring the Skewness and Kurtosis matrices of Tables 2, 3, 4 and 5, it has been inferred that for LL, LG and LLG faults in generator bus, prediction can be fruitfully done, monitoring the load side current and for the inception of these faults in load bus, fault analysis can be done monitoring the source current. Features have been extracted from these matrices and has been presented in succeeding section for more specific identification of these faults in the system.

6 Feature Extraction

Analysis on the Skewness and Kurtosis coefficients from Tables 2, 3, 4 and 5 depicts significant change in the generator and load bus current's MRA of DWT decomposition levels, which has been consolidated to develop a pictographic feature extraction from the Skewness and Kurtosis matrices. The feature extraction from the above tables is presented in Tables 6 and 7.

Monitoring Tables 6 and 7, it is clear that for LG, LL and LLG faults in source and load buses of the network, the Skewness and Kurtosis signatures are distinctly different and if these patterns can be monitored, the type of faults in the network can be identified and the location of these faults in the system can be ascertained.

7 Conclusion

Identification and localization of various faults in a system is of utmost importance in successful running of a power utility system. The motivation of this work was thus to determine the presence of LG, LL and LLG faults in the system as well as to determine the zone of inception of these faults in the system, monitoring the Skewness and Kurtosis signatures of the MRA of DWT decomposition level of the source and load side currents at normal and at fault. The features extracted, from the patterns generated, distinctively determine the occurrence of these faults in the system. Also monitoring the load and source side spectrums, the faulty zone can be identified. This analysis can be extended for the identification of other type of faults in simulated as well as real time systems, which may occur in large scale or may be incipient in nature.

Table 6 Feature extraction from source and load side currents at normal and fault in generator bus

Case study		Feature extraction from Skewness monitoring	Feature extraction from Kurtosis monitoring
Source side	Approximate values	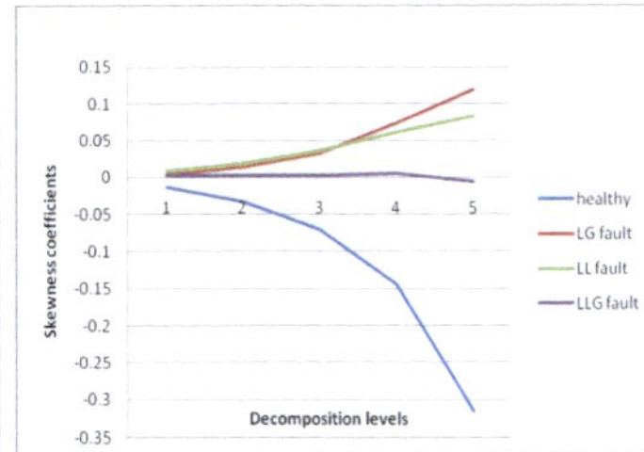	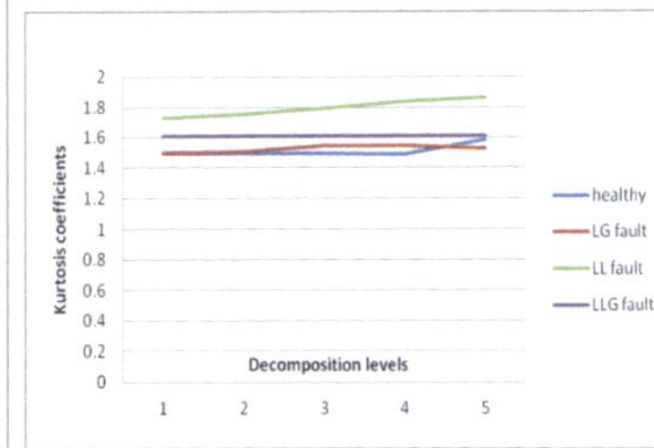
	Detailed values	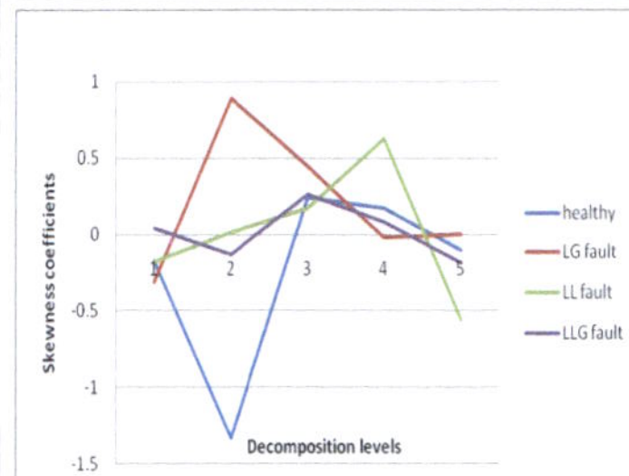	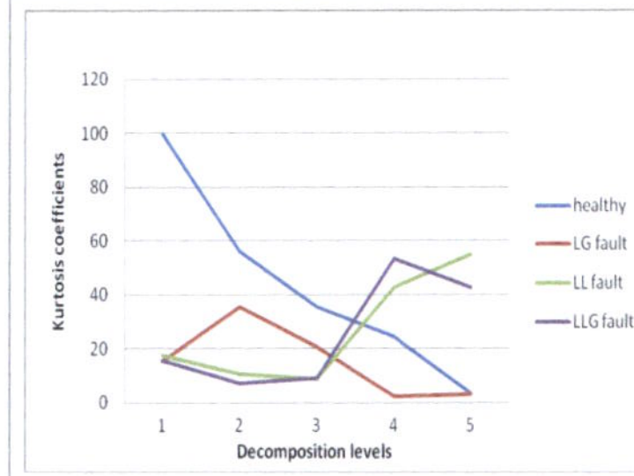

(continued)

Table 6 (continued)

Case study		Feature extraction from Skewness monitoring	Feature extraction from Kurtosis monitoring
Load side	Approximate values	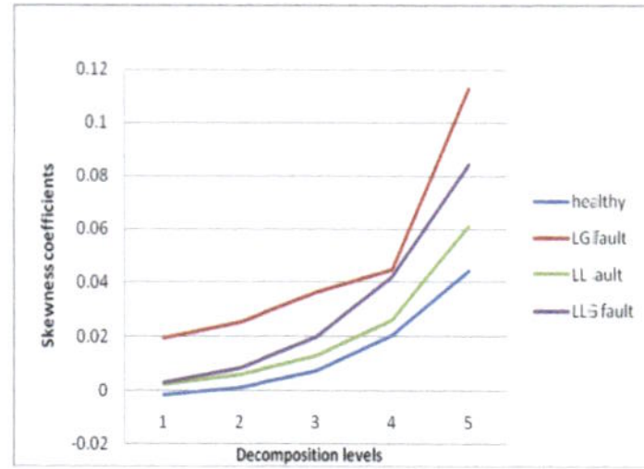	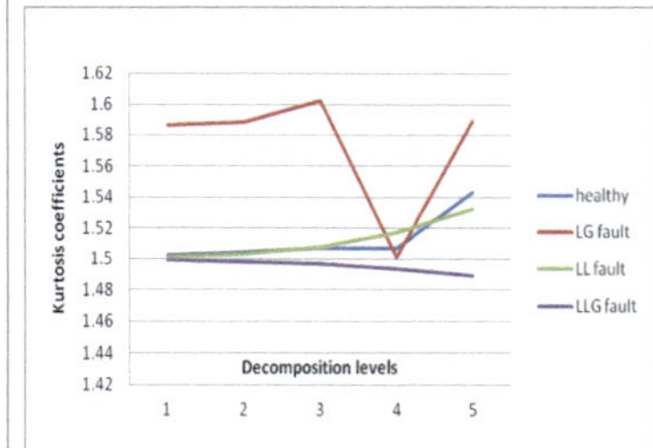
	Detailed values	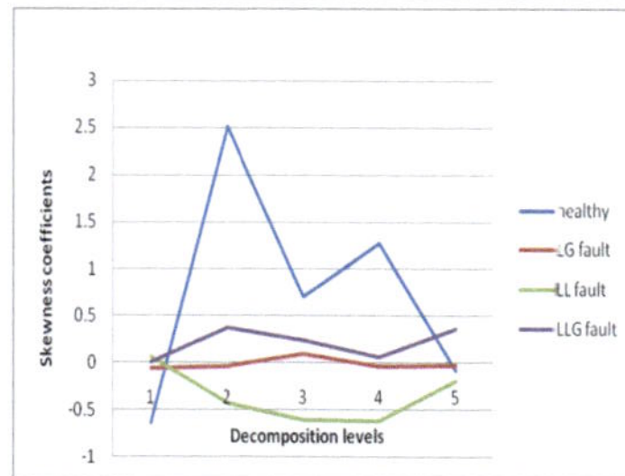	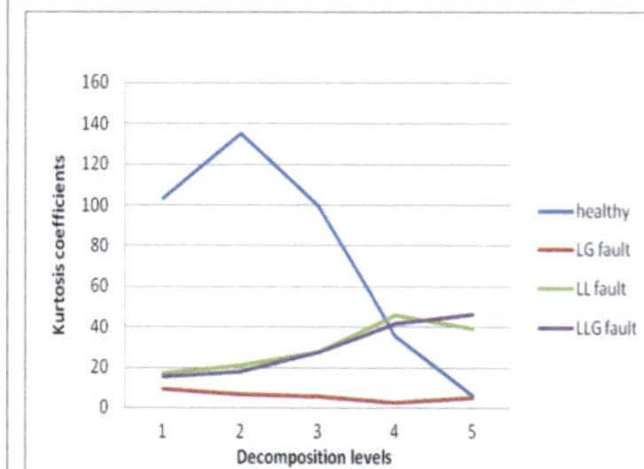

Table 7 Feature extraction from source and load side currents at normal and fault in load bus

Case study		Feature extraction from Skewness monitoring	Feature extraction from Kurtosis monitoring
Source side	Approximate values	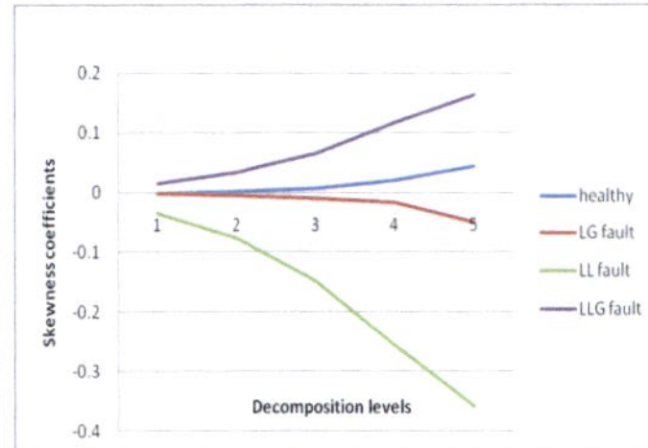	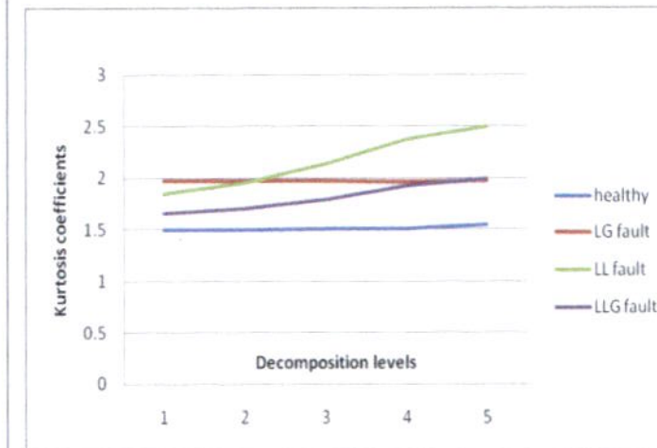
	Detailed values	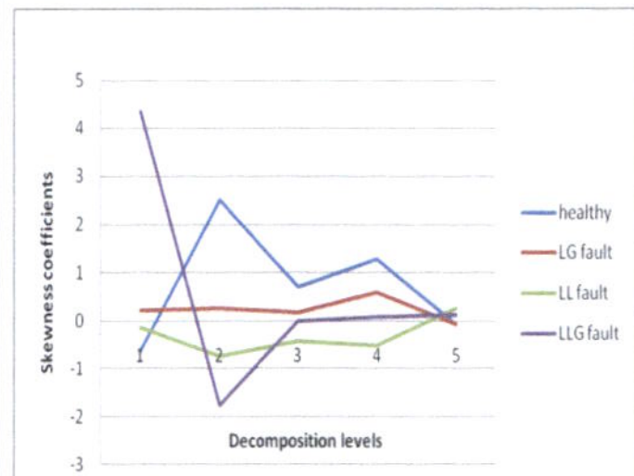	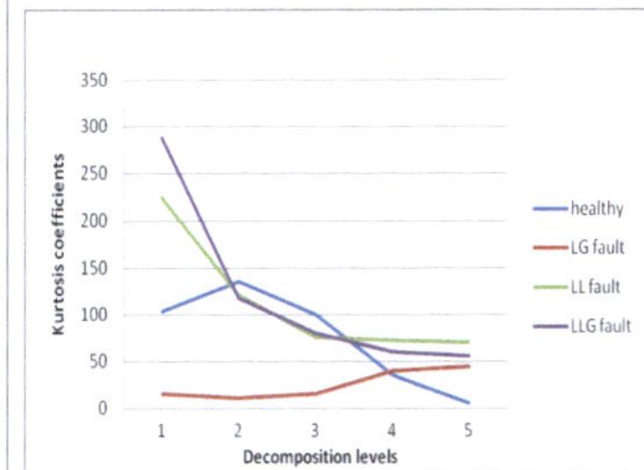

(continued)

Table 7 (continued)

<table>
<tr><th>Case study</th><th></th><th>Feature extraction from Skewness monitoring</th><th>Feature extraction from Kurtosis monitoring</th></tr>
<tr><td>Load side</td><td>Approximate values</td><td>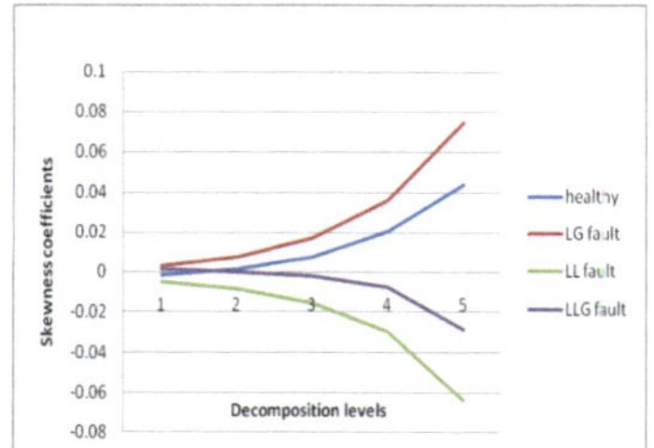
</td><td>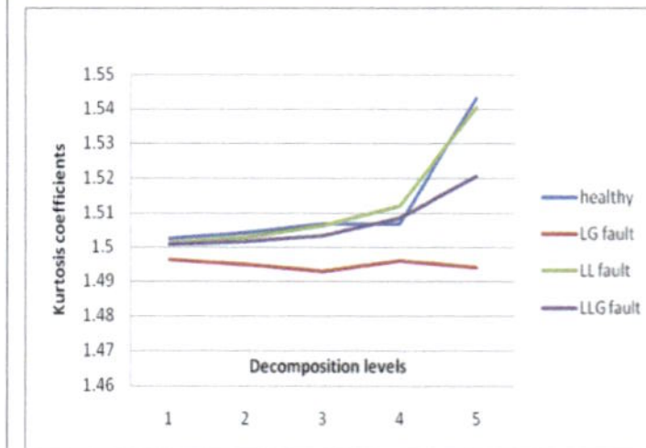
</td></tr>
<tr><td></td><td>Detailed values</td><td>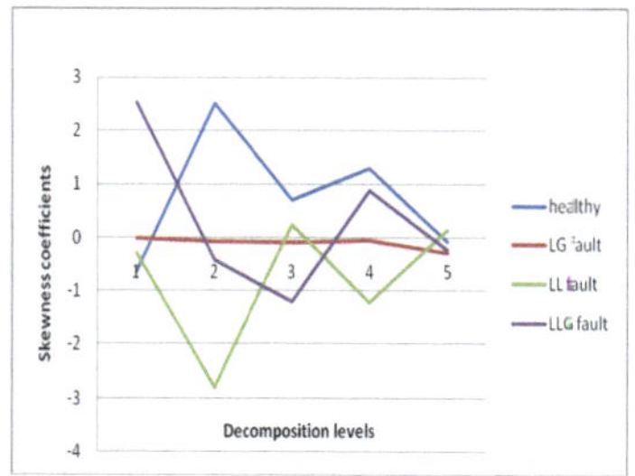
</td><td>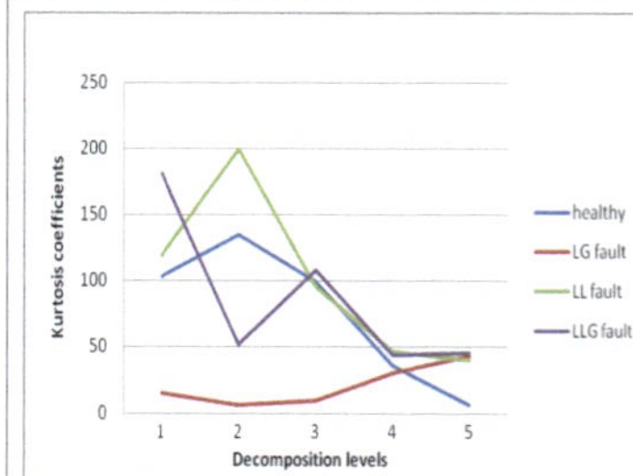
</td></tr>
</table>

References

1. B.H. Khan, *Non-Conventional Energy Resources*. McGraw Hill Education (India) Private Limited, ISBN: 978-0-07-014276-3 (2013)
2. Z. Alnasir, M. Kazerani, in *A dump load-less standalone wind energy conversion system supplying a generic load*. Electrical Power and energy Conference (EPEC), (2016) https://doi.org/10.1109/epec.2016.7771693
3. M. Pang, Y. Shi, W. Wang, X. Yuan, in *A method for optimal sizing hybrid energy storage system for smoothing fluctuations of wind power*. Power and Energy Engineering Conference (APPEEC), (2016). https://doi.org/10.1109/appeec.2016.7779913
4. Z. Hameed, Y.S. Hong, Y.M. Cho, S.H. Ahn, C.K. Song, Condition monitoring and fault detection of wind turbines and related algorithms: a review. Renew. Sustain. Energy Rev. **13**(1), 1–39 (2009)
5. H.M.K. Al-Masri, M. Ehsani, in *Impact of wind turbine modeling on a renewable energy system*. North American Power symposium (NAPS), (2016). https://doi.org/10.1109/naps.2016.7747870
6. H. Bitaraf, S. Rahman, in *Optimal operation of energy storage to minimize wind spillage and mitigate wind power forecast errors*. Power and Energy Society General Meeting (PESGM), (2016). https://doi.org/10.1109/pesgm.2016.7741550
7. S. Alnaser, L.F. Ochoa, in *Optimal sizing and control of energy storage in wind power rich distribution networks*. Power and energy Society General Meeting (PESGM), (2016). https://doi.org/10.1109/pesgm.2016.7741202
8. F. Pozo, Y. Vidal, L. Acho, in *Wind turbine fault detection through principal component analysis and multivariate statistical inference*. 8th European workshop on Structural Health Monitoring (EWSHM) (2016)
9. X. Zhang, Q. Zhang, S. Zhao, R. Ferrari, M. M. Polycarpou, T. Parisini, in *Fault Detection and Isolation of the Wind Turbine Benchmark: an Estimation-based Approach*. 18th IFAC World Congress, pp. 8295–8300 (2011)
10. S. M. M. Aval, A. Ahadi, "Wind Turbine Fault Diagnosis Techniques and Related Algorithms", International Journal of Renewable Energy Research, vol. 6, N° 1, 2016
11. K. Kim, G. Parthasarathy, O. Uluyol, W. Foslien, S. Sheng, P. Fleming, in *Use of SCADA Data for Failure Detection in Wind Turbines*. Energy Sustainable and Fuel Cell Conference, pp. 1–9 (2011)
12. M.-J. Poggi, G. Oppenheim, Y. Misiti, M. Misiti, Wavelet Toolbox™ User's Guide (R2012a). MATLAB®, Mathworks
13. S. Karmakar, S. Chattopadhyay, M. Mitra, S. Semgupta, *Induction Motor Fault Diagnosis-approach through Current Signature Analysis*. (Springer, Berlin). https://doi.org/10.1007/978-981-10-0624-1

FFT Based Harmonic Assessment of Line to Ground Fault in 14 Bus Microgrid System

Sagnik Datta, Surajit Chattopadhyay and Arabinda Das

1 Introduction

A microgrid is a small-scale power grid which operates independently or in conjunction with the main electrical grid of that area. Microgrid effectively cuts down the dependency on the main electrical grid and also improves the overall reliability of the electrical power system. Recently rapid development of "green energy" and storage devices are taking place throughout the world. Microgrids implemented in line with the concept of distributed generation are emerging as an effective form of power management. Alike the conventional power grid, effective fault identification and isolation is of paramount importance in microgrid system. In order to do so, the location of the fault has to be precisely found out.

A lot of research works have been going on in this regard. Hooshyar et al. [1] has carried out fault type classification in Microgrids including photovoltaic DGs. Short-circuit fault analysis on microgrid has been done by Bayindir et al. [2]. DC short circuit fault analysis has been done and protection of ring type DC microgrid has been developed by Yu et al. [3]. A short-circuit current calculation method has been introduced by Lai et al. [4] for low-voltage DC microgrid. Suppression strategy has been introduced by Zha et al. [5] for short-circuit current in loop-type DC microgrid. Petrea et al. [6] presented factors influencing a micro-grid recovery process following a short-circuit. Park et al. [7] developed DC Ring-Bus Microgrid Fault

S. Datta (✉)
SKF Group of Institution, Hooghly, West Bengal, India
e-mail: sagnik.ee@gmail.com

S. Chattopadhyay
Ghani Khan Choudhury Institute of Engineering and Technology, Malda, India
e-mail: surajitchattopadhyay@gmail.com

A. Das
Jadavpur University, Kolkata, India
e-mail: arbinda.das@ju.in

S. Chattopadhyay et al. (eds.), *Modelling and Simulation in Science, Technology and Engineering Mathematics*, Advances in Intelligent Systems and Computing 749,
https://doi.org/10.1007/978-3-319-74808-5_8

Protection and Identification of Fault Location. Fault detection and isolation have been done by Park et al. [8] in Low-Voltage DC-Bus Microgrid System. Modeling and real-time simulation of an AC microgrid have been done by Sahoo et al. [9] with solar photovoltaic system. Modeling and reliability assessment have been carried out by Quevedo et al. [10] of microgrids including renewable distributed generation. Reliability assessment of a microgrid distribution system has also been carried out Tuffaha and AlMuhaini [11] with PV and storage. Photovoltaic power generation system low voltage ride has been analysed by Jin et al. [12] through control during asymmetric fault. Integrated Fault Location and Power-Quality Analysis have been done by Bíscaro et al. [13] in Electric Power Distribution Systems. Analysis of two fault locators considering operation variations of the power distribution systems has been done by Ramírez-Ramírez et al. [14]. Ehrenbenberger [15] has introduced fault analysis of Smart Grid Power System employing Simultaneous Faults Method. Zhu and Zhang [16] introduced a novel control strategy of DC microgrid under unbalanced grid voltage. FFT and wavelet decomposition based harmonics assessments have been observed using current signature analysis for fault diagnosis [17–20].

This paper aims to reveal a harmonics assessment based technique of fault location identification for Line to ground (LG) faults occurring at different load buses of a 14 bus microgrid system which is working in conjunction with main electrical grid. Fast Fourier Transform (FFT) based total harmonic distortion (THD) analysis of outgoing currents from the different generator buses are examined when LG fault occurs at different load buses.

2 Microgrid Modelling

Single line diagram of the IEEE standard 14 bus microgridsystem is shown in Fig. 1. In Fig. 1, it can be seen that there are four sources connected to the system. Bulk power generator is present at generator bus G13, PV cell at G12, diesel generator 1(DG 1) at G8 and diesel generator 2 (DG 2) at G3. Five different kinds of loads are present. A non-linear load is present at bus B10, furnace at B6, battery charging system at B1, linear load at B2 and linear load 1 at B3. Π section lines are joining one bus to the other and these sections are 1 km in length.

3 Fault Simulation

LG faults are made to occur at different load buses and THD values of outgoing currents from the different generator buses are measured. At first, the entire system is kept healthy and the THD values of outgoing currents from the different generator buses are recorded. Afterwards LG fault is made to occur on one specific load bus while leaving the other load buses healthy and THD values of outgoing currents from all the different generator buses are monitored. Like this all the load buses

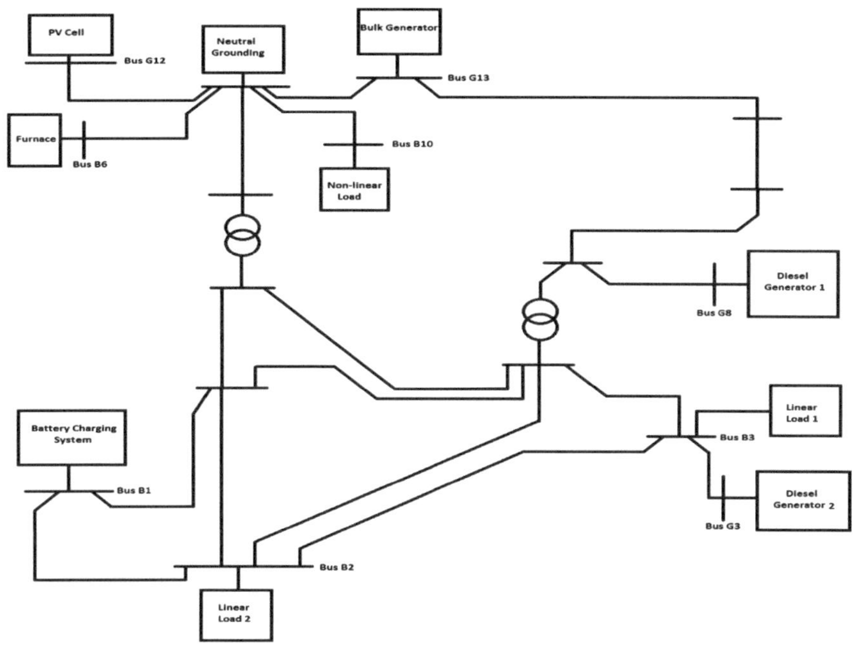

Fig. 1 Single line diagram of IEEE standard 14 bus microgrid system

are considered separately and in all cases distortion in outgoing currents from the different generator buses are assessed through the THD values of those currents. THD values are obtained by FFT analysis of the above mentioned current waveforms. Total simulation time is set at 0.8 s. Fault duration is kept within 0.2–0.4 s. 40 cycles of the outgoing currents from the different generator buses are considered for FFT analysis with start time of 0.15 s and maximum sampling frequency of 1000 Hz.

4 Harmonic Assessment and Results

Results obtained from the FFT analysis are provided in the Tables 1, 2, 3, 4 and 5. Data provided in the following tables present harmonic contents of outgoing currents from different generator buses, when LG fault occurs at different load buses.

5 Observation

Based on the results given above, THD values of outgoing currents from one specific generator bus for LG fault at different load buses is ascertained. This process is

Table 1 Harmonic content at different generator bus outgoing currents for LG fault in B10 (non-linear load)

Gen. bus	Fund. (%)	Values are in % of fundamental							Total THD (in %)
		DC component	2nd order	3rd order	5th order	7th order	9th order	11th order	
G13 (Bulk Gen)	100%	0.32	0.01	0.03	0.02	0.00	0.00	0.00	0.37
G12 (PV)	100%	21.93	0.45	1.14	0.87	0.44	0.24	0.08	15.8
G8 (DG 1)	100%	1.84	0.27	0.32	0.14	0.01	0.04	0.02	4.4
G3 (DG 2)	100%	1.5	0.2	0.25	0.11	0.01	0.01	0.01	3.38

Table 2 Harmonic content at different generator bus outgoing currents for LG fault in B6 (furnace)

Gen. bus	Fund. (%)	Values are in % of Fundamental							Total THD (in %)
		DC component	2nd order	3rd order	5th order	7th order	9th order	11th order	
G13 (Bulk Gen)	100	0.35	0.04	0.01	0.01	0.01	0.00	0.00	0.43
G12 (PV)	100	55.51	3.22	3.55	0.78	0.32	0.27	0.10	48.7
G8 (DG 1)	100	7.58	0.41	0.45	0.08	0.02	0.02	0.02	6.08
G3 (DG 2)	100	7.1	0.5	0.56	0.05	0.02	0.02	0.01	7.53

Table 3 Harmonic content at different generator bus outgoing currents for LG fault in B1 (battery storage)

Gen. bus	Fund. (%)	Values are in % of fundamental							Total THD (in %)
		DC component	2nd order	3rd order	5th order	7th order	9th order	11th order	
G13 (Bulk Gen)	100	0.37	0.01	0.04	0.03	0.01	0.01	0.00	0.53
G12 (PV)	100	21.16	0.25	1.42	1.19	0.68	0.39	0.08	20.28
G8 (DG 1)	100	0.47	0.24	0.27	0.21	0.06	0.04	0.04	4.26
G3 (DG 2)	100	0.35	0.2	0.25	0.19	0.06	0.04	0.02	3.8

Table 4 Harmonic content at different generator bus outgoing currents for LG fault in B2 (linear load)

Gen. bus	Fund. (%)	Values are in % of fundamental							Total THD (in %)
		DC component	2nd order	3rd order	5th order	7th order	9th order	11th order	
G13 (Bulk Gen)	100	0.37	0.01	0.04	0.03	0.01	0.01	0.00	0.53
G12 (PV)	100	21.16	0.25	1.42	1.19	0.68	0.39	0.08	20.28
G8 (DG 1)	100	0.47	0.24	0.27	0.21	0.06	0.04	0.04	4.26
G3 (DG 2)	100	0.35	0.2	0.25	0.19	0.06	0.04	0.02	3.8

Table 5 Harmonic content at different generator bus outgoing currents for LG fault in B3 (linear load 1)

Gen. bus	Fund.	Values are in % of Fundamental							Total THD (in %)
		DC component	2nd order	3rd order	5th order	7th order	9th order	11th order	
G13 (Bulk Gen)	100%	0.37	0.01	0.04	0.03	0.01	0.01	0.00	0.53
G12 (PV)	100%	21.16	0.25	1.40	1.19	0.68	0.38	0.08	20.28
G8 (DG 1)	100%	0.46	0.24	0.27	0.21	0.06	0.04	0.04	4.26
G3 (DG 2)	100%	0.35	0.2	0.25	0.19	0.06	0.04	0.02	3.8

Table 6 THD values of outgoing current of bulk generator bus (G13)

Site of fault	THD values (in %)
B10 (non linear load)	0.37
B6 (furnace)	0.43
B1 (battery storage)	0.53
B2 (linear load)	0.53
B3 (linear load 1)	0.53

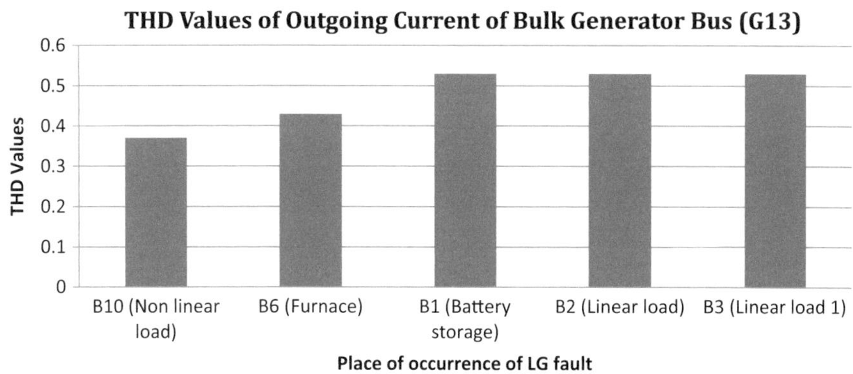

Fig. 2 THD values of outgoing current of bulk generator bus (G13) for LG fault in the load buses

repeated for all the generator buses. One by one those values have been presented as follows:

a. Bulk Generator Bus (G13)

THD values of outgoing currents from bulk generator bus (G13) for LG fault at different load buses have been considered and given in Table 6.

As seen from Fig. 2, THD values of the outgoing current of Bulk Generator Bus (G13) are much lower in comparison with the other Generator buses as shown in Tables 8 and 9.

b. PV Cell Bus (G12)

THD values of outgoing currents from PV cell bus (G12) for LG fault at different load buses have been considered and given in Table 7.

From Fig. 3, it has been observed that THD values of the outgoing current of PV Cell Bus (G12) are much higher in comparison with the other Generator buses as shown in Tables 7 and 9. When LG fault occurs at the bus (B6) connected with the furnace, outgoing current waveform of PV Cell Bus (G12) gets considerably distorted and contains high values of THD as shown in Fig. 2.

Table 7 THD values of outgoing current of PV cell bus (G12)

Site of fault	THD values (in %)
B10 (non linear load)	15.8
B6 (furnace)	48.7
B1 (battery storage)	20.3
B2 (linear load)	20.3
B3 (linear load 1)	20.3

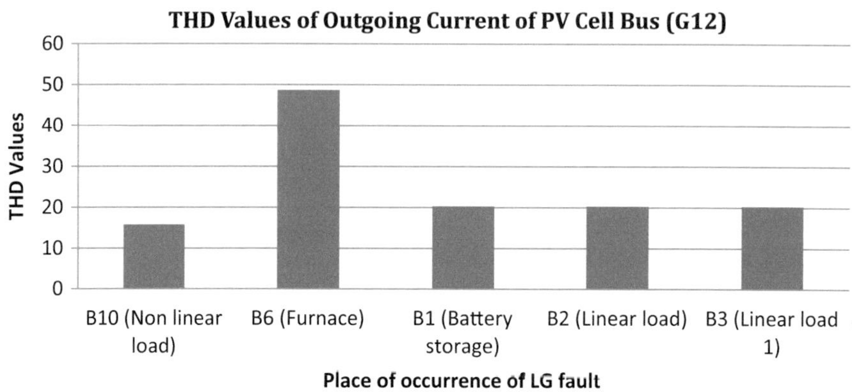

Fig. 3 THD values of outgoing current of PV cell bus (G12) for LG fault in the load buses

Table 8 THD values of outgoing current of 'diesel generator 1' bus (G8)

Site of fault	THD values (in %)
B10 (non linear load)	4.4
B6 (furnace)	6.08
B1 (battery storage)	4.3
B2 (linear load)	4.3
B3 (linear load 1)	4.3

c. 'Diesel Generator 1' Bus (G8)

THD values of outgoing currents from 'Diesel Generator 1' bus for LG fault at different load buses have been considered and given in Table 8.

Figure 4, shows that THD value of the outgoing current of 'Diesel Generator 1' (G8) is the highest for LG fault in Bus 10, where Non-linear load is connected.

d. 'Diesel Generator 2' Bus (G8)

THD values of outgoing currents from 'Diesel Generator 2' bus for LG fault at different load buses have been considered and given in Table 9.

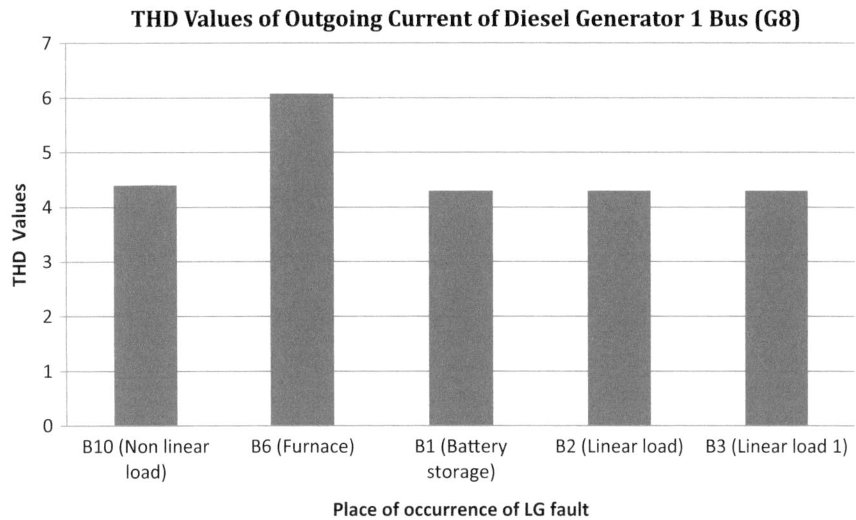

Fig. 4 THD values of outgoing current of 'diesel generator 1' (G8) for LG fault in the load buses

Table 9 THD values of outgoing current of 'diesel generator 1' bus (G8)

Site of fault	THD values (in %)
B10 (non linear load)	3.38
B6 (furnace)	7.53
B1 (battery storage)	3.8
B2 (linear load)	3.8
B3 (linear load 1)	3.8

6 Rule Set

A rule set has been developed based upon the above results and observations.

- Outgoing currents of the bulk generator bus is the least affected by the LG faults occurring in different load buses of the microgrid system shown in Fig. 1. For LG fault in Non-linear load bus THD value is less than 0.4%, furnace bus THD value is more than 0.4%. For LG faults in buses where battery storage unit and linear loads are connected (B1, B2 and B3 respectively), maximum amount of distortion is occurring in the current waveform which results into THD value in excess of 0.5%.
- Outgoing currents of the PV cell bus is the most affected among all the generator buses. For LG fault in furnace bus (B6), current waveform is most distorted with a THD value close to 50%. THD value is exactly 20% for LG faults in buses where battery storage unit and linear loads are connected (B1, B2 and B3 respectively), while THD value stays below 20% mark for LG fault in non-linear load bus.

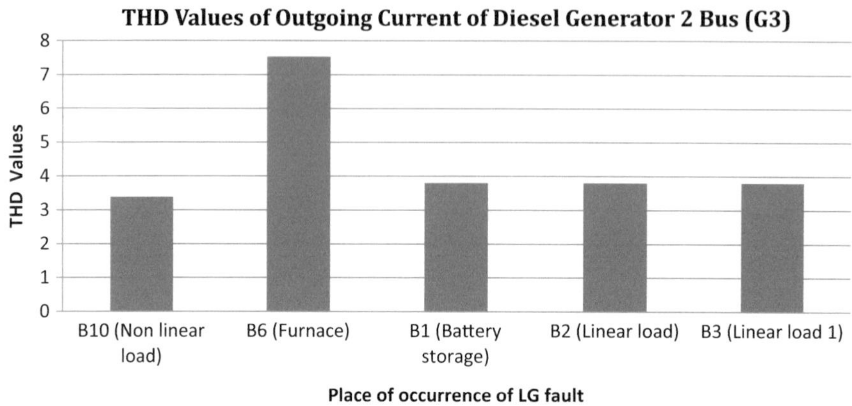

Fig. 5 THD values of outgoing current of 'diesel generator 2' (G3) for LG fault in the load buses

- Outgoing current of 'Diesel generator 1' bus is the most affected in case of an LG fault at furnace bus (B6) with THD value above 6%. For LG fault in non-linear load bus (B10) THD value is 4.4%. In case of LG faults in other three load buses THD value stays fixed at 4.3%.
- Outgoing current of 'Diesel generator 2' bus is also the most affected in case of an LG fault at furnace bus (B6) with THD value of 7.5%. Comparing Figs. 4 and 5 it is observed that among the two diesel generators, diesel generator 2 is the worst affected when LG fault takes place at furnace bus (B6). For LG fault in non-linear load bus (B10) THD value is 3.38%. In case of LG faults in other three load buses THD value stays fixed at 3.8%.

7 Specific Outcome

Due to LG faults at different load buses the outgoing current of PV cell bus is the most affected and an LG fault in furnace bus creates the most distortion. It has also been observed that amount of distortion in a specific generator bus currents does not depend on the distance of the load bus at which LG fault is taking place, rather it is dependent upon the nature of the load which is connected to the load bus where LG fault is occurring.

8 Conclusion

In this paper LG faults in a microgrid system has been assessed by FFT based THD analysis. Different THD values in the outgoing currents of the generator buses have been observed for LG fault in different load buses. From the THD values, a rule set has been developed which would be helpful for identification of location of the LG fault.

References

1. A. Hooshyar, E.F. El-Saadany, M. Sanaye-Pasand, Fault type classification in microgrids including photovoltaic DGs. IEEE Trans. Smart Grid **7**(5), 2218–2229 (2016). INSPEC Accession Number: 16249142
2. R. Bayindir, E. Irmak, F. Issi, N. Guler, in *Short-circuit fault analysis on microgrid*. 2015 International Conference on Renewable Energy Research and Applications (ICRERA), pp. 1248–1252 (2015). https://doi.org/10.1109/icrera.2015.7418608
3. M. Yu, Y. Wang, L. Zhang, Z. Zhang, in *DC short circuit fault analysis and protection of ring type DC microgrid*. IEEE 8th International Power Electronics and Motion Control Conference (IPEMC-ECCE Asia), pp. 1694–1700 (2016). https://doi.org/10.1109/ipemc.2016.7512549
4. X. Lai, F. Liu, K. Deng, Q. Gao, X. Zha, in *A short-circuit current calculation method for low-voltage DC microgrid*. International Power Electronics and Application Conference and Exposition, pp. 365–371 (2014). https://doi.org/10.1109/peac.2014.7037883
5. X. Zha, H. Ning, X. Lai, Y. Huang, F. Liu, in *Suppression strategy for short-circuit current in loop-type DC microgrid*. IEEE Energy Conversion Congress and Exposition (ECCE), pp 758–764 (2014). https://doi.org/10.1109/ecce.2014.6953472
6. I.C. Petrea, M. Corneliu, in *Factors influencing a micro-grid recovery process following a short-circuit*. 10th International Conference on Environment and Electrical Engineering, pp. 1–4 (2011). https://doi.org/10.1109/eeeic.2011.5874772
7. J.-D. Park, J. Candelaria, L. Ma, K. Dunn, DC ring-bus microgrid fault protection and identification of fault location. IEEE Trans. Power Deliv. **28**(4), 2574–2584 (2013). https://doi.org/10.1109/tpwrd.2013.2267750
8. J.-D. Park, J. Candelaria, Fault detection and isolation in low-voltage DC-bus microgrid system. IEEE Trans. Power Deliv. **8**(2), 779–787 (2013). https://doi.org/10.1109/tpwrd.2013.2243478
9. S.K. Sahoo, A.K. Sinha, N.K. Kishore, in *Modeling and real-time simulation of an AC microgrid with solar photovoltaic system*. Annual IEEE India Conference (INDICON), pp. 1–6 (2015). https://doi.org/10.1109/indicon.2015.7443619
10. P.M. de Quevedo, J. Contreras, A. Mazza, G. Chicco, R. Porumb, in *Modeling and reliability assessment of microgrids including renewable distributed generation*. IEEE 16th International Conference on Environment and Electrical Engineering (EEEIC), pp. 1–6 (2016), https://doi.org/10.1109/eeeic.2016.7555659
11. T. Tuffaha, M. AlMuhaini, in *Reliability assessment of a microgrid distribution system with pv and storage*. International Symposium on Smart Electric Distribution Systems and Technologies (EDST), pp. 195–199 (2015). https://doi.org/10.1109/sedst.2015.7315206
12. W. Jin, Z. Jinhua, H. Zhiguo, in *Analysis of photovoltaic power generation system low voltage ride through control during asymmetric fault*. China International Conference on Electricity Distribution (CICED), pp. 1–4 (2016). https://doi.org/10.1109/ciced.2016.7575950
13. A.A.P. Bíscaro, R.A.F. Pereira, M. Kezunovic, J.R.S. Mantovani, Integrated fault location and power-quality analysis in electric power distribution systems. IEEE Trans. Power Deliv. **31**(2), 428–436 (2016). https://doi.org/10.1109/tpwrd.2015.2464098
14. J. Ramírez-Ramírez; S. Pérez-Londoño; J. Mora-Flórez, in *Analysis of two fault locators considering operation variations of the power distribution systems*. IEEE 6th Latin American Symposium on Circuits & Systems (LASCAS), pp. 1–4 (2015). https://doi.org/10.1109/lascas.2015.7250470

15. J. Ehrenbenberger, in *Fault analysis of Smart Grid Power System employing Simultaneous Faults Method*. Proceedings of the 2014 15th International Scientific Conference on Electric Power Engineering (EPE), pp. 95–100 (2014). https://doi.org/10.1109/epe.2014.6839449
16. X. Zhu, Y. Zhang, in *Control strategy of DC microgrid under unbalanced grid voltage*. IEEE 8th International Power Electronics and Motion Control Conference (IPEMC-ECCE Asia), pp. 1725–1731 (2016). https://doi.org/10.1109/ipemc.2016.7512554
17. S. Chattopadhyay, A. Chattopadhyaya, S. Sengupta, in *Measurement of harmonic distortion and Skewness of stator current of induction motor at crawling in Clarke plane*. IET Science, Measurement and Technology, pp. 1–9. March, 2014
18. S. Chattopadhyay, A. Chattopadhyaya, S. Sengupta, Harmonic power distortion measurement in park plane. Measurement **51**, 197–205 (2014)
19. A. Chattopadhyaya, A. Ghosh, S. Chattopadhyay and S. Sengupta, Stator current harmonic assessment of induction motor for fault diagnosis. Int. J. Electron. Commun. Technol. **4**. ISSUE-SPL1, ISSN: 2230–7109 (Online), 2230–9543 (Print), Jan–Mar 2013
20. A. Chattopadhyaya, S. Banerjee, S. Chattopadhyay, Assessment of discrimination between inrush and fault current in a power transformer. Can. J. Technol. Innov. **1**, 187–196 (2014)

Harmonics Assessment Based Symmetrical Fault Diagnosis in PV Array Based Microgrid System

Tapash Kr. Das, Surajit Chattopadhyay and Arabinda Das

1 Introduction

With the increase of power demand different types of non-conventional energy resources are being used. Among different types of non-conventional energy resources, solar PV array based power supply has become popular in domestic as well as in small Industrial applications. To cope with the ever increasing power demand, microgrid systems are becoming more popular in day by day.

A lot of research works are going on in this regard. Anwar et al. (2013) performed detail harmonics assessment for micro grid system [1]. Power quality at voltage source converter based micro grid operation [2] has been analyzed by Dhar et al. (2015). Kumar and Zare (2015) has done performance analysis for low voltage microgrid distribution networks connected with power electronics system [3]. Wang and Yaz (2016) performed detail analysis of smart power grid synchronization with fault tolerant nonlinear estimation [4], where computer simulation techniques have demonstrated that the proposed fault tolerant extended Kalman filter (FTEKF) provides more accurate voltage synchronization results than the extended Kalman filter (EKF). Rashid et al. (2015) introduced transient stability enhancement of doubly fed induction machine-based wind generator by bridge-type fault current limiter [5] for microgrid power quality improvement, where various simulations were carried out in Matlab/Simulink environment to demonstrate the effectiveness of the BFCL and its

T. Kr. Das (✉) · S. Chattopadhyay
Department of Electrical Engineering, GKCIET (Under MHRD, Govt. of India), Malda, West Bengal, India
e-mail: p_tapash_das@yahoo.co.in

S. Chattopadhyay
e-mail: surajitchattopadhyay@gmail.com

A. Das
Department Electrical Engineering Department, Jadavpur University, Kolkata, India
e-mail: adas_ee_ju@gmail.com

S. Chattopadhyay et al. (eds.), *Modelling and Simulation in Science, Technology and Engineering Mathematics*, Advances in Intelligent Systems and Computing 749,
https://doi.org/10.1007/978-3-319-74808-5_9

performance is compared with that of the series dynamic braking resistor (SDBR). Chen et al. (2015) modeled doubly fed induction generator wind turbine systems subject to recurring symmetrical grid faults [6]. In this attempt, the performance of the doubly fed induction generator (DFIG) wind turbine system under recurring symmetrical grid faults is analyzed. Chen et al. (2014) observed the nontechnical loss and outage detection using fractional-order self-synchronization error-based fuzzy Petri nets in micro-distribution systems [7], where different computer simulations were carried out using an IEEE 30-bus power system and medium-scale micro-distribution systems to show the effectiveness of this proposed method. Mathematical morphology-based islanding detection for distributed generation has been introduced [8] by Farhan and Swarup (2016) where basic MM operators like dilate erode difference filter (DEDF) has been used to operate on three-phase voltage and current signals on target DG location. Attempt has been taken to track the islanding condition from non-islanding condition, a new operator called the MM ratio index (MM RI) computed is used for distributed generation. A GPS- based control framework for accurate current sharing and power quality improvement in Microgrids [9] has been introduced by Golsorkhi et al. (2016) to improve the current sharing accuracy at high loading conditions. A special technique for symmetrical and asymmetrical low-voltage ride through of doubly-fed induction generator wind turbines using gate controlled series capacitor [10] has been observed in detail by Mohammadpour et al. (2015), where extensive time-domain simulations using MATLAB/SIMULINK were performed to validate the effectiveness of this methods during grid faults. Rashad et al. (2016) described control methodology of inverter used in standalone micro-grid system [11]. Harmonic mitigation [12] of power distribution network in minigrid has been studied by Sabu and George (2016). Sreekumar et al. (2015) introduced a new virtual harmonic impedance scheme for harmonic power sharing in an islanded microgrid [13], where a control strategy employs negative virtual harmonic impedance to compensate the effect of line impedance on harmonic power. Zhu et al. (2015) introduced novel technique for virtual damping flux-based LVRT Control for DFIG-Based Wind Turbine [14] where the simulation has been carried out and verified with a 2-MW DFIG in MATLAB/Simulink environment to smooth the electromagnetic torque and minimized different grid faults. Taj et al. (2015) introduced an adaptive neuro-fuzzy controlled-flywheel energy storage system [15] for transient stability enhancement. In recent years many mathematical tools have been introduced for harmonics assessment [16] and they have been found very effective in fault assessment [17].

However very few works are found on harmonics assessment based microgrid based power system. This has motivated to work on harmonic assessment based fault detection in microgrid system. Attempt has been taken to model a micro grid then to perform FFT based harmonics assessment for fault diagnosis in microgrid system.

2 Modeling of 400 KW PV Array Based Micro Grid

In this work, PV Array based micro Grid of 400KW system (MATLAB, version-15) has been modeled as shown in Fig. 1. Four numbers of PV array are connected in parallel. Each PV array contains 64 strings. Each string contains 5 numbers of series connected modules having following specifications: Maximum power—315.072 W, Cells per module—86, Open circuit voltage—64.6 volts, Shot circuit current—6.14 A, voltage at maximum power point—54.7 volts, current at maximum power point—5.76 A, temperature coefficient at open circuit voltage—"−0.27269" %/°C, temperature coefficient at short circuit current—"0.061694" %/°C, Light-generated current 6.1461 A, Diode saturation current 6.5043×10^{-12} A, Diode ideality factor 0.9507, Shunt resistance 430.0559 ohms, Series resistance 0.43042 ohms. Parallel combinations of PV arrays are connected with DC to DC charge controller. Average model based VSC having 3 bridge arms has been considered in Inverter unit. Inverter output is fed to three phase 400KVA, 260 V/25 kV, 60 Hz star/delta transformer. Transformer output is fed to load Bus (BUS-2). The load bus is also connected with conventional 120 kV, 2500MVA grid supply through three phase 400KVA, 260 V/25 kV, 60 Hz star/delta transformer. Load of 47 MVA, 120 kV/25 kV.2.1 MW has been applied to load Bus.

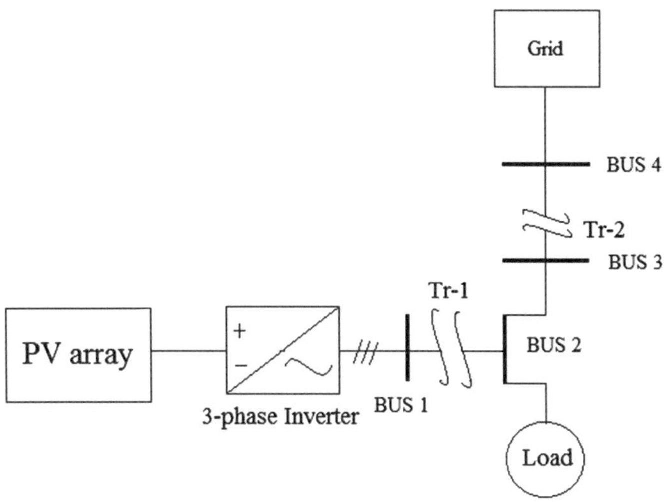

Fig. 1 Single line diagram of 400KW PV array based microgrid

3 FFT Based Harmonics Assessments of Line Current

The model described in model of Fig. 1 has been used for computer simulation. Three different current measurement units have connected at each phase output of inverter. Through these current measurement units individual phase currents have been captured. The waveform of phase currents are observed and analyzed by Fast Fourier Transform (FFT) at normal condition, LLL and LLLG faults. FFT spectrums are monitored and total harmonics distortions are measured. Continuous symmetrical fault at load bus (Bus 2) has been considered in this work.

3.1 Normal Condition

At first, line currents are captured at normal condition and FFT spectrums are generated as shown in Fig. 2a–c for phase-A, phase-B and phase-C respectively. THD at normal conditions are determined accordingly.

3.2 Fault Condition (LLL)

Then line currents are captured at LLL fault condition and FFT spectrums are generated as shown in Fig. 3a–c for phase-A, phase-B and phase-C respectively. THD at LLL fault conditions are determined accordingly.

3.3 Fault Condition (LLLG)

Then line currents are captured at LLLG fault condition and FFT spectrums are generated as shown in Fig. 4a–c for phase-A, phase-B and phase-C respectively.THD at LLLG fault conditions are determined accordingly.

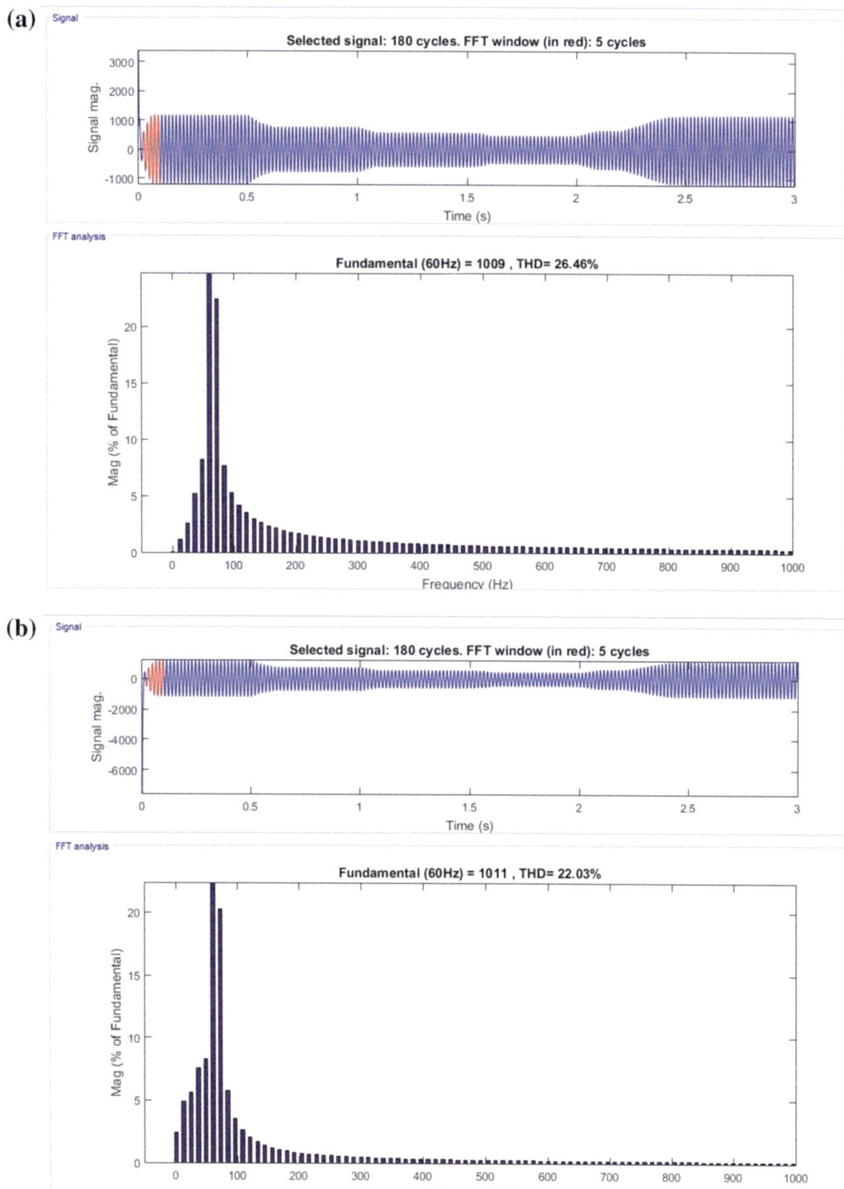

Fig. 2 Line currents and their FFT spectrum at inverter output at normal condition: **a** phase—A, **b** Phase—B and **c** Phase—C

(c)

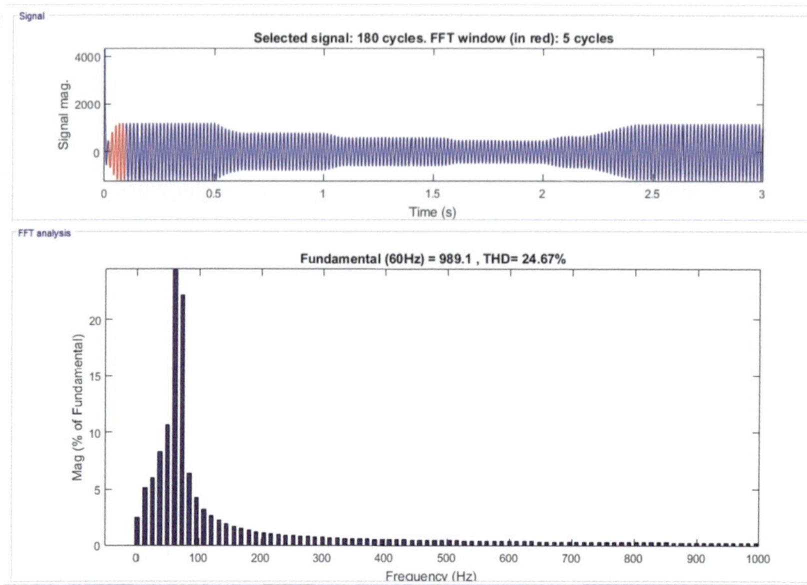

Fig. 2 (continued)

4 Comparative Study

FFT spectrums at different conditions are compared. Comparison shows significant changes in FFT spectrums of symmetrical fault conditions from that of normal condition. Also FFT spectrums at LLLG fault differ from FFT spectrums at LLL fault. The comparative results are presented in Table 1. After, FFT based spectrum comparison, THD values obtained at different conditions are compared. Maximum THD values at normal condition which reduces drastically at symmetrical fault condition. THD at LLLG and THD at LLL are found very closed to each other; however THD is found minimum at LLL fault condition as shown in Figs. 5 and 6 respectively.

5 Outcome

Specific outcome of this work is achievement of harmonics assessment based symmetrical fault diagnosis in solar PV array based micro grid system. FFT Based spectrum and THD comparison shows significant changes in those features and parameters at fault condition from that of normal condition. Also by this way significant changes in features and parameters are observed among ungrounded symmetrical fault and grounded fault conditions. THD values decrease at fault conditions and become lowest in LLL fault.

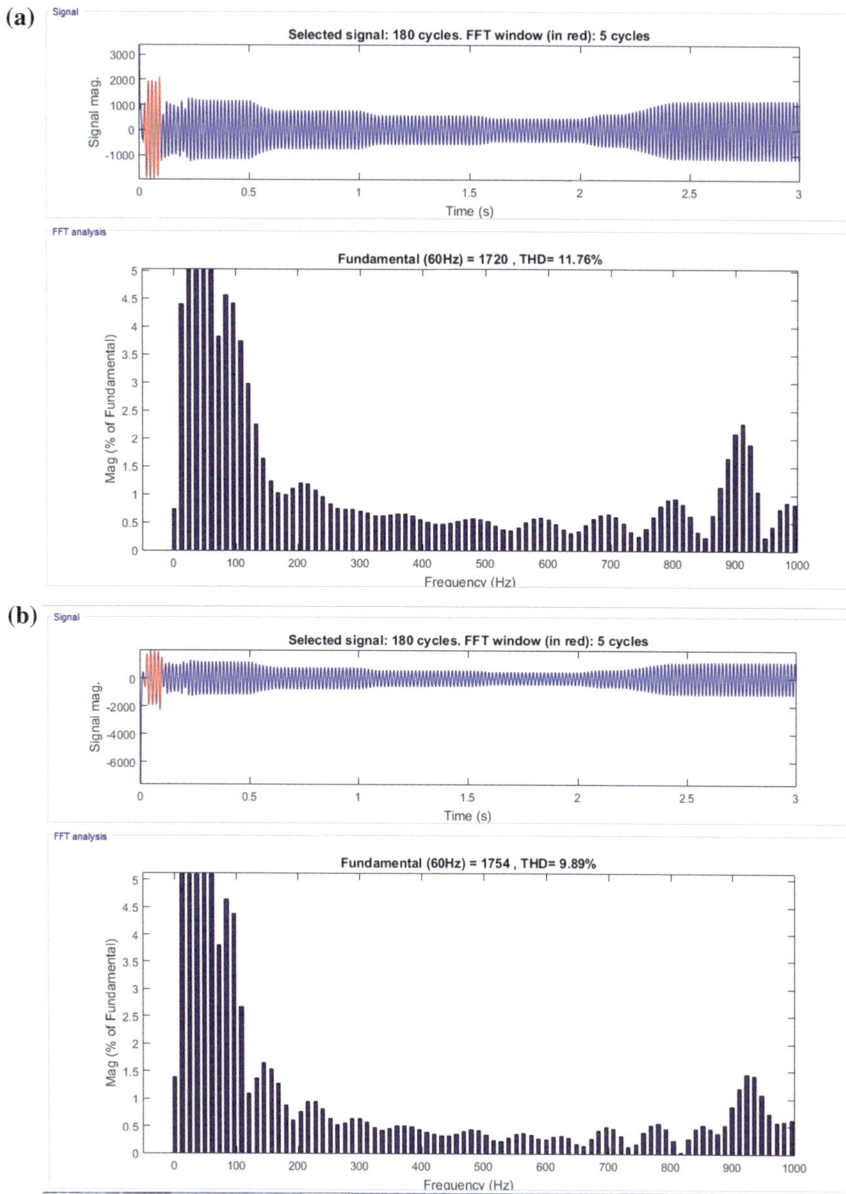

Fig. 3 Line currents and their FFT spectrum at inverter output during LLL Fault at load end: **a** phase—A, **b** Phase—B and **c** Phase—C

(c)

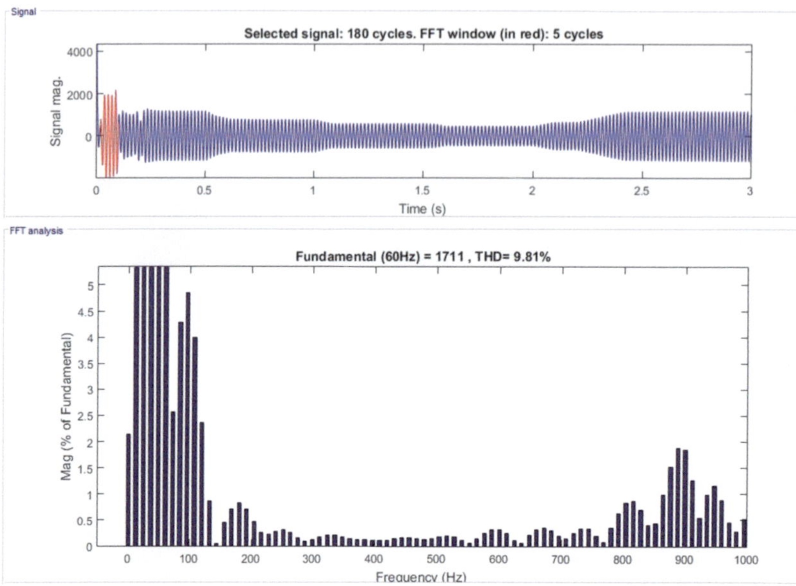

Fig. 3 (continued)

6 Conclusion

Harmonics assessment based symmetrical fault diagnosis in PV array based micro-grid system.

The paper deals with harmonics assessment based symmetrical fault diagnosis in PV array based microgrid system. This has been achieved by modeling a micro grid system consisting of 400KW PV array based power unit coupled with conventional power grid. Computer simulation performed at normal condition as well as symmetrical faults at load bus. Both grounded (LLLG) and ungrounded fault (LLL) are considered. Line currents are captured from the output currents of output system of inverter system and currents are obtained at different conditions. THD values are also determined and compared at different conditions. Based on the observation of grounded (LLLG) and ungrounded fault (LLL) of micro grid systems with respect those useful electrical parameters may be useful for synchronization, protection and performance analysis of various micro grid systems. The comparative results may be useful for symmetrical fault diagnosis and can be extended for diagnosis of other faults also.

(a)

(b)

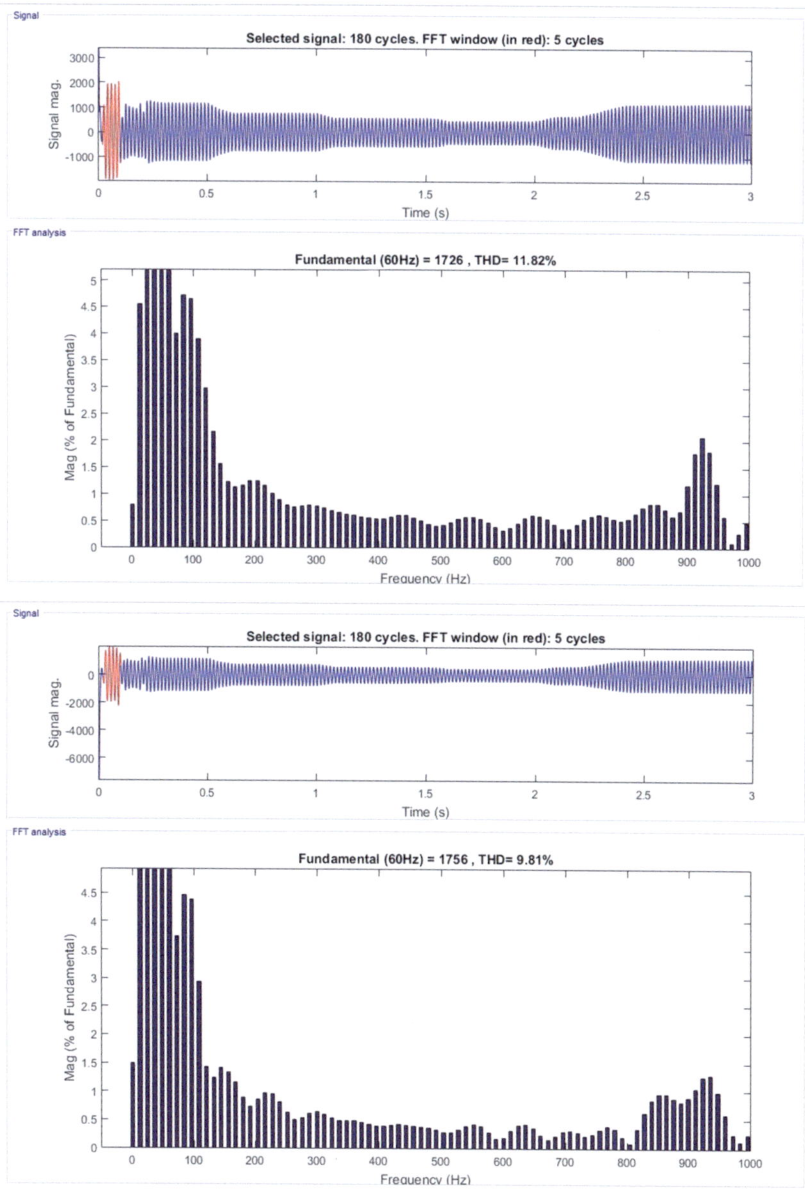

Fig. 4 Line currents and their FFT spectrum at inverter output during LLLG fault at load end: **a** phase—A, **b** Phase—B and **c** Phase—C

(c)

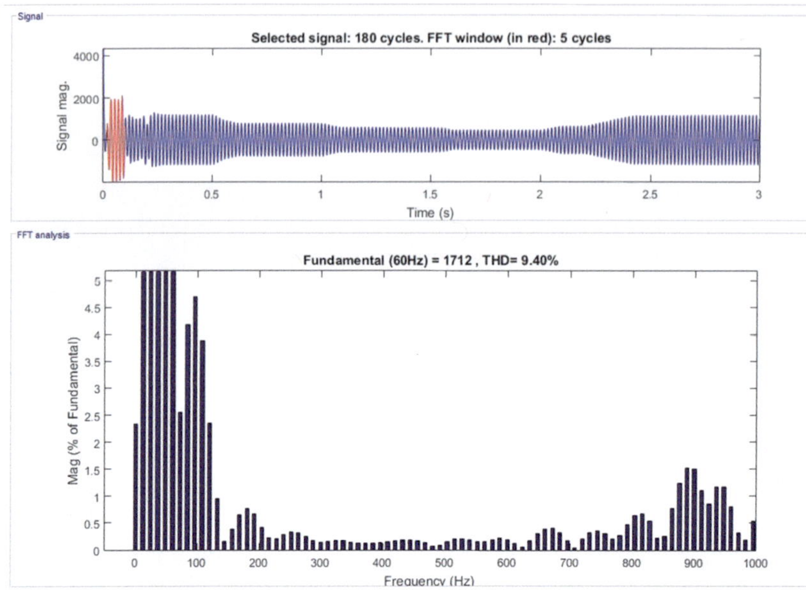

Fig. 4 (continued)

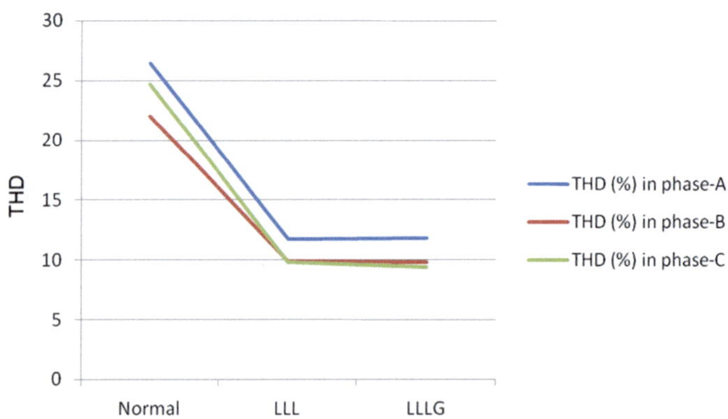

Fig. 5 THD of line currents at Phase-A, B and C during normal, LLL and LLLG fault conditions

Table 1 Result of FFT analysis at normal, LLL and LLLG fault conditions of average model 400 KW microgrid

Parameters	Phase	Normal Fig. 2a–c		LLL Fig. 3a–c		LLLG Fig. 4a–c	
SAMPLE PER CYCLE	A	333		333		333	
	B						
	C						
DC COMPONNENT	A	0.2347		12.5		13.75	
	B	24.12		24.23		26.11	
	C	24.35		36.74		39.86	
Fundamentals peak	A	1009		1720		1726	
	B	1011		1754		1756	
	C	989.1		1711		1712	
Fundamentals rms	A	713.3		1216		1220	
	B	7.5		1241		1242	
	C	699.4		1210		1210	
THD (%)	A	26.46		11.76		11.82	
	B	22.03		9.89		9.81	
	C	24.67		9.81		9.40	
Frequency (HZ)	Phase	Amplitude (%)	Angle	Amplitude (%)	Angle	Amplitude (%)	Angle
0	A	0.02	90.0°	0.73	90.0°	0.80	90.0°
	B	2.39	90.4°	1.38	90.0°	1.49	90.0°
	C	2.46	270.0°	2.15	270.0°	2.33	270.0°
60	A	100.00	69.4°	100.00	71.2°	100.00	−71.2°
	B	100.00	169.2°	100.00	167.8°	100.00	167.7°
	C	100.00	49.8°	100.00	47.3°	100.00	47.4°
180	A	1.98	5.7°	1.00	4.8°	1.16	3.5°
	B	0.95	222.1°	0.89	235.7°	0.89	224.7°
	C	1.36	160.7°	0.83	126.1°	0.77	132.4°
300	A	1.10	4.8°	0.70	9.2°	0.77	8.2°
	B	0.46	208.7°	0.63	189.2°	0.64	193.8°
	C	0.72	169.4°	0.12	131.6°	0.14	162.8°
420	A	0.77	5.2°	0.48	16.9°	0.58	10.2°
	B	0.31	202.7°	0.36	201.8°	0.41	193.2°
	C	0.49	174.1°	0.12	181.8°	0.17	182.9°
660	A	0.48	6.6°	0.45	2.7°	0.60	30.0°
	B	0.19	197.8°	0.14	174.6°	0.23	228.6°
	C	0.31	179.7°	0.32	186.3°	0.39	199.1°
780	A	0.41	7.3°	0.79	5.3°	0.54	32.3°

(continued)

Table 1 (continued)

Frequency (HZ)	Phase	Amplitude (%)	Angle	Amplitude (%)	Angle	Amplitude (%)	Angle
	B	0.16	196.4°	0.55	207.0°	0.34	238.0°
	C	0.26	181.7°	0.35	148.2°	0.28	179.0°
900	A	0.35	7.5°	2.10	−15.0°	0.17	−32.6°
	B	0.13	195.0°	0.86	104.3°	0.89	56.5°
	C	0.22	183.1°	1.84	189.8°	1.50	184.8°
960	A	0.32	9.6°	0.42	203.1°	0.59	−34.9°
	B	0.12	198.1°	0.73	−70.5°	0.59	230.9°
	C	0.21	184.5°	0.88	80.8°	0.81	97.4°

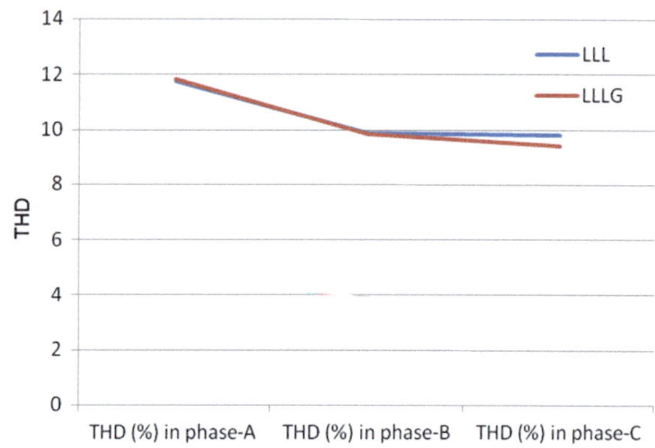

Fig. 6 THD of line currents at Phase-A, B and C during LLL and LLLG fault conditions

References

1. S. Anwar, A. Elrayyah, Y. Sozer, in *Harmonics elimination and distribution using decentralized control for microgrid applications*. Energytech, 2013 IEEE, Date of Conference: 21–23 May 2013, IEEE Xplore: 24 October 2013, Electronic ISBN: 978-1-4673-4444-9 (2013)
2. S. Dhar, P.K. Dash, Performance analysis of a new fast negative sequence power injection oriented islanding detection technique for photovoltaic based voltage source converter based micro grid operation. IET Gener. Transm. Distrib. **9**(15), 2079–2090 (2015)
3. D. Kumar, F. Zare, Harmonic analysis of grid connected power electronic systems in low voltage distribution networks. IEEE J. Emerg. Sel. Top. Power Electron. **4**(1), 70–79 (2015)
4. X. Wang, E.E. Yaz, Smart Power Grid Synchronization with Fault Tolerant Nonlinear Estimation. IEEE Trans. Power Syst. **31** (2016)
5. G. Rashid, M.H. Ali, Transient stability enhancement of doubly fed induction machine-based wind generator by bridge-type fault current limiter. IEEE Trans. Energy Convers. **30**(3), 939–947 (2015)
6. W. Chen, F. Blaabjerg, N. Zhu, M. Chen, D. Xu, Doubly fed induction generator wind turbine systems subject to recurring symmetrical grid faults. IEEE Trans. Power Electron. **31**, 1143–1160 (2015)
7. S.-J. Chen, T.-S. Zhan, C.-H. Huang, J.-L. Chen, C.-H. Lin, Nontechnical loss and outage detection using fractional-order self-synchronization error-based fuzzy petri nets in micro-distribution systems. IEEE Transactions on Smart Grid **6**(1), 411–420 (2014)

8. M.A. Farhan, K.S. Swarup, Mathematical morphology-based islanding detection for distributed generation. IET Gener. Transm. Distrib. **10**(2), 518–525 (2016)
9. M. Golsorkhi, M. Savaghebi, L. Dylan, J. Guerrero, J. Vasquez, A GPS—based control framework for accurate current sharing and power quality improvement in microgrids. IEEE Trans. Power Electron. **99**, 1–1 (2016)
10. H. Mohammadpour, S.G. Zadeh, S. Tohidi, Symmetrical and asymmetrical low-voltage ride through of doubly-fed induction generator wind turbines using gate controlled series capacitor. IET Renew. Power Gener. **9**(7, 9), 840–846 (2015)
11. R.M. Kamel, New inverter control for balancing standalone micro-grid phase voltages: a review on MG power quality improvement. Renew. Sustain. Energy Rev. **63**, 520–532 (2016)
12. B. Sabu, A. George, in *Harmonic mitigation in minigrid integrated distributed power system*. Circuit, Power and Computing Technologies (ICCPCT), 2016 International Conference on, IEEE Xplore: 04 August 2016, IEEE Xplore: 04 August 2016, INSPEC Accession Number: 16195204, 18–19 March 2016
13. P. Sreekumar, V. Khadkikar, A new virtual harmonic impedance scheme for harmonic power sharing in an islanded microgrid. IEEE Trans. Power Delivery **31**(3), 936–945 (2016)
14. R. Zhu, Z. Chen, X. Wu, F. Deng, Virtual damping flux-based LVRT control for DFIG-based wind turbine". IEEE Trans. Energy Convers. **30**(2), 714–725 (2015)
15. T.A. Taj, H.M. Hasanien, A.I. Alolah, S.M. Muyeen, Transient stability enhancement of a grid-connected wind farm using an adaptive neuro-fuzzy controlled-flywheel energy storage system. IET Renew. Power Gener. **9**, 792–800 (2015)
16. S. Chattopadhyay, M. Mitra, S. Sengupta, *Electric Power Quality*, 1st edn. (Springer, Berlin, 2011)
17. S. Chattopadhyay, S. Karmakar, M. Mitra, S. Sengupta, *Induction Motor Fault Diagnosis*, 1st edn. (Springer, Berlin, 2016)

Optimal Design of KVAr Based SVC for Improvement of Stability in Electrical Power System

Sayantan Adhikary and Sandip Chanda

1 Introduction

In modern power system, sudden change of load, faults has a substantial frequency of occurrences. A considerable amount of research of all over the globe is going on for presenting and economical reliable fast response controller to cope up with this changed environment, FACTS controller is one of them.

A method has been on developed [1] "Optimal location and sizing of static VAR compensator (SVC) based on Particle Swarm Optimization for minimization of transmission losses considering cost function. The method however is, Artificial Intelligence based and requires extensive computational time."

This method explained [2] "A novel global harmony search algorithm (NGHS) is used to determine the optimal location and size of shunt reactive power compensators such as shunt capacitors, static VAR compensators (SVCs), and static synchronous compensators (STATCOMs) in a transmission network. The algorithm though a quite efficient, requires large memory for its binary search operation."

This method approach [3] "A method to seek the optimal location of several static VAR compensators (SVCs) in a power system based on their primary function. Taking advantages of the flexible ac transmission system (FACTS) devices depends largely on how these devices are placed in the power system, namely, on their location and size."

In the work explained [4] "Advanced load flow models for the static VAR compensator (SVC) are presented in this paper. The models are incorporated into existing load flow (LF) and optimal power flow (OPF) Newton algorithms. Unlike SVC models

---

S. Adhikary (✉) · S. Chanda
Department of Electrical Engineering, Narula Institute of Technology, Kolkata, India
e-mail: sayantanaad@gmail.com

S. Chanda
e-mail: sandipee1978@gmail.com

S. Chattopadhyay et al. (eds.), *Modelling and Simulation in Science, Technology and Engineering Mathematics*, Advances in Intelligent Systems and Computing 749,
https://doi.org/10.1007/978-3-319-74808-5_10

available in open literature, the new models depart from the generator representation of the SVC and are based instead on the variable shunt susceptance concept. The work though manages to determine the optimal location of SVC but does not an enlighten, the physical design of SVC."

This method approaches [5] "Optimal location of SVC in power system based on the primary function taking advantage of faults device depends largely on how these devices are placed in power system."

This method states that [6] "Dynamic reactive power compensation is used to an increasing extended improve voltage and reactive power system. Additional takes can also be performed SVC to increase in power transmission capability."

In the work explained [7] "Power system stability enhancement via robust coordinated design of a power system stabilizer and a static VAR compensator-based stabilizer is thoroughly investigated in this paper. The coordinated design problem of robust excitation and SVC-based controllers over a wide range of loading conditions and system configurations are formulated as an optimization problem with an eigenvalue-based objective function. This work again fails to substantial the voltage level at a desired value."

In this method develop by [8] "The enhancement of power system stability properties by use of thyristor controlled series capacitors (TCSCs) and static VAR systems (SVCs). Models suitable for incorporation in dynamic simulation programs used to study angle stability are analyzed."

This method approaches [9] "Power demand has increased substantially while the expansion of power generation and transmission has been severely limited due to limited resources and environmental restrictions."

This method states that [10] "Different control techniques for damping undesirable inter area oscillation in power systems by means of power system stabilizers (PSS), static VAR compensators (SVCs), and shunt static synchronous compensators (STATCOMs)."

This method approaches [11] "The location of SVC (static VAR compensators) and other types of shunt compensation devices for voltage support is an important practical question. This paper considers a tool based on the determination of critical modes. Critical modes are computed by studying the system modes in the vicinity of the point of collapse. System participation factors for the critical mode are used to determine the most suitable sites for system reinforcement."

This method states that [12] "A novel method for optimal location of FACTS devices in a multi machine power system using Genetic Algorithm (GA). Using the proposed method, the location of FACTS controllers, their type and rated values are optimized simultaneously. Among the various FACTS controllers, Static VAR controller (SVC), Thyristor Controlled Series Compensator (TCSC) and Unified power Flow Controller (UPFC) are considered. This method again requires computational facility and also memory."

A method has been developed on [13] "A new SVC (static VAR compensation) control for damping of power system oscillations has been developed. To increase system damping an SVC uses a phase angle signal estimated from the measurement of voltage and power at the SVC location."

In this method develop by [14] "A new power system stabilizer (PSS) design for damping power system oscillations focusing on inters area modes. The input to the PSS consists of two signals. The first signal is mainly to damp the local mode in the area where PSS is located using the generator rotor speed as an input signal. The second is an additional global signal for damping inters area modes."

A method has been developed on [15] "Analysis and simulation of SVC controller have been investigated to improve the dynamic stability of power systems. Eigen-values calculated by linear system models, the impact and changes the controller parameters on the dynamic behavior of the system will be study."

From the above litterateur review it has been understood that all the controller's design for optimal locative is voltage based SVC and PSS. In this work a KVAr based SVC has been developed to limit a variation of voltage even in worst possible loading of reactive power. Simulation has been carried out in MATLAB software the result was wide encouraging and promising.

2 Theory

2.1 Use of SVC in Transmission Line

A static VAR compensator (or SVC) is an electrical device for providing fast-acting reactive power on high-voltage electricity transmission networks. SVCs are part of the Flexible AC transmission system device family, regulating voltage and stabilizing the system. The term "static" refers to the fact that the SVC has no moving parts (other than circuit breakers and disconnects, which do not move under normal SVC operation). Prior to the invention of the SVC, power factor compensation was the preserve of large rotating machines such as synchronous condensers. The SVC is an automated impedance matching device, designed to bring the system closer to unity power factor. If the power system's reactive load is capacitive (leading), the SVC will use reactors (usually in the form of Thyristor-Controlled Reactors) to consume var from the system, lowering the system voltage. Under inductive (lagging) conditions, the capacitor banks are automatically switched in, thus providing a higher system voltage. They also may be placed near high and rapidly varying loads, such as arc furnaces, where they can smooth flicker voltage. It is known that the SVCs with an auxiliary injection of a suitable signal can considerably improve the dynamic stability performance of a power system. It is observed that SVC controls can significantly influence nonlinear system behavior especially under high-stress operating conditions and increased SVC gains.

2.1.1 Principle

Typically, an SVC comprises one or more banks of fixed or switched shunt capacitors or reactors, of which at least one bank is switched by thyristors. Elements which may be used to make an SVC typically include:

- Thyristor controlled reactor (TCR), where the reactor may be air- or iron-cored
- Thyristor switched capacitor (TSC)
- Harmonic filter(s)
- Mechanically switched capacitors or reactors (switched by a circuit breaker) (Fig. 1).

By means of phase angle modulation switched by the thyristors, the reactor may be variably switched into the circuit and so provide a continuously variable MVAR injection (or absorption) to the electrical network. In this configuration, coarse voltage control is provided by the capacitors; the thyristor-controlled reactor is to provide smooth control. Smoother control and more flexibility can be provided with thyristor-controlled capacitor switching.

The thyristors are electronically controlled. Thyristors, like all semiconductors, generate heat and deionized water is commonly used to cool them. Chopping reactive load into the circuit in this manner injects undesirable odd-order harmonics and so

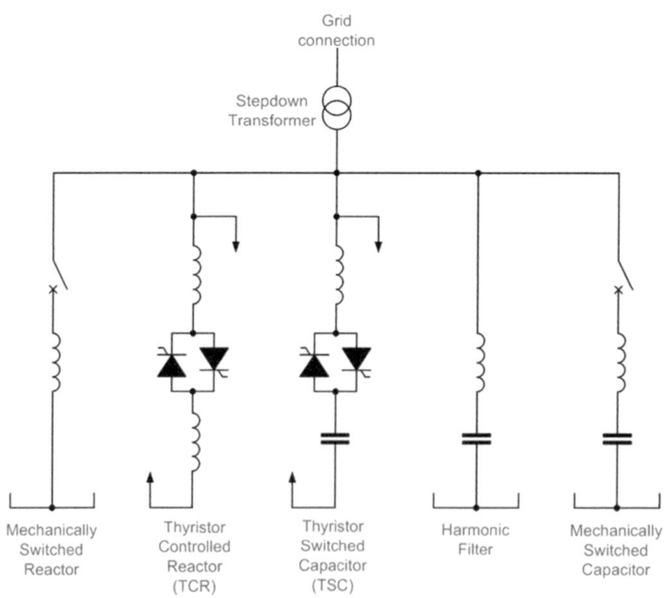

Fig. 1 One-line diagram of a typical SVC configuration; here employing a thyristor controlled reactor, a thyristor switched capacitor, a harmonic filter, a mechanically switched capacitor and a mechanically switched reactor

banks of high-power filters are usually provided to smooth the waveform. Since the filters themselves are capacitive, they also export MVARs to the power system.

More complex arrangements are practical where precise voltage regulation is required.

Voltage regulation is provided by means of a closed-loop controller. Remote supervisory control and manual adjustment of the voltage set-point are also common.

Generally, static var compensation is not done at line voltage; a bank of transformers steps the transmission voltage (for example, 230 kV) down to a much lower level (for example, 9.0 kV). This reduces the size and number of components needed in the SVC, although the conductors must be very large to handle the high currents associated with the lower voltage. In some static var compensators for industrial applications such as electric arc furnaces, where there may be an existing medium-voltage bus bar present (for example at 33 or 34.5 kV), and the static var compensator may be directly connected in order to save the cost of the transformer.

Another common connection point for SVC is on the delta tertiary winding of Y-connected auto-transformers used to connect one transmission voltage to another voltage.

The dynamic nature of the SVC lies in the use of thyristors connected in series and inverse-parallel, (forming "thyristor valves"). The disc-shaped semiconductors, usually several inches in diameter, are usually located indoors in a "valve house".

The main advantage of SVCs over simple mechanically switched compensation schemes is their near-instantaneous response to changes in the system voltage. For this reason they are often operated at close to their zero-point in order to maximize the reactive power correction they can rapidly provide when required.

They are, in general, cheaper, higher-capacity, faster and more reliable than dynamic compensation schemes such as synchronous condensers. However, static var compensators are more expensive than mechanically switched capacitors, so many system operators use a combination of the two technologies (sometimes in the same installation), using the static var compensator to provide support for fast changes and the mechanically switched capacitors to provide steady-state var.

2.2 *Traditional Operation of SVC*

2.2.1 Generation, Transmission, Distribution

In any power system, the creation, transmission, and utilization of electrical power can be separated into three areas, which traditionally determined the way in which electric utility companies had been organized.

- Generation
- Transmission
- Distribution.

Although power electronic based equipment is prevalent in each of these three areas, such as with static excitation systems for generators and Custom Power equipment in distribution systems, the focus of this paper and accompanying presentation is on transmission, i.e., moving the power from where it is generated to where it is utilized.

2.2.2 Power System Constraints

As noted in the introduction, transmission systems are being pushed closer to their stability and thermal limits while the focus on the quality of power delivered is greater than ever. The limitations of the transmission system can take many forms and may involve power transfer between areas or within a single area or region and may include one or more of the following characteristics:

- Steady-State Power Transfer Limit
- Voltage Stability Limit
- Dynamic Voltage Limit
- Transient Stability Limit
- Power System Oscillation Damping Limit
- Inadvertent Loop Flow Limit
- Thermal Limit
- Short-Circuit Current Limit
- Others.

Each transmission bottleneck or regional constraint may have one or more of these system-level problems. The key to solving these problems in the most cost-effective and coordinated manner is by thorough systems engineering analysis.

3 Proposed Methodology of SVC

See Fig. 2.

4 Development of Relation Between KVAr and SHUNT Compensation by Simulation

A 2 machine system with a load and variable susceptance support has been demonstrated in figure [3] MATLAB (Fig. 3).

The system has loaded KVAr as depicted, in table for each value of KVAr loading the variable susceptance in adjusted, to provide a voltage band of 0.9–1 pu for worst possible reactive power loading, this susceptance has been calculated and the relation

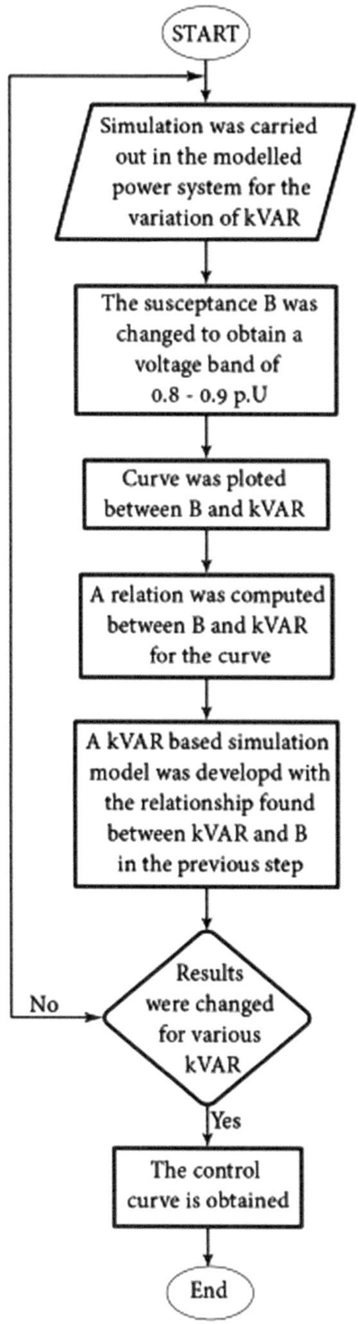

Fig. 2 Proposed methodology of SVC flow chart

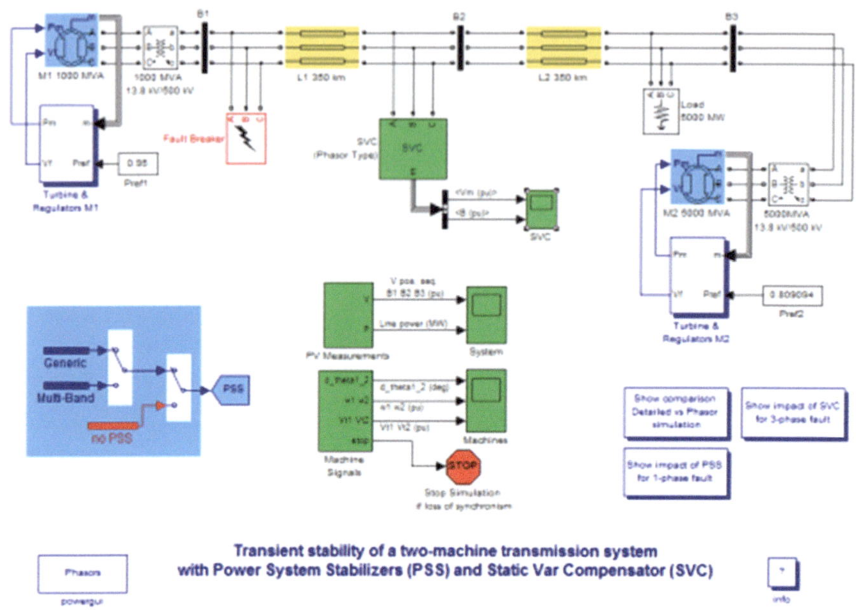

Fig. 3 Transient stability of a two-machine transmission system with power system stabilizers (PSS) and static var compensator (SVC)

between required susceptance subsequent KVAr is plotted. In figure [3] from the curve sample points were extracted and for each of the points a hyperbolic equation has been formed.

$$2.5 \times [(10)]^{\wedge}6x\_(1^{\wedge}2) + 1.2 \times [(10)]^{\wedge}6x\_1 + 1 \times [(10)]^{\wedge}6 = 15 \quad (1)$$

$$3.5 \times [(10)]^{\wedge}6x\_(2^{\wedge}2) + 3 \times [(10)]^{\wedge}6x\_2 + 2 \times [(10)]^{\wedge}6 = 20 \quad (2)$$

$$4 \times [(10)]^{\wedge}6x\_(3^{\wedge}2) + 5 \times [(10)]^{\wedge}6x\_3 + 3 \times [(10)]^{\wedge}6 = 25 \quad (3)$$

By solving the above 3 equation, the following relation has been obtained

$$63.5x\_(1^{\wedge}2) - 1.979x\_1 + 0.0822 = B$$

where, x = KVAR and, B = Susceptance

This equation becomes (Fig. 4).

$$63.5kVAR^2 - 1.97kVAR + 0.0822 = B \quad (4)$$

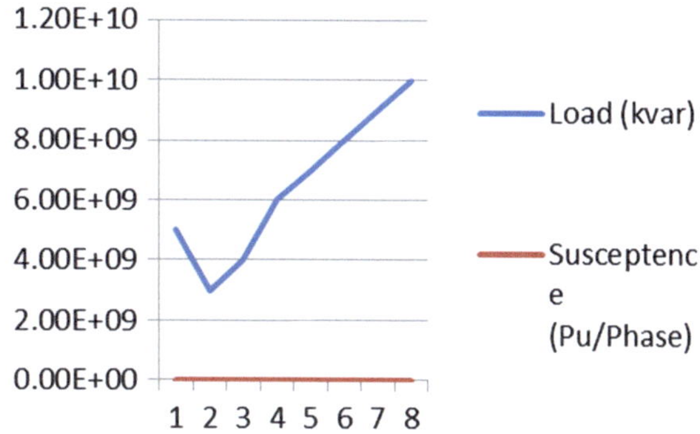

Fig. 4 Curve fitting the susceptence with respect to KVAr loading

5 Development of MATLAB Model

A Matlab based model of SVC based on equation no [4] has been developed. This model being operation, in care voltage based SVC, the range is very limited. As the span of variation is also limited (Figs. 5 and 6).

6 Description of the Blocks (with Svc and with Out Svc)

6.1 Parameters

The SVC parameters are grouped in two categories: Power Data and Control Parameters. Use the Display list box to select which group of parameters you want to visualize.

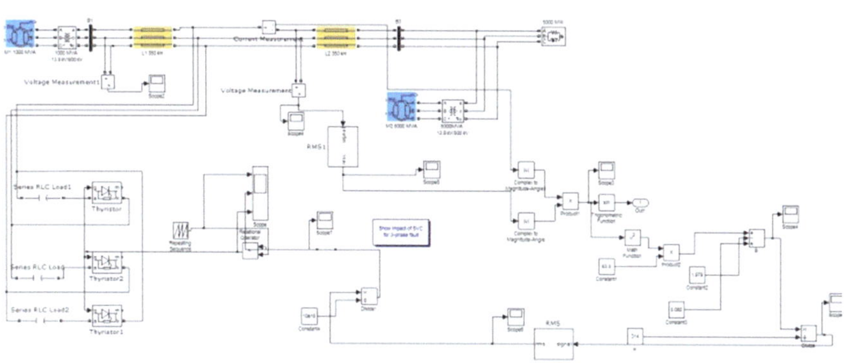

Fig. 5 Improvement of transient stability with svc diagram using MATLAB

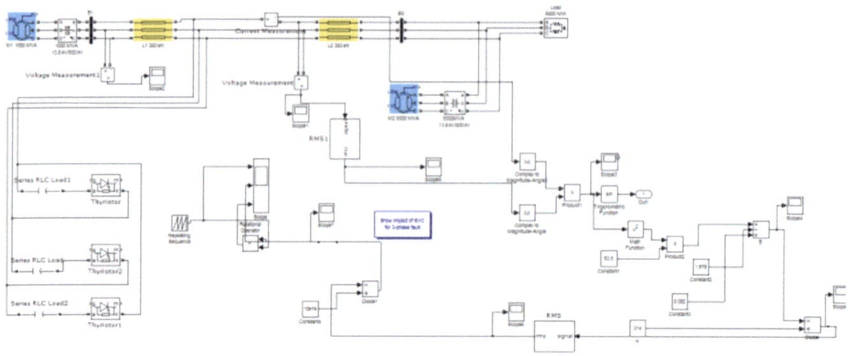

Fig. 6 Improvement of transient stability without svc diagram using MATLAB

6.2 Power Tab

Ignore negative-sequence current

The SVC is modeled by a three-wire system using two current sources. The SVC does not generate any zero-sequence current, but it can generate negative-sequence currents during unbalanced system operation. The negative sequence susceptance of the SVC is assumed to be identical to its positive-sequence value, as determined by the B value computed by the voltage regulator.

Select this box to ignore negative-sequence current. Default is selected (Fig. 7).

6.3 Nominal Voltage and Frequency

The nominal line-to-line voltage in Vrms and the nominal system frequency in hertz. Default is [500e3, 50].

6.4 Three-Phase Base Power

Three-phase base power, in VA, used to specify the following parameters in pu: droop reactance Xs, gains Kp and Ki of the voltage PI regulator, and reference susceptance Brief. This base power is also used to normalize the output B susceptance signal. Default is 200e6 (Fig. 8).

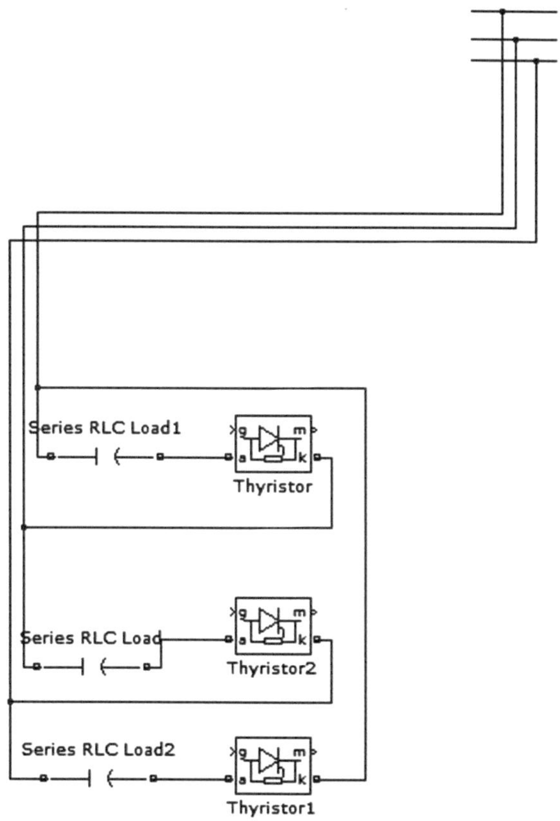

Fig. 7 Delta connected thyristor bridge

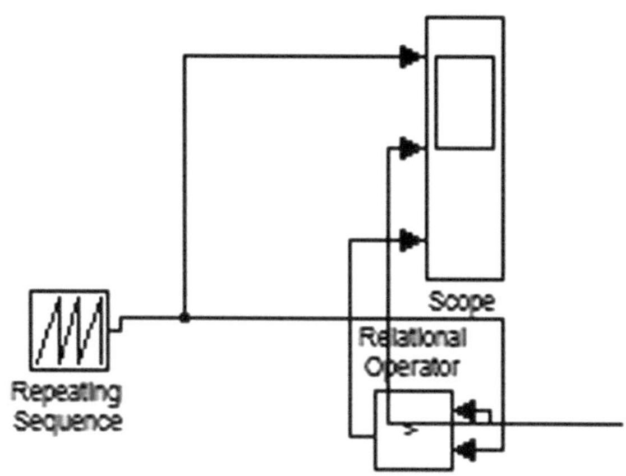

Fig. 8 Computation of PWM pulse

6.5 Reactive Power Limits

The maximum SVC reactive powers at 1 pu voltage, in vars. Enter a positive value for the capacitive reactive power Qc (var generated by the SVC) and a negative value for the inductive reactive power Ql (var absorbed by the SVC). Default is [200e6, −200e6].

Average time delay due to thyristor valves firing Average time delay simulating the non-instantaneous variation of thyristor fundamental current when the distribution unit sends a switching order to the pulse generator. Because pulses have to be synchronized with thyristor commutation voltages, this delay normally varies between 0 and 1/2 cycle. The suggested average value is 4 ms. Default is 4e-3 (Fig. 9).

6.6 Susceptance Brief

This parameter is not available when the Mode parameter is set to Voltage regulation (Fig. 10).

6.7 Reference

Susceptance, in Pu/Phase, when the SVC is operating in var control mode. Default is 0.0.

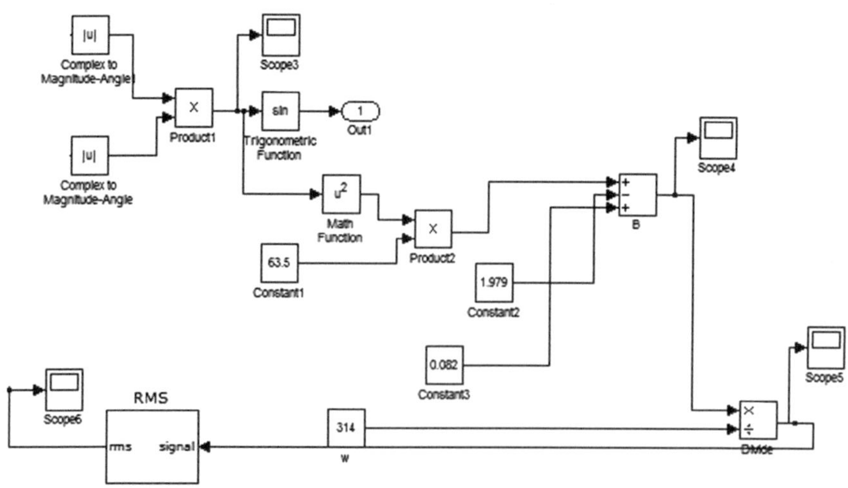

Fig. 9 KVAr sampling

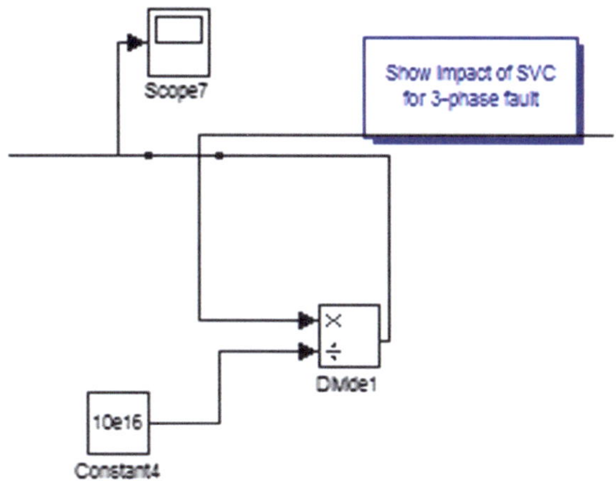

Fig. 10 Computation of requires susceptance

6.8 Simulation and Case Study

After development of improvement stability with SVC Diagram using Matlab, we get 5 simulation figures

(a) Voltage without SVC
(b) RMS Value
(c) KVAr Calculated
(d) Susceptance Calculated
(e) Carrier Signal
(f) Susceptance (RMS)
(g) PWM (Pulse Width Modulation).

1. Firstly voltage measurement block applied then we get voltage without SVC Curve.
2. Current Measurement block applied then we get RMS value.
3. We applied repeating sequence triangular value, then we get Carrier Signal, Susceptance (RMS) and PWM (Pulse Width Modulation) (Figs. A, B, C, D, E.1, E.2, and E.3).

6.9 Case Study

Comparison between with SVC and without SVC

After the development of the model a comparison between presence of develop model of SVC and without SVC has been demonstrated in the table, from this table it can be asserted that the develop model of with SVC is capable of regulating the voltage with desired level, in adverse of reactive power loading.

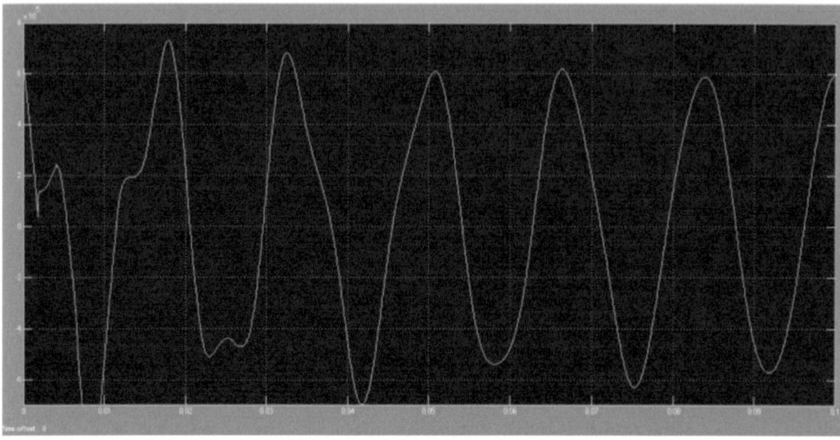

Fig. A Voltage without SVC

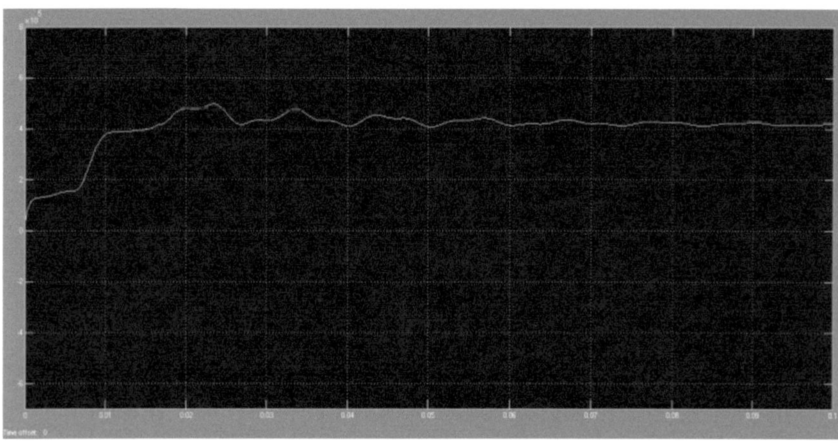

Fig. B RMS value

7 Conclusion

The work presented in this paper mainly focuses on the aspects related to Flexible AC Transmission Systems (FACTS) based controller design and assessment of their contribution to system stability improvement ensuring secure and stable operation of the power system. Most of the SVC controllers, with substantial survey has been found to be voltage based, and sub sequentially they offer less scope of making the control action in the work presented in this paper. Focuses on developing a

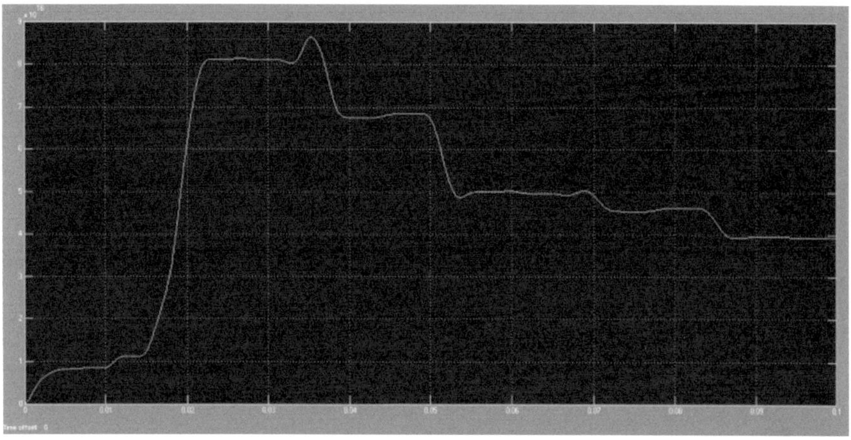

Fig. C KVAR calculated

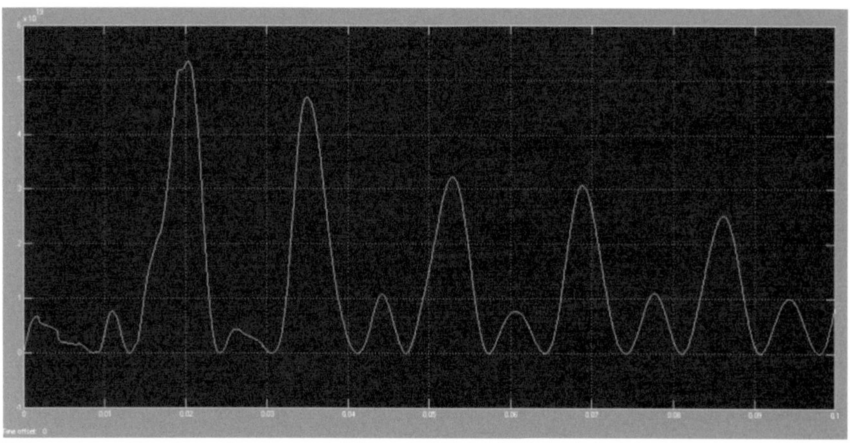

Fig. D Susceptance calculated

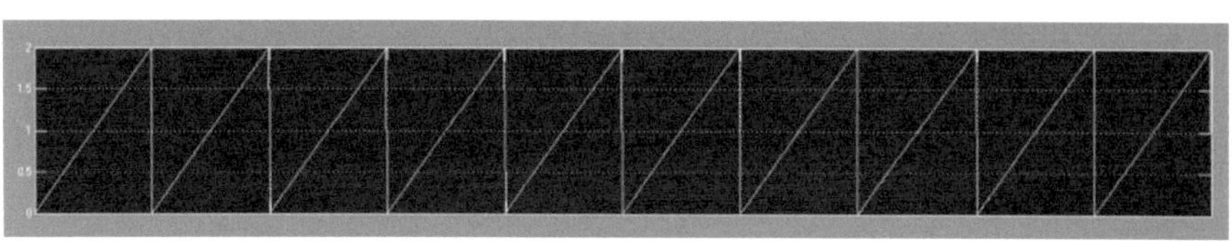

Fig. E.1 Carrier signal

KVAr based model of SVC, which without the help of in any optimization technique effectively stabilizes, the voltage profile of given power system network for adverse variation of reactive power loading. This idea may be pursuit to develop to static var compensator (SVC) for future power network.

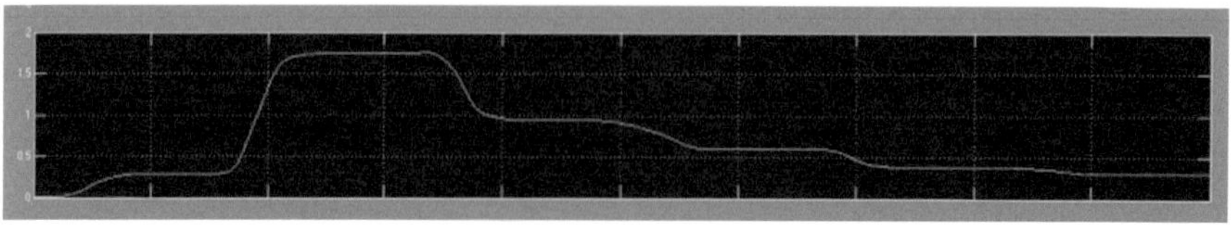

Fig. E.2 Susceptance (RMS)

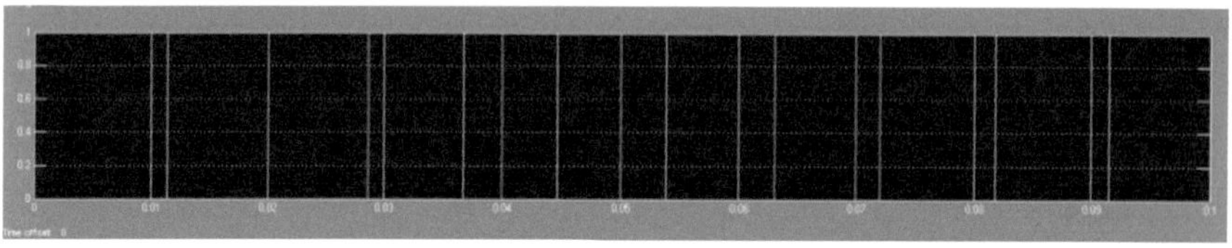

Fig. E.3 PWM

Load (KVAR)	With SVC line voltage	With out SVC line voltage
4000e6	$0.6 * 10^6$	$6 * 10^5$
5000e6	$6 * 10^5$	$7 * 10^5$
7000e6	$6.5 * 10^5$	$8 * 10^5$
8000e6	$7 * 10^5$	$8 * 10^5$
9000e6	$8 * 10^5$	$8 * 10^5$
10000e6	$9 * 10^5$	$8.25 * 10^5$
20000e6	$10 * 10^5$	$8.9 * 10^5$
30000e6	$10 * 10^5$	$9 * 10^5$
40000e6	$12 * 10^5$	$9 * 10^5$
50000e6	$12 * 10^5$	$10 * 10^5$

References

1. S.A. Jumaat, H. Mokhlis, in *Optimal Location and Sizing of SVC Using Particle Swarm Optimization Technique*. Dec 2011
2. R. Sirjani, A. Mohamed, in *Optimal allocation of shunt Var compensatorsin power systems using a novel global harmony search algorithm*. Dec 2012
3. M.M. Farsangi, H. Nezamabadi, in *Placement of SVCs and selection of stabilizing signals in power systems*. Aug 2007
4. J.A.P. Filho, H. Pinto, in *Advanced SVC models for Newton-Raphson load flow and Newton optimal power flow studies*, vol. 15. no. 1, Feb 2000
5. E.Z. Zhou, in *Application of static Var compensator to increase power system damping*. June 1982
6. M.A. Abido, Y.L. Abdel-Magid, in *Coordinated design of a PSS and an SVC based controller to enhance power system stability*. June 2003
7. S. Gerbex, R. Cherkaoui, J. Germone, in *Optimal location of FACTS device to enhance power system security*. June 2003

8. M. Noroozian, I. Hiskens, in *A robust control strategy of shunt and series reactive compensator to damp electromechanical oscillations*. Oct 2001
9. R. Sirjani, A. Mohamed, in *optimal allocation of shunt Var compensator in power system using a novel global harmony search algorithm*. Dec 2012
10. C.A. Canizares, J. Reeve, in *Comparison of PSS, SVC and statcom controllers for damping power system oscillations*. Aug 2003
11. Y. Mansoar, W. Xu, in *SVC Placement using critical modes of voltage stability*. May 1994
12. S. Gerbex, A.J. Gevaond, in *Optimal location of multi-machine system SVC using genetic algorithm*. June 2015
13. E. Larch, L. Xu, in *Advanced SVC control for damping power system oscillations*. vol. 3. May 1993
14. E. Acha, C.R. Fuerte, in *Advanced SVC models for Newton Raphson load flow And Newton optimal power flow studies*. June 2003
15. M.E. About, A.A. Sallam, *Damping controllers design for power system oscillation using global signals*. June 1996

An Improved Reactive Power Compensation Scheme for Unbalanced Four Wire System with Low Harmonic Injection Using SVC

Sankar Das, Debashis Chatterjee and Swapan K. Goswami

1 Introduction

The loads in power distribution system are generally single-phase loads supplied from a Δ/Y three-phase transformer with grounded neutral [1]. The other commonly used loads are nonlinear loads, single-phase and three-phase rectifiers, power-electronics-based equipment etc. The increased use of these types of loads generates various power quality problems in the distribution network. These are poor voltage regulation, high reactive power demand, harmonic currents, unbalanced load, excessive neutral current, etc. [2, 3]. The unequal load current due to asymmetrical load contains positive, negative, and zero sequence component. It will increase system losses and can also be harmful on industrial machines and generators. Many techniques are suggested for load balancing as well as neutral current compensation along with load harmonic elimination for three-phase four-wire distribution system.

Pulse width modulation (PWM) based switching compensator, known as 'active power filters' [4, 5], or 'power conditioner', as reactive power compensator, or both of them as hybrid devices can be applied to diminish the power quality problems effectively. It includes distribution static compensator (DSTATCOM) [6, 7] for solving power quality problem in current, dynamic voltage restorer (DVR) for compensating power quality problem in voltage, and unified power-quality conditioner (UPQC) for both current and voltage power quality problem. However, all of these techniques increase system losses; implementation cost and requires complex control strategy [8].

S. Das (✉)
Department of Electrical Engineering, Government College of Engineering and Textile Technology, Berhampore, Murshidabad, WB, India
e-mail: sankar05ju@gmail.com

D. Chatterjee · S. K. Goswami
Department of Electrical Engineering, Jadavpur University, Kolkata, WB, India
e-mail: debashisju@yahoo.com

S. Chattopadhyay et al. (eds.), *Modelling and Simulation in Science, Technology and Engineering Mathematics*, Advances in Intelligent Systems and Computing 749,
https://doi.org/10.1007/978-3-319-74808-5_11

The SVC based compensation scheme can also be used to solve power quality problems in four-wire distribution system. The combination of Y-SVC and Δ-SVC can be used to mitigate both ZPS and NPS currents simultaneously [9, 10] as single SVC configuration can not serve both the purposes. However, the operation of TCR releases significant odd harmonic currents [11] into the supply system, whereas FC or TSC amplifies the harmonic currents generated from TCR and the other nonlinear load. Thus, a combination of reactive power compensator using SVC along with passive filter [12] or combination of shunt passive filter in series with an active power filter topology [13–16] have been proposed to solve power quality problem. However, these schemes require additional investment [17] and space to accommodate filtering stages. Thus, improvement of switching schemes of SVC is also investigated to minimize harmonic generation internally [18–20] without using additional filter. The load compensation as well as source power factor improvement with minimum line harmonic injection is studied for three-phase three-wire system [21]. However, there are no suitable schemes for load balancing, neutral current compensation and source power factor correction with minimum line harmonic injection for three-phase four-wire distribution system.

In this paper, reactive power compensation is achieved by using a combined Δ-SVC and Y-SVC. At the same, the minimum harmonic injection of SVC is realized by optimizing switching function of SVC. Thus, the proposed scheme removes additional filter requirement. Both the SVCs use TSC-TCR in the proposed system modeling. The switching function of TCR is optimized using gravitational search algorithm [22, 23] and the optimized switching angles are computed offline at close interval of modulation indices which can be expressed as the ratio between fundamental component of reactor voltage to the rated load voltage. The corresponding reactive power drawn by TCR is calculated based on computed switching angles. These switching angles corresponding to minimum injected harmonics along with reactive power compensation are stored in the processor memory as a function of modulation index for online application. Different simulation results on a practical system are presented to validate the proposed concept.

2 Proposed Compensation Model Using Symmetrical Component Approach

The Fig. 1 shows schematic diagram of a Y-SVC and Δ-SVC connected to three-phase four-wire distribution system for reactive power compensation with minimum line harmonic injection. The subscript "x" denotes phase a, b and c. The I_x^L is the line current; V_x^S is the source voltage; V_x^L is the grid voltage; I_x^{TY}, I_x^{CY}, $I_x^{T\Delta}$ and $I_x^{C\Delta}$ are the Y-TCR, the Y-TSC, the Δ-TCR and the Δ-TSC current respectively; Z_x is the line impedance and Z_n is the neutral impedance. The distribution system is assumed to be a constant balanced voltage source and equal line impedances. For an unbalanced three-phase loads, the unbalanced distribution line currents causes

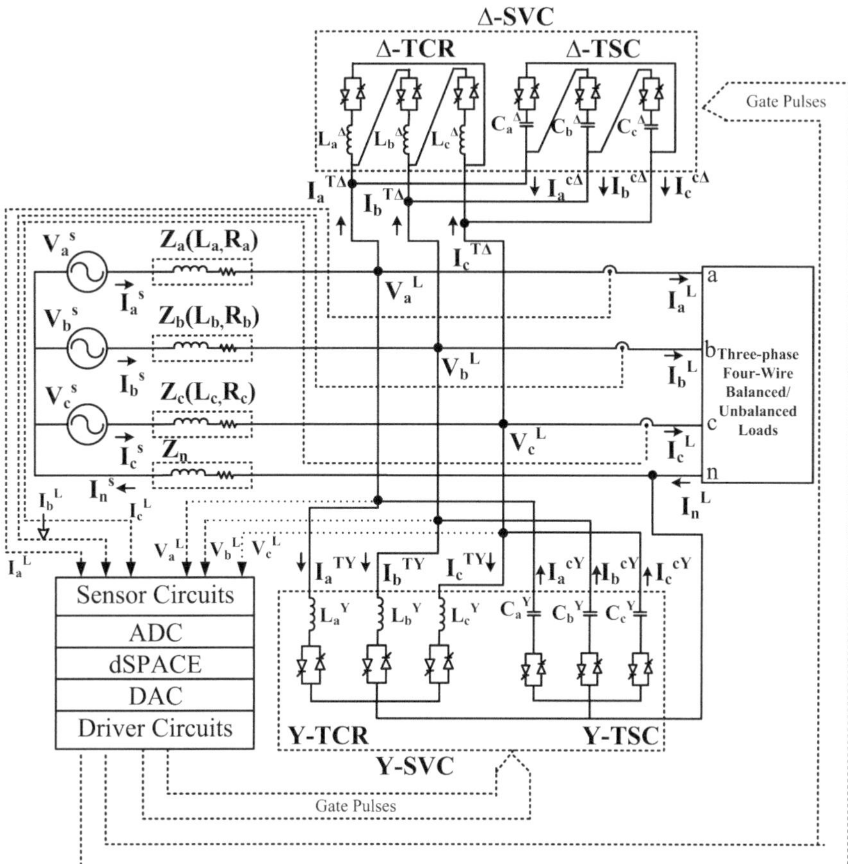

Fig. 1 Representation of distribution substation with Y and Δ-type SVC

unequal line voltage drops which make load bus voltages to be unbalanced. In order to compensate line currents and improve source power factor, SVCs are placed at the load bus to generate or absorb unbalanced reactive power. The unbalanced reactive power combined with load demand makes balanced load to the supply system. The phase-wise unbalanced loads are $P_{La} + jQ_{La}$, $P_{Lb} + jQ_{Lb}$ and $P_{Lc} + jQ_{Lc}$ while the phase-wise load seen by the source after compensation is $P_s + jQ_s$. The phase-wise reactive power absorbed by Y-TCR is Q_a^{TY}, jQ_b^{TY} and jQ_c^{TY} while for Δ-TCR is $jQ_{ab}^{T\Delta}$, $jQ_{bc}^{T\Delta}$ and $jQ_{ca}^{T\Delta}$. The phase-wise reactive power generation by Y-TSC is Q_a^{CY}, jQ_b^{CY} and jQ_c^{CY} while by Δ-TSC is $jQ_{ab}^{C\Delta}$, $jQ_{bc}^{C\Delta}$ and $jQ_{ca}^{C\Delta}$. The prefix T in all the variables denotes TCR quantities, where as prefix S, C and L denotes the source, TSC and load quantities respectively.

The compensation requirements for neutral current compensation, load balancing and source power factor improvement combining are,

$$\begin{cases} \mathrm{Re}\left[I_0^L\right] + \mathrm{Re}\left[I_0^{TY}\right] = \mathrm{Re}\left[I_0^{CY}\right] \\ \mathrm{Im}\left[I_0^L\right] + \mathrm{Im}\left[I_0^{TY}\right] = \mathrm{Im}\left[I_0^{CY}\right] \\ \mathrm{Re}\left[I_2^L\right] + \mathrm{Re}\left[I_2^{TY}\right] + \mathrm{Re}\left[I_2^{T\Delta}\right] = \mathrm{Re}\left[I_0^{CY}\right] + \mathrm{Re}\left[I_0^{C\Delta}\right] \\ \mathrm{Im}\left[I_2^L\right] + \mathrm{Im}\left[I_2^{TY}\right] + \mathrm{Im}\left[I_2^{T\Delta}\right] = \mathrm{Im}\left[I_0^{CY}\right] + \mathrm{Im}\left[I_0^{C\Delta}\right] \\ \mathrm{Im}\left[I_1^L\right] + \mathrm{Im}\left[I_1^{TY}\right] = \mathrm{Im}\left[I_1^{CY}\right] + \mathrm{Im}\left[I_1^S\right] \end{cases} \tag{1}$$

where I_0, I_1 and I_2 denotes the zero, positive and negative sequence current component respectively. The (1) has infinite solutions due to six unknowns with five constraints. Thus an additional constraint is considered to compute a unique solution. The additional constraint with a consideration that Δ-SVC doesn't generate imaginary part of positive sequence currents can be expressed as,

$$\mathrm{Im}\left[I_1^{T\Delta}\right] = \mathrm{Im}\left[I_1^{C\Delta}\right] \tag{2}$$

The compensating reactive power required by each phase of Δ-TCR and Y-TCR for load balancing, neutral current compensation and power factor correction can be expressed in terms of load power by solving (1) and (2) after substitution of each sequence component. The per phase compensating reactive power required by Δ-TCR can be calculated as,

$$\begin{cases} Q_{ab}^{T\Delta} = \frac{2(P_{La} - P_{Lb})}{\sqrt{3}} - Q_{ab}^{C\Delta} \\ Q_{bc}^{T\Delta} = \frac{2(P_{Lb} - P_{Lc})}{\sqrt{3}} - Q_{bc}^{C\Delta} \\ Q_{ca}^{T\Delta} = \frac{2(P_{Lc} - P_{La})}{\sqrt{3}} - Q_{ca}^{C\Delta} \end{cases} \tag{3}$$

Similarly, the per phase compensating reactive power required by Y-TCR can be calculated as,

$$\begin{cases} Q_a^{TY} = \left(Q_{La} - Q_a^{CY} - Q_s\right) + \frac{P_{Lb} - P_{Lc}}{\sqrt{3}} \\ Q_b^{TY} = \left(Q_{Lb} - Q_b^{CY} - Q_s\right) + \frac{P_{Lc} - P_{La}}{\sqrt{3}} \\ Q_c^{TY} = \left(Q_{Lc} - Q_c^{CY} - Q_s\right) + \frac{P_{La} - P_{Lb}}{\sqrt{3}} \end{cases} \tag{4}$$

Now, it is required to find out the appropriate switching angles corresponding to compensating reactive power. The per phase reactive power absorbed by TCR can be controlled independently by changing the firing delay angle of individual phases of TCR. For any delay angle α of a particular phase, the reactive power absorbed by Δ-TCR (Q_Δ) and Y-TCR (Q_Y) can be calculated [33] as,

$$\begin{cases} Q_{\Delta} = \left[\frac{2\pi - 2\alpha - \sin 2\alpha}{\pi x^0}\right] 3V_1^2 \\ Q_Y = \left[\frac{2\pi - 2\alpha - \sin 2\alpha}{\pi x^0}\right] V_1^2 \end{cases} \tag{5}$$

where x^0 is the reactance for full conduction of thyristor $(\alpha = 0^0)$ and V_1 is the per phase fundamental component of TCR voltage. Similarly, the per phase reactive power absorbed by Y-TCR can be calculated for any delay angle α and V_1. The per phase fundamental component of TCR voltage V_1 also depends on delay angle α. The amplitude of reactor voltage fundamental component (for $n = 1$) in terms of switching angle α can be expressed as,

$$V_1 = \frac{2}{\pi}\left[(\pi - \alpha) + \frac{1}{2}\sin(2\alpha)\right]. \tag{6}$$

Thus for required reactive power $Q_{ab}^{T\Delta}$ of Δ-TCR calculated from (3), corresponding switching angle (α) can be obtained from (5) using (6). The direct solution of (5) to obtain α requires a suitable numerical technique to be applied which can result in multiple values of α with different THDs. Thus a heuristic search based method is necessary to obtain the optimum value of α for minimum reactor voltage THD. Similar equations can be written to find α for the other two phases. The same procedure can be used to compute appropriate angle α for Y-TCR.

2.1 Control Scheme

The Fig. 2 shows control schematic for load balancing, neutral current compensation and power factor correction with minimum line harmonic injection in a three-phase four-wire system using SVC. In the control scheme, the reactive power requirement for the individual phases are calculated using (3) and (4) with a consideration of set power factor and set reactive power of TSC (Q_{TSC}). The zero crossing detector (ZCD) is used for detecting the zero crossing of input signal.

In the proposed scheme, the harmonic minimization (HM) from reactor voltage is realized by computing those values of α for solution of (5) which results in lower reactor voltage THD. Thus in this paper, a GSA based technique is used for off-line computation of the switching angles (α_{off}) as a function of modulation index (m_d^*) for individual phases. Then these computed switching angles are used to calculate reactive power (Q_{off}) absorbed by TCR in each phase using (5). The modulation indices, corresponding switching angles for optimum THD and phase wise VAr absorption are stored in the processor memory for load balancing, neutral current compensation and power factor correction with minimum reactor voltage THD.

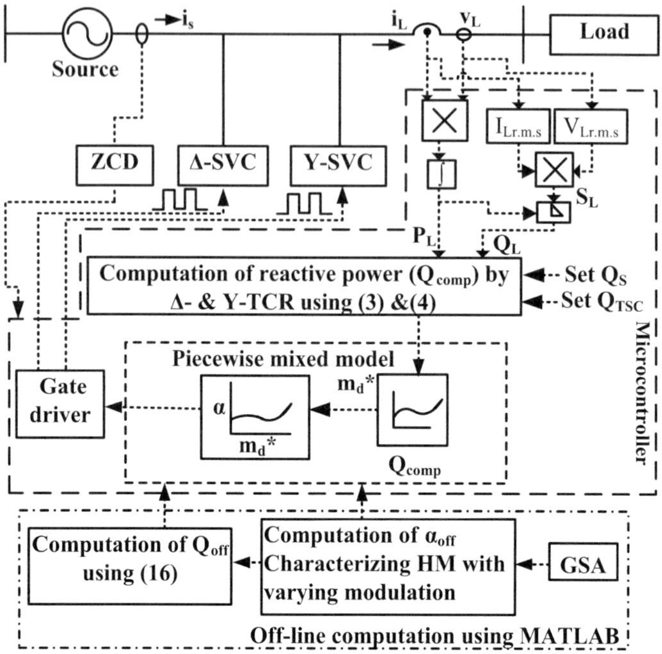

Fig. 2 Control schematic for the proposed compensation scheme

3 Proposed GSA Based Harmonic Minimization

In the proposed technique, solution of (5) for optimum switching angle is obtained through GSA based optimization technique considering minimum reactor voltage THD. The configuration of per phase thyristor controlled reactor (TCR) consisting of a reactor (L) connected in series with two anti parallel thyristors (T_1, T_2) is shown in Fig. 3a. The reactor voltage (V_{TCR}) is shown in Fig. 3b. The general expression for amplitude of nth odd harmonic for $n > 1$ for the reactor voltage, shown in Fig. 3b, is given by,

$$V_n = \frac{2}{\pi}\left[\frac{\sin(n+1)\alpha}{(n+1)} - \frac{\sin(n-1)\alpha}{(n-1)}\right]. \tag{7}$$

For a three-phase balanced system the triple n harmonics will be absent in the line and thus these are not considered in the present problem. Thus possible values of n are $n = 6i \pm 1$ ($i = 1, 2, 3, \ldots$).

Mathematically the optimization problem can be formulated as,

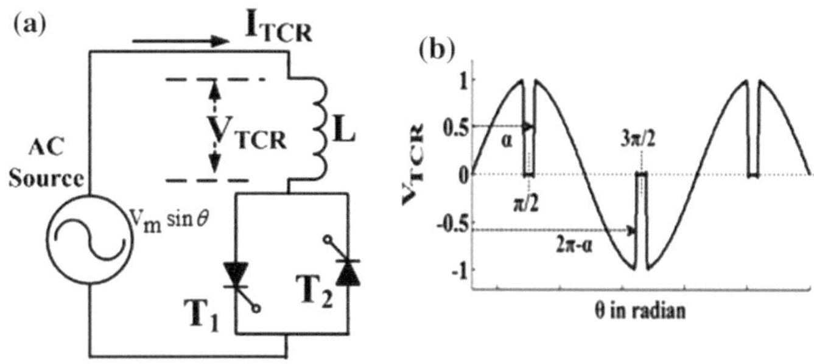

Fig. 3 Single-phase thyristor controlled reactor **a** configuration, **b** voltage across reactor

$$\begin{cases} V_1 = M \\ V_n \leq \in_n \\ \text{Subjected to} \\ \frac{\pi}{2} \leq \alpha \leq \pi. \end{cases} \tag{8}$$

where $V_1, \ldots, V_n$ are in per unit and M is the desired amplitude of the fundamental component of reactor voltage to rated load voltage which is also known as modulation index and $\in_n$ is the allowable limits of all individual harmonics and n up to 31st order.

The proposed objective function $f(\alpha)$ for the GSA satisfying (8) can be defined as,

$$f(\alpha) = K_1|V_1 - M|^2 + \sum\nolimits_{n=5,7,11,..}^{31} K_n|V_n - \in_n|^2. \tag{9}$$

where $n = 6i \pm 1(i = 1, 2, 3, \ldots)$

In (9), the coefficient K_1 needs to have larger value than K_5 to K_n for giving more priority to maintain fundamental component, at the same time K_5 to K_n are adjusted to descending order such that more priority is given to reduce lower order harmonics. Moreover, all the coefficients need to be properly adjusted so that GSA can perform nonbiased optimization. Trial and error method [23] is used until a good balance is found. For this problem, $K_1 = 257$, $K_5 = 51$, $K_7 = 12$, $K_{11} \ldots K_{31} = 5$. For each harmonic component, the $\in_n$ is selected as 0.03 according to IEEE std 519-1992.

The GSA, developed by Rashedi et al. in 2009 is the recent meta-heuristic searching algorithm [22]. It is based on Newton's law of gravity and motion. In this algorithm, agents are considered as objects and the performance of agents are measured by their masses. Hence, all these agents attract each other by a gravity force, and this force causes a global movement of all agents towards the agents with heavier masses. The heavier masses have better fitness value. Thus they describe good optimal solution to the problem and move more slowly than lighter ones. In GSA, each mass has four identifications: its position, its inertial mass, its active and passive gravitational mass [22, 23]. The position of the mass is equivalent to a solution of the problem

Table 1 Parameters of the simulation system

System parameters	Value
Line impedance (per phase)	(0.02+j 0.07) Ω
Load: a-phase	(20 + j12) kVA
b-phase	(25 + j13) kVA
c-phase	(30 + j15) kVA

and its gravitational and inertial masses are calculated by using a fitness function. Thus each mass represent a solution of the problem and the algorithm is navigated by properly adjusting the gravitational and inertia masses.

In the present problem, each solution (position) is composed of the switching angle α in half cycle. To start the algorithm, initial populations (number of agents) of switching angles are randomly generated with satisfying the constraint equation $\left(\frac{\pi}{2} \leq \alpha < \pi\right)$ for the chosen number of population. Then velocities of the agents are calculated and their next positions are updated. The other parameters of the algorithm such as gravitational constant, masses and acceleration are updated at each iteration cycle and the algorithm is terminated if it satisfies the maximum number of iterations. To get an optimal solution using GSA, the optimum settings of different input parameters are to be needed. Different trials have been performed for optimum values of input parameters. Based on these trials, the following input parameters are found to be best for optimal performance of the current problem: $G_0 = 100, \gamma = 20, T = 1000, N = \text{no. of agents} = 30$. Where the initial value of gravitational constant is G_0, γ is the user-specified constant for gravitational constant and T is the maximum number of iterations. The basic flowchart of GSA is shown in Fig. 4.

4 Simulation Results

The proposed scheme has been modeled and simulated using MATLAB and its Simulink and SimPower System toolboxes for a three-phase four-wire system. A 6.6 kV/415 V, 200 kVA distribution substation feeding a variable load is considered for simulation purpose. Thyristor switched capacitor-thyristor controlled reactor (TSC-TCR) type of SVC with 10 kVAr capacity of TCR and TSC that can vary reactive power between 0 and 30 kVAr through five steps (0, 5, 10, 20 and 30 kVAr) per phase is chosen. The line and load parameters are listed in Table 1.

The switching angles are computed at closed interval of modulation indices using GSA with minimum reactor voltage harmonics. The computed switching angles are used to calculate phase wise reactive power absorbed by TCR for each modulation index using (5). The computed switching angle α for minimum reactor voltage THD and the corresponding per phase reactive power consumption by Δ-TCR with mod-

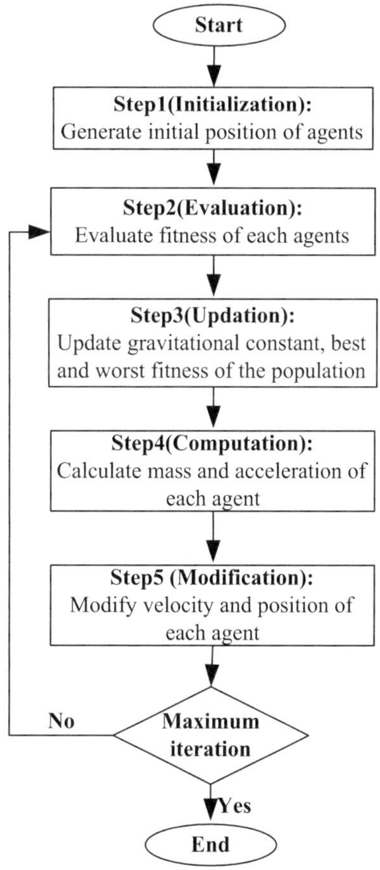

Fig. 4 Flowchart of the GSA algorithm

ulation index (m_d) are shown Fig. 5a, b respectively. In the proposed scheme, these two curves are stored in the processor memory for on-line applications.

4.1 Dynamic Response of the System Without and with Proposed Compensator

The performance of the proposed scheme for load balancing, along with neutral current compensation and source power factor improvement of a three-phase four-wire unbalanced load with low line harmonic injection is shown in Fig. 6.

At 0.03 s, the SVCs are switched into the line. The voltages $\left(V^L\right)$ at the point of common coupling (PCC), unbalanced load currents $\left(I^L\right)$, balanced source currents $\left(I^S\right)$, Δ-SVC currents $\left(I^\Delta\right)$, Y-SVC currents I^Y, source neutral current $\left(I_n^S\right)$, load neutral current I_n^L, balanced active power taken from source (P_S) and balanced

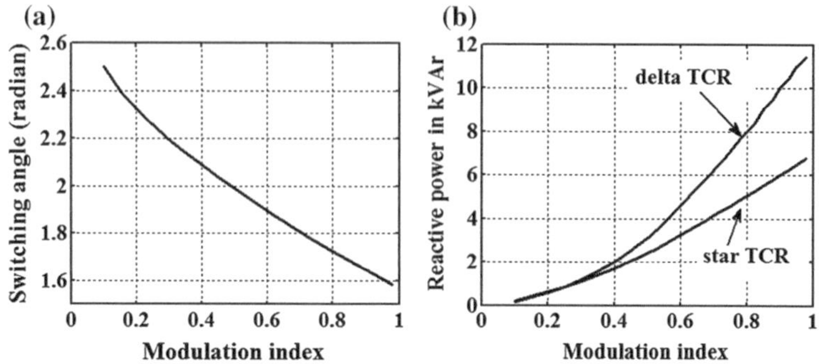

Fig. 5 Variation of **a** optimum switching angle for minimum reactor voltage THD, and **b** per phase reactive power consumption with modulation index (m_d)

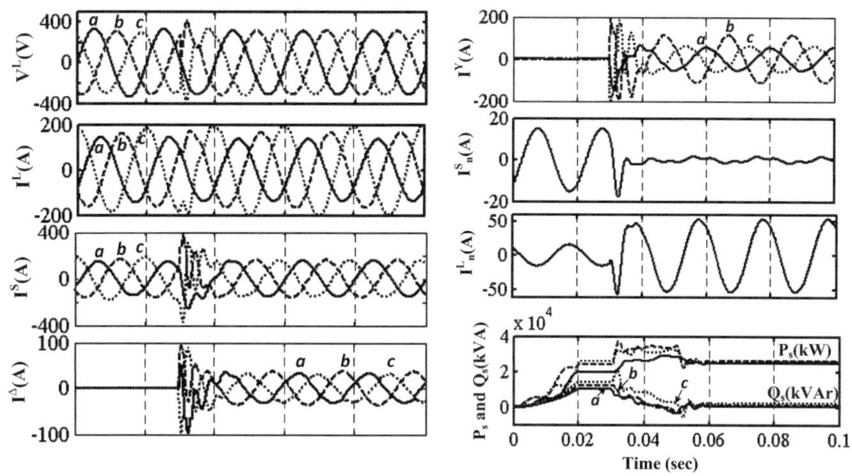

Fig. 6 Dynamic response of proposed scheme for load balancing, neutral current compensation and power factor correction when the SVCs are switched into the line at t = 0.03 s

reactive power drawn from source (Q_S) are demonstrated in Fig. 6. It is noticed that for an unbalanced three-phase load currents $\left(I^L\right)$ the compensated source currents $\left(I^S\right)$ become balanced and the source neutral current $\left(I_n^S\right)$ is maintained at nearly zero after the SVC is switched into the line at t = 0.03 s. The active (P_S) and reactive power (Q_S) seen by the source after compensation become balanced through proposed scheme. Also it can improve the source power factor by reducing the reactive power drawn from source. The harmonic spectrum of the a-phase source current (I_a^S) in steady state condition is shown in Fig. 7 which justifies the low harmonic injection by the proposed scheme

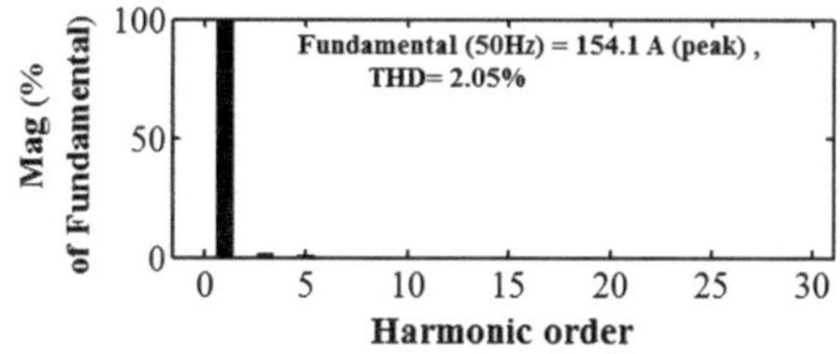

Fig. 7 Harmonic spectrum of the a-phase source current in steady condition

4.2 Dynamic Response of the System Under Sudden Load Change

The performance of the proposed scheme has also been studied under a sudden change of load condition. Initially a balanced load of (10 + j5) kVA per phase was connected to the system, then at t = 0.1 s, three-phase linear load is changed to two-phase and again to single-phase load at 0.2 s. These loads are reconnected again at 0.3 s. The PCC voltage $\left(V^L\right)$, unbalanced load currents $\left(I^L\right)$, balanced source currents $\left(I^S\right)$, Δ-SVC currents $\left(I^\Delta\right)$, Y-SVC currents I^Y, source neutral current $\left(I_n^S\right)$, load neutral current I_n^L, balanced active power taken from source (P_S) and balanced reactive power drawn from source (Q_S) under varying loads are demonstrated in Fig. 8a. From Fig. 8a it can be observed that the source current is balanced before and after the changeover which justify the effectiveness of the proposed method. Moreover, THD of the source current is within permissible limit as shown in Fig. 8b–d during different load conditions.

(a) Dynamic response of proposed scheme when load is changed from three-phase to two-phase at t = 0.1 s, to single-phase at t = 0.2 s and again reconnected to three-phase at t = 0.3 s.
(b) Harmonic spectrum of the a-phase source current during three-phase load
(c) Harmonic spectrum of the a-phase source current during two-phase load
(d) Harmonic spectrum of the a-phase source current during single-phase load.

5 Conclusion

An improved switching scheme for load balancing, neutral current compensation and source power factor improvement with minimum possible line harmonic injection without using external filter has been proposed in this paper. The proposed scheme has been implemented using TSC-TCR based combined Δ-SVC and Y-SVC. The Y-SVC is used for neutral current compensation and source power factor improvement whereas Δ-SVC is used for load balancing. The switching angles for TCR compensation are calculated by optimizing the switching function of TCR using GSA with varying modulation indices. These computed switching angles for minimum line harmonic injection along with phase-wise compensating reactive power requirement are stored in the processor memory for on-line application. From the

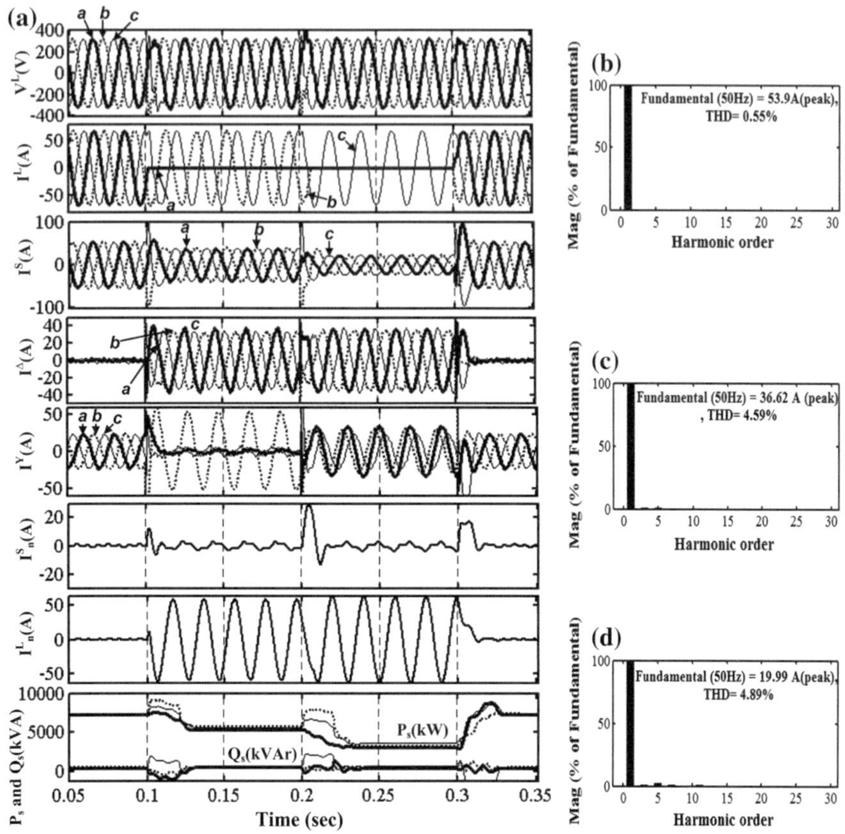

Fig. 8 Dynamic simulation results and harmonic spectrum of source current

simulation results, it is verified that the proposed SVC switching results in load balancing, neutral current compensation and source power factor improvement with reduced harmonic injection to source.

References

1. L.S. Czarnecki, P.M. Haley, Unbalanced power in four-wire systems and its reactive compensation. IEEE Trans. Power Del. **30**(1), 53–63 (2015)
2. S. Zeliang, X. Shaofeng, L. Qunzhan, Single-phase back-to-back converter for active power balancing, reactive power compensation, and harmonic filtering in traction power system. IEEE Trans. Power Electron. **26**(2), 334–343 (2011)
3. A. Bueno, J.M. Aller, J.A. Restrepo, R. Harley, T.G. Habetler, Harmonic and unbalanced compensation based on direct power control for electric railway systems. IEEE Trans. Power Electron. **28**(12), 5823–5831 (2013)

4. Q.N. Trinh, H.H. Lee, An advanced current control strategy for three-phase shunt active power filters. IEEE Trans. Ind. Electron. **60**(12), 5400–5410 (2013)
5. P. Kanjiya, V. Khadkikar, H.H. Zeineldin, A noniterative optimized algorithm for shunt active power filter under distorted and unbalanced supply voltages. IEEE Trans. Ind. Electron. **60**(12), 5376–5390 (2013)
6. S.R. Arya, R. Niwas, K.K. Bhalla, B. Singh, A. Chandra, K. Al-Haddad, Power quality improvement in isolated distributed power generating system using DSTATCOM. IEEE Trans. Ind. Appl. **51**(6), 4766–4774 (2015)
7. M. Mangaraj, A.K. Panda, An efficient control algorithm based DSTATCOM for power conditioning, in *2015 International Conference on Industrial Instrumentation and Control*, May 28–30, 2015, pp. 1069–1073
8. L. Wang, C.-S. Lam, M.-C. Wong, Design of a thyristor controlled LC compensator for dynamic reactive power compensation in smart grid. IEEE Trans. Smart Grid **8**(1), 409–417 (2017)
9. S.Y. Lee, C.J. Wu, On-line reactive power compensation schemes for unbalanced three phase four wire distribution feeders. IEEE Trans. Power Del. **8**(4), 1958–1965 (1993)
10. S.Y. Lee, C.J. Wu, Reactive power compensation and load balancing for unbalanced three-phase four-wire system by a combined system of an SVC and a series active filter. IEEE Proc.-Electr. Power Appl. **147**(6), 563–571 (1993)
11. J.R.C. Orillaza, A.R. Wood, Harmonic state-space model of a controlled TCR. IEEE Trans. Power Del. **28**(1), 197–205 (2013)
12. R. Mahanty, Large value AC capacitor for harmonic filtering and reactive power compensation. IET Gener. Transm. Distri. **2**(6), 876–891 (2008)
13. A. Luo, Z. Shuai, W. Zhu, Z. John Shen, Combined system for harmonic suppression and reactive power compensation. IEEE Trans. Ind. Electron. **56**(2), 418–428 (2009)
14. A. Luo, S. Peng, C. Wu, Z. Shuai, Power electronic hybrid system for load balancing compensation and frequency-selective harmonic suppression. IEEE Trans. Ind. Electron. **59**(2), 723–732 (2012)
15. A. Hamadi, S. Rahmani, K. Al-Haddad, A hybrid passive filter configuration for VAR control and harmonic compensation. IEEE Trans. Ind. Electron. **57**(7), 2419–2434 (2010)
16. S. Rahmani, A. Hamadi, K. Al-Haddad, A combination of shunt hybrid power filter and thyristor-controlled reactor for power quality. IEEE Trans. Ind. Electron. **57**(7), 2152–2164 (2014)
17. D.B. Kulkarni, G.R. Udupi, ANN-based SVC switching at distribution level for minimal-injected harmonics. IEEE Trans. Power Del. **25**(3), 1978–1985 (2010)
18. S. Das, D. Chatterjee, K. Swapan Goswami, A gravitational search algorithm based SVC switching function optimization technique for minimum harmonic injection. Electr. Power Compon. Syst. **43**(20), 2297–2310 (2015)
19. D. Thukaram, H.P. Khincha, B. RaviKumar, Harmonic minimization in the operation of static VAR compensators for unbalanced reactive power compensation, in *International Conference on Power System Technology-POWERCON 2004*, November 2004
20. A. Rajapakse, A. Puangpairoj, S. Chirarattananon, D. Thukaram, Harmonic minimizing neural network SVC controller for compensating unbalanced fluctuating loads, in *10th International Conference on Harmonics and Quality of Power 2002*, vol. 2, 6–9 Oct 2002, pp. 403–408
21. S. Das, D. Chatterjee, S.K. Goswami, A GSA based modified SVC switching scheme for load balancing and source power factor improvement. IEEE Trans. Power Del. **31**(5), 2072–2082 (2015)
22. E. Rashedi, H. Nezamabadi-pour, S. Saryazdi, GSA: a gravitational search algorithm. Inf. Sci. **179**(13), 2232–2248 (2009)
23. P.K. Roy, Solution of unit commitment problem using gravitational search algorithm. Electr. Power Energy Syst. **53**, 85–94 (2013). (Elsevier)

A Comprehensive Review on Distribution System

Anirban Chowdhury, Ranjit Roy, Kamal Krishna Mandal and S. Mandal

1 Introduction

The significant features of DS are its radial nature and high R/X ratio. These features make the analysis of DS different from a transmission system. The performance parameters of a DS are active power loss and voltage profile. The conventional LF techniques for transmission systems cannot be adopted for DS because of their inability to converge. Researchers have developed special/modified LF methods for analysis of DS. The above mentioned performance parameters of DS can be improved network reconfiguration which is switching combination of the distribution feeders. Overloading of DS is another issue which mal-triggers the protective devices and it can be handled by network reconfiguration as well. The reliability and cleanliness of a DS can be enhanced by RES based DG. The intermittent nature of power from DG can be improved by high capacity network storage elements. DR is a measure taken by power distributors to prevent situations like abnormally high load demand or power outages by sending explicit requests to the customers or by providing incentives

A. Chowdhury · R. Roy (✉)
Department of Electrical Engineering, Dr. Sudhir Chandra Sur Degree Engineering College, Kolkata 700074, India
e-mail: rroy.svnit@gmail.com

A. Chowdhury
e-mail: anirbanee2015@gmail.com

K. K. Mandal
Department of Power Engineering, Jadavpur University, SaltLake Campus, Sector III, Kolkata 700098, India
e-mail: kkm567@yahoo.co.in

S. Mandal
Department of Electrical Engineering, Jadavpur University, Kolkata 700032, India
e-mail: shm.here@gmail.com

S. Chattopadhyay et al. (eds.), *Modelling and Simulation in Science, Technology and Engineering Mathematics*, Advances in Intelligent Systems and Computing 749, https://doi.org/10.1007/978-3-319-74808-5_12

which will make the customers change their load usage patterns. In the last few years, research on DS have been presented under four sections namely, distribution system reconfiguration, load flow studies on distribution system, impact of renewable energy on distribution systems and effect of demand response on distribution system.

2 Distribution Network Reconfiguration

2.1 Introduction to Distribution System Reconfiguration

The configuration of most DS is radial to carry out their protection. In DS, each feeder is connected to various types of loads viz. commercial, industrial and residential. The patterns of daily load demands for the said types of load are different having different instants of occurrences of peak values. It is a problem if the DS gets overloaded as it will lead to disconnection of branches caused by actuation of the protective devices and voltage collapse. Overloading of the DS can cause problems starting from the disconnection of branches due to triggering of protective devices and can lead to a general voltage collapse resulting in financial losses for customers and the utilities. The line losses are significant in DS due to high resistance to reactance ratio. The main aim of the researchers is to investigate every possible means to reduce distribution power loss and maintain voltages within the specified limits.

In order to improve the performance of a radial distribution networks, reconfiguration is an effective solution. Reconfiguration is done by switching operation, either manually or automatically, so that the power losses are reduced resulting in enhancement of system security, power quality and reduction in network overloading. Two types of switching operations are possible, either opening sectionalizing (normally closed) switches or closing tie (normally open) switches of the network. The switching operation should ensure transfer of power to all the connected loads maintaining the radial nature of the network. The distribution network reconfiguration problem has been solved by researchers by adopting several techniques.

2.2 A Review on Distribution System Reconfiguration

Active power loss minimization by reconfiguration using several variants of Ant Colony Optimization (ACO) has been implemented widely by several researchers. ACO in Hyper-Cube framework (ACO-HC) has been implemented by Abdelaziz et al. [1] by applying a modified local heuristic approach and a standard state transition rule. Alemohammad et al. [2] have presented a model for seasonal reconfiguration of actual distribution network and it has been solved by Genetic Algorithm (GA). A multi-objective algorithm have been implemented by Alonso et al. [3] to reduce the network power losses which improves the reliability index using artificial immune systems technique (AIS) and applying graph theory considerations to improve computational performance and pareto-dominance rules. Amanulla et al. [4] proposed a binary particle swarm optimization-based search algorithm (BPSO) to find the optimal status of the switches and its effectiveness is demonstrated on a IEEE 33-bus and a IEEE 123-bus radial DS. A new method based on shuffled frog leaping algorithm (SFLA) by Arandian et al. [5], aimed reduction of network power losses and control of power generation from DGs, has been successfully tested on 33-bus and 69-bus test systems. A bi-level optimization procedure is developed by Arasteh et al. [6] using PSO, considering DR for sensitivity analysis in order to cover the effects of uncertain parameters on a distribution network and its performance has been tested on IEEE 33-bus standard test system. A novel adaptive fuzzy-based parallel GA is proposed by Asrari et al. [7] that employs the concept of parallel computing in identifying the optimal configuration of a dc distribution network. Bahrami et al. [8] worked on hybrid Big Bang-Big Crunch optimization (HBB-BC) algorithm, having faster rate of convergence and high efficiency, to solve the single -objective reconfiguration of the functions of the problem such as system average interruption frequency index, system average interruption duration index, average energy not supplied, in DS. A fuzzy multi-objective approach based reconfiguration of distribution networks have been presented by Banerjee et al. [9] considering different types of load. Bayat et al. [10] proposed a heuristic approach based on uniform voltage distribution based constructive reconfiguration algorithm (UVDA) for simultaneous DG placement, sizing and network reconfiguration. The above algorithm is applied by Bayat [11] for optimal reconfiguration of large- scale distribution networks and it was tested with various practical distribution networks varies from 16-bus system with 3 tie-switches up to 835-bus system with 146 tie-switches. Capitanescu et al. [12] explored how the DG penetration capacity of DS can be increased by both static and dynamic reconfiguration of network under thermal and voltage constraints by solving a non-linear, mixed-integer and multi-period optimal power flow problem (MP-OPF). Improvement of network reliability and reduction of network power losses based on an enhanced GA has been presented by Duan et al. [13]. Silva et al. [14] proposed a heuristic algorithm for electrical DS reconfiguration based on movement of firefly towards preys or partners where the insects positions in the space correspond to the positions of the switches in the electrical system. Oliveira et al. [15] as well as Souza et al. [16] have implemented bio-inspired metaheuristic AIS on network reconfiguration satisfying operational and network constraints, considering different load levels. Pons and Repetto [17] presented a topological reconfiguration procedure for maximizing local consumption of renewable energy in (Italian) active distribution networks. Niknam et al. [18] presented Honey Bee Mating Optimization (HBMO) approach to investigate the Distribution Feeder Reconfiguration (DFR) problem satisfying the operating limits and constraints. An efficient Modified bacterial foraging optimization algorithm (MBFOA) has been applied by Naveen et al. [19] on network reconfiguration to reduce active power losses at IEEE 16, 33 and 69 bus systems. Nguyen et al. [20] have implemented the cuckoo search algorithm (CSA) on network reconfiguration problems and it proved to be efficient and promising. Back tracking search algorithm (BSA) on network reconfiguration problems have been proposed by

Nguyen et al. [21] and its effectiveness was tested on 69- node distribution network system. Network reconfiguration based on minimum spanning tree algorithm(MST) have been tested on 33- bus, 69-bus and a real 210-bus MV utility DS by Li et al. [22] and the results were found to be very effective. Network reconfiguration based on a mixed- integer second-order conic programming (MISOCP) model has been developed by López et al. [23] and applied on 136 node DS considering the minimization of active power losses and reliability constraints. To solve the reconfiguration problem of radial DS, a scatter search, which is a metaheuristic-based algorithm, is proposed by Rupolo and Mantovani [24].

3 Load Flow Studies on Distribution System

3.1 Load Flow Approach in DS Due to a Different Topology

LF studies are performed to determine the parameters of steady-state line power flow and connected load. They provide guidance for proper planning, operation, control and optimization of power system. LF analysis helps to verify whether all the operational constraints including line voltage limits are satisfied. It is one of the most frequently carried out study by power utilities and are essential for power system planning, operation, optimization and control. It is required to explore different arrangements necessary to maintain the required voltage profile and to minimize the system losses. LF is also used as a sub problem like contingency analysis of a system. The bus voltage magnitudes of a distribution network, their phase angles, active and reactive power flows of different lines and the transmission power losses are determined from load flow studies. Some of the basic LF algorithms were developed such as Newton Raphson (NR), Gauss Seidel (GS) and were applied to the transmission network. In DS, these methods may become inefficient due to its radial nature, high resistance to reactance ratio, load unbalance etc. Thus, the LF analysis becomes complex in case of DS and fail to converge using the techniques in case of transmission systems. In the past, many approaches for DS load-flow analyses have been developed by the researchers. With respect to the nature of generation and load, there are two types of LF, probabilistic (PLF) and deterministic (DLF). In PLF, the analysis takes care of stochastic or statistical uncertainties with generation or load while in DLF the natures are taken to be consistent. Some of the works on LF in the recent years are stated below.

3.2 A Review on Load Flow Methods on Distribution Systems

Melhorn and Dimitrovski [25] proposed a three phase PLF in radial and weakly meshed distribution network for both balanced and unbalanced conditions without explicitly using Y-bus matrix and it was applied on IEEE 123 node and 13 node test feeder. A novel direct method to LF has been developed by Singh et al. [26] which saves computation time and power. Wang et al. [27] presented the explicit conditions for existence of a unique LF solution for distribution networks having generic topology. Murari et al. [28] developed a LF solution for IEEE 33 DS using matrix method. The steady-state analysis and working of an electrical DS using multi-linear probabilistic Monte Carlo (MC) simulation technique has been proposed by Carpinelli et al. [29] with integration of photo-voltaic (PV) and wind (WD) power generation. Harmonic LF in electrical distribution network has been analyzed using a fuzzy-based Monte Carlo simulation technique has been developed by Šošić et al. [30] in order to identify the weak zones of the network leading to power quality problems. PLF of unbalanced power in a DS using a point estimate method has been developed and analyzed by Delgado and Domínguez-Navarro [31], considering penetration of PV and WD sources. A new, simple and efficient LF algorithm for weakly meshed DS has been presented by Li et al. [32], using power flow variables as active and reactive powers. Ruiz-Rodriguez et al. [33] developed a hybrid modified algorithm combining jumping frog and PSO (JFPSO) and PLF in three phase network based on the MC simulation for reducing voltage unbalance in DS with PV generators. A PLF for DS with uncertain PV generation has been presented by Kabir et al. [34] using Latin Hypercube Sampling with Cholesky Decomposition (LHS-CD) in order to quantify the overvoltage issues. Voltage stability analysis of unbalanced DS has been performed by Abdel-Akher [35] using backward/forward (BF) sweep LF analysis method with secant predictor. PLF for three phase networks using binary SFLA with technical constraints has been proposed by Gomez-Gonzalez et al. [36] which handled the voltage regulation problem of a PV-connected grid within a small number of iterations. Khan et al. [37] proposed a novel LF algorithm for different radial DS which employs only three recursive equations devoid of any complex parameters and it proved to be computationally efficient and faster than other existing methods. Kocar et al. [38] tested and compared three LF solution algorithms using the modified-augmented- nodal-analysis (MANA) formulation on IEEE 8500-node distribution test feeder by means of a regulator tap-control strategy. A PLF-based approach regarding the effects of RES on the voltage quality in a DS has been proposed by Sexauer and Mohagheghi [39].

4 Impact of Renewable Energy on Distribution System

4.1 Renewable Energy—An Alternative for a Clean and Reliable Distribution System

Fossil fuel based power generation became a practice for hundreds of years. The environmental hazards related to emissions, especially CO_2, from a conventional power plant are increasing day by day. Customers and utilities have widely accepted clean and green renewable energy based generators viz., WT, PV system and fuel-cells, among others as alternate sources of energy. Distributed Generation (DG) based on renewable energy is one of the most promising solutions to the problem of high greenhouse gas emissions and research efforts across the globe are being put into this topic. But for its successful implementation, a number of challenges need to be faced. As a DS is passive, allocating a DG to it means addition of a new dynamic element to the system which needs to undergo a stability analysis. Integration of renewable distributed energy resources (DER) has the ability to affect the operation of DS by affecting the equipment reliability and customer power quality. To increase reliability, Battery Energy Storage Systems (BESS) are incorporated to the distribution network to mitigate the intermittent nature of renewable energy generation. The impact of renewable energy on DS has been a preferred area of research these years.

4.2 A Review on Renewable Energy Based Distribution Systems

Abdullah et al. [40] worked on the integration of distribution network with renewable DG which revealed the influence power output from renewable energy based generators on time varying load demand. Self-consumption and storage of power by consumers generated from PV micro grid are best means to keep the voltage levels within specified limits as suggested by Camilo et al. [41]. Optimal allocation of renewable energy based DGs in unbalanced IEEE 37-node feeder DS using Big Bang- Big crunch method has been tested by Abdelaziz et al. [42]. Kayal and Chanda [43] proposed integration of photovoltaic (PV) array, wind turbine (WT) and capacitor bank in distribution network, a sustainable way to meet the ever increasing load demand. A novel dynamic energy management strategy in integrating large-scale renewable energy sources (RES) with the distribution network has been developed by Lv and Ai [44]. The benefits of customers correlated with harvesting of renewable energy has been shown through a multi-level optimization approach for DS planning by Zeng et al. [45]. Jiang et al. [46] proposed a synchrophasor based auxiliary controller to increase voltage stability of a DS with distributed WT generators. Optimal scheduling of renewable energy integrated distribution network for BESS operation with plug in electric vehicles (PEV) have been shown by Yang et al. [47], in order to minimize active power loss, voltage fluctuation and cost of electricity. A novel stochastic programming model for active and reactive power scheduling in DS with renewable energy resources and their influence on the daily Volt/Var control (VVC) is presented by Samimi and Kazemi [48]. A novel hybrid approach to allocate RES in DS is proposed by Singh and Parida [49] and it has been demonstrated on 15-node radial DS and 69-node mesh DS. Reconfiguration problem of distribution network has been investigated by Taghi et al. [50] to improve power quality, reliability and reduce power loss by placement of solar-cell and wind turbine. Fluctuations in the magnitude of voltages at different nodes in the DS with RES have been predicted by a mathematical model developed by Iyer et al. [51]. An interval optimization based day-ahead scheduling scheme for RE management considering renewable RE uncertainties in smart DS has been presented by Chen et al. [52] and it has been tested on 33-node and 119-node systems resulting in lower active power losses and improved voltage profile. The effects of EV with vehicle to grid connectivity capability on renewable energy integrated distribution have been analyzed by Fathabadi [53]. Nijhuis et al. [54] have shown the impacts of the renewable energy and ICT driven energy transition on distribution networks and it made the energy system to be more sustainable. Tsiftsis et al. [55] suggested that the efficiency of power distribution network (PDN) can be enhanced by deployment of wireless sensor network in dispersed RES.

5 Effect of Demand Response on Distribution System

5.1 Demand Response and Its Objective

DR, an outcome of demand side management (DSM), is the change in power consumption of an electric utility customer to match power demand with supply. It is not possible to throttle the power output from the supply-side like taking generating units on/off-line or importing power from other utilities at all times due to time consumption and high expense. The main objective of DR is to manage the load side demand meticulously instead of changing the supply side power level. Utilities may send signal to the customers to cut out some of the unnecessary load during peak load hours in a number of ways like making the per unit cost of electricity cheaper in the off peak load hours than the peak load hours or by smart metering through which explicit requests or price change are intimated to the customers. In a broader sense, DR encourages the electricity customers to shift their electricity usage pattern during the peak load hours or power crisis.

5.2 A Review on Performance of Distribution System with Demand Response

Nunna and Doolla [56] presented an agent based intelligent management system to facilitate power trading among micro grids and encourage customers to participate in DR. An analytical study is reported by Homaee et al. [57] to show the impact of DR on voltages profile of DS. Zeng et al. [58] presented an integrated methodology that accounts renewable DG and DR as options for clean and sustainable planning of distribution network. A novel voltage sensitivity matrix based voltage control in a real time environment using DR in an automated DS has been presented by Zakariazadeh et al. [59]. Venkatesan et al. [60] proposed a model for DR by utilizing consumer levels of rationality and behavior for different scenarios and applied it to IEEE 8500-node test feeder resulting in improvement of voltage profile. Degefa et al. [61] showed how DR can be an integral part for smart grid planning in terms of reliability, contingency and improvement of voltage profile. Williams et al. [62] have designed a self-regulating, smart, wind power integrated DS where the LF fluctuations are controlled by self-regulating air-source heat pump (HP) cycling. Mistry and Roy [63] presented the combined effects of DR program, wind generator, as a renewable energy source and network reconfiguration on distribution network.

6 Conclusion

In this paper, research in the last few years on four topics on DS has been presented namely distribution network reconfiguration, LF techniques, integration of renewable energy based generators with or without BESS and effect of DR on DS. It is very important for a researcher to know the fore mentioned areas of DS, willing to carry out research in this field. The research works presented above aimed at analysis and/or improvement of voltage profile, efficiency, reliability and contingency of a DS. These parameters are performance indicators of a DS which needs to be improved so that transmitted power can reach to the consumer end both safely and efficiently in a cleaner way.

References

1. A.Y. Abdelaziz, R.A. Osama, S.M. El-Khodary, Reconfiguration of distribution systems for loss reduction using the hyper-cube ant colony optimisation algorithm. IET Gener. Transm. Distrib. **6**(2), 176–187 (2012)
2. S. Hamidreza Alemohammad, E. Mashhour, M. Saniei, A market-based method for reconfiguration of distribution network. Electr. Power Syst. Res. **125**, 15–22 (2015)
3. F.R. Alonso, D.Q. Oliveira, A.C. Zambroni de Souza, Artificial immune systems optimization approach for multiobjective distribution system reconfiguration. IEEE Trans. Power Syst. **30**(2), 840–847 (2014)
4. B. Amanulla, S. Chakrabarti, S.N. Singh, Reconfiguration of power distribution systems considering reliability and power loss. IEEE Trans. Power Delivery **27**(2), 918–926 (2012)
5. B. Arandian, R.A. Hooshmand, E. Gholipour, Decreasing activity cost of a distribution system company by reconfiguration and power generation control of DGs based on shuffled frog leaping algorithm. Electr. Power Energy Syst. **61**, 48–55 (2014)
6. H. Arasteh, M. Sadegh Sepasian, V. Vahidinasab, An aggregated model for coordinated planning and reconfiguration of electric distribution networks. Energy **94**, 786–798 (2016)
7. A. Asrari, S. Lotfifard, M. Ansari, Reconfiguration of smart distribution systems with time varying loads using parallel computing. IEEE Trans. Smart Grid **7**, 1–11 (2016)
8. M. Bahrami, M. Mahdi Mahmoodi, M. Sedighizadeh, Network reconfiguration to improve reliability indices and efficiency in distribution systems with an efficient hybrid big bang-big crunch algorithm systems. IJCMES **2**(1), 11–17 (2016)

9. S. Banerjee, C.K. Chanda, D. Das, Reconfiguration of distribution networks based on fuzzy multiobjective approach by considering loads of different types. J. Inst. Eng. (India): Series B **94**(1), 29–42 (2013)
10. A. Bayat, A. Bagheri, R. Noroozian, Optimal siting and sizing of distributed generation accompanied by reconfiguration of distribution networks for maximum loss reduction by using a new UVDA-based heuristic method. Electr. Power Energy Syst. **77**, 360–371 (2016)
11. A. Bayat, Uniform voltage distribution based constructive algorithm for optimal reconfiguration of electric distribution networks. Electr. Power Syst. Res. **104**, 146–155 (2013)
12. Florin Capitanescu, Luis F. Ochoa, Harag Margossian, Nikos D. Hatziargyriou, Assessing the potential of network reconfiguration to improve distributed generation hosting capacity in active distribution systems. IEEE Trans. Power Syst. **30**(1), 346–356 (2014)
13. Dong-Li Duan, Xiao-Dong Ling, Wu Xiao-Yue, Bin Zhong, Reconfiguration of distribution network for loss reduction and reliability improvement based on an enhanced genetic algorithm. Int. J. Electr. Power Energy Syst. **64**, 88–95 (2015)
14. L.I. Silva, E.A. Belati, I.C. Silva Junior, Heuristic algorithm for electrical distribution systems reconfiguration based on firefly movement equation. IEEE Latin Am. Trans. **14**(2), 752–758 (2016)
15. L.W. de Oliveira, S. da Felipe Seta, J. de Edimar Oliveira, Optimal reconfiguration of distribution systems with representation of uncertainties through interval analysis. Electr. Power Energy Syst. **83**, 382–391 (2016)
16. S.S.F. Souza, R. Romero, J. Pereir, J.T. Saraiva, Artificial immune algorithm applied to distribution system reconfiguration with variable demand. Electr. Power Energy Syst. **82**, 561–568 (2016)
17. E. Pons, M. Repetto, A topological reconfiguration procedure for maximising local consumption of renewable energy in (Italian) active distribution networks. Int. J. Sustain. Energy, Taylor & Francis **36**, 1–14 (2016). ISSN: 1478-6451
18. T. Niknam, A. Kavousi Fard, A. Seifi, Distribution feeder reconfiguration considering fuel cell/wind/ photovoltaic power plants. Renew. Energy **37**, 213–225 (2011)
19. S. Naveen, K. Sathish Kumar, K. Rajalakshmi, Distribution system reconfiguration for loss minimization using modified bacterial foraging optimization algorithm. Electr. Power Energy Syst. **69**, 90–97 (2014)
20. T. Thanh Nguyen, A. Viet Truong, T. Anh Phung, A novel method based on adaptive cuckoo search for optimal network reconfiguration and distributed generation allocation in distribution network. Electr. Power Energy Syst. **78**, 801–815 (2015)
21. T. Thanh Nguyen, H. Ngoc Pham, A. Viet Truong, T. Anh Phung, T. Trung Nguyen, A backtracking search algorithm for distribution network reconfiguration problem, chapter AETA 2015. Recent Adv. Electr. Eng. Relat. Sci. **371**, 223–234 (2016)
22. Hongwei Li, Wenjin Mao, Anan Zhang, Chao Li, An improved distribution network reconfiguration method based on minimum spanning tree algorithm and heuristic rules. Electr. Power Energy Syst. **82**, 466–473 (2016)
23. J. Camilo López, M. Lavorato, F. John Franco, J. Marcos Rider, Robust optimisation applied to the reconfiguration of distribution systems with reliability constraints. IET Gener. Transm. Distrib. **10**(4), 917–927 (2015)
24. D. Rupolo, J.R.S. Mantovani, Reconfiguration of radial electric power distribution system via a scatter search algorithm. IEEE Latin Am. Trans. **13**(4), 1022–1028 (2015)
25. A.C. Melhorn, A. Dimitrovski, Three-phase probabilistic load flow in radial and meshed distribution networks. IET Gener Transm. Distrib. **9**(16), 2743–2750 (2015)
26. N. Singh, R. Bala, F. Sahib, A novel method of load flow for radial distribution network to minimise losses. IJEDR **4**(1), 7–10 (2016). ISSN: 2321-9939
27. C. Wang, A. Bernstein, J.-Y. Le Boudec, M. Paolone, *Explicit Conditions on Existence and Uniqueness of Load-Flow Solutions in Distribution Networks*, arXiv:1602.08372v1 [math.OC], (Submitted on 26 Feb 2016)
28. K. Murari, S. Ghosh, N. Singh, Explicit conditions on existence and uniqueness of load-flow solutions in distribution networks. Int. Electr. Eng. J. (IEEJ) **6**(1), 1711–1715 (2015). ISSN 2078 -2365

29. G. Carpinelli, P. Caramia, P. Varilone, Multi-linear Monte Carlo simulation method for probabilistic load flow of distribution systems with wind and photovoltaic generation systems. Renew. Energy **76**, 283–295 (2014)
30. D. Šošić, M. Žarković, G. Dobric, Fuzzy-based Monte Carlo simulation for harmonic load flow in distribution networks. IET Gener. Transm. Distrib. **9**(3), 267–275 (2014)
31. C. Delgado, J.A. Domínguez-Navarro, Point estimate method for probabilistic load flow of an unbalanced power distribution system with correlated wind and solar sources. Electr. Power Energy Syst. **61**, 267–278 (2014)
32. H. Li, A. Zhang, X. Shen, X. Jin, A load flow method for weakly meshed distribution networks using powers as flow variables. Electr. Power Energy Syst. **58**, 291–299 (2014)
33. F.J. Ruiz-Rodriguez, F. Jurado, M. Gomez-Gonzalez, A hybrid method combining JFPSO and probabilistic three-phase load flow for improving unbalanced voltages in distribution systems with photovoltaic generators. Electr. Eng. **96**(3), 275–286 (2014). Springer
34. M.N. Kabir, Y. Mishra, R.C. Bansal, Probabilistic load flow for distribution systems with uncertain PV generation. Appl. Energy **163**, 343–351 (2015)
35. M. Abdel-Akher, Voltage stability analysis of unbalanced distribution systems using backward/forward sweep load-flow analysis method with secant predictor. IET Gener. Transm. Distrib. **7**(3), 309–317 (2012)
36. M. Gomez-Gonzalez, F.J. Ruiz-Rodriguez, F. Jurado, A binary SFLA for probabilistic three-phase load flow in unbalanced distribution systems with technical constraints. Electr. Power Energy Syst. **48**, 48–57 (2012)
37. N.A. Khan, S. Ghosh, S.P. Ghoshal, Testing of a novel load flow algorithm for different radial distribution systems. Electr. Eng. **95**(3), 229–240 (2012)
38. I. Kocar, J. Mahseredjian, U. Karaagac, G. Soykan, O. Saad, Multiphase load-flow solution for large-scale distribution systems using MANA. IEEE Trans. Power Delivery **29**(2), 908–915 (2014)
39. J.M. Sexauer, S. Mohagheghi, Voltage quality assessment in a distribution system with distributed generation—a probabilistic load flow approach. IEEE Trans. Power Delivery **28**(3), 1652–1662 (2013)
40. M.A. Abdullah, A.P. Agalgaonkar, K.M. Muttaqi, Assessment of energy supply and continuity of service in distribution network with renewable distributed generation. Appl. Energy **113**, 1015–1026 (2013)
41. F.M. Camilo, R. Castro, M.E. Almeida, V.F. Pires, Self-consumption and storage as a way to facilitate the integration of renewable energy in low voltage distribution networks. IET Gener. Transm. Distrib. **10**(7), 1741–1748 (2016)
42. A.Y. Abdelaziz, Y.G. Hegazy, W. El-Khattam, M.M. Othman, Optimal allocation of stochastically dependent renewable energy based distributed generators in unbalanced distribution networks. Electr. Power Syst. Res. **119**, 34–44 (2014)
43. P. Kayal, C.K. Chanda, Strategic approach for reinforcement of intermittent renewable energy sources and capacitor bank for sustainable electric power distribution system. Electr. Power Energy Syst. **83**, 335–351 (2016)
44. Tianguang Lv, Qian Ai, Interactive energy management of networked microgrids-based active distribution system considering large-scale integration of renewable energy resources. Appl. Energy **163**, 408–422 (2015)
45. B. Zeng, J. Wen, J. Shi, J. Zhang, Y. Zhang, A multi-level approach to active distribution system planning for efficient renewable energy harvesting in a deregulated environment. Energy **96**, 614–624 (2016)
46. H. Jiang, Y. Zhang, J.J. Zhang, D. Wenzhong Gao, E. Muljadi, Synchrophasor-based auxiliary controller to enhance the voltage stability of a distribution system with high renewable energy penetration. IEEE Trans. Smart Grid **6**(4), 2107–2115 (2015)
47. Y. Yang, W. Zhang, J. Jiang, M. Huang, L. Niu, Optimal scheduling of a battery energy storage system with electric vehicles' auxiliary for a distribution network with renewable energy integration. Energies **8**, 10718–10735 (2015). ISSN: 1996-1073

48. A. Samimi, A. Kazemi, Scenario-based stochastic programming for Volt/Var control in distribution systems with renewable energy sources. IETE Tech. Rev. **33**, 1–13 (2016). ISSN 0256–4602
49. A.K. Singh, S.K. Parida, A novel hybrid approach to allocate renewable energy sources in distribution system. Sustainable Energy Technol. Assessments **10**, 1–11 (2015)
50. M. Taghi, B. BaghmisheZad, H. Nasir Aghdam, Raising power quality and improving reliability by distribution network reconfiguration in the presence of renewable energy sources. J. Artif. Intell. Electr. Eng. **4**(14), 29–36 (2015)
51. S.V. Iyer, B. Wu, Y. Li, B. Singh, A mathematical model to predict voltage fluctuations in a distribution system with renewable energy sources. De Gruyter, Int. J. Emerg. Electr. Power Syst. (IJEEPS) **6**(6), 549–557 (2015)
52. C. Chen, F. Wang, B. Zhou, K. Wing Chan, Y. Cao, Y. Tan, An interval optimization based day-ahead scheduling scheme for renewable energy management in smart distribution systems. Energy Convers. Manag. **106**, 584–596 (2015)
53. H. Fathabadi, Utilization of electric vehicles and renewable energy sources used as distributed generators for improving characteristics of electric power distribution systems. Energy, 1–11, June 2015 (In Press)
54. M. Nijhuis, M. Gibescu, J.F.G. Cobben, Assessment of the impacts of the renewable energy and ICT driven energy transition on distribution networks. Renew. Sustain. Energy Rev. **52**, 1003–1014 (2015)
55. T.A. Tsiftsis, P.C. Sofotasios, N. Papanikolaou, M.I. Loupis, Deployment of wireless sensor network in dispersed renewable energy sources for increasing efficiency of power distribution networks. J. Mod. Power Syst. Clean Energy (MPCE) **3**(4), 610–618 (2015)
56. H.S.V.S. Kumar Nunna, S. Doolla, Demand response in smart distribution system with multiple microgrids. IEEE Trans. Smart Grid **3**(4), 1641–1649 (2012)
57. Omid Homaee, Alireza Zakariazadeh, Shahram Jadid, Real time voltage control using emergency demand response in distribution system by integrating advanced metering infrastructure. J. Renew. Sustain. Energy **6**(033145), 1–16 (2014)
58. B. Zeng, J. Zhang, X. Yang, J. Wang, J. Dong, Y. Zhang, Integrated planning for transition to low-carbon distribution system with renewable energy generation and demand response. IEEE Trans. Power Syst. **29**(3), 1153–1165 (2013)
59. A. Zakariazadeh, O. Homaee, S. Jadid, P. Siano, A new approach for real time voltage control using demand response in an automated distribution system. Appl. Energy **117**, 157–166 (2013)
60. N. Venkatesan, J. Solanki, S. Khushalani Solanki, Residential demand response model and impact on voltage profile and losses of an electric distribution network. Appl. Energy **96**, 84–91 (2011)
61. A. Safdarian, M.Z. Degefa, M. Lehtonen, M. Fotuhi-Firuzabad, Distribution network reliability improvements in presence of demand response. IET Gener. Transm. Distrib. **8**(12), 2027–2035 (2014)
62. T. Williams, D. Wang, C. Crawford, N. Djilali, Integrating renewable energy using a smart distribution system: Potential of self-regulating demand response. Renew. Energy **52**, 46–56 (2012)
63. D. Khyati Mistry, R. Roy, Impact of demand response program in wind integrated distribution network. Electr. Power Syst. Res. **108**, 269–281 (2013)

Solution of Multi-objective Combined Economic Emission Load Dispatch Using Krill Herd Algorithm with Constraints

D. Maity, M. Chatterjee, S. Banerjee and C. K. Chanda

Nomenclature

$F_i(P_i)$	Fuel cost function
$E_i(P_i)$	Emission cost function
P_i	Output power of generator i
N	Number of generators
a_i, b_i, c_i, d_i, e_i	Cost coefficients of unit i
$\alpha_i, \beta_i, \gamma_i$	Emission cost co-efficient
$P_i^{\min}$	Minimum operating limits of generator i
$P_i^{\max}$	Maximum operating limits of generator i
P_D	Load demand
P_L	Transmission losses
B_{ij}	Transmission loss co-efficient
P_N	Output of Nth generator
M_i	Change in movement due to induction

D. Maity (✉)
Electrical Engineering Department, Netaji Subhash Engineering College Garia, Kolkata, West Bengal, India
e-mail: deblina14@gmail.com

M. Chatterjee · S. Banerjee
Electrical Engineering Department, Dr. B. C. Roy Engineering College, Durgapur, West Bengal, India
e-mail: madhuragungun@gmail.com

S. Banerjee
e-mail: sumit_9999@rediffmail.com

C. K. Chanda
Electrical Engineering Department, IIEST, Shibpur, Howrah, West Bengal, India
e-mail: ckc_math@yahoo.co.in

S. Chattopadhyay et al. (eds.), *Modelling and Simulation in Science, Technology and Engineering Mathematics*, Advances in Intelligent Systems and Computing 749,
https://doi.org/10.1007/978-3-319-74808-5_13

X_i	Foraging action
D_i	Random diffusion of each individual
$M_i^{\max}$	The maximum is induced speed
α_i	Swarm density effect
ω_n	Inertia weight
M_i^{Previous}	Previous induced speed
Vf	Foraging speed
ω_x	Inertia weight of the foraging motion
β_i	Current food location
X_i^{Previous}	Previous speed
$RD\max$	Maximum diffusion speed
σ	Random directional vector
TV	Total number of the variables
UL and LL	Upper limit and the lower limit of the individual

1 Introduction

In electrical engineering mainly in power system many real life optimal problem are flattering very complex and difficult. It is very time consuming to solve these problems using conventional iterative techniques. But time cost is very important. That's why a new optimization technique i.e. krill herd algorithm is required to solve complex optimal problem like economic load dispatch (ELD) and economic emission load dispatch (EELD).

The cost function of economic load dispatch is represented by quadratic function. To find minimum fuel cost lambda iteration method is used in [1]. Problems of economic load dispatch including transmission losses are solved using dynamic programming method [2]. Basu proposed optimization technique based on artificial bee colony for solving ED problem including transmission losses, multiple fuels etc. [3].

Teaching learning based optimization technique is newly developed population based algorithm based on relationship between teacher and learners in a class [4, 5]. It has the ability to obtain convergence characteristics in relatively faster computation time to genetic algorithms [6], particle swarm optimization techniques (PSO) [7] and artificial bee colony (ABC) [8]. A new modified particle swarm optimization technique has been proposed to solve EELD problem in [9]. Gaurav Prasad et al. [10] proposed artificial bee colony optimization to solve ELD problem with considering generator constraints. It does not include emission dispatch problem. Y. Sonmez et al. [11] presented the same optimization techniques to solve the multi-EELD problem by the penalty factor approach.

In [12] also, same algorithm regarding clonal selection is proposed to find the solution of DED problem for generating units with VPL effect. A new optimization technique ABC-PSO has been proposed to solve combined EELD problem in [13].

A modified teaching learning based optimization techniques based on bare bones has been proposed to solve minimization algebraic problem in [14].

KH algorithm is proposed in this paper with various constraints. Here transmission losses are not included. Section 2 describes overview of EELD problem, Sect. 3 discusses KH and TLBO technique, Sect. 4 discuss the steps of implementation of KH on EELD problems, Sect. 5 presents the result done by simulation and Sect. 6 presents the conclusion of this paper.

2 Overview of Economic Emission Load Dispatch Problems

A. **Fuel and Emission Cost Function**

The objective of EELD is to minimize fuel and emission cost satisfying load demand. The cost function of EELD is quadratic nature. It is indicated by equation no. 1.

$$F_i(P_i) = a_i P_i^2 + b_i P_i + c_i \tag{1}$$

$$E_i(P_i) = \alpha_i P_i^2 + \beta_i P_i + \gamma_i + \eta_i * \exp(\delta_i * P_i) \tag{2}$$

$$\mathrm{Min} \cdot [F_i(P_i) + E_i(P_i)]$$

B. **Constraints**

ELD has many constraints. It has two types (I) Constraints described by equality nature (II) Constraints described by inequality nature.

(I) Constraints with equality

The generated power of each generator should be equals to summation of load demand and transmission losses.

$$\sum_{i=1}^{N} P_i(t) = Load\ Demand + Transmission\ Losses \tag{3}$$

(II) *Constraints with inequality*

The generator's output should operate in operating bounds.

$$P_i^{\min} \le P_i \le P_i^{\max} \tag{4}$$

Here $P_i^{\min}$ and $P_i^{\max}$ are the min and max operating limits of generator i.

C. *VPL (valve point loading) effect on ELD*

The effect of VPL on ELD is non-linear. When load is changed, the cost equation of ELD is represented in (5).

$$F = \min\left(\sum_{i=1}^{N} F_i\left(P_i\right)\right) = \min\left(\sum_{i=1}^{N} a_i P_i^2 + b_i P_i + c_i + \left|e_i * \sin\left\{f_i * \left(P_i^{\min} - P_i\right)\right\}\right|\right) \quad (5)$$

where a_i, b_i, c_i, d_i, e_i are the cost coefficients of unit i.

$P_i^{\min}$ Minimum generated power of unit i.

3 Overview of Krill Herd Algorithm and Teaching Learning Based Optimization Algorithm

- *Krill Herd Algorithm*

  In Krill herd (KH) algorithm the objective is the higher population size of the krill individual and searching process of food i.e. higher krill density which means that to achieve higher food density that leads to the optimal solution.

  The fitness function of each krill individual is defined as its distances from food and highest density of the swarm.

  Three essential actions considered to determine the time-dependent position of an individual krill are (i) movement induced by other krill individuals, (ii) foraging activity, and (iii) random diffusion.

  i. *Initialization*

     Reduction in the population size from the food location because of predator attack affects the objective value, this step is termed as initialization process. Position of each individual depends upon three function viz. (i) change in movement due to induction. (ii) Foraging action (iii) diffusion.

     So, in the n- dimensional space the Lagrangian model is defined as

$$\frac{dy_i}{dx} = M_i + X_i + D_i \quad (6)$$

  ii. *Change In Movement Due To The Induction*

     Due to other individual effect each individual try to keep optimal density. The direction is effected by local population size, target population size and repulsive population size. Thus, for each individual it is given as:

$$M_i^{Current} = M_i^{\max} \alpha_i + \omega_n M_i^{previous} \quad (7)$$

  iii. *Foraging Motion*

It is obtained by the mean of two parameters. Optimal solution position and the previous result i.e.

$$X_f = V_f \beta_{isolution} + \omega_x X_i^{\text{Previous}} \quad (8)$$

iv. *Random Diffusion*
It is given as

$$RD_i = RD_{\text{max}} \sigma \quad (9)$$

v. *Movement Process*
Using the result obtained from the different parameters, the position vector is given as:

$$Z_t(t + \Delta t) = Z_i(t) + \Delta t \frac{dz_i}{dt} \quad (10)$$

vi. *Crossover Operators*
In this operation the gene of an individual at next process is produced from the previous one i.e. in this operation the gene of an individual at next process is produced from the previous one i.e.

$$Z_i^{G+1} = \begin{cases} Z_{ji}^G & if\ random\ number < C_R \\ Z_i^{G+1} & else \end{cases} \quad (11)$$

vii. *Mutation Operation*
Mutation operators is given as

$$\begin{aligned} &Z_t = \{Z_{best} + \mu(Z_1 - Z_m); \quad if\ random\ number\ < M_\mu \\ &else \\ &Z_t = Z_t \end{aligned} \quad (12)$$

- *Teaching Learning Based Optimization Algorithm*
  In TLBO, population is randomly initialized within their limits. TLBO is separated also two parts.

i. Teacher Phase
ii. Learner Phase

i. *Teacher Phase*

The mean parameter of each subject of the learners in the class at generation g is given as

$$M^g = \left[m_1^g, m_2^g, \ldots, m_j^g, \ldots, m_D^g \right] \quad (13)$$

To get a new population set of learners a vector is formed using (14)

$$Xnew_i^g = X_i^g + rand\left(X_{Teacher}^g - T_F * M^g\right) \tag{14}$$

T_F is the teaching factor. Value can be either 1 or 2.

If $Xnew_i^g$ is found to be better than X_i^g in generation g, than it replaces on X_i^g otherwise it remains X_i^g.

ii. *Learner Phase*

In learner phase the students can increase their knowledge by interaction of students or sharing their knowledge. For a learner X_i^g, randomly select another learner X_r^g as $i \neq r$. To set a new vector in learner phase Eqs. (15) and (16) is to be understood.

$$Xnew_i^g = X_i^g + rand * \left(X_i^g - X_r^g\right) if\ f\left(X_i^g\right) < f\left(X_r^g\right) \tag{15}$$

$$Xnew_i^g = X_i^g + rand * \left(X_r^g - X_i^g\right) if\ f\left(X_i^g\right) > f\left(X_r^g\right) \tag{16}$$

When the stopping criteria is satisfied and means after completion of all iteration, optimum result is got.

4 Implementation of KH Algorithm to Economic-Emission Load Dispatch Problem

The steps are following

1. Initialize the Fitness Function i.e. Total cost function from the individual cost function of the various generating stations.
2. Input the Fuel cost Functions, MW limits of the generating units and the total power demand.
3. Perform change in movement due to induction when indicates minimum fuel cost.
4. Calculate mean of two parameters.
5. Movement process also is performed. For each vector of active power the value of the fitness function is calculated.
6. Crossover and mutation operation is performed.
7. By comparing two vectors new initialized vectors are formed.
8. When the stopping criteria that means when MAXIT iteration is completed, then the algorithm is stop.

5 Simulation Results with Discussions

The KH and TLBO algorithm has been applied on two cases. Case one is 6 unit systems [9] and case two is 10 unit systems [13]. MATLAB 7.01 is used for develop the program for obtaining the results.

Case 1: System including six generators (6 unit system)
The proposed two algorithm i.e. KH and TLBO has been applied on six unit systems. The coefficients of costs and limits of power generation are taken from [9]. Here emission is considered. Power generation limits are also included. Table 1 shows the solution of conventional ELD i.e. optimal power allocation for finding minimum fuel cost using KH and TLBO compared with PSO [9] shows the better convergence characteristics in proposed algorithms. The optimal allocation of generators for getting minimum emission cost is shown in Table 2. The graph between no. of iterations and fuel cost in $/hr for load of 700 MW using KH and TLBO algorithm is shown in Figs. 1 and 2 respectively. The graph between no. of iterations and emission cost in $/hr for load of 700 MW using KH and TLBO algorithm is shown in Figs. 3 and 4 respectively.

Table 1 Optimal allocation of power corresponding minimum fuel cost for 6 generator system for (700 MW) load demand

Power output	KH	TLBO	PSO [9]
P1(MW)	20.9519	10.3888	30.712
P2(MW)	12.9622	15.8524	18.681
P3(MW)	102.6855	146.2914	130.568
P4(MW)	108.4169	154.3727	134.288
P5(MW)	249.4204	275.0659	206.088
P6(MW)	205.5631	298.0289	198.252
Fuel cost($/hr)	36,022	45,553	1663066.3

Table 2 Optimal allocation of power corresponding minimum combined cost for 6 generator system for (700 MW) load demand

Power output	KH	TLBO	PSO [9]
P1(MW)	78.7743	44.5437	80.3178
P2(MW)	65.1203	92.2446	83.4732
P3(MW)	112.0194	104.3898	111.0704
P4(MW)	109.7429	115.9925	116.6904
P5(MW)	164.1777	172.3243	157.919
P6(MW)	170.1654	170.5051	167.0772
Emission cost ($/hr.)	422.2757	425.0248	432.048

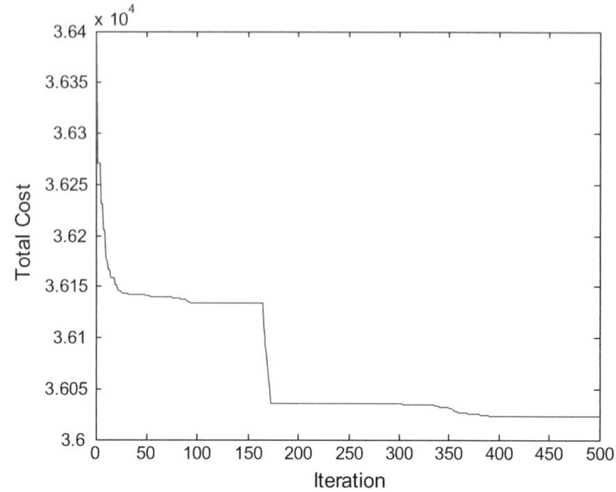

Fig. 1 Convergence characteristics of fuel cost versus iteration for six unit systems using KH algorithm

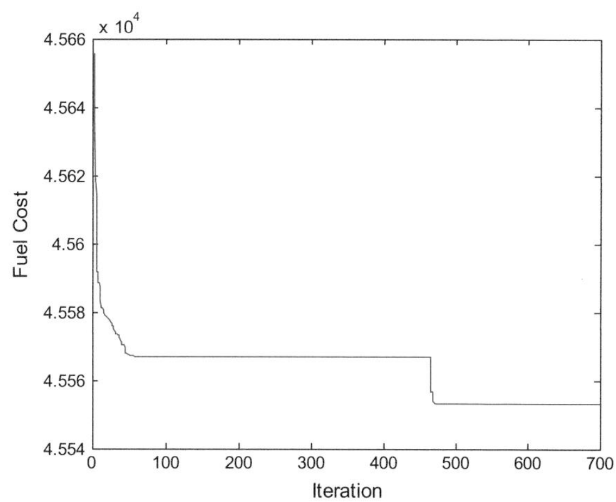

Fig. 2 Convergence characteristics of fuel cost versus iteration for six unit systems using TLBO

Case 2: System including ten generators (10 unit system)

The proposed two algorithm i.e. KH algorithm and TLBO has been applied on ten unit systems. The coefficients of costs and limits of power generation are taken from [13]. Here valve point loading effects and emission is considered. Power generation limits are also included. Table 3 shows the solution of combined EELD i.e. optimal power allocation for finding minimum fuel and emission cost. The graph between no. of iterations and cost in $/hr for load of 2000 MW using KH and TLBO algorithm is shown in Figs. 5 and 6 respectively.

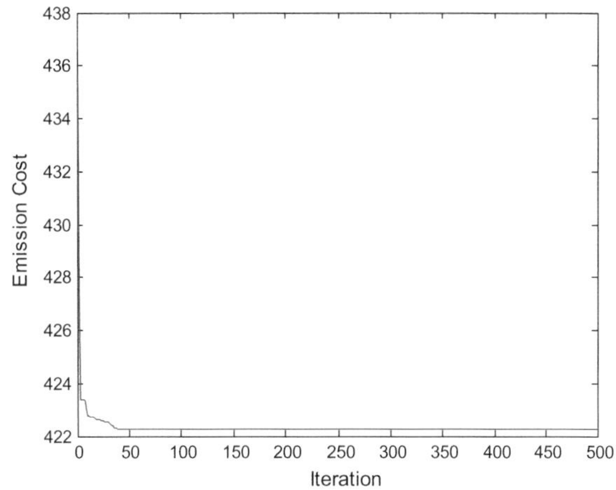

Fig. 3 Convergence characteristics of emission cost versus iteration for six unit systems using KH algorithm

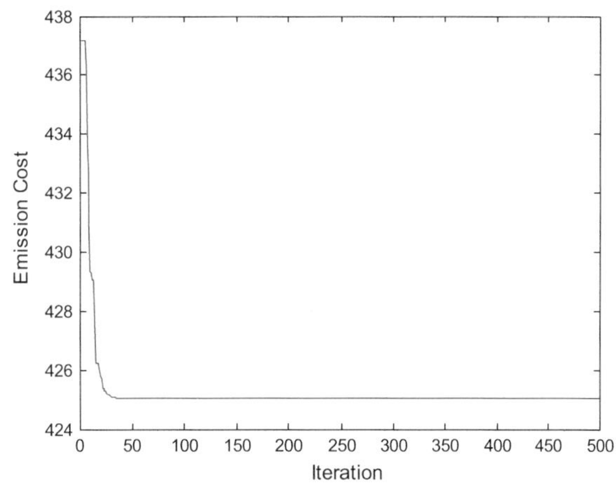

Fig. 4 Convergence characteristics of emission cost versus iteration for six unit systems using TLBO

6 Conclusion

The proposed optimization techniques based on interaction of teacher and students i.e. KH and TLBO algorithm has been successfully applied on linear and non-linear economic emission load dispatch problems. Here transmission losses are not included. The obtained results from proposed algorithms have better convergence characteristics compared to other optimization techniques. So in a word it is very efficient population based method to find optimum results in EELD problems.

Table 3 Optimal allocation of power corresponding minimum fuel cost for 10 generator system for (2000 MW) load demand

Unit power output	KH	TLBO	ABC_PSO [13]	DE [15]	NSGA-II [15]	SPEA-2 [15]
$P_1(MW)$	**41.5470**	30.8835	55	54.9487	51.9515	52.9761
$P_2(MW)$	**73.7694**	79.2830	80	74.5821	67.2584	72.813
$P_3(MW)$	**111.4648**	96.6791	81.14	79.4294	73.6879	78.1128
$P_4(MW)$	**71.1654**	130.0000	84.216	80.6875	91.3554	83.6088
$P_5(MW)$	**76.0255**	103.0483	138.3377	136.8551	134.0522	137.2432
$P_6(MW)$	**88.1139**	70.000	167.5086	172.6393	174.9504	172.9188
$P_7(MW)$	**293.9184**	297.4562	296.8338	283.8233	289.435	287.2023
$P_8(MW)$	**335.9943**	294.0677	311.5824	316.3407	314.0556	326.4023
$P_9(MW)$	**466.2090**	436.3312	420.3363	448.5923	455.6978	448.8814
$P_{10}(MW)$	**441.7923**	462.2510	449.1598	436.4287	431.8054	423.9025
Fuel cost($/h)	**108,470**	**109,880**	**113,420**	**113,480**	**113,540**	**113,520**
Emission cost (1b/h)	**5785.3**	**945.6188**	**4120.1**	**4124.9**	**4130.2**	**4109.1**

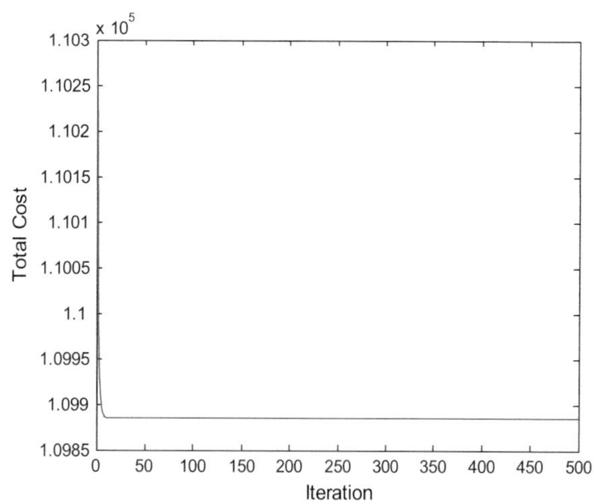

Fig. 5 Convergence behavior of fuel cost and iteration of ten unit systems of load demand 2000 MW with VPL effect using KH algorithm

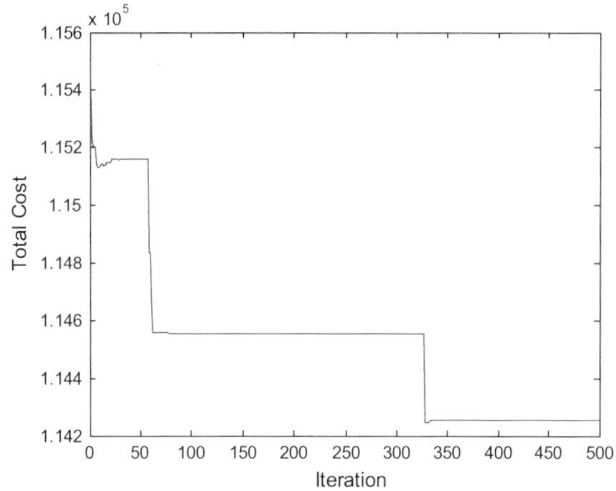

Fig. 6 Convergence behavior of combined cost and iteration of ten unit systems of load demand 2000 MW with VPL effect using TLBO

References

1. H. Saadat, *Power System Analysis* (Tata McGraw Hill Publishing Company, New Delhi, 2002)
2. Z.X. Liang, J.D. Glover, A zoom feature for a dynamic programming solution to economic dispatch including transmission losses. IEEE Trans. Power Syst. **7**(2), 544–550 (1992)
3. M. Basu, Artificial bee colony optimization for multi-area economic dispatch. Electr. Power Energy Syst. **49**, 181–187 (2013)
4. R.V. Rao, V.J. Savsani, D.P. Vakharia, Teaching-learning-based optimization: a novel method for constrained mechanical design optimization problems. CAD Comput. Aided Des. **43**(3), 303–315 (2011)
5. R.V. Rao, V.J. Savsani, D.P. Vakharia, Teaching-learning-based optimization: an optimization method for continuous non-linear large scale problems. Inf. Sci. **183**(1), 1–15 (2012)
6. D.E. Goldberg, *Genetic Algorithms in Search Optimization and Machine Learning* (Addison-Wesley, Reading, MA, USA, 1989)
7. J. Kennedy, R. Eberhart, Particle swarm optimization, in *Proceedings of the IEEE International Conference on Neural Networks*, December 1995, pp. 1942–1948
8. D. Karaboga, B. Basturk, On the performance of artificial bee colony (ABC) algorithm. Appl. Soft Comput. J. **8**(1), 687–697 (2008)
9. G. Anurag, K.K. Swarnka, K. Wadhwani, Combined economic emission dispatch problem using particle swarm optimization. Int. J. Comput. Appl. (0975–8887) **49**(6), 1–6 (2012)
10. G.P. Dixit, H.M. Dubey, M. Pandit, B.K. Panigrahi, Economic load dispatch using artificial bee colony optimization. Int. J. Adv. Electron. Eng. **1**(1), 119–124 (2011)
11. Y. Sonmez, Multi-objective environmental/economic dispatch solution with penalty factor using artificial bee colony algorithm. Sci. Res. Essays **6**(13), 2824–2831 (2011)
12. S. Chakraborty, T. Senjyu, A. Yona, A.Y. Saber, T. Funabashi, Solving economic load dispatch problem with valve-point effects using a hybrid quantum mechanics inspired particle swarm optimization. IET Gener. Transm. Distrib. **5**(10), 1042–1052 (2011)
13. E.D. Manteaw, N.A. Odero, Combined economic and emission dispatch solution using ABC_PSO hybrid algorithm with valve point loading effect. Int. J. Sci. Res. Publ. **2**(12), 1–9 (2012)
14. F. Zou, L. Wang, X. Hei, D. Chen, Q. Jiang, H. LI, Bare-bones teaching-learning-based optimization. Sci. World J. **2014**, 1–17 (2014). Article ID 136920
15. K. Basu, Economic environmental dispatch using multi-objective differential evolution. Appl. Soft Comput. **11**, 2845–2853 (2011)

Classification of Crossover Faults and Determining Their Location in a Double Circuit Power Transmission System with Multiple Sources

Nabamita Roy

1 Introduction

Faults in overhead transmission lines are more likely to happen due to lightning, falling trees and insulators breakdown. The electrical power is transmitted either by single circuit system or double circuit system. Short circuit faults are quite common and in a double circuit system there remains a scope of crossover short-circuit in which two phases of different circuits are involved. Identification of such faults and determining their location is a challenging task.

A scheme of determination of fault location for a double circuit compensated transmission lines has been proposed in [1] where the location has been estimated by using Discrete Wavelet Transform (DWT) and KNN with less than 1% error. A new approach of fault classification has been presented in [2] for EHV transmission lines using Rough Membership Neural network (RMNN). DWT has been used for feature extraction and a comparative analysis has been shown between RMNN and BPNN to establish that RMNN is faster and more accurate than BPNN as a classifier. The fault location has not been determined here. A hybrid method of ANN and DWT has been suggested in [3] for identification of faulty section and obtaining its location in a distribution network. The proposed method in this paper has been tested on a IEEE system but the effect of noise on the features extracted has not been discussed here. The paper [4] proposes an approach by combining independent component analysis (ICA) with travelling wave (TW) theory and Support Vector Machine (SVM) for fault analysis of HV Transmission lines. This method gives better performance in presence of noise. A new technique for fault location on transmission lines using only voltage measurements obtained from Wide Area Measurement Systems (WAMS) and the

N. Roy (✉)
Electrical Engineering Department, MCKV
Institute of Engineering, Liluah, Howrah,
West Bengal, India
e-mail: roynab@gmail.com

S. Chattopadhyay et al. (eds.), *Modelling and Simulation in Science, Technology and Engineering Mathematics*, Advances in Intelligent Systems and Computing 749,
https://doi.org/10.1007/978-3-319-74808-5_14

network bus admittance matrix has been reported in [5]. Fault classification has also been included in this paper using the same technique. In [6] inter-circuit shunt faults and cross-country faults in a double circuit system have been identified and classified using DWT and SVM. In this paper, the method of determining fault location and the effect of noise has not been discussed. A hybrid framework has been designed in [7] for classifying and locating short circuit faults in power transmission lines and this framework consists of a proposed two-stage finite impulse response (FIR) filter, four support vector machines (SVMs), and eleven support vector regressions (SVRs). SVM has been also applied in [8] for fault classification in a long transmission line in which the features have been selected using wavelet packet transform.

In this paper, a method is proposed for identification of the type of fault and obtaining its corresponding location in a double circuit system. ANN has been involved here in which PNN is used for fault classification and BPNN for obtaining the fault location. The input features of both the PNN and BPNN have been obtained from DST of the current signals measured at any one terminal of the network. All the faults have been simulated in MATLAB environment. The scope of this paper is limited to the simulation of only double line short—circuit faults.

The rest of the paper is organized as follows. The simulation of faults in a double circuit network is described in Sect. 2. DST is briefly discussed in Sect. 3. The features needed for fault analysis and the method of regression applied on the features have been described in Sect. 4. Section 5 explains the method of fault classification and obtaining its location. The effect of noise has also been studied in this section.

2 Simulation of Faults and the System Under Study

A 400 kV, 50 Hz, 3-phase double circuit power system network is simulated using the Simpower Toolbox of MATLAB-7 and is shown in Fig. 1. The length of each transmission line is 300 km. The double circuit network has similar sources at both its ends.

System parameters:

Generator 1, 2: Voltage = 400 kV, Impedance of generator = (0.2 + j4.49) Ω, X/R ratio = 22.45.

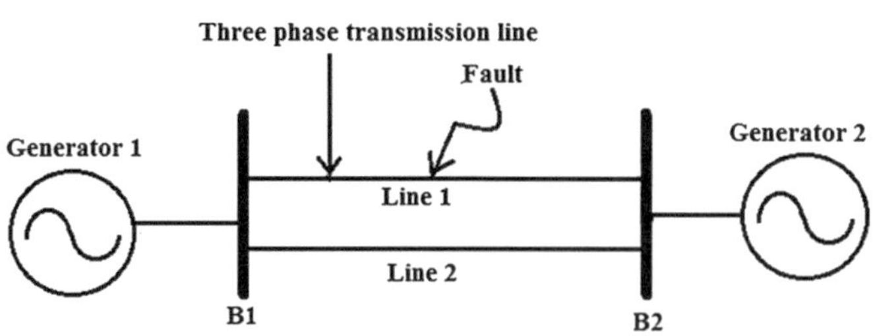

Fig. 1 Single line diagram of three phase double circuit network

Each Transmission Line: Length = 300 km, R_1 = 0.02336 Ω/km, R_2 = 0.02336 Ω/km, R_0 = 0.38848 Ω/km, L_1 = 0.95106 mH/km, L_2 = 0.95106 mH/km, L_0 = 3.25083 mH/km, C_1 = 12.37 nF/km, C_2 = 12.37 nF/km, C_0 = 8.45 nF/km.

All the signals have been simulated with a sampling time of 78.28 μs. The time period of simulation in MATLAB has been taken up to 0.04 s. The sampling frequency is 12.8 kHz.

Crossover two phase short-circuit faults have been simulated in the following way as given below:

Double Line (L-L) Faults:

A_1A_2: Phase A of Line 1, A_1 shorted to phase A of line 2, A_2
A_1B_2: Phase A_1 of Line 1, A_1 shorted to phase B of line 2, B_2
A_1C_2: Phase A_1 of Line 1, A_1 shorted to phase C of line 2, C_2
B_1A_2: Phase B_1 of Line 1, B_1 shorted to phase A of line 2, B_2
B_1B_2: Phase B_1 of Line 1, B_1 shorted to phase B of line 2, B_2
B_1C_2: Phase B_1 of Line 1, B_1 shorted to phase C of line 2, C_2
C_1A_2: Phase C of Line 1, C_1 shorted to phase A of line 2, A_2
C_1B_2: Phase C of Line 1, C_1 shorted to phase B of line 2, B_2
C_1C_2: Phase C of Line 1, C_1 shorted to phase C of line 2, C_2

All the faults have been initiated at 19 different locations starting from B1, each being 10 km apart. The fault resistances considered for the simulation are from the range of 0–100Ω in steps of 20Ω. Fault inception angle is considered to be 0°. The total number of fault simulations made in this system is 9 × 19 × 6 = 1026.

3 Discrete S-Transform (DST)

An electrical signal h(t) can be expressed in discrete form as h(kT), k = 0, 1, …, N − 1 and T is the sampling time interval, [9, 10].

The discrete Fourier transform of h(kT) is obtained as,

$$H\left[\frac{n}{NT}\right] = \frac{1}{N}\sum_{k=1}^{N-1} h(kT)e^{\frac{-i2\pi k}{N}} \tag{1}$$

where n = 0, 1, …, N − 1.

Using (4), the ST of a discrete time series is obtained by letting $f \to n/NT$ and $\tau \to jT$ as

$$S\left[jT, \frac{n}{NT}\right] = \sum_{m=0}^{N-1} H\left[\frac{m+n}{NT}\right]G(m,n)e^{i2\pi mj/N} \tag{2}$$

and $G(m,n) = e^{-2\pi^2 m^2/n^2}$, $n \neq 0$ where j, m = 0, 1, 2, …, N − 1 and n = 1, 2, …, N − 1.

A complex matrix (S-matrix) is produced from Eq. (2). The rows of the matrix represent frequencies and the columns signify times. The absolute value of the S-matrix gives the amplitude of the ST spectrum. Hence, each column of the matrix can be considered as the local spectrum at any point and time. The amplitude of the signal at different frequency resolutions remains unaffected in the matrix.

DST of a given signal provides the privilege of obtaining both the amplitude and phase informations at any point of time and at any frequency. These informations remain almost unaffected in presence of noise. Hence, in the present paper DST has been used for feature extraction. However, only the amplitude matrix has been used here for obtaining the features.

4 Feature Extraction from the S-Matrix and Regression

The waveforms of the current signals measured at the busbar B1 after simulation in Fig. 1 have been shown in Figs. 2 and 3 respectively.

The Fig. 4 shows the plot of magnitude of a parameter XB1totalphA with respect to time which has been obtained from S-matrix of a current signal corresponding to a A_1B_2 fault occurring at a distance 100 km from B1 and B2. XB1totalphA has been calculated by summing up each column of S-matrix.

The feature XareaB1 has been obtained by calculating the area under the curve plotted in Fig. 4. The other feature XreaB2 has been obtained in a similar way.

The variation of feature XareaB1 at different fault locations has been shown in Fig. 5. The curves of Fig. 5 demonstrate an irregular pattern of the feature XareaB1

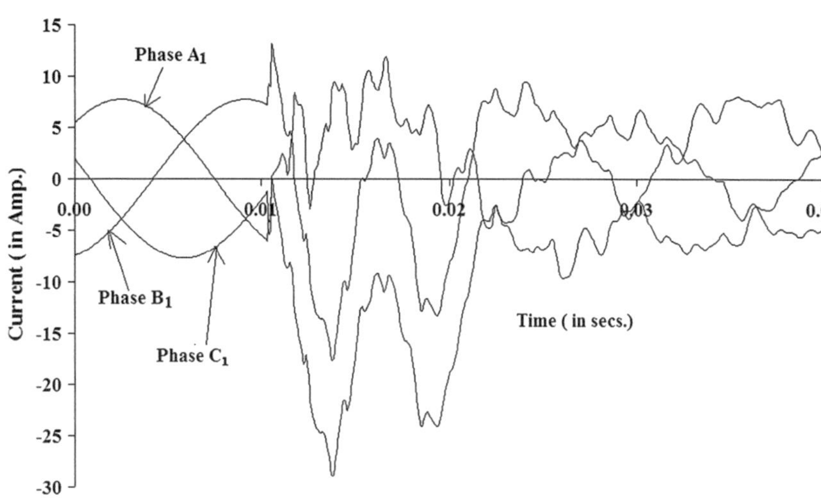

Fig. 2 Current waveforms of the three phases A_1, B_1, C_1 for a A_1B_2 type of fault at 100 km from B_1 with $R_F = 0\ \Omega$ and fault inception angle $= 0°$

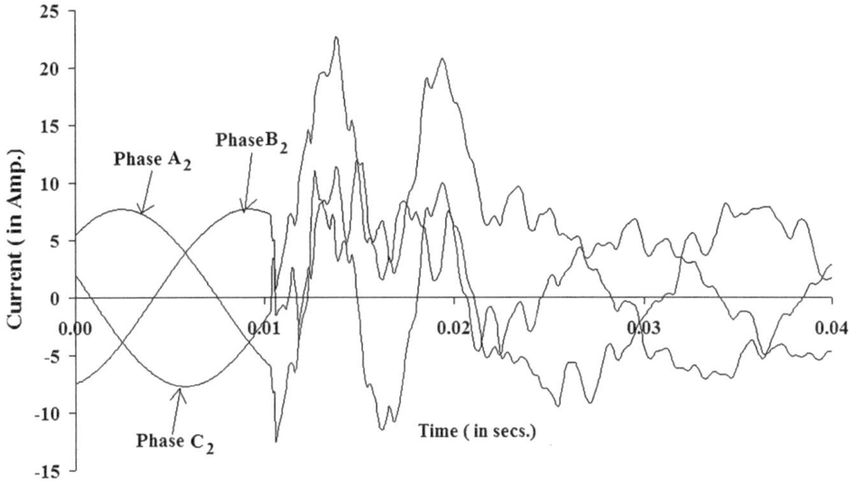

Fig. 3 Current waveforms of the three phases A_2, B_2, C_2 for a A_1B_2 type of fault at 100 km from B_1 with $R_F = 0\ \Omega$ and fault inception angle $= 0°$

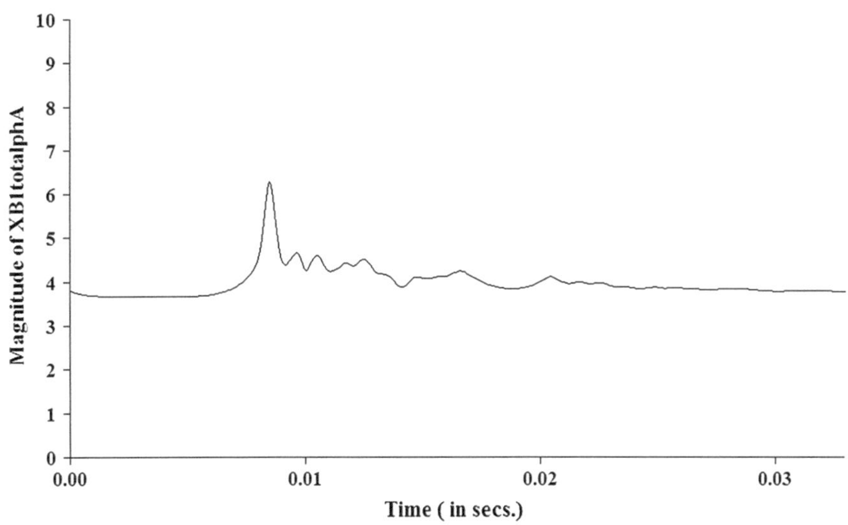

Fig. 4 Profile of magnitude of the parameter XB_1totalphA

with respect to distance of fault location. It is difficult to get a satisfactory result if such features are used for training a neural network. Henceforth, polynomial regression has been adopted to obtain a suitable pattern of the features that can be trained by a neural network to produce satisfactory results.

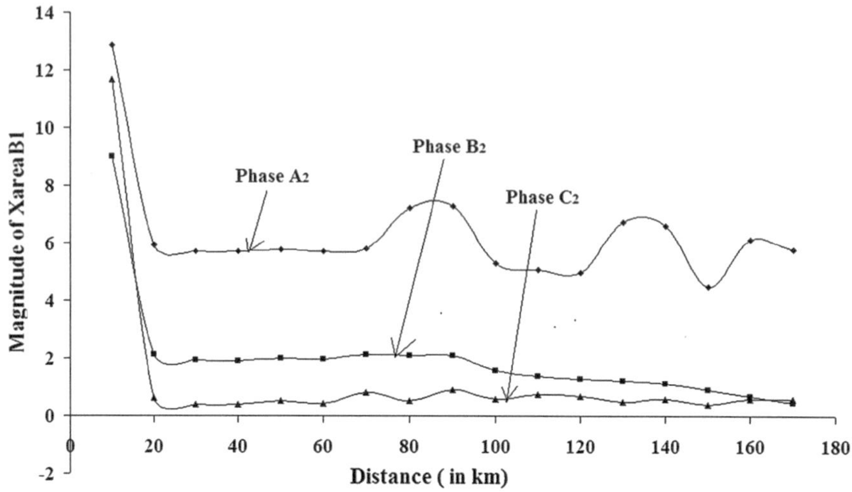

Fig. 5 Profile of the change in magnitude of the feature XareaB1 for different fault locations

4.1 *Polynomial Regression*

In case of Linear Regression Models, polynomial models represent a special case of the linear models. Polynomial models are simple. They have the ability and familiarity in their properties in following the data trends with reasonable flexibility. These models are unaffected by any changes in the location and scale of the data. The model should be selected in such a way that it should provide simple description of overall data trends and make reasonably accurate predictions.

In the present work, polynomial regression is applied on the features XareaB1 and XareaB2 by programming in MATLAB. The profiles of the features after regression have been shown in Figs. 6 and 7 respectively.

From Figs. 6 and 7 it is evident that after regression the features XareaB1 and XareaB2 follow a regular pattern with change in fault location and the regressed features can be conveniently used for training a neural network to obtain an acceptable output from an unknown input parameter.

5 Fault Classification and Determination of Fault Location

Different ANN architectures are suitable for a varied range of tasks. Selection of a particular network depends on the type of problem. PNN is widely used for the task of classification. In the present work, PNN has been used for fault classification. The architecture of the PNN used for fault classification involves two hidden layers. The working function of the first hidden layer is radial basis transfer function, and

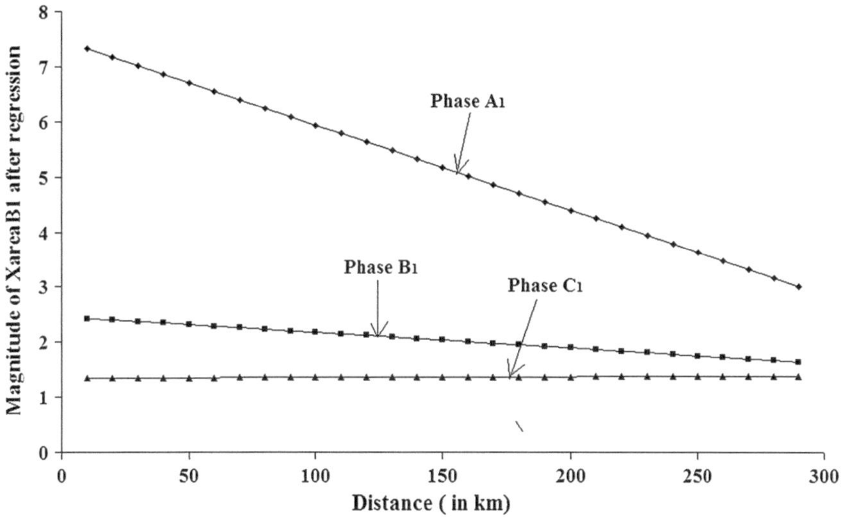

Fig. 6 Profile of the change in magnitude of the feature XareaB1 after regression

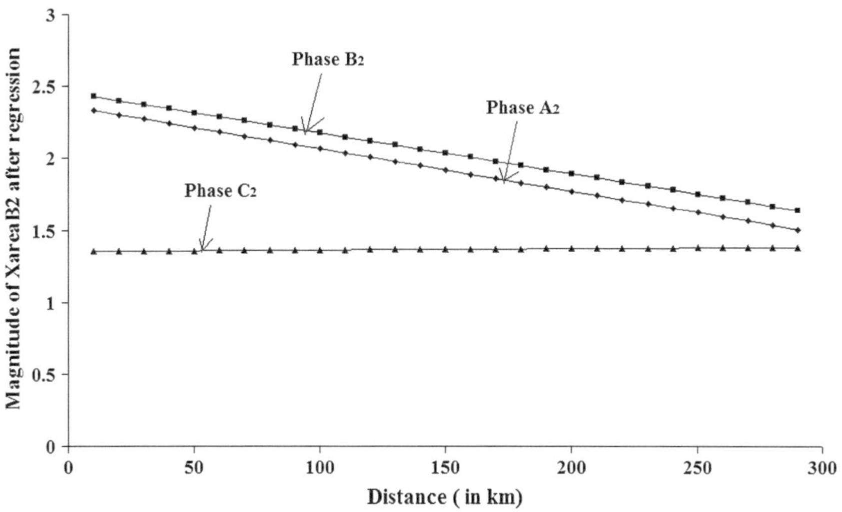

Fig. 7 Profile of the change in magnitude of the feature XareaB2 after regression

that of the second hidden layer is competitive transfer function, [11]. PNN has many advantages. Its training process is fast and it has an inbuilt parallel structure that has the highest scope of converging to an optimal classifier as the dimension of the representative training set enhances. There is a huge scope of adding or removing the training samples without involving vigorous retraining of PNN.

BPNN is relatively suitable for function approximation problems. The network is trained by input vectors and the corresponding target vectors until a function is approximated. Standard backpropagation is based on Widrow-Hoff learning rule [12] which is a gradient descent algorithm. The term *backpropagation* means the way in which the gradient is calculated for nonlinear multilayer networks. Reasonably satisfactory answers are obtained from a properly trained backpropagation network when unknown inputs are given to it. The unknown input features should be similar to the input vectors used for training so that the output feature obtained from a BPNN corresponding to the new input feature is quite close to the correct output. In this way, BPNN is generalised to be trained on a representative set of input/target pairs to produce satisfactory results without the need for training on all possible input/output pairs.

5.1 Fault Classification

The features XareaB1 and XareaB2 of the six phases have been used as input parameters. The regressed features of 10 current signals of each phase are used for training and the rest are used for testing purpose. The output of the PNN is summarised in Table 1. The average of correct predictions is 98.7%.

5.2 Determination of Fault Location

The fault location has been obtained from a BPNN. The BPNN in this paper is a 2-layer feed-forward network which consists of only one hidden layer and an output layer. The input layer consists of the input vector. In this paper, the elements of the

Table 1 Results of fault classification from the PNN

Type of fault	PNN output	% of correct predictions
A_1A_2	1	97.8
A_1B_2	2	98.9
A_1C_2	3	98.5
B_1A_2	4	97.6
B_1B_2	5	99.5
B_1C_2	6	98.7
C_1C_2	7	98.5
C_1B_2	8	98.6
C_1A_2	9	99.6
No fault	10	99.6

Table 2 Fault location in case of A_1B_2 type of fault with fault resistance, $R_F = 0\ \Omega$

Actual fault location (km)	BPNN output (km)	% error
20	21.23	1.59
40	39.36	−1.60
60	60.34	0.57
80	81.87	2.34
100	102.69	2.69
120	120.46	0.38
140	140.58	0.42
160	159.49	−0.32
180	180.85	0.47
200	200.10	0.05
220	220.57	0.26
240	241.36	0.57
260	259.06	−0.36
280	281.53	0.55

input vector are the features of 10 current signals obtained from DST. The number of neurons for the hidden layer is taken to be 100 and the transfer function used is Tan-Sigmoid. Only one neuron has been used for the output layer and the transfer function used is Purelin. The BPNN architecture has been initially tested with 60, 80, 100 and 120 neurons. It was observed that the selection of 100 neurons gives the output with satisfactory accuracy and speed. If the number of neurons is lesser than 100 then the accuracy is affected and if the number is greater than 100 then the speed of convergence of training becomes higher with a small improvement in the accuracy.

Once the short-circuited phases have been identified from the PNN either the parameter XareaB1 or XareaB2 of any one of the faulty phases has been used as the input feature. In case of A_1B_2 type of fault, the features XareaB1 obtained corresponding to phase A_1 for different fault locations has been used for training the BPNN. The features of 10 current signals have been used for training and the rest are used for testing purpose. Levenberg–Marquardt (LM) algorithm has been used for training the BPNN. The percentage error is calculated during estimation of fault location as shown in Table 2 and according to Eq. (3). Table 2 show the results of fault location obtained from the BPNN in case of a A_1B_2 type of fault.

$$\frac{BPNN\ output - Actual\ Fault\ Location}{Actual\ Fault\ Location} \times 100 \tag{3}$$

From Fig. 8 it is evident that within 4 epochs convergence is reached and the result is obtained fast corresponding to a new input parameter. The maximum error achieved in determination of fault location is 2.69%.

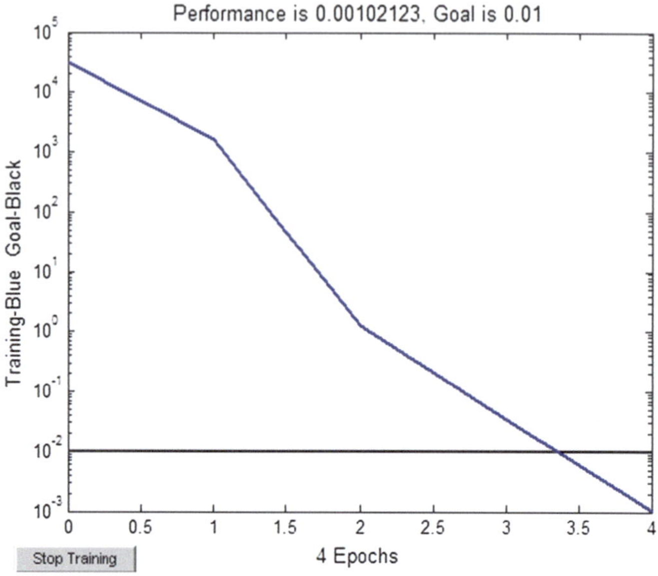

Fig. 8 Training of BPNN for obtaining fault location by LM method

5.3 Implementation of Noisy Signals

The current signals obtained at the Bus B1 from simulation have been impregnated with 20 dB white Gaussian noise by programming in MATLAB. As an illustration, the magnitude of the feature XareaB1 has been calculated for noisy current signals and the same has been plotted with respect to the signals without noise as shown in Fig. 9. The type of fault considered in Fig. 9 is A_1B_2 with fault resistance being 0 Ω. The results of classification and estimation of fault location have been given in Tables 3 and 4. The average of correct classifications from PNN is 98.6% and the maximum error achieved in obtaining fault location is 4.62%.

6 Conclusion

The selection of features is an important part that determines how effectively and accurately the faults can be classified. In this paper, only the current signals of one terminal of the network have been used for extracting features. Six features are needed for the six lines to identify the affected phases from PNN. Only one feature of the affected phase is needed for obtaining fault location from the BPNN. The technique of regression applied on the features obtained from S-matrix has produced quite accurate and fast results. The faults have been simulated at different locations with variation in fault resistance. The effect of noise has also been studied. The average

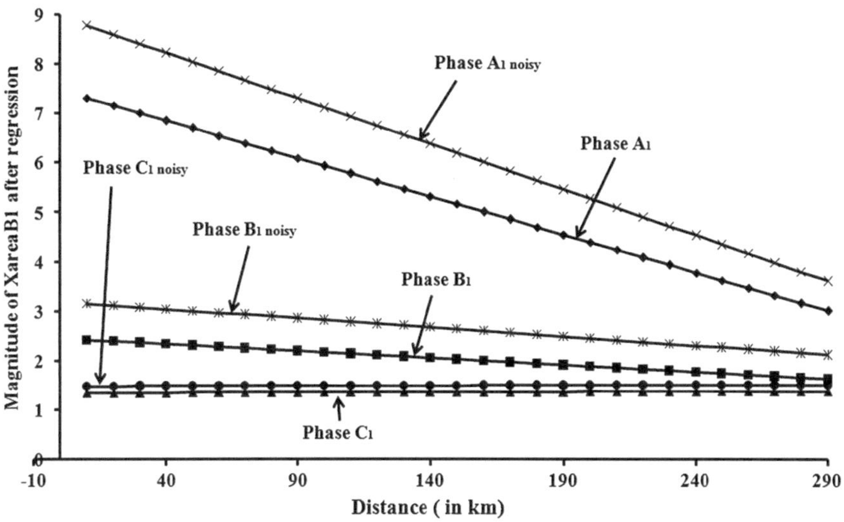

Fig. 9 Magnitude of the regressed feature XareaB1 for noisy signals and signals without noise

Table 3 Results of fault classification from the PNN with noisy current signals

Type of fault	PNN output	% of correct predictions
A_1A_2	1	97.7
A_1B_2	2	98.7
A_1C_2	3	98.9
B_1A_2	4	97.7
B_1B_2	5	99.4
B_1C_2	6	98.6
C_1C_2	7	98.3
C_1B_2	8	98.4
C_1A_2	9	99.5
No fault	10	99.5

percentage of correct classifications from the PNN is 98.7% without noise and 98.6% in presence of noise. The faults have been located by the BPNN with a maximum error of 2.69% without noise and 4.62% in presence of noise. The results indicate that the proposed method of fault classification and estimation of fault location can be effectively implemented for other systems as well.

The present work can be further extended for the analysis of Line-Ground faults, Double-line Ground Faults and Three phase short-circuit faults in which phases of different circuits are also involved.

Table 4 Fault location in case of A_1B_2 type of fault with fault resistance, $R_F = 0\ \Omega$

Actual fault location (km)	BPNN output (km)	% error
20	19.82	−0.90
40	39.87	−0.33
60	62.77	4.62
80	81.56	1.95
100	98.96	−1.04
120	119.56	−0.37
140	141.22	0.87
160	160.56	0.35
180	179.45	−0.31
200	203.56	1.78
220	223.55	1.61
240	240.88	0.37
260	262.87	1.10
280	279.76	−0.09

References

1. A. Swetapadma, P. Mishra, A. Yadav, A.Y. Abdelaziz, A non-unit protection scheme for double circuit series capacitor compensated transmission lines. Electr. Power Syst. Res. **148**, 311–325 (2017). https://doi.org/10.1016/j.epsr.2017.04.002
2. Z. He, S. Lin, Y. Deng, X. Li, Q. Qian, A rough membership neural network approach for fault classification in transmission lines. Int. J. Electr. Power Energy Syst. **61**, 429–439 (2014). https://doi.org/10.1016/j.ijepes.2014.03.027
3. A.C. Adewole, R. Tzoneva, S. Behardien, Distribution network fault section identification and fault location using wavelet entropy and neural networks. Appl. Soft Comput. **46**, 296–306 (2016). https://doi.org/10.1016/j.asoc.2016.05.013
4. A.R. Almeida, O.M. Almeida, B.F.S. Junior, L.H.S.C. Barreto, A.K. Barros, ICA feature extraction for the location and classification of faults in high-voltage transmission lines. Electr. Power Syst. Res. **148**, 254–263 (2017). https://doi.org/10.1016/j.epsr.2017.03.030
5. S. Das, S.P. Singh, B.K. Panigrahi, Transmission line fault detection and location using wide area measurements. Electr. Power Syst. Res. **151**, 96–105 (2017). https://doi.org/10.1016/j.epsr.2017.05.025
6. A. Swetapadma, A. Yadav, Directional relaying using support vector machine for double circuit transmission lines including cross-country and inter-circuit faults. Int. J. Electr. Power Energy Syst. **81**, 254–264 (2016). https://doi.org/10.1016/j.ijepes.2016.02.034
7. H. Fathabadi, Novel filter based ANN approach for short-circuit faults detection, classification and location in power transmission lines. Int. J. Electr. Power Energy Syst. **74**, 374–383 (2016). https://doi.org/10.1016/j.ijepes.2015.08.005
8. P. Ray, D.P. Mishra, Support vector machine based fault classification and location of a long transmission line. Eng. Sci. Technol. Int. J. **19**(3), 1368–1380 (2016). https://doi.org/10.1016/j.jestch.2016.04.001
9. F. Zhao, R. Yang, Localization of the complex spectrum: S-transform. IEEE Trans. On Signal Processing **44**(4), 998–1001 (1996)
10. R.G. Stockwell, S-transform analysis of gravity wave activity from a small scale network of airglow imagers, PhD thesis, University of Western Ontario, London, Ontario, Canada, 1999

11. S. Mishra, C.N. Bhende, B.K. Panigrahi, Detection and classification of power quality disturbances using S-transform and probabilistic neural network. IEEE Trans. Power Delivery **23**(1), 280–287 (2008)
12. H. Demuth, M. Beale, *Neural Network Toolbox User's Guide, Version 4*, *copyright 1992-2001*. The MathWorks, Inc

Optimal Value of Excitation of Self-excited Induction Generators by Simulated Annealing

Writwik Balow, Arabinda Das, Amarnath Sanyal and Raju Basak

1 Introduction

Generally generating stations are located far away from rural areas. Sometimes it may not be possible to deliver power to the rural areas, because of the distance and cost involved in generation and transmission. Researchers are currently giving stress on dispersed generation using alternative energy sources to generate energy at a cheap rate. Wind is found to be the best alternative among all the energy sources available in rural areas. The fuel cost is nil, however, the source is of variable energy. Under the circumstances, Induction Generator has been found to be the most effective machine used to generate energy while driven by wind turbine. Self-excited induction Generator shows better performance for generating energy from the renewable energy sources.

W. Balow
Electrical Engineering Department,
Ideal Institute of Engineering, Kalyani, India
e-mail: writwik.balow@gmail.com

A. Das
Electrical Engineering Department,
Jadavpur University, Kolkata, India
e-mail: adas_ee_ju@yahoo.com

A. Sanyal
Electrical Engineering Department,
Calcutta Institute of Engineering and Management, Kolkata, India
e-mail: ansanyal@yahoo.com

R. Basak (✉)
GEP Department, University Claude Bernard Lyon1,
Villeurbanne, France
e-mail: basak.raju@yahoo.com

S. Chattopadhyay et al. (eds.), *Modelling and Simulation in Science, Technology and Engineering Mathematics*, Advances in Intelligent Systems and Computing 749,
https://doi.org/10.1007/978-3-319-74808-5_15

When induction motor runs at super-synchronous speed i.e. at a negative slip, it is called an induction generator [1, 2]. In this mode, it converts mechanical energy of the wind turbine into electrical energy. In this case, reactive power is consumed by the motor instead of being generated as in a synchronous machine. While the machine is driven at super-synchronous speed, it starts delivering active power. The output variables like voltage, frequency, and load are highly influenced by the amount of excitation of the capacitor bank.

There are two kinds of operations for a Squirrel cage induction generator. The generator running at super-synchronous speed may be connected to grid, or a capacitor bank may be connected in parallel with the system [3–5]. The configuration of Self-excited Induction Generator (SEIG) is shown in Fig. 1 and it is used to develop an objective function for excitation of the capacitor. A case study is made for four machines having rating 4, 7.5, 15, 37 kW whose excitation is optimized by the process of Simulated Annealing (SA). The final result has been presented graphically to show the variation in results.

2 Excitation of Self-excited Induction Generator (SEIG)

Induction Generator is excited initially by residual magnetism of the core which feeds the capacitor bank—the voltage builds up as in a self-excited D.C generator. The Capacitor bank is connected in parallel with the system to supply the reactive power [6].

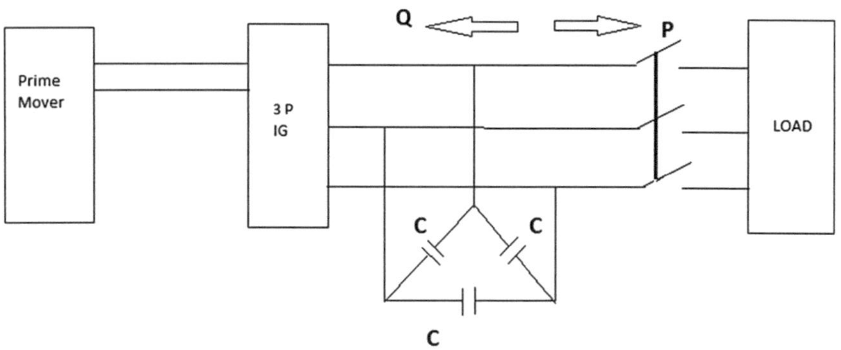

Fig. 1 Self-excited induction generator (SEIG) connected with load

3 Simulated Annealing

There are many types of optimization technique—the nature of the problem is also diverse. Sometimes it becomes quite difficult to choose the proper optimization scheme for a particular problem. The choice depends on the nature of the objective function. It is found that the excitation function of self excited induction generator is non-linear by nature [7, 8]. An attempt has been made for global search by Simulated Annealing within an area widened by lower and upper limits of variables. Choice of number of variables, number of generators, bounds of the variables, constraint-functions etc. have been chosen step by step for SA and then with proper coding in Mat lab the function have been optimized. Flow Chart for algorithm is shown in Fig. 2.

Annealing is a process used to crystallize metals. When a metal is strongly heated, the atoms reach high energy level and are set to motion. The cooling process helps the atoms to reach the equilibrium condition at minimum energy. The expression of probability is given by $P(\Delta E) = e^{(-\Delta E/KT)}$, where T is temperature and K is Boltzmann's constant. The function, which is to be optimized, is started with a high temperature and then it is slowly cooled down to its global optima. $e^{(-\Delta E/KT)}$ is calculated and a random number 'r' is generated between 0 and 1. If $r \leq e^{(-\Delta E/KT)}$ then it is saved, otherwise discarded. Thereafter we move to the next step.

4 Variables

Self-excited Induction Generator is largely influenced by its excitation—if it is not excited properly its output voltage cannot build up. Here three variables have been chosen and then the objective function (i.e. excitation of the capacitor) has been framed with these variables. Variables have been chosen in a way such that, they can governs the objective function i.e. load, frequency and speed. The variables are allowed to vary within a certain range—the range in p.u. is given below:

$$x_1 = \text{Load} \quad [0.5 \leq x_1 \leq 1.3]$$
$$x_2 = \text{Frequency} \quad [1.3 \leq x_2 \leq 1.6]$$
$$x_3 = \text{Speed} \quad [0.5 \leq x_3 \leq 1.3]$$

The output voltage is selected as inequality constraint: $[g_i(x) - x_2 \cdot k \geq 0]$

5 Objective Function

The equivalent circuit of the Induction generator is shown in Fig. 3.

A function is derived from the equivalent circuit and is taken as an objective function of excitation. Capacitive reactance is taken, which is to be optimized and expressed in terms of load resistance, output frequency and speed. The function can be expressed as:

$$\text{Minimize} = [Ax_1x_2^2 + Bx_1x_2]/[Cx_1 + Dx_2(x_2 - x_3) + E] \quad (1)$$

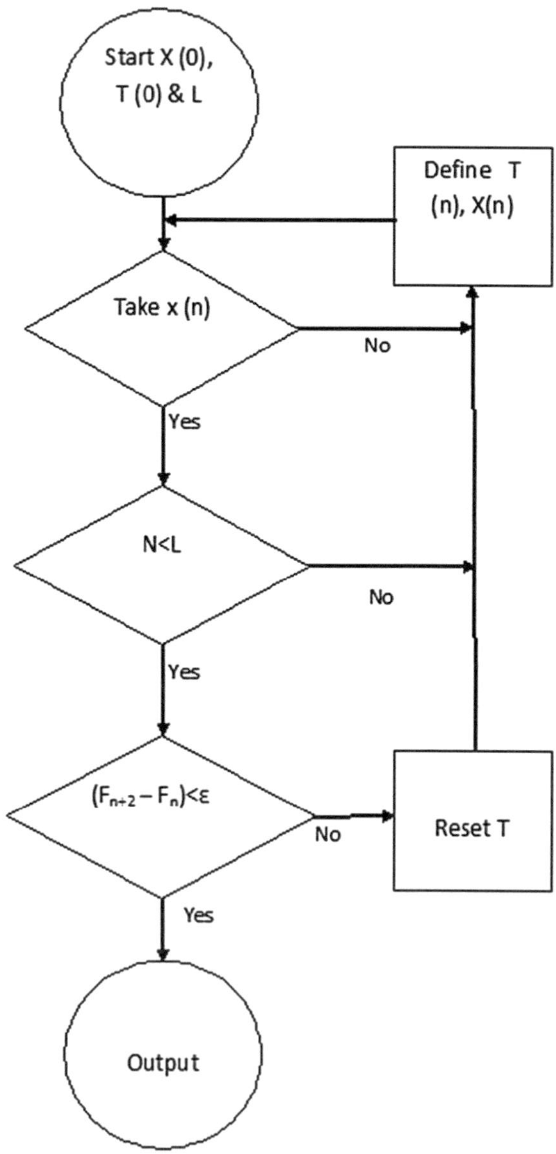

Fig. 2 Flowchart

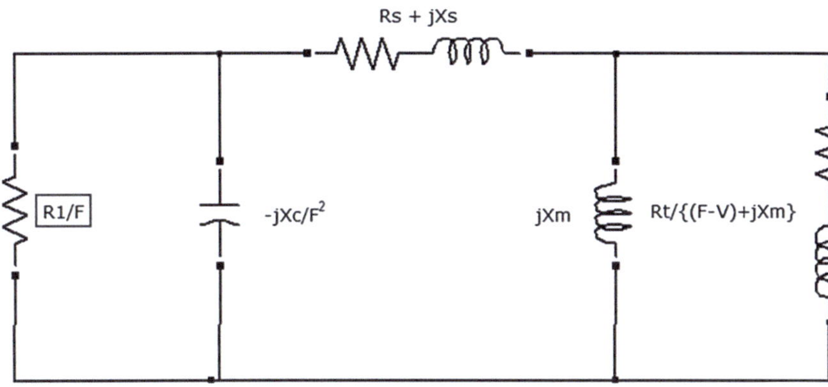

Fig. 3 Equivalent circuit of induction generator

Generated Voltage is as follows:

$$V_g = (1.6 - 0.36x_m)x_2 \quad (2)$$

Three Variables are bound with their limits. Simulated Annealing is applied to reach the optimum, under the constraints, e.g. the output voltage should not be below a certain limit [9, 10]. Optimality is reached in about thousands of iterations. First hundred convergence data for 4 kW machine is shown in Fig. 4 and last thirty-six convergence data is shown in Table 1.

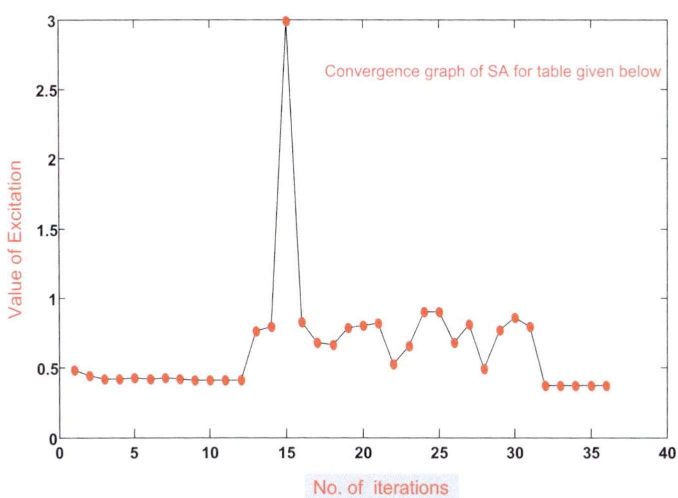

Fig. 4 Graph of convergence

Table 1 First 36 convergence data of the above graph

No.	F(x)	No.	F(x)	No.	F(x)
1	0.4869	13	0.7596	25	0.9002
2	0.4437	14	0.7961	26	0.6821
3	0.4194	15	0.9895	27	0.8132
4	0.419	16	0.8294	28	0.4901
5	0.4268	17	0.6782	29	0.7673
6	0.417	18	0.668	30	0.8608
7	0.424	19	0.7871	31	0.7977
8	0.417	20	0.8031	32	0.36812
9	0.4134	21	0.8215	33	0.3683
10	0.4135	22	0.5242	34	0.36883
11	0.4134	23	0.6525	35	0.36883
12	0.4134	24	0.903	36	0.36883

Table 2 Equivalent circuit parameters

Machines rating (kW) in p.u.	R_E	R_r	L_{LS}	L_{Lr}	L_m
4.0	0.0351	0.0348	0.0458	0.045	1.352
7.5	0.0346	0.0347	0.0448	0.044	1.827
15	0.0201	0.0206	0.0291	0.029	1.89
37	0.0190	0.0116	0.0526	0.052	1.97

6 Case Study

Four machines are taken for case study having ratings of 4, 7.5, 15 and 37 kW respectively. The parameters of the machines are shown in Table 2 and the constant of objective function is shown in Appendix.

Optimal value of the excitation is determined by the method of simulated annealing for each machine with their corresponding value of the variables in p.u. Simulated annealing works well, as an optimizing tool, for global search within the search space created by the bounds of the variables [11, 12]. The optimality is reached through two thousand iterations without violating the constraints [13].

7 Conclusion

The output parameters are shown in Table 3, which are capacitive reactance, and generated voltage per phase in p.u for different machines. Optimized results are obtained by varying three function variables namely, load (0.5–1.3 p.u), frequency

Table 3 Optimized output

Machine rating (kW) (p.u)	X_c	V_g
4.0	0.3688	1.782
7.5	0.3500	1.720
15	0.4270	1.680
37	0.5779	1.620

(1.3–1.6 p.u) and speed (0.5–1.3 p.u) [14, 15]. The generated voltage is determined in each case by using Eq. (2).

To show the variation of excitation and output voltage per phase, two graphs have been plotted and shown in Fig. 5 with different rating of the machine. The results for 4 kW machine give the minimum excitation of 0.3688 p.u.

Figure 5 shows two lines, blue and red in colour. The red colour represents the variation of output voltage in p.u. against rating of the machines in kW*10 scale.

The graph shows that output voltage firstly decreases and then almost remains constant with the machine rating. The blue line shows the variation of excitation in p.u. against rating, which is almost flat initially and then increases when the rating increases.

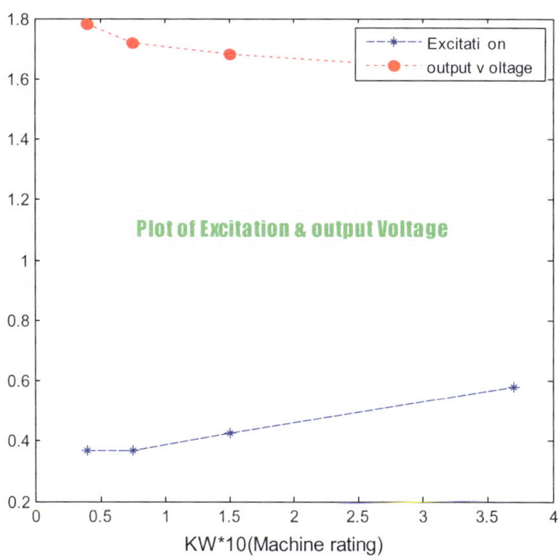

Fig. 5 Excitation and frequency with m/c rating for optimal excitation

Appendix: Constants of Objective Function for 4 kW Machine

A = 0.1010604, B = −0.050717, C = 0.03513, D = 0.2564, E = 0.001714, K = 1.0642 p.u, Xm = 1.35 p.u

References

1. S. Vadhera, K. Sandhu, Genetic algorithm toolbox based investigation of terminal voltage and frequency of self excited induction generator. Int. J. Adv. Eng. Appl. **1**(1), 243–250 (2010)
2. R.C. Bansal, Three phase self-excited induction generators-An overview. IEEE Trans. Energy Convers. **20**(2), 292–299 (2005)
3. L. Sridhar, B. Singh, C.S. Jha, B.P. Singh, S.S. Murthy, Selection of capacitors for the self-regulated short shunt self-excited generator. IEEE Trans. Energy Convers. **10**(1), 10–17 (1995)
4. B.I.J. Nagrath, D.P. Kothari, *Electrical Machines*, 2nd edn. (Tata McGraw-Hill, New York, 1997)
5. S.K. Jain, J.D. Sharma, S.P. Singh, Transient performance of three-phase self-excited induction generator during balanced and unbalanced faults. IEE Proc. Gener. Trans. Distrib. **149**(1), 50–57 (2002)
6. T.F. Chan, Analysis of self-excited induction generators using an iterative method. IEEE Trans. Energy Convers. **10**(3), 502–507 (1995)
7. E.D. Besant, F.M. Potter, Capacitor excitation for induction motors. AIEE Trans. **54**, 540–545 (1935)
8. M.H. Haque, A novel method of evaluating performance characteristics of a self-excited induction generator. IEEE Trans. EC **24**(2), 358–365 (2009)
9. L.A. Alolah, M.A. Alkanhal, Optimization-based steady state analysis of three phase self-excited induction generator. IEEE Trans. on Energy Conversion **15**(1), 61–65 (2000)
10. R. Basak, H. Yahoui, N. Siauve, Study of optimal excitation of self-excited Induction generators by genetic algorithm. IJSRD-Int. J. Sci. Res. Dev. **4**(12), 590–593 (2017)
11. L. Wang, J.Y. Su, Dynamic performance of an isolated self-excited induction generator under various loading conditions. IEEE Trans. **EC-14**(1), 93–100 (1999)
12. A.H. Al-Bahrani, N.H. Malik, Voltage control of parallel operated self-excited induction generators. IEEE Trans. **EC-8**(2), 236–242 (1993)
13. A. Nejmi, Y. Zidani, M. Naciri, Investigation on the self excited induction generator provided with a hydraulic regulator, in *FIER* (2002), pp. 494–499
14. R.C. Bansal, D.P. Kothari, T.S. Bhatti, Induction generator for isolated hybrid power system applications: A review, in *Proceedings of 24th National Renewable Energy Conversion*, Bombay, India, Nov. 30/Dec. 2, 2000, pp. 462–467
15. C. Grantham, F. Rahman, D. Seyoum, A regulated self-excited induction generator for use in a remote area power supply. Int. J. Renewable Energy Eng. **2**(1), 234–239 (2000)

Different Setting of Unified Power Flow Controller (UPFC) and Its Effect on Performance of Distance Relay

Rajib Sadhu and P. S. Bhowmik

1 Introduction

For efficient power supply from generating station to load-centers with high reliability, inter-connection of grids is not solely sufficient to provide satisfactory transmission line capacity. Again the available capacity of the line is confined within a limit due to the cost of transmission, line losses and various other economic and environmental factors. These factors also restrict up-gradation of network by construction of new transmission lines. Thus to keep an optimum balance between quality and reliability of service, new technologies should be welcomed. Advancement in the field of Power Electronics and Semiconductor devices has made power engineers to propose FACTS (Flexible AC Transmission System) as an alternative solution. It will help in controlling power as well as enhance the usability of available and planned lines. Installation of FACTS devices adds more complexity to the network and also affects the performances of distance relay and other protective devices in the network. Under a fault condition, transients superimposed on the power frequency voltage and the current waveforms can significantly differ from a system without FACTS devices.

The Unified Power Flow Controller is the most versatile FACTS device and it consists of two voltage source converters, using gate turn-off (GTO) thyristor valves. Out these converters, one is shunt type (STATCOM) and other is series type (SSSC), and they are connected through a common dc storage capacitor. SSSC provides the

R. Sadhu (✉)
Department of Electrical Engineering,
University Institute of Technology, Bardhaman, West Bengal, India
e-mail: rajib_sadhu@rediffmail.com

P. S. Bhowmik
Department of Electrical Engineering,
National Institute of Technology, Durgapur, West Bengal, India
e-mail: psbhowmik@ee.nitdgp.ac.in

S. Chattopadhyay et al. (eds.), *Modelling and Simulation in Science, Technology and Engineering Mathematics*, Advances in Intelligent Systems and Computing 749,
https://doi.org/10.1007/978-3-319-74808-5_16

main function of the UPFC by injecting an ac voltage with controllable magnitude and phase angle. This injected voltage can be considered essentially as a synchronous ac voltage source. The real and reactive power flow can be controlled by this type of arrangement. The basic function of STATCOM is to supply or absorb the real power demanded by SSSC at the common dc link. STATCOM can also generate or absorb controllable reactive power, if it is desired, and thereby it can provide independent shunt reactive compensation for the line [1].

Amongst some of the research work carried over impact of various FACTS device on distance relays, the study carried out in [2–6] shows the effect of STATCOM on distance relay under different fault conditions, fault locations and system configurations.

The impacts of SSSC on measured impedance at relay point for different fault location are discussed in [7–9].

Zhou et al. in [10] have presented analytical and simulation results of the application of distance relays for the protection of transmission systems employing FACTS device such as the unified power flow controller (UPFC) and how the performance of distance relay influenced by the UPFC for single-phase-earth and phase-phase faults.

Apparent impedance calculation procedure for digital distance relay of transmission line involving UPFC and its effects on the trip boundaries for the locations are observed in [11–15]. A new cross-differential protection algorithm for parallel transmission lines including UPFC in one of the lines is presented in [16]. Moravej et al. in [17] analyzes the distance relay performance during power swing conditions for an uncompensated and compensated transmission line with a UPFC.

All the studies discussed so far clearly show that the performance of distance relays is greatly affected by STATCOM and SSSC operating simultaneously (UPFC) or individually. At the time of fault voltage and current injection of these devices will affect both the steady and transient components of voltage and current. For this apparent impedance of system without FACTS devices and with FACTS devices seen by a conventional distance relay will be different.

2 Proposed Model

2.1 Model of Transmission Line with UPFC

An uncompensated 132 kV transmission line of a typical power system network is taken for simulation study, corresponding single line diagram of the model with VSC based FACTS device like UPFC is shown in the Fig. 1. Two 200 km parallel 132-kV transmission lines with two 6000-MVA short-circuit levels (SCLs) sources and the angle difference between them is 20°.

The two lines have same line parameters. The line positive and negative sequence impedance is $0.0255 + j0.352\ \Omega/\text{km}$. The line zero sequence impedance is $0.3864 +$

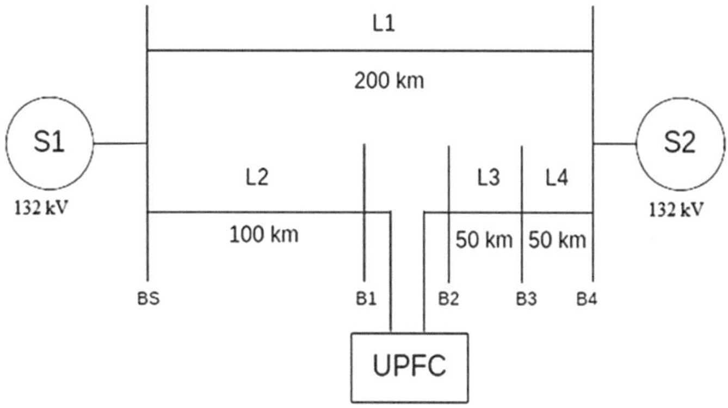

Fig. 1 Single line diagram

$j1.5556\ \Omega$/km. Shunt connected converter of UPFC named as STATCOM, it uses one 48-pulse voltage source converter which connects with two 2000 μF series DC capacitors. The shunt connected convert injects or consumes reactive power from the transmission line to regulate the voltage at the connecting point. The series part of UPFC or SSSC injects an almost sinusoidal voltage of variable magnitude and angle, in series with the transmission line to regulate the power flow through the transmission line.

2.2 *STATCOM and SSSC Model Using 48-Pulse VSC*

Static synchronous compensators (STATCOM) using Multi-pulse converters generally based on elementary six-pulse GTO—VSC (gate turn off based voltage source converter). Practically, a quasi-harmonic 48-pulse (4 × 12pulse) configuration is used with the phase angle control algorithm employing proportional and integral (PI) control methodology with a phase displacement of 7.5°. It can be shown that the fundamental component of the output voltage of a quasi-48 pulse converter is given by.

$$E_{1,48} = 4\frac{2\sqrt{2}}{\pi} \cdot V_{dc} \cos\left(\frac{\pi}{24}\right) \cos\left(\frac{\pi}{48}\right) \tag{1}$$

The harmonic component of order h is given by,

$$\boldsymbol{E}_{h,48} = 4\frac{2\sqrt{2}}{\pi} \cdot \boldsymbol{V}_{dc} \cos\left(\frac{h\pi}{24}\right) \cos\left(\frac{h\pi}{48}\right); \boldsymbol{h} = 12\boldsymbol{k} \pm 1, \boldsymbol{k} = 1, 2, 3, \ldots \tag{2}$$

It is observed that the DC bus (V_{dc}) is connected to the four 3-phase inverters. The four voltages generated by the inverters are applied to secondary windings of four zig-zag phase-shifting transformers connected in Wye (Y) or Delta (D). The four transformer primary windings are connected in series and the converter pulse patterns are phase shifted so that the four voltage fundamental components sum in phase on the primary side.

Each 3-level inverter generates three square-wave voltages which can be $+V_{dc}$, 0, $-V_{dc}$. The duration of the $+V_{dc}$ or $-V_{dc}$ level can be adjusted between 0° and 180° by varying the conduction angle (σ) of the Firing Pulse Generator. Each inverter uses a Three-Level Bridge block where specified power electronic devices are GTOs. Each leg of the inverter uses 3 ideal switches to obtain the 3 voltage levels ($+V_{dc}$, 0, $-V_{dc}$).

Except for the 23rd and 25th harmonics, this transformer arrangement neutralizes all odd harmonics up to the 45th harmonic. Y and D transformer secondary cancel harmonics $5+12n$ (5, 17, 29, 41, ...) and $7+12n$ (7, 19, 31, 43, ...). In addition, the 15° phase shift between the two groups of transformers allows cancellation of harmonics $11+24n$ (11, 35, ...) and $13+24n$ (13, 37, ...).Considering that all $3n$ harmonics are not transmitted by the transformers (delta and ungrounded Y), the first harmonics that are not cancelled by the transformers are therefore the 23rd, 25th, 47th and 49th harmonics. By choosing the appropriate conduction angle for the three-level inverter ($\sigma = 172.5°$), the 23rd and 25th harmonics can be minimized. The first significant harmonics generated by the inverter will then be 47th and 49th.

SSSC model employing 48-pluse GTO based VSC is same as STATCOM only change is there for variable amplitude of injected voltage conduction angle is not fixed.

2.3 Apparent Impedance Calculation for Distance Relay

This calculation is generally based on symmetrical component transformation by the use of power frequency components of current and voltage signals, which is measured at relay point. In this calculation, certain assumptions are performed beforehand such as signal acquisition, pre-processing, and sequence component calculations.

When a single phase to ground fault occurs on the transmission line and the distance between the fault point and the relay point is p × L, then at the time of fault the positive, negative and zero sequence networks of the system are as shown in Fig. 2.

$$V_1 = I_1 0.5Z_1 + V_{ij1} + I_{l1}(p - 0.5)Z_1 + R_f I_{f1} \tag{3}$$

$$V_2 = I_2 0.5Z_1 + V_{ij2} + I_{l2}(p - 0.5)Z_1 + R_f I_{f2} \tag{4}$$

$$V_0 = I_0 0.5Z_0 + V_{ij0} + I_{l0}(p - 0.5)Z_0 + R_f I_{f0} \tag{5}$$

$$I_{l1} = I_1 + I_{ij1} \tag{6}$$

$$I_{l2} = I_2 + I_{ij2} \tag{7}$$

$$I_{l0} = I_0 + I_{ij0} \tag{8}$$

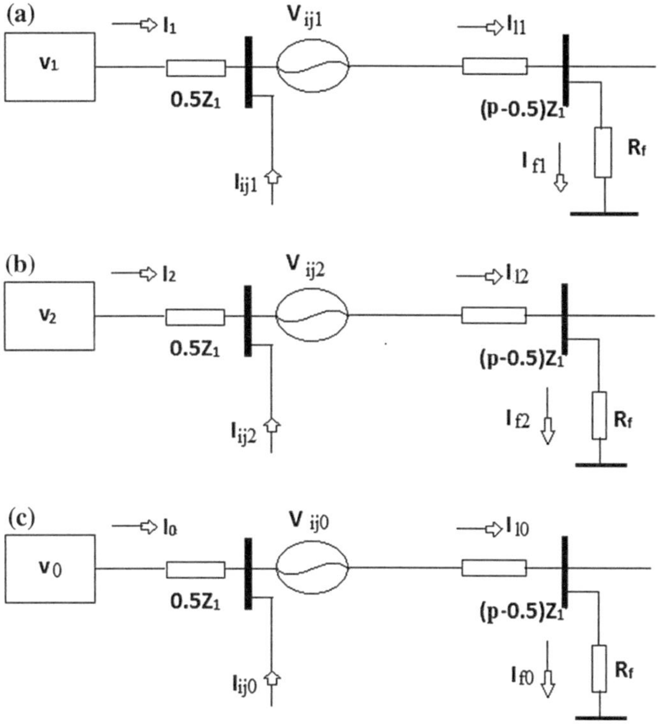

Fig. 2 **a** Positive sequence network. **b** Negative sequence network. **c** Zero sequence network of the system from the relay location to fault

where, V_1, V_2, V_0 are the sequence phase voltages at the relay location. $V_{\text{ij1}}, V_{\text{ij2}}, V_{\text{ij0}}$ are series sequence phase voltages injected by SSSC. I_1, I_2, I_0 are sequence phase currents at the relay location. I_{11}, I_{12}, I_{10} are sequence phase currents in transmission line. I_{f1}, I_{f2}, I_{f0} are sequence fault currents. I_{ij1}, I_{ij2}, I_{ij0} are shunt sequence phase currents injected by SATCOM. Z_1, Z_0 are sequence impedance of the transmission line. p is per-unit distance of a fault from the relay location. From above, the voltage at the relay point can be derived as

$$V = pIZ_1 + pI_0(Z_0 - Z_1) + I_{\text{ij}}(p - 0.5)Z_1 + (p - 0.5)I_{\text{ij0}}(Z_0 - Z_1) + V_{ij} + R_f I_{\text{f}} \tag{9}$$

where,

$$V = V_1 + V_2 + V_0 \tag{10}$$

$$I = I_1 + I_2 + I_0 \tag{11}$$

$$I_{\text{ij}} = I_{ij1} + I_{ij2} + I_{\text{ij0}} \tag{12}$$

$$V_{ij} = V_{ij1} + V_{ij2} + V_{ij0} \tag{13}$$

In the transmission system without UPFC, for a single phase-to-ground fault, the apparent impedance of distance relay can be calculated using the equation

$$Z = \frac{V}{I + \frac{Z_0 - Z_1}{Z_1} \times I_0} = \frac{V}{I_r} \tag{14}$$

where, V, I phase voltage and current at relay point. I_0 is zero sequence phase current. I_r is the relaying current.

If this conventional distance relay is applied to the transmission system with UPFC, the apparent impedance seen by this relay can be expressed as,

$$Z = pZ_1 + \frac{I_{ij}}{I_r}(p - 0.5)Z_1 + \frac{I_{ij0}}{I_r}(p - 0.5)(Z_0 - Z_1) + \frac{V_{ij}}{I_r} + \frac{I_f}{I_r}R_f \tag{15}$$

In practice, one side of the shunt transformer is in delta connection, and thus there is no zero sequence current injected by STATCOM, so, $I_{ij0} = 0$, then the equation can be rewritten as

$$Z = pZ_1 + \frac{I_{ij}}{I_r}(p - 0.5)Z_1 + \frac{V_{ij}}{I_r} + \frac{I_f}{I_r}R_f. \tag{16}$$

From the above, it is observed that the impact of UPFC on the apparent impedance, can be divided two parts; one results from the shunt current STATCOM and another is the impact of the series voltage injected by the SSSC; the last part of the apparent impedance is due to the fault resistance.

Now, if the UPFC working as STATCOM only, the apparent impedance seen by relay is given by,

In general, one side of shunt transformer has a delta connection, as a result there is no zero sequence current injected by STATCOM, so $I_{ij0} = 0$. So, the modified equation is

$$Z = pZ_1 + \frac{I_{ij}}{I_r}(p - 0.5\,Z_1) + \frac{I_f}{I_r}R_f \tag{17}$$

When solid single phase to ground fault occurs, then the equation become,

$$Z = pZ_1 + \frac{I_{ij}}{I_r}(p - 0.5\,Z_1) \tag{18}$$

For SSSC, the apparent impedance seen by relay is given by,

$$Z = pZ_1 + \frac{V_{ij}}{I_r} + \frac{I_f}{I_r}R_f \tag{19}$$

2.4 Relay Model

The flowchart shown in Fig. 3 gives the procedure to calculate the apparent impedance for drawing apparent impedance trajectory. The relay point three phase current and voltage phasors are sent to current and potential transformers (CT and PT) respectively where they are scaled down to acceptable voltage levels for the subsequent process. 'Signal Conditioner' converts these signals into a form that can be converted to digital values. 'Sample and Hold circuit' samples time varying analog signals and holds the instantaneous sample values constant during conversion period of 'Analog to Digital Converter' (ADC). ADC converts these samples to equivalent numerical values and outputs in binary. In 'Fourier Transform', fast Fourier transform (FFT) is used to extract fundamental frequency components from the post-fault relaying signals and remove dc offset from signals. The output is used for 'Apparent Impedance calculation' Z. Real (*R*) and imaginary (*X*) components of *Z* are obtained and used for drawing apparent impedance trajectory.

3 Result Analysis

Simulation is carried out with step length of 0.02 ms. A single phase to ground (SLG) fault is introduced at 150 km distance from source 1 in the second transmission line.

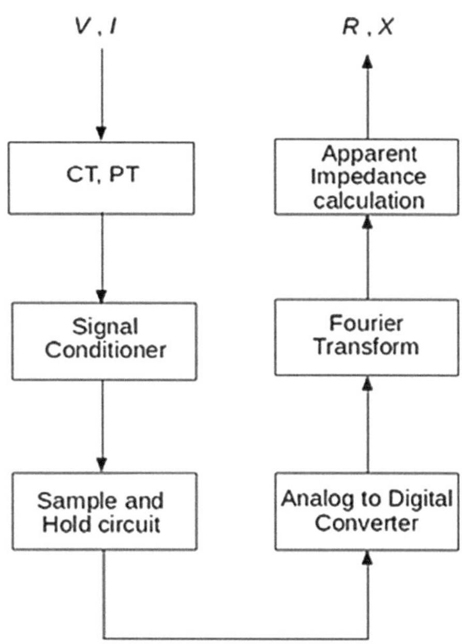

Fig. 3 Flowchart of distance relay

The UPFC is connected between buses B_1 and B_2 and it can act in STATCOM, SSSC or UPFC mode. The distance relay is connected after bus B_S.

3.1 Statcom Results

UPFC is operated in STATCOM mode for voltage control with $V_{ref} = 1$ pu. STATCOM adjusts the three phase voltages and currents in such a way that line voltage remains constant at 1 pu. Also during fault, the line voltages have remained close to V_{ref} due to lagging current injection on healthy phases and leading current injection on faulty phase by the device, reducing voltages for higher current of healthy phases and increasing voltage for reduced current of faulty phase. In this manner it controls system voltage.

Figure 4 shows when the system is working under no-load with STATCOM, the reactive power oscillates near zero axis with slight negative average value, employing STATCOM is exchanging reactive power to maintain constant voltage. During fault, the reactive power nearly doubles (80 MVAr) system reactive power (40 MVAr) indicating STATCOM lagging power injection to maintain line constant voltages.

For introduction STATCOM two things happened. They are,

1. Voltage control—The voltage of the power system remains almost constant at 1.0 pu and phase displacement between all the buses is reduced. These things are observed even during fault.

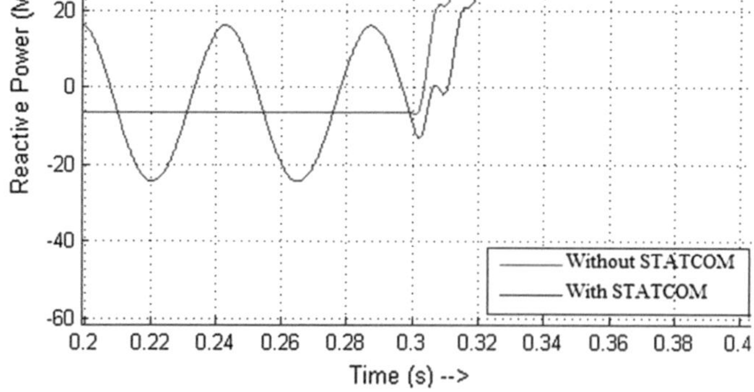

Fig. 4 Reactive power injunction by STATCOM

2. Reactive power control—Due to inclusion of STATCOM, at no fault system reactive power oscillates around uncompensated system no fault reactive power implying STATCOM power exchange to maintain constant line voltage. Also it injects extra power during fault to keep up bus voltage at reference value.

From Figs. 5 and 6 it is observed that during fault, apparent resistance decreases and apparent reactance increases for STATCOM. The apparent resistance starts from higher value (~40 ohms) for STATCOM than uncompensated line (~30 ohms). Also apparent reactance steady value is greater (~60 ohms) from uncompensated system (~40 ohms).

Figure 7 shows the apparent impedance trajectory of the uncompensated system and STATCOM compensated system and relay characteristics. Clearly, the really detects fault in case of uncompensated system but with STATCOM the R-X trajectory goes outside the circle, thereby under-reaching the relay and it does not trip. So with STATCOM, relay settings should be adjusted.

3.2 SSSC Results

By selecting SSSC mode from UPFC settings (controlling series connected GTO converter) and choosing SSSC injected voltage V_{ij} in p.u results show comparatively least effect than STATCOM in Figs. 5, 6 and 7 respectively.

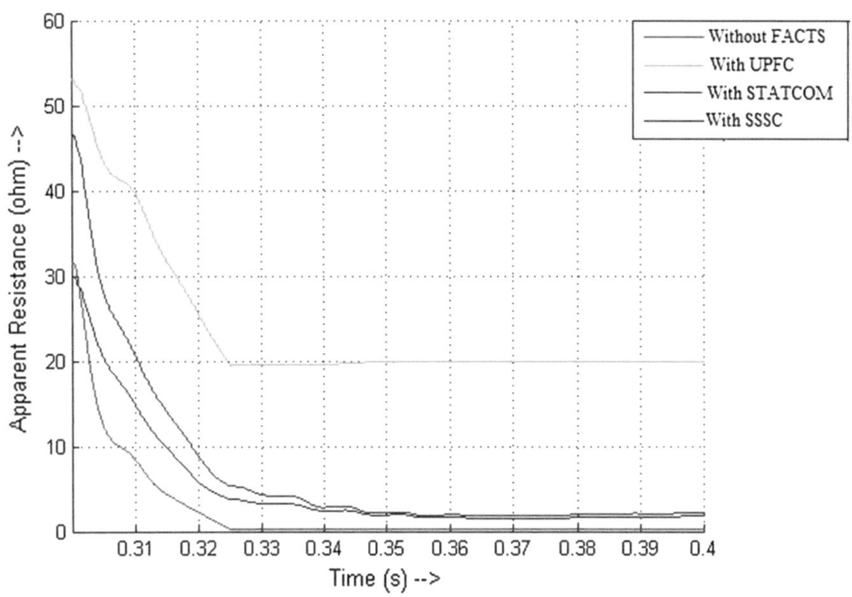

Fig. 5 Apparent resistance trajectory

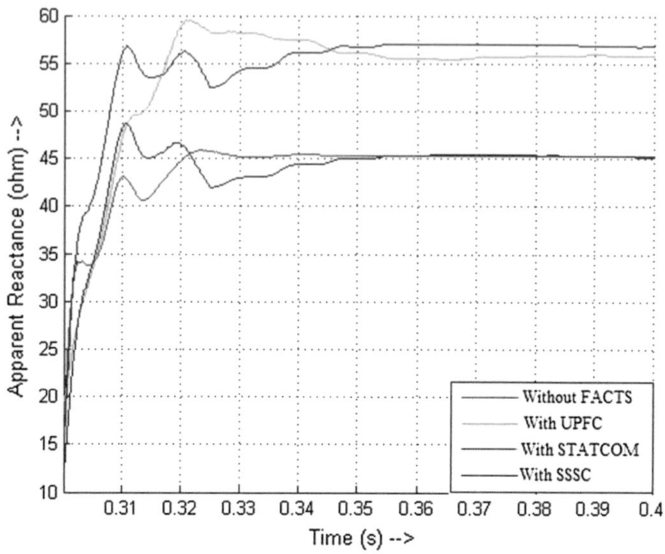

Fig. 6 Apparent reactance trajectory

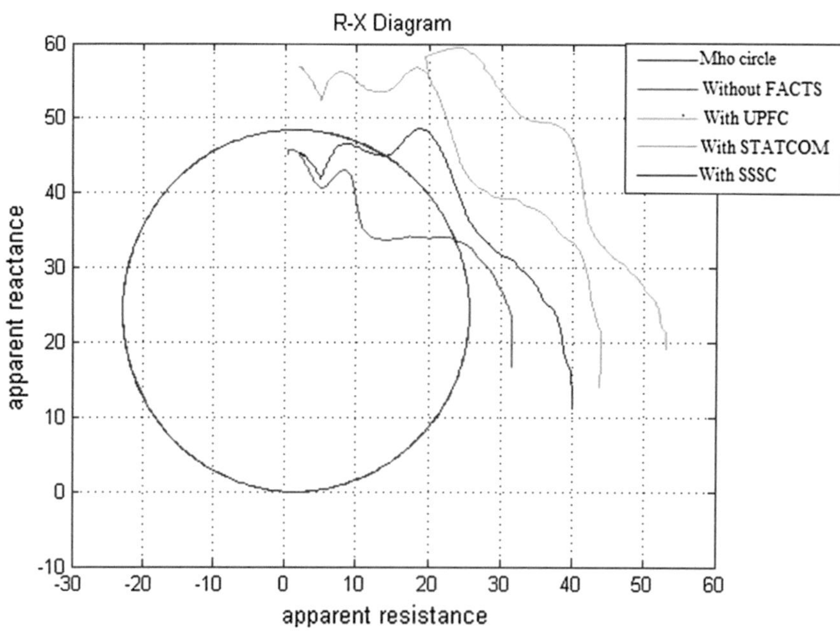

Fig. 7 Mho-relay characteristics and apparent impedance trajectory

3.3 UPFC Results

In UPFC mode by controlling both shunt and series connected GTO converters simultaneously and setting UPFC active power reference P_{ref} and reactive power reference Q_{ref} in p.u within a given boundary, UPFC can control both active and reactive power flow also the voltage of the system. Figures 5 and 6 shows UPFC has more adverse effect on distance protection relay operation than STATCOM or SSSC. Also in Fig. 7 for UPFC, the impedance trajectory goes very much outside of the circle and relay under-reach occurs more.

4 Conclusion

STATCOM, SSSC and UPFC always help to improve power system efficiency by controlling system current, voltage and power. Though they help the power system but they always have harmful effect on reliability of the distance protection schemes. STATCOM in voltage control mode, keep the bus voltage nearly constant by reducing the faulted phase current during fault. SSSC can transfer desired active power through the line under fault condition in voltage injection mode. Incorporation of UPFC with power (active and reactive) flow control, desired (within a limited range) active and reactive power can be transferred through line with medium varying system conditions (for faults reactive power becomes uncontrollable). As a result the apparent resistances for STATCOM and UPFC connected systems start from higher value than uncompensated system, for SSSC apparent resistance starts from nearly same value as in uncompensated system but attain slightly higher value than uncompensated system in steady state for STATCOM and SSSC, and very high value for UPFC. Apparent reactances for these compensated networks start from nearly same value as in uncompensated system but attain higher and very higher steady state values for STATCOM and UPFC respectively and more or less same value for SSSC. Within UPFC, STATCOM and it-self have greater effect on apparent resistance and reactance hence it is observed in impedance trajectory more than SSSC. Employment of these FACTS devices causes distance relay to under-reach in this study and the relay settings have to be reconsidered. The results thus obtained significantly show that the performance and characteristics of a distance relay under fault or normal conditions significantly depend upon the presence of FACTS devices, their type and control parameters setting. Hence, study of the effects of FACTS devices laid upon the performance of distance relays is of paramount importance for ensuring satisfactory and reliable operation of the system.

References

1. L. Gyugyi, Unified power-flow control concept for flexible AC transmission systems. IEE Proc. C—Gener. Transm. Distrib. **139**(4), 323–331 (1992)
2. W.H. Zhang, S.J. Lee, M.S. Choi, S. Oda, Considerations on distance relay setting for transmission line with STATCOM, in *IEEE Power and Energy Society General Meeting*, July 2010
3. K. El-Arroudi, G. Joos, D.T. McGillis, Operation of impedance protection relays with the STATCOM. IEEE Trans. Power Delivery **17**(2), 381–387 (2002)
4. T.S. Sidhu, R.K. Varma, P.K. Gangadharan, F.A. Albasri, G.R. Ortiz, Performance of distance relays on shunt—FACTS compensated transmission lines. IEEE Trans. Power Delivery **20**(3), 1837–1845 (2005)
5. F.A. Albasri, T.S. Sidhu, R.K. Varma, Performance comparison of distance protection schemes for shunt-FACTS compensated transmission lines. IEEE Trans. Power Delivery **22**(4), 2116–2125 (2007)
6. M. Khederzadeh, The impact of FACTS device on digital multifunctional protective relays, in *Proceedings of IEEE/PES Transmission and Distribution Conference and Exhibition 2002*, vol. 3 (Asia Pacific, 2002), pp. 2043–2048
7. A. Kazemi, S. Jamali, H. Shateri, Comparing impacts of SSSC and STATCOM on measured impedance at relaying point, in *IEEE Power and Energy Society General Meeting*, 2009
8. H. Abdollahzadeh, B. Mozafari, A. Tavighi, J. Marti, Impact of shunt capacitance of a SSSC-compensated transmission line on performance of distance relays, in *IEEE Power and Energy Society General Meeting*, 2013
9. M. Khederzadeh, A. Ghorbani, A. Salemnia, Impact of SSSC on the digital distance relaying, *IEEE Power and Energy Society General Meeting*, 2009
10. X. Zhou, H. Wang, R.K. Aggarwal, P. Beaumont, Performance evaluation of a distance relay as applied to a transmission system with UPFC. IEEE Trans. Power Delivery **21**(3), 1137–1147 (2006)
11. P.K Dash, A.K Pradhan, G. Panda, A.C Liew, Digital protection of power transmission lines in the presence of series connected facts devices, in *IEEE Power Engineering Society Winter Meeting*, pp. 1967–1972, 2000
12. P.K. Dash, A.K. Pradhan, G. Panda, Distance protection in the presence of unified power flow controller. Electric Power Syst. Res. **54**(3), 189–198 (2000)
13. M. Khederzadeh, UPFC operating characteristics impact on transmission line distance protection, in *IEEE Power and Energy Society General Meeting-Conversion and Delivery of Electrical Energy in 21st Century*, 2008
14. S. Jamali, A. Kazemi, H. Shateri, Effects of UPFC on measured impedance by distance relay in double-circuit lines, *IEEE Power and Energy Society General Meeting*, 2009
15. S. Jamali, A. Kazemi, H. Shateri, Comparing series and shunt reactive power compensation via UPFC from distance relay point of view, in *Transmission and Distribution Conference and Exposition*, 2010
16. L.N. Tripathy, P.K. Dash, S.R. Samantaray, A new cross-differential protection scheme for parallel transmission lines including UPFC. IEEE Trans. Power Delivery **29**(4), 1822–1830 (2013)
17. Z. Moravej, M. Pazoki, M. Khederzadeh, Impact of UPFC on power swing characteristic and distance relay behavior. IEEE Trans. Power Delivery **29**(1), 261–268 (2014)

Assessment of Discrimination Between Fault and Inrush Condition of Power Transformer by Radar Analysis and Wavelet Transform Based Kurtosis and Skewness Analysis

Sushil Paul, Shantanu Kr Das, Aveek Chattopadhyaya and Surajit Chattopadhyay

1 Introduction

In ever increasing power system scenario power transformer plays a vital role for proper and reliable operation of power system. Being one of the most important components of power system, power transformer needs adequate protection for its proper operation. When a transformer in unloaded or lightly loaded condition is connected to a power supply, then a large transient current may appear due to flux asymmetries and saturation in the core of the transformer which is known as inrush current [1]. Inrush current decays very fastly for few cycles then it varies slowly. Inrush current may take 4–6 s to subside. There are some factors which affect magnitude and duration of inrush current, like (i) residual flux in the transformer (ii) type of magnetic material which is used in the core (iii) size of power system (iv) size of transformer (v) switching instant of energization of the transformer [1]. Inrush current of transformer may be divided into three categories: energization inrush, recovery inrush, sympathetic inrush [1]. Simple model has been proposed to sim-

S. Paul (✉) · S. K. Das · A. Chattopadhyaya
Department of Electrical Engineering, SKFGI, MAKAUT, Kolkata, West Bengal, India
e-mail: sushilpaul28@gmail.com

S. K. Das
e-mail: shantanu.das95@gmail.com

A. Chattopadhyaya
e-mail: aveek_chatterjee40@yahoo.com

S. Chattopadhyay
Electrical Engineering Department,
Ghani Khan Chaoudhury Institute of Engineering and
Technology (Under Ministry of HRD, Government of India),
Malda, West Bengal, India
e-mail: surajitchattopadhyay@gmail.com

S. Chattopadhyay et al. (eds.), *Modelling and Simulation in Science, Technology and Engineering Mathematics*, Advances in Intelligent Systems and Computing 749,
https://doi.org/10.1007/978-3-319-74808-5_17

ulate the magnetizing inrush current of transformers using real-time measurements and then simulation data used for harmonic analysis [2]. Different techniques have been proposed by different researchers to assess the inrush current. Hidden Markov Model (HMM) based method proposed in [3], to detect inrush current of power transformer. Three factors like the conventional second harmonic content, the decaying DC time constant, and the ratio between the fundamental component and the first peak magnitude based dynamic filter has been proposed to identify inrush current and fault current of power transformer [4]. Correlation coefficient between the sampling waveform based method proposed in [5] to discriminate between inrush current and fault current of power transformer. Inrush current plays a vital role in differential protection of power transformer. Phase angle difference between primary and secondary current based method proposed by the authors to avoid unwanted tripping of differential protection on magnetizing inrush current [6]. In [7], authors proposed a method distinguish between inrush currents and internal faults based on the differential current gradient. Different signal processing techniques and soft computing techniques used by the researchers to assess the inrush current of power transformer. Wavelet Transform (WT) based feature has been extracted in [8], to identify the inrush and fault current of power transformer. Median Absolute Deviation (MAD) of wavelet coefficients based method proposed by the researchers to distinguish of different nature of currents of power transformer where five level of decomposition have been considered in DWT decomposition [9]. WT and correlation coefficient based method proposed in [10] to discriminate between inrush and fault current. S-transform based technique used as another signal processing based technique to assess inrush and fault current [11, 12], where in [12] probabilistic neural network used as a classifier for classification of inrush and fault current. Slantlet transform (S-transform) and Artificial Neural Network (ANN) based method used for classification of over current and inrush current of power system where ANN used as a classifier [13]. Back propagation algorithm based ANN has been used as a tool to discriminate between inrush and fault current of power transformer [14]. Multi resolution analysis and space vector analysis based method proposed in [15] for solution of dilemma of fault and inrush current of power transformer. WT and PNN based method proposed as another technique to assess inrush current [16], where EMTP simulation has been used to simulate inrush current along with other transients current for this purpose. Fuzzy and neuro fuzzy based approach have been proposed by the researchers to distinguish inrush current from fault current in power system [17, 18]. To analyze the abnormal condition of electrical systems some techniques based on Clarke and Park plane have been proposed by the researchers [19–23]. Radar analysis, FFT and THD based approach proposed in [24] to discriminate between inrush and fault current of power transformer.

None of the research works, inrush and fault condition of power transformer has been assessed by the CWT, Radar analysis and DWT based skewness, kurtosis, rms and mean value analysis. For this reason an attempt has been made to discriminate between inrush and fault condition of power transformer based on CWT, Radar analysis and DWT based skewness, kurtosis, rms and mean value analysis. Different feature patterns have been observed in CWT and Radar analysis and DWT based different parameters values have been noted of primary current of power transformer in different conditions form there inrush and fault conditions of power transformer have been assessed.

2 Model for Simulation of Inrush and Fault Current of Power Transformer

Two MATLAB models [25] have been prepared to simulate inrush and inrush with fault condition in a three phase power transformer which are depicted in Figs. 1 and 2 respectively. A three phase two winding 6.3 MVA, 33/11 kV, 50 Hz power transformer has been connected to three phase source of 33 kV through a three phase circuit breaker. Primary current values of three phase transformer have been stored in workspace after the sampling. All the three phases, which were initially open, were closed at a transition time of 0.1 s by the three phase breaker. Sampling time was taken as 50e-6 s (available in MATLAB) for the analysis. Using two models three different conditions have been created which are, normal condition, inrush condition and short circuit fault with inrush condition. In all the cases primary side current of the transformer have been used for assessment of inrush condition of power transformer.

3 Wavelet Transform (WT) Analysis

To analyse non-stationery signal in better ways WT was introduced [26, 27]. WT is used to get better time frequency representation from a non-stationery signal which was the limitation of Fourier Transform (FT) and Short Time Fourier Transform (STFT). Different signals aspects like trends, breakdown points, discontinuities etc.

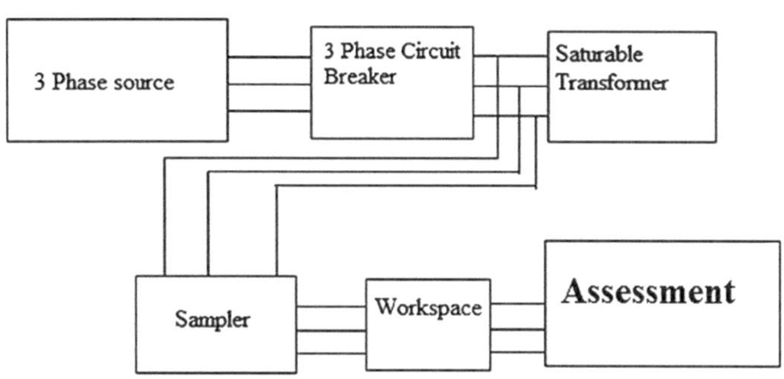

Fig. 1 MATLAB model for simulation—of inrush current of transformer

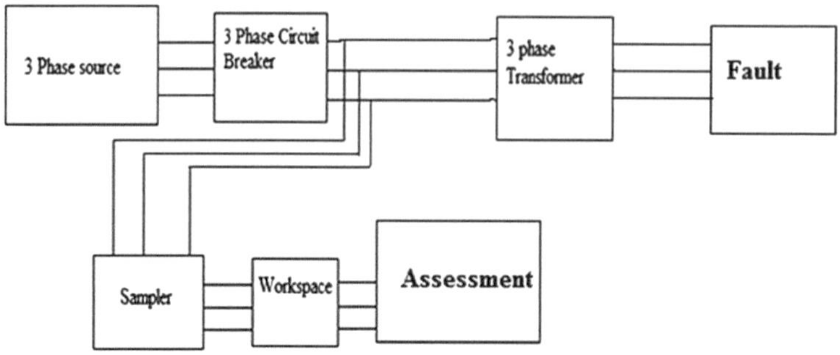

Fig. 2 MATLAB model for—simulation of fault current of transformer

can be analysed by WT from a particular signal. It can be classified as (i) Continuous Wavelet Transform (CWT) (ii) Discrete Wavelet Transform (DWT).

Continuous Wavelet Transform (CWT)

The formula of CWT, which is used to achieve time frequency representation from a signal $\boldsymbol{x}(\boldsymbol{t})$ is defined as [26, 27],

$$\mathbf{X}_{\mathbf{WT}(\tau,\, s)} = \frac{1}{\sqrt{|s|}} \int x(t) \cdot \psi^{*}\left(\frac{t-\tau}{s}\right) dt \tag{1}$$

The transformed signal $\mathbf{X}_{\mathbf{WT}\,(\tau,\, s)}$ is a function of the translation parameter τ and the scale parameter s. The mother wavelet is denoted as ψ (t) and the *(asterisk) indicates the complex conjugate which is used in case of a complex wavelet. For the CWT analysis, signal can be discretized arbitrarily without violating the Nyquist criterion.

Discrete Wavelet Transform (DWT)

Calculation of wavelet coefficients at every possible scale is a fair amount of work and it generates lots of data not only that, the computation of CWT may consume significant amount of time and resources depending on the resolution required. In DWT [26, 27], the signal which is to be analysed is passed through filters with different cut off frequencies at different scales. In this work 'db4' is used as the mother wavelet because it is compactly supported in time frame and this mother wavelet is used to detect sudden jump or notch in the signal. Some parameters like skewness, kurtosis, rms and mean values of approximation (approximate) coefficients have been found out in all the cases after decomposed the signal by 'db4' based DWT.

4 Assessment of Inrush, Normal and Fault with Inrush Condition of Power Transformer

Different conditions of power transformer have been assessed by CWT, Radar analysis and DWT based parameter analysis which is given below.

4.1 Results and Observation of Continuous Wavelet Transform (CWT)

Figures 3, 4, 5 and 6 are used to depict the result of CWT of R-phase primary current of power transformer in different conditions. In short circuit conditions CWT result of R phase current is almost same where as in inrush and normal condition it is different in nature. Different critical areas have been observed in different CWT results and observing the feature pattern of the CWT results, normal, inrush and short circuit conditions of power transformer can be discriminated properly.

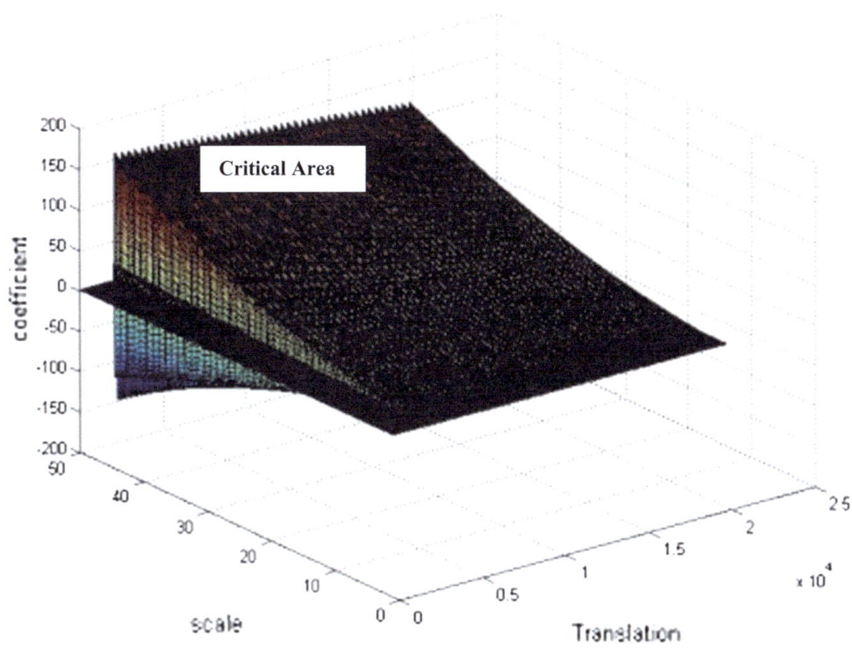

Fig. 3 CWT of R-phase primary current under inrush condition

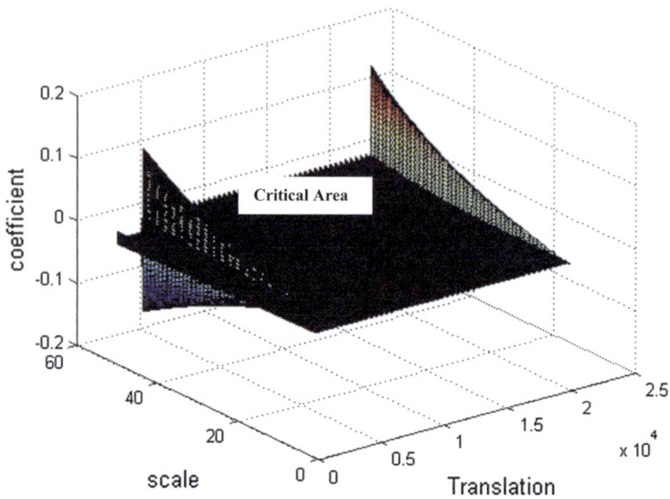

Fig. 4 CWT of R-phase primary current under normal condition

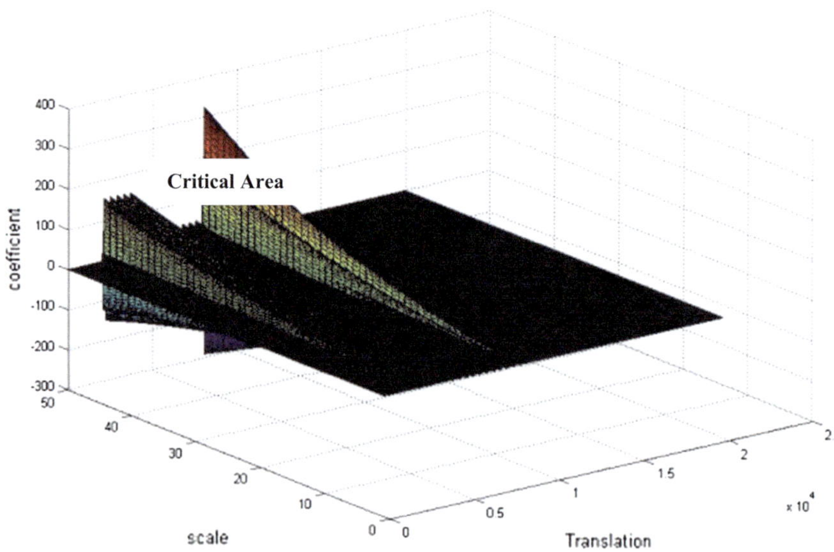

Fig. 5 CWT of R-phase primary current under inrush with short circuit fault (L-L-L) condition

4.2 Discrete Wavelet Transform (DWT)

The main disadvantage of CWT is that, it generates lots of data which sometimes very cumbersome to properly handle it. For this reason in this work DWT based parameter analysis has been done to discriminate normal, inrush and fault conditions of power transformer which is very easy to implement to detect and discriminate

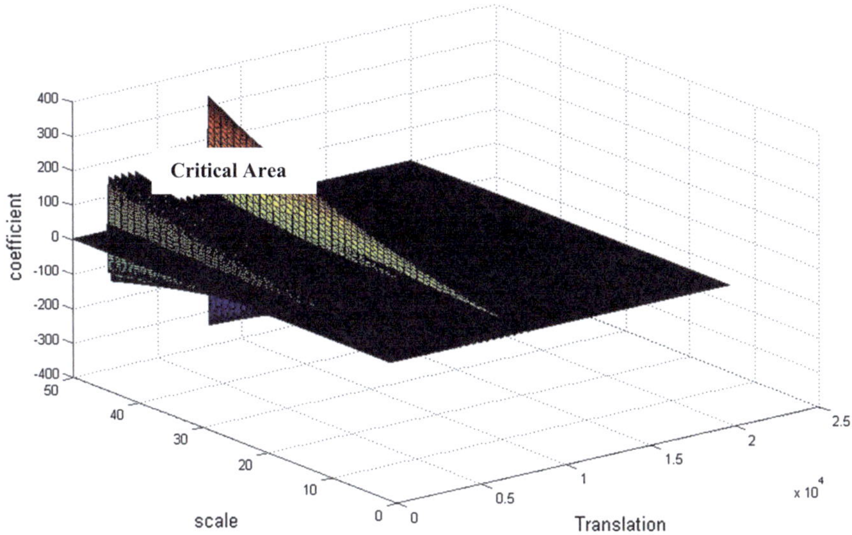

Fig. 6 CWT of R-phase primary current under inrush with short circuit fault (L-G) condition

those conditions. Kurtosis, skewness, rms and mean values have been found out of R phase primary current in different conditions to detect different conditions of power transformer. Skewness [28] can be mathematically defined as the averaged cubed deviation from the mean divided by the standard deviation cubed where as kurtosis [28] is used as an indicator in distribution analysis as a sign of flattening or "peakedness" of a distribution.

4.2.1 Assessments of DWT Based Parameter Analysis of R Phase Current

Figure 7 is used to depict the results of DWT based kurtosis values of approximate coefficients in normal, inrush and short circuit fault conditions of power transformer. In this figure constant and clear differences of kurtosis values have been observed in three different conditions where maximum difference have been observed from DWT decomposition level 8–9.

Maximum and constant differences have been observed of DWT based mean values of approximate coefficients in all those mentioned conditions which are shown in Fig. 8.

Figure 9 is used to show the results of DWT based rms values of approximate coefficients in normal, inrush and short circuit fault conditions, where differences are maximum up to DWT decomposition level 7 then it is decreasing in nature.

Figure 10 depicts the result of DWT based skewness values of approximate coefficients of R phase primary current in different conditions. One distinct feature has

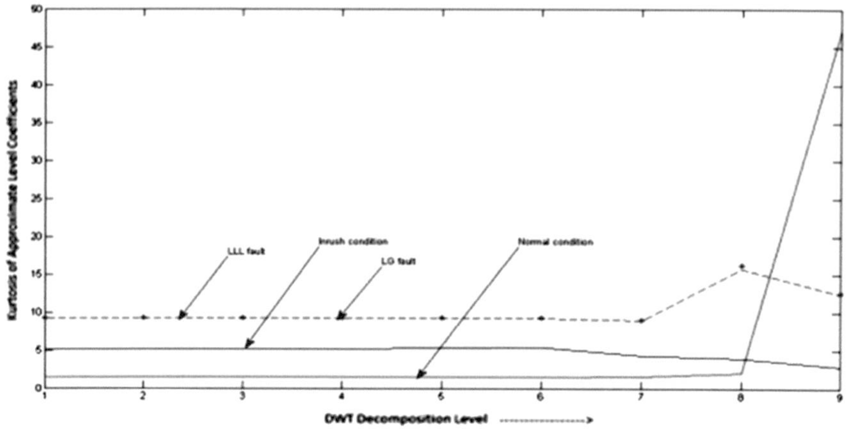

Fig. 7 Kurtosis values of approximate coefficients for normal, inrush, LLL fault and LG fault condition of R phase current

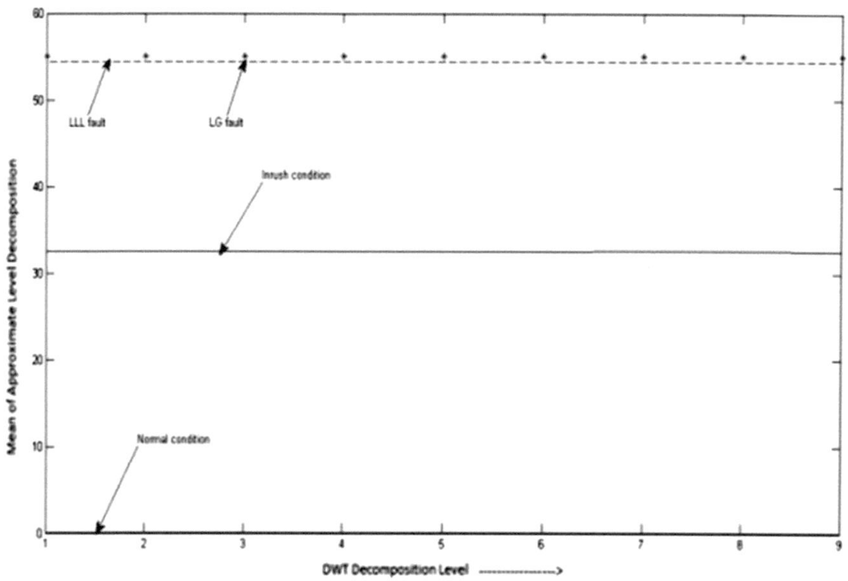

Fig. 8 Mean values of approximate coefficients for normal, inrush, LLL fault and LG fault condition of R phase current

been observed that, up to DWT decomposition level 6 skewness values of R phase current in inrush and fault conditions is same then it is increasing in nature where as maximum difference of skewness values in three conditions have been observed in DWT decomposition level 9.

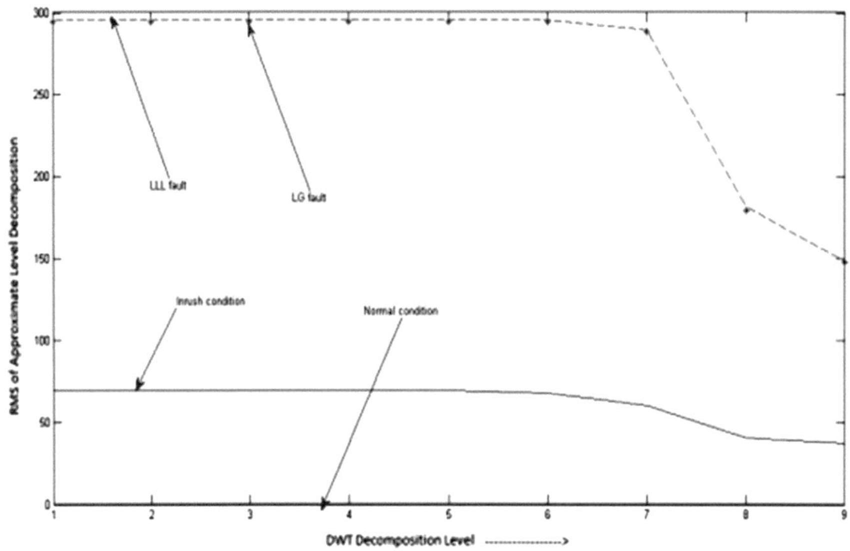

Fig. 9 Root mean square values of approximate coefficients for normal, inrush, LLL fault and LG fault condition of R phase current

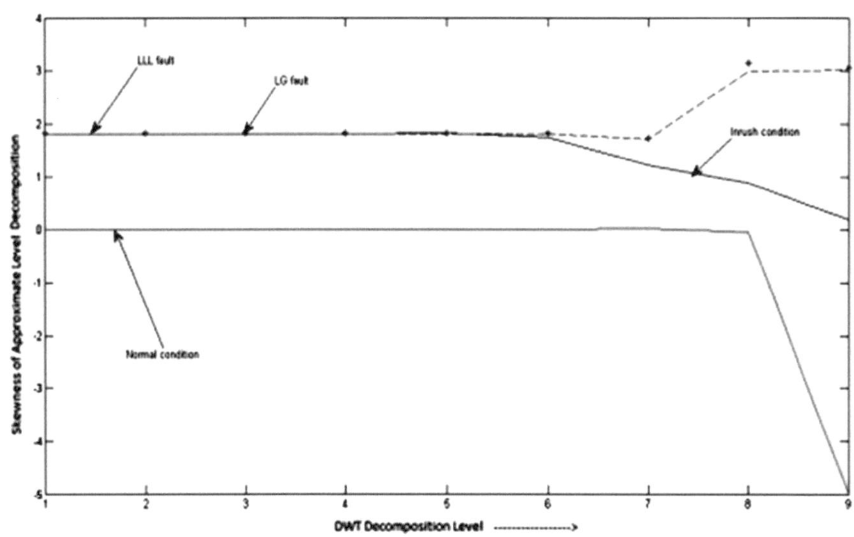

Fig. 10 Skewness values of approximate coefficients normal, inrush, LLL fault and LG fault condition of R phase current

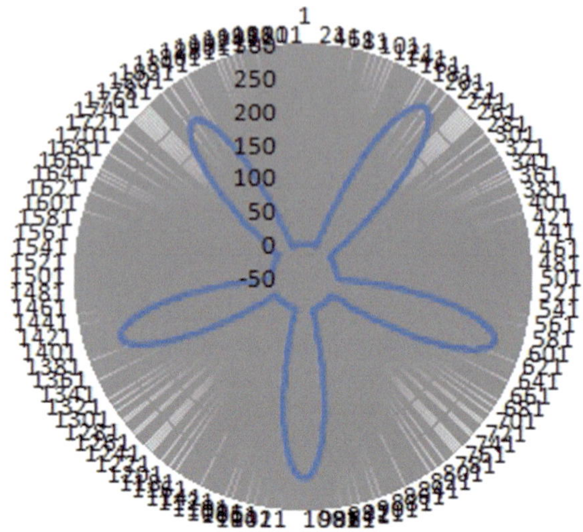

Fig. 11 Radar chart of R phase primary current in inrush condition

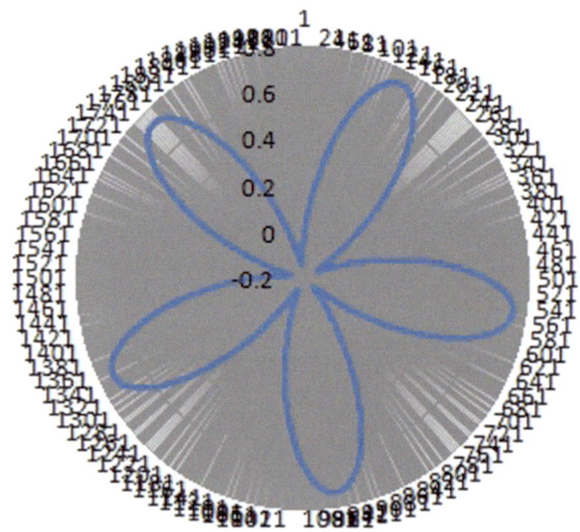

Fig. 12 Radar chart for R phase primary current in normal condition

4.3 Radar Analysis

Radar analysis has been done of R phase current at inrush, normal and fault conditions after taking the primary currents of power transformer which are shown in Figs. 11, 12 and 13, where clear difference of pictorial representation has been observed in those figures; from there inrush normal and fault conditions can be discriminated properly.

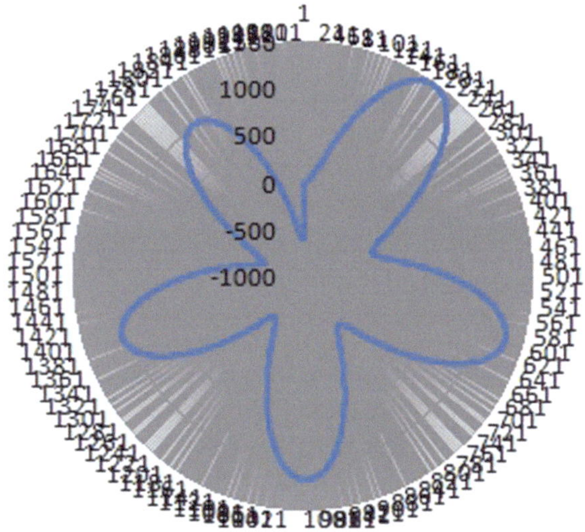

Fig. 13 Radar chart for R phase primary current in different fault (L-L-L and L-G) conditions

5 Algorithm of Assessment of Different Conditions of Power Transformer

An algorithm for assessment of different conditions of power transformer has been made as follows which can be implemented in numerical protection of power transformer:

(a) Step down the three phase primary currents of power transformer through current transformer
(b) Sample them at proper sampling frequency
(c) Capture the sampled values through data acquisition system
(d) Apply CWT and Radar analysis on the captured signal
(e) Determine skewness, kurtosis, rms and mean values of approximation (approximate) coefficients from DWT decomposition levels (up to 9th level).
(f) Diagnose the results to assess different conditions of power transformer.

6 Specific Outcome

Normal condition, Inrush condition and short circuit fault condition of power transformer have been assessed by Radar analysis, CWT and DWT based skewness, kurtosis, rms and mean value analysis of approximation coefficients based technique. Different patterns have been observed in CWT and Radar analysis of primary current of power transformer in different conditions and different parameters values have been noted in DWT based parameter analysis of 'R' phase primary current of power transformer from where all these three conditions of power transformer have been assessed properly.

7 Conclusion

In this paper, inrush, fault and normal condition of power transformer have been discriminated by Radar analysis, CWT and DWT based skewness, kurtosis, rms and mean value analysis based techniques. Different patterns have been observed for all those conditions of power transformer in Radar analysis and CWT based techniques from there different conditions of power transformer have been discriminated properly. DWT based skewness, kurtosis, mean and rms values also calculated to assess fault, inrush and normal condition of power transformer where approximation coefficients of DWT has been used for this purpose. Using all these parameters, different conditions of power transformer have been assessed properly which can be implemented for numerical protection of power transformer in real time applications.

References

1. S.S. Sunil, *Switchgear Protection and Power Systems: Theory, Practice and Solved Problems*, 13th edn. (Khanna Publishers, Delhi, 2008)
2. C.L. Cheng, C.E. Lin, C.L. Huangand, J.C. Yeh, A simple model for transformer inrush current calculation and harmonic analysis. Electr. Power Syst. Res. **24**(2), 153–163 (1992)
3. X. Ma, J. Shi, A new method for discrimination between fault and magnetizing inrush current using HMM. Electr. Power Syst. Res. **56**(1), 43–49 (2000). https://doi.org/10.1016/S0378-7796(00)00099-7
4. A.K. Al-othman, K.M. El-naggar, A new digital dynamic algorithm for detection of magnetizing inrush current in transformers. Electr. Power Compon. Syst. **37**, 355–372 (2009). https://doi.org/10.1080/15325000802548699. Taylor & Francis Group, LLC
5. D.Q. Bi, X.A. Zhang, H.H. Yang, G.W. Yu, X.H. Wang, W.J. Wang, Correlation analysis of waveforms in nonsaturation zone-based method to identify the magnetizing inrush in transformer. IEEE Trans. Power Delivery **22**(3), 1380–1385 (2007). https://doi.org/10.1109/ICPST.2006.321623
6. A. Hosny, V.K. Sood, Transformer differential protection with phase angle difference based inrush restraint. Electr. Power Syst. Res. **115**, 57–64 (2014). https://doi.org/10.1016/j.epsr.2014.03.027
7. R.J.N. Alencara, U.H. Bezerrab, A.M.D. Ferreira, A method to identify inrush currents in power transformers protection based on the differential current gradient. Electr. Power Syst. Res. **111**, 78–84 (2014)
8. P.L. Mao, R.K. Aggarwal, A wavelet transform based decision making logic method for discrimination between internal faults and inrush currents in power transformers. Electr. Power Energy Syst. **22**, 389–395 (2000). https://doi.org/10.1016/S0142-0615(00)00013-2
9. A.A.H. Eldin, M.A. Refaey, A novel algorithm for discrimination between inrush current and internal faults in power transformer differential protection based on discrete wavelet transform. Electr. Power Syst. Res. **81**, 19–24 (2010)
10. M. Rasoulpoor, M. Banejad, A correlation based method for discrimination between inrush and short circuit currents in differential protection of power transformer using discrete wavelet trans-

form: theory, simulation and experimental validation. Electr. Power Energy Syst. **51**, 168–177 (2013)
11. Z. Moravej, A.A. Abdoos, M.S. Pasand, Power transformer protection using improved S-transform. Electr. Power Compon. Syst. Taylor & Francis Group, LLC **39**, 1151–1174 (2011)
12. Z. Moravej, A.A. Abdoos, M.S. Pasand, A new approach based on S-transform for discrimination and classification of inrush current from internal fault currents using probabilistic neural network. Electr. Power Compon. Syst. Taylor & Francis Group, LLC **38**, 1194–1210 (2010)
13. A. Chatterjee, M. Maitra, S.K. Goswami, Classification of overcurrent and inrush current for power system reliability using Slantlet transform and artificial neural network. Expert Syst. Appl. **36**, 2391–2399 (2009)
14. M. Şengül, S. Öztürk, H.B. Çetinkaya, T. Erfidan, *New phenemenon on power transformers and fault identification using artificial neural networks* (Springer, Berlin, Heidelberg, 2006), pp. 767–776
15. P. Arboleya, G. Díaz, J. Gómez-Aleixandre, C. González-Morán, A solution to the dilemma inrush/fault in transformer differential relaying using MRA and Wavelets. Electr. Power Compon. Syst. Taylor & Francis Group, LLC **34**, 285–301 (2007)
16. G. Mokryani, P. Siano, A. Piccolo, Inrush current detection based on wavelet transform and probabilistic neural network, in *SPEEDAM 2010, International Symposium on Power Electronics, Electrical Drives, Automation and Motion*, pp. 62–67, 2010
17. F. Zhalefar, M.S. Pasand, A new fuzzy-logic-based extended blocking scheme for differential protection of power transformers. Electr. Power Compon. Syst. Taylor & Francis Group, LLC **38**, 675–694 (2010)
18. A. Esmaeilian, M. Mohseninezhad, M. Khanabadi, M. Doostizadeh, A novel technique to identify inrush current based on adaptive neuro fuzzy, *10th International Conference on Environment and Electrical Engineering (EEEIC),* IEEE Xplore: 13 June 2011. https://doi.org/10.1109/eeeic.2011.5874743
19. S. Chattopadhyay, M. Mitra, S. Sengupta, Harmonic analysis in a three-phase system using park transformation technique. Int. J. Model. Measur. Control General Phys. Electr. Appl., AMSE, Series–A, Modeling-A **80**(3), 42–58 (2007)
20. S. Chattopadhyay, M. Mitra, S. Sengupta, *Electric Power Quality*, 1st edn. (Springer, Berlin, 2010). ISBN 978-94-007-0634-7
21. S. Chattopadhyay, M. Mitra, S. Sengupta, Area based approach for three phase power quality assessment in clarke plane. J. Electr. Syst. **4**(3), 60–76 (2008). ISSN: 1112-5209
22. S. Chattopadhyay, S. Karmakar, M. Mitra, S. Sengupta, Assessment of crawling of an induction motor by stator current Concordia analysis. IET Electron. Lett. **48**(14), 841–842 (2012)
23. S. Chattopadhyay, M. Mitra, S. Sengupta, Area based approach in power quality assessment. Int. J. Power Manage. Electron. **2008**, 6 (2008). ID-147359, ISSN: 16876679
24. A. Chattopadhyaya, S. Banerjee, S. Chattopadhyay, Assessment of discrimination between inrush and fault current in a power transformer. Can. J. Technol. Innov. **1**, 187–196 (2014)
25. MATLAB Software Version7.7 (MATLAB® 7.7)
26. The Wavelet Tutorial by Robi Polikar
27. A. Chattopadhyaya, S. Chattopadhyay, J.N. Bera, S. Sengupta, Wavelet decomposition based skewness and kurtosis analysis for assessment of stator current harmonics in a PWM—fed induction motor drive in single phasing condition. AMSE J. 2016-Series: Adv. B **59**(1), 1–14 (2016)
28. Engineering Statistics Handbook, in *NIST/SEMATECH e-Handbook of Statistical Methods, NIST*. Retrieved 18 Mar 2012

SCADA Based Real Time Reactive Power Compensation Scheme for Assessment and Improvement of Voltage Stability in Power System

Kabir Chakraborty and Arghyadeep Majumder

1 Introduction

In present day, the topic voltage stability is taking a great collapses places around the world. The recent power networks are undergrounding frequent modifications and introducing extra complexity in the power networks from operation, stability, control and protection point of view to meet up the ever-growing electrical consumer requirement. The main difficulty which is linked with such a stressed network is voltage instability [1]. In recent years a great deal of effort has been devoted to analyse voltage stability of power network [2–6].

An electrical power network is supposed to go into a situation of voltage instability when a disturbance results an unmanageable drop in voltage profile of load buses. The cause behind this is the inability of the system to meet the increased reactive power demand. Due to the lack of adequate reactive power in power networks when the system experiences huge load demand and/or serious contingencies the voltage instability occurs. During voltage instability, magnitude of some load bus voltages decreases slowly and afterward quickly reaches the voltage collapse point. The major voltage collapse occurrences are believe to be connected to heavily loaded systems when necessary quantity of real and reactive power are not obtainable to preserve standard voltage magnitudes of the network buses.

In this paper, a method for real time SCADA system has been suggested for reactive compensation scheme in power system to assess and improve the voltage stability. The system consists of measuring instruments for data acquisition, simulation software for supervisory control and FACTS devices for reactive power compensation.

K. Chakraborty · A. Majumder (✉)
Department of Electrical Engineering, Tripura Institute of Technology, Narsingarh, Tripura, India
e-mail: arghyadeep.majumder@gmail.com

K. Chakraborty
e-mail: kabir_jishu@rediffmail.com

S. Chattopadhyay et al. (eds.), *Modelling and Simulation in Science, Technology and Engineering Mathematics*, Advances in Intelligent Systems and Computing 749,
https://doi.org/10.1007/978-3-319-74808-5_18

This system has been applied to a standard power network and load flow solution of the network is obtained. Weakest segment of the power system has been find out by dV/dQ indicator [2] values. Integrated Voltage Stability Indicator values has been find out and compared with reactive power for different system conditions to evaluate the voltage instability of power network. In addition, the operation of the system has been shown with closed loop feedback algorithm for real time application.

2 Concept of Proposed Methodology

In Fig. 1 the measuring instruments of RTUs [3] are connected to the transmission network. The data acquisition are done through this section. For real time implementation, real time data are needed from the measuring instruments but generally, these data are not in Per Unit values. Generally, all the parameters are like line data, bus data, impedances, resistance, reactance, half line charging etc. are calculated in Per Unit values because Per Unit values do not change when they are referred to one side of a transformer to other side of transformer.

$$\text{Per Unit Value} = \text{Actual value}/\text{Base Value} \quad (1)$$

This can be a major advantage as because in a large power system huge numbers of transformers are interconnected. That is why, to make complex power system calculation more convenient all parameters are expressed in the same units irrespective of their ratings.

Real time data can be converted to Per Unit Value through simulation using above formula in Eq. (1). This information then can be fed to the ECC (Energy Control Centre) for load flow solution purpose. For analysis purpose, the data is considered for a particular instant of standard IEEE 6-bus power system. To solve any power system problem load flow solution [4] must be solved and this is solved by simulation.

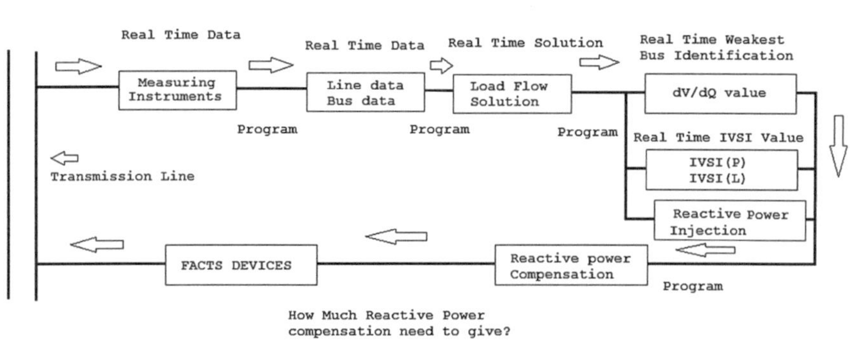

Fig. 1 Real time SCADA system

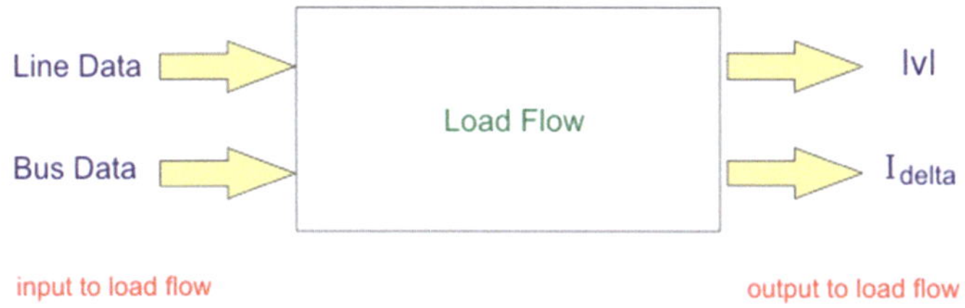

Fig. 2 Basic load flow diagram

The load flow in a power system has been shown in Fig. 2. Basic load flow expression based on N-R method is given as

$$\begin{bmatrix} \Delta P \\ \Delta Q \end{bmatrix} = \begin{bmatrix} J_1 & J_2 \\ J_3 & J_4 \end{bmatrix} \begin{bmatrix} \Delta \partial \\ \Delta |V| \end{bmatrix} \quad (2)$$

The value of ∂V/∂Q i.e. the variation of voltage with respect to reactive component is highest for weakest bus of the system. So, it is required to find the highest value of [∂V/∂Q] from J4 elements in the jacobian matrix.

Now, multi-bus electrical power network can be symbolized by an correspondent two-bus system comprising of one slack bus having bus voltage magnitude Vs and generated power ($P_g + jQ_g$) is supplied along with one load bus having bus voltage magnitude Vr and load ($P_{load}+jQ_{load}$) is connected to this bus [1]. The line connected these two buses having equivalent impedance Z_{eq}. The active and reactive power losses of the equivalent system are given by Eqs. (3) and (4)

$$P_{loss} = \frac{R_{eq}\left(P_g^2 + Q_g^2\right)}{V_s^2} \quad (3)$$

$$Q_{loss} = \frac{X_{eq}\left(P_g^2 + Q_g^2\right)}{V_s^2} \quad (4)$$

Integrated voltage stability indicator (IVSI) has been used for the detection of weaker segment of the network using the equivalent system methodology. Based on these quantities maximum transferred power, i.e., maximum values of reactive power and the maximum values real power and reactive power loss of the transmission line, the expressions for integrated voltage stability indicators can expressed in the Eqs. (5) and (6) as

$$IVSI(P) = \frac{P_r}{P_{r(\max)}} = \frac{Q_r}{Q_{r(\max)}} \quad (5)$$

$$IVSI(L) = \frac{P_l}{P_{l(\max)}} = \frac{Q_l}{Q_{l(\max)}} \quad (6)$$

Since, voltage collapse is considered imminent when the value of IVSI is near or equal to 1, the which means that smaller the value of IVSI more healthy is the system state [1].

3 Simulation

A standard IEEE 6-bus power system has been taken for simulation purpose. The main objective lies with Integrated Voltage Stability Indicator (IVSI) & dV/dQ indicator values for formulation of the real time application. Load flow simulation results for a particular instant are given as follows in Table 1.

To locate the weakest bus [6] in the network, the Jacobian Matrix (J) is computed and [∂V/∂Q] value for all the load buses are calculated and shown in Table 2.

From this simulation, it is obtained that the ∂V/∂Q value is the highest for bus no. 5 whose value is equal to 0.0653022. Therefore, bus no. 5 is the weakest bus of the network [7] for that particular instant. The value of IVSI obtained from this simulation is shown in the Table 3.

From this simulation, it is obtained that the IVSI value is closer to the 1. Bus number 5 is the weakest bus of the system as the value of IVSI calculated for bus number 5 in the IEEE 6-bus power system is greatest. Therefore, the system is imminent to voltage collapse [8].

CASE STUDY-1: Change of ∂V/∂Q values and IVSI values with respect to reactive compensation given to the weakest bus shown in Fig. 3.

Table 1 Load flow solution

Bus	Voltage	Angle	Active power	Reactive power
No.	Per unit	Radian	Per unit	Per unit
1	1.050	0.000	0.015	−0.485
2	1.080	−0.609	0.010	0.046
3	1.080	0.576	0.018	−0.137
4	1.076	0.469	−0.014	−0.006
5	1.083	−0.644	−0.012	−0.004
6	1.084	−0.658	−0.009	−0.004

Table 2 ∂V/∂Q values for weakest bus identification

Bus no.	4	5	6
∂V/∂Q Value	0.0635343	0.0653022	0.054249

Table 3 IVSI values

IVSI (power)	IVSI (loss)
0.744932	0.488930

For Condition-1 where no reactive compensation is provided [9], the IVSI value (power) is quite high. In Condition-2 a little amount of reactive power (0.3 pu) is injected in the weakest bus. so the IVSI value (power) has been reduced. Further reactive power injection is increased to 0.6 pu and the IVSI value (power) has been decreased significantly as shown in condition 3. The system condition after further increase of reactive compensation (equal or more than 0.6 pu) to the weakest is as shown in condition 4.

CASE STUDY-2: Change of ∂V/∂Q values and IVSI values with respect to reactive load connected to the weakest bus shown in Fig. 4.

In condition-1 small reactive load connected to the weakest bus. ∂V/∂Q values & IVSI values (power) are very less. In condition-2 a larger reactive load (0.4 pu) is linked to the weakest bus, now ∂V/∂Q value is increased little bit but IVSI value (power) has been increased significantly. By connecting a large load (0.8 pu), as shown in condition 3, to the weakest bus the voltage collapse is considered to be more imminent because IVSI value (power) is very closer to the unity (i.e. 0.67).

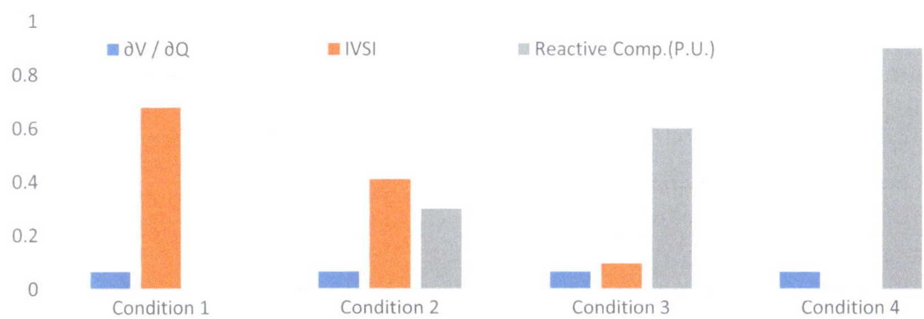

Fig. 3 Change of ∂V/∂Q value and IVSI value with respect to reactive compensation given to the weakest bus

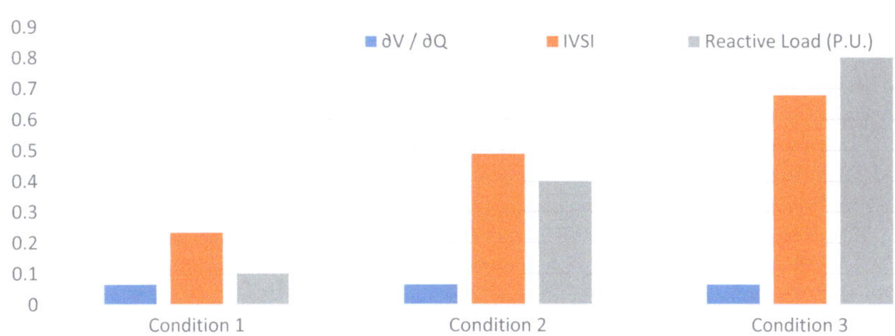

Fig. 4 Change of ∂V/∂Q value and IVSI value with respect to reactive load connected to the weakest bus

Table 4 IVSI values before compensation

∂V/∂Q	IVSI	Reactive compensation
0.0653022	0.744932	0.00

Table 5 IVSI values after compensation

∂V/∂Q	IVSI	Reactive compensation
0.0652831	0.33631	0.25

4 Closed Loop Feedback Algorithm

From the above simulation results and case studies, it can be seen that reactive compensation is correlated with ∂V/∂Q values and IVSI values. To apply in the real time system it is required to use closed loop feedback algorithm, which is shown in Fig. 1.

The reference point for IVSI value is considered as 0.35 that means if in real time IVSI is equal or less then the reference value then the system will be considered as in the safe state in terms of voltage stability [10]. The simulated result for a particular instant show that the real time IVSI value as given in Table 4 is much higher than the reference IVSI value 0.35.

Therefore, a new simulated program is required which will now calculate the reactive compensation needed (supplied by FACTS devices) [11] for the system to lower down the real time IVSI value equal or less than the reference IVSI value 0.35.

From Table 5, it can be observed that the reactive compensation given here is 0.25 in per unit [12] and indicator values are calculated by the simulated program. Therefore, the new real time IVSI value is 0.33631, which is below or lower than the reference IVSI value. This way, the system will continuously track the real time data and simulated calculation; will try to keep real time IVSI value closer to the reference IVSI value.

5 Conclusion

A SCADA based real-time reactive power compensation scheme has been presented in the paper to assess and improve the stability in terms of bus voltage in multi-bus power system using closed loop feedback algorithm.

Load flow solution and weakest segment of the electrical power network is obtained. Integrated Voltage Stability Indicator (IVSI) value, which indicate the system voltage stability is obtained and it has been reduced by reactive power compensation given to the weakest bus of standard IEEE 6- bus system. The operation of close loop feedback algorithm has been shown in this paper with IVSI values. Some case studies are presented with IVSI and reactive compensation values. The performance of the system depends upon data acquisition speed, software processing, calculation and operation of FACTS devices.

This SCADA system will be very much helpful in ECC (Energy Control Centre) for real time voltage stability control because ECC operator has to monitor only parameter i.e. to keep real time IVSI value closer to the reference IVSI value. These will ease the ECC operator stress and make voltage control operation more reliable.

References

1. K. Chakraborty, A. Chakrabarti, *Soft Computing Techniques in Voltage Security Analysis* (Springer, Berlin, 2015)
2. K. Chakraborty, S.D. Biswas, An offline simulation method to identify the weakest bus and its voltage stability margin in a multibus power network, in *International Conference on Modelling and Simulation, MS 2007*, India, December 3–5, 2007
3. A. De, K. Chakraborty, A. Chakrabarti, Classification of power system voltage stability conditions using Kohonen's self-organising feature map and learning vector quantization. Euro. Trans. Electr. Power **22**(3), 412–420 (2012)
4. P. Srikanth, O. Rajendra, A. Yesuraj, M. Tilak, K. Raja, Load flow analysis of IEEE 14 bus system using Matlab. Int. J. Eng. Res. Technol. **2**(5), 149–155 (2013)
5. K. Chakraborty, A. Chakraborty, A. De, Integrated voltage stability indicator based assessment of voltage stability in a power system and application of ann. Iranian J. Electr. Comput. Eng. **10**(2), 85–92 (2011)
6. K. Chakraborty, B. Saha, S. Das, A method for improving voltage stability of a multi-bus power system using network reconfiguration method. Int. J. Electr. Eng. **8**(1), 91–102 (2015). ISSN 0974-2158
7. P. Gao, L. Shi, L. Yao, Multi-criteria integrated voltage stability index for weak buses identification, in *Transmission & Distribution Conference and Exposition: Asia and Pacific*, ISBN: 978-1-4244-5230-9, 2009
8. M.S.S. Danish, A. Yona, T. Senjyu, A review of voltage stability assessment techniques with an improved voltage stability indicator. Int. J. Emerging Electr. Power Syst. **16**(2), 107–115 (2015)
9. P. Roy, P. Bera, S. Halder, P.K. Das, Reactive power sensitivity index based voltage stability analysis to a real system. Int. J. Electron. Commun. Technol. (IJECT) **4**(1), 167–169 (2013)
10. K. Chakraborty, A. De, A. Chakraborty, Assessment of voltage security in a multi-bus power system using artificial neural network and voltage stability indicators. J. Electr. Syst. **6**(4), 517–529 (2010)
11. A.K. Mohanty, Power system stability improvement using facts devices. Int. J. Mod. Eng. Res. (IJMER) **1**(2), 666–672 (2011)
12. S. Dudhe, Reactive power compensation techniques in transmission lines. Int. J. Recent Innovation Trends Comput. Commun. **3**(5), 3224–3226 (2015). ISSN: 2321-8169

Part III
Energy

Solar Photovoltaic Power Supply to Utility Grid and Its Synchronization

Sonalika Dutta, Soumya Kanti Bandyopadhyay and Tapas Kumar Sengupta

1 Introduction

SPV roof top system is widely used in the world, employ as clean technology to reduce CO_2 emission. Whenever SPV roof top system is connected to grid for supply power in grid, it needs a grid connected inverter to couple with grid. In this paper discusses about multistage grid connected inverter due to the SPV generated voltage level is low as compare to the grid voltage. Grid connected inverter is also named as grid tie inverter (GTI). The applications are in net metering, dual metering, SPV without use battery storage system. The multistage inverter consists with two stage converter (a boost converter, a dc-dc converter with high frequency transformer) and a grid connected inverter. A LCCL filter is connected between GTI and utility grid to reduce harmonic distortion. By using PLL control technique of GTI, reduce complexity and it more reliable for synchronization. The Grid connected inverter is nothing but an H- Bridge single phase VSI inverter but its control mechanism is differ from traditional inverter. The GTI and utility grid are synchronized by a special type Phase locked Loop (PLL) and is designed and verified by PSIM software.

PSIM is a platform for engineering simulation and design; for research, development and application in various sectors i.e. power supply and generation, Nonconventional generation, motor drives, power conversion and control systems. T. Lesster has first developed the PSIM software in 1994, PSIM is capable to develop power

S. Dutta (✉) · S. K. Bandyopadhyay · T. K. Sengupta
Department of Electrical Engineering,
Supreme Knowledge Foundation Group of Institutions,
MAKAUT, Kolkata, India
e-mail: sonalika.dutta.35@gmail.com

S. K. Bandyopadhyay
e-mail: soumya.bandyopadhyay@skf.edu.in

T. K. Sengupta
e-mail: tksg1948@yahoo.co.in

S. Chattopadhyay et al. (eds.), *Modelling and Simulation in Science, Technology and Engineering Mathematics*, Advances in Intelligent Systems and Computing 749,
https://doi.org/10.1007/978-3-319-74808-5_19

electronics simulation and design for various power electronic applications. Here the circuit is developed PSIM 9.0 version which is shown in Fig. 1 and the graphical results of this simulation shown in Figs. 3, 4, 5, 6, 7, 8, 9, 10 and 11. Moreover by reducing components harmonic distortion is reduced, the size of LCCL filter is reduced (Fig. 2).

2 DC-DC Converter

Multistage GTI topology is used for low voltage (12 V) SPV generation. First stage is voltage increased by a boost converter. In second stage a dc-dc converter with high frequency transformer is use to increase voltage further step. Now the circuit detail is discussed in below.

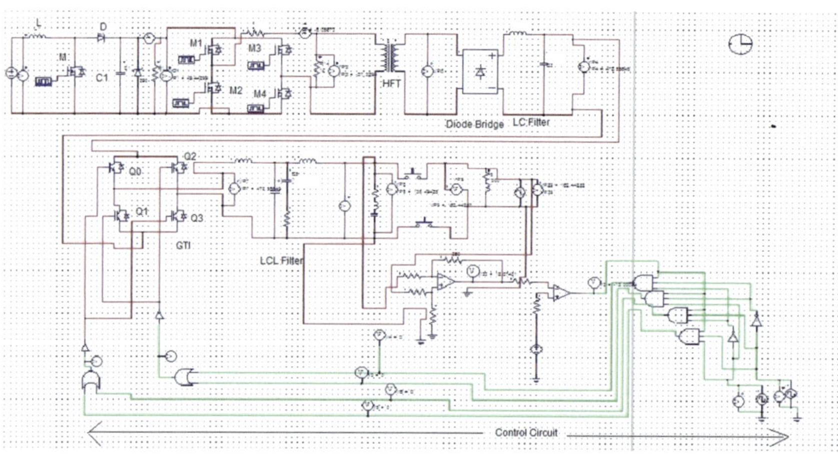

Fig. 1 Simulation model of synchronized solar photo voltaic power supply to utility grid

Fig. 2 Simulation model Subtractor circuit

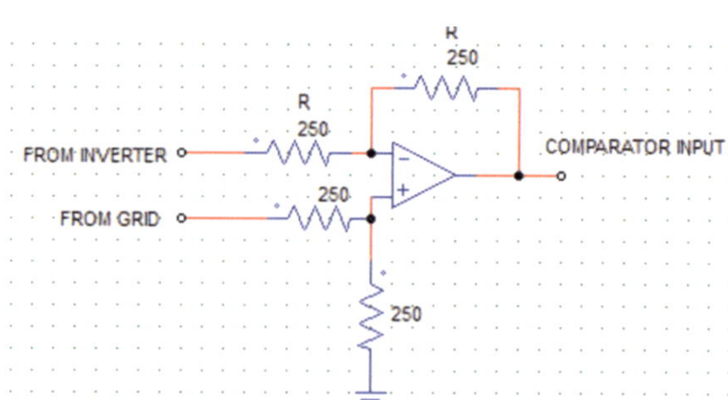

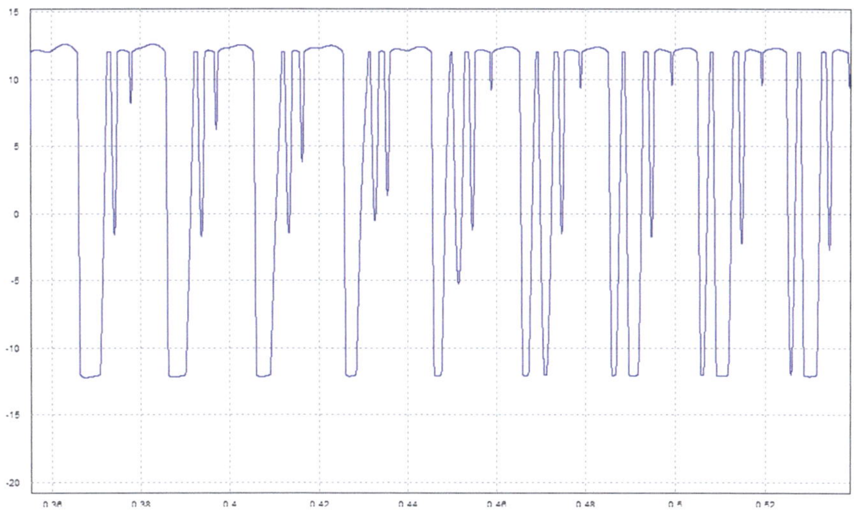

Fig. 3 Output waveform of subtractor circuit

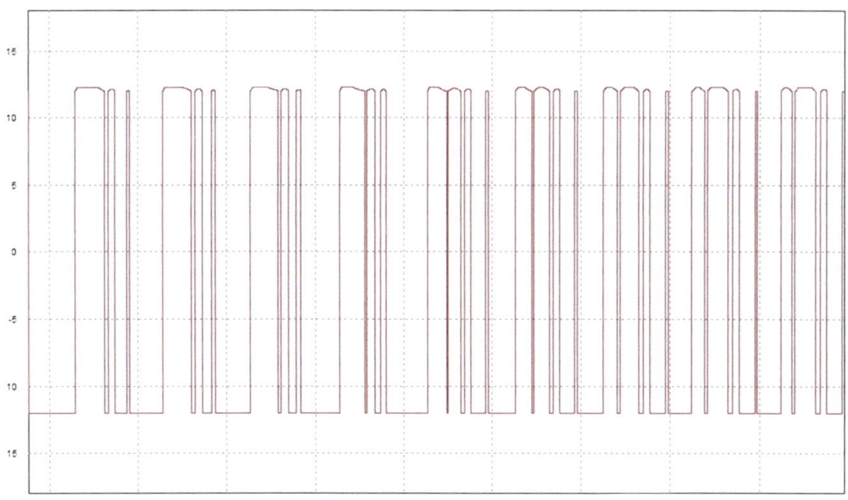

Fig. 4 Output waveform of comparator circuit

2.1 Boost Converter

The boost converter is a dc–dc step up converter. This converter is desired in here to reduce turns ratio of transformer in next stage, otherwise leakage reactance increase in transformer and switching of MOSFETs are affected [1]. The converter consist of two solid state devices one is a MOSFET(M) switch and other is diode and energy

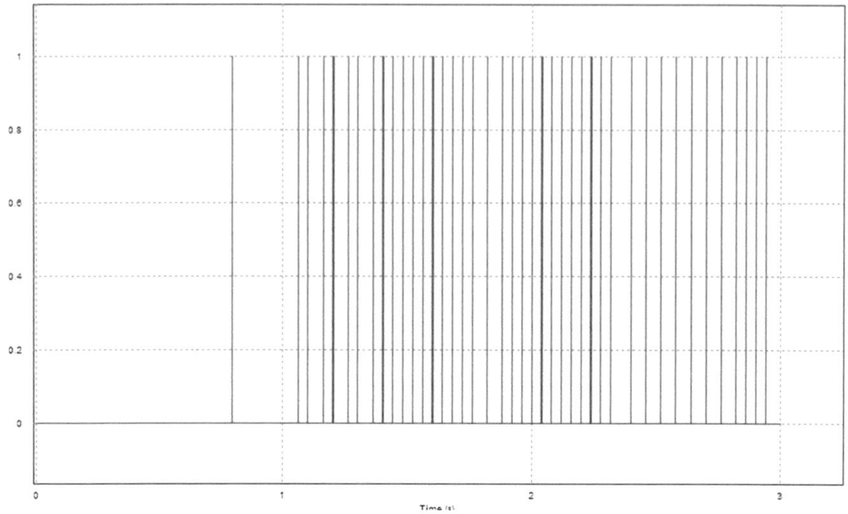

Fig. 5 Output wave form of demultiplexer for gate pulse of IGBT 1

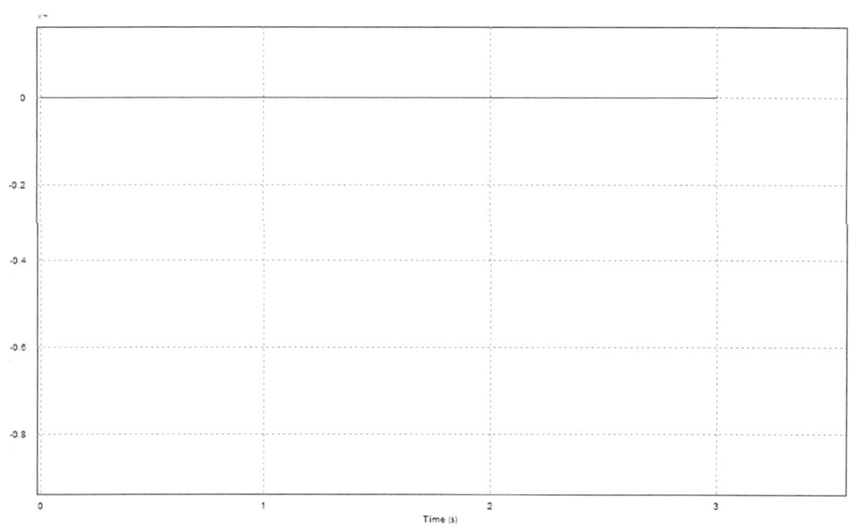

Fig. 6 Output waveform of demultiplexer for gate pulse of IGBT 2

storage passive elements i.e. an Inductor and a filter capacitor. The filter capacitor reduces the ripple output voltage. To get a steady state voltage applies a Zener diode across voltage output.

Boost converter operates in two modes,

At mode 1, $t = t_1$, inductor L charging at switch M is on. At that condition inductor current raises initial value I_1 to final value I_2.

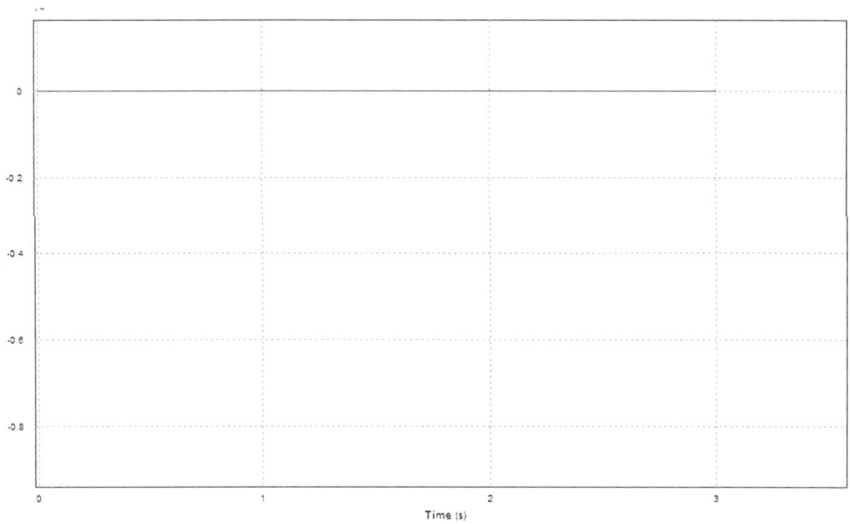

Fig. 7 Output wave form of demultiplexer for gate pulse of IGBT 4

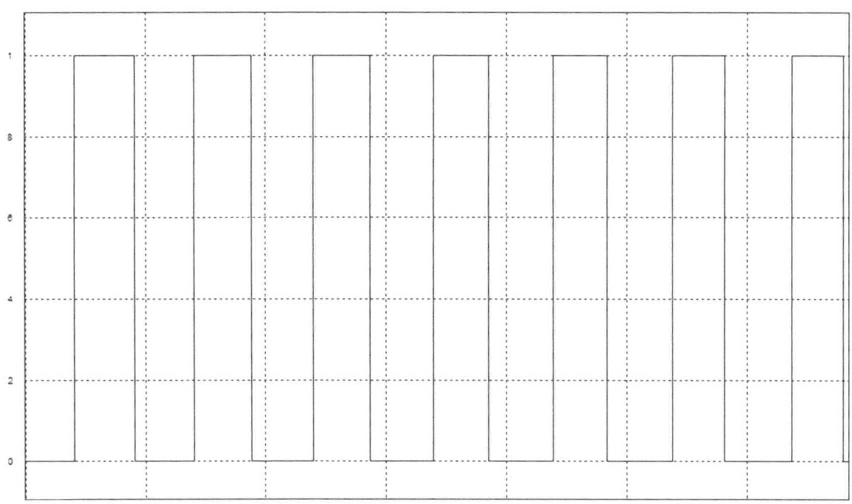

Fig. 8 Output wave form of demultiplexer for gate pulse of IGBT 3

$$V_s = L\Delta I/t_1 \tag{1}$$

$$t_1 = \frac{\Delta I L}{Vs} \tag{2}$$

At mode 2, $t = t_2$, L discharge still MOSFET switch is open next at cycle. In this cycle,

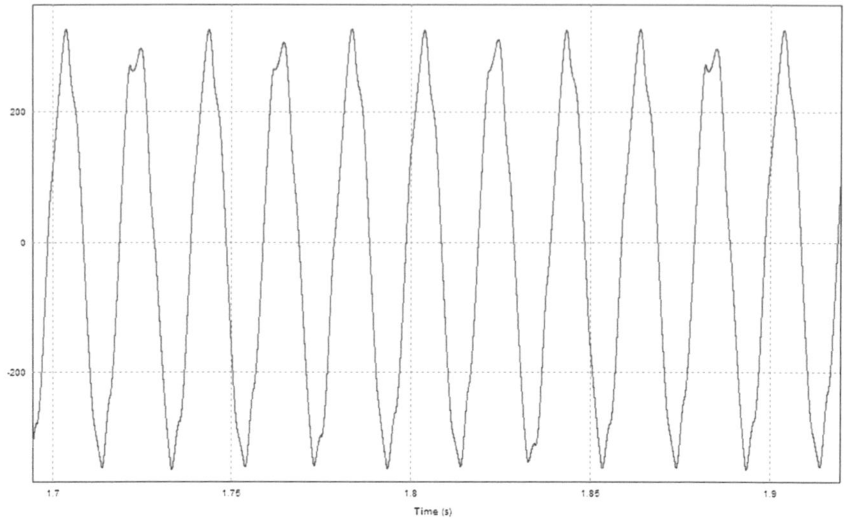

Fig. 9 Output waveform of grid tie inverter

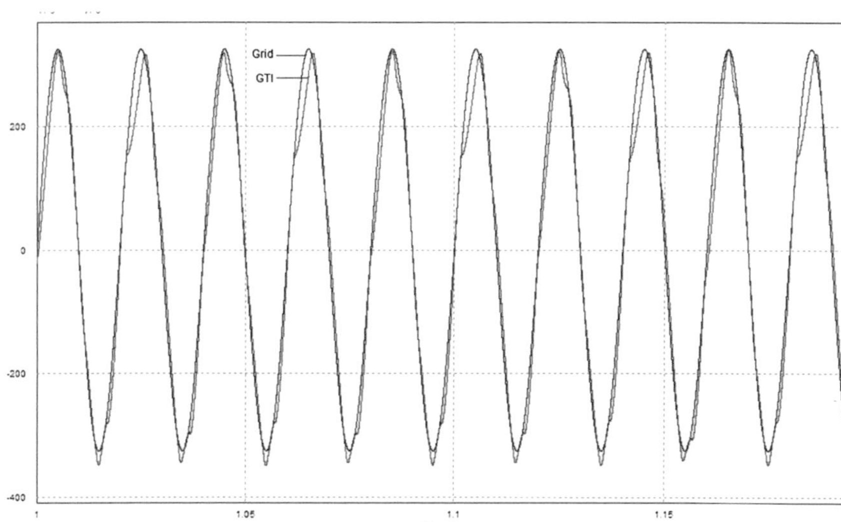

Fig. 10 Output wave forms of grid tie inverter and utility grid

$$t_2 = \frac{\Delta I L}{V_a - V_s} \tag{3}$$

The switching period T is found by from adding Eqs. (2) and (3)

$$T = t_1 + t_2 = \frac{\Delta I L V_a}{V_s (V_a - V_s)} \text{ and , } T = \frac{1}{f}$$

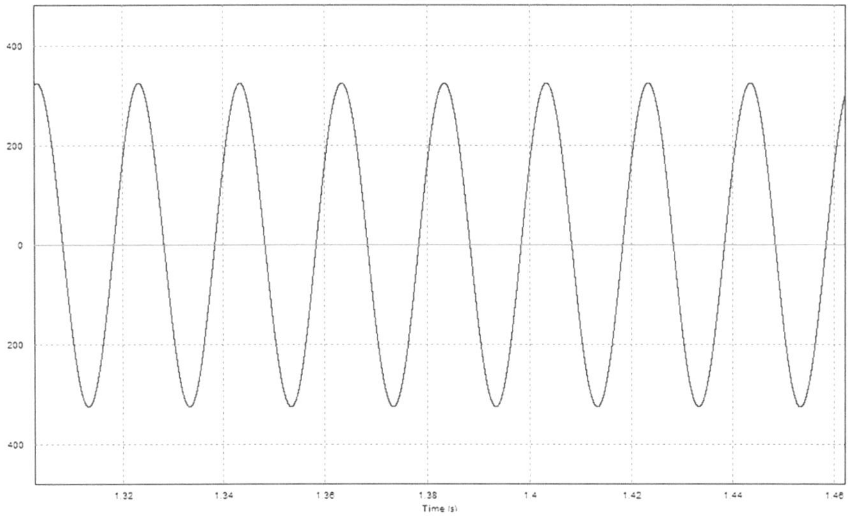

Fig. 11 Output waveforms of grid tie inverter and utility grid at when grid couple switches are connected

From this we get switching frequency (f) of MOSFET switch in boost converter.

2.2 *Full Bridge High Frequency Converter*

The second stage of conversion is implied through a full bridge converter, a high frequency (HF) step up transformer and a diode bridge rectifier. The full bridge converter is H-bridge PWM inverter with faster MOSFET switches. The Switching frequency high for matching the frequency of the HF transformer. The HF transformer use for reduce the size, higher order harmonics and cost [2]. The output voltage of transformer followed by the equation which is given below

$$V_{out} = 2\frac{N_2}{N_1}DV_{dc} \tag{4}$$

Here D is the duty cycle of Switching element (here use MOSFET), $\frac{N_2}{N_1}$ the turn ratio of transformer [3].

The ac output voltage fed in a full bridge diode rectifier to get dc regulated voltage. A LC filter use to reduce ripple from dc output voltage.

3 Grid Connected Inverter with Filter

The interfacing between a SPV system and utility grid is by a Grid tie inverter. The GTI play an important role when dc power resource is connected to an ac grid. After analyzing all criteria for synchronization, the GTI connect with grid. The DC voltage fed to inverter input from Diode Bridge rectifier. The basic difference from conventional VSI inverter from GTI is in its control mechanism. The inverter consists of four IGBT switches alien in two limb of inverter. Two switches of each limb is operated at a time. The line commuted inverter is operated with grid reference [4] which is consist of a close loop control system. The PWM gate pulses of IGBT switches are generated by using an analog—digital combination circuit of PLL. These pulses are controlled the inverter for grid synchronization. After match the frequency, voltage amplitude and phase angle the inverter is switched to grid [5] as per IEEE standard 1547.2 [6]. In between inverter and grid a LCCL filter is placed to reduce harmonic contain and improve power quality of inverter output. The Simulation model is shown in Fig. 1.

4 Control Technique of GTI Switching Devices

The most important part of this paper is the control technique of GTI. It is different form PLL control of grid tie inverter. The analog and digital circuits generate the pulses with reference of grid values. These pulses are the PWM pulse feed to gates of IGBT switches as per need and directs by demultiplexer selector switches. The control circuit consists of a subtrator, a comparator and a demultiplexer with two OR gates.

4.1 Subtractor

The inputs of subtractor fed from grid and inverter. Grid and inverter phases are synchronized (locked) here. The subtractor's one input (+) voltage fed from grid and other input (-) take voltage from Grid connected inverter output.

If all external resistors(R) have equal value, the output voltage is derived by using 'superposition theorem' [7].

First we take $V_2 = 0$, V_1 is only input source of op-amp, the circuit is now like non inverting amplifier.

The output voltage

$$V_{01} = V_1/2(1 + \mathrm{R}/\mathrm{R}) = V_1 \tag{5}$$

Similarly, the output V_{02} due to input V_2 only, $V_1 = 0$, can be express as inverting amplifier i.e.

$$V_{02} = -V_2 \tag{6}$$

Then the output voltage

$$V_0 = V_{01} + V_{02} = V_1 - V_2 \tag{7}$$

V_1 is the grid voltage and V_2 is the inverter output voltage.

4.2 Comparator

The comparator has a reference dc input and other is from phase detector (subtractor). In here the comparator use with open loop mode to get high open loop gain [8]. In here V_{ref} takes (12 V DC) in one input (+) and in another input (-) take a signal from the subtractor output. Any minimum change in voltage the comparator gets signal [9] and error is signified by this.

$$V_{out} = Av_0\left(V^+ - V^-\right)$$

In open loop mode the amplifier voltage gain is nearly equal to Av_0. This V_{out} is the input of digital pulse generator, PWM Pulse control the firing angle of IGBTs of GTI.

4.3 Demultiplexer

The gates of IGBT switches of GTI are got pulse from 1 line to 4 lines demultiplexer [10]. Demultiplexer consists of two select inputs, one data input and four outputs. The select inputs are select which output (AND Gate) of Demultiplexer is active at a time among the four output switches. The four outputs are generating pulses for each gate of IGBT switches. The data input is here the pulse from comparator. Any change in grid value sense by the subtractor and in demultiplex through the comparator. The IGBTs of GTI are also getting this effect through gates switching. The firing angle is controlled by phase angle control by using of PLL. So it is the model based close loop control system of GTI.

Output waveforms of PLL control circuit, Grid Tie Inverter and Utility Grid are shown in Fig. 3.

5 Latest Trends and Scope of Future Developments

In present trend the SPV power is supplied to utility grid as three phase or single phase system. For grid synchronization the GTI is controlled by using various technologies. PLL is a popular control technology among these. The single phase GTI already has four types of PLL control [11].

This paper is proposed a new type of PLL for synchronizes GTI and utility grid. After taking grid value i.e. voltage amplitude, frequency and phase angle the PLL is senses and send signal to inverter, then the inverter produce sine wave and achieve synchronization criteria.

To improve power quality further may be used of H_5 structure [12] inverter in future development. In case of grid contingency, isolation will be required and for resynchronization different control circuit will be designed and adopted.

6 Conclusion

In this paper is monitoring and synchronizing of Grid tie inverter by use a new variation of PLL. This synchronization experiment has been developed by simulation model of PSIM-software. Here the isolation of low pass filter [11] the PLL system is more robust in configuration and cost is also reduced. The LCCL filter components size is reduced by using digital signal of control circuit which reduced harmonic distortion. The compactness of overall circuit which maintains power quality and by using line commutated inverter the complexity and cost are also reduced.

Acknowledgements The authors would like to thank Electrical Department, Supreme Knowledge Foundation Group of Institutions, MAKAUT, Kolkata, India, for coordination and support.

References

1. A Grid Tie Inverter For Solar Systems, solar.smps.us, 17.08.2015
2. Design of High Frequency Pulse Transformer www.electrical4u.com, 26.03.2017
3. A. Singh, V.S. Jabir, *Voltage Fed Full Bridge DC-DC and DC-AC Converter for High-Frequency Inverter Using C2000*. (Texas Instruments, Texas, Application Report) June 2015
4. Line-commutated Inverter, definedterm.com, 10.03.2017
5. U. Solanke Tirupati, A. Kulkarni Anant, Effective microgrid synchronization in islanded mode: controlled input/output PI-Fuzzy-PI algorithm. Int. J. Comput. Appl. (0975–8887) **75**(16), 39–45 (2013)
6. Standard for interconnecting Distributed Resources with Electric Power System, IEEE standard 1547.2, 2008
7. D. De, K. Prasad Ghotok, *Basic Electronics* (Pearson Education, London, 2012)
8. Op amp Comparator, www.electronics-tutorials.ws, 22.03.2017
9. D. Roy Choudhury, S.B. Jain, *Linear Integrated Circuits*, 3rd edn. (New age international publishers, London, 2007)

10. A. Anand Kumar, *Fundamentals of Digital Circuits* (Prentice-Hall of India Pvt. Ltd, Delhi, 2007)
11. Y. Yang, F. Blaabjerg, Synchronization in single-phase grid-connected photovoltaic systems under grid faults, in *3rd IEEE International Symposium on PEDG Conference*, Aalborg, Denmark, June 2012, pp. 476–482, 2012
12. R. Teodorescu, M. Liserre, P. Rodriguez, *Grid Converters for Photovoltaic and Wind Power Systems* (Wiley, Hoboken, 2011)

Optimum Sizing and Economic Analysis of Renewable Energy System Integration into a Micro-Grid for an Academic Institution—A Case Study

Nithya Saiprasad, Akhtar Kalam and Aladin Zayegh

1 Introduction

World energy demand has been estimated to be greater than 800EJ by 2050. For this estimation, with the present scenario of escalating oil prices when considered, renewable energy could promise to be an alternate option as an energy resource [1]. Alternately, the global concern towards pollution and global warming has supported this cause. In recent years, there has been much technical advancement in renewable energy systems (RES) including the storage units. Many countries have been striving to reach their renewable energy target towards the global energy contribution; Australia being one among them.

Australia is the world's 9th largest energy producer using coal and the largest exporter of uranium [2]. In its share of renewable energy generation, Australia's renewable energy contribution is far too minimal for the abundance of natural resources it possesses. Despite the fact of the volatility of the conventional energy market, this cheaper environmentally unfriendly energy has been dominant in the energy market. Although several studies conducted on Australia being 100% renewable have given negative results [3, 4]. However, pondering renewable energy being a part of the modern grid has equally been dealt with [5–10].

N. Saiprasad (✉) · A. Zayegh
College of Engineering and Science, Victoria University,
Melbourne, VIC 3011, Australia
e-mail: nithya.saiprasad@live.vu.edu.au

A. Zayegh
e-mail: Aladin.Zayegh@vu.edu.au

A. Kalam
Smart Energy Research Unit, College of Engineering and Science,
Victoria University, Melbourne, VIC 3011, Australia
e-mail: Akhtar.Kalam@vu.edu.au

S. Chattopadhyay et al. (eds.), *Modelling and Simulation in Science, Technology and Engineering Mathematics*, Advances in Intelligent Systems and Computing 749,
https://doi.org/10.1007/978-3-319-74808-5_20

Designing and optimizing a micro-grid and analyzing their economic and environmental impacts have been the template of this study. Similar studies have been conducted using Solar cells or Photo-voltaic (PV), wind turbines, fuel cells (used either as an energy source or as a storage unit) for isolated villages, islands, wind farms, resorts [5, 8, 11–20]. The current study is aiming at integrating renewable energy like PV and wind turbine connected to the grid for Victoria University located at the St Albans campus in Melbourne, Australia. The location map is shown in Fig. 1.

To design and optimize any micro-grid, it is significant to understand and study the load requirement of the desired location. This crucial step during the design of a micro-grid should not terminate in underestimating or overestimating the consumption, either of which could result in unmet load or oversized setup respectively. Various methods have been used to optimize a micro-grid including genetic algorithm and swarm optimization techniques. However, many software have been used in such studies like MATLAB/SIMULINK, HOMER etc. [21].

HOMER (Hybrid Optimization of Multiple Energy Resources) is a software that was initially created by the National Renewable Energy Laboratory and now marketed by a company called HOMER Energy. HOMER consists of 3 main modules Simulation, Optimization, and Sensitivity Analysis. The crucial task lies in the architecture of the micro-grid setup for the load demand of the university with least cost demand and greater efficiency. The aforementioned problem has been studied using HOMER software which designs and optimizes the setup with least Net Present Cost (NPC) of the system [22]. The study conducted also includes the environmental impact of the designed system by analyzing the amount of harmful gases they emit to the environment.

2 System Description

The RES designed here, considers the total cost of the system which includes the total capital cost and the maintenance cost. The architecture of this system consists of PV arrays, wind turbine, controller, batteries and grid support. To minimize the cost of the system and meet the load demand, HOMER defines a few terminologies which are deciding factors for the suggested model [22].

They are expressed as follows:

a. Net Present Cost (NPC): Net Present Cost determines the profitability of the project, which is the total net present value of the component subtracted by the (income) profit it incurs for the complete lifetime of the project.

$$NPC = \sum \left(Total\ Cash\ flow / (1 + Interest\ rate)^{(Project\ Life\ time)} \right) - Initial\ Investment \quad (1)$$

b. Annualized cost of the system (ACS): Annualized cost is that cost of the set up when factored equally over the entire lifetime of the project considered.

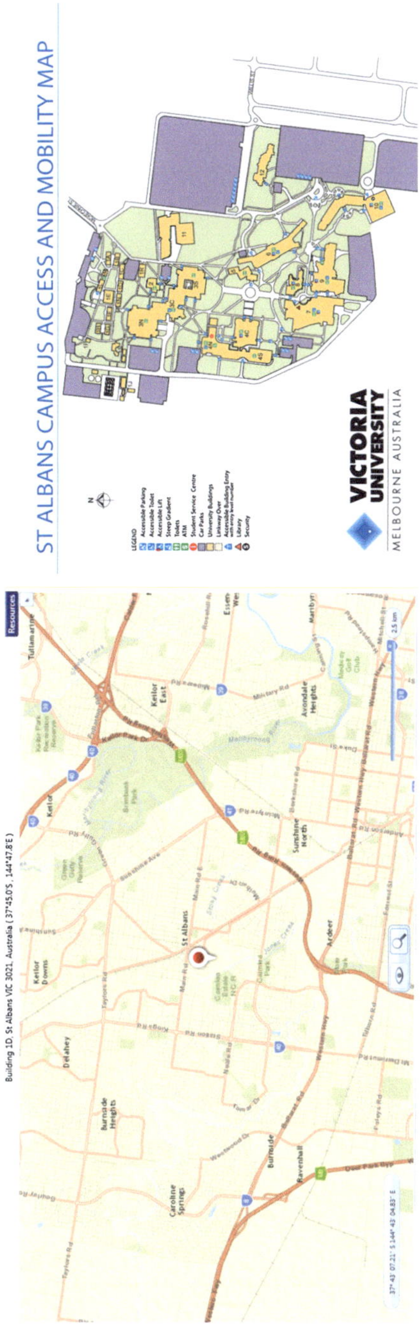

Fig. 1 Victoria university St. Albans location map from Google and campus access map

$$ACS = (Cost\ of\ the\ Project \times Discount\ rate) / \left(1 - (1 + Discount\ rate)^{-Project\ lifetime}\right) \quad (2)$$

c. Levelised Cost of Energy (COE): It is the average cost of useful electrical energy produced by the system. To calculate the levelised cost of energy, HOMER divides the annualized cost of producing electricity (the total annualized cost minus the cost of serving the thermal load) by the total electric load served, using the following equation:

$$COE = (Total\ annual\ electricity\ production) / (Load\ Served\ by\ the\ system) \quad (3)$$

d. Renewable Energy penetration (REP): It is the amount of renewable energy that serves the load annually

$$REP = (Power\ produced\ from\ renewable\ energy) / (Total\ electrical\ load\ served) \quad (4)$$

To design and optimize the RES into the grid for this study it is necessary to identify the sensitive variables along with evaluating their electricity load profile, solar irradiation and wind energy which are introduced in this section.

2.1 Solar Radiation Data

The solar radiation data has been analyzed from the National Renewable Energy Laboratory (NREL) data for St. Albans, Melbourne. This data is used to design the RES to integrate into a micro-grid to meet the load demand. Figure 2 shows the average solar radiation at the given place is 4.13 kWh/m2/day. Clearness index for the same location was used to design the micro-grid setup using HOMER.

2.2 Wind Resource

The wind resource data has been analyzed from NASA Surface meteorology for the desired location which provides monthly averaged values of wind speed at 50 m above the Earth's surface over a period of 10 years (July 1983–June 1993). Figure 3 shows the wind distribution at the desired location with an average wind speed of 4.53 m/s.

However, since the data seem to have been collected till June 1993, to understand the wind speed over recent years was also considered from Bureau of Meteorology, Australian Government. The site details closest to the university were found out to

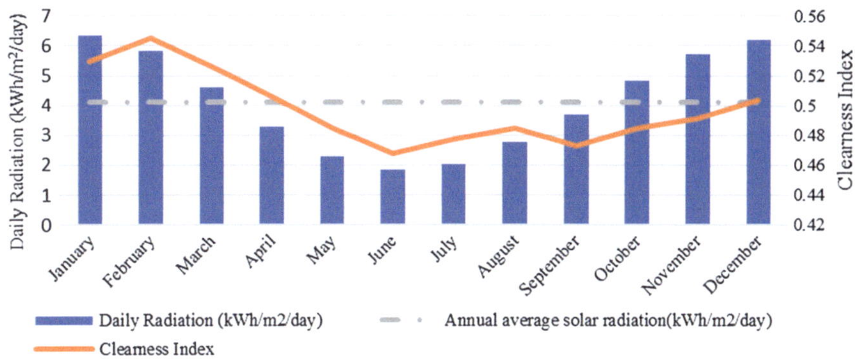

Fig. 2 Daily solar radiation and clearness index for the desired location

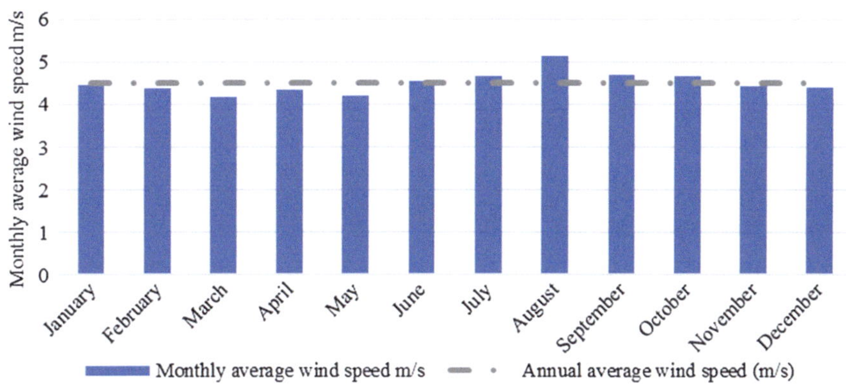

Fig. 3 Average wind speed for a year at desired location

be Melbourne airport (lat 37.67 °S, long 144.83 °E) and the mean 9 am wind speed statistics from 1970 to 2010 was 5.28 m/s. It is noted that the data measured was at an elevation of 113 m.

To use the above information in the simulation the wind speed was evaluated for 50 m (similar to NASA surface meteorology data) using Power law of wind profile given by Eq. (5).

$$(u/u_r = (z/z_r)^{\alpha}) \tag{5}$$

where u is the wind speed at a height z and u_r is the known wind speed at a reference height. From Eq. (5), the wind speed at a height of 50 m using the data measured from Bureau of Meteorology, is measured as 4.708 m/s with the power law exponent factor (α) to be 0.14.

Table 1 Sensitive variables used as boundary conditions in simulation

Inflation period (%)	2.5
	3.5
	5
Discount rate (%)	6.7
	3.5
	8
Lifetime of the project (years)	15
	25
Feed in tariffs	\$0.05/kWh
	\$0.03/kWh
	\$0.1/kWh
Electricity price	\$0.226512/kWh
	\$0.5/kWh

2.3 Electrical Load Analysis

The Electric power consumption of the university was studied using their electricity bill procured for one year. The average electricity consumption is 11091.27 kWh/d. A few variables reflect on the economics of the system, they are: inflation period, discount rate, lifetime of the project, feed in tariffs, electricity price. These were considered as the sensitive variables or the boundary conditions in the analysis and their values are shown in Table 1.

3 HOMER Simulation Model

The simulated model shown in Fig. 4 considers integration of Solar cells or PV and wind turbine into the grid. Wind energy and solar energy complement each other as distributed energy resources in the micro-grid. The fact of their energy production benefits and drawbacks has resulted in studying such renewable energy systems penetration into the grids. The intent to use a grid supported system instead of battery is its resilience and the fact that the presence of battery would escalate the cost of the setup which is already high due to the presence of wind turbine. Supplementing the above criteria, excess of power generation from this RES can be fed into the grid to acquire an additional profit in the form of energy sell back through Feed-in Tariffs (FiTs). The presence of converter in the system is to converts DC source output from PV to AC.

For the HOMER simulation, the size of the PV and converters was scaled for a definite range of numbers whilst the number of wind turbines to be integrated was varied between 1 and 10 and the details are provided in Table 2.

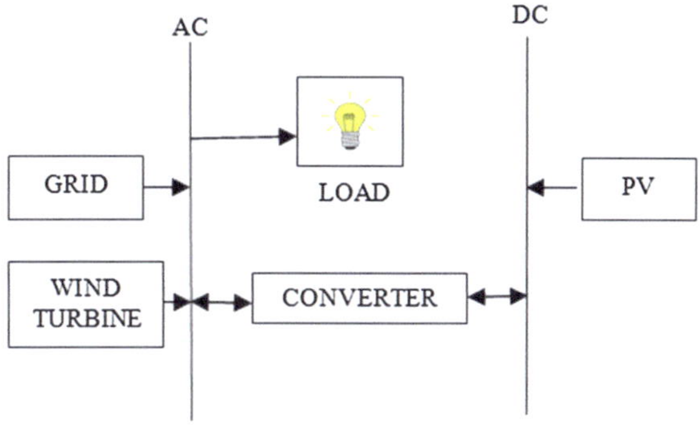

Fig. 4 Schematic diagram of the micro-grid considered

Table 2 Component details considered in the analysis

Component	Size	Details	Capital cost ($)	Operational and maintenance cost ($/year)
PV (1 kW)	$0 \leq 10^7$	Generic flat plate	680	10
Converter	$0 \leq 10^7$	Generic system converter	240	–
Generic 1.5 MW wind turbine (G1500)	1.5 MW of Quantity 1–5	Rated capacity 1500 kW, hub height 80 m	3,900,000	39,000

4 Results and Discussion

The setup for simulation considers PV and Wind turbines as RES. HOMER simulates a set of values having least NPC, considering the sensitive variables, and optimizing the size of the system. However, when the current Discount rate of 6.7% and current inflation rate of 3.5% and a sell back of $0.03/kWh was considered [23, 24]. HOMER optimized the size of the RES having least NPC, the results are shown in Table 3.

The architecture of the above model considered are about 2400–3200 kW of PV. A single 1.5 MW wind turbine scaled for different wind speeds integrated into a grid through converters ranging from 1400 to 1600 kW. The lifetime for the project and turbine lifetime considered are 15 and 25 years. The smallest architecture for the RES is about 2.4 MW PV and one 1.5 MW wind turbine connected to the grid through converter of about 1.4 MW with a wind speeds 5.3 m/s and project lifetime of 15 years and turbine lifetime of 15 and 25 years.

The above architecture of the RES from Table 3 has NPC ranging between $7 M and $12 M. The renewable energy penetration for these systems with an average of 83% is tabulated in Table 4. The unmet load from renewable energy of less than 20% is bought from the grid, while the excess of renewable energy being sold using the RES for a year converts to a profit or revenue.

Table 3 Architecture details of optimized model by homer

Architecture						
Project lifetime (years)	1.5 MW wind turbine lifetime (years)	Wind speed scaled average (m/s)	PV (kW)	1.5 MW wind turbine	Grid (kW)	Converter (kW)
15	25	5.277778	2441.406	1	999999	1424.154
15	25	4.708428	2644.857	1	999999	1424.154
15	15	5.277778	2441.406	1	999999	1424.154
15	15	4.708428	2644.857	1	999999	1424.154
25	25	5.277778	2644.857	1	999999	1424.154
25	25	4.708428	3255.208	1	999999	1627.604
25	15	5.277778	2644.857	1	999999	1424.154
25	15	4.708428	3255.208	1	999999	1627.604

Table 4 Energy and economics details of the optimized model

Energy and economics						
COE ($)	NPC ($)	Operating cost ($)	Initial capital ($)	Renewable energy fraction (%)	Energy purchased in kWh (percent-age)	Excess energy sold in kWh (percent-age)
0.087219	7016055	93945.19	5901953	84.94	1021381 (14.1%)	2734872 (40.3%)
0.101916	7651731	135881.8	6040300	81.82	1150687 (16.6%)	2282596 (36.1%)
0.099501	8004071	177258.3	5901953	84.942	1021381 (14.1%)	2734872 (40.3%)
0.115076	8639745	219194.9	6040300	81.82	1150687 (16.6%)	2282596 (36.1%)
0.081688	9729412	214030.5	6040300	85.62	993156 (13.2%)	2861718 (41.4%)
0.090606	1.07E+07	241373	6504167	84.32	1070182 (14%)	2780436 (40.7%)
0.097329	1.16E+07	322105.6	6040300	85.62	993156 (13.2%)	2861718 (41.4%)
0.106432	1.25E+07	349448	6504167	84.32	1070182 (14%)	2780436 (40.7%)

Table 5 Annual energy production details of the micro-grid sources considered

Energy production	kWh/yr	Percentage
Generic flat plate PV	3,537,357	51.08
Generic 1.5 MW wind turbine	2,236,621	32.3
Grid purchases	1,150,687	16.62
Total	6,924,664	100

Comparing the results of Tables 3 and 4, it is observed that there are two architectures of RES with a project lifetime of 15 years and wind speed 4.7 m/s, size of PV and converter is 2645 and 1424 kW respectively. However, the lifetime of 1.5 MW generic wind turbine are 15 and 25 years. These two architectures of RES have about 82% of renewable energy fraction, energy purchased from the grid and energy sold to the grid are 1150687 and 2282596 kWh respectively. The architecture of the system with the 1.5 MW wind turbine of 15 years have larger NPC, COE and smaller renewable energy penetration of about 81.8% compared to the system discussed earlier. This larger value of NPC and COE is due to the performance of 1.5 MW wind turbine for the lower wind velocity of about 4.7 m/s. It is also observed that for RES consisting of 1.5 MW wind turbine performing at velocity of 4.7 m/s result in smaller energy purchased or sold compared to the system operating with wind velocity of 5.3 m/s.

When monthly average electric production is considered for the above discussed RES, PV and Wind turbine contributed the major share of energy to reach the load demand as shown in Fig. 5. The maximum energy produced by the RES are during the months when solar energy radiation and wind energy are at their maximum.

Table 5 summarizes the annual energy production details of the micro-grid sources considered. 51% of energy contribution is by PV and 32% of energy production is from Generic 1.5 MW wind turbine. Total Grid purchase of about 17% is noted. The contribution of grid energy is mainly when the PV and wind turbine is not able to meet the load requirement when there is not enough sunlight or wind.

Figure 6 discusses the toxic gas emissions of the winning system. The data shows it illustrates the net toxic gas of carbon dioxide being maximum compared to Sulphur dioxide and Nitrogen dioxide.

5 Conclusion

In this project, we considered integrating RES like generic flat plate solar PV, wind turbine optimized according to the sensitive values and HOMER presented a list of values according to the least NPC. However, with the present condition of discount rate, sell back price for 15 years considered, and NPC of $8.64 M resulted with 82% of renewable energy penetration. The negative emission of Carbon dioxide explains the fact that the energy sold is greater than the energy purchased through the grid.

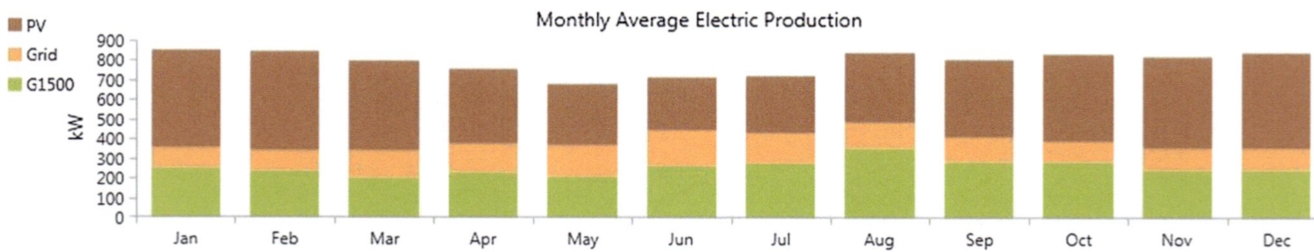

Fig. 5 Monthly average electric production of the desired model

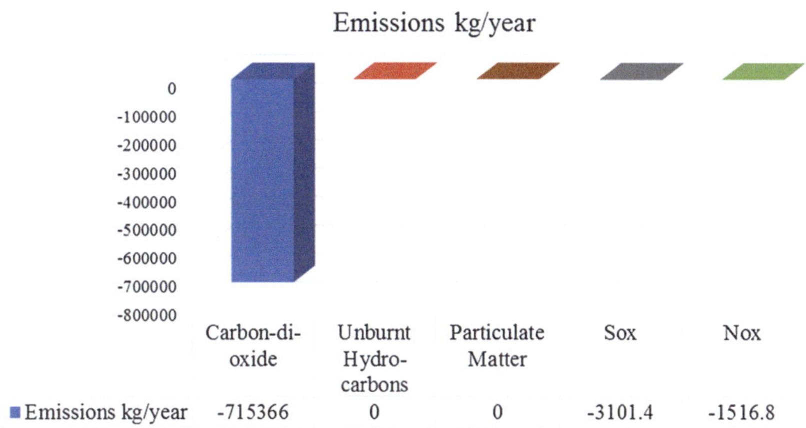

Fig. 6 Toxic gas emissions of the model considered

This system proves to be environmental friendly when the toxic gas emissions are considered. Further considerations on Small-scale Technology Certificates (STC) and other Government aided subsidies along with the solar energy installation costs are not considered in our current studies [25]. However, it could lead to future research work.

Acknowledgements The authors wish to acknowledge Anil Chaudhary from Greenova Solutions Pty Ltd, St Albans, VIC 3021 for helping us in providing the current market prices of the RES used in our study.

References

1. P. Moriarty, D. Honnery, What is the global potential for renewable energy? Renew. Sustain. Energy Rev. **16**, 244–252 (2012)
2. C.E. Council, Clean Energy Australia Report 2015. Available: https://www.cleanenergycouncil.org.au/policy-advocacy/reports/clean-energy-australia-report.html. Accessed 11 Oct 2016
3. T. Trainer, Can Australia run on renewable energy? The negative case. Energy Policy **50**, 306–314 (2012)
4. T. Trainer, Can renewables etc. solve the greenhouse problem? The negative case. Energy Policy **38**, 4107–4114 (2010)
5. R. Velo, L. Osorio, M.D. Fernández, M.R. Rodríguez, An economic analysis of a stand-alone and grid-connected cattle farm. Renew. Sustain. Energy Rev. **39**, 883–890 (2014)
6. L. Zhang, N. Gari, L.V. Hmurcik, Energy management in a microgrid with distributed energy resources. Energy Convers. Manag. **78**, 297–305 (2014)
7. X. Guan, Z. Xu, Q.S. Jia, Energy-efficient buildings facilitated by microgrid. IEEE Trans. Smart Grid **1**, 243–252 (2010)
8. G.J. Dalton, D.A. Lockington, T.E. Baldock, Feasibility analysis of renewable energy supply options for a grid-connected large hotel. Renew. Energy **34**, 955–964 (2009)

9. D.P. Kaundinya, P. Balachandra, N. Ravindranath, Grid-connected versus stand-alone energy systems for decentralized power—a review of literature. Renew. Sustain. Energy Rev. **13**, 2041–2050 (2009)
10. E. Mashhour, S.M. Moghaddas-Tafreshi, Integration of distributed energy resources into low voltage grid: a market-based multiperiod optimization model. Electr. Power Syst. Res. **80**, 473–480 (2010)
11. S.R. Pradhan, P.P. Bhuyan, S.K. Sahoo, G.S. Prasad, Design of standalone hybrid biomass & PV system of an off-grid house in a remote area. Int. J. Eng. Res. Appl. **3**, 433–437 (2013)
12. L. Olatomiwa, S. Mekhilef, A.S.N. Huda, O.S. Ohunakin, Economic evaluation of hybrid energy systems for rural electrification in six geo-political zones of Nigeria. Renew. Energy **83**, 435–446 (2015)
13. A.B. Kanase-Patil, R.P. Saini, M.P. Sharma, Integrated renewable energy systems for off grid rural electrification of remote area. Renew. Energy **35**, 1342–1349 (2010)
14. S. Wencong, Y. Zhiyong, C. Mo-Yuen, Microgrid planning and operation: solar energy and wind energy, in *IEEE PES General Meeting*, 2010, pp. 1–7
15. H. Dagdougui, R. Minciardi, A. Ouammi, M. Robba, R. Sacile, Modeling and optimization of a hybrid system for the energy supply of a "Green" building. Energy Convers. Manag. **64**, 351–363 (2012)
16. B.U. Kansara, B.R. Parekh, Modelling and simulation of distributed generation system using HOMER software, in *2011 International Conference on Recent Advancements in Electrical, Electronics and Control Engineering (ICONRAEeCE)*, 2011, pp. 328–332
17. S. Mizani, A. Yazdani, Optimal design and operation of a grid-connected microgrid, in *Electrical Power & Energy Conference (EPEC)*, IEEE, 2009, pp. 1–6
18. H. Rui, S.H. Low, U. Topcu, K.M. Chandy, C.R. Clarke, Optimal design of hybrid energy system with PV/wind turbine/storage: a case study, in *IEEE International Conference on Smart Grid Communications (SmartGridComm)*, pp. 511–516, 2011
19. B. Zhao, X. Zhang, P. Li, K. Wang, M. Xue, C. Wang, Optimal sizing, operating strategy and operational experience of a stand-alone microgrid on Dongfushan Island. Appl. Energy **113**, 1656–1666 (2014)
20. C. Marnay, G. Venkataramanan, M. Stadler, A.S. Siddiqui, R. Firestone, B. Chandran, Optimal technology selection and operation of commercial-building microgrids. IEEE Trans. Power Syst. **23**, 975–982 (2008)
21. S. Sinha, S.S. Chandel, Review of software tools for hybrid renewable energy systems. Renew. Sustain. Energy Rev. **32**, 192–205 (2014)
22. Homer Energy. Available: http://www.homerenergy.com/software.html. Accessed 10 Oct 2016
23. W. Australia, Business loans. Available: http://www.westpac.com.au/business-banking/business-loans/business-loans-interest-rate/, Accessed 12 Oct 2016
24. Australia Inflation Rate. Available: http://www.tradingeconomics.com/australia/inflation-cpi. Accessed 13 Oct 2016
25. I. MacGill, Electricity market design for facilitating the integration of wind energy: experience and prospects with the Australian national electricity market. Energy Policy **38**, 3180–3191 (2010)

Modelling and Simulation of Solar Cell Under Variable Irradiance and Load Demand

Payel Ghosh and Palash Kumar Kundu

1 Introduction

The breakneck depletion of fossil fuels taken together with the overloading of the atmosphere (with global warming emissions) due to human activity has shifted the focus towards the exploration of more abundant and benign energy resources over the past few decades. The tried-and-true technique of energy production from renewable energy resources (viz. the wind, solar, geothermal, hydroelectric, and biomass) is not only more sustainable but requires very less maintenance, causes considerably less noise pollution, effectively less or no production of the greenhouse or net carbon emissions as compared to traditional generators and hence render minimal impact on the environment.

The life-giving Sun is a cornucopia of energy which is harvested by photovoltaic and solar thermal technologies to produce electricity. A number of solar cells (the fundamental block of PV systems) are assembled, wired and sealed together in an environmentally protective laminate to form PV Modules. In order to meet the power requirements in terms of voltage and current one or more photovoltaic modules are connected in series and parallel (or a combination of both) to form a PV array—in parallel to increase current and in series to produce a higher voltage [1]. The power output of PV system varies from kilowatt range in residential applications increasing to the megawatt range, in utilities. Domestic installation of PV array is typically done on the rooftop where partial shading of the cells from neighboring structures or trees is often ineludible [2]. Hence the sum of the individually rated power of each module is, however, more than the total power in such an array [3]. Earlier studies

P. Ghosh (✉)
Department of Electrical Engineering, Meghnad Saha Institute of Technology, Kolkata, India
e-mail: payel4ever@gmail.com

P. K. Kundu
Department of Electrical Engineering, Jadavpur University, Kolkata, India
e-mail: palashm.kushi@gmail.com

S. Chattopadhyay et al. (eds.), *Modelling and Simulation in Science, Technology and Engineering Mathematics*, Advances in Intelligent Systems and Computing 749,
https://doi.org/10.1007/978-3-319-74808-5_21

assume that this decrease in the array output is proportional to the shaded area i.e. reduction in solar irradiance causing the solar cells being unevenly illuminated, thus introducing the concept of shading factor. This concept may be true for a single cell, but the decrease in power at the module or array level is often far from linearity with the shaded portion [2]. The reverse bias of the shaded cells makes it act as a load thus draining power from other fully illuminated cells [1]. Apart from reduced output, if all the cells are not equally illuminated, hot spot problem may arise causing the system to be irreversibly damaged. In order to maintain continuity of supply and to meet the power demand even in shading condition switching of cells is necessary. Switching of cells uses the idea of two or more cells operating in parallel or series mode (as desired) whenever due to decreased ambient irradiance the output from a respective cell is not enough for the load. This article aims to show the switching of cells at various irradiance levels and also under various load demand using SIMULINK models and embedded MATLAB function.

2 Photovoltaic System

2.1 *Single Diode Model*

A p-n junction when illuminated acts as a solar cell. A solar cell is basically a current source connected in parallel with a diode. However, the model changes taking into account the non-ideality factors—especially the parasitic series and shunt resistances. It generates current when illuminated. However, it acts as a diode i.e. the solar cell is an inactive device resulting in zero voltage and current during darkness. This section briefly describes the single diode model of a solar cell taking into account the effects of ambient irradiance and temperature and the associated equations are:

The Shockley diode equation can be stated as:

$$I_D = I_0\left(e^{\frac{qV_D}{nKT}} - 1\right) \tag{1}$$

where I_D is the diode current (in Ampere), I_0 is the reverse saturation bias current (or scale current in Ampere) of the diode corresponding to working temperature T (in Kelvin). I_0 is not constant for any given device but varies widely with T. For every 10 °C temperature rise, I_0 doubles itself. q is the charge of an electron equal to 1.602×10^{-19} C, V_D is the voltage across the diode (in Volts), n is the ideality factor (also termed as the quality factor or sometimes emission coefficient) of the diode typically varying between 1 and 2, however, it can be more based on the fabrication process and semiconductor material. It is set to 1 for an ideal diode. K is the Boltzmann Constant equal to 1.38×10^{-23} J/K.

The diode equation can also be expressed as:

$$I_D = I_0\left(e^{\frac{V_D}{nV_T}} - 1\right) \tag{2}$$

where V_T is the thermal voltage equal to Boltzmann constant times temperature of p-n junction whole divided by the elementary charge of an electron and is denoted as

$$V_T = \frac{KT}{q}$$

It is approximately 25.85 mV at 300 K.

Under ideal conditions, the output current (in Ampere), $I = I_L - I_D$ neglecting the parasitic series and shunt resistances. I_L is the photon current (in Ampere) corresponding to a particular irradiance level and given temperature, varying directly with irradiance level. Thus,

$$I = I_L - I_0\left(e^{\frac{qV_D}{nKT}} - 1\right) \quad (3)$$

The series resistance R_S (in Ohms) is the equivalent resistance in contacts, metal grids as well as the resistance encountered (internal losses) by the current flow in the p-n layers of the semiconductor material. Shunt resistance R_{SH} (in Ohms) corresponds to the leakage current of the p-n junction. Hence the expression for output current, I, corresponds to:

With series resistance R_S,

$$I = I_L - I_0\left(e^{\frac{q(V+IR_S)}{nKT}} - 1\right) \quad (4)$$

With series resistance R_S and shunt resistance R_{SH} (Fig. 1)

$$I = I_L - I_D - I_{R_{SH}} \quad (5)$$

$$I = I_L - I_0\left(e^{\frac{q(V+IR_S)}{nKT}} - 1\right) - \left(\frac{V + IR_S}{R_{SH}}\right) \quad (6)$$

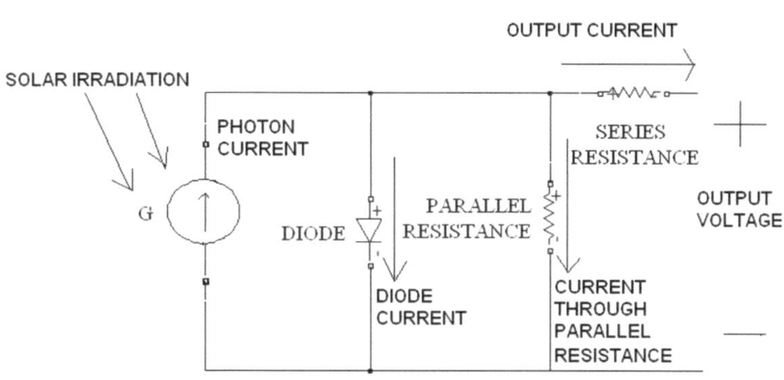

Fig. 1 Model of a solar cell with equivalent series resistance and shunt resistance

However $R_S = 0$ in an ideal solar cell [4]. In this paper, R_{SH} is neglected hence considering a moderately complex model with series resistance only [5]. The other equations involved can be listed as:

$$I_L(T_1) = I_{SC}(T_{1,NOM})\frac{G}{G_{NOM}} \tag{7}$$

where G_{NOM} and $T_{1,NOM}$ are the values of suns and temperature at standard test condition (i.e. $G_{NOM} = 1000\ \text{W/m}^2$, $T_{1,NOM} = 25\ °\text{C}$).

$$I_L = I_L(T_1) + K_0(T - T_1) \tag{8}$$

$$K_0 = \frac{I_{SC}(T_2) - I_{SC}(T_1)}{(T_2 - T_1)} \tag{9}$$

$$I_0(T_1) = \frac{I_{SC}(T_1)}{\left(e^{\frac{qV_{OC}(T_1)}{nKT_1}} - 1\right)} \tag{10}$$

$$I_0 = I_0(T_1) \times \left(\frac{T}{T_1}\right)^{\frac{3}{n}} e^{\frac{qV_g(T_1)}{nK\left(\frac{1}{T} - \frac{1}{T_1}\right)}} \tag{11}$$

$$X_V = I_0(T_1)\frac{q}{nKT_1}e^{\frac{qV_{OC}(T_1)}{nKT_1}} - \frac{1}{X_V} \tag{12}$$

$$R_S = -\frac{dV}{dI_{V_{OC}}} - \frac{1}{X_V} \tag{13}$$

[6] where T_1 is the normalized temperature (=25 °C STC) in Kelvin, V_{OC} is the open circuit voltage in Volts, G is the number of Suns in Watt/metre^2 (1 Sun = 1000 W/m^2), K_0 is current/temperature coefficient in Ampere/Kelvin [A/K], V_g is the voltage of the Crystalline Silicon ($V_g = 1.12$ and 1.75 eV for Amorphous Silicon) in Electron volt [eV], dV/dI_{Voc} is the dV/dI coefficient at V_{OC}.

The basic parameters characterizing the solar cell are:

(I) **Short circuit current (I_{SC})**: I_{SC} is the maximum value of current (roughly equal to the photon current for very small values of series parasitic resistance) of the solar cell under short circuit conditions i.e. zero voltage appearing across the terminals.

(II) **Open circuit voltage (V_{OC})**: V_{OC} is the maximum voltage under open circuit (zero current) condition. Neglecting the parasitic resistances,

$$I = I_L - I_0\left(e^{\frac{qV_D}{nKT}} - 1\right).$$

Under open circuit conditions when $I = 0$, $V_D = V_{OC}$ and

$$V_{OC} = \frac{nKT}{q}\ln\left(\frac{I_L}{I_0} + 1\right).$$

The equation clearly indicates that V_{OC} is widely controlled by the dark saturation current.

(III) **Maximum power point** (MPP): I_{SC} and V_{OC} do not occur simultaneously and hence maximum output power, P_{MAX} that can be delivered to the connected load by the PV cell is not equal to I_{SC} X V_{OC} rather P_{MAX} is the product of I_{MAX} and V_{MAX} (Current and Voltage corresponding to Maximum Power Point), which are much less than I_{SC} and V_{OC} respectively.

(IV) **Efficiency of PV cell (η)**: Efficiency is the ratio of output of PV cell i.e. the maximum current times the maximum voltage (at MPP) to the input light power and is denoted as

$$\text{Efficiency} = \frac{P_{OUT}}{P_{IN}} = \frac{P_{MAX}}{G} = \frac{I_{MAX}V_{MAX}}{GA} = \frac{I_{MAX}V_{MAX}}{1000A}$$

where G is the ambient irradiation taken as 1000 W/m^2 at Standard test conditions and A corresponds to exposed PV cell area. Efficiency ranges are: 6%-amorphous silicon-based solar cell to 42.8% with multiple cells: 14–19% for commercially available multi-crystalline solar cells and widely depends on several critical factors like temperature, irradiance, shading, snow, etc.

(V) **Fill Factor** (FF): The product of current and voltage corresponding to the maximum power point ($I_{MAX}V_{MAX}$) divided by the product of short circuit current, I_{SC} times the open circuit voltage, V_{OC} is termed as Fill Factor. The idea about the cell quality is conveyed by the fill factor which typically ranges between 0.7 and 0.8 for good cells.

$$FF = \frac{I_{MAX}V_{MAX}}{I_{SC}V_{OC}}.$$

2.2 *Solar Cell Module and Array Model*

(I) **Series**

In order to increase the module voltage, N solar cells are connected in series and the module output voltage is given by $V_{OUT} = V_1 + V_2 + V_3 + V_4 + \cdots + V_N$.N, the number of cells to be connected in series, is decided according to the voltage demand by the load. Some examples are shown of possible series combinations:

(A) *Similar Solar cells in Series*: Using the same three 2 V/1 A solar cells in series, the output voltage is 6 V (2+2+2) at the same rated current of 1A (Fig. 2a).

(B) *Solar cells in Series with different Voltage*: Three solar cells of different voltage rating are connected in series (2 V/1 A, 3 V/1 A, and 4 V/1 A) yielding the same amperage of 1 A but an augmented voltage of 9 V (2+3 + 4) (Fig. 2b).

(C) *Solar cells in Series with different Voltage and Current*: Three solar cells (2 V/1 A, 3 V/2 A, and 4 V/3 A) are connected in series resulting in a voltage jump of 9 V (2+3+4) but the current is restricted to the lowest rating of the module i.e. 1 A here (Fig. 2c).

(II) **Parallel**

Connecting N solar cells in parallel increases the output current of the module which is given by $I_{OUT} = I_1 + I_2 + I_3 + I_4 + \cdots + I_N$ N, the number of cells to be connected in parallel is decided according to the current demand by the load. Some examples are shown of possible parallel combinations:

(A) *Similar Solar cells in Parallel*: Using the same three 2 V/1 A solar cells in parallel, the output current is increased to 3 A (1+1+1) at the same rated voltage of 2 V (Fig. 3a).

(B) Solar cells in Parallel with different voltage and current: Three solar cells (3 V/1 A, 5 V/3 A, and 7 V/4 A) are connected in parallel resulting in an increase in current equal to 8 A (1+3+4) but the module voltage is restricted to the lowest rated i.e. 3 V here (Fig. 3b).

In order to increase both module voltage and the current series-parallel combination is preferred. In the photovoltaic module with N_P cells branches in parallel and N_S cells in series, total shunt resistance in Ohm is equal to,

$$R_{SH,MODULE} = \left(\frac{N_P}{N_S}\right) R_{SH,CELL}$$

where $R_{SH,CELL}$ corresponds to shunt resistance in one photovoltaic cell, Ohm.

Total series resistance is Ohm is given by,

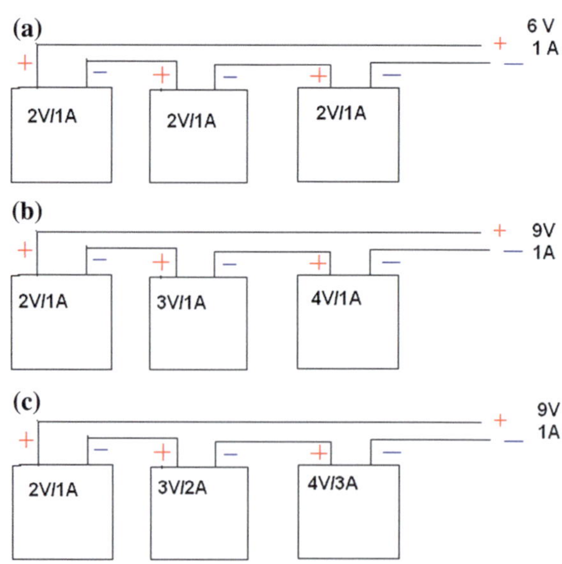

Fig. 2 **a** Similar solar cells in series **b** Solar cells in series with different voltage **c** Solar cells in series with different voltage and current

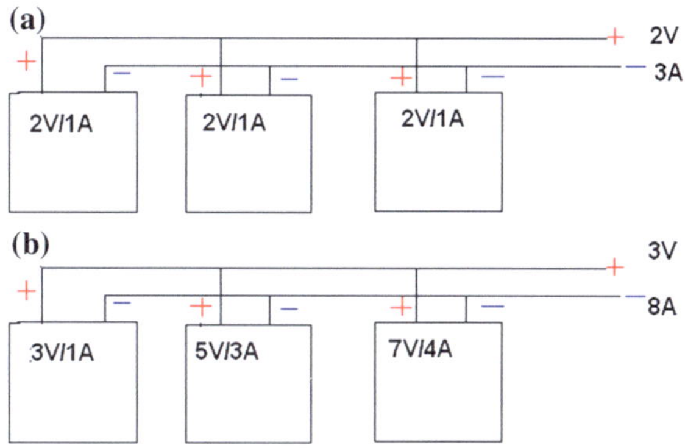

Fig. 3 **a** Similar solar cells in parallel **b** Solar cells in parallel with different voltage and current

$$R_{S,MODULE} = \left(\frac{N_S}{N_P}\right) R_{S,CELL}$$

where $R_{S,CELL}$ corresponds to series resistance in one photovoltaic cell, Ohm.

Therefore for a module with N_P and N_S, we will add $R_{SH,MODULE}$ and $R_{S,MODULE}$ instead of R_S and R_{SH} in Eq. (6) of Single Diode Model section. In order to find the specifications of the module the equations already stated in the previous section will be applied and finally, the characteristic curves (I-V and P-V) are obtained according to values of ambient temperature and irradiance [7, 8].

Total short circuit current in the module (in Ampere) is

$$I_{SC,MODULE} = (N_P) I_{SC,CELL}$$

where $I_{SC,CELL}$ is the short circuit current of one photovoltaic cell, in Ampere. The open circuit voltage of the photovoltaic module (in Volts) is

$$V_{OC,MODULE} = (N_S) V_{OC,CELL}$$

where $V_{OC,CELL}$ is the open circuit voltage of one photovoltaic cell, in Volts [9].

Modules in a PV system are typically connected to form arrays. With M_P parallel branches each with M_S modules in series, V^A is the applied voltage at the terminals of the array and the array current, I^A is denoted by

$$I^A = \sum_{i=0}^{M_P} I_i$$

where A correspond to the branches number. But $I^A = M_P I^M$ if it is assumed that the ambient irradiation is same on all the identical modules (Fig. 4).

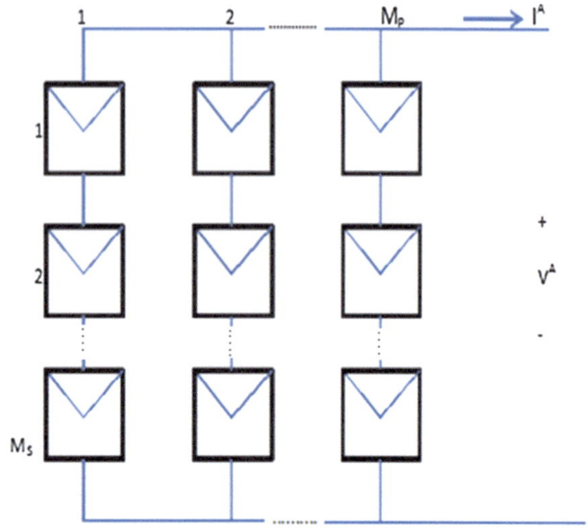

Fig. 4 Solar cell array with M_P parallel branches, with M_S modules in series in each branch

2.3 *Characteristic Curves of Solar Cell*

Solar cell I-V and P-V characteristic curves are the input-output analysis of the cell which helps in determining the cell output and solar efficiency. The I-V Curve is a plot of all possible values of output current corresponding to each voltage levels exhibiting an inverse relationship (i.e. the current decreases from a maximum value to zero as we sweep the voltage from zero to its maximum value). In any DC electrical circuit, Power (P) in Watts (W) = the Current (I) in Amperes (A) X the Voltage (V) in Volts (V). Thus the P-V curve is the measure of the output power (product of current and voltage from I-V curve) corresponding to respective voltage levels (Fig. 5).

2.4 *Effect of Ambient Irradiance and Temperature*

The variation of solar irradiation and temperature throughout the day results in different characteristic curves. At fixed temperature, with increasing solar irradiance the maximum power point varies as both the short-circuit current and the open-circuit voltage increase. I_{SC} exhibits a linear variation as more electron-hole pairs are formed but V_{OC} increases marginally with the increase in irradiance.

The rate of photon generation increases with the increase in temperature which in turn rapidly increases the reverse saturation current and thus the band gap is reduced. Although this leads to marginal changes in current, the voltage undergoes major changes (roughly around −0.35%/°C or −2.2 mV/°C). Thus temperature acts like a negative factor adversely affecting solar output. Thus with regards to both irradiance and temperature, it can be inferred that temperatures between 26 and 30 °C coupled

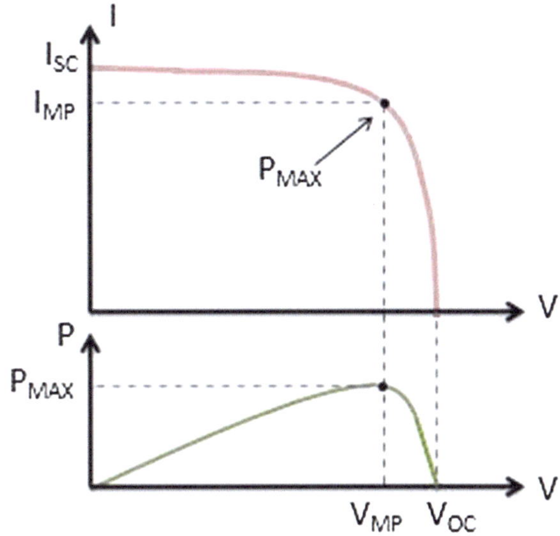

Fig. 5 Ideal IV and PV curve

with high irradiance are necessary for high panel output on sunny days with low temperature [10].

3 Results

The characteristic parameters of SUNPOWER A-300 solar cell are used as a reference (Table 1). The four ranges of irradiance used are obtained by dividing the maximum irradiance (approximate) of Kolkata equal to 0.27 Suns into equal ranges and respective maximum power output are evaluated in each case for a single cell and also when switching takes place.

This paper widely explains the switching of cells due to varying irradiance and load demand. The same is implemented by MATLAB script and is used as an embedded MATLAB function in SIMULINK. The logic as per which the switching of cells takes is as follows:

(I) For an irradiance of $a \leq 0.0675$ Suns, 4 solar cells will be operating in parallel.
(II) For an irradiance of value in the range, 0.0675 Suns < a ≤ 0.135 Suns, if user defined load demand is less than or equal to 0.5533 W, 3 solar cells will be operating in parallel. Whereas if user defined load demand is greater than 0.5533 W, 4 solar cells will be operating in parallel.
(III) For an irradiance of 0.135 Suns < a ≤ 0.2025 Suns, if user defined load demand is less than or equal to 0.7554 W, 2 solar cells will be operating in parallel. If 0.7554 W < user defined load demand ≤ 1.331 W, 3 solar cells will be operating

Table 1 Typical electrical performance of SUNPOWER A-300 solar cell [a](mono crystalline silicon)

Parameter	Symbol	Value
Open circuit voltage	V_{OC}	0.665 V
Short circuit current	I_{SC}	5.75 A
Maximum power voltage	V_{MAX}	0.560 V
Maximum power current	I_{MAX}	5.35 A
Rated power	P_{RATED}	3.0 W
Efficiency	η	20.0% minimum
Temperature coefficient of voltage	−1.9 mV/°C	
Temperature coefficient of power	−0.38%/°C	

[a]Data are given at STC: Illumination 1000 W/m^2, Temperature: 25 °C and spectrum of light AM 1.5 [11]

in parallel. When user defined load demand is greater than 1.331 W, 4 solar cells will be operating in parallel.

(IV) For an irradiance of a > 0.2025 Suns, if user defined load demand is less than or equal to 0.5602 W, only 1 cell will operate. If 0.5602 W < user defined load demand ≤1.1410 W, 2 solar cells will be operating in parallel. If 1.1410 W < user defined load demand ≤ 1.7115 W, 3 solar cells will be operating in parallel. When user defined load demand is greater than 1.7115 W, 4 solar cells will be operating in parallel.

The SIMULINK model (Fig. 6a) operates on the above-stated logic which takes irradiance and load demand as input (shown in the subsystem Fig. 6b, c). The individual scopes in the

SIMULINK model give the respective I-V and P-V curves under varying irradiance and load demand. In the plots, the blue color is used for single cell operation and red for multiple cell operation.

3.1 Analysis with Various Irradiance and Load Demand Values

Some of the possible cases are:

(a) Input Irradiance: 0.0675 Suns, User Defined Load Demand: 0.50 W, Results: Number of cells operating in parallel: 4, 1 cell: P_{MAX} = 0.1842 W and I_{SC} = 0.4056 A. 4 cells: P_{MAX} = 0.7366 W and I_{SC} = 1.6225 A (Fig. 7).

(b) Input Irradiance: 0.05 Suns, User Defined Load Demand: 0.7 W, Results: Number of cells operating in parallel: 4, 1 cell: P_{MAX} = 0.1360 W and I_{SC} = 0.3050 A.

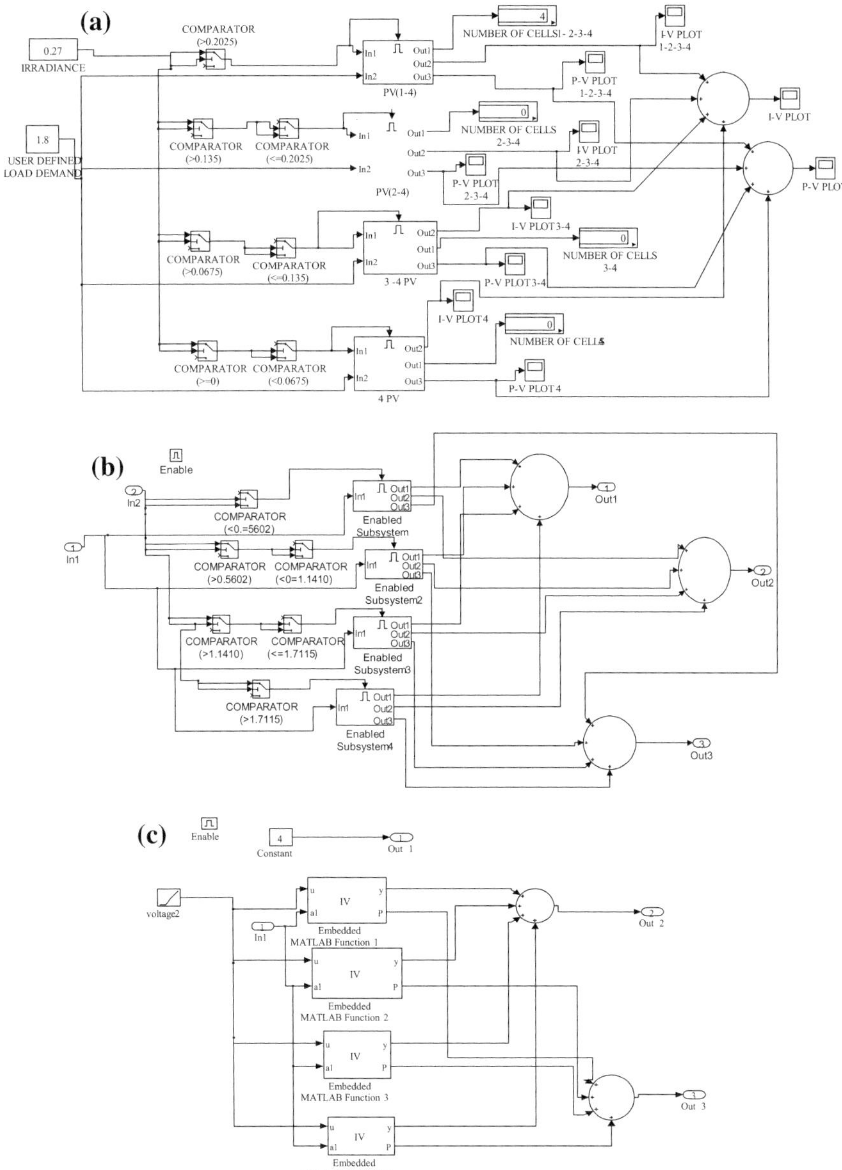

Fig. 6 **a** SIMULINK model. **b** Subsystem model. **c** Subsystem model

4 cells: $P_{MAX} = 0.544$ W and $I_{SC} = 1.22$ A. Here even the parallel operation of 4 cells fails to supply the load demand and calls for more cells to be connected in parallel.

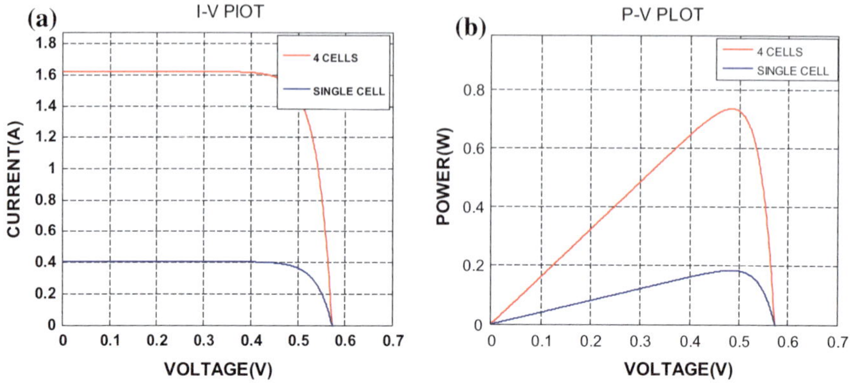

Fig. 7 I-V and P-V curves under irradiance 0.0675 Suns and load demand 0.50 W

(c) Input Irradiance: 0.135 Suns, User Defined Load Demand: 0.55 W, Results: Number of cells operating in parallel: 3, 1 cell: $P_{MAX} = 0.3748$ W and $I_{SC} = 0.7937$ A. 3 cells: $P_{MAX} = 1.124$ W and $I_{SC} = 2.381$ A ((Fig. 8).
(d) Input Irradiance: 0.135 Suns, User Defined Load Demand: 0.7 W, Results: Number of cells operating in parallel: 4, 1 cell: $P_{MAX} = 0.3748$ W and $I_{SC} = 0.7937$ A. 4 cells: $P_{MAX} = 1.499$ W and $I_{SC} = 3.175$ A (Fig. 9).
(e) Input Irradiance: 0.140 Suns, User Defined Load Demand: 0.75 W, Results: Number of cells operating in parallel: 2, 1 cell: $P_{MAX} = 0.3892$ W and $I_{SC} = 0.8225$ A. 2 cells: $P_{MAX} = 0.7784$ W and $I_{SC} = 1.645$ A (Fig. 10).
(f) Input Irradiance: 0.2025 Suns, User Defined Load Demand: 1.3 W, Results: Number of cells operating in parallel: 3, 1 cell: $P_{MAX} = 0.5702$ W and $I_{SC} = 1.1819$ A. 3 cells: $P_{MAX} = 1.7106$ W and $I_{SC} = 3.545$ A (Fig. 11).

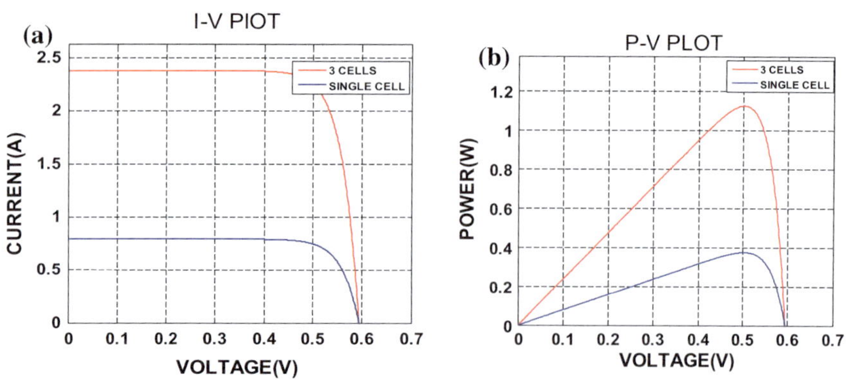

Fig. 8 I-V and P-V curves under irradiance 0.135 Suns and load demand 0.55 W

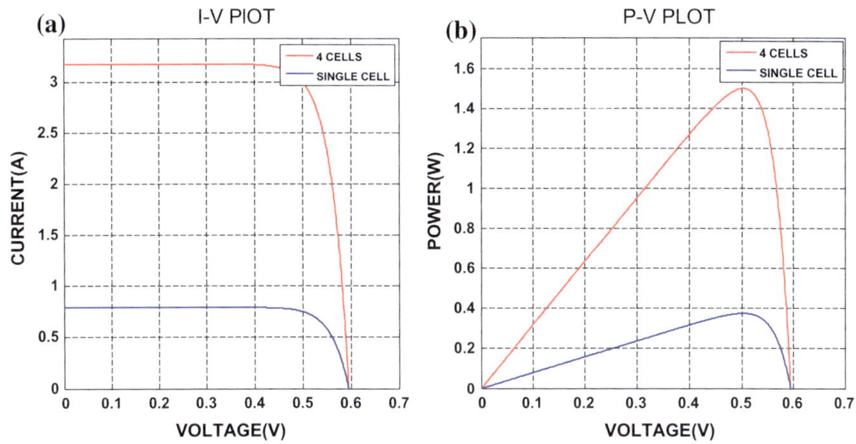

Fig. 9 I-V and P-V curves under irradiance 0.135 Suns and load demand 0.7 W

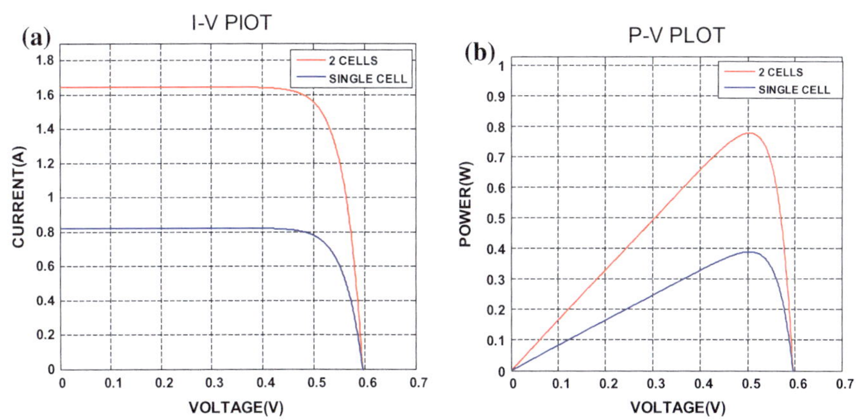

Fig. 10 I-V and P-V curves under irradiance 0.140 Suns and load demand 0.75 W

(g) Input Irradiance: 0.2025 Suns, User Defined Load Demand: 2 W, Results: Number of cells operating in parallel: 4, 1 cell: $P_{MAX} = 0.5702$ W and $I_{SC} = 1.1819$ A. 4 cells: $P_{MAX} = 2.2808$ W and $I_{SC} = 4.7276$ A (Fig. 12).

(h) Input Irradiance: 0.27 Suns, User Defined Load Demand: 0.5 W, Results: Number of cells operating in parallel: 1, 1 cell operates to meet the load demand of $P_{MAX} = 0.7682$ W and $I_{SC} = 1.57$ A.

(i) Input Irradiance: 0.27 Suns, User Defined Load Demand: 0.6 W, Results: Number of cells operating in parallel: 2, 1 cell: $P_{MAX} = 0.7682$ W and $I_{SC} = 1.57$ A. 2 cells: $P_{MAX} = 1.5364$ W and $I_{SC} = 3.14$ A.

(j) Input Irradiance: 0.27 Suns, User Defined Load Demand: 1.5 W, Results: Number of cells operating in parallel: 3, 1 cell: $P_{MAX} = 0.7682$ W and $I_{SC} = 1.57$ A. 3 cells: $P_{MAX} = 2.3046$ W and $I_{SC} = 4.71$ A.

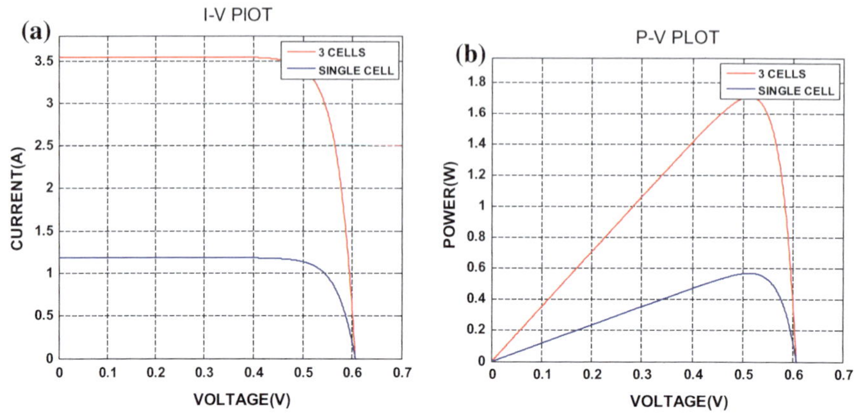

Fig. 11 I-V and P-V curves under irradiance 0.2025 Suns and load demand 1.3 W

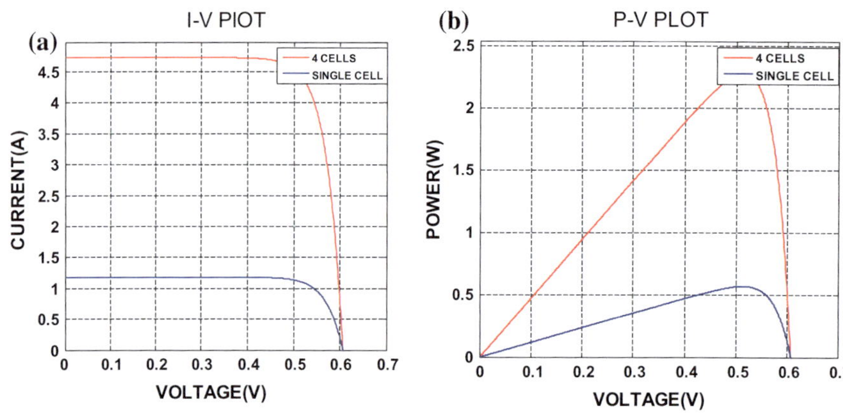

Fig. 12 I-V and P-V curves under irradiance 0.2025 Suns and load demand 2 W

(k) Input Irradiance: 0.27 Suns, User Defined Load Demand: 1.8 W, Results: Number of cells operating in parallel: 4, 1 cell: $P_{MAX} = 0.7682$ W and $I_{SC} = 1.57$ A. 4 cells: $P_{MAX} = 3.0782$ W and $I_{SC} = 6.28$ A.

4 Conclusion

Switching of cells according to irradiance and load demand is shown with the help of SIMULINK model which uses embedded MATLAB function. The I-V and P-V plots for some possible cases are shown and the logic of execution is clearly stated in the Results section. There is a restriction of maximum power output that can be delivered by connecting all cells in parallel and whenever it is less than the user defined load input, it calls for more number of parallel cells to be operated to meet the load demand.

Acknowledgements Authors would like to thank Jadavpur University for support to do this work.

References

1. B. Alsayid, S. Alsadi, J. Jallad, M. Dradi, Partial shading of PV system simulation with experimental results. Smart Grid Renew. Energy **4**(6), 429–435 (2013)
2. D. Sera, Baghzouz, Y, in *On the Impact of Partial Shading on PV Output Power*. Proceedings of RES'08 WSEAS Press (2008)
3. A. Chaudhary, S. Gupta, D. Pande, F. Mahfooz, G. Varshney, Effect of partial shading on characteristics of PV panel using Simscape. Int. J. Eng. Res. Appl. (IJERA) **5**(10), 85–89 (2015)
4. CIGRE TF38.01.10, in *Modeling New Forms of Generation and Storage*, Nov 2000
5. J.A. Gow, C.D. Manning, Development of a photovoltaic array model for use in power electronics simulation studies. IEE Proc. Electr. Power Appl. **146**(2), 193–200 (1999)
6. F.M. González-Longatt, in *Model of Photovoltaic Module in Matlab*™, 2DO Congreso Iberoamericano de estudiantes de ingeneria electrica, electronica y Computacion (II CIBELEC 2005)
7. C.-S. Tu, H.-L. Tsai, Y.-J. Su, in *Development of Generalized Photovoltaic Model Using MATLAB/SIMULINK*, in the World Congress on Engineering and Computer Science 2008, San Francisco, USA, 2008
8. M.G. Villalva, J.R. Gazoli, Comprehensive approach to modeling and simulation of photovoltaic arrays. IEEE Trans. Power Electron. **24**, 1198–1208 (2009)
9. A. Chaouachi, R.M. Kamel, K. Nagasaka, A novel multi-model neuro-fuzzy-based MPPT for three-phase grid-connected photovoltaic system. Sol. Energy **2010**(84), 2219–2229 (2010)
10. P. Arjyadhara, S.M. Ali, J. Chitralekha, Analysis of solar PV cell performance with changing irradiance and temperature. Int. J. Eng. Comput. Sci. **2**(1), 214–220 (2013)
11. SunPower Corporation. Document Number 70-0006 Rev 02, 2003

Power Management of Non-conventional Energy Sources Connected to Local Grid

Siddhartha Singh and Biswarup Basak

1 Introduction

Solar and wind have huge amount of potential for the energy requirement of the world. If 14% of earth's surface is installed with conventional onshore wind turbines of 80-m tower height, renewable power generated will be around 70 TW (five times of global power consumption). Another renewable resource is SP having incredible potential to meet the world energy demand only by installing solar panels with efficiency of 8% around 0.22% of earth's surface. Roughly 85,000 TW of SP is available on earth's surface. In order to utilise the SP and WP directly across the TPL, it should be connected to the GS which can fill the gap in TPL and generated Power. But without power management it is uneconomical, so the Power management of NCES connected to GS is necessary for proper operation.

2 Simple Block Diagram

The aim is to build up a power network for any island system or some small area network whose power can be managed. As the name suggests the power management of non-conventional energy sources, is an attempt to control the power of two means i.e. the wind generation and the solar generation. From Fig. 1 we can see that the three phase GS, WP generation and SP generation are connected in parallel to feed a common load.

S. Singh (✉) · B. Basak
Electrical Engineering, IIEST, Shibpur, Howrah 711103, India
e-mail: singhsiddhartha1811@gmail.com

B. Basak
e-mail: biswarup_basak@yahoo.com

S. Chattopadhyay et al. (eds.), *Modelling and Simulation in Science, Technology and Engineering Mathematics*, Advances in Intelligent Systems and Computing 749,
https://doi.org/10.1007/978-3-319-74808-5_22

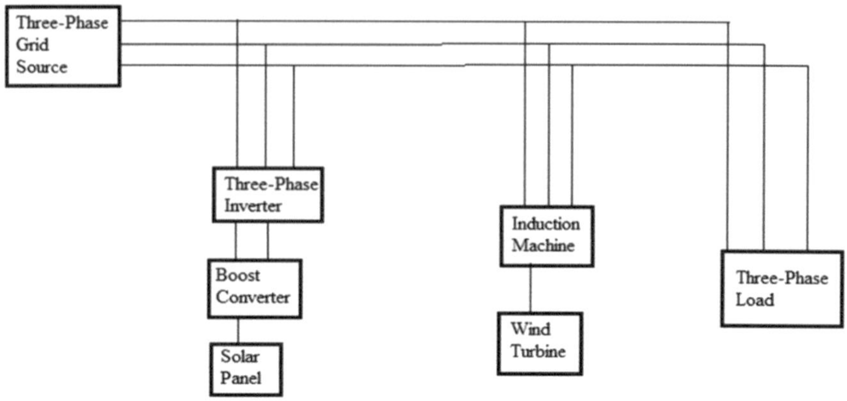

Fig. 1 Simple block diagram

Grid power can not be controlled directly so the WP and SP are tracked and controlled to produce its rated power. WP controlled by blade pitch angle and wind speed to generate rated electrical power with the help of induction machine and supplied to the grid. SP generation is depending on irradiance of sunlight. The output voltage of solar panel is boosted to higher suitable value using boost converter and then converted in ac with the help of three phase inverter so that it can be fed to the grid directly without any transformer. Whether if the WP and SP are not enough to supply the load than in the case, remaining load is being fed by the GS. So, indirectly the GS power supply dependency is also controlled and hence we are capable of reducing the consumption of power generation by means of conventional methods. By this approach the Renewable resources and non-renewable resources can be implied together to feed the same load.

3 Interconnected Overall Simulink Model

See Fig. 2.

3.1 Induction Machine Model

IM Simulink model is comprised of three blocks voltage transformation block, IM d-q equation block and current transformation block. Voltage transformation block is used to convert three phase a-b-c to d-q voltages, following by the IM block in d-q equations is developed and the output current is again transformed from d-q to a-b-c by Current transformation block. All the equations used are shown below.

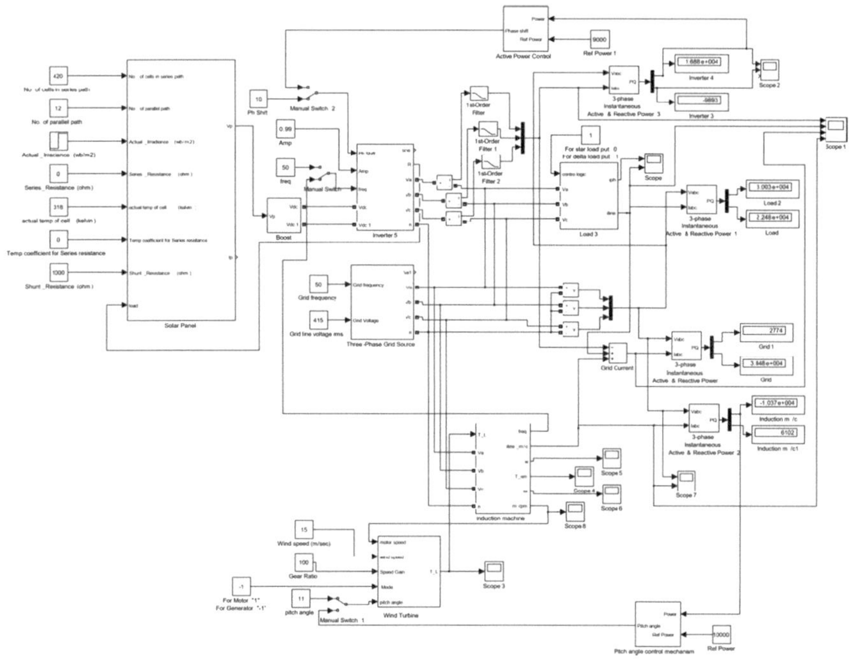

Fig. 2 Interconnected overall network

The equations used for voltage a-b-c to d-q transformations:

$$V_a = V_m * \sin(\omega t) \tag{1}$$

$$V_b = V_m * \sin\left(\omega t - 120^0\right) \tag{2}$$

$$V_c = V_m * \sin\left(\omega t + 120^0\right) \tag{3}$$

The parks transformation (a-b-c to d-q) is defined as follows:

$$V_d = \frac{2}{3}(V_a * \sin(\theta) + V_b * \sin(\theta - 120^\circ) + V_c * \sin(\theta + 120^\circ)) \tag{4}$$

$$V_q = \frac{2}{3}\left(V_a * \cos(\theta) + V_b * \cos(\theta - 120^0) + V_c * \cos(\theta + 120^0)\right) \tag{5}$$

The equations used for induction machine d-q model [1]:
Stator circuit equations:

$$V_{ds} = r_s * i_{ds} + p\lambda_{ds} - \omega_e \lambda_{qs} \tag{6}$$

$$V_{qs} = r_s * i_{qs} + p\lambda_{qs} + \omega_e \lambda_{ds} \tag{7}$$

where $p = d/dt$ *operator*.

Rotor circuit equations:

$$V'_{dr} = r'_r * i'_{dr} + p\lambda'_{dr} - (\omega_e - \omega_r)\lambda'_{qr} \tag{8}$$

$$V'_{qr} = r'_r * i'_{qr} + p\lambda'_{qr} + (\omega_e - \omega_r)\lambda'_{dr} \tag{9}$$

Flux linkage expressions:

$$\lambda_{ds} = L_{ls} * i_{ds} + L_m * (i_{ds} + i'_{dr}) \tag{10}$$

$$\lambda'_{dr} = L'_{lr} * i'_{dr} + L_m * (i_{ds} + i'_{dr}) \tag{11}$$

$$\lambda_{qs} = L_{ls} * i_{qs} + L_m * (i_{qs} + i'_{qr}) \tag{12}$$

$$\lambda'_{qr} = L'_{lr} * i'_{qr} + L_m * (i_{qs} + i'_{qr}) \tag{13}$$

$$\lambda_{dm} = L_m * (i_{ds} + i'_{dr}) \tag{14}$$

$$\lambda_{qm} = L_m * (i_{qs} + i'_{qr}) \tag{15}$$

$$i_{ds} = \frac{\lambda_{ds} * (L'_{lr} + L_m) - L_m * \lambda'_{dr}}{L_{ls} * L'_{lr} + (L_{ls} + L'_{lr}) * L_m} \tag{16}$$

$$i_{qs} = \frac{\lambda_{qs} * (L'_{lr} + L_m) - L_m * \lambda'_{qr}}{L_{ls} * L'_{lr} + (L_{ls} + L'_{lr}) * L_m} \tag{17}$$

$$i'_{dr} = \frac{\lambda'_{dr} * (L'_{lr} + L_m) - L_m * \lambda_{ds}}{L_{ls} * L'_{lr} + (L_{ls} + L'_{lr}) * L_m} \tag{18}$$

$$i'_{qr} = \frac{\lambda'_{qr} * (L'_{lr} + L_m) - L_m * \lambda_{qs}}{L_{ls} * L'_{lr} + (L_{ls} + L'_{lr}) * L_m} \tag{19}$$

where $L_m = \frac{3}{2} * M$

M = Stator to rotor mutual inductance in phase variable mode.

$$T_e = P * L_m(i_{qs} * i'_{dr} - i_{ds} * i'_{qr}) \tag{20}$$

$$\omega_r = \int \frac{P}{2J}(T_e - T_l)dt \tag{21}$$

Based on Eqs. (6)–(21) the IM model is developed.

The inverse parks transformation equations used for current (d-q to a-b-c) transformations:

$$i_a = i_{ds} * \sin(\theta) + i_{qs} * \cos(\theta) \tag{22}$$

$$i_b = i_{ds} * \sin(\theta - 120^0) + i_{qs} * \cos(\theta - 120^0) \tag{23}$$

$$i_c = i_{ds} * \sin(\theta + 120^0) + i_{qs} * \cos(\theta + 120^0) \tag{24}$$

3.2 Turbine Model

A wind turbine generates electricity from moving air with the help of an alternator or generator. Energy of moving wind is transferred to the spinning shaft, which is mechanically coupled with the alternator. This shaft power depends on wind speed and swept area of turbine blades, is given as: [2]

$$P_s = 0.5C_P(\lambda, \beta)\rho A v^3 \tag{25}$$

where

$$\lambda = \frac{Rn\pi}{30v} \tag{26}$$

The Eq. (27) is the performance coefficient of wind turbine [2], is;

$$C_P = c_1\left(\frac{c_2}{\lambda_i} - c_3\beta - c_4\right)e^{-\frac{c_5}{\lambda_i}} + c_6\lambda \tag{27}$$

where

$$\frac{1}{\lambda_i} = \frac{1}{\lambda + 0.08\beta} - \frac{0.035}{\beta^3 + 1} \tag{28}$$

$c_1 = 0.5176$ $c_2 = 116$ $c_3 = 0.4$ $c_4 = 5$ $c_5 = 21$ $c_6 = 0.0068$.

Based on above equations a Turbine Simulink model is generated to provide shaft power as an output.

The parameters provided as input are:

- v=wind speed (m/s)
- n=rotor speed (rpm)
- β=blade pitch angle (degree).

3.3 Solar Panel Model

The schematic of a single-diode solar cell is shown in Fig. 3. A light generated current source, is connected in parallel with an ideal diode, together with a parallel and series resistance respectively [3].

The Current I is given by:

$$I = I_L - I_D - I_{sh} \tag{29}$$

The Voltage across diode is given by:

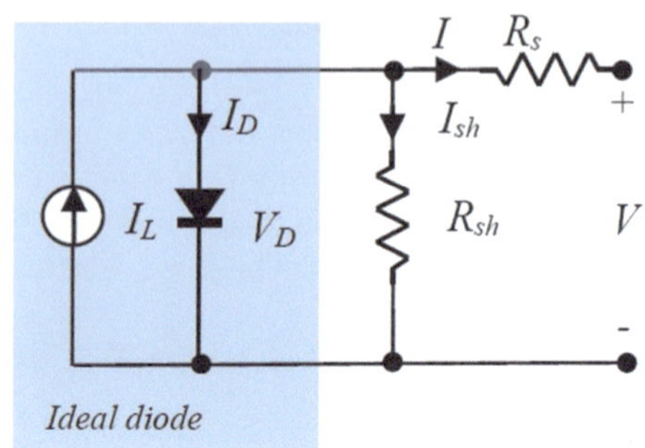

Fig. 3 Single diode solar cell

$$V_D = V + I R_S \tag{30}$$

The diode current by employing Shockley diode equation

$$I_D = I_o \left\{ e^{\frac{V_D}{\eta V_T}} - 1 \right\} \tag{31}$$

where

$$V_T = \frac{kT}{q}$$

The shunt current is given by

$$I_{sh} = \frac{V_D}{R_{sh}} \tag{32}$$

The load current is defined as

$$I_L = \left(\frac{\phi}{\phi^*} \right) \left\{ I_{SC} + \alpha_{Isc} \left(T_C^* - T_C \right) \right\} \tag{33}$$

The Short Circuit current is provided by

$$I_{SC} = I_{SC}^* \left(\frac{\phi}{\phi^*} \right) \left\{ 1 + \alpha_{Isc} \left(T_C - T_C^* \right) \right\} \tag{34}$$

The Open Circuit Voltage is provided by

$$V_{OC} = V_{OC}^* \left\{ 1 + \beta_{Voc} \left(T_C - T_C^* \right) \right\} \tag{35}$$

By employing all of the above equations from Eq. (29) to Eq. (35) a single solar cell model is generated in Simulink platform which gives output voltage and output current. The series and parallel combination of solar cell is used to generate solar panel model. Actual irradiance is a variable which can vary in the range of 1000–2000 W per square meter.

3.4 Boost Converter Model

A step-up dc-dc boost converter is having four components: inductor, electronic switch, diode and output capacitor. The converter is capable of operating in two different modes depending on the length of switching period and energy storing capacity [4].

The Boost converter is designed by using component modelling as seen in Fig. 4 PI-controller and closed loop feedback system are used to limit the output voltage at 600 volts. The values of component parameters are defined by following equations:

$$\text{Capacitance} \quad C_{out} = \frac{D * Timeperiod * I_{dcload}}{\Delta V_{ppload}} \tag{36}$$

$$\text{Inductance} \quad L = \frac{V_i * Timeperiod * D}{\Delta i_{pp}} \tag{37}$$

Specification of Boost Converters:

$P_{rated} = 10$ kW
Input Voltage = 200 V
Output Voltage = 600 V.

Depending on above parameters the calculated values of capacitance, inductance and load resistance is coming out to be as follows

Load = 25 Ω
$C_{out} = 192$ μF
L = 1.68 mH.

For PI controller the values used are

$K_p = 0.01$
$K_i = 5$.

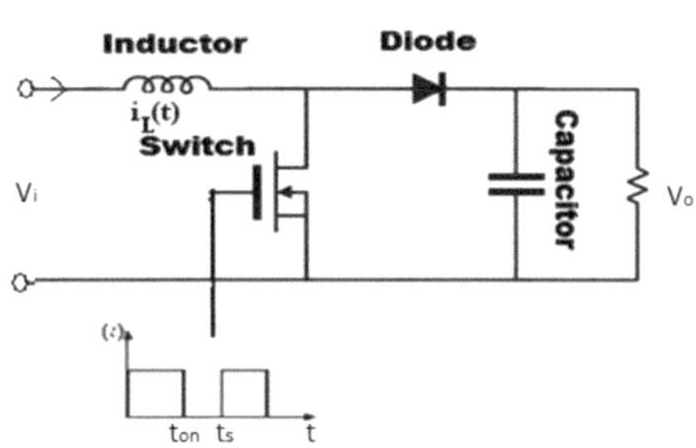

Fig. 4 Step-up dc-dc converter

3.5 Three Phase Inverter Model

The job of inverter is to convert DC input into AC output. Sinusoidal Pulse Width Modulation (SPWM) technique is used as PWM technique in the three-phase inverter model to control the inverter output voltage and output frequency [5].

The SPWM signals are generated by comparing reference sine wave with the carrier triangular wave. The two signals are compared by using relational operators, whenever the value of reference sine wave is more than the carrier triangular wave, the output is set to +1 else 0. In this way, a switching pulse signal is generated. Reference signals are

$$V_{Sa} = \sin(\omega t) \tag{38}$$

$$V_{Sb} = \sin(\omega t - 120°) \tag{39}$$

$$V_{Sc} = \sin(\omega t + 120) \tag{40}$$

Q_1 and Q_4 are generated by using V_{sa}
Q_3 and Q_6 are generated by using V_{sb}
Q_5 and Q_2 are generated by using V_{sc}.

The line to neutral voltages of the three-phase inverter are defined as:

$$V_{an} = V \sin(\omega t) \tag{41}$$

$$V_{bn} = V \sin(\omega t - 120°) \tag{42}$$

$$V_{cn} = V \sin(\omega t + 120°) \tag{43}$$

And the line voltages are found from

$$V_{ab} = V_{an} - V_{bn} \tag{44}$$

$$V_{bc} = V_{bn} - V_{cn} \tag{45}$$

$$V_{ca} = V_{cn} - V_{an} \tag{46}$$

3.6 Voltage Grid and Load Model

In this a small voltage grid is designed along with three phase load which is having logic to get connected in star and delta format of its own. The voltage of grid is 415-line voltage and 50 Hz frequency. The Load value is R = 11.022 Ω and L = 26.27 mH with power factor of 0.8 lagging.

4 Operation Results of Model

The Model needs a startup which is described as the without Active Power control mode and once it reaches complete startup than after switching it is brought in Active Power Control Mode.

4.1 Operation of Model Without Active Power Control

In above Fig. 1 the Actual Irradiance Parameter is shifting from 1000 to 1200 (W/m^2) at 0.05 s of simulation time. Initially all the three manual switches are at constant side. The simulation is started and output of boost converter and induction machine speed is kept in consideration. The boost converter output during starting is shown in Fig. 5.

Similarly, we have to wait for the induction machine speed is to reach above synchronous speed i.e. 1500 rpm. Under this consideration, the steady state output results for 30 kW 0.8 pf lagging Delta connected load are as follows:

- Boost converter output line voltage = 600 V
- Boost converter output current = varying in between 0 and 20 A
- Active Power at terminals of three phase VSI = 9330–9450 W
- Reactive Power at terminals of three phase VSI = −4980 to −5150 VAR
- Induction machine speed = 1580 rpm
- Induction machine Active Power = 10.4 kW
- Induction machine Reactive Power = −6112 VAR
- Grid Active Power = 10.22 kW
- Grid Reactive Power = 33.64 kVAR

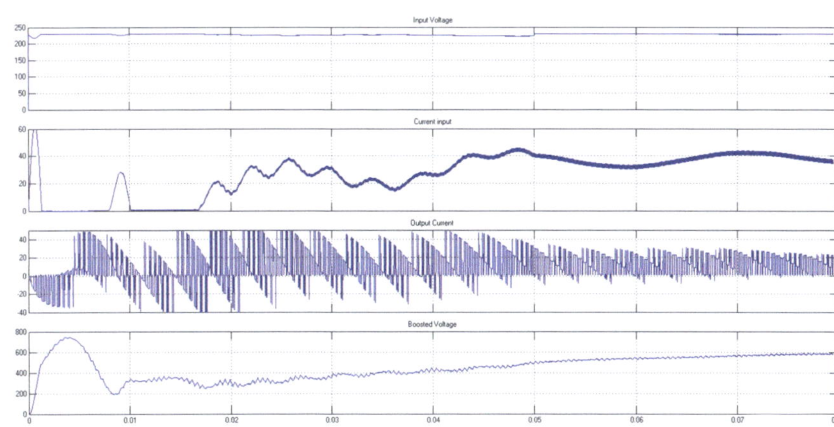

Fig. 5 Boost converter output during starting (y-axis) versus time (x-axis) (s)

- Load Active Power = 30.03 kW
- Load Reactive Power = 22.48 kVAR.

4.2 Operation of Model with Active Power Control

The steady state is reached and corresponding output values are noted down. At this all the manual switches are thrown back to feedback mode and wait for steady state results. The switching point leads to a transition which is visible in all the graph plots. Boost converter output is used to show the behavior of transition.

Figure 6 shows how switching transition is affecting the boost converter results from earlier steady state reached value. From Fig. 7 it is clearly visible that the boost converter trying to reach its steady state again in this condition.

In Fig. 2 the manual switches are used for connecting the feedback loop. There are three manual switches in the model, one switch is used to transfer the frequency from open loop to closed loop and the other two switches are for Active Power control method of induction machine and inverter.

$K_p = 1$ and $K_i = 30$ is used for induction machine active power control, whereas $K_p = 0.7$ and $K_i = 10$ is used for inverter active power control.

The overall steady state results for 30 kW 0.8 pf lagging Delta connected load are as follows:

- Boost converter output line voltage = 600 V
- Boost converter output current = varying in between 0 and 20 A.
- Active Power at terminals of three phase VSI = 8900–9020 W
- Reactive Power at terminals of three phase VSI = −5020 to −5140 VAR
- Induction machine speed = 1570 rpm

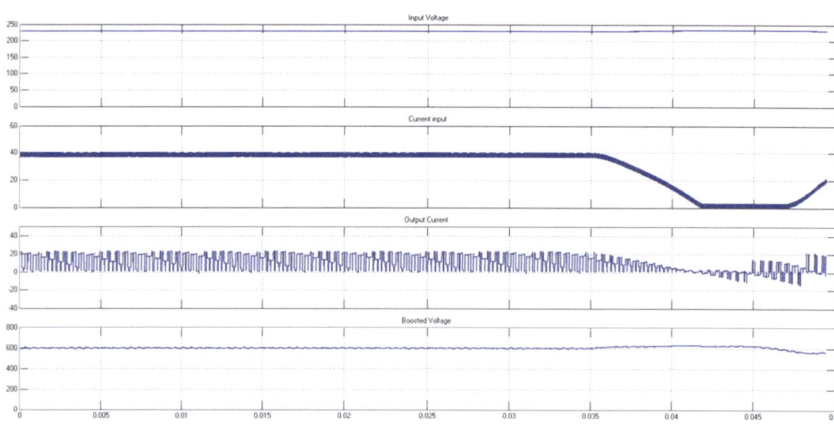

Fig. 6 Switching transition in Boost converter results (y-axis) versus time (x-axis) (s)

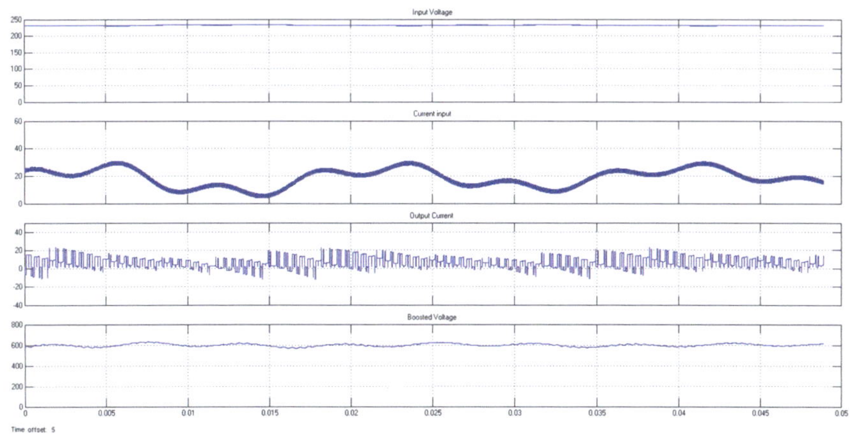

Fig. 7 Boost converter trying to attain steady state (y-axis) versus time (x-axis) (s)

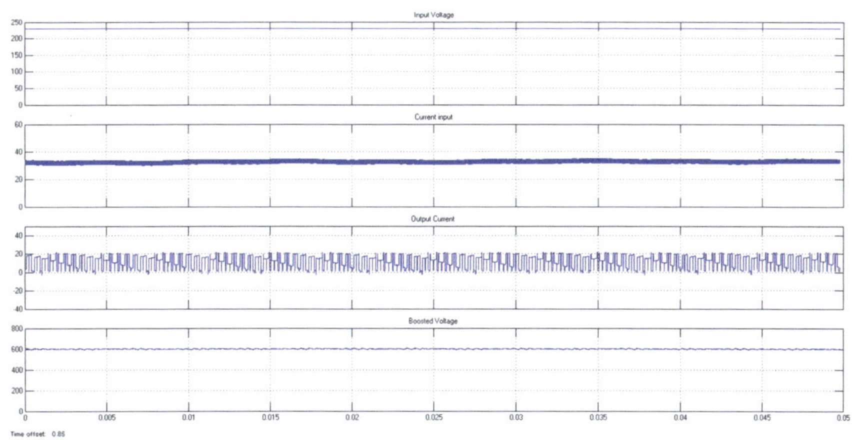

Fig. 8 Steady state input and output of boost converter (y-axis) after switching versus time (x-axis) (s)

- Induction machine Active Power = 10 kW
- Induction machine Reactive Power = −5967 VAR
- Grid Active Power = 11.10 kW
- Grid Reactive Power = 33.58 kVAR
- Load Active Power = 30.03 kW
- Load Reactive Power = 22.48 kVAR (Figs. 8, 9, and 10).

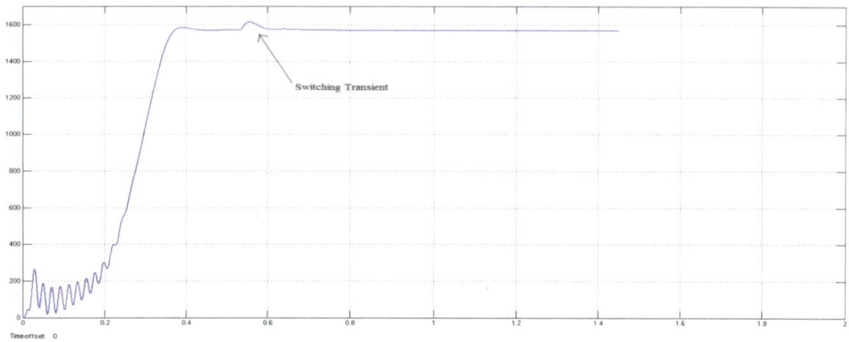

Fig. 9 Induction machine speed Nm (rpm) along y-axis versus time (s) on x-axis

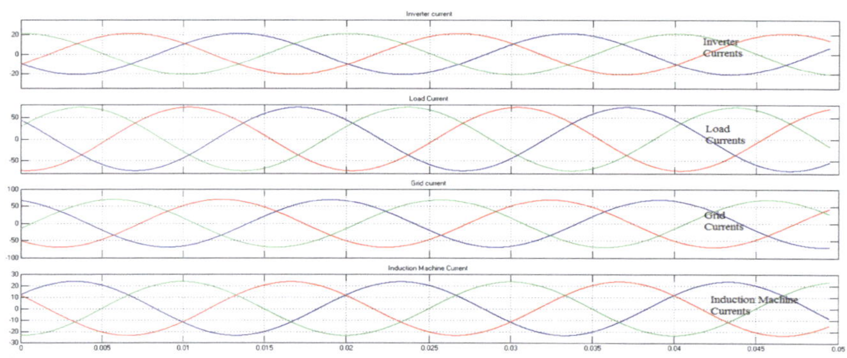

Fig. 10 Line current (red), (blue), (green) (Amps) respectively on y-axis versus time (s) on x-axis

5 Conclusion

For Power management of NCES in smart grid the various individual models are developed working efficiently in standalone cases.

The models completed are:

1. Induction Machine
2. Wind Turbine
3. Solar cell and Solar panel
4. Boost Converter
5. Three phase VSI
6. Voltage Grid and Load.

In complete interconnected model, the same common load under steady state condition is being fed by three sources:

1. Induction Machine
2. Solar Panel via Boost converter followed by Three phase VSI
3. Grid.

The output Active power of Induction Machine and Solar Panel side (i.e. at Three phase VSI terminals) is controlled by the PI controller method. Induction Machine Active Power control is Providing nearly perfect result, whereas the Solar Panel Side Active Power control is providing good result although a very small variation in power is observable but still the results are widely acceptable.

Hereby it can be said that the Power Management of Non-Conventional Energy sources when connected to a local grid is performed with good result.

References

1. G. Renukadevi, K. Rajambal, "Generalized d-q model of n-phase induction motor drive", World Academy of Science, Engineering and Technology. Int. J. Electr. Comput. Energ. Electron. Commun. Eng. **6**(9), 1066–1075 (2012)
2. A.B. Cultura II, Z.M. Salameh, in *Modeling and Simulation of a Wind Turbine-Generator System*, http://www.ieee.org, IEEE 2011
3. D. Stefan, in *Matlab/Simulink Solar Cell Model Based on Electrical Parameters at Only One Operating Condition*, 18th International Conference on System Theory, Control and Computing, Sinaia, Romania, pp. 709–714, 17–19 Oct. 2014
4. M.R. Dave, K.C. Dave, Analysis of boost converter using PI control algorithms. Int. J. Eng. Trends Technol. **3**(2), 71–73 (2012). http://www.internationaljournalssrg.org
5. N.I. Raju, Md Shahinur Islam, A.A. Uddin, Sinusoidal PWM signal generation technique for three phase voltage source inverter with analog circuit & simulation of PWM inverter for standalone load & micro-grid system. Int. J. Renew. Energy Res. **3**(3), 647–658 (2013)

Smart Coordination Approach for Power Management with PEV Based on Real Time Pricing

Purbasha Singha, Debanjan Ghosh, Sayan Koley, Rishiraj Sarkar and Sawan Sen

1 Introduction

A restructured electrical grid that integrates the behaviours of suppliers and consumers using modern information and communications technology, to improve the efficiency, reliability and sustainability of the production of power and distribute the same to the consumer, is called smart grid. With the application of these new technologies peak power prices were effectively averaged out and passed on to all power market participants equally. In Smart Grid scenario, inclusion of demand response (DR) may be found out to be a significant input to power network operations for achieving operational excellence. The unprecedented and growing concerns over environmental issues from traditional fossil-fired power stations turned the engineering minds towards efficient use of large amounts of renewable energy. The impediments of high generation cost and intermittent nature of wind power and solar power needed more advanced control systems to facilitate the connection to the grid. Moreover, to improve reliability and to respond to natural disaster or malicious sabotage, the Smart Grid is designed such that it can mend the fault by itself which means it has got

P. Singha (✉) · D. Ghosh · S. Koley · R. Sarkar · S. Sen
Electrical Engineering, Kalyani Government Engineering College,
Nadia, West Bengal, India
e-mail: purbasha.singha95@gmail.com

D. Ghosh
e-mail: dbnjnghosh1996@gmail.com

S. Koley
e-mail: sayanko1000@gmail.com

R. Sarkar
e-mail: rishiraj1000sarkar@gmail.com

S. Sen
e-mail: sawansen@gmail.com

S. Chattopadhyay et al. (eds.), *Modelling and Simulation in Science, Technology and Engineering Mathematics*, Advances in Intelligent Systems and Computing 749,
https://doi.org/10.1007/978-3-319-74808-5_23

self healing property [1]. Proper installation of information network integrated with Smart Grid can appear to be an invaluable resource for regulation of the operational condition of the system [2, 3]. Over the last decade, Plug-in Electric Vehicles (PEV) has taken the transportation sector to reach a new level. The concept of PEV has been very popular as dependence on petroleum is to be reduced for nature's sake. Preliminary studies indicate that PEVs will become the key solution of the electricity industry in the near future as pollution-free alternatives to the conventional petroleum based transportation. The high acceptability of PEVs in electrical market have significant impacts on power market especially at the distribution level [4, 5]. In the quest of optimizing the utilization of these new resources, researchers in the recent past have been proposing indigenous methodologies and solution algorithm. The popularity of PEV is favourable utilization of Off-peak periods of power networks that is to charge the batteries to supply for the vehicles during peak loading periods when the price of electricity is high. But PEV is also capable of supplying electricity to grid when demand curve is high. While the power system planners heavily rely upon the methodologies in [6, 7] that introduced a distinctive work where operating conditions were optimized with a coordination methodology of PEV charging. Plug-In Vehicles in a Smart Grid has been depicted in [8, 9]. A prediction based charging method of PEVs was depicted in [10] where dynamic price information was utilized to produce optimum charging schedule of PEVs. Ref. [11] designed an OPF to produce an optimum schedule of charging of PEVs considering the operating condition but the methodology could not minimize generation cost by maximizing load catering. The advantages of battery storage were utilized in [12] for optimal operation of micro-grids that is optimizing both generation and loading schedule but it could not assure a standard of operating conditions. All these methods pointed towards effective utilization of DR to optimize the PEV and Smart Grid operation. Ref. [13] A framework of future transmission grid is portrayed while [14–16] identifying the challenges faced with implementation of DR in distribution of the existing grid. During the inclusion of renewable energy sources, immense difficulty to system optimization is posed by the unparalleled sporadic nature and cost curve. Refs. [17–20] proposed optimization methodologies to intensify the operational status of power grid endorsing a particular renewable energy source. Effective deployment of chargeable batteries has been depicted for optimization of intermittent or renewable energy resources. The literature survey has shown that PEVs can be made to work in conjunction with demand response, intermittent renewable energy sources and it is also capable of standardizing the conditions prevailing during operation in terms of voltage and loss profile of the system. Moreover it is efficient in peak load flattening to reduce the dynamics of electricity price [21–24]. Social welfare has been taken on ground as an objective but applied less importance on enduring operating conditions of the system or the load shedding technique whereas [25, 26] some new approaches have been considered. PEVs schedule of charging and discharging has been planned in such a way that energy can be uprooted from grid to vehicle or from vehicle to grid respectively. A no of other authors have adopted different methodologies for charging profiles of PEV using recorded vehicle usage data.

Thus, the above literature review necessitates of an algorithm that ensures a standard parametric operational state with a goal to minimize the electricity price with favourable load catering without contravening the price equilibrium of the market by harmonizing multiple charging and discharging of PEVs, that can greatly affect grid repeatability, security and performance. The proposed algorithm in this paper not only integrates the DR and generation characteristics but also involves price sensitivity of voltage profile, line loss, transmission congestion and load curtailment to accomplish all the features of Smart Grid. To compare the solution obtained, a standard OPF [27] has been embraced. To ratify the power management infrastructure and distribution network loss minimization, the smart coordination strategy is incorporated on modified IEEE 30 bus system that consists of a mix of residential, commercial and industrial customers penetrated with PEVs.

2 Theory

An electrical grid that comprises a variety of functioning and energy measures with higher order of reliability. Researchers designed charging and discharging schedule of PEVs under smart grid, in which one category of research was focused on charging and discharging schedule based on the information about the present state of the power grid and the second category is the one in which charging or discharging schedule is based on forecasted estimates of the state of the grid. Unlike classic grids, here local sub networks generate more power than it is consuming. System Operator will be able to reach everywhere of network to maintain optimized operating conditions under the worst possible states of the system with the help of deployment of new resources. In this context ISO will be able to identify the state variable creating imbalance. The price change response of the state variables of modern power markets incorporating PEVs have been elaborated in the following sections.

2.1 Electric Vehicles and Distribution Networks

PEVs are turning out for being the most innovative and environment friendly alternative of only traditional fuel based automobiles. The term Plug-in relates to the electric power storage system in built in these kind of vehicles to accumulate electrical power from power outlet. Scarcity of fossil fuel and their corresponding price hike are forcing the development of more of these kinds of vehicles [28–30]. But exploitation of PEV in improper way has a great impact on power distribution network. If simultaneous charging of the vehicles would be accomplished in a small geographic area, the increased demand caused due to charging could cause major troubles for the utility. While PEVs are charged all along at system peak, it could result in supply shortages or create an urge for large new investments in expansion of capacity of generation and setting up new generation plants, congestion problem at distribution level for most utilities will be the most noted concern due to PEVs charged all along. Thus, PEVs must be able to communicate bi-directionally with the grid where the grid can inform PEV about its constraints, generation capabilities, load catering priorities, and the price at which it can afford to sell power to PEVs and the PEVs need to express its requirements time of charging and price it can to buy power from the grid.

2.2 *State Space Modeling of Power Network Including PEVs*

If numbers of generators = N and Numbers of PEV loads = M and Number of other loads = m, the price sensitive state variables of a power network can be written as,

$$x = [P_{G1} P_{G2} .. P_{GN} L_{p1} L_{p2} \ldots L_M L_1 L_2 \ldots L_m P_c T_L V_{\text{min}} P_{l\text{max}}]^T \quad (1)$$

where P_c = load retrenched, T_L = Total line loss, V_{min} = minimum bus voltage, $P_{l\text{max}}$ = maximum line flow.

As the market players are connected with ISO through smart metering, on receipt of the price information at the nth hour, the participants either change or persevere with their propositions best suitable to them. Under the fencing of social welfare the ISO determines the price dependent state variables $x(n+1)$ at the $(n+1)$th hour based on the price of nth hour. Mathematically

$$x(n+1) = f(p) = Ax(n) + Bu(n) \quad (2)$$

where A is the sensitivity matrix and B is the contingency or state modification matrix, which will arise only when participant, characteristics are reoriented by either deliberately or inadvertently.

3 Problem Statement

The aim of power management is to make the total power losses least and energy transferred to the PEVs highest over time duration [1, T] maintaining other operating condition. Thus, the multi-objective function considering overall social welfare may be defined as follows:

$$\max \Lambda = \sum_{k=1}^{T} C(d_k) + \max \sum E_D - C(glv) \quad (3)$$

where function $C(\cdot)$ is strictly increasing and convex. Convexity of the above cost function causes heavier penalty on larger instantaneous power losses, which is important in alleviating power loss values. PEVs over time duration [1, T] is represented by $\sum E_D$ and finally, $C(glv)$ represents the operational limit violation constrained generation cost function. Mathematically, these terms can be expressed as follows:

$$C(d_k) = \sum_{k=1}^{M+m} a_k P_{d(M+m)k}^2 + b_k P_{d(M+m)k} + c_k \tag{4}$$

$$C(glv) = \left(\sum_{i=1}^{n_g} a_i P_{gi}^2 + b_i P_{gi} + c_i + P_c.PF_1 + T_L.PF_2 + V_{\text{min}}.PF_3 + P_{l\text{max}}.PF_4 \right) \tag{5}$$

Subject to the power balance equations at all buses and other constraints are as follows:

$$P_{gi}^{\text{min}} \le P_{gi}^0 \le P_{gi}^{\text{max}} \quad \forall i \in N \tag{6}$$

$$Q_{gi}^{\text{min}} \le Q_{gi}^0 \le Q_{gi}^{\text{max}} \quad \forall i \in N \tag{7}$$

$$\Delta P_{gi}^{\text{min}} \le \Delta P_{gi}^j \le \Delta P_{gi}^{\text{max}} \quad \forall i \in N \tag{8}$$

$$\Delta Q_{gi}^{\text{min}} \le \Delta Q_{gi}^j \le \Delta Q_{gi}^{\text{max}} \quad \forall i \in N \tag{9}$$

$$\left| V_k^{\text{min}} \right| \le \left| V_k \right| \le \left| V_k^{\text{max}} \right| \quad \forall k \in M + m \tag{10}$$

and PEV charging constraint for every $t \in \{1, \ldots\ldots T\}$,

$$\sum_{t=1}^{T} P_{Ek}[t] = C_k \quad \forall k \in N, t \in \{1, \ldots T\}$$

a_i, b_i, c_i are cost coefficients of the ith generating station, P_{gi} is generation of the ith generator, a_k, b_k are bid coefficients of jth consumer, $P_{d(M+m)k}$ denotes the power demand of the M numbers of PEV loads and m number of other loads, PF_1, PF_2, PF_3 are the penalties for operational limit violation set by ISO, C_k is the capacity of PEV battery connected to bus k.

4 Simulation and Result

As the proposed methodology claims that apposite charging and discharging schedule of PEVs may lead to an operating condition which overall is beneficial for all the power market participants, the case study has been performed in modified IEEE 30 bus system having both generator characteristics and price responsive demand characteristics as input to cost optimization algorithm. This modified system has six GENCOs connected to buses 1, 2, 5, 8, 11, 13 and 24 Load Despatch Centres (LDCs) with which PEV may be connected for charging or discharging as per consumers' requirement. Aside from the elastic/inelastic PEV loads, each LDCs excluding GENCOs, $k \in \{1, 2, \ldots, 24\}$ is connected to an inelastic load as well. The system description has been tabulated in Table 1.

To check the efficacy of the proposed methodology for maintaining the standard operational constraints with an aim of sustaining both individual and social welfare maximum by smart coordination approach of PEV with inelastic power demand of consumer, without violating the price equilibrium of the power market, the simulations have been performed in different steps. In the first step, different LDCs have been grouped (HIG and LIG) through price depending zone-clustering algorithm. In next step, generation-demand schedule has been prepared with PEV charging-discharging plan. With this charging and discharging schedule of PEV, market demand can be flattened during peak time and can be enhanced during off peak time to maintain the base demand. This has been shown in step 3 of the simulation. Charging and discharging of PEV is not only significant to fulfil the purpose of a vehicle, but maintaining appropriate charging-discharging schedule of PEV has assisted to maintain the price equilibrium of the power market. In step 4 of simulation, it can be shown that suitable charging-discharging schedule of PEV leads toward the equilibrium of market demand as well as price of electricity. Once more, to prepare a charging-discharging schedule of PEV, it is not efficient to prove the adequacy of proposed methodology without judging its efficiency for maintaining standard operational constraints of the power network. Hence, inspections of various operational standards (making transmission line loss less, maintaining voltage profile, etc.) are the important parts of step 5 of simulation.

4.1 Clustering of LDCs Through Price Depending Zone-Clustering Algorithm

The proposed algorithm depends on non-linear relation between system demand and price of electricity, which may vary throughout the whole day. Higher Income Group (HIG) consists of LDCs where with minimal change of demand, the change of cost

Table 1 System description

Serial no.	Specifications	Provision
1.	LDCs	24
2.	PEV	1000
3.	GENCOs	6
4.	Maximum elastic/inelastic demand (MW)	283.4
5.	Maximum reactive demand (MVAr)	126

is more than Lower Income Group (LIG). This non-linear relation between system demand and electricity price depends on the fact that, the consumers from HIG are willing to pay more for maintaining the reliability and quality of power. It has been shown from the following graph that 17 LDCs may be assembled in HIG group and remaining LDCs are under LIG group in projected system (Fig. 1).

4.2 Preparation of Generation-Demand Schedule with PEV Charging-Discharging Plan

Now here, the generation-demand map has been shown for demand-price equilibrium condition, depending on that different LDCs can plan their offering price to have their inelastic demand with PEV charging and discharging. But in Smart Grid arena, price of electricity is not fixed; it fluctuates all over the day-night depending on demand pattern. Following graph shows the power market inelastic demand plan throughout a whole day, where total 24 h have been subdivided into four sectors depending on variable price schedule.

From (Fig. 2), ISO can fix up the PEV charging-discharging schedule to maintain the market equilibrium, which can be broadcasted to the consumer through smart meter. Consequently the willing participants of power market can reschedule their demand to enjoy quality power at lowest possible price. Thus the proposed algorithm for smart coordination has got a short-term forecasting module and an optimization section. This short term module remits information about the number of PEVs in the parking garage which are ready for charging during off peak time or supply power to consumer during peak time in the course of stored power and the objective of the optimization module is to make the energy delivered to PEV maximum and satisfy optimization conditions for the PEV incorporating operational constraints which is connected to the power grid and customer demands.

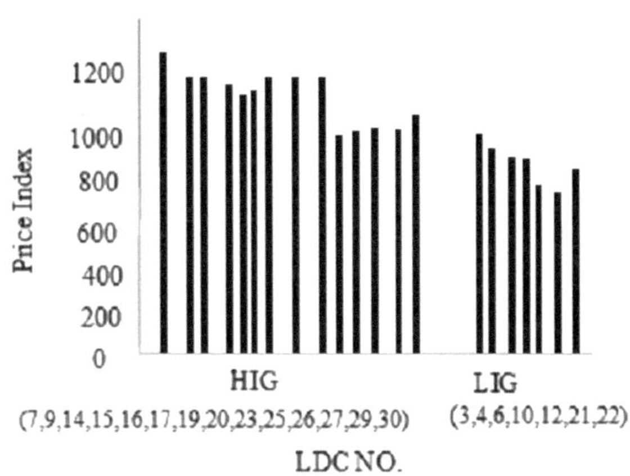

Fig. 1 Zone clustering of LDCs

4.3 Demand Response of the Power Market with PEV Charging and Discharging Schedule

From the above demand schedule (Fig. 2) it is apparently clear that, during 4 PM to 10 PM, the market demand is high (peak load condition). Thus it becomes very essential to make PEV ready for discharging when peak loading condition prevails, which in turn reduces the demand peak to base demand. Again, the cost of electricity does not remain same throughout the whole day (it will be high during peak demand and low during off peak time under Smart Grid arena), so it is desirable to charge the PEV during low cost time duration (off peak state: 10 am to 4 pm) after reaching the office when the purpose of vehicle is over for few hours. Similarly, during evening time period (4 pm to 10 pm), the cost of electricity will be high; hence it is desirable to use stored energy in PEV, which in turn flatten the peak demand in one side and other side it will reduce the expenses of power consumers. The competence of the proposed algorithm is to heal the power market by its own under the rearrangement of operating conditions particularly during peak periods, when the available generation is most costly (Fig. 3a, b).

4.4 Cost Evaluation to Estimate the Economical Benefits of Power Consumer

With the base generation schedule of GENCOs, the performance of the proposed algorithm has been observed to be remarkably superior in sustaining the operating conditions within safe limit, owing to the fact that the proposed algorithm not

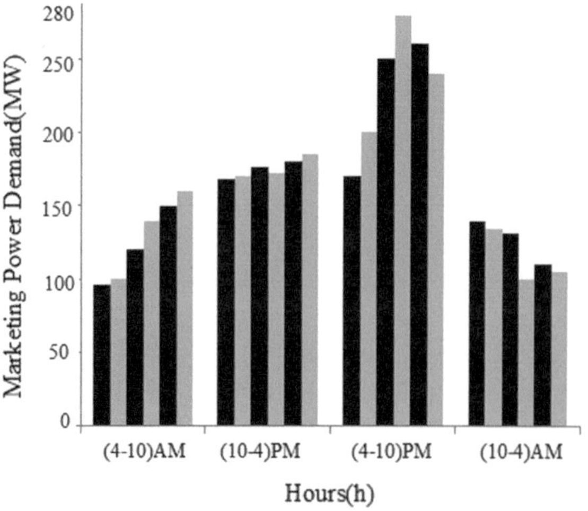

Fig. 2 Power market demand for a whole day

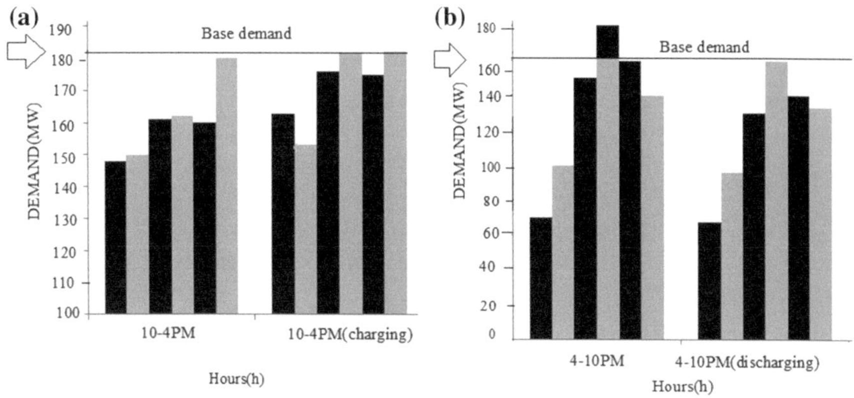

Fig. 3 **a** Enhancement of load line with proper charging of PEVs. **b** Flattening of load line with proper use of stored energy of PEVs

only manages generation with optimal cost but also manages loads of PEVs with efficiency to distribute maximum demand at minimum cost in the most favourable way, maintaining all operating constraints. Table 2 shows the economical benefits of power consumer with proposed methodology for proper charging and discharging plan of PEVs. Projected algorithm efficiently administers both circumstances of power market, where cost of electricity is independent of demand and dependent of demand.

With economical benefits of power consumers, this algorithm as well can be used to manage PEV as supplier of power (other than GENCOs) to consumers during peak load time and enhance the load line during off-peak time. Following graph (Fig. 4) shows that change of electricity cost during off peak time as well as peak time under variable demand-price condition.

Table 2 Economical benefits of power consumers

Savings during discharging	4–5 PM	5–6 PM	6–7 PM	7–8 PM	8–9 PM	9–10 PM
Fixed price (Rs.)	7.31	10.79	71.72	72.12	72.43	25.5
Variable price (Rs.)	27.1	39.9	257.2	258.1	259.4	93.7

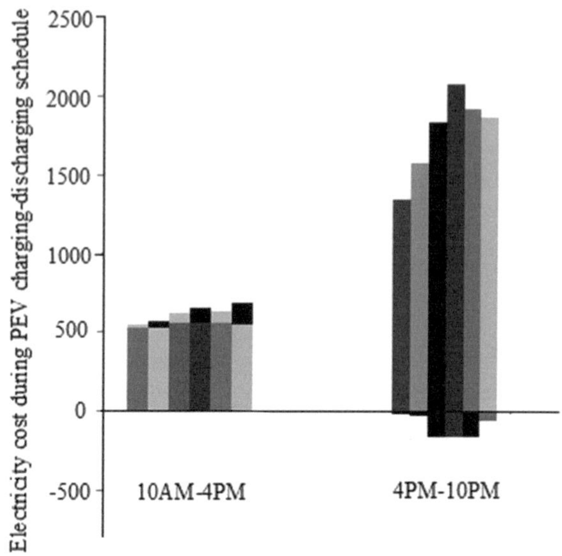

Fig. 4 Change of electricity cost due to inclusion of PEVs in market demand

4.5 Applicability of Proposed Algorithm to Maintain Standard Operational Constraints

Planning of charging-discharging schedule of PEV, maintaining load line close to base load, sustaining power market price equilibrium are not sufficient to prove the competence of proposed methodology without evaluating its efficiency for maintaining standard operational constraints of the power network. Therefore, the optimization section also monitors the distribution network losses and voltages of all affected LDCs considering all other operational constraints. Thus, following studies have been carried out to formalize the power management infrastructure and distribution network loss minimization (Fig. 5).

5 Conclusion

The approachability to smart metering communications has obtained profound changes in power grid operation. Here this paper presents a methodology to illustrate optimal and efficient operations of Smart Grid with the presence of PEVs. The model effectively regulates the charging-discharging schedule of PEVs to reach prolific solutions negotiating with generator characteristics, demand response of shiftable loads, PEVs, voltage stability and transmission loss limits. Simulation results convincingly shows that the application of proposed methodology palliate network constraints while regaling peak demand and reducing the cost of electricity even under worst possible states of the system operation. Comparing with the conven-

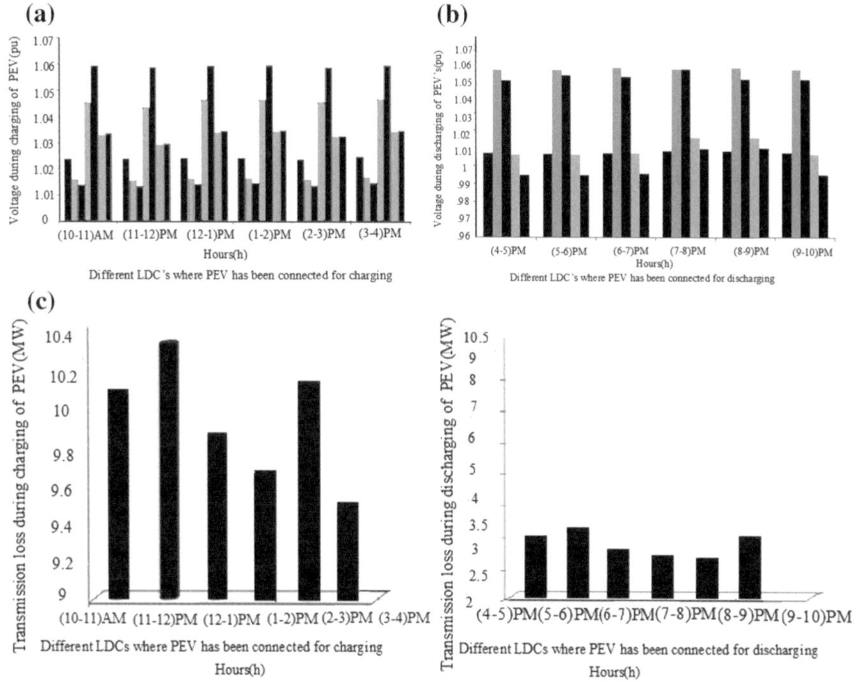

Fig. 5 **a** Voltage pattern of affected LDCs during charging of PEVs. **b** Voltage pattern of affected LDCs during discharging of PEVs. **c** Comparison of transmission loss pattern during charging-discharging schedule of PEVs

tional system functioning based on only GENCOs scheduling between highest and lowest possible generation, the proposed algorithm implements consumer or PEVs charging-discharging schedule with a smart communication facility where not only GENCOs but also power consumer with PEV characteristics can take part to change their position for their own and social benefit.

References

1. S. Sen, S. Chanda, S. Sengupta, A. De, Demand response governed swarm intelligent grid scheduling framework for social welfare. Int. J. Electr. Power Energy Syst., Elsevier **78**, 783–792 (2016)
2. R. Farrokh, I. Ali, Demand response as a market resource under the smart grid paradigm. IEEE Trans. Smart Grid **1**(1), 110–116 (2000)
3. A. Ipakchi, in *Demand Side and Distributed Resource Management—A Transactive Solution*. IEEE Power and Energy Society General Meeting (2011)
4. L. Fernandez, T. Roman, R. Cossent, C. Domingo, P. Frias, Assessment of the impact of plug-in electric vehicles on distribution networks. IEEE Trans. Power Syst. **26**(1), 206–213

5. J. Lopes, F. Soares, P. Almeida, Integration of electric vehicles in the electric power system. Proc. IEEE **99**(1), 168–183 (2011)
6. N. Ramasamy, *Computer-aided system analysis, Marcel Dekkar* (INC, New York, 2002)
7. D. Sara, S. Amir et al., Real-time coordination of plug-in electric vehicle charging in smart grids to minimize power losses and improve voltage profile. IEEE Trans. Smart Grid **2**(3) (2011)
8. B. Terry, in *Energy Efficiency, Demand Response and PHEVs and the Smart Grid: Pulling It All Together*, Summer Seminar (2010)
9. M. Erol-Kantarci, in *Decentralized demand side management of Plug-in hybrid vehicles in a Smart Grid*, Summer Seminar (2010)
10. T. Hussein, in *Prediction-based charging of PHEVs from the smart grid with dynamic pricing*, 1st IEEE Workshop on Smart Grid Networking Infrastructure (Colorado, SGNI, 2010)
11. S. Sojoudi et al., in *Optimal Charging of Plug-In Hybrid Electric Vehicles in Smart Grids*, Power Engineering Society, IEEE General Meeting—PES (2010)
12. E. Sortomme, M.A. El-Sharkawi, in *Optimal Power Flow for a System of Microgrids with Controllable Loads and Battery Storage*, Power Systems Conference and Exposition, 2009. PSCE '09
13. L. Fangxing, Q. Wei et al., Smart transmission grid: vision and framework. IEEE Trans. Smart Grid **1**(2) (2010)
14. M. Jose, M. Nelson, R. Ilya, Demand response and distribution grid operations: opportunities and challenges. IEEE Trans. Smart Grid **1**(2) (2010)
15. A.-H. Mohsenian-Rad, V.W.S. Wong et al., Autonomous demand-side management based on game-theoretic energy consumption scheduling for the future smart grid. IEEE Trans. Smart Grid **1**(3) (2010)
16. K. Pouyan, M. Hassan, A. Hassan, in *Load profile reformation through demand response programs using smart grid*, Modern Electric Power Systems 2010 (MEPS'10), Wroclaw, Poland
17. R. Yann, B. Seddik et al., Optimal power flow management for grid connected PV systems with batteries. IEEE Trans. Sustain. Energy **2**(3) (2011)
18. W.D.J. King, C.S. Özveren, D.A. Bradley et al., in *Economic Load Dispatch Optimization of Renewable Energy in Power System Using Genetic Algorithm*, Power Tech 2007, Lausanne
19. G. Martin, A. Göran, Optimal power flow of multiple energy carriers. IEEE Trans. Power Systems **22**(1) (2007)
20. C. Carlo, C. Costantino, S. Pierluigi, Combined operations of renewable energy systems and responsive demand in a smart grid, IEEE Trans. Sustain. Energy **2**(4) (2011)
21. K. Arman, A. Anuradha, in *The Effect of a Smart Meter on Congestion and Stability in a Power Market*, 49th IEEE Conference on Decision and Control (Atlanta, GA, USA, 2010) pp. 194–199
22. S. Pedram, A.-H. Mohsenian-Rad et al., in *Optimal real-time pricing algorithm based on utility maximization for smart grid*, 1st IEEE International Conference on Smart Grid Communication (Maryland, USA, 2010)
23. M.O. Buygi, G. Balzer et al., Market-based transmission expansion planning. IEEE Trans. Power Syst. **19**(4) (2004)
24. T.S.P. Fernandes et al., Load shedding strategies using optimal load flow with relaxation of restrictions. IEEE Trans. Power Syst. **23**(2), 712–718 (2008)
25. G.M. Huang, N.C. Nair, in *An OPF Based Algorithm to Evaluate Load Curtailment Incorporating Voltage Stability Margin Criterion*, www.pserc.wisc.edu. Accessed Nov. 2001
26. F. Kjetil et al., in *Joint State-Space Model for Electricity Spot and Futures Prices*, Norwegian Computing Centre (2002)
27. C. Florin, G. Mevludin et al., in *Interior-Point Based Algorithms for the Solution of Optimal Power Flow Problems*, Electric Power Systems Research (Elsevier, New York, 2007) pp. 508–517
28. C. Romain, M. Samir, Electrical vehicles in the smart grid: a mean field game analysis. IEEE J. Sel. Areas Commun. **30**(6), 1086–1096 (2012)

29. C. Gerkensmeyer et al., in *Technical Challenges of Plug in Hybrid Electric Vehicles and Impacts to the US Power System: Distribution System Analysis*, Prepared for the U.S. Department of Energy under Contract DE-AC05-76RL01830, 2010
30. S. Rajakaruna, F. Shahnia, A. Ghosh, in *Plug in Electrical Vehicles in Smart Grids* (Springer Nature, Berlin, 2015)

Fault Analysis in Grid Connected Solar Photovoltaic System

Nirjhar Saha, Atanu Maji, Subhra Mukherjee and Niladri Mukherjee

1 Introduction

In hugely increasing power demand scenario solar energy plays a vital role. In solar energy system, photovoltaic (PV) array is used to convert the solar energy to electrical energy. The output from the photovoltaic (PV) array is dependent on the weather conditions. To get the maximum power output various maximum power point tracking (MPPT) algorithm has been developed throughout the years. Connecting the PV array to the grid is a challenging part as in order to connect it needs synchronization to the grid. After connecting to the grid it is very essential to analyze the effect of different types of faults on the system to enhance the reliability of the system. It is very essential to detect the different types of faults as early as possible and from the protection point of view it is very much obvious that, the protection system should be able to detect different faults in the system properly.

Detail simulation diagram of grid connected photovoltaic system has been developed using PSCAD, a transient software package [1]. The detail modeling of inverter to connect the PV system to the grid and the control of active and reactive power has been discussed in [2]. F. Liu, Y. Kang and Y. Zhang proposed an effective maximum power point tracking (MPPT) algorithm named as perturb and observe (P and O)

N. Saha (✉) · S. Mukherjee
Electrical Engineering Department, NIT, MAKAUT, Kolkata, West Bengal, India
e-mail: nirjharsaha94@gmail.com

S. Mukherjee
e-mail: subhra.reek@gmail.com

A. Maji · N. Mukherjee
Electrical Engineering Department, SKFGI, MAKAUT, Kolkata, West Bengal, India
e-mail: atanu.maji@skf.edu.in

N. Mukherjee
e-mail: niladrimukherjee88@gmail.com

S. Chattopadhyay et al. (eds.), *Modelling and Simulation in Science, Technology and Engineering Mathematics*, Advances in Intelligent Systems and Computing 749,
https://doi.org/10.1007/978-3-319-74808-5_24

and using this method performance of the system has been evaluated and compared with the other methods [3]. Similarly another MPPT method, incremental conductance (IC) has also been employed with variable step size for more accuracy and good adaptability with the rapid changing weather conditions [4]. Different types of unbalanced faults have been simulated and their effects in the grid system have been analyzed [5]. Another method to detect the fault condition has been developed taking the power loss on DC side which is used as an indicator [6], where an automatic supervision system has been used for reliable discrimination. A control scheme has been proposed for grid faults with unbalanced grid voltage by controlling various control parameters and sensing the current harmonics distortion [7]. Fault analysis of three phase grid connected solar system has been discussed in [8], where synchronously-rotating *d-q* reference theory is used to model the converter. Three-phase grid-connected Photovoltaic (PV) system based on Current Source Inverter (CSI) has been developed in [9] and fault analysis has been done. Dynamic modeling based grid connected hybrid system proposed in [10], where hybrid system consists of photovoltaic (PV) and wind energy system. Voltage and current signature analysis based different techniques already proposed by the researchers to analyze the faults in electrical systems where different planes (Clarke and Park plane), different signal processing and soft computing techniques have been used for this purpose [11–17].

None of the research works, fault condition of grid connected solar PV system has been assessed by the THD inter harmonics groups and DWT based skewness, kurtosis, rms and mean value analysis. For this reason an attempt has been made to assess the fault condition of grid connected solar PV system by THD, inter harmonics and DWT based skewness, kurtosis, rms and mean value analysis. In this work different short circuit faults are considered in grid side and inverter output side current (system current) is used for fault analysis which is shown in Fig. 1.

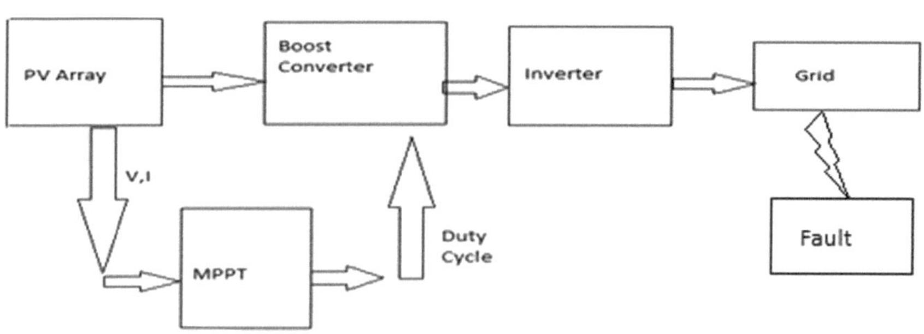

Fig. 1 Block diagram of grid connected solar PV system

2 System Description

The simulink model which is shown in Fig. 1 is used for fault analysis. A 250 kW solar array is considered for simulation. The power output from solar array is fed to the boost converter, whose duty cycle is controlled by the maximum power point tracking (MPPT) block. The controlled dc output is fed to the inverter which converts the dc voltage to ac voltage (250 V) which is stepped up by a transformer and ultimately it was connected to 25 kV grid. Different short circuit faults are considered on the grid side and all the analysis has been done based on 'R' phase system current or inverter output side current. Fast Fourier transform (FFT) based THD, inter harmonics calculations and Discrete Wavelet transform (DWT) based parameter analysis has been done on the captured 'R' phase system current for fault analysis.

3 Mathematical Techniques Used for Fault Analysis

3.1 Total Harmonic Distortion (THD)

The Total Harmonic Distortion (THD) of a signal is the measurement of the harmonic distortion which is present in the signal and it can be defined as the ratio of the square root of the sum of the currents of all harmonic components to the current of the fundamental frequency. It is used for linearization the power quality of electric power systems. The current waveform of 'R' phase has been taken in normal condition and as well as in different fault conditions then using Fast Fourier Transform (FFT), the total harmonic distortion (THD) has been calculated. Table 1 shows the THD values of 'R' phase current signal in different conditions from where normal and different fault conditions can be assessed properly. Following formula is used to calculate the THD of 'R'phase current [18].

$$THD = \frac{\sqrt{(I_2^2 + I_3^2 + I_4^2 + \ldots\ldots I_n^2)}}{I_1} \times 100 \quad (1)$$

Table 1 THD results of 'R' phase current in different conditions

Conditions	dc component in %	2nd harmonics in %	3rd harmonics in %	4th harmonics in %	5th harmonics in %	6th harmonics in %	7th harmonics in %	THD
Normal	0.01	0.34	0.04	0.03	0.28	0.01	0.01	0.443509
L-G	13.29	0.55	0.26	0.07	0.06	0.06	0.03	0.618951
L-L-L-G	11.66	0.44	0.1	0.11	0.08	0.01	0.04	0.473075
L-L	13.87	0.48	0.22	0.09	0.04	0.04	0.03	0.539444
L-L-G	13.85	0.47	0.21	0.09	0.04	0.03	0.03	0.525833

where

I_1 Current in fundamental frequency
I_n Current harmonics. n = 2, 3… n.

3.2 Inter-harmonics Group Analysis

For further analysis of the signals the magnitude of inter harmonics has been observed and the inter harmonics group has been calculated using the following formula.

$$IG_k = \sqrt{\sum\nolimits_{i=1}^{11} I_{k*n+5i}} \times 100 \quad (2)$$

where

IG Inter-harmonic Group
k Group number, k = 0, 1, 2, 3…
n fundamental frequency of the signal

In this work up to second inter-harmonics group (i.e. k = 2) of 'R' phase system current signal has been considered for fault analysis and the results has been given in Table 2.

3.3 Discrete Wavelet Transform (DWT) Analysis

Using Wavelet Transform (WT) better time frequency representation can be achieved from non-stationery signals which were the short-comings of other signal processing techniques like, FFT, STFT etc. Continuous Wavelet Transform (CWT) and Discrete Wavelet Transform (DWT) are the two classifications of WT [15]. Due to some computational difficulties of CWT; in this work DWT is considered as a technique to assess the transient conditions occurring due to different types of fault. In DWT, signal is passing through high pass and low pass filter bank in each decomposition

Table 2 Inter-harmonics group analysis of 'R' phase current in different conditions

Inter-harmonics group	Normal condition	L-G condition	L-L-L-G condition	L-L condition	L-L-G condition
IG0	0.722011	20.09749	20.1727	25.33755	25.76861
IG1	0.822496	6.186186	5.271243	13.05052	12.27778
IG2	0.091104	0.78905	0.744983	0.94016	0.855161

level, to get different frequency band present in the signal. Here, nine (9) levels of DWT decomposition and 'db4' mother wavelet is considered for DWT analysis.

For assessment of different conditions DWT has been done on 'R' phase current in all the conditions. In DWT up to level nine is considered in all the cases. Mean, r.m.s, Skewness and kurtosis parameters have also been calculated in all these decomposition levels to detect and discriminate the fault and normal conditions in grid connected solar PV system.

3.4 *Skewness*

Skewness can be mathematically defined as the averaged cubed deviation from the mean divided by the standard deviation cubed. If the result of the computation is greater than zero, the distribution is positively skewed. If it's less than zero, it's negatively skewed and equal to zero means it's symmetric [19].

3.5 *Kurtosis*

Kurtosis refers to the degree of peak in a distribution. More peak than normal means that a distribution also has fatter tails and that there are lesser chances of extreme outcomes compared to a normal distribution. The kurtosis formula measures the degree of peak [19].

4 Results of Fault Analysis Using Different Techniques

See Fig. 2.

DWT Analysis

In this analysis, at first 'R' phase current signal has been decomposed up to DWT level nine by 'db4' based mother wavelet and then rms, mean, skewness and kurtosis has been calculated for detail level and approximation (approximate) level coefficients in different conditions which is shown from Figs. 3, 4, 5, and 6.

Results of Mean Values

Results of rms Values

Results of Skewness Values

Results of Kurtosis Values

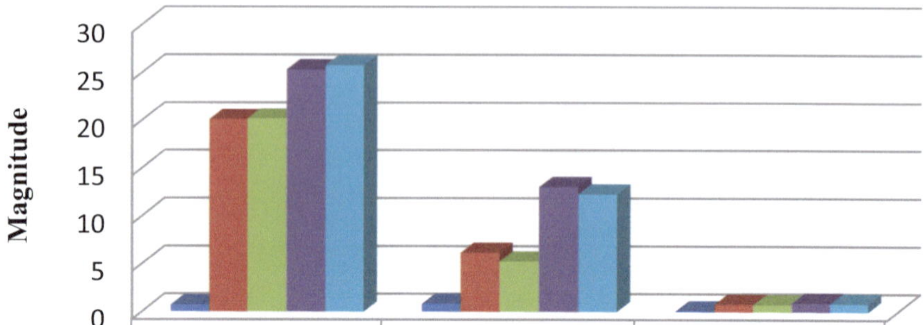

	IG0	IG1	IG2
Normal	0.72201108	0.822496201	0.091104336
LG	20.09749487	6.186186224	0.789050062
LLLG	20.17270185	5.271242738	0.744983221
LL	25.33754526	13.05051723	0.940159561
LLG	25.76861075	12.27778074	0.855160804

Fig. 2 Pictorial representation of inter-harmonics group analysis of 'R' phase current in different conditions

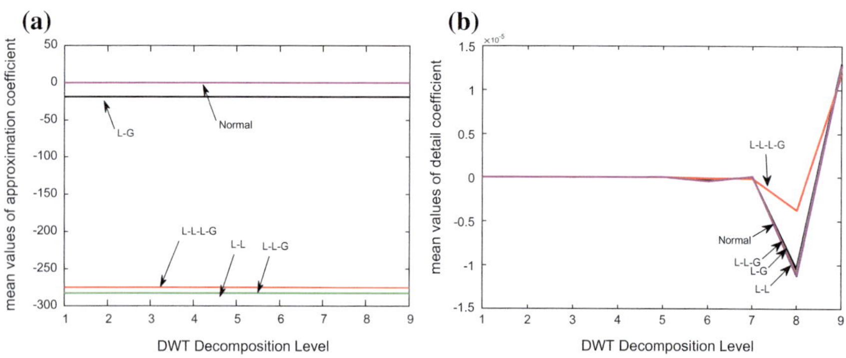

Fig. 3 Mean values of **a** approximation coefficients and **b** detail coefficients versus DWT decomposition levels

5 Observations

From the FFT based THD results (Table 1), different fault conditions and normal condition can be classified properly because THD is found minimum at normal condition and is higher (different values) at different fault conditions.

In this work, three inter harmonics group has been calculated of 'R' phase current signals in different conditions which is given in Table 2. In this results, maximum inter-harmonics group values has been observed in L-L and L-L-G fault conditions and minimum inter-harmonics group values has been observed in normal condition.

Figure 3a, b is used to depict the results of mean values of approximation and detail coefficients respectively in different DWT decomposition levels at different conditions. In Fig. 3a differences between mean values in different conditions is constant up to level nine and in Fig. 3b difference is maximum at DWT decomposition level eight (8).

In Fig. 4a, a constant difference has been observed between rms values of approximation coefficients in all those mentioned conditions and in Fig. 4b, maximum difference has been observed between rms values of detail coefficients at DWT decomposition level nine but for both the cases minimum values of both the coefficients has been observed in normal condition.

Figure 5a is used to depict the skewness values of approximate coefficients with respect to different DWT decomposition levels, where the differences of skewness

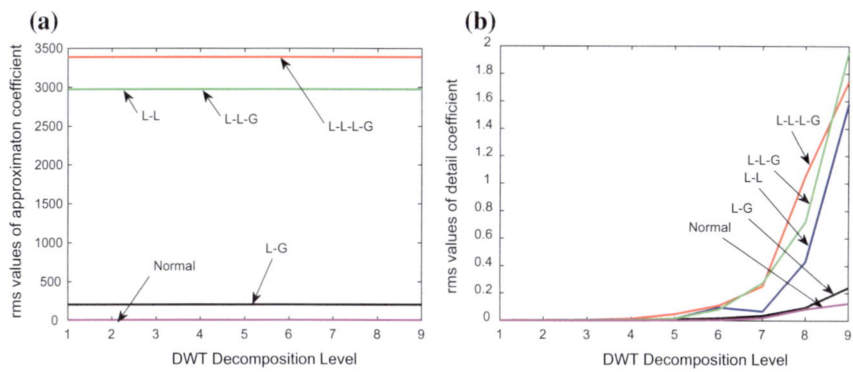

Fig. 4 RMS values of **a** approximation coefficients and **b** detail coefficients versus DWT decomposition levels

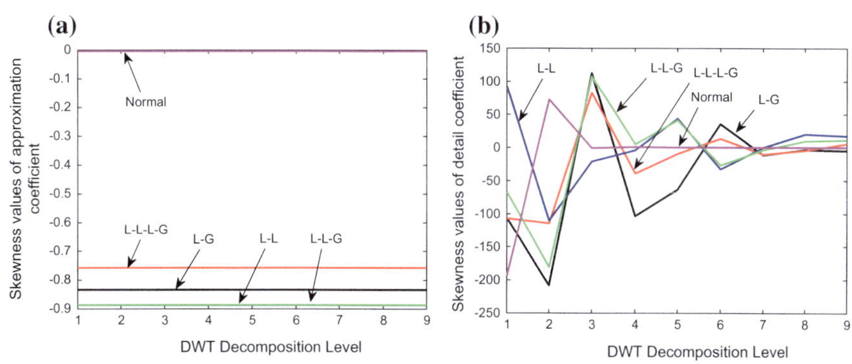

Fig. 5 Skewness values of **a** approximation coefficients and **b** detail coefficients versus DWT decomposition levels

values between normal condition and different fault conditions are almost constant and maximum negative skewness values has been observed in L-L, L-L-G conditions. In Fig. 5b zig-zag differences of skewness values of detail coefficients has been observed in different conditions, where the difference is maximum at DWT decomposition level one (1).

Figure 6a, b is used to show the kurtosis values of approximation and detail coefficients respectively in different conditions with respect to DWT decomposition levels. In Fig. 6a, constant differences has been observed where as in Fig. 6b decreasing differences has been observed up to DWT level six (6) though maximum difference among all those conditions has been observed at DWT decomposition level one (1).

6 Algorithm of Assessment of Different Conditions of Grid Connected Solar PV System

An algorithm for assessment of different conditions of grid connected solar PV system has been made as follows which can be implemented in numerical protection of grid connected solar PV system:

(a) Step down the any phase system current of solar PV side.
(b) Sample it at proper sampling frequency.
(c) Capture the sampled values through data acquisition system.
(d) Calculate THD, inter-harmonics group of the captured signal.
(e) Determine skewness, kurtosis, rms and mean values of approximation (approximate) and detail coefficients at DWT decomposition levels (up to 9th level).
(f) Diagnose the results to assess different conditions of grid connected solar PV system.

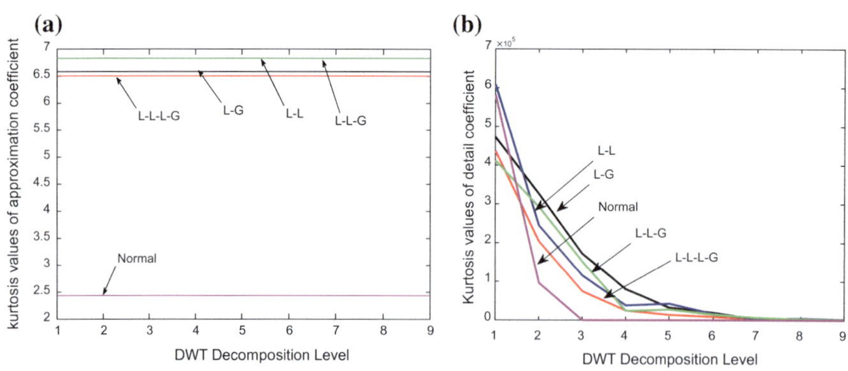

Fig. 6 Kurtosis values of **a** approximation coefficients and **b** detail coefficients versus DWT decomposition levels

7 Specific Outcome

In this paper, Normal condition and different short circuit fault conditions of grid connected solar PV system has been analyzed by THD, inter-harmonics group and DWT based parameter analysis. Minimum THD value has been recorded at normal conditions and maximum THD values has been recorded at L-G fault condition. In case of inter-harmonics group analysis, minimum inter-harmonics has been observed at normal condition where as maximum inter-harmonics has been observed at L-G and L-L-G fault conditions. In DWT based parameter analysis maximum and constant differences among parameter values have been observed for approximation coefficients in different conditions where as in detail coefficients it is different for different cases.

8 Conclusion

In this work, Normal condition and different short circuit fault conditions of grid connected solar PV system has been analyzed by THD, inter-harmonics group and DWT based parameter analysis. Incremental conductance (IC) is used here to extract maximum power from solar PV system which has been fed to the grid and then different fault conditions at grid side have been analyzed. Different THD and different inter-harmonics group values has been observed at different conditions which can be used to assess the fault conditions of grid connected solar PV system. DWT based skewness, kurtosis, mean and rms values also calculated in different conditions for approximation and detail coefficients where maximum differences of these parameters values have been observed for approximation coefficients which can be used further to analyze the fault conditions in grid connected solar PV system. It can be also used for numerical protection of microgrid system.

References

1. S.K. Kim, H. Jeon, C.H. Cho, E.S. Kim, J.B. Ahn, Modeling and simulation of a grid-connected PV generation system for electromagnetic transient analysis. Elsevier J. Solar Energy **83**(5), 664–678 (2009)
2. T.I. Maris, St. Kourtesi, L. Ekonomou, G.P. Fotis, Modeling of a single-phase photovoltaic inverter. Elsevier J. Solar Energy Mater. Solar Cells **91**(18), 1713–1725 (2007)
3. F. Liu, Y. Kang, Y. Zhang, S. Duan, in *Comparison of P&O and Hill Climbing MPPT Methods for Grid-Connected PV Converter*. IEEE Conference on Industrial Electronics and Applications, 2008 (ICIEA, 2008) pp. 804–807
4. F. Liu, S. Duan, F. Liu, B. Liu, Y. Kang, A variable step size INC MPPT method for PV systems. IEEE Trans. Industr. Electron. **55**(7), 2622–2628 (2008)
5. L.W.A. Badi, Z. Zakaria, A.H.M. Nordin, R.F. Mustapa, in *Unbalanced Fault Analysis in Grid-Connected PV System*, IEEE International Conference Power & Energy (2014) pp. 360–365

6. A. Chouder, S. Silvestre, Automatic supervision and fault detection of PV systems based on power losses analysis. Elsevier J. Energy Convers. Manage., 1929–1937 (2010)
7. M. Castilla, J. Miret, J.L. Sosa, J. Matas, L.G.D. Vicuña, Grid-fault control scheme for three-phase photovoltaic inverters with adjustable power quality characteristics. IEEE Trans. Power Electron. **25**(12), 2930–2940 (2010)
8. P.K. Hota, B. Panda, Fault analysis of grid connected photovoltaic system. Am. J. Electr. Power Energy Syst. **5**, 35–44 (2016). https://doi.org/10.11648/j.epes.20160504.12
9. S.S. Kashyap, K.S. Verma, Fault analysis of grid connected current source inverter based PV system. Int. J. Eng. Res. Gen. Sci. **3**(2), 966–981 (2015). ISSN 2091-2730
10. D.M. Naik, D.S. Reddy, Dr. T. Devaraju, Dynamic modeling, control and simulation of a wind and PV hybrid system for grid connected application using MATLAB. Int. J. Eng. Res. Appl. **4**(7)(Version 6), 131–139 (2014). ISSN: 2248-9622
11. S. Chattopadhyay, M. Mitra, S. Sengupta, in *Electric Power Quality*, 1st edn. (Springer, Berlin, 2010) ISBN: 978-94-007-0634-7
12. S. Chattopadhyay, M Mitra, S Sengupta, Area based approach for three phase power quality assessment in clarke plane. J. Electr. Syst. **4**(3), 60–76 (2008). ISSN: 1112-5209
13. S. Chattopadhyay, S. Karmakar, M. Mitra, S. Sengupta, Assessment of crawling of an induction motor by stator current Concordia analysis. IET Electron. Lett. **48**(14), 841–842 (2012)
14. S. Chattopadhyay, M. Mitra, S Sengupta, Area based approach in power quality assessment. Int. J. Power Manage. Electron. ID-147359 (2008). ISSN: 16876679
15. A. Chattopadhyaya, S. Chattopadhyay, J.N. Bera, S. Sengupta, Wavelet decomposition based skewness and kurtosis analysis for assessment of stator current harmonics in a PWM—fed induction motor drive in single phasing condition. AMSE J. 2016-Ser.: Adv. B **59**(1), 1–14 (2016)
16. S. Chattopadhyay, A. Chattopadhyaya, S. Sengupta, Harmonic power distortion measurement in park plane. Elsevier Measurement **51**, 197–205 (2014)
17. S. Chattopadhyay, A. Chattopadhyaya, S. Sengupta, Analysis of stator current of induction motor used in transport system at single phasing by measuring phase angle, symmetrical components, skewness, kurtosis and harmonic distortion in park plane. IET Electr. Syst. Trans., 1–8 (2013)
18. A. Chattopadhyaya, S. Banerjee, S. Chattopadhyay, Assessment of discrimination between inrush and fault current in a power transformer. Can. J. Technol. Innov. **1**, 187–196 (2014)
19. Engineering Statistics Handbook, NIST/SEMATECH e-Handbook of Statistical Methods, NIST, Retrieved 18 Mar. 2012. MATLAB® 7.7

Sub-harmonics Based String Fault Assessment in Solar PV Arrays

Tapash Kr. Das, Ayan Banik, Surajit Chattopadhyay and Arabinda Das

1 Introduction

Application of photo voltaic (PV) system as energy resource in microgrid is becoming more popular. Modeling and simulation of such system have become important area in such system. One of the major issues in such system is monitoring of photo voltaic system and its fault diagnosis. An extensive research work is progressing to study the various aspects of PV array connected inverter system. Kuitche et al. (2014) analyzed in detail about a new dominant failure mechanism for field-aged crystalline silicon photo voltaic modules during extreme weather conditions [1] where, encapsulate discoloration of PV modules were noticed during desert condition. In this attempt, the several risk priority number based quantitative measures and sizable datasheet have been developed for dominant failure modes. New leakage current detection technique for solar photo voltaic modules [2] has been introduced by Dhere et al. (2014) where, different tests under high voltage condition have been performed and characteristics of leakage current for poly crystalline modules dependent on construction, material and different climatic condition have been observed. Hariharan et al. (2016)

T. Kr. Das (✉) · S. Chattopadhyay
Department of Electrical Engineering, GKCIET (Under MHRD, Government of India), Malda, West Bengal, India
e-mail: p_tapash_das@yahoo.co.in

S. Chattopadhyay
e-mail: surajitchattopadhyay@gmail.com

A. Banik
Department of Electrical Engineering,
Cooch Behar Government Engineering College, Cooch Behar, West Bengal, India
e-mail: ayanbanik97@gmail.com

A. Das
Department of Electrical Engineering, Jadavpur University, Kolkata, India
e-mail: adas_ee_ju@gmail.com

S. Chattopadhyay et al. (eds.), *Modelling and Simulation in Science, Technology and Engineering Mathematics*, Advances in Intelligent Systems and Computing 749,
https://doi.org/10.1007/978-3-319-74808-5_25

developed a whole new partial shading based fault investigation system for photo voltaic array [3]. In this attempt, to improve the reliability and efficiency under different irradiation conditions various experiments have been carried out where, photo voltaic array has been classified into three categories. A new optimization method for solar PV system to increase the output under irregular irradiation [4] has been introduced by Obane et al. (2012) where, to minimize losses the reconnection of PV field has been observed. Yi et al. (2017) presented a DC short circuit fault identification technique under low irradiance conditions for photovoltaic systems where, multi-resolution signal decomposition and fuzzy inference based schemes have been used [5]. A new logic based intelligent fault detection system in solar power electronic converter [6] has been developed by Chen and Bazzi (2017). In this attempt the proposed model has been experimentally validated and tested under the environment of different computer simulations and hardware. Hejri and Mokhtari (2017) introduced a new parameterization scheme of solar Photo voltaic cells and module [7]. Cárdenas et al. (2017) proposed another new single diode based parameter extraction technique of solar photo voltaic module [8]. Irradiance and temperature dependence of photovoltaic modules in solar photo voltaic system [9] has been analyzed in detail by Sauer et al. (2015) where an optimized model relative to standard model has been demonstrated for improving the energy yield prediction. A new luminescent materials based technique for photovoltaic modules [10] has been developed by Klampaftis et al. (2015) where, both color and graphical design were used in photo voltaic production technologies to improve the efficiency. Manganiello et al. (2015) reviewed in detail about aging and mismatching mechanisms occurred in photovoltaic modules [11]. In this attempt, cause and effect of aging and mismatching mechanisms of PV cells have been discussed.

A whole new cost effective MOSFET based RC circuit driven educational tool to identify different characteristics of photo voltaic modules [12] has been designed by Dos Santos et al. (2017) where, different data and curves were captured and displayed with the help of digital oscilloscope. In this attempt, estimation of eradiation, maximum power point and temperature of PV module have been analyzed. In some recent work, harmonic assessment based fault diagnosis approaches [13–16] have also been introduced. However, very few works have been found to utilize subharmonics component in monitoring and fault assessment of solar PV arrays.

In this work, an attempt has been taken to study the effect of disconnection of few strings in solar arrays. Based on subharmonic assessment, an algorithm has been proposed for such string fault monitoring and has been validated by simulation.

2 Data Acquisition

PV array based microgrid system has been considered and has been modeled as shown in Fig. 1. Four PV arrays connected in parallel each consisting of 64 strings are used as energy resource. Combination of PV arrays is connected to three phase average model based voltage source converter (VSC) through DC to DC charge controller.

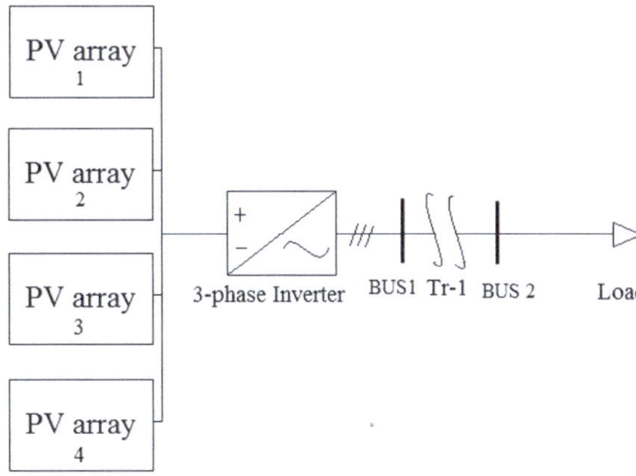

Fig. 1 PV arrays connected with inverter unit

Inverter output is connected to load through three phase 400 KVA, 260 V/25 kV, 60 Hz star/delta transformer.

Current signals from the three phase inverter output terminal have been captured. Data capturing has been carried out both at normal condition and string fault. In simulation current measurement blocks have been connected in series with each phase of inverter output. From the captured current signals various frequencies present in the waveform have been measured for further assessment.

3 Fault Simulation

At normal condition four arrays have been used; each having 64 numbers of strings. All-together 256 numbers of strings has been used in normal condition. Then faults have been created in certain portion of the string. Gradually the numbers of faulty strings have been increased from 0 to 25%. For each condition current signal is captured and harmonic assessment has been done out on starting transient. Inverter output current at normal condition (0% String fault) has been shown Fig. 2 and inverter output current at 9.38% string fault has been shown in Fig. 3.

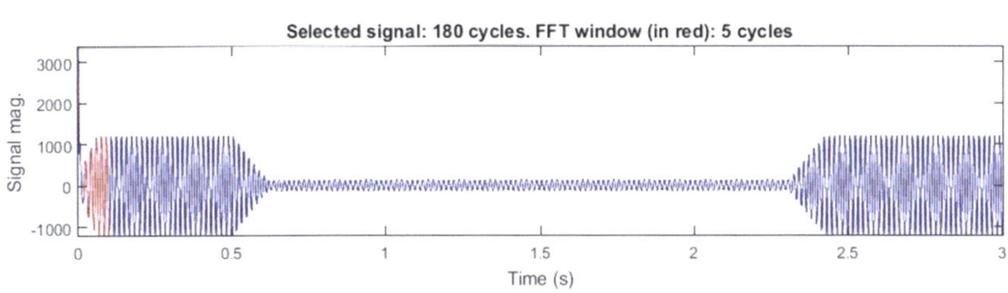

Fig. 2 Inverter output current at normal condition (0% string fault)

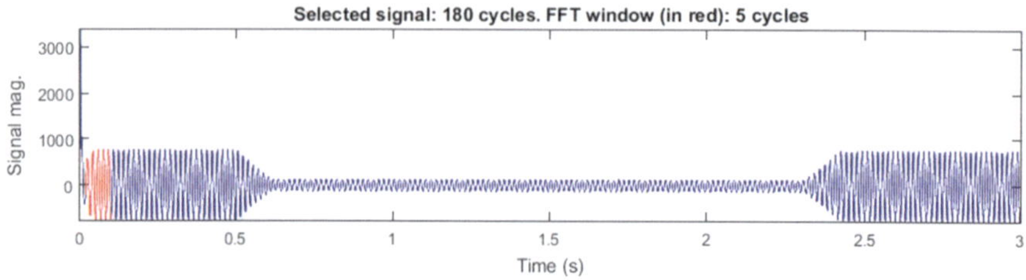

Fig. 3 Inverter output current at 9.38% string fault condition

4 Sub-harmonics Feature Extraction

Sub-harmonics component with 12 Hz interval have been measured from the inverter output current signal both at normal condition as well as string fault conditions. String fault has been considered up to 25% considering 256 numbers of total strings. String fault refers disconnection of faulty strings or inactiveness of faulty strings in contributing energy to the microgrid system through inverter input terminals. The magnitudes of sub-harmonic components present in the inverter output current at different conditions have been presented in Table 1.

Magnitudes of sub-harmonic components versus frequency of sub-harmonic components of have been shown in Fig. 4 and their comparative bar chart representation has been shown Fig. 5. The Fig. 4 shows that sub-harmonic components having frequencies of 12 and 36 Hz are decreasing gradually with the increase in percentage of fault. The other two components remain almost constant.

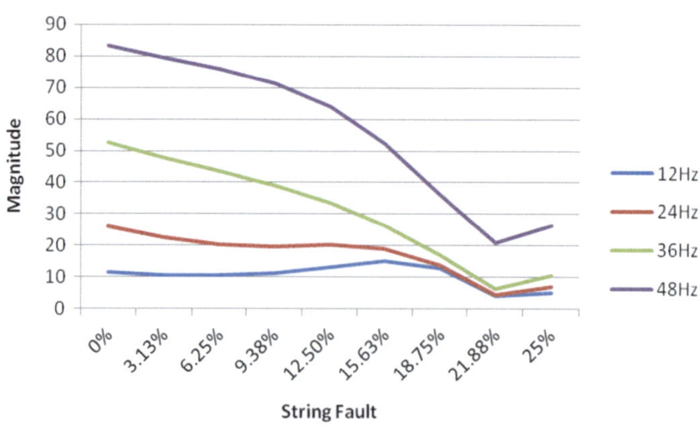

Fig. 4 Magnitude of sub-harmonics versus percentage of string fault

Table 1 Magnitudes of sub-harmonic components present in inverter output current at different conditions

Percentage string fault (%)	0	3.125	6.25	9.375	12.5	15.625	18.75	21.875	25
Frequency (Hz)	Magnitude of sub-harmonics components								
12	11.5	10.47	10.38	11.12	12.98	15.04	12.74	3.94	5.03
24	25.9	22.5	20.16	19.61	20.12	18.92	13.86	4.44	6.91
36	52.46	47.52	43.4	38.9	33.46	26.31	17.05	6.27	10.34
48	83.25	79.3	75.87	71.28	63.9	52.34	36.24	20.99	26.36

5 Curve Selection and Linier Approximation

Among the four relations shown in Figs. 4 and 5, Amplitude (A) versus string fault(S_F) corresponding to sub-harmonic component having frequency 48 and 36 Hz are very close to linear nature and therefore these two natures have selected for further assessment.

Linear approximation of these two natures has been done as shown in Fig. 6. Mathematical equation of these two approximate linearized natures has been determined as follows:

$$\text{for 48Hz, } A_{48} = -2.751\ S_F + 86 \quad (1)$$

$$\text{for 36 Hz, } A_{36} = -1.981\ S_F + 55 \quad (2)$$

from Eq. (1) string fault(S_F) can be written as

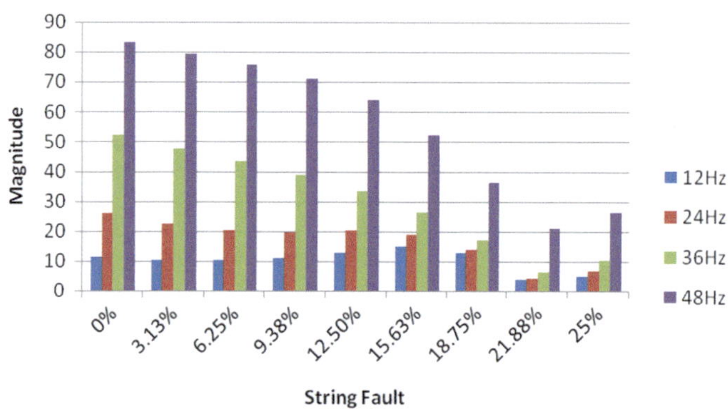

Fig. 5 Sub-harmonics amplitude comparison

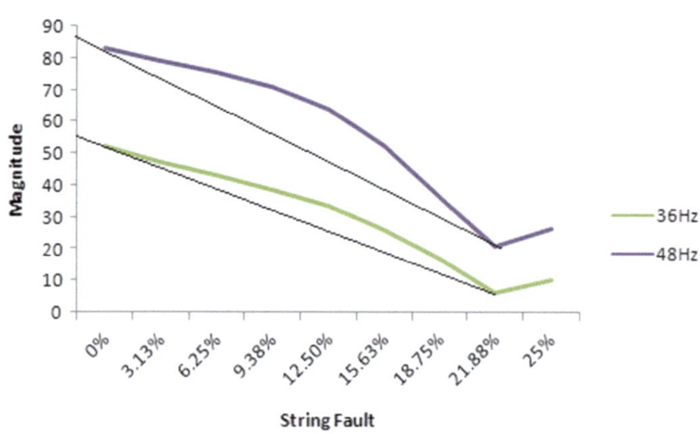

Fig. 6 Linear approximate nature of selected sub-harmonic components

$$S_F = \frac{86 - A_{48}}{2.751} \tag{3}$$

from Eq. (2) string fault(S_F) can be written as

$$S_F = \frac{55 - A_{36}}{1.981} \tag{4}$$

Equations (3) and (4) have been used for assessment of string faults.

6 Algorithm and Validation

6.1 Algorithm

Based on curve selection and linear approximation mathematical expressions have been established by Eqs. (3) and (4) among amplitudes of selected sub-harmonics components and percentage of string fault. Using these equations an algorithm proposed as follows:

i. Capture inverter output current
ii. Select starting transient
iii. Determine sub-harmonic component
iv. Assess string fault using Eqs. (3) and (4).

6.2 Validation

The above algorithm has been validated in the same type of system. As in practical cases percentage of string faults remains low at initial stage, validation has been done for very small percentage of string fault conditions (considering 5 and 10 numbers faulty strings). First amplitude of 48 and 36 Hz components of inverter output currents corresponding to 5 and 10 numbers faulty strings have been determined and then percentage of string faults have been calculated using Eqs. (3) and (4). The results obtained in validation have been presented in Table 2.

Comparative study of the results obtained in Table 2 shows very similarities with the actual. Among 48 and 36 Hz the result obtained from the equation using 36 Hz of sub-harmonic component is closer to the actual than other.

Table 2 Percentage of string fault obtained in validation

Known number of faulty string	Amplitude of sub-harmonic components	Percentage of string fault
5 (1.95%)	$A_{36} = 51.56$	1.736
	$A_{48} = 82.54$	1.276
10 (3.91%)	$A_{36} = 50.71$	2.165
	$A_{48} = 81.36$	1.712

6.3 Discussion

Accuracy of Eq. (3) is better than that of Eq. (4) this is because original nature of amplitude versus percentage of string fault for 36 Hz sub-harmonic component is more nearer to linear nature. The accuracy of the proposed algorithm can be improved by using polynomial expression of the natures of sub-harmonic components. It may be noted that the best frequency component is system dependent and may vary in other system. However, the present approach may be an effective means of monitoring string fault. Use of starting transient reduces the effect of loading in the relative amount of different sub-harmonic components.

7 Conclusion

The work has presented monitoring of string fault in PV arrays. An attempt has been taken to monitor the string fault which refers to disconnection of few strings connected with PV arrays feeding energy to the inverter. Waveforms of inverter output currents have been captured and monitored. Staring transient of inverter output current has been selected and sub-harmonic components have been extracted. Specific relations have been observed among for amplitudes of selected frequencies versus string faults. Frequency selection has been carried out and linear approximations have been done in the relation between amplitude versus percentage of string faults. At the end, algorithm has been proposed for monitoring the number of faulty strings and has been validated; limitation as well as utility has been highlighted.

References

1. J.M. Kuitche, R. Pan, G. TamizhMani, Investigation of dominant failure mode(s) for field-aged crystalline silicon PV modules under desert climatic conditions. IEEE J. Photovoltaics **4**(3), 814–826 (2014)
2. N.G. Dhere, N.S. Shiradkar, E. Schneller, Evolution of leakage current paths in MC-Si PV modules from leading manufacturers undergoing high-voltage bias testing. IEEE J. Photovoltaics **4**(2), 654–658 (2014)

3. R. Hariharan, M. Chakkarapani, G. Saravana Ilango, C. Nagamani, A method to detect photovoltaic array faults and partial shading in systems. IEEE J. Photovoltaics **6**(5), 1278–1285 (2016)
4. H. Obane, K. Okajima, T. Oozeki, T. Ishii, PV system with reconnection to improve output under nonuniform illumination. IEEE J. Photovoltaics **2**(3), 341–347 (2012)
5. Z. Yi, A.H. Etemadi, Fault detection for photovoltaic systems based on multi-resolution signal decomposition and fuzzy inference systems. IEEE Trans. Smart Grid **8**(3), 1274–1283 (2016)
6. W. Chen, A.M. Bazzi, Logic-based methods for intelligent fault diagnosis and recovery in power electronics. IEEE Trans. Power Electron. **32**(32), 5573–5589 (2017)
7. M. Hejri, H. Mokhtari, On the comprehensive parameterization of the photovoltaic (PV) cells and modules. IEEE J. Photovoltaics **7**(1), 250–258 (2017)
8. A.A. Cárdenas, M. Carrasco, F. Mancilla-David, A. Street, R. Cárdenas, Experimental parameter extraction in the single-diode photovoltaic model via a reduced-space search. IEEE Trans. Ind. Electron. **64**(2), 1468–1476 (2017)
9. K.J. Sauer, T. Roessler, C.W. Hansen, Modeling the irradiance and temperature dependence of photovoltaic modules in PVsyst. IEEE J. Photovoltaics **5**(1), 152–158 (2015)
10. E. Klampaftis, D. Ross, G. Kocher-Oberlehner, B.S. Richards, Integration of color and graphical design for photovoltaic modules using luminescent materials. IEEE J. Photovoltaics **5**(2), 584–590 (2015)
11. P. Manganiello, M. Balato, M. Vitelli, A survey on mismatching and aging of PV modules: the closed loop. IEEE Trans. Ind. Electron. **62**(11), 7276–7286 (2015)
12. R.L. Dos Santos, J.S. Ferreira, G.E. Martins Jr., K.C.A. de Souza, Low cost educational tool to trace the curves PV modules. IEEE Latin Am. Trans. **15**(8), 1392–1399 (2017)
13. S. Chattopadhyay, M. Mitra, S. Sengupta, Power quality analysis in park plane, lectures on modeling and simulation. AMSE **7**(1)(No. 1-2-3), 8–15 (2006), ISSN: 1961-5086
14. S. Chattopadhyay, M. Mitra, S. Sengupta, Intelligent power quality monitoring in an electric power system, lectures on modeling and simulation, AMSE **5**(2)(No. 1-2-3), 146–156 (2004), ISSN: 1961-5086
15. A. Chattopadhyaya, S. Chattopadhyay, J.N. Bera, S. Sengupta, Wavelet decomposition based skewness and kurtosis analysis for assessment of stator current harmonics in a PWM—fed induction motor drive during single phasing condition. AMSE J. 2016 Ser.: Adv. B **59**(1), 1–14 (2016)
16. S. Chattopadyay, A. Chattopadhyaya, S. Sengupta, Measurement of harmonic distortion and skewness of stator current of induction motor at crawling in Clarke plane, IET Sci. Measur. Technol. **8**(6), 528–536 (2014). ISSN: 1751-8822, https://doi.org/10.1049/iet-smt.2013.0082

Part IV
Control Techniques

Design of Bacterial Foraging Optimization Algorithm Based Adaptive Sliding Mode Controller for Inverted Pendulum

Rajeev Ranjan Pathak and Anindita Sengupta

1 Introduction

Traditional Sliding Mode Control, also called variable structure systems (VSS), was designed for the systems with uncertainties due to its excellent robustness. But it also presents some drawbacks such as undesired large control authority and chattering problem. These drawbacks were overcome by Slotine and Sastry [1] by replacing control switching at a fixed sliding surface by a smooth control interpolation in a boundary layer neighboring a time varying sliding surface. This eliminated the excitations of high-frequency un-modeled dynamics and leads to an explicit trade-off between the model uncertainty and controller tracking performance. In recent years a large amount of research into the area of sliding control has been developed. Fernanderz and Hedrick [2] generalized the sliding mode approach to a larger class of multivariable systems. Shyu and Tsai [3] introduced a multiple sliding surfaces in the sliding controller design to eliminate control chattering. In Oh and Khalil [4] a VSS controller with a high-gain observer was designed as a function of the state estimates to ensure attractiveness of sliding manifold. Won and Hedrick [5] developed a 'multiple-surface' sliding control for a class of single-input single-output (SISO) non-linear systems whose uncertainties do not satisfy the matching condition. Edwards and Spurgeon [6] presented a controller/observer pair based on sliding mode ideas, provided robust output tracking of a reference signal. Chen and Toshio [7, 8] used the transfer function method based on the VSS theory to estimate the disturbance. The estimated disturbance was employed to construct a VSS-type state observer. Further, the estimated disturbance and the state observer were applied to a

R. R. Pathak (✉) · A. Sengupta
Electrical Engineering Department, IIEST, Shibpur, Howarh, India
e-mail: rajeevpathak27@yahoo.in

A. Sengupta
e-mail: aninsen2002@yahoo.com

S. Chattopadhyay et al. (eds.), *Modelling and Simulation in Science, Technology and Engineering Mathematics*, Advances in Intelligent Systems and Computing 749,
https://doi.org/10.1007/978-3-319-74808-5_26

controller to place the desired stable poles and to cancel the disturbance [8]. A common assumption of these studies was that the bounds of model uncertainties must be known. If the bounds of model uncertainties were unknown, traditional sliding control schemes would not work.

In this work, a new Adaptive Sliding Control scheme is proposed for non-linear systems with unknown bound time varying uncertainties, Sliding-Mode Control theory and adaptive control scheme has been used to solve the tracking reference input for non-linear systems with bounded unknown parameters and external disturbances. First, a non linear system has been linearized using feedback linearization technique with the assumption of system parameters which are known a priori. Second, the structure of adaptive sliding-mode controller has been developed using the so-called certainty equivalent principle of adaptive control theory. At this point we consider appropriate techniques to eliminate the effect of approximation errors and disturbances to achieve a desired control objective. The gain of Adaptive SMC has been optimized using BFOA. The proposed method is not only robust against approximate errors, disturbances and un-modeled dynamics, but also guarantees the overall system stability.

2 The Principle of Bacterial Foraging Optimization Algorithm

Bacteria Foraging Optimization Algorithm (BFOA), proposed by Passino [9], is a new comer to the family of nature-inspired optimization algorithms. BFOA [10] is developed by simulating the bacterial foraging process of human E. coli, which contains three steps: chemotaxis, reproduction and finally Elimination and Dispersal operation.

2.1 *Chemotaxis Operation*

This process simulates the movement of an E. coli cell through swimming and tumbling via flagella. Biologically an E. coli bacterium can move in two different ways [11]. It can swim for a period of time in the same direction or it may tumble and alternate between these two modes of operation for the entire lifetime. Suppose $\theta^i(j, k, l)$ represents ith bacterium at jth chemotaxis, kth reproductive and lth elimination-dispersal step and $C(i)$ is the size of the step taken in the random direction specified by the tumble (run length unit). Then in computational chemotaxis the movement of the bacterium may be represented by

$$\theta^i(j+1, k, l) = \theta^i(j, k, l) + C(i)\frac{\Delta(i)}{\sqrt{\Delta^T(i)\Delta(i)}} \tag{1}$$

where Δ indicates a vector in the random direction whose elements lie in $[-1, 1]$.

2.2 Reproduction

The least healthy bacteria eventually die when each of the healthier bacteria (which yielding lower value of the objective function) asexually split into two bacteria, which are then placed in the same location. This keeps the swarm size constant.

2.3 Elimination and Dispersal Operation

Gradual or sudden changes in the local environment where a bacterium population lives may occur due to various reasons. Events can occur such that all the bacteria in a region are killed or a group is dispersed into a new part of the environment. For example, a significant local rise of temperature may kill a group of bacteria that are currently in a region with a high concentration of nutrient gradients. Events can take place in such a fashion that all the bacteria in a region are killed or a group is dispersed into a new location. Over long periods of time, such events had spread various types of bacteria into every part of our environment from our intestines to hot springs and underground environments. To simulate this phenomenon in BFOA some bacteria are liquidated at random with a very small probability while the new replacements are randomly initialized over the search space. Elimination and dispersal events have the effect of possibly destroying chemotaxis progress, but they also have the effect of assisting in chemotaxis, since dispersal may place the bacteria near good food sources. From a broad perspective, elimination and dispersal are parts of the population-level long-distance motile behavior.

3 The Fundamental Principle of Adaptive Sliding Mode Controller

Sliding mode control, or SMC, is a non-linear control method that alters the dynamics of a system by application of a discontinuous control signal that forces the system to “slide” along a sliding surface (as designed by the designer). The control law of SMC is based on state feedback control law, but it is not a continuous function of time. Instead, it can switch from one continuous structure to another based on the current position in the state space. Hence, sliding mode control is a variable structure control method. The main advantages of this approach are:

(A) While the system is on the sliding manifold it behaves as a reduced order system with respect to the original plant.

(B) The dynamic of the system, while in sliding mode is insensitive to model uncertainties and disturbances.

Adaptive control [12] is an appealing approach for controlling uncertain dynamic systems. In principle, the systems can be uncertain in terms of its dynamic structure or its parameters. So far, however, Adaptive control can only deal with parameter-uncertain systems. Furthermore, existing adaptation methods generally require linear parameterization of the control law or the system dynamics. Systematic theories exist on how to design adaptive controllers for general linear systems. If the full state is available, adaptive control design and implementation is quite simple. If only output feedback is available, however, adaptive control design is much more involved because of the need to introduce dynamics into the controller. Some classes of nonlinear systems can also be adaptively controlled (Fig. 1).

4 Design of Adaptive Sliding Mode Controller Based on BFOA

4.1 Design of BFOA Encoding

Adaptive Sliding Mode controller is an algorithm based on the estimation of the information of the past, present and future. The control system is mainly combined with Adaptive Sliding Mode Controller and the optimal gain of controller has been obtained using BFOA encoding. Schematic diagram of Inverted Pendulum is shown in Fig. 2.

State space model of lab based inverted pendulum is given by

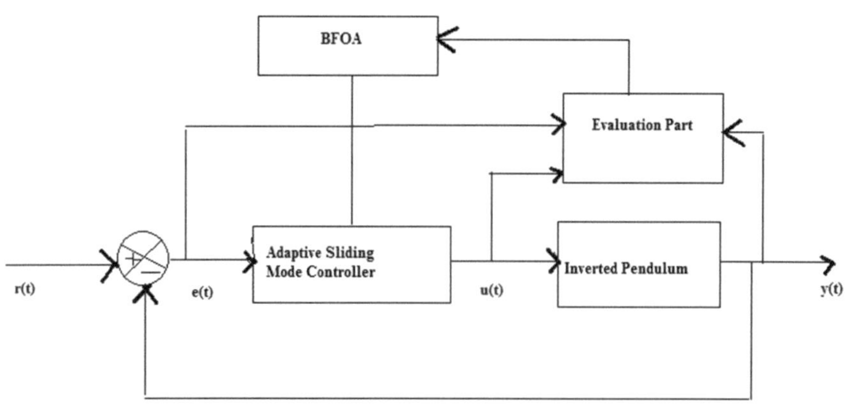

Fig. 1 Adaptive sliding mode controller based on BFOA

Fig. 2 Schematic diagram of inverted pendulum

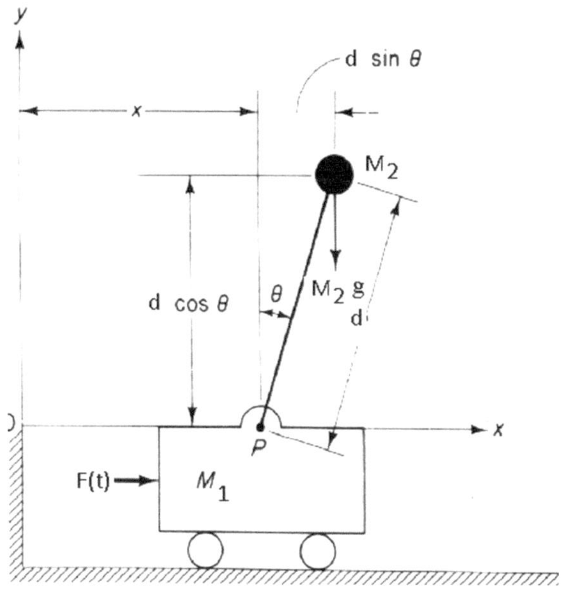

$$\begin{bmatrix} \dot{x}_1 \\ \dot{x}_2 \\ \dot{x}_3 \\ \dot{x}_4 \end{bmatrix} = \begin{bmatrix} 0 & 1 & 0 & 0 \\ 0 & 0 & -0.97 & 0 \\ 0 & 0 & 0 & 1 \\ 0 & 0 & 21.57 & 0 \end{bmatrix} \begin{bmatrix} x_1 \\ x_2 \\ x_3 \\ x_4 \end{bmatrix} + \begin{bmatrix} 0 \\ 0.9124 \\ 0 \\ -1.824 \end{bmatrix} u(t)$$

$$Y = \begin{bmatrix} 0 & 0 & 1 & 0 \end{bmatrix} \begin{bmatrix} x_1 \\ x_2 \\ x_3 \\ x_4 \end{bmatrix} \tag{2}$$

where

Linear displacement of cart	$(x) = x_1$
Linear velocity of cart	$\dot{x} = x_2$
Angular position of pendulum	$\theta = x_3$
Angular velocity of pendulum	$\dot{\theta} = x_4.$

Adaptive Sliding Mode Controller [13, 14] designed for inverted pendulum is given below

$$u_{eq} = \frac{1}{1.824(21.57 + 3\lambda^2)}\left[\left(465.27 + 64.71\lambda^2\right)x_3 + \left(64.71\lambda + \lambda^3\right)x_4\right] \tag{3}$$

where u_{eq} is the equivalent control law of ASMC, and λ is gain of ASMC, our main objective is to find optimal gain λ using BFOA technique

That is the objective function, means the index of the fitness value which is calculated by using of input variables, output variables and intermediate variables. Although different control systems have different objective fitness function and even the control systems have changed dramatically, the algorithm has little change relatively. Even the transfer function has been changed, gain of controller can be obtained easily, this is the fundamental principle of Adaptive Sliding Mode controller which can be realized.

4.2 Design of Fitness Function

Fitness function is obtained from objective function of control system. It is the key performance index which determines the stability of control system [15]. The objective function of Adaptive Sliding Mode controller can be determined based on the principle of ASMC as:

$$J = 0.3IAE + 0.7\,ITAE = 0.3\int_0^{\infty} |e(t)| + 0.7\int_0^{\infty} t|e(t)| \tag{4}$$

Fitness function is F = 1/J.

Chemotaxis operation is the main part of algorithm, which determines the algorithm's local search ability and direction. The individual of the population is good or not is determined by the index of fitness functions' value. If a bacterium obtains a better fitness function value at a new location, the information of the position will be stored, and then the bacterium continues to swim. Otherwise the storage value is not changed until up to the maximum number of steps Ns.

Reproduction and elimination–dispersal operations are to speed up the convergence of the control system, and also to avoid local optimum. The sorting method is used in this research work with elimination dispersion probability 0.5.

4.3 Flow Chart of BFOA Encoding

See Fig. 3 and Table 1.

5 Experimental Simulation and Results Analysis

A Simulation design of Adaptive Sliding Mode Controller has been framed for lab based Inverted Pendulum for tracking angular position of Inverted Pendulum having unknown parameters and unbound disturbances. The gain of ASMC has been optimized using BFOA and its value is 102.1 for given objective function. The no

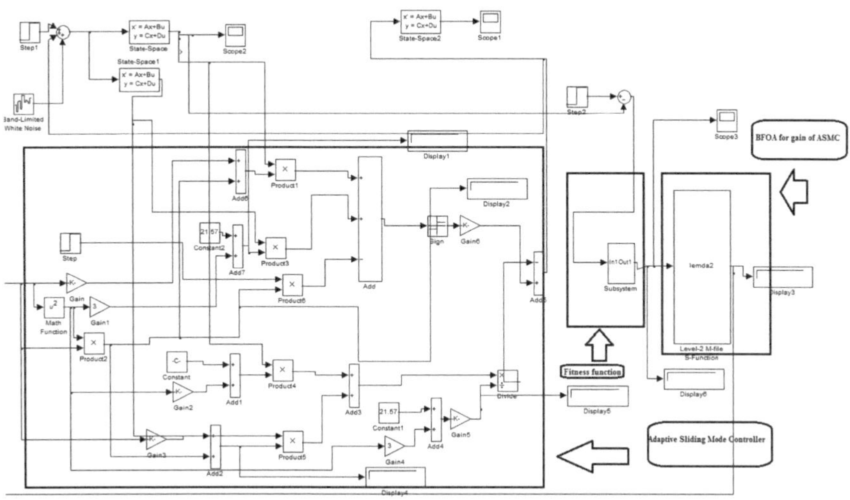

Fig. 3 Simulink model of inverted pendulum with sliding mode controller

Table 1 Control parameters of bacterial foraging optimization algorithm

Sl No.	Parameter	Values
1	Number of bacterium, S	30
2	Number of chemotaxis steps, Nc	10
3	Number of reproduction steps, Nre	4
4	Number of elimination-dispersal steps, Ned	4
5	Elimination dispersal probability, Ped	0.5
6	Size of step C(i)	0.1

bacteria taken for simulation is 30 and it has been found that, the simulation process slows down when the no of bacteria is increased (Fig. 4).

The Inverted Pendulum is found to be highly unstable non linear system. In that case initially PID Controller was used to track the reference angular position of pendulum. But it was found that in terms of external unbound disturbance, the PID could not track the desired reference signal. The same tracking response is shown in Fig. 5.

Finally AMSC was applied to the Inverted Pendulum which could track the angular reference signal even in presence of unbound external disturbance and unknown Parameter uncertainties which is shown in Fig. 6.

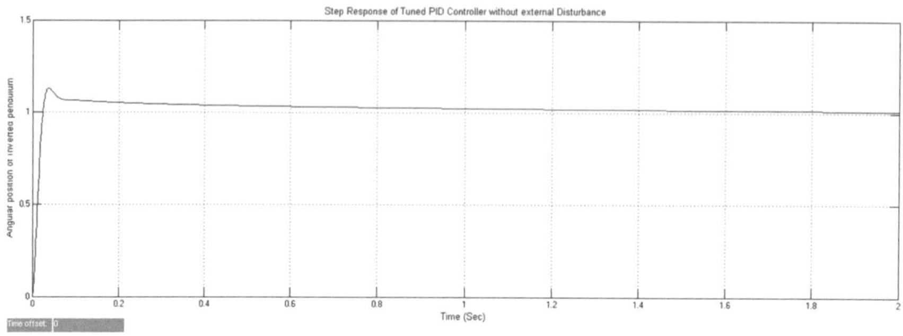

Fig. 4 Tracking response of angular position of inverted pendulum with PID controller without external unbounded disturbances

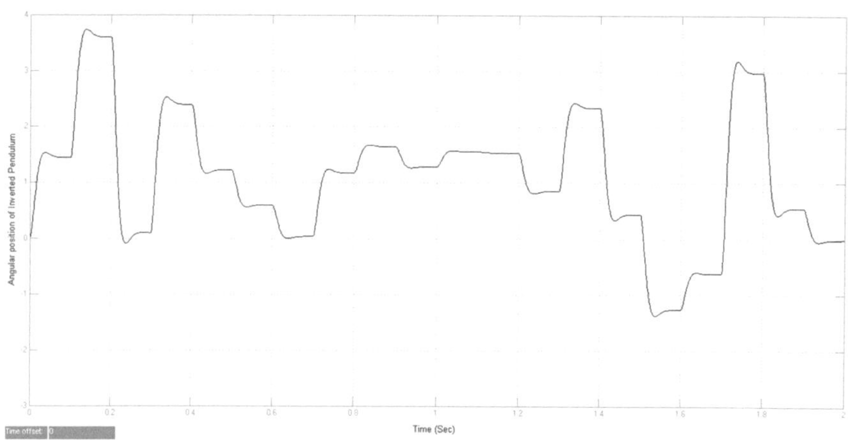

Fig. 5 Tracking response of angular position of inverted pendulum with PID controller with external disturbances

6 Comparison of Different Controller Performances

See Table 2.

7 Conclusion

According to the above principle and simulation results, the parameter of different control system can be easily proposed by just changing initialization parameter by using BFOA. Adaptive Sliding Mode Controller shows good robustness properties in presence of known external disturbances as compared to that shown by conventional PID controller.

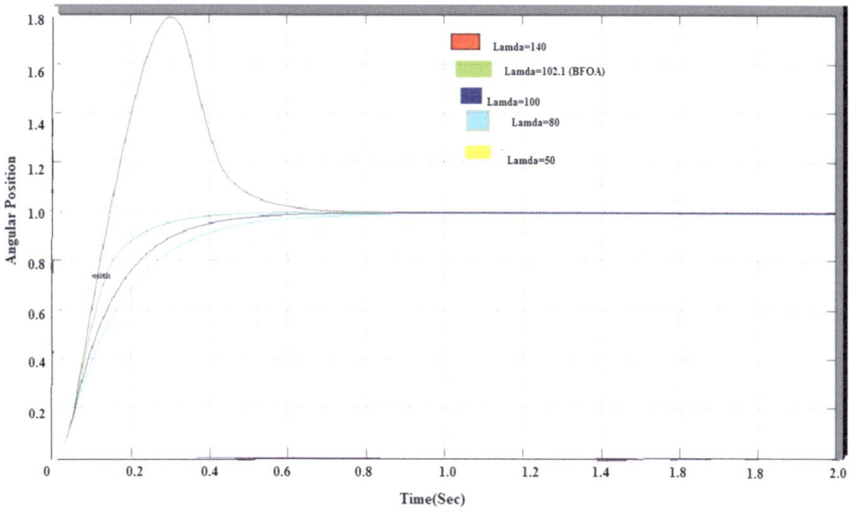

Fig. 6 Tracking response of inverted pendulum with adaptive sliding mode controller different value of lamda

Table 2 Comparison of different controllers for time = 2 s

Different controller	Gain of controller	J = 0.3IAE + 0.7ITAE		Remark
		Without disturbance	With disturbance	
Adaptive sliding mode Controller with different gain	Lamda = 20	0.3434	0.4434	
	Lamda = 50	0.09342	0.09942	
	Lamda = 80	0.057	0.097	
	Lamda = 100	0.04827	0.05827	
	Lamda = 130	0.0474	0.0574	
	Lamda = 140	0.05278	0.05278	
BFOA based ASMC	**Lamda = 102.1**	**0.04385**	**0.04352**	**Minimum value of J among all controller**

References

1. J.J.E. Slotine, S.S. Sastry, Tracking control of non-linear systems using sliding surfaces, with application to robot manipulators. Int. J. Control **38**, 465–492 (1983)
2. B. Fernanderz, J.K. Hedrick, Control of multivariable non-linear systems by the sliding mode method. Int. J. Control **46**, 1019–1040 (1987)
3. K.K. Shyu, Y.W. Tsai, C.F. Yung, A modified variable structure controller. Automatica **28**, 1209–1213 (1992)

4. S. Oh, H.K. Khalil, Output feedback stabilization using variable structure control. Int. J. Control **62**, 831–848 (1995)
5. M. Won, J.K. Hedrick, Multiple-surface sliding control of a class of uncertain non-linear systems. Int. J. Control **64**, 693–706 (1996)
6. C. Edwards, S. Spurgeon, Robust output tracking using a sliding-mode controller/observer scheme. Int. J. Control **64**, 967–983 (1996)
7. X. Chen, T. Fukuda, Variable structure system theory based disturbance identifications. Int. J. Control **68**, 373–384 (1997)
8. C. Xinkai, F. Toshio, Variable structure system theory based disturbance identification and its applications. Trans. Int. J. Control **68**(2), 373–384 (1997)
9. K.M. Passino, Biomimicry of bacterial foraging for distributed optimization and control. IEEE Control Syst. Mag., 52–67 (2002)
10. S. Das, A. Biswas, S. Dasgupta, in *Bacterial Foraging Optimization Algorithm: Theoretical Fundamental Analysis and Application*, Foundations of Computer Intel (2009), pp. 23–55
11. S. Dasgupta, S. Das, A. Abraham, Adaptive computational chemotaxis in bacterial foraging optimization. IEEE Trans. Evol. Comput., 919–941 (2009)
12. S.S. Sastry, A. Isidori, Adaptive control of linearizable systems. IEEE Trans. Autom. Control, 1123–1131 (1989)
13. A. Dhar, A. Sengupta, in *Sliding Mode Control Algorithm with Adaptive Gain and Implementation on Inverted Pendulum System*, IET International Summit MFIIS-2015 (2015), pp. 8–13
14. A. Sen, A. Sengupta, in *Parameter Selection Strategy for Robust Sliding Mode Controller and Its Implementation in Real Time System*, IEEE International Conference on Power Electronics, Intelligent Control and Energy Systems(ICPEICES) (2016), pp. 1–6
15. W. Guozhong, in *Application of Adaptive PID Controller Based on Bacterial Foraging Optimization Algorithm*, 25th Chinese Control and Decision Conference (2013), pp. 2353–2356

Design of Sliding Mode Excitation Controller to Improve Transient Stability of a Power System

Asim Halder, Debasish Mondal and Manas kr. Bera

1 Introduction

The stability problem due to poor damping of the electromechanical oscillations is the intrinsic dynamic characteristics of an electrical power system. Power transfer in a single machine power system (SMIB) or in a large interconnected power systems may be limited causing this poor damping of the electrical machines. To improve the power transfer capability of SMIB or interconnected power systems, damping controllers are generally used. The most economical and common damping controllers are the power system stabilizers (PSSs) [1]. PSSs are the type of feedback controllers installed at generators and they use locally available signals as their inputs. Though PSSs served the purpose partially and lot of research work has been done for the same, but still there is a scope to improve transient stability of the power systems.

To improve the small signal stability of the power systems a nonlinear control scheme [2] has been proposed for Thyristor Controlled Series Capacitor (TCSC) to damped out the electromechnical oscillation of the power system. In this method the affine nonlinear model of a SMIB system with TCSC controller has been linearized with the help of feedback linearization and transformed to a linear system. Then, the control law for TCSC has been derived from that linear model. A different approach

A. Halder (✉)
Applied Electronics and Instrumentation Engineering, Haldia Institute of Technology, M.A.K.A.U.T., Haldia, Kolkata, India
e-mail: asim_calcutta@yahoo.com

D. Mondal
Electrical Engineering, RCC Institute of Information Technology, M.A.K.A.U.T., Kolkata, India
e-mail: mondald12@yahoo.in

M. kr. Bera
Electronics and Instrumentation Engineering, National Institute of Technology (NIT), Silchar, India
e-mail: manas.bera@gmail.com

S. Chattopadhyay et al. (eds.), *Modelling and Simulation in Science, Technology and Engineering Mathematics*, Advances in Intelligent Systems and Computing 749,
https://doi.org/10.1007/978-3-319-74808-5_27

of exact linearization process of nonlinear model of power system has been reported in [3]. Here, at first optimal parameters are introduced to the system and then power system is exactly linearized without any approximation. In [4] a nonlinear control strategy has been proposed for the Static Synchronous Series Compensator (SSSC) in which the control strategy has been derived from the Lyapunov's theory. A different approach, know as Zero Dynamic Design, has been proposed in [5] where the system is partially linearizable. Here, it has been proposed that if the system is not fully or exactly linearizable, then the zero dynamic design approach is the effective and simpler one to derive control law.

To improve the transient stability of power systems, coordinated control schemes has also been proposed by some researchers. A coordinated excitation and governor control system approach has been presented in [6]. This method shows that the coordination control system between nonlinear robust excitation control (NREC) and governor power system stabilizer (GPSS) in multimachine power system in order to mitigate transient oscillations. Another approach of coordinated control design has been depicted in [7], in which a coordinated H_∞ controller is designed for excitation and governor of hydroturbo-generator sets on the basis of port-controlled Hamiltonian (PCH) method. In [8] a TCSC controller along with a robust nonlinear generator excitation controller has been designed as a nonlinear coordinated controller to enhance the transient stability of power systems. A probabilistic approach has been proposed in [9] to design coordinate controller. Here, the parameters of power system stabilizers (PSSs) and static VAR compensator (SVC) has been coordinated and optimized to improve the probabilistic small-signal stability of power systems with large-scale wind generation. In [10] another FACTS device, Static Synchronous Compensator (STATCOM), has been considered to design nonlinear coordinated STATCOM and excitation controller to mitigate transient oscillations of SMIB system.

It has been observed that most of the research work of the design of conventional or nonlinear damping controllers are based on the information of the operating conditions of the system and proper tuning of the parameters of the controllers. These methods have serious limitations like slugiesh convergence property, complex computation process etc. Some research works based on global optimization methods like Artificial Neural Network (ANN), Genetic Algorithm (GA), Particle Swarm Optimization (PSO) and Evolutionary Programming (EP) has been reported in [11–14]. These methods showed reliable optimization of several power system problems.

This paper presents the design of a nonlinear excitation controller based on Super twisting Sliding Mode Control along with robust exact differentiator to damp the transient stability as well as to reject external disturbances of an SMIB system. Further, its performance has been compared with Zero Dynamic Design of nonlinear excitation controller. The simulation results predict that the proposed nonlinear controller is more advantageous and effective. To the best of author's knowledge the proposed work has not been explored in the existing literature.

The whole paper is organized as follows; Sect. 2 describes the design process of excitation controller which consists of three sub-sections. The Sect. 2.1 describes the affine form of nonlinear model of SMIB system and the estimation of relative degree has been described in Sect. 2.2. The derivation of control law based on STSMC has been presented in Sect. 2.3. Section 3 describes the design of excitation controller based on Zero Dynamic Design Approach. Finally, the performances of nonlinear controller has been compared with the Zero Dynamic Design based controller in Sect. 4.

2 Design of Nonlinear Excitation Controller Based on STSMC

2.1 *Affine Model of an SMIB System*

To develop the nonlinear control scheme, a simple third-order power system model of Single Machine Infinite Bus (SMIB) system is adopted in Fig. 1. The basic equations of motion of a SMIB power system can be written as [15];

$$\dot{E}_q' = -\frac{1}{T_d'}E_q' + \frac{1}{T_{do}}\frac{x_d - x_d'}{x_{d\sum}'}V_s\cos\delta + \frac{1}{T_{do}}V_f \tag{1}$$

$$\dot{\omega}(t) = \frac{\omega_s}{H}\left(P_M - \frac{D(\omega(t) - \omega_s)}{\omega_s} - P_e\right) \tag{2}$$

$$\dot{\delta}(t) = \omega(t) - \omega_s \tag{3}$$

where, V_f is the excitation control voltage, δ: rotor angle of the machine, ω: machine speed. P_M and $P_e = \frac{E_q' V_s \sin\delta}{X_{d\sum}'}$ are the mechanical and electrical power of the generator in p.u. respectively. H is the inertia constant and D is the damping coefficient. ω_s: synchronous speed. $T_d' = T_{do}\frac{x_{d\sum}'}{x_{d\sum}}$ = time constant of the field winding for closed stator circuit. $x_{d\sum} = x_d + x_L + x_T$ and $x_{d\sum}' = x_d' + x_L + x_T$. x_d and x_d' are direct-axis synchronous and transient reactance respectively. x_T, x_L, E_q' have their usual meanings.

The Eqs. (1)–(3) can be rewritten in an affine nonlinear from as follows;

$$\dot{X} = f(X) + g(X)u \tag{4}$$

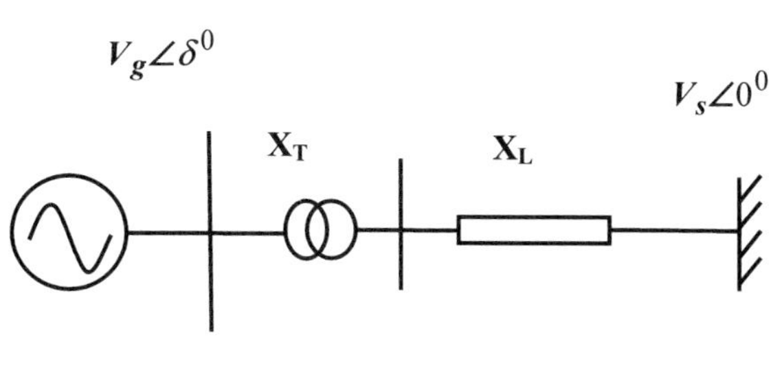

Fig. 1 Simplified model of SMIB system

$$y = h(X) \tag{5}$$

where, $X = \left[E'_q\ \omega\ \delta\right]^T$; with the initial condition $X(0) = X_0$; $X_0 = \left[E'_{q0}\ \omega_0\ \delta_0\right]^T$;

$$g(X) = \left[\frac{1}{T_{do}}\ 0\ 0\right]^T \text{ and } f(X) = \begin{bmatrix} -\frac{1}{T'_d} + \frac{1}{T_{do}} \frac{x_d - x'_d}{x'_{d\Sigma}} V_s \cos\delta \\ \frac{\omega_0}{H} P_M - \frac{D}{H}(\omega - \omega_0) - \frac{\omega_0}{H} \frac{E'_q v_s}{x'_{d\Sigma}} \sin\delta \\ (\omega - \omega_0) \end{bmatrix}$$

Now, choosing the rotor angle 'δ' as the output function i.e., $h(X) = \delta$, the relative degree of the test system, which is an important factor to calculate nonlinear control law, is derived in the following sub-section.

2.2 Estimation of Relative Degree of the Test System

To design a nonlinear feedback controller, one of the important factors is the calculation of the relative degree of the system. The standard procedure of the relative degree estimation is given in [15]. According to this procedure, if the lie derivatives of the output function $h(X)$ i.e. $L_f^{r-1}h(X)$ along a vector field $g(X)$ is not equal to zero in a neighbourhood $\hat{R}$ $i.e.\left(L_g L_f^{r-1} h(X) \neq 0\right)$, then the system has the relative degree 'r' in the neighbourhood $\hat{R}$. Therefore, by choosing output function $y = h(X) = \delta$, the relative degree of the system expressed by the Eqs. (1)–(3) can be derived as follows;

$$L_f^{1-1}h(X) = L_f^0 h(X) = \omega \tag{6}$$

Therefore,

$$L_g L_f^0 h(X) = \frac{\partial h(X)}{\partial X} g(X) = 0 \tag{7}$$

For $r{=}2$, $L_f^{2-1}h(X) = L_f h(X) = \frac{\partial h(X)}{\partial X} f(X) = \left[\frac{\omega_0}{H} P_M - \frac{D}{H}(\omega - \omega_0) - \frac{\omega_0}{H} \frac{E'_q V_s}{x'_{d\Sigma}} \sin\delta\right]$
Hence,

$$L_g L_f^{2-1} h(X) = L_g L_f h(X) = -\frac{1}{T_{do}} \frac{\omega_0 V_s}{H x'_{d\Sigma}} \sin\delta \neq 0 \tag{8}$$

Therefore, the relative degree of the test system is 'r' = 2. The design of sliding surface followed by the derivation of the control law is given in the next sub-section. It is to be noted that the conventional sliding (1-sliding) mode can only be achieved with relative degree '1', while the second-order sliding (2-sliding) mode requires relative degree '2' with respect to discontinuous control.

2.3 Estimation of Control Law by Super Twisting Sliding Mode Control (STSMC)

The detail theory of design of conventional sliding mode and Super Twisting Sliding Mode Control (STSMC) is given in [16]. According to this theory, under uncertain condition, the classical sliding modes provide robust and highly accurate solution for a wide range of control problem. However, there are two main restrictions present. First, the constraint to be made zero in conventional sliding mode control must have the relative degree one, which means that the control needs to explicitly appear in the first derivative of the constraint. Thus, one has to find an appropriate constraint. Second, the chattering effect may cause unacceptable practical complications if the control has any physical sense. These restrictions may be solved by STSMC. Therefore, to derive the control law using STSMC, the output of the system in (5) can be represented as constraint and represented as follows;

$$y(t) = e(t) = \delta - \delta_0 \tag{9}$$

where 'δ' and 'δ_0' are the instantaneous and initial value of the rotor angle respectively.

The nonlinear constraint function for an affine nonlinear system given in (4) and (5), usually chosen as;

$$\sigma = C_1\dot{e} + C_0 e \tag{10}$$

where, C_1 and C_0 are real valued constant. The Eq. (10) also known as the sliding surface of the above control strategy.

Now for $t \to \infty$, if the error $e(t) \to 0$ is considered. Then (10) can be modified to;

$$C_1\dot{e} + C_0 e = 0$$

or,

$$e(t) = \exp\left(-\frac{C_0}{C_1}t\right)e(0) \tag{11}$$

Here, 'e' is the error signal which is the deviation of the rotor angle of the machine from equilibrium position. The aim is to minimize the error, i.e. $\underset{t\to\infty}{Lt}\ e(t) = 0$.

Now, the time derivative of (10) is as follows;

$$\dot{\sigma} = C_1\ddot{e} + C_0\dot{e} \tag{12}$$

Combining (2), (3) and (12) the time derivative of constraint function can be written as;

$$\dot{\sigma} = C_1\left[\frac{\omega_s}{H}\left(P_M - \frac{D(\omega-\omega_s)}{\omega_s} - \frac{E_q' V_s \sin\delta}{X_d' \sum}\right)\right] + C_0(\omega - \omega_s) \tag{13}$$

In the next step the method of design of robust exact differentiator has been described which is required for the design of constraint function or the sliding surface represented in Eq. (10).

In order to design the robust exact differentiator as shown in (Fig. 2) an auxiliary system is considered as;

$$\dot{Z}_0 = v \tag{14}$$

In the robust exact differentiator,
$\sigma_0 = -e(t) + Z_0$ or $\dot{\sigma}_0 = -\dot{e}(t) + \dot{Z}_0$ or $\dot{\sigma}_0 = -\dot{e}(t) + v$
and the task is to keep $\sigma_0 = 0$.

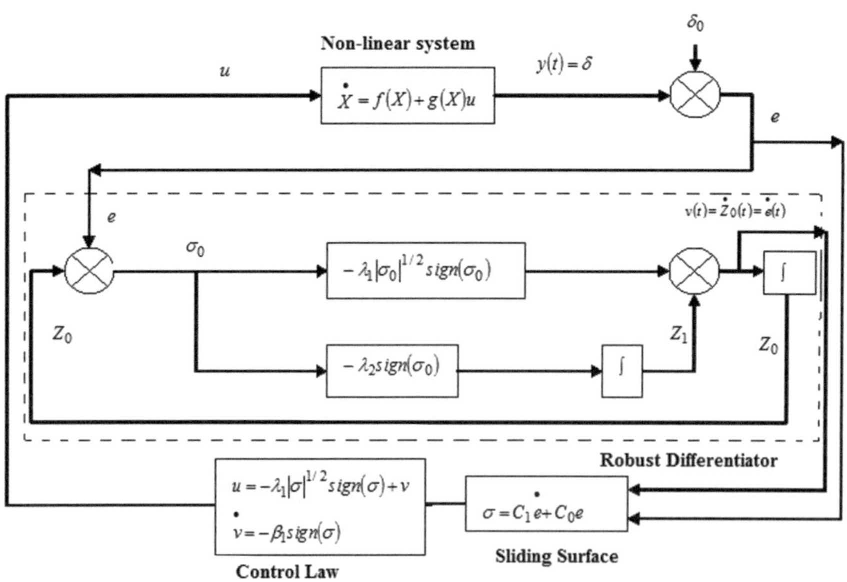

Fig. 2 Block diagram representation of the nonlinear system with STSMC controller

According to 2-sliding mode control [16], $\sigma_0 = \dot{\sigma}_0 = 0$. Therefore, and

$$\left. \begin{aligned} e(t) &= z_0 \\ \dot{Z}_0 &= \dot{e}(t) = v \end{aligned} \right\} \tag{15}$$

Thus, the resulting form of the robust exact differentiator is;

$$\dot{Z}_0 = v = -\lambda_1 |Z_0 - e(t)|^{\frac{1}{2}} sign(Z_0 - e(t)) + Z_1 \tag{16}$$

and $\dot{Z}_1 = -\lambda_2 sign(Z_0 - e(t))$

where, $\lambda_1 = L^{1/2}$ and $\lambda_2 = 2L$, L is any positive integer here.

Finally, constraint function (10) can be designed by combining (9)–(15) and the super twisting sliding mode control law is obtained as;

$$\left. \begin{aligned} u &= -\alpha_1 |\sigma|^{1/2} sign(\sigma) + v \\ \dot{v} &= \beta_1 sign(\sigma) \end{aligned} \right\} \tag{17}$$

where, α_1 and β_1 are real valued constant. The phenomenon of chattering can be minimized by proper selection of the values of α_1 and β_1.

The Eq. (17) will be used in section-4 for simulation and performance evaluation of the STSMC controller.

3 Zero Dynamic Design of Excitation Controller

The theory of zero dynamics design of nonlinear controller [15] is applicable for a system which has the relative degree (r) less than the order of the system (n). From Eq. (8) it can be seen that the relative degree of the system ($r = 2$) is less than the order of the system ($n = 3$). Therefore, the theory of zero dynamics design is applied to derive the control law.

Now, the system described by (4) and (5) has to be linearized by the nonlinear coordinate transformation. The process of mapping from X space to Z space is given as follows;

$$Z_1 = y = h(X) = L_f^{1-1} h(X) \tag{18}$$

then the time derivative of Eq. (18) it can be written as

$$\dot{Z}_1 = \frac{\partial h(X)}{\partial X} \dot{X} \tag{19}$$

Now substituting Eq. (4) in (19) for $\dot{X}$, we have

$$\dot{Z}_1 = \frac{\partial h(X)}{\partial X} f(X) + \frac{\partial h(X)}{\partial X} g(X)u = L_f^{2-1} h(X) + L_g L_f^{1-1} h(X)u$$

As $L_g L_f^{1-1} h(X) = 0$, then the above equation can be transformed as

$$\dot{Z}_1 = L_f f(X) = Z_2 \tag{20}$$

$$\dot{Z}_2 = L_f^2 h\big(\varphi^{-1}(Z)\big) + L_g L_f^{2-1} h\big(\varphi^{-1}(Z)\big)u \tag{21}$$

$$\dot{Z}_3 = L_f \varphi_3\big(\varphi^{-1}(Z)\big) \tag{22}$$

considering the fact that $\dot{Z}_r = L_f^r h(X) + L_g L_f^{r-1} h(X)u$ for the first r equations and $\dot{Z}_n = L_f \varphi_n\big(\varphi^{-1}(Z)\big)$ for the rest (n-r) equations.

Hence, the coordinate transformation $Z = \varphi(X)$ can be written as;

$$Z_1 = \varphi_1(X) = \delta$$

$$Z_2 = \varphi_2(X) = \frac{\omega_0}{H} P_M - \frac{D}{H}(\omega - \omega_0) - \frac{\omega_0}{H} \frac{E_q' V_s}{x_{d\Sigma}'} \sin\delta$$

$$Z_3 = -\frac{D}{H}\dot{\omega} - \frac{\omega_0}{H} \frac{\dot{E}_q' V_s}{x_{d\Sigma}'} \sin\delta$$

This coordinate transformation is valid if the Jacobian Matrix at $X = X_0$, $J_\varphi = \left.\frac{\partial\varphi(X)}{\partial X}\right|_{X=X_0}$ is non singular and the function $\varphi_3(X)$ satisfies $L_g \varphi_3(X) = 0$.

The key idea of the zero dynamics design approach is that at any time 't', the deviation of the output response of the system from its equilibrium point should be zero. Therefore, the output response can be expressed as;

$$y(t) = h(X(t)) = 0 \quad 0 \le t \le \infty$$

As $y(t) = h(X) = Z_1(t)$ has been set to zero at any time. Therefore, Eq. (18) can be written as $Z_1(t) = 0$. So, the first r components of coordinates Z are $\begin{bmatrix} Z_1 & Z_2 \end{bmatrix}^T = 0$ for all $t \ge 0$ and there exists

$$\dot{Z}_2(t) = 0 \tag{23}$$

Under such condition, from (21) the control input 'u' can be computed as follows;

$$0 = L_f^2 h\big(\varphi^{-1}(Z)\big) + L_g L_f^{2-1} h\big(\varphi^{-1}(Z)\big)u \tag{24}$$

$$u = -\frac{L_f^2 h\big(\varphi^{-1}(Z)\big)}{L_g L_f h\big(\varphi^{-1}(Z)\big)} = -\frac{L_f^2 h(X)}{L_g L_f h(X)} \tag{25}$$

The term $L_f^2 h(X)$ can be computed as;

$$\begin{aligned} L_f^2 h(X) = {} & \frac{\omega_s}{H}\frac{E_q' V_s}{T_d' x_{d\sum}'}\sin\delta - \frac{\omega_0}{H}\frac{V_s^2}{x_{d\sum}'^2}\frac{1}{T_{do}}\left(x_d - x_d'\right)\sin\delta\,\cos\delta - \frac{D\omega_s}{H^2}P_M \\ & + \frac{D^2}{H^2}(\omega - \omega_s) + \frac{D\omega_s E_q' V_s}{H^2 x_{d\sum}'}\sin\delta \\ & - \frac{\omega_0 E_q' V_s}{H x_{d\sum}'}(\omega - \omega_s)\cos\delta \end{aligned} \quad (26)$$

Thus, on substituting $L_f^2 h(X)$ from (26) and $L_g L_f h(X)$ from (8) in the Eq. (25), the nonlinear excitation control law *via* zero dynamic design becomes;

$$u = V_f = E_q - \frac{T_{do}E_q'}{P_e}\left(Q_e + \frac{V_s^2}{x_{d\sum}'}\right)\Delta\omega - \frac{T_{do}E_q'}{P_e}\frac{D}{H}\dot{\omega} \quad (27)$$

where, $E_q = \frac{x_{d\sum}}{x_{d\sum}'}E_q' - \left(x - x_d'\right)\frac{V_s}{x_{d\sum}'}\cos\delta$; $Q_e = \frac{V_s E_q'}{x_{d\sum}'}\cos\delta - \frac{V_s^2}{x_{d\sum}'}$ and $\Delta\omega = \omega - \omega_s$ are the basic electrical equations of an SMIB power system. The Eq. (27) will be used for simulation and performance analysis in the following section. The block-diagram of the SMIB system with non-linear excitation controller is presented in the Fig. 3.

4 Performance Analysis

The dynamic behaviour of electromechanical oscillations of the SMIB test system (Fig. 3) with proposed nonlinear STSMC controller is investigated here under three-phase-to-earth fault scenario for a simulation time 7 s. The three-phase-to-earth fault is applied in the generator bus at $t = 1.0$ s. and the fault is cleared at $t = 1.2$ s successfully. It has been observed that the rotor angle (Fig. 4) of the machine

Fig. 3 SMIB system with nonlinear excitation controller based on zero dynamic design

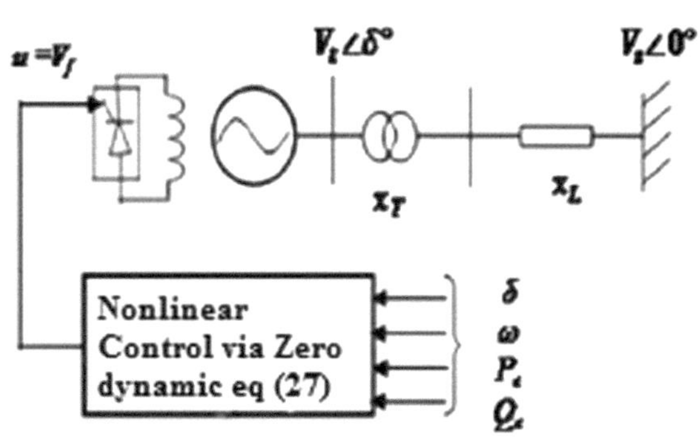

increases exponentially after the occurrence of fault. It is further observed that with the application of proposed nonlinear STSMC controller the rotor angle reaches to its equilibrium position very quickly in comparison to the zero dynamic design of nonlinear controller. Moreover, from the simulation, it has been observed that there is a considerable steady state error in the response of rotor angle applying zero dynamic design of nonlinear controller. The dynamic behaviour of the machine speed is shown in Fig. 5. Here, the occurrence of fault, fault clearance time and the total simulation time is considered as that of the case of rotor angle. The machine speed increases monotonically after the occurrence of fault at $t = 1.0$ s. It has been found that at post fault clearance condition the machine speed reaches to its equilibrium position immediately and becomes asymptotically stable. The simulation results for rotor angle and machine speed presented in the Figs. 4 and 5 are signifying that the transient stability of the test SMIB system is substantially improved applying STSMC controller compared to the controller designed through zero dynamic approach. In Fig. 4 after fault clearance the system becomes stable at around 1.3 s whereas using zero dynamic design of nonlinear controller not only the oscillation continues but also there is a considerable steady state error remains.

In study of machine speed response in Fig. 5, the machine speed of the system becomes stable and settled quickly after clearance of the fault when nonlinear STSMC controller is applied. But for use of the zero dynamic design of nonlinear controller, the oscillation of the system continues and it appears from the figure that it requires larger time more than specified simulation time (7 s) to attain stable operating condition.

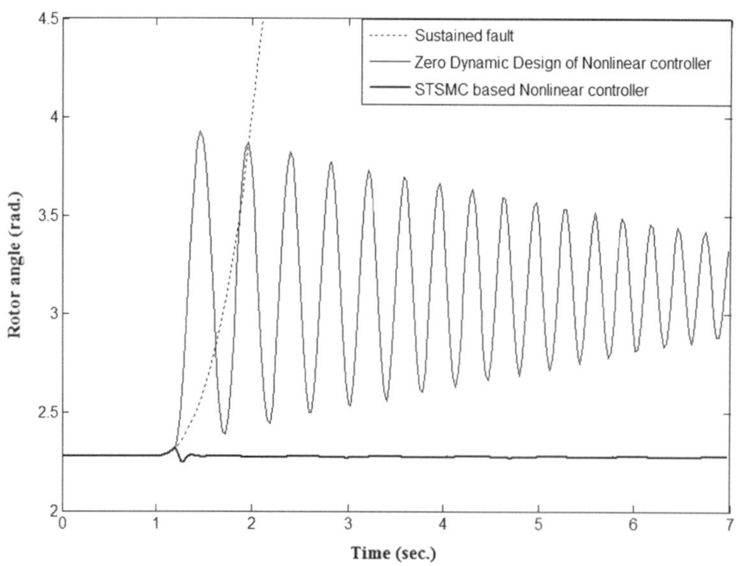

Fig. 4 Dynamic response of rotor angle of SMIB system

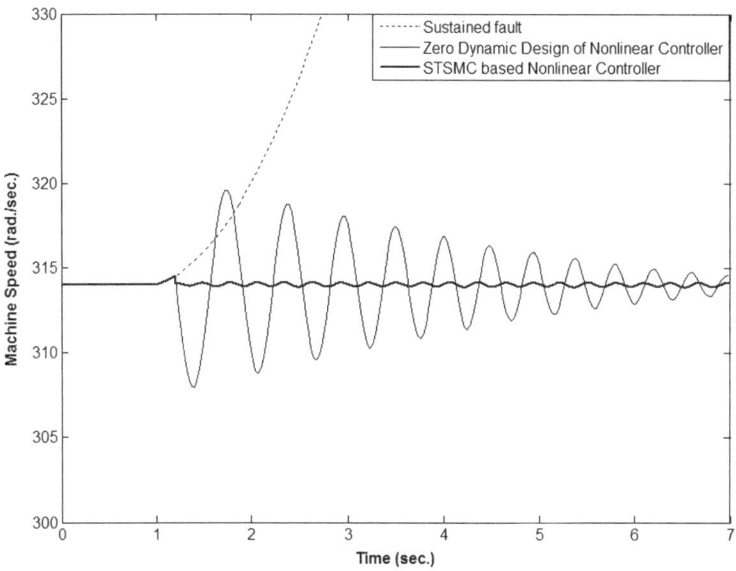

Fig. 5 Dynamic response of machine speed of SMIB system

5 Conclusion

This paper presents a novel approach to design a nonlinear Variable Structure (VS) based sliding mode excitation control for a Single Machine Infinite Bus (SMIB) power system. The Super Twisting Sliding Mode Control (STSMC) has been utilized to formulate the nonlinear control law. The performance of the propose nonlinear controller has been compared with the performance of zero dynamic design based nonlinear excitation controller. The obtained simulation results confirm that the proposed nonlinear STSMC is more effective than the zero dynamic design of nonlinear controller. Moreover, in the proposed controller design method there is no need of linearization of nonlinear model of the system. The control law can be derived directly from the nonlinear model of the system where as in zero dynamic design method the nonlinear system must be linearize to derive control law. Furthermore, in zero dynamic design approach the stability analysis of internal states of the system is essential which introduces further mathematical complexity. Thus it is possible to conclude that the proposed nonlinear control strategy have good performance and is superior than the zero dynamic design approach in mitigating transient stability in a SMIB power system. The present approach of nonlinear controller design can also be implemented in multimachine power systems.

Appendix

Parameters of the SMIB Test System

$H = 2.37$ s; $D = 0.0$; $R_s = 0.0$ pu; $R_e = 0.02$ pu; $T_d = 5.90$ s; $\omega_s = 314$ rad/sec; $X_d = 1.70$ pu; $X'_d = 0.245$ pu; $X_e = 0.7$ pu; $X_q = 1.64$ pu; $V_{inf} = 1.00\ \angle 0^o$ pu; $V_t = 1.72\ \angle 19.31^o$ pu. $\delta = 130$ deg.

References

1. P. Kundur, M. Klein, G.J. Rogers, M.S. Zywno, Application of power system stabilizer for enhancement of overall system stability. IEEE Trans. Power Syst. **4**(2), 614–626 (1989)
2. D. Jing, X. Lei, in *A Nonlinear TCSC Control Strategy for Power System Stability Enhancement*, Proceedings of the 5th International Conference on Advances in Power System Control, Operation and Management, AF'SCOM 2000, Hong Kong, vol. 2, pp. 576–581, Oct. 2000
3. S. Kawomoto, in *Nonlinear Control of One-Machine Power System by Exact Linearization Using Optimal Parameters*, Transmission and Distribution Conference and Exhibition 2002: Asia Pacific. IEEE/PES, vol. 1, pp. 201–205, Oct. 2002
4. P. Kumkratug, Nonlinear control design of series FACTS devices for damping power system oscillation. Am. J. Appl. Sci. **8**(2), 124–128 (2011)
5. M.A. Mahmud, M.J. Hossain, H.R. Pota, Nonlinear excitation controller for power systems using zero dynamic design approach. North Am. Power Symp. (NAPS) **1**, 1–6 (2011)
6. W. Zhang, F. Xu, W. Hu, M. Li, W. Ge, Z. Wang, in *Research of Coordination Control System between Nonlinear Robust Excitation Control and Governor Power System Stabilizer in Multimachine Power System*, IEEE International Conference on Power System Technology, Auckland (POWERCON), pp. 1–5, 2012
7. S. Mei, F. Liu, Y. Chen, Q. Lu, Co-ordinated H_∞ control of excitation and governor of hydroturbo-generator sets: a Hamiltonian approach. Int. J. Robust Nonlinear Control **14**(9–10), 807–832 (2004)
8. Y. Wang, Y.L. Tan, G. Guo, Robust nonlinear co-ordinated excitation and TCSC control for power systems. IEE Proc. Gen. Transm. Distrib. **149**(3), 367–372 (2002)
9. X.Y. Bian, Y. Geng, K.L. Lo, Y. Fu, Q.B. Zhou, Coordination of PSSs and SVC damping controller to improved probabilistic small-signal stability of power system with wind farm integration. IEEE Trans. Power Syst. **31**(3), 2371–2382 (2016)
10. L. Gu, J. Wang, Nonlinear coordinated control design of excitation and STATCOM of power systems. Electr. Power Syst. Res. **77**, 788–796 (2007)
11. H.E.A. Talat, A. Abdennour, A.A. Al-Sulaiman, Design and experimental investigation of a decentralized GA-optimized neuro-fuzzy power system stabilizer. Int. J. Electr. Power Energy Syst. **32**(7), 751–759 (2010)
12. N. Magaji, M.W. Mustafa, Z.B. Muda, in *Power system damping using GA based fuzzy controlled SVC device*, IEEE Region 10 Conference Singapore, TENCON, pp. 1–7, 2009
13. R. Poli, J. Kennedy, T. Blackwell, Partical swarm optimization: an overview. Swarm Intell. **1**(1), 33–57 (2007)
14. S. Jalilzadeh, M.R. Tirtashi, M. Sadeghi, PSS and SVC controller design by chaos and PSO algorithms to enhancing the power system stability. Word Acad. Sci. Eng. Technol. **67**, 689–693 (2010)
15. Q. Lu, Y. Sun, S. Mei, *Nonlinear control systems and power system dynamics* (Kluwer Academic Publishers, Norwel, 2001)
16. Y. Shtessel, C. Edwards, L. Fridman, A. Levant, in *Sliding Mode Control and Observation* (Springer, Berlin, 2014)

Modelling of an Optimum Fuzzy Logic Controller Using Genetic Algorithm

Piyali Ganguly, Akhtar Kalam and Aladin Zayegh

1 Introduction

Fuzzy logic has emerged as one of the active areas of research, particularly in control applications. It is a very powerful method of reasoning when mathematical models are not available and input data are imprecise [1]. Nowadays, the T-S fuzzy modelling technique is becoming powerful engineering tool for modelling and control of complex nonlinear dynamic systems [2]. T-S fuzzy models represent fuzzy dynamic models or fuzzy systems which are described by fuzzy if-then rules that can give a local linear representation of the non-linear system [3]. For the reason that it employs the linear model in the consequent part, conventional linear system theory can be applied for the system analysis and synthesis easily, The methods for learning T-S fuzzy models from data are based on the idea of consecutive structure and parameter identification [2].

A major drawback of FLC is that the tuning process becomes more difficult and very time consuming when the number of the system inputs and outputs are increased. Different evolutionary algorithms regarding tuning the membership function parameters of FLC have been studied in the literature [4]. Genetic Algorithm (GA) is very efficient and systematic technique to optimise the FLC [5]. Genetic algorithms (GA), which are adopted from the principle of biological evolution, are efficient search techniques that manipulate the coding representing a parameter set to reach a near optimal solution [6].

P. Ganguly (✉) · A. Kalam · A. Zayegh
Department of Engineering and Science, Victoria University, Melbourne, Australia
e-mail: Piyali.ganguly@live.vu.edu.au

A. Kalam
e-mail: Akhtar.Kalam@vu.edu.au

A. Zayegh
e-mail: aladin.zayegh@vu.edu.au

S. Chattopadhyay et al. (eds.), *Modelling and Simulation in Science, Technology and Engineering Mathematics*, Advances in Intelligent Systems and Computing 749,
https://doi.org/10.1007/978-3-319-74808-5_28

In this present work, a T-S type FLC is optimised using GA to control the liquid level of a tank. The performance of the GA optimised FLC is compared with that of the conventional PID controller. The MATLAB/SIMULINK software forms part of the modelling and design tools employed in this research.

2 Fuzzy Logic Controller (FLC)

Recently fuzzy control techniques have been applied to many industrial processes. FLCs are rule-based systems which are useful in the context of complex ill-defined processes, especially those which can be controlled by a skilled human operator without knowledge of their underlying dynamics [7, 8]. Fuzzy control system a control mechanism based on fuzzy set theory. As per the fuzzy theory and logic, a decision is made by mainly three operations: fuzzification process, an inference engine for rule base and defuzzification process [9]. It is a mathematical system that analyses analogue input values in terms of logical variables that takes a continuous value between 0 and 1, unlike classical logic and operates on discrete values of either 1 or 0. It utilises the expert knowledge of an experienced user to design the knowledge base of the controller. Figure 1 represents the configuration of fuzzy logic.

3 Takagi-Sugeno Type Fuzzy Logic Controller

The T-S fuzzy model is a system described by fuzzy IF-THEN rules which can give local linear representation of the nonlinear system by decomposing the whole input space into several partial fuzzy spaces and representing each output space with a linear equation [2]. This FLC model is capable of approximating a wide class of nonlinear systems.

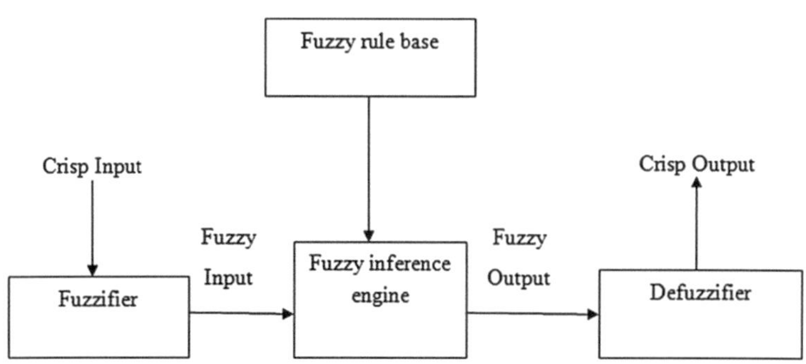

Fig. 1 Configuration of fuzzy logic

Takagi-Sugeno model (for short TS model) consists of a set of if-then rules, where the rule premises are expressed by fuzzy sets and the rule consequents are considered to be mostly linear functions of input variables. Therefore, we can formulate the model as follows [10]

R_i: **If** x_i is A_{i1} **and** … x_n is A_{in} **then**

$$g_i = p_{i1}x_1 + \cdots p_{in}x_n + p_{i(n+1)}, \quad i = 1, \ldots M, \tag{1}$$

where $x = [x_1, \ldots x_n]^T$ is the vector of input variables, g_i the output variable, R_i i-th fuzzy rule,

$A_{i1}, \ldots A_{in}$ fuzzy sets defined in the premise space by membership functions $\mu_{Aij}(x_j)$ and p_{i1} …, $p_{i(n+1)}$ consequent parameters. The output y of the model is given as a weighted mean of the individual fuzzy rule contributions:

$$y = \frac{\sum_{i=1}^{M} \beta_i g_i}{\sum_{i=1}^{M} \beta_i}, \tag{2}$$

where β_i denotes the degree of fulfilment of the *i*-th rule:

$$\beta_i = \prod_{j=1}^{n} A_{ij}(x_j), i = 1, \ldots, M, \tag{3}$$

where $A_{ij}(x_j)$ is the membership of input variable x_j in the fuzzy set A_{ij}.

Generally, the selections of fuzzy rules and premise inputs and consequent parameters are made by trial and error method or using some clustering methods. In this work, the p_{i1} …, $p_{i(n+1)}$ (consequent parameters) of a T-S type FLC are optimized using GA to control the liquid level of a tank.

4 Genetic Algorithm

Genetic algorithms are search algorithms based on the mechanics of natural selection, natural genetics [11]. They combine survival of the fittest among string structures with a structured yet randomised information exchange to form search algorithms. Genetic algorithm searches the solution space of a function through the use of simulated evolution, i.e., the survival of the fittest strategy. In general, the fittest individuals of any population tend to reproduce and survive to the next generation, thus improving successive generations [12].

The genetic algorithms developed by Holland [12] is to simulate the natural evolution process that operates on chromosomes [5]. The simple genetic algorithm that yields satisfactory results in many practical problems. The central theme of research on genetic algorithm has been robustness, if the artificial systems can be more robust, a costly redesign can be reduced or eliminated. If a higher level of adoption can be achieved, existing systems can perform longer and better.

Any simple GA is composed of three operators [5]

1. Reproduction
2. Crossing over
3. Mutation.

GA can be summarized in Fig. 2.

5 Modelling of Liquid Level System

For the present work, the liquid level system shown in Fig. 3, has been considered [13]. The system includes one inflow and one outflow. The objective of this work is to design a heuristic fuzzy logic controller capable of driving the liquid level in the tank to a given set point. In the present work the level of the liquid will be controlled by adjusting only the input or inlet valve.

The system variables are defined as follows:

- Q_{in}: the rate of flow of liquid into the tank at time t, m^3/s.
- Q_{out}: the rate of flow of liquid out of the tank at time t, m^3/s.
- q_i: small deviation of inflow rate from its steady-state value, m^3/s.
- q_o: small deviation of outflow rate from its steady-state value, m^3/s.
- H: the steady state height, m.
- h: small deviation of head from its steady-state value, m.
- R: the resistance for the liquid flow out, s/m^2.
- A: the cross-sectional area, m^2.

The present design assumes that angular position of the value equal to the rate of flow in liquid or in other words, ratio between the steady-state flow rate (before any change has occurred) and the rate of flow in liquid is equal to one.

(1) Supply a population P_0 of N individuals and respective function values.
(2) *i* - 1
(3) *P'_i - Selection_function (P_i-1)*
(4) *P'_i - Reproduction_function(P'_i)*
(5) *evaluate (P_i)*
(6) *i - i + 1*
(7) Repeat step 3 until termination
(8) Print out best solution found

Fig. 2 GA optimization algorithm

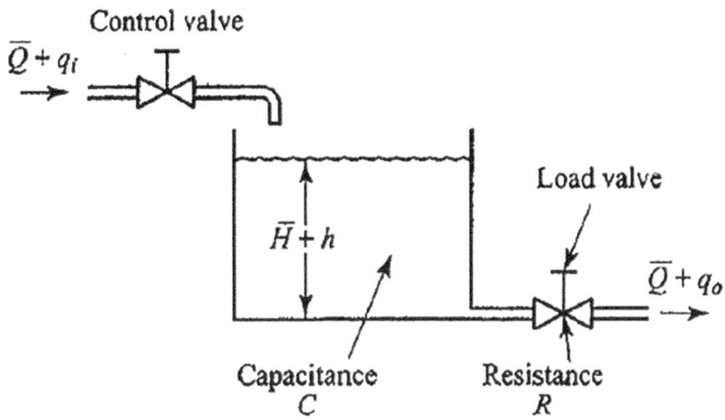

Fig. 3 Liquid level system [13]

The flow rate is considered as turbulent flow rate in this work, not laminar flow rate, hence the resistance R for the liquid flow consider as R_t.

$$Q = k \times \sqrt{H} \tag{4}$$

Where K is coefficient. The resistance R_t for turbulent flow, which can be obtained as:

$$R_t = \frac{dH}{dQ} \tag{5}$$

From Eq. (4) it can be written as:

$$dQ = \frac{k}{2\sqrt{H}}dH \tag{6}$$

$$\frac{dH}{dQ} = \frac{2\sqrt{H}}{k} = \frac{2\sqrt{H}\sqrt{H}}{Q} = \frac{2H}{Q} \tag{7}$$

Since $R_t = \frac{2H}{Q}$

The value of turbulent-flow R_t depend on the flow rate and the head. By using the turbulent flow resistance, the relationship between Q & H can be expressed as:

$$Q = \frac{2H}{R_t} \tag{8}$$

The difference between the inflow and the outflow during the small time interval dt is equal to the additional amount stored in the tank which can be represented as:

$$Adt = (q_i - q_o)dt \tag{9}$$

From the definition of resistance, the relationship between q_o and h is:

$$q_o = \frac{h}{R} \tag{10}$$

for a constant value of R, the differential equation for this system becomes

$$RC\frac{dh}{dt} + h = Rq_i \tag{11}$$

RC is the time constant of the system. Taking the Laplace transforms of both sides of Eq. (11) assuming zero initial condition:

$$(RCs + 1)H(s) = RQ_i(s) \tag{12}$$

where $H(s) = L[h(t)]$ and $Q_i(s) = L[q_i(t)]$

Considering Q_i as input and H as output, the transfer function can be represented as:

$$\frac{H(s)}{Q_i(s)} = \frac{R}{RCs + 1} \tag{13}$$

Considering H = 1 m, k = 1 and $C = 0{:}5\ \text{m}^2$

$$R = \frac{2H}{Q} \tag{14}$$

$$\text{Q} = \text{K}\sqrt{H} = 1 \ \text{ and } \ R = \frac{2H}{k\sqrt{H}} = 2$$

Hence the transfer function is:

$$\frac{H(s)}{Q(s)} = \frac{2}{s + 1} \tag{15}$$

6 Fuzzy Logic Controller Design

The T-S type FLC is implemented using MATLAB/SIMULINK environment. The block diagram of the control system is shown in Fig. 4. The simulated controller has two input variable: e(t) (error) which is the difference between the set value and the process value, $\frac{de(t)}{dt}$, (rate)is the differential of $e(t)$ and an output variable which is the control signal of the actuator [14]. Triangular shaped built-in membership functions have been used in the Sugeno-type fuzzy algorithm for the input variables. The membership functions of the input variables are shown in Figs. 5 and 6.

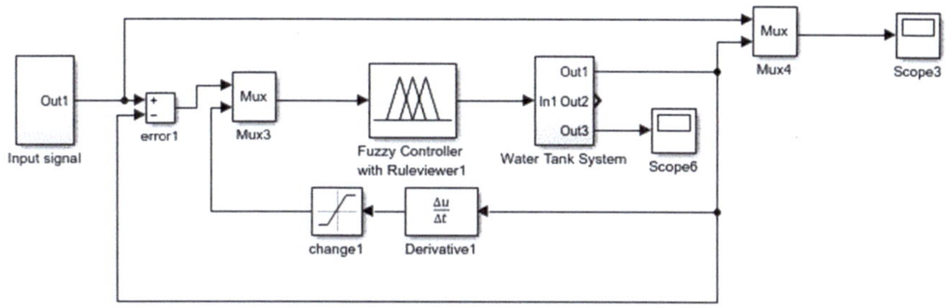

Fig. 4 The SIMULINK block diagram of the liquid level control system

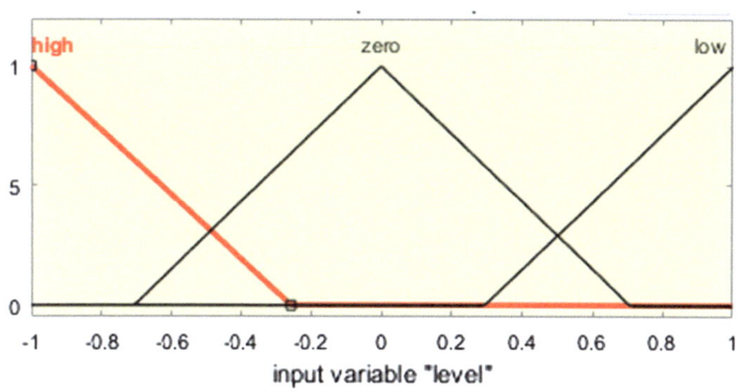

Fig. 5 Membership functions of the input variable 'error'

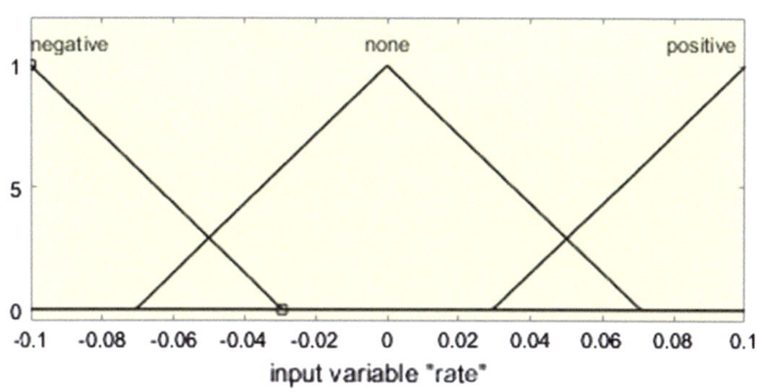

Fig. 6 Membership function of the input variable 'rate'

Output membership functions have three constant parameters. The surface of the FLC before optimisation is represented in Fig. 7.

The fuzzy controller output is applied to the actuator in order to control of liquid level. Then, the flow of liquid has been adjusted to obtain the desired level. The FLC has five rules. The output of the T-S type FLC can be calculated using Eqs. (1)–(3). The parameters of the T-S type FLC is optimized using GA within a specified limit.

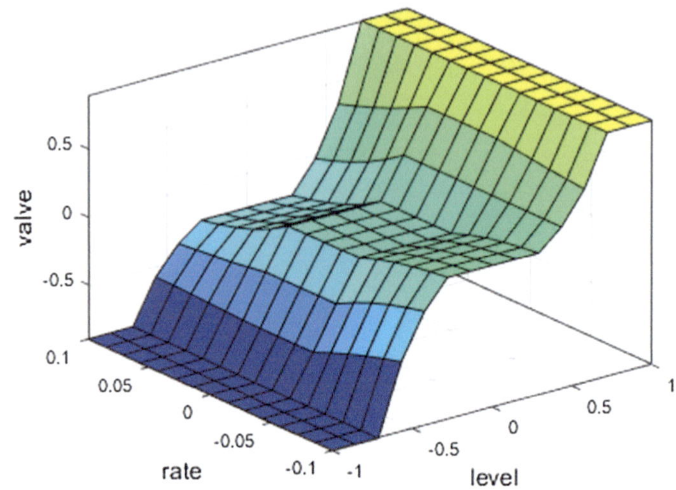

Fig. 7 The control surface of the un-optimized FLC

The optimization is done using MATLAB programing. The programme optimises the 15 variables (population) (for 5 rules using 3 variables at a time used for the present problem) with the help of GA. This programme continues until 200 generations, and gets the optimized result. The evaluation function (fitness function) used here can be represented by Balochian and Ebrahimi [14]:

$$\text{Fitness} = \text{sse}(y - \hat{y}) + 100 \times \text{Overshoot}$$

where 'sse' is sum squared error performance function, y is reference input and $\hat{y}$ is output of the Simulink model. The optimized values of the parameters of the T-S type FLC are fed into the controller to get the optimized controller for the system.

7 Simulation and Results

Figure 8 shows the output (liquid level) of the system using un-optimized FLC against the reference input. Figure 9 represents the system output using PID controller against a set input and Fig. 10 represents the output of the system with optimized FLC against the reference input. In first two cases, the water level of the tank follows the input signal with a significant error in case of both maximum and minimum specified limit. It can be observed that with the optimised FLC, the error is reduced to a great extent. The optimised FLC can maintain the water level to the specified maximum value. It can also track the specified minimum level with a minimum error.

Hence, it can be stated that comparison between the control results obtained from the un-optimized FLC, PID controller and optimized FLC clearly shows that

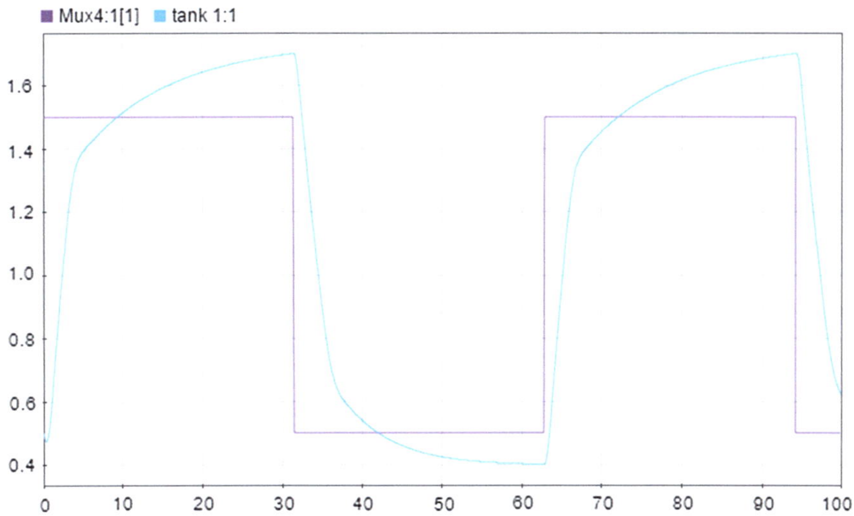

Fig. 8 Liquid level of the tank with un-optimized FLC. Purple line represents the set input and the blue line shows the actual liquid level

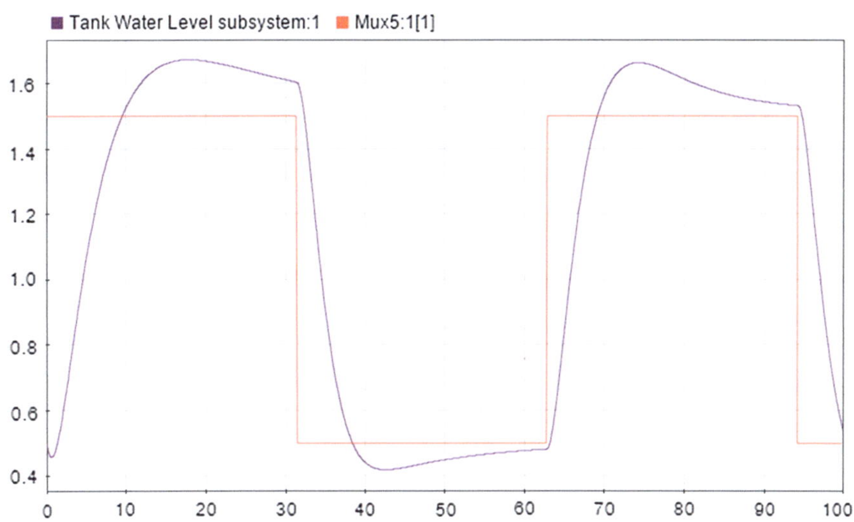

Fig. 9 Liquid level of the tank with PID controller. Orange line represents the set input and the purple line shows the actual liquid level

optimized FLC has more accurate and acceptable results. Figure 11 represents the control surface of the optimum FLC.

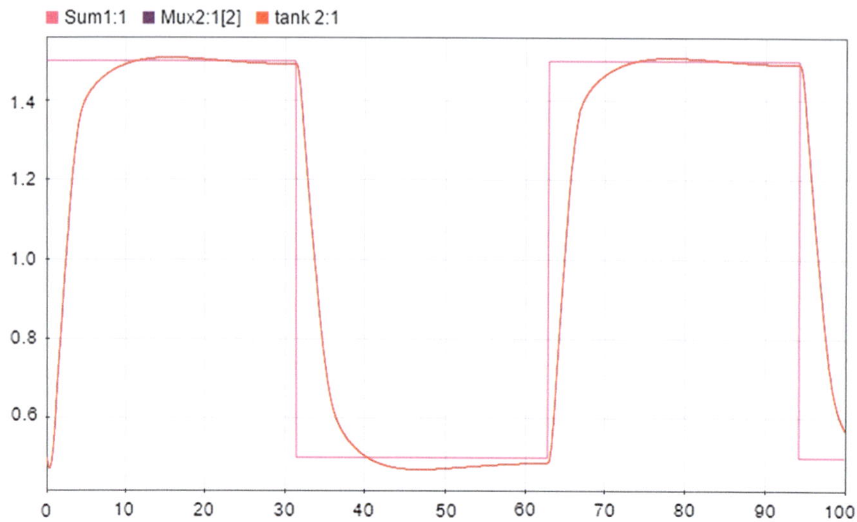

Fig. 10 Liquid level of the tank with optimized FLC. Pink line represents the set input and the red line shows the actual liquid level

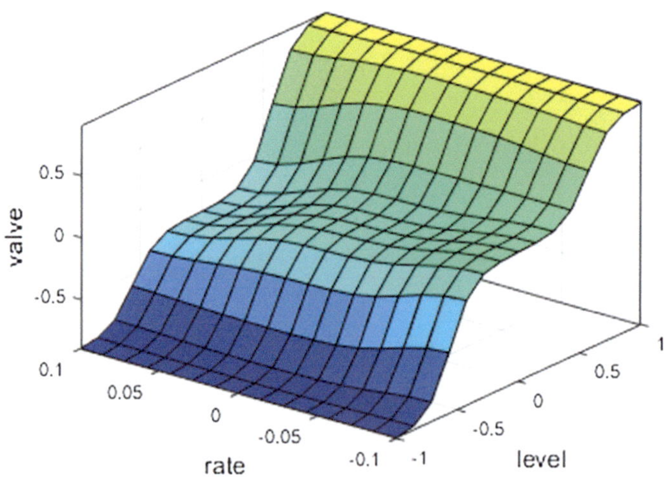

Fig. 11 Control surface of the optimized FLC

8 Conclusion

In this paper, an efficient and effective tuning approach-based on Genetic Algorithm (GA) is presented to obtain the optimal Takagi-Sugeno type fuzzy controller parameters to control the liquid level of a tank. The results obtained clearly shows that one can improve the performance of fuzzy controller by optimizing the parameters.

Simulation results show clearly that the optimized controller have better performance compared with un-optimised FLC or conventional PID controller. GA could tune up fuzzy controller parameters promptly with uppermost level of accuracy.

References

1. D. Su, K. Ren, J. Luo, C. He, L. Wang, X. Zhang, Programmed and simulation of the fuzzy control list in fuzzy control, in IEEE/WCICA, 2010, pp. 1935–1940
2. H. Du, N. Zhang, Application of evolving Takagi-Sugeno fuzzy model to nonlinear system identification. Appl. Soft Comput. **8**(1), 676–686 (2008)
3. M. Männle, FTSM-fast Takagi-Sugeno fuzzy modeling, in *Fault Detection, Supervision and Safety for Technical Processes (SAFEPROCESS'00)* (Budapest, Romaina, 2000), pp. 663–668
4. Y.-C. Chiou, L.W. Lan, Genetic fuzzy logic controller: an iterative evolution algorithm with new encoding method. Fuzzy Sets Syst. **152**(3), 617–635 (2005)
5. K.C. Ng, Y. Li, Design of sophisticated fuzzy logic controllers using genetic algorithms, in *Fuzzy Systems. Proceedings of the IEEE World Congress on Computational Intelligence*, 1994
6. S. Khan et al., Design and implementation of an optimal fuzzy logic controller using genetic algorithm. J. Comput. Sci. **4(**10), 799–806 (2008)
7. S. Balochian, E. Ebrahimi, Parameter optimization via cuckoo optimization algorithm of fuzzy controller for liquid level control. J. Eng. (2013)
8. Francisco Herrera, Manuel Lozano, Jose L. Verdegay, Tuning fuzzy logic controllers by genetic algorithms. Int. J. Approximate Reasoning **12**(3-4), 299–315 (1995)
9. L.A. Zadeh, Outline of a new approach to the analysis of complex systems and decision processes. IEEE Trans. Syst. Man Cybern. **1**, 28–44 (1973)
10. H. Roubos, S. Magne, Compact fuzzy models and classifiers through model reduction and evolutionary optimization, in *The Practical Handbook of Genetic Algorithms: Applications* (Chapman & Hall/CRC 2001)
11. D.E. Goldberg, *Genetic algorithms in search, optimization and machine learning* (Addison-Wesley, Reading, 1989)
12. J.H. Holland, *Adaptation in natural and artificial systems* (University of Michigan Press, Ann Arbor, 1975)
13. E.A. Elayan, Design of heuristic fuzzy logic controller for liquid level control, in *International Conference on Intelligent System Modelling and Simulation*, 2014
14. S. Balochian, E. Ebrahimi, Parameter optimization via cuckoo optimization algorithm of fuzzy controller for liquid level control. J. Eng. 2013

Evolutionary Smith Predictor for Control of Time-Delay Systems

Neelbrata Roy, Anindita Sengupta and Ashoke Sutradhar

1 Introduction

In most cases, plants having time delay can't be controlled efficaciously using Proportional plus Integral plus Derivative controller as the lag in phase added by the delay term tries to make the closed-loop system unstable. By decreasing the controller gain the stability problem can be addressed even though the output becomes very nebbish in nature [1, 2]. An exclusive approach for the control of systems accompanying a certified time-delay was developed by Otto. J. Smith in the 1950s by balancing the delayed response using input entities stored over a time frame $[t - t_l]$ and evaluates the response of the system employing a calculated plant prototype [3].

This philosophy of compensating the time delay was extended for unstable plants using finite-time integrals of the delayed input values thereby avoiding unstable pole-zero cancellations that may occur in Smith's controller [4].

Different versions of designs on Smith predictor has been developed after the invention led by Smith, followed by Astrom et al., Matusek et al., Majhi et al., Liu et al. and Kaya [1–6].

Liu et al. proposed an efficient design of smith predictor [1]. It consists of a set point tracking controller and disturbance estimator. The estimator constants are the gain's of the traditional PID controller and the parameter of setpoint tracking controller is λ_c and k_c respectively that are being dealt in the Liu's model of smith

N. Roy (✉) · A. Sengupta · A. Sutradhar
Department of Electrical Engineering, IIEST, Shibpur,
Howrah 711103, India
e-mail: neelbrataroy.rs2016@ee.iiests.ac.in

A. Sengupta
e-mail: aninsen@ee.iiests.ac.in

A. Sutradhar
e-mail: as@ee.iiests.ac.in

S. Chattopadhyay et al. (eds.), *Modelling and Simulation in Science, Technology and Engineering Mathematics*, Advances in Intelligent Systems and Computing 749,
https://doi.org/10.1007/978-3-319-74808-5_29

predictor. Two examples had been discussed here, an unstable plant [1, 2] and another stable plant [7–11]. The accomplishment of the closed-loop system realized in this study on the plants discussed in section two as well as in paper [1] is better than the outcomes reported in [1, 2] and hence the design procedure was applied for other plants also [7–11]. The Optimization technique used are PSO [12, 13], ACO [14], HSO [15], and GWO [16]. The results have overshoot within 5%, which is quite acceptable for practical systems [17, 18].

2 System Identification and Mathematical Model

The system considered for testing is a single tank system. The mathematical modeling is discussed briefly below. The single tank system has a uniform cross-sectional area A, a flow resistance R is attached to the bottom this can be a valve or a pipe. The layout of Liquid Level system (LLS) is shown in Fig. 1.

The rate of flow of fluid through the resistance is related to the head of tank 'h' by linear relationship given by [7–11].

$$q_0(t) = \frac{h(t)}{R}$$

Thus we can say the difference in flow rate is directly proportional to height and inversely proportional to the flow resistance that can be mathematically stated using the following differential equation.

$$q_i(t) - q_0(t) = A\frac{dh(t)}{dt}$$

Using the above equations,

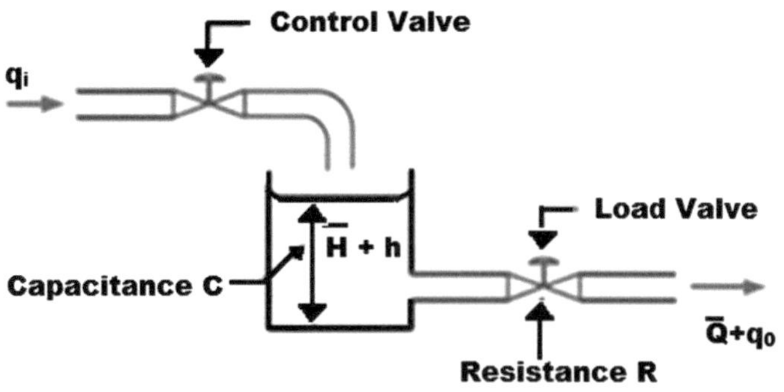

Fig. 1 Single tank system

$$q_i(t) - \frac{h(t)}{R} = A\frac{dh(t)}{dt}$$

Thus the system transfer function is given by

$$\frac{H(s)}{Q_i(s)} = \frac{R}{1+Ts}$$

From experimental datasheet of the instrument, it was found that $R=5380$ and $T=204$. Therefore overall transfer function of the system is

$$\frac{H(s)}{Q_i(s)} = \frac{5380}{1+204s}$$

The delay of the LLS is approximately 2s multiplying the additional delay to the transfer function stated above the overall TF of the LLS is written as

$$\frac{H(s)}{Q_i(s)} = \frac{5380}{1+204s}\mathrm{e}^{-2s}$$

And the close loop Transfer Function of the system was found to be

$$G(s) = \frac{5380\mathrm{e}^{-2s}}{1+204s+5380\mathrm{e}^{-2s}}$$

It can easily be inferred from the step response of the system that the system is unstable as illustrated in Fig. 2.

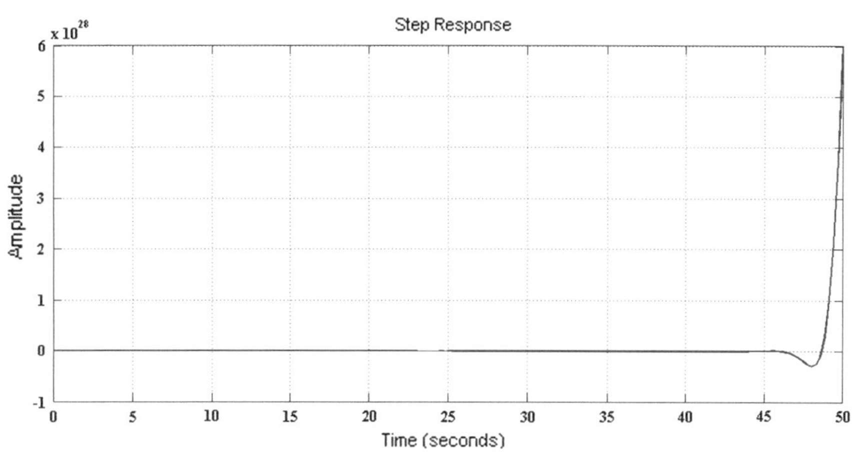

Fig. 2 Closed loop response of STS

2.1 An Introduction of Smith Predictor Scheme for Processes with Dead Time

The paper [1] details the smith predictor scheme that controls a class of unstable system with dead time generally called FODT system; the schema results in a lucid control structure and refine tuning capability for reference inputs along with load perturbations. The control system scheme is shown below [1] (Fig. 3).

Looking to the above scheme and literature cited the disturbance estimator and setpoint tracking controller parameters namely K_p, $K_i K_d$ and λ are revised using all the optimization procedure stated above keeping the objective function ITAE [12–16].

3 Ant Colony Optimization Technique

It's seen in nature that ants move in a random fashion and upon finding food return to their respective colony while putting down pheromone trails. If another ant discovers the same path, they do not keep traveling at random, but instead, they follow the laid trail, retracing and enriching it if they eventually find food in the same location.

Over time, however, the pheromone trail fades away, thus making feeble the time taken by pheromone to evaporate is equivalent to a summation of go and return time of the ant. Therefore the ants move over the same path more frequently, and thus the pheromone density becomes higher on shorter paths than the longer ones. The procedure is explained below.

STEP 1: Initially a random number of ants visits a random number of nodes and ramblingly moves by the edges moving from one node to other node.
STEP 2: Based on constraints all ants build his own solution.

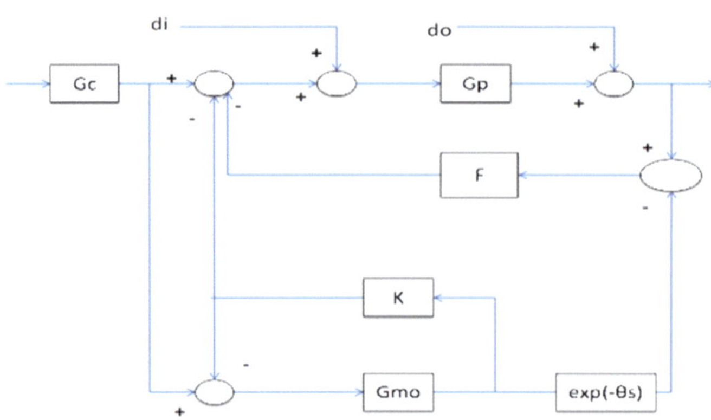

Fig. 3 Smith predictor scheme. *Courtesy* Liu et al. [1]

STEP 3: The next node is selected based on the pheromones accumulated at edge of each node.
STEP 4: The probability of choosing next node 'j' from current node 'i' is given by:

$$P_{ij}^{k}(t) = \frac{[\tau_{ij}^{\alpha}][\eta_{ij}^{\beta}]}{\sum [\tau_{il}^{\alpha}][\eta_{il}^{\beta}]}$$

where

τ_{ij}	symbolizes the pheromone concentration with the edges
η_{ij}	Symbolizes the heuristic information which is given by
$\eta_{ij} = \frac{1}{d_{ij}}$; d_{ij}	is the distance between two nodes.
$\alpha = 0$	If nodes closer to each other are chosen.
$\beta = 0$	If probability depends on pheromone intensification.

STEP 5: if all nodes are traversed then, pheromone yoked with each node is amended by:

$$\tau_{ij}(t+1) = (1-\rho).\tau_{ij}(t) + \sum_{k=1}^{m} \Delta\tau_{ij}^{k} \forall (i, j)$$

$\Delta\tau_{ij}^{k}(t)$ is the quantity of pheromone would be supplemented in the route k traversed by a ant, and ρ is assumed to be vanishing rate of pheromone

$$\Delta\tau_{ij}^{k}(t) = \begin{cases} \dfrac{1}{L_k} \\ 0 \end{cases}$$

$L_k L_k$, is the net distance of path k. Now, as is inversely equivalent to $\Delta\tau_{ij}^{k}$ therefore with the increase $\Delta\tau_{ij}^{k}$ increase in $\tau_{ij}(t+1)$ is obvious. Thus the quantity of pheromones yoked with that particular path is more and hence quintessential solution of the cost function is obtained [14].

4 Particle Swarm Optimization Technique

It is a nature-based optimization technique. It continuously tries to improve a candidate solution with regard to a given measure of quality i.e. the cost function. It elucidates a problem by having a populace of candidate solutions, these particles percolate around in the search-space according to simple mathematical formulae over the particle's position and velocity. The movement of particles can be altered by its local best-known position and all of them try to move towards the best-known positions in the search-territory. The better positions are updated by other particles.

Thus the swarm is expected to move toward the best solutions [12, 13]. The algorithm is detailed as follows.

STEP 1: The *numbers* of particles, total no of iterations, size of swarm, the locus of intervals of the maximum and minimum bound of the parameters are being initialized.
STEP 2: Assessment of the function ITAE (standard cost function).

(a) The particles initialized are hied to the standard cost function.
(b) The functional value obtained is assigned as the pioneer *Pbest* of the swarm. The pick of the liter among *Pbest's* is known as *gbest.*

STEP 3: Until the cessation requirement is reached the looping is to be continued (amendment of velocity and position of particles).

(a) The velocity of the particle is updated using the following formula as shown below

$$V_1(t+1) = V_1(t) + c_1(P_1 - x_1(t))R_1 + c_2(g - x_1(t))R_2$$

here the amended velocity $v_i(t+1)$ is, P_i the *Pbest*, g is the *gbest*, c_1 is the self-learning rate, c_2 is the global erudition coefficient, these values are any two random numbers between zero and four respectively, R_1 and R_2 are diagonal matrices generated using unifrand command the range is in between [0,1] respectively.
(b) The up gradation of the position of the particle in the swarm done using the following formula

$$x_i(t+1) = x_i(t) + v_i(t+1)$$

(c) The amended particles are again passed to the cost function.
(d) The *Pbest* and *gbest* are amended as follows

$$\text{If } G(x_i(t+1)) > G(p_i) \text{ then } p_i = x_i(t+1)$$
$$\text{If } G(x_i(t+1)) > G(g_i) \text{ then } g_i = x_i(t+1)$$

.

Step 4: As the process ends gbest gives the prime estimate of the disturbance estimator parameters and setpoint tracking controller parameter [12, 13, 18].

5 Harmony Search Algorithm

Harmony search is a music-based metaheuristic optimization algorithm. It is seen that the aim of music is to achieve for a sublime position of harmony. The venture to get the harmony in music is as equivalent to achieve optimality in an optimization problem. A music virtuoso always intends to produce a piece of music with perfect harmony. The process goes similarly for an optimization process to find the best solution available for the problem under the given cost functions and restricted to constraints and other inequality. The procedure of harmony search is shown below [15].

STEP 1: The HM (Harmony Memory) is generated. This generated matrix has the solutions for the problem discussed in Sect. 2.1. The content of the matrix is generated randomly.
STEP 2: A set of solution is created from the HM $[x_1', x_2', \ldots x_n']$ and each component of the solution is based on harmony memory (HM) considering rate (HMCR). The HMCR is the probability of selecting a component from the HM. The solution is further muted according to pitch adjustment rate (PAR). The PAR decides the probability of a candidate from the HM to be metamorphosed. Thus the phenomenon is similar to the production of the progeny.
STEP 3: Update the HM. The new solution from Step 2 is evaluated. If it yields a better fitness than that of the worst member in the HM, it will replace that one. Otherwise, it checks out.
STEP 4: Repeat Step 2 to Step 3 until the maximal number of iterations, is met [15].

6 Grey Wolf Optimization

The GWO algorithm adopts the leadership level and hunting strategy of grey wolves and is proposed by Mirjalili et al. in 2014. Four types of grey wolves such as alpha, beta, delta, and omega are employed for simulating the leadership levels. In addition, three main steps of hunting, searching for prey, encircling prey, and attacking prey, are implemented to perform optimization. The algorithm is as follows [16].

STEP 1: The Pack of grey wolves (prospective solutions), 'N' and a guess for no of required iteration, 'N_{iter}' is being entered by the end user for the cost optimization problem.
STEP 2: With, x_α, the initial pack position, the value of $f(x_\alpha)$ is calculated.
STEP 3: Based on the fitness value the grey wolves are categorized as α, β, δ.
STEP 5: The looping process is started until cessation condition is met.
STEP 6: Another looping process is started for all successive wolf j attached to the pack.
STEP 7: The place of the wolf is amended by the following relationship:

$$\vec{x}(t+1) = \frac{\vec{x_1} + \vec{x_2} + \vec{x_3}}{3}$$

This marks the end of step 7.

STEP 8: Amend the trade-off parameter 'a' using the formula a $= 2 - t\frac{2}{MaxIter}$, A $=$ $2ar_1 - a$, C $= 2r_2a$, the parameter 'a' is the deciding factor between investigation and profiteering between the pack. The parameter A and C are vectors that represent the position or solutions of the prey or the problem cast.
STEP 9: The cost function is again examined with this value of position, β, and δ are updated. This Marks the End of Step 5. α, β are subjected as the best solution for control problem cast [16, 18].

7 Results and Discussions

On simulating the above four optimization algorithms, the comparative performances of the unstable system [1] have been illustrated in Fig. 4.

Example 1. Consider the unstable time delay process as in [1].

$$G_p(s) = \frac{4e^{-2s}}{4s - 1}$$

For the above plant the set point tracking controller and disturbance estimator designed is as follows.

$$G_c(s) = \frac{s + 0.75}{2s + 1}$$

Keeping the tuning Parameter $\lambda_c = 2$ and

$$F(s) = 0.5186 + \frac{1}{32.7873s} + 0.4s$$

A unit step input is added at t=0 and an inverse step load disturbance with magnitude 0.1 is added to the process input at t=30 s [1].

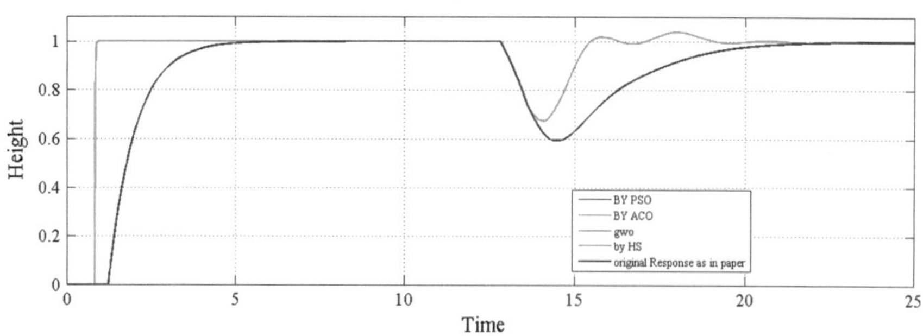

Fig. 4 Comparison of responses of four tuning algorithm for Smith predictors

Table 1 Computed estimator/controller parameter values for the unstable system

	Name of algorithm / Estimator/controller parameters	Ant colony algorithm	Particle swarm optimization	Harmony search algorithm	Grey wolf algorithm
$G_p(s) = \frac{4e^{-2s}}{4s-1}$	K_p	0.61274	0.6529	0.61274	0.6529
	K_i	0.045864	0.0941	0.0941	0.0941
	K_d	0.69508	0.6392	0.69508	0.6392
	λ_c	0.36121	0.0305	0.0305	0.0305

The computed values of the estimator/controller parameters in different optimization methods have been given in Table 1.

The above algorithms had also been tested for the single tank system [7–11] with delay. The comparative performances for controllers optimized using four above algorithms have been shown in Fig. 5 and the corresponding values of the computed estimator/controller parameters have been shown in Table 2.

8 Conclusion

This paper adduces a tuning procedure of smith predictor based on Liu et al. [1] model for time-delay systems using four different evolutionary algorithms. The optimization process tuned the disturbance estimator and setpoint tracking controller minimizing

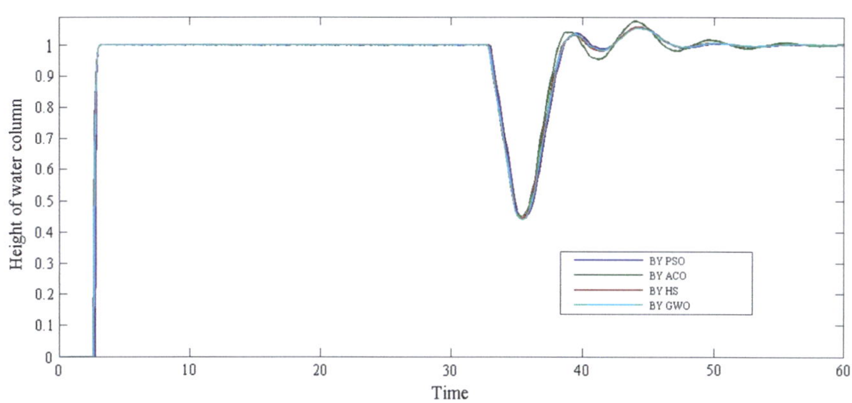

Fig. 5 Response Smith predictors for a single tank system with time delay

Table 2 Computed estimator/controller parameter values for stable liquid level system

	Name of algorithm	Ant colony algorithm	Particle swarm optimization	Harmony Search algorithm	Grey wolf algorithm
	Estimator/controller parameters				
$G_p(s) = \frac{5380}{1+204s}e^{-2s}$	K_p	0.0253	0.0242	0.0244	0.024259
	K_i	0.0063	0.0059	0.0061	0.0059498
	K_d	0.0243	0.0221	0.0226	0.0221
	λ_c	0.0714	0.0890	0.0786	0.078584

the performance index ITAE. The responses for all the cases show improved rise time and disturbance rejection than that reported by Liu et al.

The obtained outcome had improvised the reported results even though a little peak is observed but it is within 5% of the desired result so it is quite acceptable. Further development in the stabilizing controller and setpoint tracking controller may improve the results in terms of settling time and disturbance rejection.

References

1. T. Liu, Y.Z. Cai, D.Y. Gu, W.D. Zhang, New modified Smith predictor scheme for integrating and unstable processes with time delay, IEE Proc.-Control Theory Appl. **152**(2) (2005)
2. Ibrahim Kaya, A new Smith predictor and controller for control of processes with long dead time. ISA Trans. **42**, 101–110 (2003)
3. O.J.M. Smith, A controller to overcome dead time. ISA J. **6**(2), 28–33 (1959)
4. S. Majhi, D.P. Atherton, Obtaining controller parameters for a new Smith predictor using auto-tuning. Automatica **36**, 1651–1658 (2000)
5. M.R. Matausek, A.D. Micic, On the modified Smith predictor for controlling a process with an integrator and long dead-time. IEEE Trans. Autom. Control **44**(8), 1603–1606 (1999)
6. K.J. Astrom, C.C. Hang, B.C. Lim, A new Smith predictor for controlling a process with an integrator and long dead time. IEEE Trans. Autom. Control **39**, 343–345 (1994)
7. R. Paul, A. Sengupta, Discrete wavelet packet transform based controller for liquid level system and its performance analysis. Measurement **97**, 226–233 (2017)
8. S. Sen, S. Chakraborty, A. Sutradhar, Estimation of vehicle yaw rate and lateral motion for dynamic stability control using unscented Kalman filtering (UKF) approach. IET Digital Library, MFIIS-2015
9. U. Mondal, A. Sengupta, Rajeev R. Pathak, Servomechanism for periodic reference input: discrete wavelet transform-based repetitive controller. Trans. Inst. Meas. Control **38**(1), 14–22 (2016)
10. R. Paul, A. Sengupta, R.R. Pathak, Wavelet-based denoising technique for the liquid level system. Measurement **46**(6), 1979–1994 (2013)
11. U. Mondal, A. Sengupta, A. Roy, Repetitive controller: an advanced s servomechanism for periodic reference input. Int. J. Dyn. Control **4**(4), 428–437 (2016)
12. M. Clerc, Standard particle swarm optimization. *HAL open access archive, 2012*

13. R. Baskarane, T. Sendhil Kumar, Hybrid optimization for multiobjective multicast routing. IJRAT **2**(3) (2014)
14. M. Dorigo, T. Stültze, *Ant Colony Optimization* (MIT Press, 2004). p. 12
15. Z.W. Geem, *Music-Inspired Harmony Search Algorithm Theory and Applications* (Springer, Berlin, 2009)
16. S. Mirjalili, S.M. Mirjalili, A. Lewis, Grey wolf optimizer. Adv. Eng. Softw. **69**, 46–61 (2014)
17. IEEE Guide for Identification, Testing, and Evaluation of the Dynamic Performance of Excitation Control Systems, IEEE Standard 421.2-1990
18. N. Roy, A. Sengupta, A. Sutradhar, A comparison between bio-inspired and music-inspired smith predictor for control of time-delay systems. 2017 IEEE CALCON (2017)

On-line Adaptation of Parameter Uncertainties of a Practical Plant Employing L_1 Adaptive Controller

Roshni Maiti, Kaushik Das Sharma and Gautam Sarkar

1 Introduction

Adaptive controllers are introduced to handle system with uncertainties, time varying disturbances and nonlinearities. Different types of adaptive controllers such as model reference adaptive sliding mode control [1], fuzzy adaptive controller [2, 3] are used to control different types of systems. Adaptive controller [4] estimates the uncertainties present in the system and adapt them to produce control signal. Error between the output of the reference model and system is used to produce adaptation law and the controller reduces the error asymptotically is commonly known as direct adaptive control whereas when the system parameters are dynamically estimated to produce adaptive law and control signal is called indirect adaptive control [5].

Conventional model reference adaptive controller (MRAC) [6] has some disadvantages like slow transient performance, dependency upon systems input etc. To cope up with such problem in the year of 2006 Cao and Hovakimyan introduces a novel adaptive controller named as L_1 adaptive controller [7] inserting a filter to eliminate the high frequency introduces into the control signal due to high adaptation rate. They guarantee high robustness and stability with quick transient performance [7].

Different systems such as aircraft [8, 9], single-link armed robot [7] are controlled using L_1 adaptive controller in simulation environment efficiently. It is very much important to validate a controller in real life experimentation. Now in this paper the effectiveness of this method is tested in real time environment. A dc motor is

R. Maiti (✉) · K. Das Sharma · G. Sarkar
Department of Applied Physics, University of Calcutta, Kolkata, India
e-mail: roshni.maiti@gmail.com

K. Das Sharma
e-mail: kdassharma@gmail.com

G. Sarkar
e-mail: gautamgs2010@yahoo.in

S. Chattopadhyay et al. (eds.), *Modelling and Simulation in Science, Technology and Engineering Mathematics*, Advances in Intelligent Systems and Computing 749,
https://doi.org/10.1007/978-3-319-74808-5_30

tested with different trajectories to justify the quick transient performance with high robustness in real time. L_1 adaptive controller at first estimates the uncertainties, unknown constant present in the system and then adapt those and produce control signal [10, 11]. Rate of adaptation is made high to give swift transient performance. Robustness is also guaranteed by use of low pass filter and projection operator. To obtain non-adaptive parameters of L_1 adaptive controller a well known stochastic optimization technique named as particle swarm optimization (PSO) [12, 13] is used. Adaptive parameters are adapted online continuously in each time step to give desired results.

There are a very few literature where L_1 adaptive controller is tested online. Maalouf et al. [11] test L_1 adaptive controller in real time experiment to control an AC-ROV submarine which is an under-actuated underwater vehicle. They augment L_1 adaptive controller with proportional integral (PI) controller to drastically reduce the tracking time lags. Here in this proposed method no other controller is required in conjunction with L_1 adaptive controller to get proper controlling parameters. In simulation environment PSO is giving the optimal parameter setting which is capable of controlling systems without using any further with L_1 adaptive controller. This naturally reduces the online computational burden as well as computational time which are of great importance.

The paper is arranged as follows. Second section lights up on L_1 adaptive controller architecture. Third section describes the L_1 adaptive controller implementation in online mode with two sub section as inclusion of model uncertainties, online L_1 adaptive controller. Experimental results are tabulated and depicted in section four. Discussion and conclusion ends up the paper in fifth section.

2 Problem Formulation

2.1 L*1* *Adaptive Controller Architecture*

L_1 adaptive controller shows in Fig. 1 comprises of system with time varying uncertainties, disturbances, unknown constant in it, predictor, adaptive law block and controller with low pass filter. Cao and Hovamikiyan clearly describe the different parts of it [7]. State predictor predicts those uncertainties and estimates them in adaptive manner with some adaptive law derived from Lyapunov stability criteria. Control signal will produce and filtered to remove high frequency introduced due to high adaptation gain and given to the system.

The state predictor [7], described by this equation

$$\dot{\hat{\underline{x}}} = A_m\hat{\underline{x}}(t) + b(\hat{\omega}(t)u(t) + \hat{\underline{\theta}}^T(t)\underline{x}(t) + \hat{\sigma}(t)) \quad (1)$$

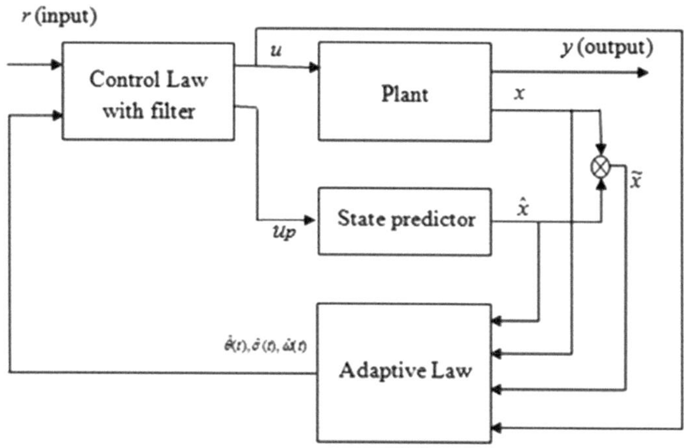

Fig. 1 Architecture of L_1 adaptive controller

gives the adaptive estimates of unknown constants, $\hat{\omega}(t) \in \Re$; uncertainties, $\underline{\hat{\theta}}(t) \in \Re^n$; time varying disturbances $\hat{\sigma}(t) \in \Re$ present in the system model described by these equations,

$$\begin{aligned} \dot{\underline{x}}(t) &= A_m \underline{x}(t) + b(\omega u(t) + \underline{\theta}^T(t)\underline{x}(t) + \sigma(t)) \\ \underline{x}(0) &= \underline{x}_0, \end{aligned} \tag{2}$$

$$y(t) = c^T \underline{x}(t) \tag{3}$$

where, A_m is the Hurwitz matrix $\in \Re^{n \times n}$. $b \in \Re^n$, $c \in \Re$ are known constant vector. $\underline{x}$ is the state vector $\in \Re^n$. u is the control input to the system $\in \Re$ and $y \in \Re$ is the output of the system. ω is the unknown constant $\in \Re$, $\underline{\theta} \in \Re^n$ is the unknown uncertainties and $\sigma(t) \in \Re$ is the time varying disturbances present in the system.

The adaptation law can be derived from the stability analysis of the system. The Lyapunov equation for this system is:

$$\begin{aligned} v\left(\underline{\tilde{x}}(t), \hat{\omega}(t), \underline{\hat{\theta}}(t)\right) &= \frac{1}{2}\underline{\tilde{x}}^T(t)P\underline{\tilde{x}}(t) + \frac{1}{2}\tilde{\omega}^T(t)\Gamma_1^{-1}\tilde{\omega}(t) \\ &+ \frac{1}{2}\underline{\tilde{\theta}}^T(t)\Gamma_2^{-1}\underline{\tilde{\theta}}(t) + \frac{1}{2}\tilde{\sigma}^T(t)\Gamma_3^{-1}\tilde{\sigma}(t) \end{aligned} \tag{4}$$

where, $\underline{\tilde{x}} = \hat{x} - x$ is the state error vector, $\tilde{\omega} = \hat{\omega} - \omega$ is the error for unknown constant term, $\underline{\tilde{\theta}} = \underline{\hat{\theta}} - \underline{\theta}$ is the error for unknown uncertainties and $\tilde{\sigma} = \hat{\sigma} - \sigma$ is the error of disturbances.

The derivative of it is:

$$\dot{v}\left(\underline{\tilde{x}}(t), \hat{\omega}(t), \underline{\hat{\theta}}(t)\right) = \frac{1}{2}\dot{\underline{\tilde{x}}}^T(t)P\underline{\tilde{x}}(t) + \frac{1}{2}\underline{\tilde{x}}^T(t)P\dot{\underline{\tilde{x}}}(t) + \tilde{\omega}^T(t)\Gamma_1^{-1}\dot{\hat{\omega}}(t)$$
$$+\underline{\tilde{\theta}}^T(t)\Gamma_2^{-1}\dot{\underline{\hat{\theta}}}(t) + \tilde{\sigma}^T(t)\Gamma_3^{-1}\dot{\hat{\sigma}}(t), \quad (5)$$

From Lyapunov stability theorem it must hold that $\dot{v} < 0$ to make the system stable.

Satisfying the above stability condition the adaptive law will be as follows:

$$\dot{\hat{\omega}}(t) = \Gamma_1 \mathrm{Proj}(\hat{\omega}(t), -\underline{\tilde{x}}^T(t)Pb\,u(t)) \quad (6)$$

$$\dot{\underline{\hat{\theta}}}(t) = \Gamma_2 \mathrm{Proj}(\underline{\hat{\theta}}(t), -\underline{x}(t)\underline{\tilde{x}}^T(t)Pb) \quad (7)$$

$$\dot{\hat{\sigma}}(t) = \Gamma_3 \mathrm{Proj}(\hat{\sigma}(t), -(\underline{\tilde{x}}^T(t)Pb)^T) \quad (8)$$

Now, the projection (Proj(,)) operator [14] will limit the values of adapted parameters. Therefore, all the terms with adaptive parameters of Eq. (5) are bounded and they belongs to some compact set $[\omega\ \underline{\theta}\ \sigma] \in [\Omega\ \underline{\Theta}\ \Sigma]$. The state error vector $Lim\,\underline{\tilde{x}} = 0$ as $t \to \infty$. Therefore, $\dot{v}$ will become <0 and the system will be stable.

Now the aim is to produce a control signal that the system can track the desired trajectory properly. The control law of L_1 adaptive controller is:

$$u(s) = -kC(s)\left[\hat{\eta}(s) - k_g r(s)\right]. \quad (9)$$

where, k is the gain, k_g is the pre filter feed forward gain, $\hat{\eta}(s)$ is the Laplace transform of $\hat{\eta}(t)$ given by,

$$\hat{\eta}(t) = \hat{\omega}(t)u(t) + \underline{\hat{\theta}}^T(t)\underline{x}(t) + \hat{\sigma}(t). \quad (10)$$

Due to high value of $\underline{\Gamma} = [\Gamma_1\ \Gamma_2\ \Gamma_3]$, high frequency signals introduce into the control channel which will affect the stability of the system. Therefore, one low pass filter

$$C(s) = \frac{f_c D(s)}{1 + f_c D(s)} \quad (11)$$

with $C(0) = 1$ is attached in the control channel to remove the high frequencies and to give high robustness with quick transient response. f_c is the cut off frequency of the filter.

3 L_1 Adaptive Controller Implementation in Online Mode

From L_1 norm condition the values of the L_1 adaptive controller parameters can be found out but that may not give optimal results. A stochastic optimization technique named as particle swarm optimization is used along with L_1 adaptive controller at the time of simulation to get optimal result which is used at the time of online adaptation. PSO is very efficient method in finding optimal parameter setting for controllers [15, 16]. At first parameter estimation has done which includes the uncertainties into the model and to cope up with those uncertainties online L_1 adaptive controller is implemented and tested successfully.

3.1 Inclusion of Parameter Uncertainties

In this paper, a DC motor is used as experimental setup as shown in Fig. 2. Voltage signal is given as the input to the motor through driver circuitry and speed of the motor is the output. A fly wheel is attached with the shaft and one permanent magnet is so arranged that it can produce load disturbances to the motor. For parameter estimation the DC motor runs in open loop environment where a pre-defined voltage input is given to the motor and output speed is recorded.

The model of the DC motor is gives as:

$$\ddot{\omega} + \left(\frac{R_a}{L_a} + \frac{B}{J}\right)\dot{\omega} + \left(\frac{BR_a}{JL_a} + \frac{K_b K_T}{JL_a}\right)\omega = \frac{KK_T}{JL_a}u(t) \tag{12}$$

where, ω is the angular speed of the motor in rad/sec. u is the dc voltage input and angular speed is taken as the output. Unknown motor parameters such as inertia (J), damping ratio (B), armature resistance (R_a), armature inductance (L_a), motor back emf coefficient (K_b), motor torque coefficient (K_T) and motor driver circuit coefficient (K) are obtained from parameter estimation of the motor. At the time of parameter estimation, uncertainties includes into the model. Let, the uncertainties are

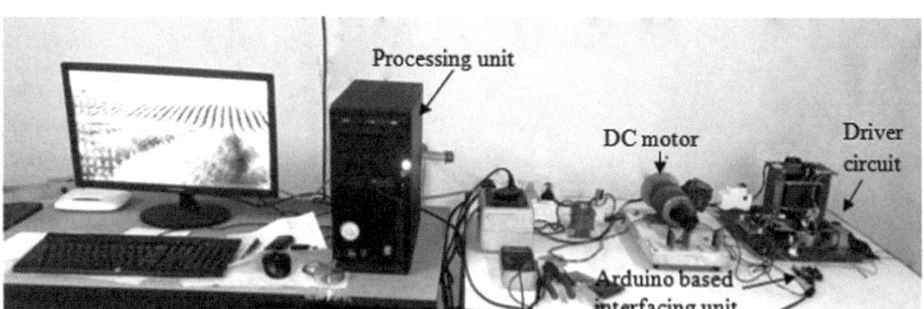

Fig. 2 DC motor experimental setup

ΔJ, ΔB, ΔR_a, ΔL_a, ΔK_b, ΔK_T and ΔK respectively for the above mentioned parameters. Now the DC motor model with uncertainties and time varying disturbances $d(t)$ can be represented as,

$$\ddot{\omega} + \left(\frac{R_a + \Delta R_a}{L_a + \Delta L_a} + \frac{B + \Delta B}{J + \Delta J}\right)\dot{\omega} + \left(\frac{(B + \Delta B)(R_a + \Delta R_a)}{(J + \Delta J)(L_a + \Delta L_a)} + \frac{(K_b + \Delta K_b)(K_T + \Delta K_T)}{(J + \Delta J)(L_a + \Delta L_a)}\right)\omega + d(t)$$
$$= \left(\frac{(K + \Delta K)(K_T + \Delta K_T)}{(J + \Delta J)(L_a + \Delta L_a)}\right)u(t) \tag{13}$$

or, $\ddot{\omega} + C_1\dot{\omega} + C_2\omega + d(t) = C_3 u(t)$

$$C_1 = \left(\frac{R_a + \Delta R_a}{L_a + \Delta L_a} + \frac{B + \Delta B}{J + \Delta J}\right),$$

$$C_2 = \left(\frac{(B + \Delta B)(R_a + \Delta R_a)}{(J + \Delta J)(L_a + \Delta L_a)} + \frac{(K_b + \Delta K_b)(K_T + \Delta K_T)}{(J + \Delta J)(L_a + \Delta L_a)}\right), C_3 = \left(\frac{(K + \Delta K)(K_T + \Delta K_T)}{(J + \Delta J)(L_a + \Delta L_a)}\right)$$

$$\begin{aligned} C_1 &= \left(\frac{R_a}{L_a} + \frac{L_a \Delta R_a - R_a \Delta L_a}{L_a(L_a + \Delta L_a)} + \frac{B}{J} + \frac{J\Delta B - B\Delta J}{J(J + \Delta J)}\right) \\ &= \left(\frac{R_a}{L_a} + \frac{B}{J} + \frac{L_a \Delta R_a - R_a \Delta L_a}{L_a(L_a + \Delta L_a)} + \frac{J\Delta B - B\Delta J}{J(J + \Delta J)}\right) \end{aligned}$$

$$\begin{aligned} C_2 &= \left(\begin{array}{l} \frac{BR_a}{JL_a} + \frac{R_a\Delta B + B\Delta R_a + \Delta B\Delta R_a - BR_a(L_a\Delta J + J\Delta L_a + \Delta J\Delta L_a)}{JL_a(J+\Delta J)(L_a+\Delta L_a)} \\ + \frac{K_bK_T}{JL_a} + \frac{K_T\Delta K_b + K_b\Delta K_T + \Delta K_b\Delta K_T - K_bK_T(L_a\Delta J + J\Delta L_a + \Delta J\Delta L_a)}{JL_a(J+\Delta J)(L_a+\Delta L_a)} \end{array}\right) \\ &= \left(\begin{array}{l} \frac{BR_a}{JL_a} + \frac{K_bK_T}{JL_a} + \frac{R_a\Delta B + B\Delta R_a + \Delta B\Delta R_a - BR_a(L_a\Delta J + J\Delta L_a + \Delta J\Delta L_a)}{JL_a(J+\Delta J)(L_a+\Delta L_a)} \\ + \frac{K_T\Delta K_b + K_b\Delta K_T + \Delta K_b\Delta K_T - K_bK_T(L_a\Delta J + J\Delta L_a + \Delta J\Delta L_a)}{JL_a(J+\Delta J)(L_a+\Delta L_a)} \end{array}\right) \end{aligned}$$

$$C_3 = \left(\frac{KK_T}{JL_a} + \frac{K_T\Delta K + K\Delta K_T + \Delta K\Delta K_T - KK_T(L_a\Delta J + J\Delta L_a + \Delta J\Delta L_a)}{JL_a(J + \Delta J)(L_a + \Delta L_a)}\right)$$

Therefore, Eq. (13) can be written as:

$$\ddot{\omega} + \left(\frac{R_a}{L_a} + \frac{B}{J} + \Delta P_1\right)\dot{\omega} + \left(\frac{BR_a}{JL_a} + \frac{K_bK_T}{JL_a} + \Delta P_2\right)\omega + d(t) = \left(\frac{KK_T}{JL_a} + \Delta P_3\right)u(t) \tag{14}$$

where, $\Delta P_1 = \frac{L_a\Delta R_a - R_a\Delta L_a}{L_a(L_a+\Delta L_a)} + \frac{J\Delta B - B\Delta J}{J(J+\Delta J)}$,

$$\begin{aligned} \Delta P_2 = &\frac{R_a\Delta B + B\Delta R_a + \Delta B\Delta R_a - BR_a(L_a\Delta J + J\Delta L_a + \Delta J\Delta L_a)}{JL_a(J + \Delta J)(L_a + \Delta L_a)} \\ &+ \frac{K_T\Delta K_b + K_b\Delta K_T + \Delta K_b\Delta K_T - K_bK_T(L_a\Delta J + J\Delta L_a + \Delta J\Delta L_a)}{JL_a(J + \Delta J)(L_a + \Delta L_a)} \end{aligned}$$ and

$\Delta P_3 = \frac{K_T \Delta K + K \Delta K_T + \Delta K \Delta K_T - K K_T (L_a \Delta J + J \Delta L_a + \Delta J \Delta L_a)}{J L_a (J + \Delta J)(L_a + \Delta L_a)}$ are the uncertainty terms with angular speed, angular acceleration and voltage input of the machine respectively. Therefore, the state space representation of the model with time varying uncertainties and disturbances become,

$$\begin{bmatrix} \dot{x}_1 \\ \dot{x}_2 \end{bmatrix} = \begin{bmatrix} 0 & 1 \\ -\left(\frac{R_a}{L_a} + \frac{B}{J} + \Delta P_1\right) & -\left(\frac{BR_a}{JL_a} + \frac{K_b K_T}{JL_a} + \Delta P_2\right) \end{bmatrix} \begin{bmatrix} x_1 \\ x_2 \end{bmatrix} + \begin{bmatrix} 0 \\ \left(\frac{KK_T}{JL_a} + \Delta P_3\right) \end{bmatrix} u + \begin{bmatrix} 0 \\ d(t) \end{bmatrix} \quad (15)$$

or,

$$\begin{bmatrix} \dot{x}_1 \\ \dot{x}_2 \end{bmatrix} = \begin{bmatrix} 0 & 1 \\ -\left(\frac{R_a}{L_a} + \frac{B}{J}\right) & -\left(\frac{BR_a}{JL_a} + \frac{K_b K_T}{JL_a}\right) \end{bmatrix} \begin{bmatrix} x_1 \\ x_2 \end{bmatrix} + \begin{bmatrix} 0 \\ 1 \end{bmatrix} \left(\left(\frac{KK_T}{JL_a} + \Delta P_3\right) u + \begin{bmatrix} -\Delta P_1 & -\Delta P_2 \end{bmatrix} \begin{bmatrix} x_1 \\ x_2 \end{bmatrix} + d(t) \right) \quad (16)$$

Comparing (2) and (16) the unknown constant, uncertainties and disturbance become,

$\omega = \left(\frac{KK_T}{JL_a} + \Delta P_3\right)$, $\underline{\theta} = \begin{bmatrix} \theta_1 & \theta_2 \end{bmatrix} = \begin{bmatrix} -\Delta P_1 & -\Delta P_2 \end{bmatrix}$ and $\sigma = d(t)$ respectively.

Equation (16) represents the dc motor model with parameter uncertainties, disturbances and unknown constant.

To estimate those parameters the DC motor is run for 10 min with time step 0.1 s in open loop condition. Variable step input with two different voltage steps is given as input to the motor, such as:

$$u_{oc}(t) = \begin{cases} 28.8 * 0.5\, u(t) & 0 < t \leq 300\,\text{s.} \\ 28.8 * 0.9\, u(t) & 300 < t \leq 600\,\text{s.} \end{cases}$$

Figure 3 shows the open loop input given to the motor. In this condition the output voltage is measured and converted back into speed in rpm which is given in Fig. 4.

Parameter estimation is done with that speed set as reference and the DC motor model is run in open loop with same input $u_{oc}(t)$ in simulation and the parameter values are estimated by help of a stochastic search algorithm, named PSO. This process is repeated 10 times with 30 particles and 200 iterations of PSO run to get minimum integral absolute error value. From those results the obtained nominal parameter values are tabulated in Table 1 and the result is given in Fig. 5.

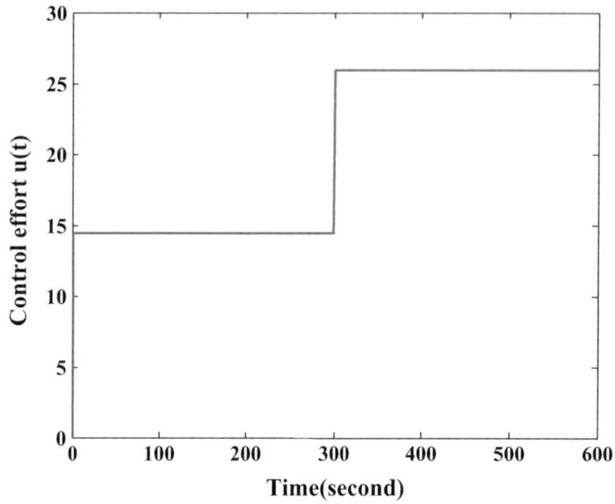

Fig. 3 Variable step input given to open loop system in real time environment

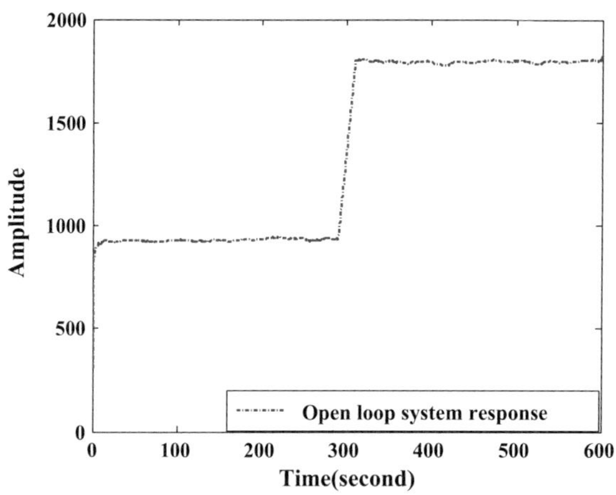

Fig. 4 Open loop system response in real time environment

Table 1 Estimated nominal parameter values of DC motor

R_a	L_a	J	B	K	K_b	K_T
0.43401	0.00073	0.00620	0.00043	0.43025	0.00329	0.00329

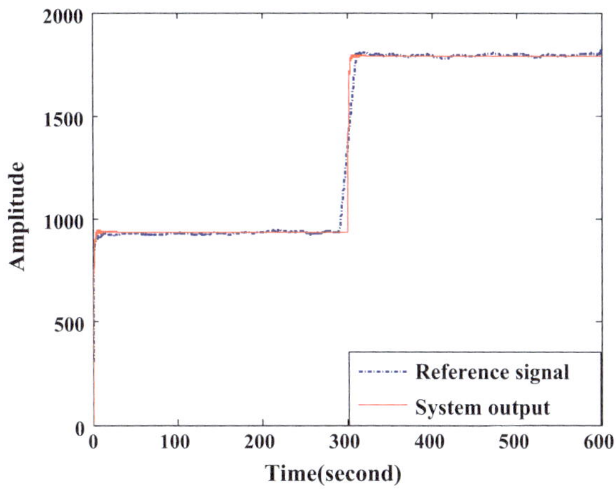

Fig. 5 Open loop system response in simulation environment

From Fig. 5 it can be clearly shown that due to parameter uncertainties and unknown constant, the real time output varies from simulation output though same input is given to the real time DC motor model (16) as well as to the simulation model (12). To produce effective control law for such system the uncertainties must be properly estimated and adapted. In this present paper, L_1 adaptive controller is run in an online manner to get proper estimation of those uncertainties and then from L_1 adaptive law those uncertainties are adapted to produce efficient control signal.

3.2 *Online* L_1 *Adaptive Controller*

Theoretically L_1 adaptive controller is robust and gives fast transient performances. It is of great importance to validate it in real time environment.

Parameters of L_1 adaptive controllers can be classified into two groups such as adaptive parameters and non-adaptive parameters. The parameter vector can be seen as: $\underline{P}_{\mathrm{L}_1} = [\omega|\underline{\theta}|\sigma\,|\underline{\Gamma}|k|k_g|f_c]$, where the adaptive parameters are: $[\,\omega|\underline{\theta}|\sigma\,]$ and non-adaptive parameters are $[\underline{\Gamma}|k|k_g|f_c]$. In this paper, first the PSO is used in an offline manner to get the non-adaptive parameters and the adaptive parameters are adapted in each time step from L_1 adaptation rule.

Table 2 Offline and online adaptation results in terms of IAE

	Integral absolute error (IAE)	
	Offline adaptation	Online adaptation
Input 1	19216.2484	18793.5851
Input 2	21852.4425	21506.4504
Input 3	32627.1874	30465.4716

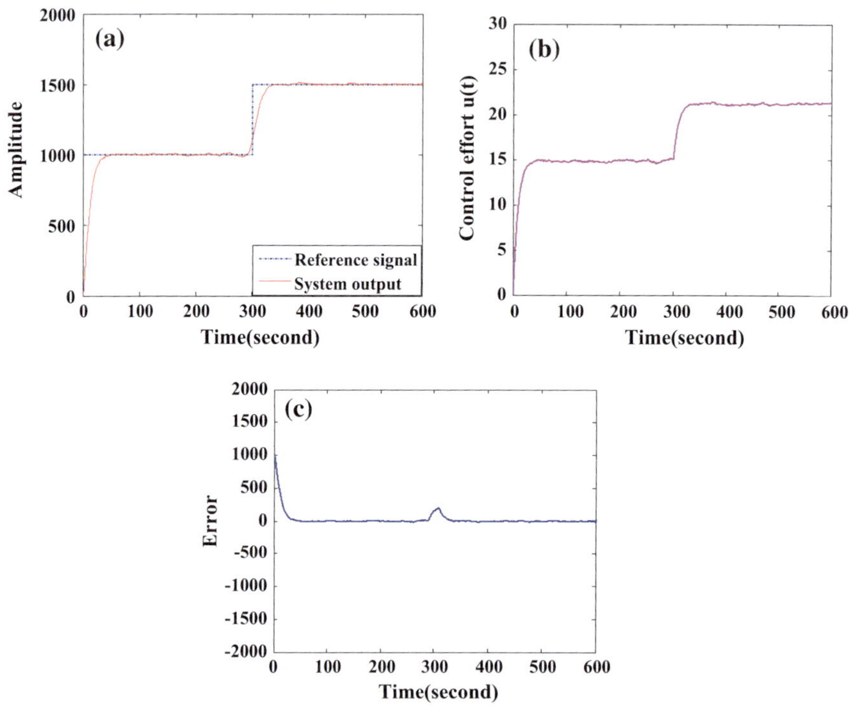

Fig. 6 Online adaptation responses of L_1 adaptive controller with input 1 for **a** system response with reference, **b** control effort, **c** error signal

With high adaptation gain the transient response is satisfactory enough, although it introduces high frequency into the control signal. Projection operator limits the values of adaptive parameters into the compact set given by $[\Omega \, \underline{\Theta} \, \Sigma]$ and the filter eliminates the high frequency contents into the bandwidth of the control channel and the system in real time gives very good results.

4 Experimental Results

The motor is run for 10 min with sampling time $\Delta t = 0.1$ second to get the results.

Two different variable step trajectories are given as reference such as:

Input 1: $r(t) = \begin{cases} 1000\,u(t)\ 0 < t \leq 300\,\text{s.} \\ 1500\,u(t)\ 300 < t \leq 600\,\text{s.} \end{cases}$

Input 2: $r(t) = \begin{cases} 1200\,u(t)\ 0 < t \leq 300\,\text{s.} \\ 1800\,u(t)\ 300 < t \leq 600\,\text{s.} \end{cases}$

Input 3: $r(t) = 1000\,u(t) \quad 0 < t \leq 600\,\text{s}$ with certain full load disturbances.

Offline Adaptation: Here in simulation, the criteria of selecting the parameters $\underline{P}_{L_1}$ of the L_1 adaptive controller is calculated from L_1 norm condition and from that mathematical calculation, some arbitrary values are chosen and adaptive parameters are adapted. For different trajectories the DC motor is then tested with those obtained non-adaptive and final values of adaptive parameters in real time.

Online Adaptation: Here in simulation, particle swarm optimization is used to select optimized parameter $\underline{P}_{L_1}$ values satisfying the L_1 norm condition and adaptive parameters are adapted following L_1 adaptive law. When this is tested in real time then optimized values are taken for non-adapted parameters and adaptive parameters are adapted spontaneously. Online adaptation of estimated uncertainties and unknown constant gives better result than offline adaptation process.

4.1 Real Time Case Study

At first the DC motor is run in offline adaptation mode using the L_1 adaptive controller parameters getting from the mathematical calculation of L_1 norm condition. Those results are given in Table 2 in terms of IAE.

Then the results are given for online L_1 adaptive controller where non-adaptive parameters are set from PSO and adaptive parameters are adapted in each step online. IAEs are tabulated in Table 2 and results are given in Figs. 6, 7 and 8 for input 1, input 2 and input 3 respectively. Figures 6a, 7a and 8a show the system output compared with reference signal whereas, Figs. 6b, 7b and 8b show the control effort required and Figs. 6c, 7c and 8c portray the error signal.

Comparing those results it can be clearly shown that the online adaptation of L_1 adaptive controller gives better results than L_1 adaptive controller implementation in offline adaptation mode.

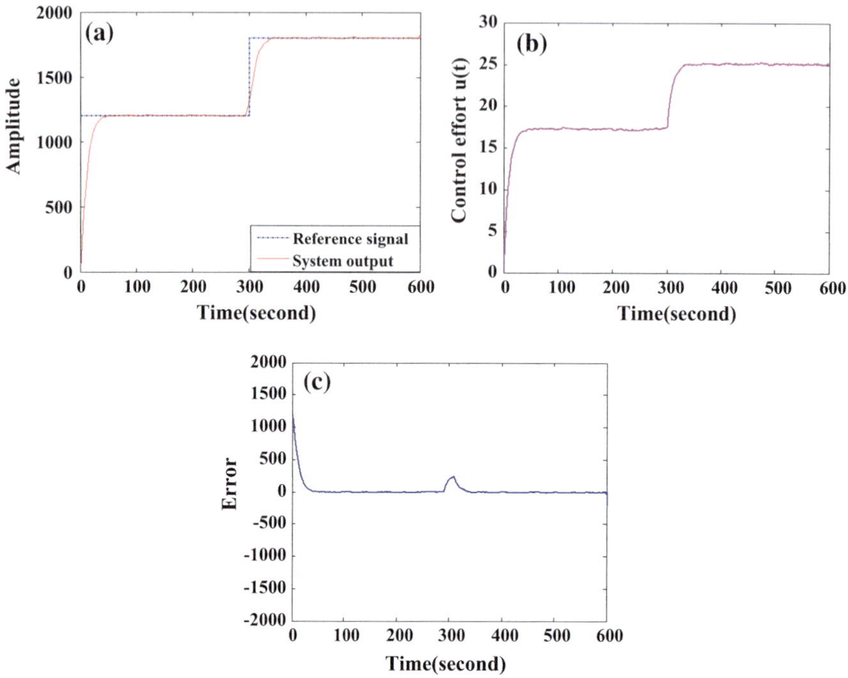

Fig. 7 Online adaptation responses of L_1 adaptive controller with input 2 for **a** system response with reference, **b** control effort, **c** error signal

5 Conclusion

In this paper on-line adaptation of L_1 adaptive controller is implemented successfully for speed control of a DC motor with unknown parameters and time varying disturbances. PSO provides the optimal parameter setting for non-adaptive parameters which give perfect selection of adaptive gain, filter parameters which are most important parameters in designing of L_1 adaptive controller. Adaptive parameters of the controller are adapted on-line to produce fast transient performance with high robustness. The proposed method is successfully applied on DC motor experimental setup and the results are compared in terms of integral absolute error to prove its effectiveness.

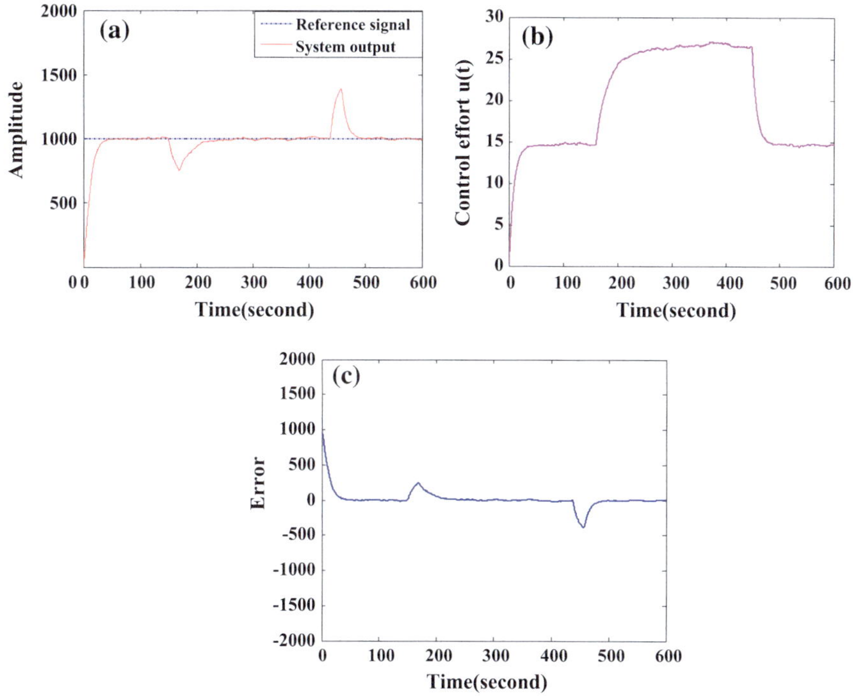

Fig. 8 Online adaptation responses of L_1 adaptive controller with input 3 for **a** system response with reference, **b** control effort, **c** error signal

References

1. M. Ganesan, D. Ezhilarasi, J. Benni, Hybrid model reference adaptive second order sliding mode controller for automatic train operation. IET Contr. Theo. & App. **11**(8), 1222–1233 (2017)
2. Amar Bechkaoui, Aissa Ameur, Slimane Bouras, Kahina Ouamrane, Diagnosis of turn short circuit fault in PMSM sliding-mode control based on adaptive fuzzy logic-2 speed controller. AMSE J. Ser. Adv. C **70**(1), 29–45 (2015)
3. A. Boulkroune, M. Tadjine, M. M'Saad, M. Farza, Design of a unified adaptive fuzzy observer for uncertain nonlinear systems. Inf. Sci. **265**, 139–153 (2014)
4. S.H. Hashemipour, N. Vasegh, A.K. Sedigh, Decentralized model reference adaptive control for interconnected time delay systems with delay in state and compensation of long delay in input by nested prediction. Int. J. Contr. Automat. Syst. **15**(3), 1198–1206 (2017)
5. L.R.L. Teixeira, J.B. Oliveira, A.D. Araujo, Stability analysis of indirect binary model reference adaptive controller for plants with relative degree one. J. Contr. Automation Elec. Syst. **26**(4), 2195–3899 (2015)
6. H.H. Nguyen, V.T. Duong, H. Yim, C. Van Ho, H.K. Kim, S.B. Kim, A model reference adaptive controller for belt conveyors of induction conveyor line in cross-belt sorting system with input saturation, in *International Conference on Advanced Engineering Theory and Applications, AETA'16: Recent Advances in Electrical Engineering & Related Science*, 2016, pp. 129–139

7. C. Cao, N. Hovakimyan, Guaranteed transient performance with L_1 adaptive controller for systems with unknown time-varying parameters and bounded disturbances: part I, in *Proceedings of the 2007 American Control Conference*, Marriott Marquis Hotel at Times Square, New York City, USA, 2007, pp. 3925–3930
8. R. Choe, E. Xargay, N. Hovakimya, L_1 adaptive control for a class of uncertain nonaffine-in-control nonlinear systems. IEEE Trans. Autom. Control **61**(3), 840–846 (2016)
9. F. Hellmundt, A. Wildschek, R. Maier, R. Osterhuber, F. Holzapfel, Comparison of L_1 adaptive augmentation strategies for a differential PI baseline controller on a longitudinal F16 aircraft model, in *Advances in Aerospace Guidance, Navigation and Control*, 2015, pp. 99–118
10. H.T. Song, T. Zhang, G.L. Zhang, L_1 adaptive controller of nonlinear reference system in presence of unmatched uncertainties. J. Central South University **23**(4), 834–840 (2016)
11. D. Maalouf, A. Chemori, V. Creuze, A new extension of the L_1 adaptive controller to drastically reduce the tracking time lags, in *NOLCOS'2013: 9th IFAC Symposium on Nonlinear Control Systems*, Toulouse, France, 2013, pp. 001–006
12. P. Biswas, R. Maiti, A. Kolay, K. Das Sharma, G. Sarkar, PSO based PID controller design for twin rotor MIMO system, in *Proceedings of International Conference on Control, Instrumentation, Energy and Communication (CIEC14)*, Kolkata, India, 2014, pp. 106–110
13. S. Kar, K. Das Sharma, M. Maitra, Gene selection from microarray gene expression data for classification of cancer subgroups employing PSO and adaptive K-nearest neighborhood technique. Exp. Sys. App. **42**(1), 612–627 (2015)
14. J. Niu, S. Ding, Y. Xing, Embedding theorems for composition of homotopy and projection operators. J. Inequalities Appl. **2015**(375), 1–14 (2015)
15. R. Maiti, K. Das Sharma, G. Sarkar, Optimal state feedback controller and observer design for twin rotor MIMO system, in *International Conference on Control, Instrumentation, Energy & Communication (CIEC16)*, Kolkata, India, 2016, pp. 30–34
16. M.M.d. Menezes, P.B.d. Araujo, D.B.d. Valle, Design of PSS and TCSC damping controller using particle swarm optimization. J. Cont. Auto. Elec. Syst. **27**(5), 554–561 (2016)

Two-Degree-of-Freedom Control of Non-minimum Phase Mechanical System

Mita Pal, Gautam Sarkar, Ranjit Kumar Barai and Tamal Roy

1 Introduction

Designing of Controller for a Non-minimum phase is a challenging tasks due to its Right-hand plane (RHP) zero dynamics. RHP zero of NMP system is responsible for large phase lag in the frequency domain and initial undershoot cannot be avoided in time domain analysis of NMP system. This special characteristic is not desirable for any appropriate system. Many Industrial processes like flexible link manipulator, aircraft, steam generator, electronic circuits etc. [1] has non-minimum phase characteristics, and it is an attractive job to handle this type of system, as undesirable phenomena is obvious in the system's dynamic response. Various methods have been developed during past decade for the control processes with non-minimum phase characteristics. QFT design method for NMP unstable system is revisited in 2001 [2]. Predictive control with PID structure [3], Filter basis function [4] has been studied for discrete type NMP system. Flight Control has been proposed by approximating non-linear NMP system by minimum phase system [5]. Recent research works on designing of a control system for NMP system have been performed using filter basis function and sliding mode learning control method [6, 7]. Input-output

M. Pal (✉) · G. Sarkar · R. K. Barai
Electrical Engineering Department, Jadavpur University,
Kolkata 700032, India
e-mail: mitapal91@gmail.com

G. Sarkar
e-mail: sgautam63@gmail.com

R. K. Barai
e-mail: ranjit.k.barai@gmail.com

T. Roy
Electrical Engineering Department, MCKV Institute of Engineering,
Liluah, Howrah 711204, India
e-mail: tamalroy77@gmail.com

S. Chattopadhyay et al. (eds.), *Modelling and Simulation in Science, Technology and Engineering Mathematics*, Advances in Intelligent Systems and Computing 749,
https://doi.org/10.1007/978-3-319-74808-5_31

linearization of the non-linear system has been tried to achieve local asymptotic stability [8] and recently, a Proportional plus Integral current-output observer based linear quadratic discrete tracker(LQDT) methodology has been theoretically applied on a discrete time NMP system [9]. One peculiar characteristic of NMP system is being experimented, where, the series combination of inverse and non-inverse transfer function model of NMP system always produce oscillatory and unbounded output response [10, 11]. It is well understood, that this phenomenon is not true for minimum phase system. This unusual nature of RHP zero dynamics inspired the researcher to design an appropriate control system for NMP system. The degree of freedom in control system engineering is the number of variables that can be manipulated or regulated in a process. In another way, it may be stated that a number of closed loop transfer functions which have to be controlled, indicates the DOF of the system [12]. The Two-degree-of-freedom (2DOF) strategy in design technique of control system has obviously been a better way to get precise control performance. The Recent development of 2DOF controller has been found in Ball and Beam system [13], where the ball can track the square wave with certain given specifications. Single input single output industrial processes with model uncertainty are automatically tuned with Model Predictive Control in 2DOF structure [14]. By surfing others literature regarding 2DOF, it is observed that 2DOF concept becomes a reliable tool for researchers. The combination of Feed-forward and feedback compensators in 2DOF structure may be an effective tool for NMP system also. Adaptive feed-forward zero phase error tracking control in 2DOF structure shows the good result for the minimum as well as NMP System [15].

Among all Adaptive Controller, Model Reference Adaptive Control (MRAC) is very much popular as it is a direct approach to force the uncertain system to obtain the desired performance [16]. Here is an added advantage for control designer that, they can choose the shape of the transient part of the output response, and it is only possible by reference model, which is an essential part of the MRAC design [17]. A Model Reference Adaptive Controlled Permanent Magnet Synchronous Motor has been simulated successfully, where time response shows optimum performance [18]. Both MIT rule and Lyapunov stability rule can apply for designing of Model Reference Adaptive Control and its efficiency has been observed by implementing on a Chemical Reactor [19], that the stability of the Controller is completely granted for Lyapunov stability theory. To improve the lateral stability of articulated heavy vehicles (AHVs), MRAC has been theoretically applied to active trailer steering (ATS) [20]. The transient performance of closed loop Model reference adaptive controlled fractional order nonlinear system improves in the sense of generating smooth system output [21]. Proportional plus Integral (PI), Proportional plus Integral plus Derivative (PID), Phase lead or Phase lag Controller may not be able to control all the poles of the higher order system independently. To circumvent these difficulties, state feedback control with arbitrary pole placement approach can solve this problem independently under certain condition, that the system must be completely state controllable [22]. Transient dynamics of piezo-actuated bimorph atomic force microscopy (AFM) probe is controlled by state feedback controller, where quality factor and the resonance frequency of the probe have been adjusted simultaneously

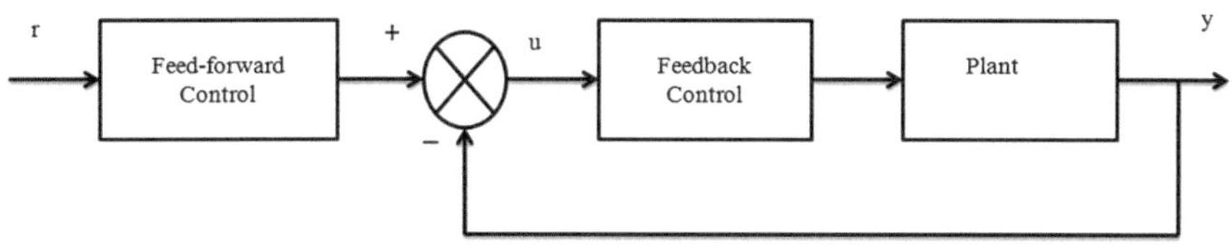

Fig. 1 Block diagram of two-degree-of-freedom control scheme

[23]. An optimal state feedback design has been considered for the classical unstable plant like Inverted Pendulum by calculating the appropriate state feedback gains using flower pollination algorithm (FPA) [24]. The efficiency of MRAC and Pole placement method encourages us to design a suitable controller for inversion based Non-minimum phase system.

The objective of this paper is to present a Two-degree-of-freedom controller for NMP system, in which inverse transfer function model becomes compensated by Model Reference Adaptive Control as feed-forward technique and non-inverse transfer function model is controlled by arbitrary pole placement method as feedback counter part of the 2DOF structure. This proposed method is verified with a practical mass spring damper system [25] theoretically and compared it with two different compensators, like PID and State Feedback Controller.

This paper is organized as follows, Sects. 2, 3, 4 and 5 represent brief descriptions of the 2DOF, MRAC, State feedback and PID controller respectively. The mathematical background of MRAC is provided by Sect. 3.1. Section 6 describes problem formulation. Description of NMP mechanical system is provided by Sect. 7. Simulation results observation and analysis of simulation results discussed in Sects. 8 and 9 respectively. The demonstration ended with concluding remarks in Sect. 10.

2 Two-Degree-of-Freedom

The number of closed loop transfer function that can be actuated independently is known as Degree-of-Freedom. 2DOF structure not only guarantees stability, it also tries to achieve the desired performances.

The 2DOF control structure consists of feed-forward and feedback controller and they are connected in various fashion. In this block diagram (Fig. 1), it is shown that output signal of feed-forward controller becomes the control input of feedback controlled plant. The two controllers are decoupled with each other. Basically, feedback control is used to stabilize the system, whereas feed-forward controller help to satisfy the desired dynamic characteristics of the given system.

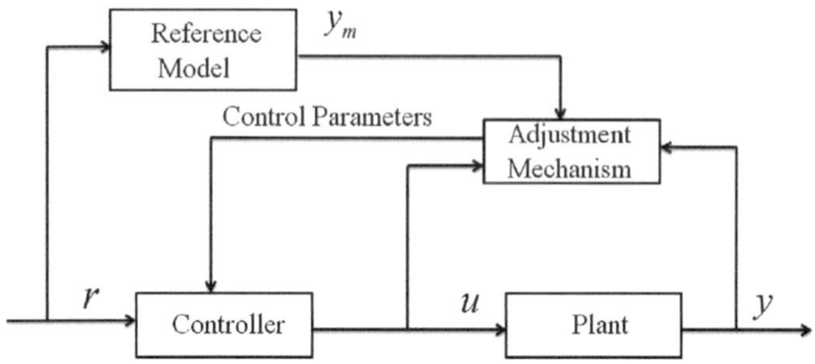

Fig. 2 Model reference adaptive system

3 Model Reference Adaptive Control

In the domain of Adaptive Control System, Model Reference Adaptive system is an effective Adaptive control scheme. Originally it was applicable for Flight Control System, where the reference model describes the desired trajectories of the Aircraft [26]. The block diagram shown in Fig. 2 has two control loops, inner loop, and outer loop. The inner loop is an ordinary feedback loop composed of Plant and Controller and the outer control loop consists of the adjustment mechanism and reference model. Outer loop calculates the control parameters on the basis of error which is produced by the difference of reference and plant model outputs. In a model reference adaptive control, the mechanism for adjusting the parameters can be obtained in two ways, by using gradient method or by using stability theory.

3.1 Design of 2DOF Controller Using Model Reference Adaptive Control

The desired plant response to a command signal is specified by means of a parametrically defined reference model. An adaptation mechanism keeps track of the process output, and reference model output, and calculates the suitable control parameters so that the difference of these outputs tends to zero. In addition to the process output, y, plant or control input, u and reference input, r may be used in adaptation mechanism.

Tracking error, e is simply the difference between the plant output, y and the reference model output, y_m refer to Fig. 3. Considering,

Reference Model:

$$\frac{d^2 y_m}{dt^2} = -a_m \frac{dy_m}{dt} + b_m r \tag{1}$$

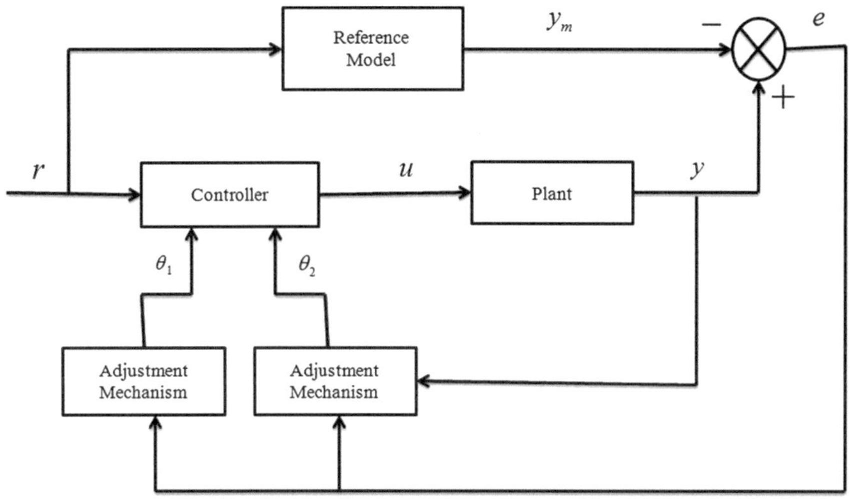

Fig. 3 Model reference adaptive control system

Plant Model:

$$\frac{d^2y}{dt^2} = -a\frac{dy}{dt} + bu \tag{2}$$

Let, Control input:

$$u = \theta_1 r - \theta_2 \frac{dy}{dt} \tag{3}$$

Tracking error:

$$e = y - y_m \tag{4}$$

By subtracting Eq. (1) from Eq. (2), we get,

$$\frac{d^2e}{dt^2} = \frac{d^2y}{dt^2} - \frac{d^2y_m}{dt^2} = -a\frac{dy}{dt} + bu - (-a_m\frac{dy_m}{dt} + b_m r) \tag{5}$$

Replacing u in Eq. (5) from Eq. (3), we get,

$$\frac{d^2e}{dt^2} = -a\frac{dy}{dt} + b(\theta_1 r - \theta_2\frac{dy}{dt}) + a_m\frac{dy_m}{dt} - b_m r \tag{6}$$

By adding and deducting $a_m \frac{dy}{dt}$ with Eq. (6), we get,

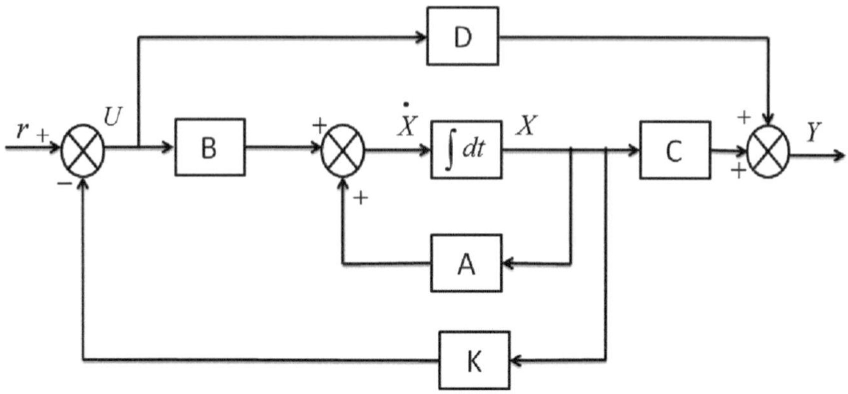

Fig. 4 Block diagram of state feedback controlled system

$$\frac{d^2e}{dt^2} = -a\frac{dy}{dt} + a_m\frac{dy}{dt} + b\theta_1 r - b\theta_2\frac{dy}{dt} - a_m\frac{dy}{dt} + a_m\frac{dy_m}{dt} - b_m r \tag{7}$$

$$\frac{d^2e}{dt^2} = -a_m(\frac{dy}{dt} - \frac{dy_m}{dt}) - (b\theta_2 + a - a_m)\frac{dy}{dt} + (b\theta_1 - b_m)r \tag{8}$$

$$\frac{d^2e}{dt^2} = -a_m\frac{de}{dt} - (b\theta_2 + a - a_m)\frac{dy}{dt} + (b\theta_1 - b_m)r \tag{9}$$

Assuming the initial values of the control parameter θ_1 and θ_2, and integrating Eq. (9) with respect to t, we get,

$$\frac{de}{dt} = -a_m e - (b\theta_2 + a - a_m)y + (b\theta_1 - b_m)r \tag{10}$$

From the error dynamics it is observed that the tracking error will go to zero if $b\theta_2 = a_m - a$ and $b\theta_1 = b_m$. The parameter adjustment rule thus achieve the goal.

4 Description of State Feedback Controller

Any System can be represented by state model, which consists of the following state equation and output equation $\dot{X} = AX + Bu$ and $Y = CX + DU$ where, X is state vector (n-vector), U is control signal (scalar), Y is output vector (scalar), A is $n \times n$ system matrix, B is $n \times 1$ input matrix, C is $1 \times n$ output matrix D is input-output coupling matrix (1×1) [27] (Fig. 4).

In Pole Placement technique, all the closed loop poles are placed at desired locations. It may be assumed that the entire state variable is measurable and are available for feedback. A necessary condition for this arbitrary pole placement technique is that, the system must be stated controllable. If the condition is satisfied, the control

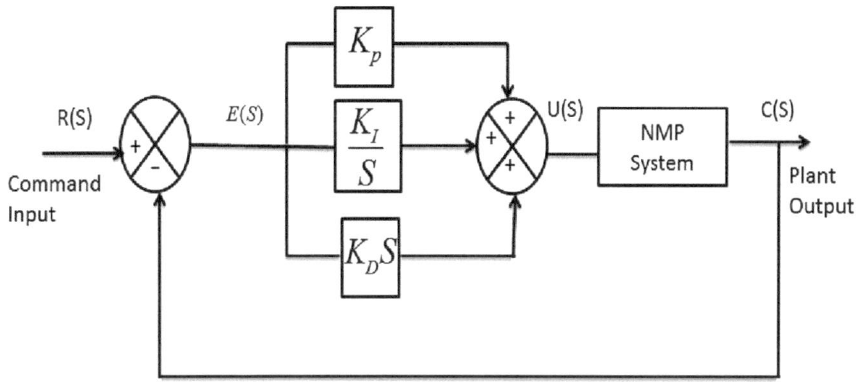

Fig. 5 Block diagram of PID controlled NMP System

signal will be $u = -KX$. It means that the control signal is determined by the instantaneous state. Such a scheme is called State feedback Control scheme and the matrix is called the state feedback gain matrix. The Eigen values of the closed loop system may be placed at any desired location through this state feedback gain matrix. The state equation will be modified to $\dot{X}(t) = (A - BK)X(t)$. If the Eigen values of the new system matrix placed at Left hand plane (LHP) of jω axis, the state model will be stable in sense of Hurwitz.

5 Description of Proportional Plus Integral Plus Derivative (PID) Control

PID Controller is a classical controller, which include three different control modes, proportional action P, an Integral action I and derivative action D in Fig. 5. All the three control modes require three independently operational amplifiers to be adjusted separately and it needs electronics circuits. The mathematical model of PID controller is given by the following equation, $u(t) = k_p e(t) + k_i \int_0^t e(\tau)d\tau + k_d \frac{de}{dt} = k_p\left(e(t) + \frac{1}{T_i}\int_0^t e(\tau)d\tau + T_d\frac{de}{dt}\right)$, where, $u(t)$ is control input, $e(t)$ is an error signal, and $k_p k_i$, k_d are the proportional, integral and derivative adjustable constant respectively. The transfer function of PID Controller is $k_p + \frac{k_i}{S} + k_d S$. To meet given performance specification of the control system, adjustable gains are selected by using a particular tuning method or by trial and error method. Ziegler-Nichols tuning method has been used here as it is very effective technique.

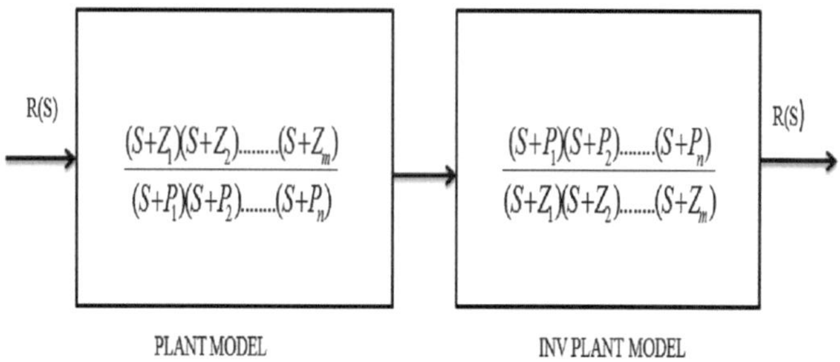

Fig. 6 Block diagram of cascaded inverse with non-inverse transfer function model of NMP system

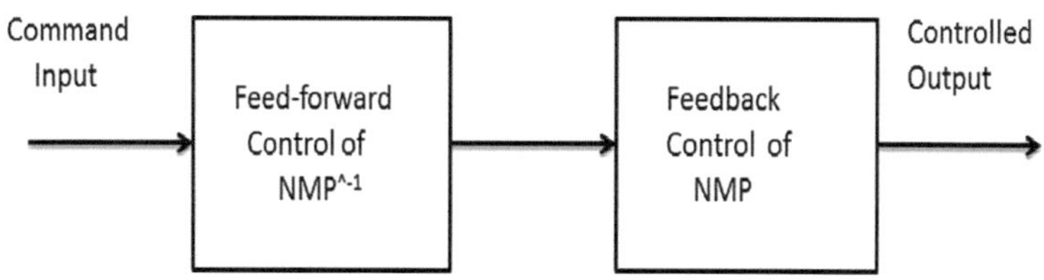

Fig. 7 Block diagram of decoupled 2DOF controlled NMP plant

6 Problem Formulation

When an inverse transfer function model connected in cascade with the original transfer function model, exact reference input trajectory must be achieved as numerator denominator canceled to each other (Fig. 6).

From the above diagram, it is needless to say, that exact reference input trajectory can be possible by canceling the numerator denominator polynomial, but it is applicable for minimum phase only where R(S) is reference input signal. But for Non-minimum phase system, actual output trajectory not only unable to meet the desired trajectory, it also produces unbounded output response (Fig. 7).

Two-degree-of-freedom consists of feed-forward and the feedback controller, where MRAC controller as feed-forward part includes the inverse transfer function model and State feedback controller as feedback controller has been taken the original transfer function model of the NMP system. A practical Non-minimum phase system has been taken for experimental simulation.

7 Mechanical System Model

Mechanical Experimental set up developed by Freeman et al. [25], where non-minimum phase component has been placed in the upper left corner of the test bed and

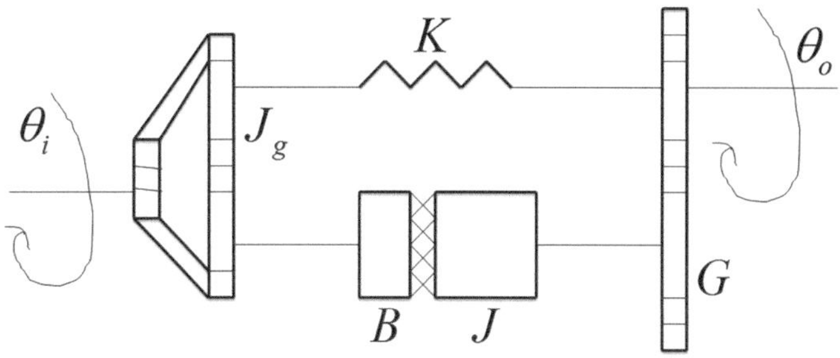

Fig. 8 Schematic representation of mechanical realization of non-minimum phase system

two more mass-spring-damper systems have been connected with the non-minimum phase system (Fig. 8).

Electrical analogous to the above mechanical system shows non-minimum phase characteristics. This NMP system consists of inertias, a damper, a torsional spring, timing belt, pulleys, and gears. In the schematic diagram [25], inertias are represented by J, J_g, G represents gear and B, K represents damping friction co-efficient and spring constant respectively. Two spring mass damper systems have been inserted before the non-minimum phase component to increase the relative degree of the entire system. The mechanical experimental set up has been formulated by the following transfer function model.

$$G(S) = \frac{123.853 * 10^4(3.5 - S)}{(S^2 + 6.5S + 42.25)(S + 45)(S + 190)} \tag{11}$$

8 Simulation Result

Unit step signal is applied to the uncontrolled NMP system, inverse transfer function model of NMP system and series connected NMP system with its inverse transfer function model. Square wave signal has been used as a set point for all the three, state feedback, PID and 2DOF controlled NMP systems. All the simulation experiments have been performed in a MatLab environment. An illustrative example for simulation is briefly described in Sect. 7.

9 Result Analysis

Here, Fig. 9 shows output response of uncontrolled NMP system, where a deep undershoot has been found. Inverse NMP system is obviously an unstable system and Fig. 10 shows it's unbounded output response. Unlike minimum phase system,

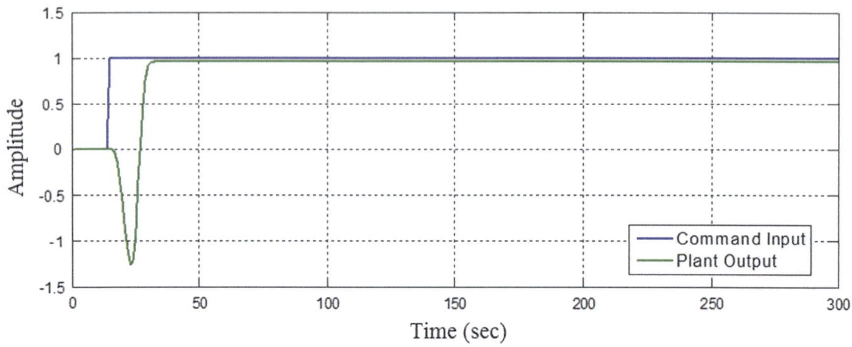

Fig. 9 Unit step response of uncontrolled NMP system

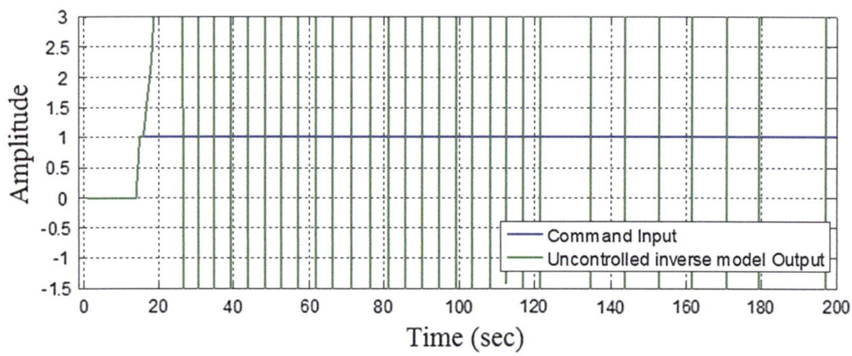

Fig. 10 Unit step response of uncontrolled inverse NMP system

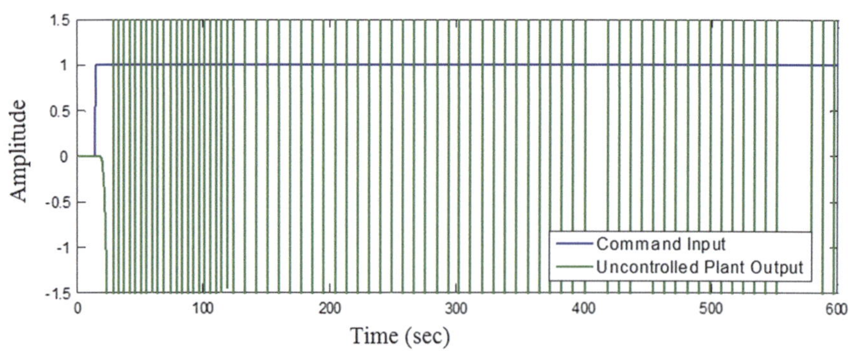

Fig. 11 Unit Step response of series connected NMP system with it's inverse transfer function model

NMP system connected in cascaded with its inverse transfer function model produce unbounded output response (Fig. 11). It is observed that Two-degree-of-freedom controlled NMP system produce far better result (Fig. 14) than PID controlled (Fig. 13)

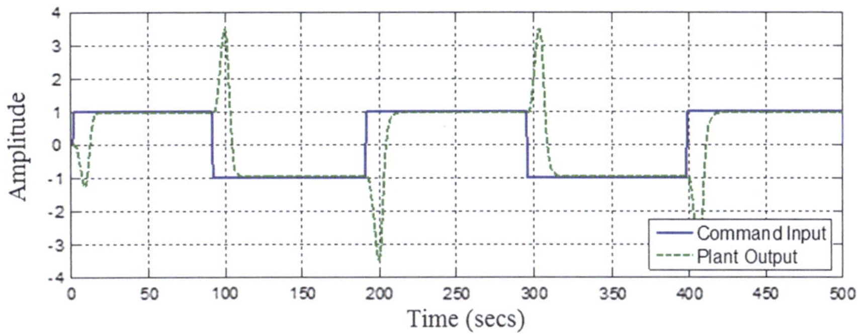

Fig. 12 State feedback controlled NMP system with square wave signal as reference input

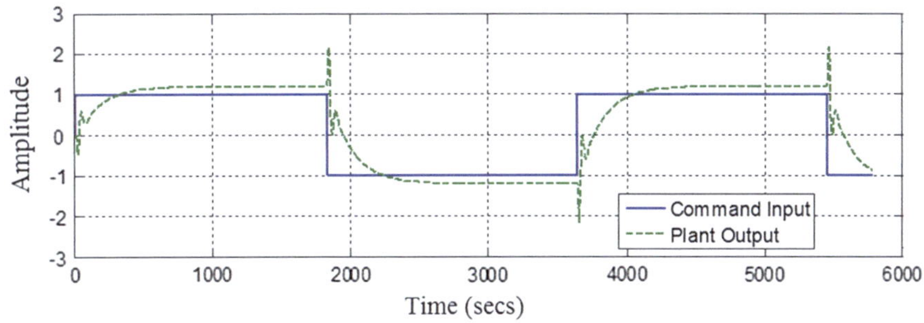

Fig. 13 PID controlled NMP system with square wave signal as reference input

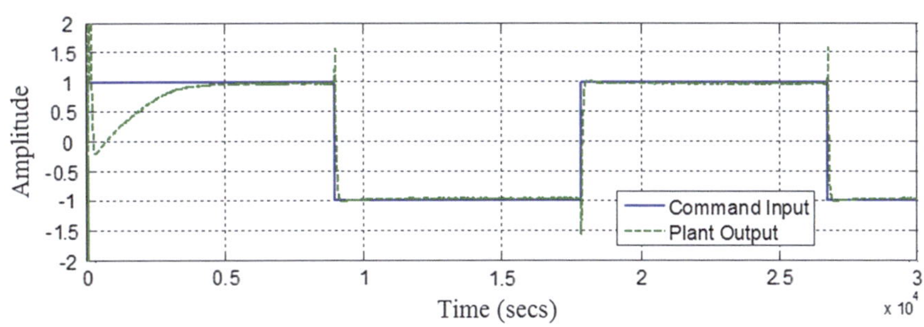

Fig. 14 2DOF controlled NMP system with square wave signal as reference input

and state feedback controlled (Fig. 12) NMP system when Square Wave signal has been applied as Reference input signal. Unit step response of state feedback, PID and 2DOF controlled NMP system have been superimposed (Fig. 15). It is found that output trajectory of 2DOF controlled NMP Plant is very near to reference input trajectory and also initial undershoot becomes completely nullified. Percentage RMS errors from the square wave responses of the 2DOF controlled NMP system with the existing compensation technique have been listed below.

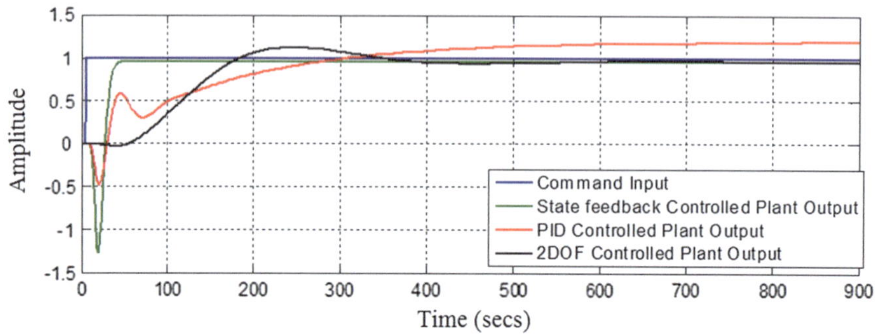

Fig. 15 Unit step responses of state feedback, PID and 2DOF controlled NMP system

Type of controller	State feedback	PID	2DOF
% error in RMS	46%	7.8%	0.79%

Comparison chart of State feedback, PID and 2DOF controlled system dynamics has been listed below.

Name of the controller	Initial undershoot	Steady state error	Rise time (sec)
State feedback	1.27	0.04	0
PID	0.47	20	6.1
2DOF	0.03	0.04	3.8

10 Conclusion

This paper presents a theoretical technique for designing a Model Reference Adaptive control (MRAC) in Two-degree-of-Freedom (2DOF) framework for a practical Mechanical System, which has Non-minimum phase (NMP) characteristics. Decoupled Feed-forward and feedback compensators are the main components of the 2DOF controller. As the inverse model connected NMP systems are not able to produce bounded output, here, inverse model NMP system is compensated by MRAC scheme as feed-forward part and non-inverse NMP system has been taken care by state feedback controller in the 2DOF control structure. The simulation study of this proposed 2DOF methodology has been shown the effectiveness of the proposed control methodology, especially, initial undershoot, which is unavoidable phenomena in NMP system becomes almost nullified. After comparing with PID control and State feedback control technique, it has been observed that proposed control method can contribute satisfactory result than other two control schemes.

References

1. F. Merrikh-Bayat, F. Bayat, Method for undershoot-less control of non-minimum phase plants based on partial cancellation of the non-minimum phase zero: application to flexible-link robots, in *Proceedings of the International Conference on Electrical and Computer Engineering (ICECE2013)*, 26–28, pp. 1–6 (2013)
2. W.-H. Chen, D.J. Balance, QFT design for uncertain non-minimum phase and unstable plants revisited. Int. J. Control **74**(9), 957–965 (2001)
3. K. Uren, G. van Schoor, in *Predictive PID Control of Non-Minimum Phase Systems*, in ed. by K. Uren, G. van Schoor, V.D. Yurkevich. ISBN 978-953-307-267-8 (2011)
4. K.S. Ramani, M. Duan, C.E. Okwudire, A.G. Ulsoy, Tracking control of linear time-invariant non-minimum phase systems using filtered basis. J. Dyn Syst Measur. Control ASME **139**, 1–11 (2017)
5. L. Benvenuti, M.D. Di Benedetto, J.W. Grizzle, Approximate output tracking for non-linear non-minimum phase system with an application to flight control. Int. J. Robust Non-linear Control. **4**, 397–414 (1994)
6. M. Duan, K.S. Ramani, C.E. Okwudire, Tracking control of non-minimum phase system using filtered basis functions: a nurbs-based approach, in *Proceedings of the ASME 2015 Dynamic Systems and Control Conference*, 28–30, October 2015, pp. 1–10
7. M.T. Do, Z. Man, J. Jin, C. Zhang, J. Zheng H. Wang, Sliding mode learning control of non-minimum phase non-linear system. Int. J. Robust Nonlinear Control **26**, 2281–2298 (2015)
8. D. Ho, Hedrick, J. Carl, Control of nonlinear non-minimum phase systems with input-output linearization, in *Proceedings of the American Control Conference (ACC)* (IEEE, 2015), pp. 4016–4023
9. J.S.-H. Tsai, F. Ebrahimzadeh, Y.-Y. Lin, S.-M. Guo, L.-S. Shieh, Y.-T. Juang, An efficient robust servo design for non-minimum phase discrete-time systems with unknown matched/mismatched input disturbances. Autom. Control Intell. Syst. **5**(2), 14–28 (2017)
10. M. Pal, G. Sarkar, R.K. Barai, T. Roy, Design of adaptive two-degree-of-freedom controller for inversion based non-minimum phase system, in *Proceedings of Michel Faraday IET International Summit—2015 (MFIIS-2015)*, Kolkata, India, vol. 1, pp. 78–83 (2015)
11. M. Pal, G. Sarkar, R.K. Barai, T. Roy, Reference input tracking of inversion-based non-minimum phase system using adaptive two-degree-of-freedom control, in *IEEE First International Conference on Control, Measurement and Instrumentation (CMI)*, pp. 508–513 (2016)
12. M. Araki, H. Taguchi, Two-degree-of-freedom PID controllers. Int. J. Control Autom. Syst. **1**(4), 401–411 (2003)
13. N.N. Abdul Aziz, M.I. Yusoff, M.I. Solihin, R. Akmeliawati, Two degrees of freedom control of a ball and beam system, in *Proceedings of the 5th International Conference on Mechatronics (ICOM'13)*, pp. 1–12 (2013)
14. N. He, D. Shi, J. Wang, M. Forbes, J. Backström, T. Chen, Automated two-degree-of-freedom model predictive control tuning. Ind. Eng. Chem. Res. **54**(43), 10811–10824 (2015)
15. R. Adnan, H. Ismail, N. Ishak, M. Tajjudin, M.H.F. Rahiman, Adaptive feed-forward zero phase error tracking control for minimum phase and non-minimum phase systems—XY table real-time application, in *IEEE Control and System Graduate Research Colloquium (ICSGRC 2012)*, pp. 359–363 (2012)
16. J.F. Quindlen, G. Chowdhary, P.H. Jonathan, Hybrid model reference adaptive control for unmatched uncertainties, in *American Control Conference (ACC)*, pp. 1125–1130 (2015)
17. M. Pal, G. Sarkar, R.K Barai, T. Roy, Model reference adaptive control of unstable system with different reference models, in *Proceedings of the International Conference on Advances in Science and Technology*, 17th to 19th March, pp. 252–256 (2017)
18. L. Yujie, C. Shaozhong, Model reference adaptive control system simulation of permanent magnet synchronous motor, in *Proceedings of IEEE Advanced Information Technology, Electronic and Automation Control Conference (IAEAC)*, pp. 498–502 (2015)

19. H. Tahersima, M. Saleh, A. Mesgarisohani, M. Tahersima, Design of stable model reference adaptive system via Lyapunov rule for control of a chemical reactor, in *Proceedings of the Australian Control Conference*, pp. 348–353 (2013)
20. Q. Wang, S. Zhu, Y. He, Model reference adaptive control for active trailer steering of articulated heavy vehicles. SAE Technical Paper, pp. 1–10 (2015)
21. Y. Wei, S. Liang, Y. Hu, Y. Wang, Composite model reference adaptive control for a class of nonlinear fractional order systems, in *Proceedings of the ASME 2015 International Design Engineering Technical Conferences and Computers and Information in Engineering Conference*, pp. 1–9 (2015)
22. D. Roy Choudhury, *Modern Control Engineering* (Prentice Hall of India Private Limited, 2005), pp. 530–645
23. B. Orun, S. Necipoglu, C. Basdogan, L. Guvenc, State feedback control for adjusting the dynamic behaviour of a piezoactuated bimorph atomic force microscopy probe, in *Proceedings of the 9th International Conference on Robotic, Vision, Signal Processing and Power Applications*, pp. 419–426 (2009)
24. D. Puangdownreong, Optimal state-feedback design for inverted pendulum system by flower pollination algorithm. Int. Rev. Automatic Control (IREACO) **9**(5) (2016)
25. C.T. Freeman, P.L. Lewin, E. Rogers, Experimental evaluation of iterative learning control algorithms for non-minimum phase plants. Int. J. Control **78**(11), 826–846 (2005)
26. K.J. Åström, B. Wittenmark, *Adaptive Control*, 2nd edn. (Dover Publications Inc), pp. 20–21
27. K. Ogata, *Modern Control Engineering*, 5th edn. (Prentice Hall), pp. 723–729

LFT Modeling of Differentially Driven Wheeled Mobile Robot

Tamal Roy, Ranjit Kumar Barai and Rajeeb Dey

1 Introduction

Popular mobile robotic systems result from the synergistic integration of mechanical systems, microelectronics and intelligent computer control [1]. As the various subsystems of control systems are interconnected in a delicate manner to serve the primary objective of precise and stable control, the overall closed-loop system is very complex. As demand increases, it becomes necessary to improve the performance of the WMR for conventional applications. It is possible to improve the performance of the WMR through the better modelling of the system. Model-based control approaches are widely accepted technique to control WMR and the ability to develop a realistic model will greatly benefit the development of advanced controllers. From the perspective of the design of an appropriate control law for WMR; such complicated model may be inconvenient due to the high degree of computational complexity. Moreover, from a wide range of control law, a particular control law may be most suitable for a particular class of systems from the perspective of tackling the nonlinearities, parameter variation, model uncertainties, noise, and disturbance. Therefore, it becomes necessary to select an appropriate mathematical model of the system for the design of a particular control law in order to obtain optimum control performance.

T. Roy (✉)
Electrical Engineering Department, MCKV Institute of Engineering,
Liluah, Howrah 711204, India
e-mail: tamalroy77@gmail.com

R. K. Barai
Electrical Engineering Department, Jadavpur University, Kolkata 700032, India
e-mail: ranjit.k.barai@gmail.com

R. Dey
Electrical Engineering Department, NIT, Silchar, Assam 788 010, India
e-mail: rajeeb.iitkgp@gmail.com

S. Chattopadhyay et al. (eds.), *Modelling and Simulation in Science, Technology and Engineering Mathematics*, Advances in Intelligent Systems and Computing 749,
https://doi.org/10.1007/978-3-319-74808-5_32

In order to achieve the intended control performance, research efforts have been directed towards the development of control-oriented modeling of a benchmark mechatronics system namely, differentially driven wheeled mobile robot (WMR). Owing to its widespread industrial applications, cost-effectiveness and availability of its small-scale laboratory-based model and highly nonlinear cross-coupled dynamics, WMR has become a test-bed for the development of robust controller among control community. Kinematic modeling of the WMR has been discussed in the numerous literature related to engineering mechanics [2] and analytical mechanics [3]. There has been growing interest among the control community researchers on dynamic modeling of WMR due to the challenge posed by this system as in this case the interaction between the outer wheel and the ground surface is the contact point with the environment and has a considerable influence in the dynamic motion of the system [4–6]. Although effort has been made to derive a systematic framework of the system include the dynamics into the overall mathematical modeling [7, 8]. However, due to the various uncertainties associated with the dynamic model of the WMR, some kinds of robustness measure in the control law are incorporated to develop a controller that is effective in the real-world situation [9]. H_∞ control strategy has been proved useful to overcome or compensate the effects of these model uncertainties [10, 11]. Thus, a modeling structure has been proposed to develop LFT based model uncertainty representation of WMR would make it convenient for the application of modern robust control technique like μ-analysis and synthesis in addition to H_∞-control.

Over the last two decades, there has been a widespread interest to design a robust controller where system model is considered to be consisting of a nominal model and a model uncertainty part [12]. There are different forms of model uncertainty representation for robust control design like: additive or multiplicative model uncertainty bounds, or parametric uncertainty regions, co-prime factor uncertainty, uncertainty representation in a feedback-like connection as in case of linear fractional transform (LFT) modeling, and the performance of the robust controller depends on the appropriate representation of the model uncertainty.

The model of nonlinear systems varies due to changes in system configuration and operating conditions. This system variation can be characterized as model uncertainties [13] and can be represented in the linearized model by expressing the system state-matrices as matrix polynomials in the uncertain parameters in the form of a linear fractional transform (LFT) [12, 14]. This LFT based model uncertainty representation of the nonlinear system is essential for the application of modern robust control technique like μ-analysis and synthesis [15] in addition to H_∞-control and H_∞-Loop Shaping [14, 16–18]. Linearization of uncertain nonlinear systems as LFT model relates each of the uncertainty with a physically meaningful parameter of the actual system [19]. Linear Fractional Transform (LFT) technique offers a unified framework for parameter identification problems [20]. In the LFT framework, a wide variety of identification problems concerning structured nonlinear systems, linear parameter varying (LPV) systems, and also the various parametric linear system model structures can be accommodated due to its general nature. Moreover, LFT framework is convenient to address the issues such as identifiability and persistence of excitation for a large class of model structures. The main aspect of the LFT-based uncertainty modeling is the generation of the lower order LFT modeling [21] but it is only applicable to those nonlinear systems where the linearization of the mathematical modeling is possible [22]. The LFT representation consists known or invariant part separated from the uncertain or variable parts which are summarized and stacked together in the "perturbation" matrix.

In view of the above discussion on the modelling aspects of WMR, the authors propose a novel methodology of uncertainty modelling in the linear fractional transformation (LFT) modelling framework, considering the nonlinear model of WMR with non-holonomic constraints that arise from constraining the wheels of the robots to roll without slipping [23]. A versatile and well-organized uncertainty modelling representation of WMR in LFT framework has never been addressed in the literature.

The rest of the paper has been organized as follows. The system dynamics of WMR is described in Sect. 2. The LFT modeling formulation of differentially driven WMR has been described in Sect. 3. Section 4 explains the frequency domain model validation of the proposed modeling structure of WMR.

2 Differentially Driven Wheeled Mobile Robot

The most popular construction of the wheel mobile robot is the two-wheeled mobile robot differentially driven by two independent DC motor shown in Fig. 1. Two DC motors are used for left and right drive wheels of the robot in a horizontal axis and one free caster wheel is used to keep the platform stable. The control of the robot depends on the angular velocities of the two drive wheels [24]. If the angular velocities of the two drive wheels are identical in values and in same relative rotational senses, the robot makes a spin motion. If the angular velocities of the two wheels are identical in values but opposite in relative rotational senses, the robot makes a linear motion and if the angular velocities of the two drive wheels are different in values, the robot makes a curve motion.

The robot configuration is represented by the position of the centre of the axis between the two wheels in the Cartesian space (x_c and y_c) and by its orientation θ (angle between the vector of the robot orientation and the abscissas) is shown in Fig. 1.

The kinematic modelling describes the relations between the derivatives of robot position and orientation and the robot linear and angular speeds, v and ω. While, the dynamic model of WMR is derived from the physics laws. All the physical parameters are taken into the consideration in the dynamic modelling of the WMR [24]. In the dynamic modelling of WMR non-holonomic constraints have been considered which results from assuming no slip between the ground and the wheels. In our present work, dynamic modelling of the wheel mobile robot is considered to formulate the control-oriented modelling in LFT modelling framework.

Fig. 1 Schematic diagram of wheeled mobile robot

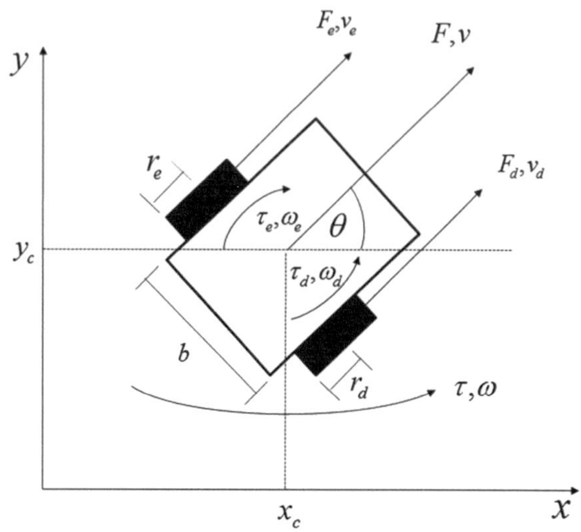

2.1 Kinematic Modeling

In kinematic modelling, the main purpose is to point out the structural properties of the WMR by introducing the concepts of the degree of mobility and degree of steribility in the variety of possible robot constructions and wheel configurations. WMR is operating on a horizontal plane and considered as the rigid body on wheels and out of these three dimensionalities of the robot chassis, two for position in the plane and one for orientation along the vertical axis, which is orthogonal to the plane.

The kinematic model of the differential-drive robot relates the forward and angular velocities (v, ω) to cartesian velocities $\left(\dot{x}, \dot{y}, \dot{\theta}\right)$ [8, 14]. The x component of the forward velocity can be expressed as

$$\dot{x} = v \cos \theta \tag{1}$$

and the y component can be expressed as

$$\dot{y} = v \sin \theta \tag{2}$$

The kinematic model can then be represented as

$$\dot{\mathbf{q}} = {}^{q}T_V.\mathbf{V} \tag{3}$$

where $\mathbf{V} = \begin{bmatrix} v \\ \omega \end{bmatrix}$, $\dot{\mathbf{q}} = \begin{bmatrix} \dot{x} \\ \dot{y} \\ \dot{\theta} \end{bmatrix}$ and ${}^{q}T_V = \begin{bmatrix} \cos\theta & 0 \\ \sin\theta & 0 \\ 0 & 1 \end{bmatrix}$

2.2 Dynamic Modelling

The parameterized dynamic model of a differential drive wheeled mobile robot is derived from the physical laws that govern the robot subsystems, including the actuator dynamics like electrical and the mechanical characteristics of the actuators, friction and robot dynamics (movement equations) [24].

The modelling of the WMR can be presented by the differential equation as given below

$$M\dot{\mathbf{V}} + B\mathbf{V} = K\mathbf{u} \tag{4}$$

where,

$$M = \begin{bmatrix} m + \frac{J_d}{r_d^2} & 0 \\ 0 & J + \frac{J_e b^2}{4r_e^2} \end{bmatrix} = \begin{bmatrix} J_1 & 0 \\ 0 & J_2 \end{bmatrix}$$

where

$$J_1 = m + \frac{J_d}{r_d^2} \text{ and } J_2 = J + \frac{J_e b^2}{4r_e^2}$$

$$B = \begin{bmatrix} \beta_1 + \frac{1}{r_d^2}(\frac{K_d^2}{R_d} + \beta_d) & 0 \\ 0 & \beta_2 + \frac{b^2}{4r_e^2}(\frac{K_e^2}{R_e} + \beta_e) \end{bmatrix} = \begin{bmatrix} \beta_1 + \alpha_1 & 0 \\ 0 & \beta_2 + \alpha_2 \end{bmatrix}$$

where,

$$\alpha_1 = \frac{1}{r_d^2}(\frac{K_d^2}{R_d} + \beta_d) \text{ and } \alpha_2 = \frac{b^2}{4r_e^2}(\frac{K_e^2}{R_e} + \beta_e)$$

and

$$K = \begin{bmatrix} \frac{K_d}{R_d r_d} & 0 \\ 0 & -\frac{K_e b}{2R_e r_e} \end{bmatrix} = \begin{bmatrix} K_1 & 0 \\ 0 & K_2 \end{bmatrix}$$

where,

$$K_1 = \frac{K_d}{R_d r_d} \text{ and } K_2 = -\frac{K_e b}{2R_e r_e}$$

The vector $V = [v, \omega]^T$ represent the linear and the angular velocity of the robot and $\mathbf{u} = \begin{bmatrix} e_r \\ e_l \end{bmatrix}$ contains the input signals mainly armature voltages applied to the right and left actuators (motors). $K \in \Re^{n\times n}$ is the matrix which transforms the applied electrical signals u into forces to the robot wheels for movement. $M \in \Re^{n\times n}$ is inertia matrix of the WMR and $B \in \Re^{n\times n}$ is the damping matrix which includes terms of viscous friction and electric resistance. Parameterized dynamic modelling of a differential-driven wheeled mobile robot is taken into the consideration for the formulation of robust control oriented modelling in LFT framework.

3 LFT Modelling of Differentially Driven Wheeled Mobile Robot

The equations of motion of WMR are obtained using the Newtonian as well as Euler formula [30,24]. The moment of inertia and the co-efficient of viscous friction of WMR are supposed to be uncertain with bounded uncertainties. These associated model uncertainties can be lumped into one single perturbation block Δ. LFT modeling of WMR consisting of the nominal model represented by known transfer function and the uncertain block Δ is represented by an unknown transfer function matrix. The dynamic modeling of WMR is shown in Fig. 4.

$$M\dot{\mathbf{V}} + B\mathbf{V} = K\mathbf{u} \tag{5}$$

$$\dot{\mathbf{V}} = -M^{-1}B\mathbf{V} + M^{-1}K\mathbf{u} \tag{6}$$

Based on the practical consideration, moments of inertia and the viscous friction coefficient are uncertain parameters within the system but we assumed that the moments of inertia are constants with possible relative error of 10% around the nominal values; similarly, the viscous friction coefficients may have with 15% relative errors around the nominal values [25]. In the uncertainty representation of the moments of inertia of WMR inverse additive perturbation representation technique is used.

Therefore, the actual moments of inertia are presented as

$$J_i = \bar{J}_i(1 + p_i\delta_{J_i})\,\text{where}\quad i = 1, 2 \tag{7}$$

$\bar{J}_i$ is the nominal value of the corresponding moment of inertia, $p_i = 0.1$ is the maximum relative uncertainty in each of these moments and $-1 \le \delta_{J_i} \le 1$.

Matrix M contained the moment of inertia J_i, then matrix M decomposed as

$$M = \overline{M} + M_p\Delta_J \tag{8}$$

where the elements of $\overline{M}$ is determine by the nominal values of the moment of inertia and Δ_J represents the uncertain part of the matrix M

$$\overline{M} = \begin{bmatrix} \bar{J}_1 & 0 \\ 0 & \bar{J}_2 \end{bmatrix}, M_p = \begin{bmatrix} \bar{J}_1 p_1 & 0 \\ 0 & \bar{J}_2 p_2 \end{bmatrix} \text{and } \Delta_J = \begin{bmatrix} \delta_{J_1} & 0 \\ 0 & \delta_{J_2} \end{bmatrix}$$

The upper LFT representation of M^{-1} matrix is defined with

$$M^{-1} = F_U(Q_J, \Delta_J) = Q_{J_{22}} + Q_{J_{21}}\Delta_J(I_2 - Q_{J_{11}}\Delta_J)^{-1}Q_{J_{12}} \tag{9}$$

The block partition matrix Q_J represented as

$$Q_J = \left[\begin{array}{c:c} Q_{J_{11}} & Q_{J_{12}} \\ \hdashline Q_{J_{21}} & Q_{J_{22}} \end{array}\right] = \left[\begin{array}{c:c} -\overline{M}^{-1}M_p & \overline{M}^{-1} \\ \hdashline -\overline{M}^{-1}M_p & \overline{M}^{-1} \end{array}\right] \tag{10}$$

The uncertainty representation of the viscous friction coefficient matrix B is express in terms of the uncertain parametric representation can be described by

$$\beta_i + \alpha_i = (\overline{\beta_i} + \bar{\alpha}_i)(1 + s_i\delta_{\beta_i}), \quad i = 1, 2 \tag{11}$$

where $\overline{\beta_i} + \bar{\alpha}_i$ is the nominal value of the corresponding viscous friction coefficients, $s_i = 0.15$ is the maximum relative uncertainty in each of these coefficient and $-1 \leq \delta_{\beta_i} \leq 1,\ i = 1, 2$.

Matrix B can be decomposed as

$$B = \overline{B} + \Delta B \tag{12}$$

where the elements $\overline{B}$ are determined by the nominal values of the viscous friction coefficient

$$\overline{B} = \begin{bmatrix} \bar{\beta}_1 + \bar{\alpha}_1 & 0 \\ 0 & \bar{\beta}_2 + \bar{\alpha}_2 \end{bmatrix}$$

The uncertainty matrix ΔB in Eq. (12) can be further decomposed as

$$\Delta B = B_f \Delta_\beta B_g \tag{13}$$

where $B_f = \begin{bmatrix} \overline{\beta_1} s_1 & 0 \\ 0 & \overline{\beta_2} s_2 \end{bmatrix}$, $\Delta_\beta = \begin{bmatrix} \delta_{\beta_1} & 0 \\ 0 & \delta_{\beta_2} \end{bmatrix}$ and $B_g = \begin{bmatrix} 1 & 0 \\ 0 & 1 \end{bmatrix}$

The upper LFT representation of the viscous friction coefficients matrix B is given by

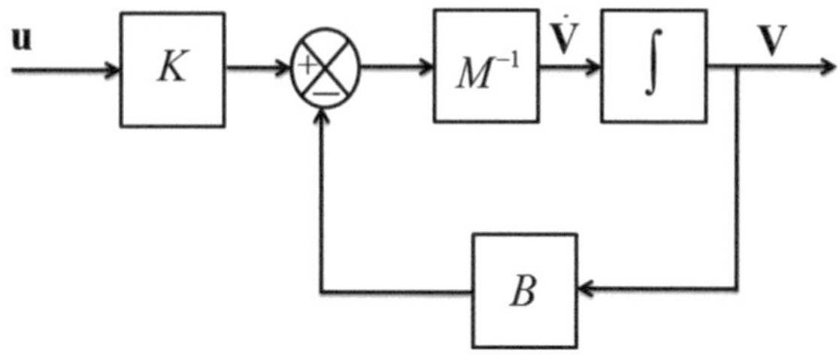

Fig. 2 Block diagram of wheeled mobile robot

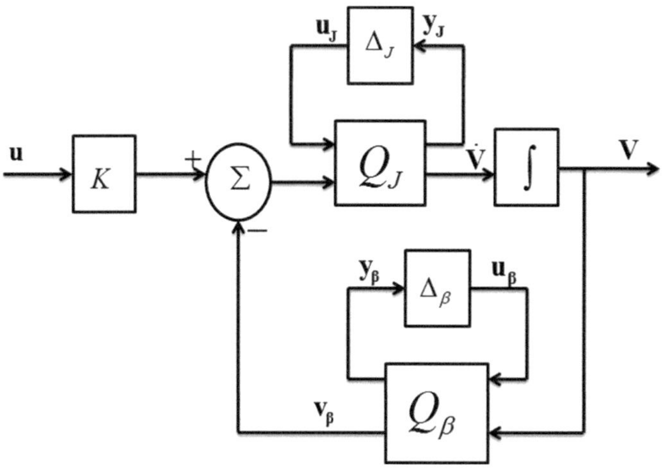

Fig. 3 Block diagram of WMR in LFT structure

$$B = F_U(Q_\beta, \Delta_\beta) = Q_{\beta_{22}} + Q_{\beta_{21}} \Delta_\beta (I_2 - Q_{\beta_{11}} \Delta_\beta)^{-1} Q_{\beta_{12}} \tag{14}$$

The block partition matrix Q_β is represented as

$$Q_\beta = \left[\begin{array}{c:c} Q_{\beta_{11}} & Q_{\beta_{12}} \\ \hdashline Q_{\beta_{21}} & Q_{\beta_{22}} \end{array}\right] = \left[\begin{array}{c:c} 0_{2\times 2} & B_g \\ \hdashline B_f & \bar{B} \end{array}\right] \tag{15}$$

Using LFT structure of matrices M^{-1} and B in (9) and (14) the block diagram of WMR in Fig. 2, now redrawn with nominal block partition matrices Q_J and Q_β and uncertainty matrices Δ_J and Δ_β as shown in Fig. 3.

Now considering u_J, u_β being the output of the uncertainty blocks Δ_J and Δ_β are fed as an input to Q_J and Q_β blocks respectively. Similarly, y_J, y_β are the outputs of Q_J and Q_β blocks that are fed as an input to the Δ_J and Δ_β blocks respectively.

Assuming $x_1 \triangleq v$ and $x_2 \triangleq \omega$.

The state vector of WMR can be defined as

$$\mathbf{X} \triangleq \begin{bmatrix} x_1 \; x_2 \end{bmatrix}^T \tag{16}$$

Next, $\dot{\mathbf{V}}$ is represent in terms of state variables as defined above

$$\dot{\mathbf{V}} = \left[\dot{\upsilon}\ \dot{\omega}\right]^T = \left[\dot{x}_1\ \dot{x}_2\right]^T \tag{17}$$

The controlled output vector is defined in terms of state variable as

$$\mathbf{y_c} \triangleq \left[x_1\ x_2\right]^T = \left[y_{c_1}\ y_{c_2}\right]^T = \left[\upsilon\ \omega\right]^T \tag{18}$$

Now, output of WMR can be expressed as

$$\mathbf{y_c} = \mathbf{V} \tag{19}$$

where $\mathbf{V} = \left[\upsilon\ \omega\right]^T$

The measured output vector of the system is represented as

$$\mathbf{y_m} = \left[y_{m_\upsilon}\ y_{m_\omega}\right]^T \tag{20}$$

It is to be noted that the output are measured using position encoder for the system under study. Considering the resolution of the encoder to measure output y_m is further written as

$$\mathbf{y_m} = \Gamma_m \mathbf{V} \tag{21}$$

The LFT representation of WMR can be represent

$$\begin{bmatrix} \dot{\mathbf{x}}_1 \\ \dot{\mathbf{x}}_2 \\ \mathbf{y_J} \\ \mathbf{y_\beta} \\ \mathbf{y_c} \\ \mathbf{y_m} \end{bmatrix} = \prod \begin{bmatrix} \mathbf{x_1} \\ \mathbf{x_2} \\ \mathbf{u_J} \\ \mathbf{u_\beta} \\ \mathbf{u} \end{bmatrix} \tag{22}$$

where

$$\prod = \begin{bmatrix} -\overline{M}^{-1}\bar{B} & -\overline{M}^{-1}M_p & -\overline{M}^{-1}B_f & -\overline{M}^{-1}K \\ -\overline{M}^{-1}\bar{B} & -\overline{M}^{-1}M_p & -\overline{M}^{-1}B_f & -\overline{M}^{-1}K \\ B_g & 0_{2\times 2} & 0_{2\times 2} & 0_{2\times 2} \\ I_{2\times 2} & 0_{2\times 2} & 0_{2\times 2} & 0_{2\times 2} \\ \Gamma_m & 0_{2\times 2} & 0_{2\times 2} & 0_{2\times 2} \end{bmatrix}_{10\times 8}$$

Input output expression of the uncertainty blocks can be represent as

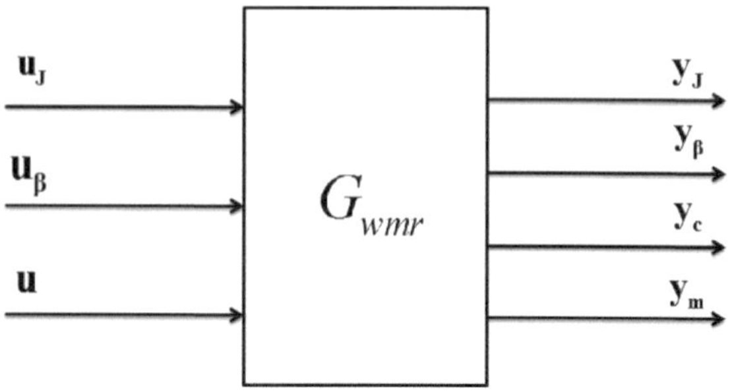

Fig. 4 Input output block diagram of the WMR

$$\begin{bmatrix} \mathbf{u_J} \\ \mathbf{u_\beta} \end{bmatrix} = \Delta \begin{bmatrix} \mathbf{y_J} \\ \mathbf{y_\beta} \end{bmatrix} \tag{23}$$

where $\Delta = \begin{bmatrix} \Delta_J & 0 \\ 0 & \Delta_\beta \end{bmatrix}_{4\times 4}$

Input-output representation of WMR is expressed as The open loop block diagram representation of LFT structure of WMR with three inputs ($\mathbf{u_J}$, $\mathbf{u_\beta}$, $\mathbf{u}$) and four outputs ($\mathbf{y_J}$, $\mathbf{y_\beta}$, $\mathbf{y_c}$, $\mathbf{y_m}$), is shown in Fig. 4.

Input-output representation of WMR in LFT framework is expressed as

$$\begin{bmatrix} \dot{\mathbf{x}} \\ \mathbf{y_J} \\ \mathbf{y_\beta} \\ \mathbf{y_c} \\ \mathbf{y_m} \end{bmatrix}_{10\times 1} = G_{wmr} \begin{bmatrix} \mathbf{x} \\ \mathbf{u_J} \\ \mathbf{u_\beta} \\ \mathbf{u} \end{bmatrix}_{8\times 1} \tag{24}$$

where

$$G_{wmr} = \left[\begin{array}{c|cc} A & B_1 & B_2 \\ \hline C_1 & D_{11} & D_{12} \\ C_2 & D_{21} & D_{22} \end{array}\right]$$

and

$$A = \left[-\overline{M}^{-1}\bar{B}\right]_{2\times 2},\ B_1 = \left[-\overline{M}^{-1}M_p\ -\overline{M}^{-1}B_f\right]_{2\times 4},\ B_2 = \left[-\overline{M}^{-1}K\right]_{2\times 2}$$

$$C_1 = \begin{bmatrix} -\overline{M}^{-1}\bar{B} \\ B_g \end{bmatrix}_{4\times 2},\ D_{11} = \begin{bmatrix} -\overline{M}^{-1}M_p & -\overline{M}^{-1}B_f \\ 0_{2\times 2} & 0_{2\times 2} \end{bmatrix}_{4\times 4},\ D_{12} = \begin{bmatrix} -\overline{M}^{-1}K \\ 0_{2\times 2} \end{bmatrix}_{4\times 2}$$

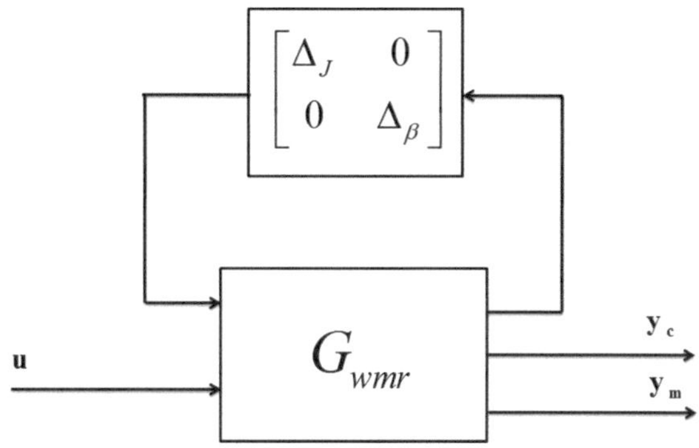

Fig. 5 LFT representation of perturbed WMR

$$C_2 = \begin{bmatrix} I_{2\times 2} \\ \Gamma_m \end{bmatrix}_{4\times 2}, D_{21} = \begin{bmatrix} 0_{2\times 2} & 0_{2\times 2} \\ 0_{2\times 2} & 0_{2\times 2} \end{bmatrix}_{4\times 4}, D_{22} = \begin{bmatrix} 0_{2\times 2} \\ 0_{2\times 2} \end{bmatrix}_{4\times 2}$$

The Upper LFT representation of the WMR with uncertainty block Δ can be represented by

$$\mathbf{y} = F_U(G_{wmr}, \Delta)\mathbf{u} \tag{25}$$

where $\Delta = \begin{bmatrix} \Delta_J & 0 \\ 0 & \Delta_\beta \end{bmatrix}$

The uncertain modeling structure of WMR can be described by an upper LFT representation with the diagonal uncertain matrix is shown in Fig. 5.

Frequency response of the perturbed open-loop WMR is shown in Fig. 6.

4 Frequency Domain Validation

The frequency domain validation of uncertainty modeling in LFT framework has been investigated in the context of robust control theory [8, 14]. The objective is to design an H_∞ controller that achieves certain performance specification and remain stable in the presence of all possible uncertainties.

4.1 Simulation Result

The (sub) optimal H_∞ control law has been implemented for the interconnected system as shown in Fig. 7. The H_∞ optimum control minimizes the $\| . \|_\infty$ norm of the nominal transfer function matrix $F_L(P, K)$ over the stabilizing controller

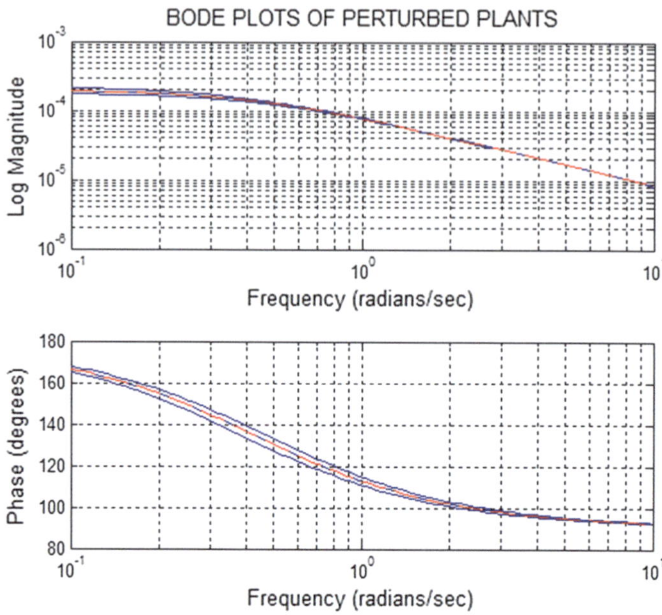

Fig. 6 Frequency response of perturbed open-loop systems

Fig. 7 Closed-loop LFTs in H_∞ design

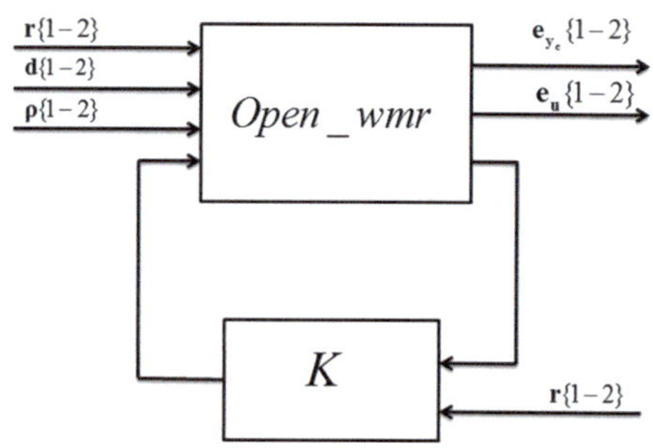

transfer matrix K, where P is the transfer function matrix of the augmented system. The nominal closed loop transfer function matrix $F_L(P, K)$ of WMR is expressed in terms of references, disturbances and noises (the signal r, d and ρ) to the weighted output e_{y_c} and e_u is shown in Fig. 7.

The interval for γ iteration is chosen between 1 and 100 with a tolerance of 0.0001. The controller of the closed loop system achieves the $\|\,.\|_\infty$ norms equal to 1.0001. The most suitable property of the controller is that it has all stable poles, which makes the proposed modeling structure of the WMR more acceptable in practice. The closed loop system with H_∞ controller achieves the robust stability and the maximum value of μ is 0.17461 is shown in Fig. 8 and also achieve the robust performance with a maximum value of the μ is 0.94528 shown in Fig. 9.

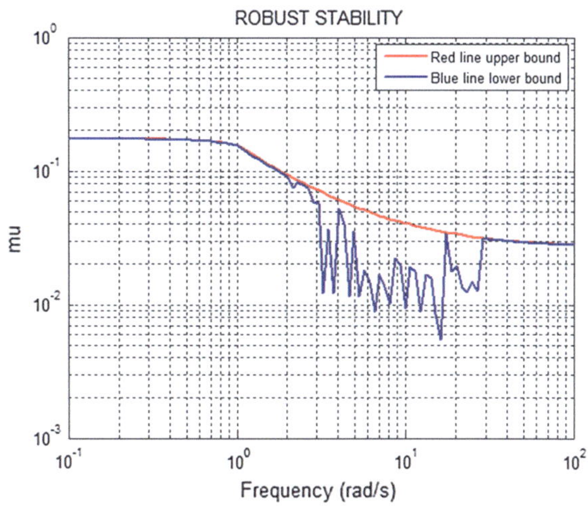

Fig. 8 Robust stability test of H_∞ controller

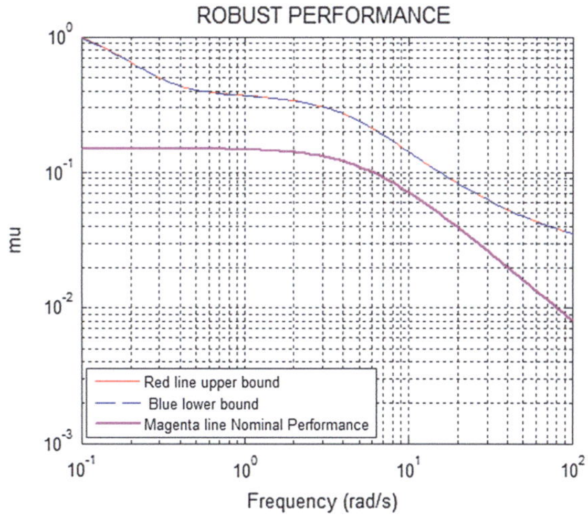

Fig. 9 Nominal and robust performance of H_∞ controller

Numerical values of the WMR are given in Table 1.

5 Conclusions

In this paper, LFT modeling of a nonlinear cross couple multivariable WMR system has been derived for implementing robust control law. A novel modeling technique, which can truly integrates control objectives of the robust control law, may be most suitable for the particular class of systems from the perspective of tackling the non-linearities, parameter uncertainties, reduction of model order, disturbances and noise.

Table 1 Numerical values of WMR

Symbol	Value	Unit	Symbol	Value	Unit
m	0.352	Kg	K_d	2.05×10^{-2}	–
b	0.075	m	K_e	2.01×10^{-2}	–
r_d	0.0251	m	J_1	0.003×10^{-6}	kg – m^2
r_e	0.0254	m	J_2	0.005×10^{-6}	kg – m^2
R_d	17.33	ohm	β_d, β_e	7.004×10^{-5}	–
R_e	17.00	ohm	β_1	0.1002	–
J	13.02×10^{-6}	kg – m^2	β_2	7.004×10^{-5}	–

A popular cross coupled nonlinear WMR system has been considered for develop a compact and manageable mathematical representation in LFT framework for implementing a particular control law in order to obtain optimum control performance. Frequency domain validation technique has been introduced for ensuring satisfactory results for validating uncertainty model of the WMR in LFT framework.

References

1. D.D. Von, T. Bretl, Kinematic and dynamic control of a wheeled mobile robot, in *Proceedings of IEEE/RSJ International Conference on Intelligent Robots and Systems*, San Diego, CA, USA, pp–4065–4070 (2007)
2. P.F. Muir, C.P. Neuman, Kinematic modeling of wheeled mobile robot, technical report. J. Robot. Syst. **4**(2), 281–340 (2007)
3. J.C. Alexander, J.H. Maddocks, On the kinematics of wheeled mobile robot. Auton. Robot Veh. 5–24 (1990)
4. Y. Zhao, S.L. Bement, Kinematics, dynamics and control of wheeled mobile robot, in *Proceedings of International Conference on Robotics and Automation*, Nice, France, pp-91–96 (1992)
5. S.N. Sidek, *Dynamic Modelling and Control of Nonholonomic Wheeled Mobile Robot Subjected to Wheel Slip*. Ph.D. Thesis, Graduate School of Vanderbilt University, Nashville, Tennessee (2008)
6. U. Kumar, N. Sukavanam, Dynamic modeling and tracking control of a four wheeled nonholonomic mobile robot, in *Proceedings of 13th National Conference on Mechanics and Machines*, NaCoMM, pp. 127–134 (2007)
7. N. Nawash, *H-Infinity Control of an Autonomous Mobile Robot*. Master of Science Thesis, Cleveland State University (2005)
8. M.L. Corradini, G. Orlando, Robust tracking control of mobile robots in the presence of uncertainties in the dynamical model. J. Robot. Syst. **18**(6), 317–323 (2001)
9. P.N. Guerra, P.J, Alsina, A.D.A. Medeiros, A.P. Araújo, Linear modeling and identification of a mobile robot with differential drive, in *International Conference on Informatics in Control, Automation and Robotics—ICINCO*, Setúbal, Portugal, pp. 263–269
10. K. Sato, M. Yanagi, K. Tsuruta, Adaptive H_∞ trajectory control of nonholonomic mobile robot with compensation of input uncertainty, in *Proceeding of IEEE International Conference on Control Applications (CCA)*, CO, USA, pp. 28–30

11. W. Jiang, W. Ge, Modeling and H_∞ robust control for mobile robot, in *Proceedings of International Conference of Robotics, Automation and Mechatronics*, Chengdu, China, pp. 1108–1112 (2008)
12. A. Marcos, D.G. Bates, I. Postlethwaite, Exact nonlinear modelling using symbolic linear fractional transformation, in *Proceedings of the 16th IFAC World Congress*, Czech Republic, 2005, vol. 16, No. 1, pp. 31–36 (2005)
13. G.E. Boukarim, J.H. Chow, Modelling of nonlinear system uncertainties using a linear fractional transformation approach, in *Proceedings of the American Control Conference*, Philadelphia, Pennsylvania, pp. 2973–2979 (1998)
14. S. Hecker, A. Varga, Generalized LFT—based representation of parametric uncertain models. Eur. J. Control **10**(4), 326–337 (2004)
15. Z. Xingfeng, Z. Zhiqiang, Uncertainty modeling and robustness analysis of flight control using μ-analysis techniques, in *The 3rd International Conference on Innovative Computing Information and Control (ICICIC'08)* (2008)
16. R. Swaminathan, K. Ramkumar, S.R. Kumar, Design of linear fractional transformation based robust control for interacting pressure tank process, in *Proceedings of IEEE International Conference on Advances in Engineering, Science and Management (ICAESM-2012)*, pp. 136–140 (2012)
17. A.J. Helmicki, C.A. Jacobson, C.N. Nett, Control oriented system identification: a worst-case/deterministic approach in H_∞. IEEE Trans. Autom. Control **36**(10), 1163–1176 (1991)
18. S. Hecker, A. Varga, Symbolic manipulation techniques for low order LFT-based parametric uncertainty modeling. Int. J. Control **79**(11), 1485–1494 (2006)
19. Y.C. Paw, G.J. Balas, Parametric uncertainty modeling for LFT model realization, in *IEEE International Conference on Computer-Aided Control System Design*, San Antonio, Texas, USA, pp. 834–839 (2008)
20. G. Wolodkin, S. Ranganz, K. Poolla, An LFT approach to parameter estimation, in *Proceedings of the American Control Conference*, Albuquerque, New Mexico pp. 2088–2092 (1997)
21. H. Pfifer and S. Hecker, Mixed parametric/unstructured LFT modeling for robust controller design, in *Proceedings of American Control Conference*, San Francisco, USA, pp. 727–232 (2001)
22. P.N. Guerra, P. J. Alsina, A.A.D. Medeiros, A.P. Araújo Jr., Linear modeling and identification of a mobile robot with differential drive, in *International Conference on Informatics in Control, Automation and Robotics—ICINCO*, Setúbal, Portugal, pp. 263–269 (2004)
23. K. Zhou, J.C. Doyle, *Essentials of Robust Control* (Prentice-Hall, New Jersey, 1998)
24. M.M. Yamamoto, D.P.F. Pedrosa, A.A.D. Medeiros, Um simulador diâmico para-mini robôs môveis com modelagem colisões, in SBAI—Simpósios Brasileiro de Automacão Inteligente, Bauru, SP, Brazil, Portuguese (2003)
25. Y.C. Paw, G.J. Balas, Parametric uncertainty modeling for LFT model realization, in *Proceeding of IEEE International Symposium on Computer-Aided Control System Design Control Systems CACSD*, pp. 834–839 (2008)

Part V
Neuro Fuzzy, Control System and Optimization

Automatic Electronic Excitation Control in a Modern Alternator

Avik Ghosh, Sourish Sanyal, Arabinda Das and Amaranth Sanyal

1 Introduction

The alternators commonly used for generation of a.c. power are of the synchronous type. The magnetic field of such alternator rotates at synchronous speed—hence the name. These generators have their armatures in the stator and the field winding in the rotor. The armatures are the power-producing parts and are of the 3-phase type. EMFs are induced in the armature conductors due to change in flux-linkage produced by rotation of the field winding. The field winding is excited by d.c. power, mostly through slip-rings and brushes. Brushless excitation is another possibility. There are various schemes for supplying d.c. power to the field winding. It may be obtained from a separate d.c. generator or an alternator-rectifier or from the stator itself through controlled rectification [1–3].

A. Ghosh (✉)
Department of Electrical Engineering, Ideal Institute of Engineering,
Kalyani, Nadia, West Bengal 741235, India
e-mail: avik_be@yahoo.com

S. Sanyal
Department of Electronics and Communication Engineering,
Techno India College, Salt Lake, Kolkata, India
e-mail: maysourish2013@gmail.com

A. Das
Department of Electrical Engineering, Jadavpur University, Kolkata
700032, India
e-mail: adas_ee_ju@yahoo.com

A. Sanyal
Department of Electrical Engineering, Calcutta Institute of Engineering
and Management, Kolkata 700040, India
e-mail: ansanyal@yahoo.co.in

S. Chattopadhyay et al. (eds.), *Modelling and Simulation in Science, Technology and Engineering Mathematics*, Advances in Intelligent Systems and Computing 749,
https://doi.org/10.1007/978-3-319-74808-5_33

The modern alternators are invariably high voltage high power units. Their voltage and the frequency must be closely controlled to maintain the quality of supply. The excitation system (AVR) is to provide controlled excitation to a synchronous machine to keep the generator terminal voltage (or at the terminal of generator-transformer) constant [4]. The reactive power generation is also controlled by the excitation system, thus controlling the power factors of the generators operating in parallel. The excitation system also has protective functions viz. it imposes limits on under-excitation and over-excitation. The frequency is controlled by the turbine-governor. It is kept within close limits (between 49 and 50.5 Hz. as per ABT) but is not held constant in our country. We do not use an automatic frequency controller (AFC) due to technical and economic reasons. However the use of electronic automatic voltage regulator (AVR) is a must.

2 Mathematical Description

From the basic voltage equation of the alternator, we get:

$$\mathbf{E} = \mathbf{V} + \mathbf{I} \cdot \mathbf{Z} \tag{1}$$

where E = induced voltage, V = terminal voltage, I = armature current, and Z = synchronous impedance

To maintain the terminal voltage constant, the induced emf and hence the field current has to be continuously adjusted with change in loading. This is the main task of the AVR. It has also additional functions e.g., proper sharing of reactive power between alternators in parallel, boosting up the excitation to ceiling under sudden disturbance etc.

There is a limit to the power that a generator can develop. For a cylindrical pole turbo-generator, the power developed and the maximum power are given as:

$$\mathrm{P} = P_{\max} \sin \delta,\ P_{\max} = (EV/X_s) \tag{2}$$

where X_s = synchronous reactance.

If the power demand exceeds P_{max}, then the machine will not be able to deliver any power and will fall out of synchronism. This is the power limit at steady state. Actually the machine will become unstable at a much lower power, if the load change be sudden or there be a sudden disturbance like short-circuit. This lower power limit is the transient stability limit which must not be exceeded in order to maintain synchronism between generators operating in parallel. This transient stability limit is enhanced by modern fast-acting exciter control which forces the excitation to its ceiling on occurrence of a fault. Thereby it increases the value of field current, I_f, hence induced emf, E and consequently maximum power, P_{max}. This is called field-forcing [5–7].

3 Evolution of the Exciter System

Therefore, we note that the exciter has multiple functions—both control and protective. The AVR is embedded in it which is the heart of the excitation control. Earlier versions of AVR were slow and sluggish and could not respond to changes immediately. Such systems employed d.c. generators as exciters and vacuum tubes, Magamps etc. for control. In a later stage came the a.c. exciters with rectifiers—either static or rotating. They were much faster. In a still later stage, solid state excitation (using high power SCRs) was brought in. In this method a part of the a.c. power generated by the alternator is rectified in a controlled manner and is applied to the field winding [8]. This happens to be the fastest. For various reasons the rotating rectifier system is also being used for many modern alternators.

4 Modern Alternators and Their Excitation

Modern turbo-generators are designed with low value of air-gap which gives rise to low short-circuit ratio (SCR). This practice reduces the requirement of excitation current and gives rise to a lighter rotor of smaller inertia. The lighter rotor obviously reduces the bearing losses, but increases the amplitude of oscillation under transients. It also increases the synchronous reactance and the operating power angle. It reduces the maximum power, the angle margin and the coupling coefficient. For example in BTPS, the length of air-gap for the 210 MW set of BHEL is 70 mm. The unsaturated value of synchronous reactance is 2.22 p.u. and inertia constant is 4.18 s. The corresponding values for the Westinghouse 89.25 MW sets are 1.6 p.u. and 7.0 s.

All modern generators are equipped with fast-acting excitation controllers and automatic voltage regulators. Such devices have radically improved the transient stability. The high value of synchronous reactance and the reduced angle margin are of no significance, now-a-days. The excitation controllers with excitation forcing facility under fault condition can bring back the generator into stable zone from a power angle of even 140°, without pole-slip. The modern excitation controllers using solid state devices are extremely fast. But their use may cause dynamic instability if the parameters of the controller are not properly tuned with the generator and the system [9].

The synchronous reactance of a generator is many times larger than its resistance. So the synchronous reactancc is largely responsible for the voltage drop. We want to closely regulate the terminal voltage of the generator (sometimes or the generator-transformer), we make use of automatic control systems—the excitation controller and the AVR. The AVR also fixes up the reactive power allocation.

The frequency also falls with loading. It can also be regulated by using Automatic Frequency Controllers (AFC). As yet, we do not use AFC in our systems for various reasons.

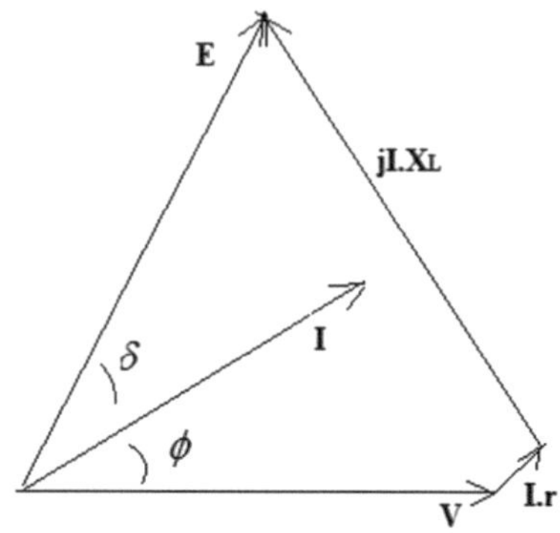

Fig. 1 Alternator with under-excitation (leading p.f.)

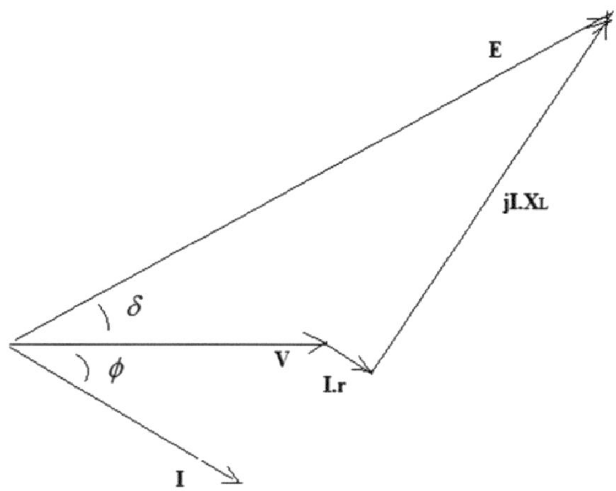

Fig. 2 Alternator with over-excitation (lagging p.f.)

5 Change of Excitation with Change in Load

The power factor of the aggregate load is an important factor. At leading power factor the field current decreases. This is called under-excitation. It reduces the stability margin. Leading p.f. may occur due to charging current of cable transmission and long distance high voltage transmission while the load is small. If the load power factor is too much lagging, the field current becomes too high. This is called over-excitation. Too much of over-excitation is not good as it increases the rotor temperature and may cause thermal injury. The phasor diagrams of alternator under over-excited and under-excited conditions are given in Figs. 1 and 2 respectively. It is observed that the induced e.m.f. must vary over wide range with change in load and its power factor if we like to keep the terminal voltage constant. That is why an excitation controller is used. Manual control is inaccurate and human response to changes in voltage and corrective action are extremely sluggish. Therefore the excitation controller must be automatic [10].

6 Requirements of the Excitation System

Functionally, the excitation system is to provide regulated direct current to the field winding of a synchronous machine. It has both control and protective functions. It has to control the terminal voltage by regulating the field current and, as and when necessary, to boost up the excitation to ceiling for system protection.

The excitation control aims at improved quality, greater reliability and higher stability margins in a generating system. It is, in essence, the heart of the generator control.

The most important requirements of excitation systems are as follows:

(a) The excitation system should ensure good damping of free and forced oscillation of small and large amplitude. Good damping increases the resultant stability, enhances the transient stability margin and reduces the voltage fluctuations at the load nodes.
(b) The excitation system should have high operational reliability. It has to choose the operating variables to be controlled and the circuit components. The excitation system should enable the generators to operate under overexcited and under-excited conditions. It must ensure single-unit and multi-unit control and proportional distribution of reactive power between synchronous machines.
(c) The excitation control should ensure high quality of voltage under steady state i.e. it must accurately maintain the voltage at system nodes to which the controlled synchronous machines are connected. This is obtained by choosing high gains for automatic excitation regulators.
(d) The automatic excitation control system should ensure steady state stability of the electrical system under all possible operating conditions including the following cases: a loaded or unloaded synchronous machine is disconnected from the system, a synchronous generator is connected to an unloaded transmission line, a synchronous machine is operating under overload and under-excited condition, normal load conditions, post fault condition.
(e) The automatic excitation control of generators connected to a long transmission line should provide maximum transmitted power equal to the maximum capacity of the line with a constant voltage at the system nodes to which large controlled synchronous machines are connected.
(f) A desired limit of transient stability must be ensured by the excitation system. It has to respond to a sudden disturbance with field forcing. The forcing during transience must conform to the instantaneous and short term capabilities of the generator. The generator capabilities in this regard are limited by factors such as:

    i. rotor insulation failure due to high field voltage
    ii. rotor heating due to high field current
    iii. stator heating due to high armature current
    iv. core-end heating during under-excited operation and
    v. heating due to excess flux (volts/Hz).

The thermal limits are time-dependent. The short-term overload capability of the generators extends from 15 to 60 s.

7 Types of Exciter

The exciters may either be of the proportional type or of the over-action type. The P-type control responds to the polarity and the magnitude of variations in the signal. But the over-action type also responds to the rate of change of the signals. The P-type control is slow and is associated with a static error. The over-excitation controllers operate more quickly and keep the voltage at generator terminals or at the terminals of the generator-transformer practically constant.

8 Change of Excitation with Load and Its P.F

The phasor diagrams of a cylindrical pole turbo-generator under both leading and lagging power factor conditions with terminal voltage constant are shown in Figs. 1 and 2. It is observed that the excitation requirement is high under lagging p.f. condition. So this condition is called over-excitation. The power angle at this condition is relatively low and hence there is enough of steady state stability margin. On the other hand the excitation requirement is low under leading p.f. condition. This condition is called under-excitation. The corresponding power angle is high and the stability margin low. The excitation requirement and the angle margin remains at intermediate values under unity p.f.

When no automatic control is incorporated the induced emf is determined by the invariable value of the field current and remains constant under changing load and fault condition (E = const). The power limit is given by:

$$P = EV \sin \delta / X_S = P_{\max} \sin \delta \tag{3}$$

With increasing load at lagging p.f., characteristic of power system, the terminal voltage will gradually fall and hence the value of P_{max}. This will not only make the quality of power supply unacceptable for the consumers but also will increase the power angle thus reducing the stability margin. But if the terminal voltage is to be kept constant by automatic devices, the value of E is to be regularly adjusted in conformity with the following equation:

$$\mathbf{E} = \mathbf{V} + \mathbf{I} \cdot \mathbf{Z} \text{ or } \quad \mathrm{E} < \delta = \mathrm{V} < 0 + \mathrm{I} < -\theta \cdot (\mathrm{r} + \mathrm{j} \cdot \mathrm{X}) \tag{4}$$

The values of induced emf required to keep the terminal voltage constant at rated value at different loads and at different power factors can be computed for an alternator by using an appropriate program. The synchronous reactance method is generally adopted, neglecting the effect of saturation. Infinite bus idealization has been made.

9 Induced Voltage at Different Loads and P.F

Bus is assumed to be infinite. All quantities are in p.u. The machine is connected to the bus through a transformer as shown in Fig. 3.

The MW-rating of the generator = 210	Transformer resistance = 0.005 (machine base)
Voltage at infinite bus = 1.0	Transformer reactance = 0.1367 (do)
Synchronous reactance = 2.225	Total line resistance = 0.0069
Armature resistance = 0.0019	Total line reactance = 2.3617

It is revealed from Table 1 that the variation of emf. is through a very wide range as the load and its p.f. changes. Leading p.f. occurs during lean period, generally it is not less than 0.95. At 70% load, the emf becomes 1.5692. At peak hours, the p.f. does not fall below 0.9 lag. At 90% load, the induced emf becomes 2.6013. So the range of variation in excitation, hence the field current, is quite large.

10 Automatic Excitation Control

From this table it is evident that the excitation must be controlled continuously to keep the terminal voltage (at generator terminal or generator-transformer terminal) constant. This is electronically done using power electronic components, at present making use of microprocessor/microcontroller or digital computer for close control. Much research work has been made in this area and bulk of papers published. Some publications in this area [11–14] have been referred to.

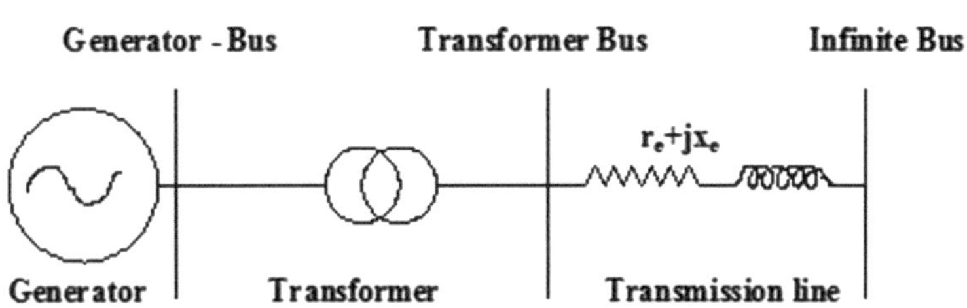

Fig. 3 Synchronous generator on infinite bus

Table 1 Induced e.m.f. versus varying load and p.f

PU load	Power factor		Induced emf	PU load	Power factor		Induced emf
0.5	0.8	Lag	1.8916 ∠28.00	0.6	0.8	Lag	2.0954 ∠30.57
	0.85		1.8481 ∠30.71		0.85		2.0484 ∠33.57
	0.9		1.7927 ∠33.90		0.9		1.9883 ∠37.11
	0.95		1.7143 ∠38.02		0.95		1.9036 ∠41.73
	1.0	Upf	1.4982 ∠47.95		1.0	Upf	1.6705 ∠53.05
	0.95	Lead	1.2448 ∠58.21		0.95	Lead	1.3987 ∠65.19
	0.9		1.1287 ∠62.68		0.9		1.2751 ∠70.69
0.7	0.8	Lag	2.2528 ∠35.91	0.8	0.8	Lag	2.5126 ∠34.43
	0.85		2.1891 ∠39.75		0.85		2.4603 ∠37.87
	0.9		2.1250 ∠42.25		0.9		2.3938 ∠41.93
	0.95		2.0994 ∠44.76		0.95		2.3001 ∠47.26
	1.0	Upf	1.8535 ∠57.18		1.0	Upf	2.0444 ∠60.54
	0.95	Lead	1.5692 ∠70.72		0.95	Lead	1.7514 ∠75.14
	0.9		1.4411 ∠76.95		0.9		1.6207 ∠81.88
0.9	0.8	Lag	2.7245 ∠35.92	1.0	0.8	Lag	2.9380 ∠37.19
	0.85		2.6703 ∠39.51		0.85		2.8821 ∠40.92
	0.9		2.6013 ∠43.77		0.9		2.8112 ∠45.34
	0.95		2.5045 ∠49.36		0.95		2.4422 ∠65.66
	1.0	Upf	2.2411 ∠63.33		1.0	Upf	2. 4422∠65.65
	0.95	Lead	1.9420 ∠78.70		0.95	Lead	2.1388 ∠81.62
	0.9		1.8098 ∠85.80		0.9		2.0058 ∠88.97

11 Conclusion

The excitation system of an alternator must effectively control the terminal voltage under varying load and at the same time enhance the system stability. It must respond quickly to a disturbance for enhancing transient stability and for modulating the generator field for enhancing dynamic stability. Modern excitation systems have almost instantaneous response along with high ceiling voltages. The high field forcing capability, along with auxiliary stabilizing signals substantially enhance the overall dynamic performance of the system. The excitation system parameters must be properly matched with the network under all quiscent conditions. Otherwise, the transient stability may be enhanced but there may be dynamic instability due to growing oscillations. The excitation parameters are tuned with the system by simulation studies.

The power angle depends very much on the load and its power factor. The task of AEC is to bring forth necessary changes in excitation to keep the voltage constant at generator terminal (or generator-transformer). As the magnitude of induced emf changes with loading conditions, P_{max} also changes. The effect of AEC on static stability is an increase of power limit. Modern AEC can bring back a rotor swinging beyond 120°–130° even by its over-action. For this reason AEC is so important.

References

1. P.M. Anderson, A.A. Fouad, *Power System Control and Stability*, 2nd edn. (IEEE Press, Wiley-Interscience, 2011)
2. M. Watanabe, Y. Mitani, H. Iki, Y. Uriu, Y. Urano, Dynamic stability assessment of customer's power systems with detailed models implementation, in *17th Power Systems Computation Conference*, Stockholm, Sweden, 22–26 Aug 2011
3. M.J. Basler, R.C. Schaefer, Understanding power-system stability. IEEE Trans. Ind. Appl. **44**(2) (2008)
4. M.J. Basler, R.C. Schaefer, Understanding power-system stability. IEEE Trans. Ind. Appl. **44**(2) (2008)
5. P. Kundur, et al., Definition and classification of power system stability. IEEE Trans. Power Syst. **19**(2) (2004)
6. M. Klein, G.J. Rogers, P. Kundur, A fundamental study of inter-area oscillations. IEEE Trans. PWRS-6(3), 914–921 (1991)
7. J. Arilagga, N.R. Watson, *Computer Modeling of Electrical Power System*, 2nd edn. (Wiley)
8. P. Kundur, *Power System Stability and Control*. The EPRI Power System Engineering Series (McGraw-Hill Inc.)
9. S. Hasan Saeed, *Automatic Control Systems (with MATLAB Programs)*, 5th edn. (S.K. Kataria and Sons)
10. Bharat Heavy electrical Ltd. (Operating manual on 210 Mw turbogenerator set)
11. R. Jamil, I. Jamil, Z. Jinquan, L. Ming, W. Y. Dong, Control and configuration of generator excitation system as current mainstream technology of power system. ISSR J. (2013)

12. A.K. Datta, M. Dubey, S. Jain, *Modelling and Simulation of Static Excitation System in Synchronous Machine Operation and Investigation of Shaft Voltage* (Hindawi Publishing Corporation) Advances in Electrical Engineering, vol. 2014, Article ID 727295
13. M. Prajapati, J. Patel, H. Chandwanic, V. Patel, Digital excitation system for synchronous generator. J. Electr. Eng
14. B. Shanngguan, Analysis of strong excitation control of A.C. synchronous generator based on excitation voltage regulation. Electrotehnica Electronica Automatica (EEA, English version), **64**(3) (2016)

Analysis of Linear Time Invariant and Time Varying Dynamic Systems via Taylor Series Using a New Recursive Algorithm

Suchismita Ghosh

1 Introduction

The Taylor series represents a function as an infinite sum of terms calculated from the values of its different order derivatives at a single point. This kind of expansion technique was formally introduced by Brook Taylor in 1715. Even after three centuries, the concept, though quite old, has much potential to be of importance in many areas of mathematics.

This paper solves dynamic differential equations or the state equations of different systems [1] using Taylor series. The method is essentially recursive algebraic formulation using both the first order and second order Taylor series [2] to solve time invariant and time varying systems. Earlier some works have been done where operational matrices were employed using Taylor series [3]. In this method, the operational matrix and subsequently matrix inversions for obtaining desired solution are avoided.

State space problem has been investigated with the framework of several orthogonal basis functions, e.g., Haar function, Walsh function [4], block pulse function [5]. Recently, triangular functions [6], hybrid functions [7] have been used extensively in this area. However, the Taylor series recursive technique seems to be much simpler and a powerful tool as well. In fact, some works on time varying system analysis, optimal control, system identification problem etc. have been done earlier using the Taylor series [8–12], But these never focused on recursion so that the computation can be much faster.

Further, integral equations have been solved by Taylor series [13], and delay systems are solved using the hybrid of block pulse functions and Taylor series

S. Ghosh (✉)
Department of Electrical Engineering, Faculty, MCKV Institute of Engineering, 243, G. T. Road, Liluah, Howrah 711204, India
e-mail: suchismita.ghosh@ymail.com

S. Chattopadhyay et al. (eds.), *Modelling and Simulation in Science, Technology and Engineering Mathematics*, Advances in Intelligent Systems and Computing 749,
https://doi.org/10.1007/978-3-319-74808-5_34

[14, 15]. In recent studies, a variety of works are found on the recursive solution using Taylor series [16–19] for the state analysis.

2 Function Approximation via Taylor Series

2.1 First Order Taylor Approximation [2]

A time function $f(t)$ defined over an interval h, $t \in (ih, (i+1)h)$, can be approximated about a point if the function exists at that point.

A first order Taylor approximation $\bar{f}1(t)$ of the function $f(t)$ [19] around a specific point μ_i is given by

$$\bar{f}1(t) \triangleq f(\mu_i) + \dot{f}(\mu_i)(t - \mu_i) \tag{1}$$

where, μ_i is a point in the $(i + 1)$-th interval with $\mu_i \in (ih,\ (i + 1)h)$ and $i = 0$, 1, 2,…, $(m - 1)$. Thus, $f(t)$ may be approximated up to the m-th interval. The time intervals in $[0, T)$ are of equal width h so that $h = \frac{T}{m}$. It should be mentioned that for the approximation to be pertinent, $(t - \mu_i)$ should be small compared to T.

Assuming $\mu_i = ih$, Eq. (1) becomes

$$\bar{f}1(t) = f(ih) + \dot{f}(ih)(t - ih) \tag{2}$$

Putting $t = (i + 1)h$ in (2),

$$\left.\begin{aligned} \bar{f}1\{(i+1)h\} &= f(ih) + h\dot{f}(ih) \\ \bar{f}1\{(i+2)h\} &= \bar{f}1\{(i+1)h\} + h\dot{\bar{f}}1\{(i+1)h\} \end{aligned}\right\} \tag{3}$$

Equation (3) provide a recursive solution of $f(t)$ using first order Taylor series. By changing the index i, approximate values of the function $f(t)$ can be obtained at m equidistant points, knowing the initial value $f(0)$ and $\dot{f}(0)$. For the very next interval $t \in ((i + 1)h, (i + 2)h)$, the Taylor approximated value $\bar{f}1\{(i + 1)h\}$ and $\dot{\bar{f}}1\{(i + 1)h\}$ are considered as the initial values for computation.

2.2 Second Order Taylor Approximation

The function $f(t)$ can also be represented using the second order Taylor series [19] around the point μ_i, knowing the function value at the point. Thus

$$\bar{f}2(t) \triangleq f(\mu_i) + \dot{f}(\mu_i)(t - \mu_i) + \frac{1}{2!}\ddot{f}(\mu_i)(t - \mu_i)^2 \tag{4}$$

where, as before, μ_i is a point in the $(i+1)$th interval with $\mu_i \in (ih, (i+1)h)$ and $i = 0, 1, 2, \ldots, (m-1)$. Thus, $f(t)$ may be approximated up to the mth intervals of equal width h within span $t \in [0, T)$ so that $h = \frac{T}{m}$. For a good approximation, $(t - \mu_i)$ should be small compared to T.

As in the previous case, if μ_i coincides with the point ih, i.e., $\mu_i = ih$, then Eq. (4) becomes

$$\bar{f}2(t) = f(ih) + \dot{f}(ih)(t - ih) + \frac{1}{2!}\ddot{f}(ih)(t - ih)^2 \tag{5}$$

Assuming $t = (i+1)h$,

$$\left.\begin{aligned} \bar{f}2\{(i+1)h\} &= f(ih) + h\dot{f}(ih) + \tfrac{h^2}{2!}\ddot{f}(ih) \\ \bar{f}2\{(i+2)h\} &= \bar{f}2\{(i+1)h\} + h\dot{\bar{f}}2\{(i+1)h\} + \tfrac{h^2}{2!}\ddot{\bar{f}}2\{(i+1)h\} \end{aligned}\right\} \tag{6}$$

Equation (6) provide a recursive solution of $f(t)$ using second order Taylor series approximation. By changing the index i, approximate values of the function $f(t)$ can be obtained at m equidistant points, knowing three initial values $f(0)$, $\dot{f}(0)$ and $\ddot{f}(0)$. For the very next interval $t \in ((i+1)h, (i+2)h)$, the Taylor approximated value $\bar{f}2\{(i+1)h\}$, $\dot{\bar{f}}2\{(i+1)h\}$ and $\ddot{\bar{f}}2\{(i+1)h\}$ are considered as the initial values for computation.

Now, state equations of different linear systems are solved using both first order and second order Taylor series based upon the recursive technique elaborated in Eqs. (3) and (6) respectively.

3 Analysis of the State via Taylor Approximation

3.1 Linear Time Invariant (LTI) System

The state equation of a linear non-homogenous time invariant (LTI) system is

$$\dot{\mathbf{x}}(t) = \mathbf{A}\mathbf{x}(t) + \mathbf{B}\mathbf{u}(t) \text{ and } \quad \mathbf{x}(0) = \mathbf{x}_0 \tag{7}$$

where, the notations are having usual significances.

Here, the state $\mathbf{x}(t)$ has to be solved knowing $\mathbf{A}$, $\mathbf{B}$, $\mathbf{u}(t)$ and the initial values of the states.

Differentiating (7), we have

$$\ddot{\mathbf{x}}(t) = \mathbf{A}\dot{\mathbf{x}}(t) + \mathbf{B}\dot{\mathbf{x}}(t) \tag{8}$$

Putting $t = ih$ in Eqs. (7) and (8)

$$\dot{\mathbf{x}}(ih) = \mathbf{A}\mathbf{x}(ih) + \mathbf{B}\mathbf{u}(ih) \tag{9}$$

$$\ddot{\mathbf{x}}(ih) = \mathbf{A}\dot{\mathbf{x}}(ih) + \mathbf{B}\dot{\mathbf{u}}(ih) \tag{10}$$

Calling the first order Taylor approximation of $\mathbf{x}(t)$ as $\bar{\mathbf{x}}1(t)$, one can write an equation similar to (9). That is

$$\dot{\bar{\mathbf{x}}}1(ih) = \mathbf{A}\bar{\mathbf{x}}1(ih) + \mathbf{B}\,\mathbf{u}(ih) \tag{11}$$

Now, to apply second order Taylor approximation, similar equations may be written following (9) and (10). Hence,

$$\dot{\bar{\mathbf{x}}}2(ih) = \mathbf{A}\bar{\mathbf{x}}2(ih) + \mathbf{B}\,\mathbf{u}(ih) \tag{12}$$

$$\ddot{\bar{\mathbf{x}}}2(ih) = \mathbf{A}\dot{\bar{\mathbf{x}}}2(ih) + \mathbf{B}\,\dot{\mathbf{u}}(ih) \tag{13}$$

where, the second order Taylor approximation of $\mathbf{x}(t)$ is considered as $\bar{\mathbf{x}}2(t)$.

Following Eq. (3), we can write the recursive equations to determine the states via first order Taylor series, as under

$$\bar{\mathbf{x}}1(h) = \bar{\mathbf{x}}1(0) + h\dot{\bar{\mathbf{x}}}1(0) \tag{14}$$

and

$$\bar{\mathbf{x}}1\{(i+1)h\} = \bar{\mathbf{x}}1(ih) + h\dot{\bar{\mathbf{x}}}1(ih) \tag{15}$$

where, $\bar{\mathbf{x}}1(0) \triangleq \mathbf{x}(0)$ and $i = 1, 2, 3, \ldots, m$.

Substituting (11) in Eqs. (14) and (15), we can write

$$\bar{\mathbf{x}}1(h) = [\mathbf{I} + \mathbf{A}h]\bar{\mathbf{x}}1(0) + h\mathbf{B}\mathbf{u}(0); \quad \text{for } i = 0, \tag{16}$$

$$\bar{\mathbf{x}}1\{(i+1)h\} = [\mathbf{I} + \mathbf{A}h]\bar{\mathbf{x}}1(ih) + h\mathbf{B}\mathbf{u}(ih); \quad \text{for any } i. \tag{17}$$

Equation (17) along with (16) provides a recursive solution of state vector $\mathbf{x}(t)$ using the first order Taylor series. The solution known at any point of the time scale is the data required for obtaining the solution at the next point, where the step size h is very small. As the initial values of the state vector $\mathbf{x}(0)$ are known, we can start the recursive process from $\mathbf{x}(0)$. Hence, knowing the input vector $\mathbf{u}(t)$ and by changing the index i up to m, the process gives a point wise solution of the state vector $\mathbf{x}(t)$.

When higher derivative terms of the states are considered, the recursive Eq. (15) becomes more and more accurate. Hence, to obtain a better approximation, second order Taylor approximation may be employed and Eqs. (14) and (15) can be modified based on Eq. (6) to determine the states. Thus

$$\bar{\mathbf{x}}2(h) = \bar{\mathbf{x}}2(0) + h\dot{\bar{\mathbf{x}}}2(0) + \frac{h^2}{2!}\ddot{\bar{\mathbf{x}}}2(0) \tag{18}$$

$$\bar{\mathbf{x}}2\{(i+1)h\} = \bar{\mathbf{x}}2(ih) + h\dot{\bar{\mathbf{x}}}2(ih) + \frac{h^2}{2!}\ddot{\bar{\mathbf{x}}}2(ih) \tag{19}$$

where, $\bar{\mathbf{x}}2(0) \triangleq \mathbf{x}(0)$ and $i = 1, 2, 3, \ldots, m$.

Substituting (12) and (13) in Eq. (19), we get

$$\bar{\mathbf{x}}2\{(i+1)\,h\} = \bar{\mathbf{x}}2(ih)\left(1 + h\mathbf{A} + \frac{h^2}{2!}\mathbf{A}^2\right) + \mathbf{u}(ih)\left(h\mathbf{B} + \frac{h^2}{2!}\mathbf{AB}\right) + \frac{h^2}{2!}\mathbf{B}\dot{\mathbf{u}}(ih)$$

So,

$$\bar{\mathbf{x}}2(h) = \bar{\mathbf{x}}2(0)\left(1 + h\mathbf{A} + \frac{h^2}{2!}\mathbf{A}^2\right) + \mathbf{u}(0)\left(h\mathbf{B} + \frac{h^2}{2!}\mathbf{AB}\right) + \frac{h^2}{2!}\mathbf{B}\dot{\mathbf{u}}(0); \quad \text{for } i = 0, \tag{20}$$

$$\bar{\mathbf{x}}2\{(i+1)h\} = \left(1 + h\mathbf{A} + \frac{h^2}{2!}\mathbf{A}^2\right)\bar{\mathbf{x}}2(ih) + \left(h\mathbf{B} + \frac{h^2}{2!}\mathbf{AB}\right)\mathbf{u}(ih) + \frac{h^2}{2!}\mathbf{B}\dot{\mathbf{u}}(ih); \quad \text{for any } i. \tag{21}$$

Thus, from Eqs. (20) and (21) knowing the initial values of the states $\mathbf{x}(0)$ and the input vector $\mathbf{u}(t)$, the solution of the states can be determined recursively by changing the index i up to any chosen m.

One particular advantage of this approach is, the step size h may be constant or it may be varied during computation dynamically.

3.2 *Linear Time Varying System*

The state equation of a linear non-homogenous time varying (LTI) system is

$$\dot{\mathbf{x}}(t) = \mathbf{A}(t)\mathbf{x}(t) + \mathbf{B}(t)\mathbf{u}(t) \text{ and } \quad \mathbf{x}(0) = \mathbf{x_0} \tag{22}$$

where, the notations are having usual significances.

Knowing the initial values of the states $\mathbf{x}(0)$, we are to determine the solution of the states $\mathbf{x}(t)$. Differentiating (22), we get

$$\ddot{\mathbf{x}}(t) = \dot{\mathbf{A}}(t)\mathbf{x}(t) + \mathbf{A}(t)\dot{\mathbf{x}}(t) + \dot{\mathbf{B}}(t)\mathbf{u}(t) + \mathbf{B}(t)\dot{\mathbf{u}}(t) \tag{23}$$

Putting $t = ih$, Eqs. (22) and (23) reduce to

$$\dot{\mathbf{x}}(ih) = \mathbf{A}(ih)\mathbf{x}(ih) + \mathbf{B}(ih)\mathbf{u}(ih) \tag{24}$$

$$\ddot{\mathbf{x}}(ih) = \dot{\mathbf{A}}(ih)\mathbf{x}(ih) + \mathbf{A}(ih)\dot{\mathbf{x}}(ih) + \dot{\mathbf{B}}(ih)\mathbf{u}(ih) + \mathbf{B}(ih)\dot{\mathbf{u}}(ih) \tag{25}$$

For first order Taylor approximation, we can modify another equation similar to (24) as

$$\dot{\bar{\mathbf{x}}}1(ih) = \mathbf{A}(ih)\bar{\mathbf{x}}1(ih) + \mathbf{B}(ih)\mathbf{u}(ih) \tag{26}$$

Again, Eqs. (24) and (25) can be modeled using second order Taylor series as,

$$\dot{\bar{\mathbf{x}}}2(ih) = \mathbf{A}(ih)\bar{\mathbf{x}}2(ih) + \mathbf{B}(ih)\mathbf{u}(ih) \tag{27}$$

$$\ddot{\bar{\mathbf{x}}}2(ih) = \dot{\mathbf{A}}(ih)\,\bar{\mathbf{x}}2(ih)+\mathbf{A}(ih)\ddot{\bar{\mathbf{x}}}2(ih) + \dot{\mathbf{B}}(ih)\mathbf{u}(ih) + \mathbf{B}(ih)\dot{\mathbf{u}}(ih) \tag{28}$$

Based on Eq. (3), the following recursive equations are formed to determine the states using first order Taylor approximation. Substituting (26) in Eqs. (14) and (15),

$$\bar{\mathbf{x}}1(h) = [\mathrm{I} + h\mathbf{A}(0)]\bar{\mathbf{x}}1(0) + h\mathbf{B}(0)\mathbf{u}(0); \quad \text{for } i = 0, \tag{29}$$

$$\bar{\mathbf{x}}1\{(i+1)h\} = [\mathrm{I} + h\mathbf{A}(ih)]\bar{\mathbf{x}}1(ih) + h\mathbf{B}(ih)\mathbf{u}(ih); \quad \text{for any } i. \tag{30}$$

Equations (29) and (30) provide a recursive solution of the state vector $\mathbf{x}(t)$ for the linear time varying system described by (22). Since initial values of the states $\mathbf{x}(0)$ are known, the recursive process can be started easily by choosing $i = 0$ in (30) or using Eq. (29). Here, the input of the system $\mathbf{u}(t)$ should be known as in the earlier case. But additionally, the time varying matrices $\mathbf{A}(t)$ and $\mathbf{B}(t)$ should also be known at various points of time as sample values.

Use of second order derivative term of the state in Eq. (15) gives a better approximation. So, substitution of (27) and (28) in Eq. (19) yields,

$$\begin{aligned}\bar{\mathbf{x}}2\{(i+1)h\} = \mathbf{x}(ih)&\left(1 + h\mathbf{A}(ih) + \frac{h^2}{2!}\dot{\mathbf{A}}(ih) + \frac{h^2}{2!}\mathbf{A}^2(ih)\right)\\ &+ \mathbf{u}(ih)\left(h\ \mathbf{B}(ih)+\frac{h^2}{2!}\mathbf{A}(ih)\mathbf{B}(ih)+\frac{h^2}{2!}\dot{\mathbf{B}}(ih)\right)\\ &+ \frac{h^2}{2!}\mathbf{B}(ih)\dot{\mathbf{u}}(ih)\end{aligned}$$

So, for $i = 0$;

$$\begin{aligned}\bar{\mathbf{x}}2(h) = \mathbf{x}(0)&\left(1 + h\mathbf{A}(0) + \frac{h^2}{2!}\dot{\mathbf{A}}(0) + \frac{h^2}{2!}\mathbf{A}^2(0)\right)\\ &+ \mathbf{u}(0)\left(h\ \mathbf{B}(0)+\frac{h^2}{2!}\mathbf{A}(0)\mathbf{B}(0)+\frac{h^2}{2!}\dot{\mathbf{B}}(0)\right)\\ &+ \frac{h^2}{2!}\mathbf{B}(0)\dot{\mathbf{u}}(0)\end{aligned} \tag{31}$$

and in general,

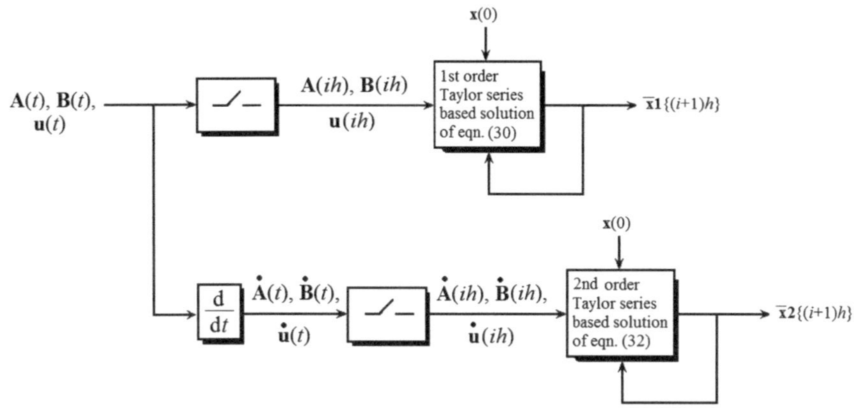

Fig. 1 Schematic block diagram of the algorithm for approximating $\mathbf{x}(t)$ via first and second order Taylor series

$$\begin{aligned}\bar{\mathbf{x}}2\{(i+1)\,h\} &= \left(1 + h\mathbf{A}(ih) + \frac{h^2}{2!}\dot{\mathbf{A}}(ih) + \frac{h^2}{2!}\mathbf{A}^2(ih)\right)\mathbf{x}(ih) \\ &+ \left(h\,\mathbf{B}(ih) + \frac{h^2}{2!}\mathbf{A}(ih)\mathbf{B}(ih) + \frac{h^2}{2!}\dot{\mathbf{B}}(ih)\right)\mathbf{u}(ih) \\ &+ \frac{h^2}{2!}\mathbf{B}(ih)\dot{\mathbf{u}}(ih)\end{aligned} \tag{32}$$

Equation (32) is the recursive equation of the state using second order Taylor series. Changing the value of i up to any integer m will give point wise solution of $\mathbf{x}(t)$. Similar to the first order Taylor approximation, in this case also, we should know the initial value of the state $\mathbf{x}(0)$, the input $\mathbf{u}(t)$ as well as the time varying matrices $\mathbf{A}(t)$ and $\mathbf{B}(t)$.

One important feature of this approach is, the step size h may be fixed or variable, which means the interval length h can be changed dynamically during recursion.

The method above mentioned can be implemented for online process for the determination of state simply by programming Eqs. (17), (21) or (30), (32) using microprocessor or microcontroller. To assist this algorithm other requirements are sampler and the differentiator. A block diagram for state analysis of time varying system is shown in Fig. 1. It is quite obvious that the time invariant system would follow the same process, but as the matrices **A** and **B** are constant in nature, so, the differentiation of those is not needed.

4 Numerical Example

(i) Consider the linear non-homogeneous time invariant system.

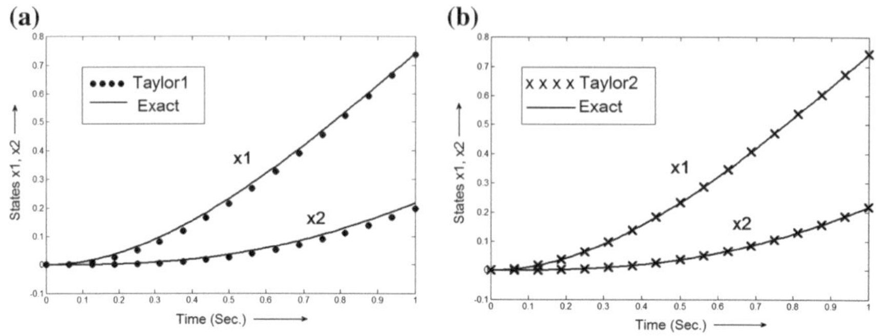

Fig. 2 Pointwise solution of the states $x1$ and $x2$ using **a** first order Taylor series (Taylor1) from Eq. (17) and **b** second order Taylor series (Taylor2) from Eq. (21) in recursive way for $T = 1$ s and $m = 16$. The exact solutions of the states $x1$ and $x2$ of Eqs. (34) and (35) are given in solid line

$$\dot{\mathbf{x}}(t) = \begin{bmatrix} -0.8 & -2.2 \\ 1 & -1 \end{bmatrix} \mathbf{x}(t) + \begin{bmatrix} 2.2 \\ 0 \end{bmatrix} \mathbf{u}(t) \tag{33}$$

and $\mathbf{x(0)} = \mathbf{x_0} = \begin{bmatrix} 0 \\ 0 \end{bmatrix}$ with unit ramp input. Applying traditional Laplace and inverse Laplace on (33), the exact solution is obtained as $\mathbf{x}(t) = L^{-1}\left[(s\mathbf{I} - \mathbf{A})^{-1}\{\mathbf{x}(0) + \mathbf{BU}(s)\}\right]$; where, $\mathbf{A} = \begin{bmatrix} -0.8 & -2.2 \\ 1 & -1 \end{bmatrix}$, $\mathbf{B} = \begin{bmatrix} 2.2 \\ 0 \end{bmatrix}$ and thereby the exact solution becomes,

$$x1(t) = \frac{11t}{15} - \frac{22\left(\cos\frac{\sqrt{219}t}{10} + \frac{34\sqrt{219}\sin\frac{\sqrt{219}t}{10}}{219}\right)}{75\exp\left(\frac{9t}{10}\right)} + \frac{22}{75}; \tag{34}$$

$$x2(t) = \frac{11t}{15} + \frac{11\left(\cos\frac{\sqrt{219}t}{10} - \frac{23\sqrt{219}\sin\frac{\sqrt{219}t}{10}}{657}\right)}{25\exp\left(\frac{9t}{10}\right)} - \frac{11}{25} \tag{35}$$

Using Eqs. (16), (17) and (20), (21) states of the system can be solved via first and second order Taylor series approximations respectively. Table 1 shows the results obtained via two methods. For computation, time T is considered to cover the range 0–1 s, with the number of steps $m = 16$. Thus, $h = T/m = 0.0625$ s.

For clarity, the solutions are also presented graphically in Fig. 2(a) and (b). For both the cases, the time interval T and the number of steps m are taken to be the same. Along with the Taylor solutions, both the figures present the exact curves of the states $x1$ and $x2$ from Eqs. (34) and (35) to judge the quality of Taylor series based results.

It is noted that, the second order Taylor approximation is much better than the first order approximation.

Table 1 Recursive solution of the state $x1$ and state $x2$ obtained via first order and second order Taylor approximation compared with the exact solution (for $T = 1$ s, $m = 16$ and $h = 0.0625$ s)

Time (s)	Pointwise solution of state $x1(t)$			Pointwise solution of state $x2(t)$		
	Exact data	Approximated data using		Exact data	Approximated data using	
		First order Taylor series	Second order Taylor series		First order Taylor series	Second order Taylor series
0	0	0	0	0	0	0
1/16	0.0042	0	0.0043	0.0001	0	0
2/16	0.0166	0.0086	0.0167	0.0007	0	0.0005
3/16	0.0366	0.0254	0.0368	0.0022	0.0005	0.0020
4/16	0.0637	0.0498	0.0641	0.0051	0.0021	0.0049
5/16	0.0974	0.0814	0.0978	0.0096	0.0051	0.0094
6/16	0.1371	0.1196	0.1376	0.0161	0.0098	0.0159
7/16	0.1822	0.1638	0.1828	0.0248	0.0167	0.0246
8/16	0.2321	0.2135	0.2329	0.0359	0.0259	0.0357
9/16	0.2864	0.2680	0.2872	0.0494	0.0376	0.0492
10/16	0.3444	0.3268	0.3453	0.0655	0.0520	0.0654
11/16	0.4056	0.3892	0.4066	0.0843	0.0692	0.0842
12/16	0.4696	0.4548	0.4706	0.1057	0.0892	0.1057
13/16	0.5357	0.5229	0.5367	0.1297	0.1120	0.1298
14/16	0.6036	0.5931	0.6046	0.1564	0.1377	0.1566
15/16	0.6728	0.6648	0.6738	0.1856	0.1662	0.1858
16/16	0.7430	0.7376	0.7440	0.2173	0.1973	0.2176

(ii) Consider the linear homogeneous time varying system [8].

$$\dot{\mathbf{x}}(t) = \begin{bmatrix} 0 & 0 \\ t & 0 \end{bmatrix} \mathbf{x}(t); \mathbf{x}(\mathbf{0}) = \mathbf{x_0} = \begin{bmatrix} 1 \\ 1 \end{bmatrix} \text{ with step input} \tag{36}$$

The exact solution is

$$x1(t) = 1 \text{ and } \quad x2(t) = 1 + \frac{1}{2}t^2 \tag{37}$$

This system is solved by using the recursive Eqs. (29)–(32). The solutions are exhibited in Table 2. For computation, a time interval of $T = 1$ s is considered and the number of steps $m = 4$, so that $h = T/m = 0.25$ s.

The first order Taylor approximation, compared with the exact solution, is shown in Fig. 3(a), while the second order Taylor approximation is shown in Fig. 3(b). The exact curves are drawn from the exact solutions provided in Eq. (37).

Table 2 Recursive solution of the state $x1$ and state $x2$ obtained via first order and second order Taylor approximation compared with the exact solution (for $T = 1$ s, $m = 4$ and $h = 0.25$ s)

Time (s)	Pointwise solution of state $x1(t)$			Pointwise solution of state $x2(t)$		
	Exact data	Approximated data using		Exact data	Approximated data using	
		First order Taylor series	Second order Taylor series		First order Taylor series	Second order Taylor series
0	1.0000	1.0000	1.0000	1.0000	1.0000	1.0000
1/4	1.0000	1.0000	1.0000	1.0313	1.0000	1.0313
2/4	1.0000	1.0000	1.0000	1.1250	1.0625	1.1250
3/4	1.0000	1.0000	1.0000	1.2813	1.1875	1.2813
4/4	1.0000	1.0000	1.0000	1.5000	1.3750	1.5000

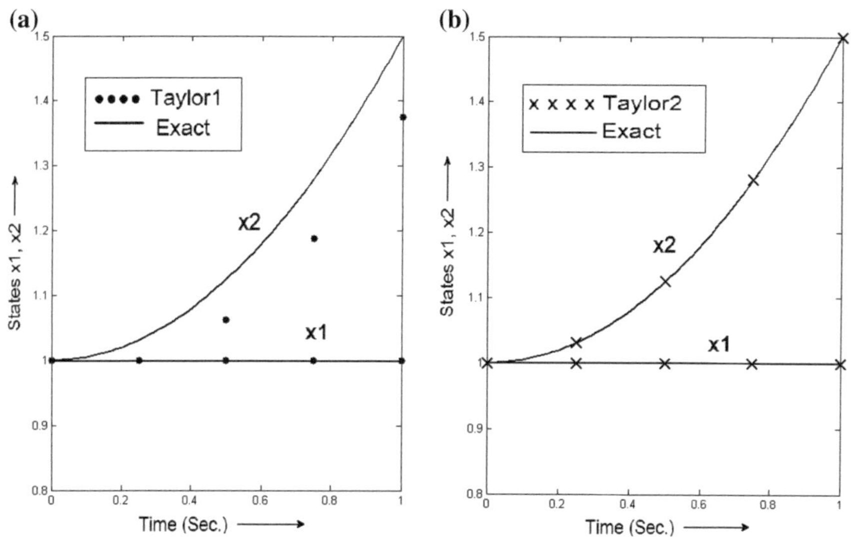

Fig. 3 Pointwise solution of the states $x1$ and $x2$ using **a** first order Taylor series (Taylor1) from Eq. (30) and **b** second order Taylor series (Taylor2) from Eq. (32) in recursive way for $T = 1$ s and $m = 4$. The exact solution in Eq. (37) of the states $x1$ and $x2$ of are shown in solid line

In Fig. 4, first order Taylor approximation results are shown with more number of steps ($m = 60$) within $T = 1$ s and reasonably match the exact results. However, the improved accuracy is achieved in lieu of increased computational burden.

(iii) Consider the linear non-homogeneous time varying system [11].

$$\dot{\mathbf{x}}(t) = \begin{bmatrix} 0 & te^{-t} \\ 0 & 0 \end{bmatrix} \mathbf{x}(t) + \begin{bmatrix} 0 \\ 1 \end{bmatrix} \mathbf{u}(t); \mathbf{x}(\mathbf{0}) = \mathbf{x}_0 = \begin{bmatrix} 1 \\ 1 \end{bmatrix} \text{ with step input} \tag{38}$$

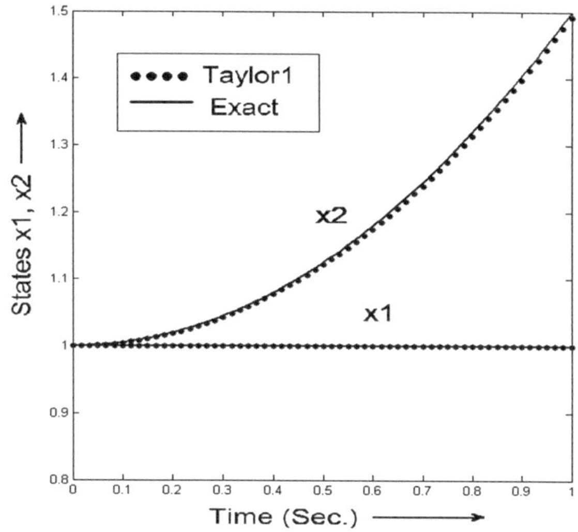

Fig. 4 The exact solutions of Eq. (37) along with solution via first order approximation (Taylor1) of Eq. (30) for $T = 1$ s and more number of steps i.e., $m = 60$. It shows better approximation than Fig. 2(a), which has less number of steps, $m = 4$

The exact solution is

$$x1(t) = 4 - \left(3 + 3t + t^2\right)e^{-t} \text{ and } \quad x2(t) = 1 + t \tag{39}$$

The time varying system given in this example is solved by using the recursive Eqs. (29)–(32). The point wise solutions are shown in Table 3 along with the exact samples obtained from (39). For computation, time interval $T = 1$ s. is considered and the number of steps $m = 10$, so that $h = T/m = 0.1$ s.

The comparison of the exact solution with the first order Taylor approximation is shown in Fig. 5(a) and the comparison with the second order Taylor approximation is presented in Fig. 5(b). In both the cases, the time interval considered is $T = 1$ s and the number of steps $m = 10$, i.e., $h = 0.1$ s.

Decreasing the step size obviously improves the accuracy of approximation with increased computational burden. This is established in Fig. 6, where first order Taylor approximation is shown with $m = 40$ and $T = 1$ s.

5 Error Estimates

In case of time invariant and time varying systems, for first order approximation using fixed intervals, the solution appreciably deviates from the exact results. But for the second order Taylor approximation using the same number of steps, the solution almost overlaps with the exact solution. The accuracy level for different step size can be measured with respect to mean integral square error (MISE). This is presented in Table 4 for three numerical examples.

Table 3 Recursive solution of the state $x1$ and state $x2$ obtained via first order and second order Taylor approximation compared with the exact solution (for $T = 1$ s, $m = 10$ and $h = 0.1$ s)

Time (s)	Pointwise solution of state $x1(t)$			Pointwise solution of state $x2(t)$		
	Exact data	Approximated data using		Exact data	Approximated data using	
		First order Taylor series	Second order Taylor series		First order Taylor series	Second order Taylor series
0	1.0000	1.0000	1.0000	1.0000	1.0000	1.0000
1/10	1.0050	1.0000	1.0050	1.1000	1.1000	1.1000
2/10	1.0198	1.0100	1.0199	1.2000	1.2000	1.2000
3/10	1.0441	1.0296	1.0443	1.3000	1.3000	1.3000
4/10	1.0774	1.0585	1.0777	1.4000	1.4000	1.4000
5/10	1.1190	1.0960	1.1194	1.5000	1.5000	1.5000
6/10	1.1681	1.1415	1.1686	1.6000	1.6000	1.6000
7/10	1.2241	1.1942	1.2247	1.7000	1.7000	1.7000
8/10	1.2861	1.2533	1.2868	1.8000	1.8000	1.8000
9/10	1.3532	1.3180	1.3541	1.9000	1.9000	1.9000
10/10	1.4248	1.3875	1.4259	2.0000	2.0000	2.0000

Table 4 Error estimates based on mean integral square error (MISE) for three numerical examples with a specified number of recursions (m) and length of interval (h), using first and second order Taylor series

Examples along with number of recursion points (m) and length of intervals (h)	For states $\mathbf{x}(t)$	Approximation error (MISE) using	
		First order Taylor series	Second order Taylor series
Example (i) $m = 16$; $h = 0.0625$ s	State $x1(t)$	$10^{-3} \times 0.1864$	$10^{-5} \times 0.1108$
	State $x2(t)$	$10^{-3} \times 0.1393$	$10^{-5} \times 0.0045$
Example (ii) $m = 4$; $h = 0.25$ s	State $x1(t)$	0	0
	State $x2(t)$	0.0046	$10^{-4} \times 0.3255$
Example (iii) $m = 10$; $h = 0.1$ s	State $x1(t)$	$10^{-3} \times 0.5581$	$10^{-5} \times 0.1163$
	State $x2(t)$	0	0

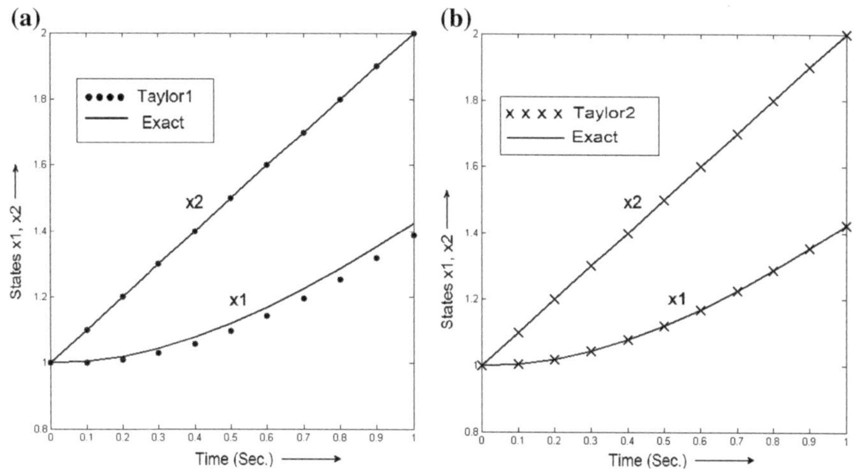

Fig. 5 Pointwise recursive solution of the states $x1$ and $x2$ using **a** first order Taylor series (Taylor1) from Eq. (30) and **b** second order Taylor series (Taylor2) from Eq. (32) for $T = 1$ s and $m = 10$. The exact solution in Eq. (39) of the states $x1$ and $x2$ of are shown in solid line

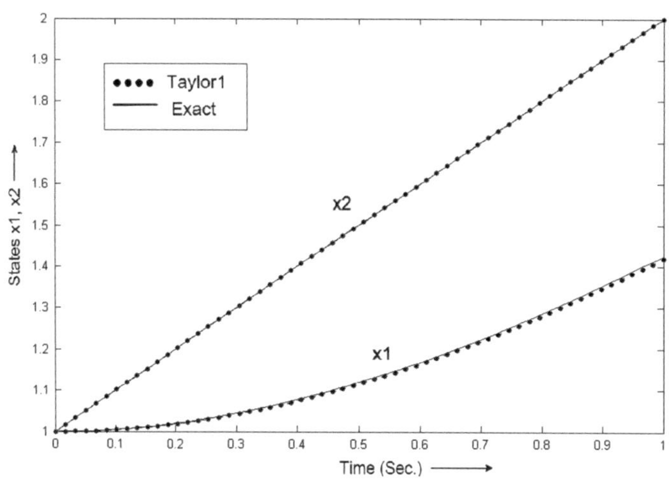

Fig. 6 The exact solutions of Eq. (39) along with solution via first order approximation (Taylor1) of Eq. (30) for $T = 1$ s and more number of steps i.e., $m = 40$. It shows an approximation better than Fig. 4a, which has less number of steps, $m = 10$

As expected, it shows, the second order Taylor approximation is a stronger tool than the first order approximation.

6 Conclusion

A simple recursive method for solving the states of linear time invariant as well as time varying control systems using Taylor series expansion have been presented. Here, the solution which results in every iteration, changes into the new initial value for the next iteration. The solutions are compared with the exact samples using different step sizes. It is shown that, to reach the efficiency of first order Taylor series based method like second order Taylor approximation, requirement of iteration increases, which in turn increases computational burden. Four tables and eight figures are presented to compare the results in quantitative threadbare manner. Comparison with the exact solutions shows this recursive approach to be attractive and quite reliable as well.

In some papers [9–11, 13–15], where the state is analyzed using the Taylor series, is a way which needs operational matrix of integration. This matrix has a fixed dimension that depends on the length of interval h or number of interval m, which has to be set prior the computation of state, whereas, this approach avoids this kind of matrix. Hence during computation, one can easily change h or m and thereby perform a dynamic computation.

Being the algorithm recursive in nature, avoids the complexity of handling large matrices and its inversion too unlike traditional approaches. The memory requirement and time consumption during computation is reduced considerably in this context.

References

1. Lawrence Perko, *Differential Equations and Dynamical Systems*, 3rd edn. (Springer, Berlin, 2006)
2. M. Krasnov, A. Kiselev, G. Makarenko, E. Shikin, *Mathematical Analysis for Engineers*, vol. 2 (Mir Publishers, Moscow, 1990)
3. M.R. Eslahchi, M. Dehghan, Application of Taylor series in obtaining the orthogonal operational matrix. Comput. Math Appl. **61**(9), 2596–2604 (2011)
4. K.G. Beauchamp, *Applications of Walsh and Related Functions, with an Introduction to Sequency Theory* (Academic Press, London, 1984)
5. J.H. Jiang, W. Schaufelberger, *Block Pulse Functions and their Application in Control System* (LNCIS – 179, Springer, Berlin, 1992)
6. A. Deb, G. Sarkar, A. Sengupta, *Triangular Orthogonal Functions for the Analysis of Continuous time Systems* (Anthem Press, London, 2011)
7. A. Deb, G. Sarkar, A. Ganguly, A. Biswas, Approximation, integration and differentiation of time functions using a set of orthogonal hybrid functions (HF) and their application to solution of first order differential equations. Appl. Math. Comput. **218**(9), 4731–4759 (2012)
8. M.J. Tsai, C.K. Chen, Analysis of linear time-varying systems by shifted legendre polynomials. J. Franklin Inst. **318**(4), 275–282 (1984)
9. Ching-Yu Yang and Cha'o-Kuang Chen, Analysis and parameter identification of time-delay systems via Taylor series. Int. J. Syst. Sci. **18**(7), 1347–1353 (1987)
10. P.N. Paraskevopoulos, A.S. Tsirikos, K.G. Arvanitis, New Taylor series approach to state space analysis and optimal control of linear systems. J. Optim. Theory Appl. **71**(2), 315–325 (1991)
11. M.H. Perng, An effective approach to the optimal-control problem for time-varying linear systems via Taylor series. Int. J. Control **44**(5), 1225–1231 (1986)

12. S. Ghosh, A. Deb, G. Sarkar, A new recursive method for solving state equations using Taylor series. Int. J. Electr. Electron. Comput. Eng. **1**(2), 22–27 (2012)
13. S.C. Tsay, T.T. Lee, Solutions of integral equations via Taylor series. Int. J. Control **44**(3), 701–709 (1986)
14. H.R. Marzban, M. Razzaghi, Analysis of time-delay systems via hybrid of block pulse functions and Taylor series. J. Vibr. Control **11**(12), 1455–1468 (2005)
15. H.R. Marzban, M. Razzaghi, Solution of multi-delay systems using hybrid of block pulse functions and Taylor series. J. Sound Vibr. **292**, 954–963 (2006)
16. S. Ghosh, A. Deb, G. Sarkar, A new recursive method for solving state equations using Taylor series. Int. J. Electr. Electron. Comput. Eng. **1**(2), 22–27 (2012)
17. S. Ghosh, A. Deb, G. Sarkar, Recursive solution of systems modeled by state space and multi-delay systems via first and second order Taylor series, in *2nd International Conference on Control, Instrumentation, Energy & Communication (CIEC 16)*, January 28–30, Kolkata, pp. 55–59 (2016)
18. S. Ghosh, A. Deb, G. Sarkar, Analysis of singular systems via Taylor series using a recursive algorithm, in *2016 IEEE First International Conference on Control, Measurement and Instrumentation (CMI)*, Jadavpur University, pp. 356–360 (2016)
19. S. Ghosh, A. Deb, G. Sarkar, Taylor series approach for function approximation using ‘estimated’ higher derivatives. Appl. Math. Comput. **284**, 89–101 (2016)

Severity and Location Detection of Three Phase Induction Motor Stator Fault Using Sample Shifting Technique and Adaptive Neuro Fuzzy Inference System

S. Samanta, J. N. Bera and G. Sarkar

1 Introduction

In industry, Induction motors are most commonly used due to various mechanical and economical reasons like low cost, reasonably small size, ruggedness, low maintenance, smooth operation etc. When in operation, various stresses are imparted on the induction motor which may lead to incipient failures/faults in the motor. If these incipient faults are not detected in an early stage, a permanent or damage of the motor could be the result. Different kinds of machine faults are studied in literature [1, 2] like stator inter turn fault, broken rotor bar fault, bearing fault, unbalanced stator and rotor parameters, eccentricity fault etc. So, early detection of faults is a serious issue to reduce consequential damage, increase life span of machine, reduces spare parts inventories and avoid breakdown. Implementation of an efficient condition monitoring scheme can provide warning and predict the fault at early stages.

According to an IEEE and Electric Power Research Institute motor reliability study [3], 37% of the induction motor failure is due to stator fault. In most cases, a stator fault starts as a turn to turn (inter turn) and finally grows as coil-to-coil, phase-to-phase or phase-to-ground failures, and ultimately causes motor breakdown. In modern system, due to the advancement in the condition monitoring technologies,

S. Samanta (✉)
Electrical Engineering Department, MCKV Institute of Engineering,
243 G. T. Road (N), Liluah, Howrah 711204, India
e-mail: sudeep0809@gmail.com

J. N. Bera · G. Sarkar
Applied Physics Department, University College of Science and Technology,
92, A.P.C Road, Kolkata 700009, India
e-mail: jitendrabera@rediffmail.com

G. Sarkar
e-mail: gautamgs@yahoo.co.in

S. Chattopadhyay et al. (eds.), *Modelling and Simulation in Science, Technology and Engineering Mathematics*, Advances in Intelligent Systems and Computing 749,
https://doi.org/10.1007/978-3-319-74808-5_35

automation in diagnosis process are being implemented with the adoption of different soft computing techniques [4–6].

Recently different kinds of soft computing techniques like expert system, neural network, fuzzy logic, adaptive neural fuzzy inference system, genetic algorithm etc. are being employed to assist the diagnostic task with correct interpretation of the fault condition. These techniques become popular over other conventional techniques as they are easy to extend and modify, improved performance, with increased precision and accuracy of the monitoring system.

In [7, 8] Fuzzy logic based motor condition monitoring and fault diagnosis is presented. Filippetti et al. [9] has introduced an expert system based comprehensive study about the application of artificial intelligence along with Fuzzy logic in machine monitoring and fault diagnosis. In [10], stator three phase rms values of currents and the variance are used as the input to the fuzzy logic system for fault diagnosis of induction motor. Bouzid [11] has suggested a feed forward multi layer perceptron neural network approach for automatically detection of an inter-turn short circuit fault in the stator windings of an induction motor. Tan and Huo [12] have introduced a generic neuro-fuzzy model based approach for rotor broken bar fault detection of an induction motor. This approach overcomes the practical limitations of model based strategies because it reduces the amount of experimental data needed to design the fault detector. Ballal et al. [13] have approached an adaptive neural fuzzy inference system for the detection of inter-turn insulation and bearing faults in induction motor.

The sequence components [14, 15] are used as very important tool for fault detection in electrical system. In normal operating condition, the negative sequence component of current is nearly equal to zero. But, during stator inter turn fault in any phase, this component of current increases to a considerable amount due to the flow of huge current through the corresponding phase causing an unbalance in the stator current.

In this proposed paper, Sample Shifting Technique (SST) is utilized for the evaluation of sequence components based on three phase current samples only. The intricacy of SST is discussed briefly in next section. An ANFIS based fault detection strategy is then developed based on the various sequence based index for detection of fault location and fault severity. The proposed work is simulated by modeling a three phase induction motor with inter turn fault condition in MATLAB. The proposed diagnosis method is also tested on a real three phase motor hardware to justify its effectiveness.

2 Materials and Methods

2.1 Sequence Component Evaluation Using SST

Based on Fortescue's theorem, any unbalance quantity can be resolved to three independent sequence components namely, Positive, Negative and Zero Sequence component. In a three phase system, the positive and negative sequence component of currents can be calculated as [16, 17]:

$$\begin{bmatrix} I_{a1} \\ I_{a2} \end{bmatrix} = \left[A^{-1}\right] \begin{bmatrix} I_a \\ I_b \\ I_c \end{bmatrix} \quad (1)$$

where, $[A] = \begin{bmatrix} 1 & a^2 & a \\ 1 & a & a^2 \end{bmatrix}$ and $\left[A^{-1}\right] = \frac{1}{3}\begin{bmatrix} 1 & a & a^2 \\ 1 & a^2 & a \end{bmatrix}$

The operator $a = 1\llcorner 120°$ and $a^2 = 1\llcorner 40°$, I_{a1} and I_{a2} are positive and negative sequence components of current respectively.

In this paper, Sample Shifting Technique (SST) [18] is used to calculate the sequence components. Sample Shifting Technique stands on the fact that shifting of any sinusoidal wave at any angle actually means shifting the time of occurrence of instantaneous amplitudes of the sinusoidal wave by that angle. So, very easily a shifted sine wave can be generated from its original waveform, only by rearranging its sample values. Let us consider, a sine wave has 360 samples per cycle (Sampled at 1° interval), then only by shifting 121st to 360th samples of original wave to 1st to 240th position and 1st to 120th samples of original wave to 241th to 360th position, we can generate its 120° shifted sine wave.

For calculation of Positive and Negative sequence components using Sample Shifting Technique, it is considered that the multiplication with operator 'a' to a sine wave is equivalent to 120° sample shifting from its original samples and multiplication with operator 'a^2' is equivalent to 240° sample shifting from its original samples. So, to evaluate current sequence component in a three phase system using SST, Eqs. (2)–(3) are utilized.

$$\begin{aligned} I_1(n) &= \frac{1}{3}[I_a(n) + I_b(n + \frac{N}{3}) + I_c(n + \frac{2N}{3})] \quad \text{for, } 1 \le \text{n} \le \text{N}/3; \\ &= \frac{1}{3}[I_a(n) + I_b(n + \frac{N}{3}) + I_c(n - \frac{N}{3})] \quad \text{for, N}/3 < \text{n} \le 2\text{N}/3; \\ &= \frac{1}{3}[I_a(n) + I_b(n - \frac{2N}{3}) + I_c(n - \frac{N}{3})] \quad \text{for, 2N}/3 < \text{n} \le \text{N}; \end{aligned} \quad (2)$$

$$\begin{aligned} I_2(n) &= \frac{1}{3}[I_a(n) + I_b(n + \frac{2N}{3}) + I_c(n + \frac{N}{3})] \quad \text{for, } 1 \le \text{n} \le \text{N}/3; \\ &= \frac{1}{3}[I_a(n) + I_b(n - \frac{N}{3}) + I_c(n + \frac{N}{3})] \quad \text{for, N}/3 < \text{n} \le 2\text{N}/3; \\ &= \frac{1}{3}[I_a(n) + I_b(n - \frac{N}{3}) + I_c(n - \frac{2N}{3})] \quad \text{for, 2N}/3 < \text{n} \le \text{N}; \end{aligned} \quad (3)$$

where, N is total no. of samples in a cycle of the sinusoidal wave.

2.2 *Formulation of Fault Detection Methodologies*

For detection of stator inter turn fault, the magnitude and phase angle of negative sequence component are used. In healthy condition, the magnitude of negative sequence component is approximately zero. But, when inter turn fault occurs, due to current unbalance negative sequence component are generated. As the fault severity goes on increasing, magnitude of negative sequence component also increases. Again, depending on the different fault location, different phase relationship between positive and negative sequence component are introduced. On the basis of these two characteristics, the fault location and severity are detected by introducing two unique parameters, defined as Sequence Component Amplitude Index (SCAI) and Sequence Component Phase Index (SCPI). SCAI is used to detect the severity of fault whereas the faulty phase is detected using SCPI parameter. The formulations SCAI and SCPI are described as follows.

2.2.1 Formulation of SCAI

Sequence Component Amplitude Index (SCAI) is defined as per unit change in magnitude of negative sequence components with respect to positive sequence component.

$$SCAI = (I_1 - I_2)/I_1 \tag{4}$$

Here, I_1 is the magnitude of positive sequence component and I_2 is the magnitude of negative sequence component.

2.2.2 Formulation of SCPI

Sequence Component Phase Index (SCPI) is defined as per unit change in phase angle of negative sequence components with respect to positive sequence component.

$$SCPI = (\phi_1 - \phi_2)/120 \tag{5}$$

Here, Φ_1 is the phase angle of positive sequence component and Φ_2 is the phase angle of negative sequence component.

2.3 Adaptive Neuro Fuzzy Inference System (ANFIS) for Fault Diagnosis

Though, Artificial Neural Network (ANN) is capable for machine condition monitoring and fault diagnosis using an inexpensive and reliable procedure, it does not provide heuristic reasoning about fault detection process. On the other hand, Fuzzy logic can easily provide heuristic reasoning but fails to provide exact solution. So, by merging the good features of ANN and Fuzzy logic, a simple noninvasive fault detection technique ANFIS [19] has been developed. ANFIS become popular over other fault detection technique because of its knowledge extraction feasibility, domain partitioning, rule structuring and modifications [20]. ANFIS based fault detection system has a fuzzy inference system along with five layers feed-forward network. Knowledge is extracted in terms of membership functions and Takagi Sugeno type Fuzzy if then rule. By using hybrid learning procedure, ANFIS can construct an input—output in the form of Takagi-Sugeno type if-then rules.

In this paper, a suitable ANFIS program as written in MATLAB for interturn fault detection purposes. Fault detection system is prepared with two input parameters namely SCAI and SCPI. The main objective of ANFIS based fault diagnostic system is to learn the relationships between the fault signature under different load conditions (ANFIS inputs) and the corresponding operating condition (ANFIS outputs) which will be able to provide fault information correctly. In Fig. 1, The ANFIS based fault diagnostic system is given.

2.4 Preparation of Suitable Training Data Set for ANFIS

In this work, a suitable training data set for ANFIS is prepared for collecting three phase current samples with stator interturn fault at different phases under different loading conditions such as no load, 5 Nm load, 10 Nm load. Here, eight different

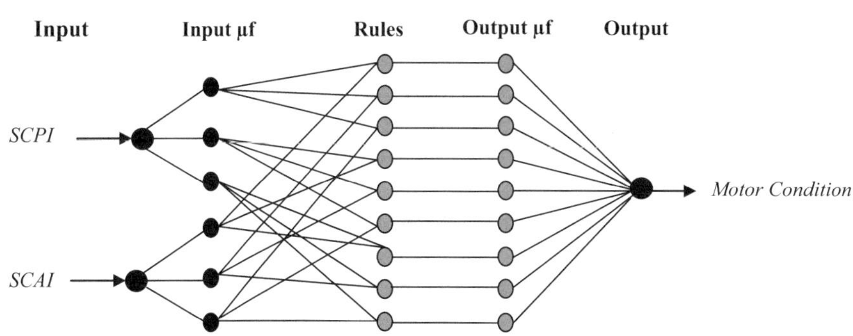

Fig. 1 ANFIS based fault diagnostic system

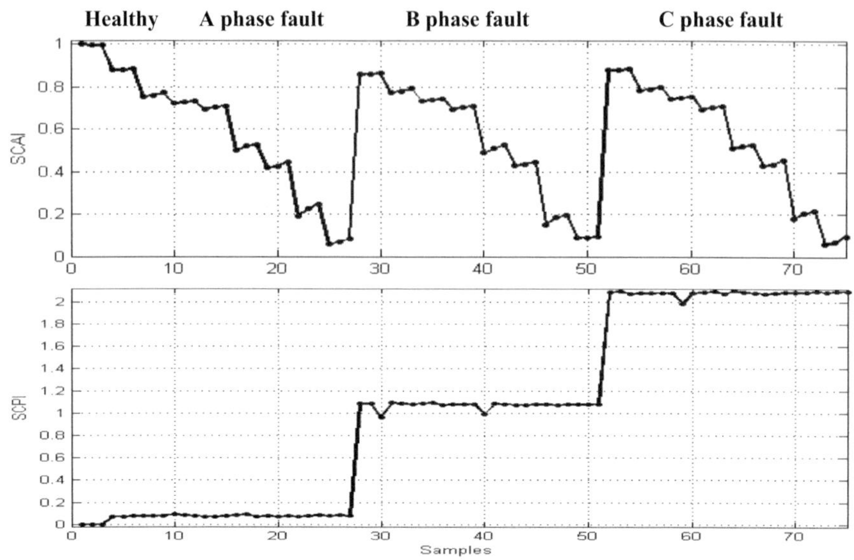

Fig. 2 Simulated training input data set

parentages of inter turn fault are considered like 2, 5, 8, 10, 12, 15, 18 and 20%. For all conditions, SCAI and SCPI are calculated, which are used for training. Thus for three load conditions, total ($3 \times 8 = 24$) numbers of inputs are considered. Thus for three phase fault, 72 (24×3) number and 3 number for healthy condition, all together 75 number of training and testing data patterns are generated for two inputs, shown in Fig. 2.

3 MATLAB Simulation

3.1 Simulink Model of Three Phase Induction Motor

In this paper, 3-Φ Squirrel cage Induction motor model is developed in MATLAB/SIMULINK by assuming that identical three phase stator windings with equivalent no. of turns N_s and resistance r_s, stator windings are symmetrically distributed and displaced from each other at angle 120°.

The rotor windings are also considered as three identical sinusoidal distributed windings with equivalent no. of turns N_r and resistance r_r and displaced at 120° with each other. The uniform air gap is also considered in between stator and rotor.

Considering all of these, stator and rotor voltage equations of the 3-Φ induction motor can be derived as:

$$V_{abcs} = r_s i_{abcs} + \frac{d}{dt}\psi_{abcs} \tag{6}$$

$$V_{abcr} = r_r i_{abcr} + \frac{d}{dt}\psi_{abcr} \tag{7}$$

For, squirrel cage induction motor $V_{abcr} = 0$.

Where, Stator resistance, $r_s = \begin{bmatrix} r_s & 0 & 0 \\ 0 & r_s & 0 \\ 0 & 0 & r_s \end{bmatrix}$ and rotor resistance $r_r = \begin{bmatrix} r_r & 0 & 0 \\ 0 & r_r & 0 \\ 0 & 0 & r_r \end{bmatrix}$

In above equations, stator and rotor voltages are

$$V_{abcs} = \begin{bmatrix} V_{as} \\ V_{bs} \\ V_{cs} \end{bmatrix}, \quad V_{abcr} = \begin{bmatrix} V_{ar} \\ V_{br} \\ V_{cr} \end{bmatrix}$$

stator and rotor currents are

$$I_{abcs} = \begin{bmatrix} I_{as} \\ I_{bs} \\ I_{cs} \end{bmatrix}, \quad I_{abcr} = \begin{bmatrix} I_{ar} \\ I_{br} \\ I_{cr} \end{bmatrix}$$

Flux linkage equations can be expressed as,

$$\begin{bmatrix} \psi_{abcs} \\ \psi_{abcr} \end{bmatrix} = \begin{bmatrix} L_s & L_{sr} \\ L_{sr}^T & L_r \end{bmatrix} \begin{bmatrix} i_{abcs} \\ i_{abcr} \end{bmatrix} \tag{8}$$

Here, stator winding inductance matrix L_s, rotor winding inductance matrix L_r and stator to rotor mutual inductance matrix L_{sr} can be expressed as,

$$L_s = \begin{bmatrix} L_{ls} + L_{ms} & -\frac{1}{2}L_{ms} & -\frac{1}{2}L_{ms} \\ -\frac{1}{2}L_{ms} & L_{ls} + L_{ms} & -\frac{1}{2}L_{ms} \\ -\frac{1}{2}L_{ms} & -\frac{1}{2}L_{ms} & L_{ls} + L_{ms} \end{bmatrix} \text{ and } Lr = \begin{bmatrix} L_{lr} + L_{mr} & -\frac{1}{2}L_{ms} & -\frac{1}{2}L_{ms} \\ -\frac{1}{2}L_{ms} & L_{lr} + L_{mr} & -\frac{1}{2}L_{ms} \\ -\frac{1}{2}L_{ms} & -\frac{1}{2}L_{ms} & L_{lr} + L_{mr} \end{bmatrix}$$

where, $L_{ls}, L_{lr}, L_{ms}, L_{mr}$ = stator and rotor leakage and magnetizing inductance.

Considering, electrical angular velocity ω_r and displacement θ_r,

$$Lsr(\theta_r) = \begin{bmatrix} \cos\theta_r & \cos(\theta_r + \frac{2\pi}{3}) & \cos(\theta_r - \frac{2\pi}{3}) \\ \cos(\theta_r - \frac{2\pi}{3}) & \cos\theta_r & \cos(\theta_r + \frac{2\pi}{3}) \\ \cos(\theta_r + \frac{2\pi}{3}) & \cos(\theta_r - \frac{2\pi}{3}) & \cos\theta_r \end{bmatrix} \tag{9}$$

The electromagnetic torque produced by a three phase Induction motor can be derived from energy conversion principles [21],

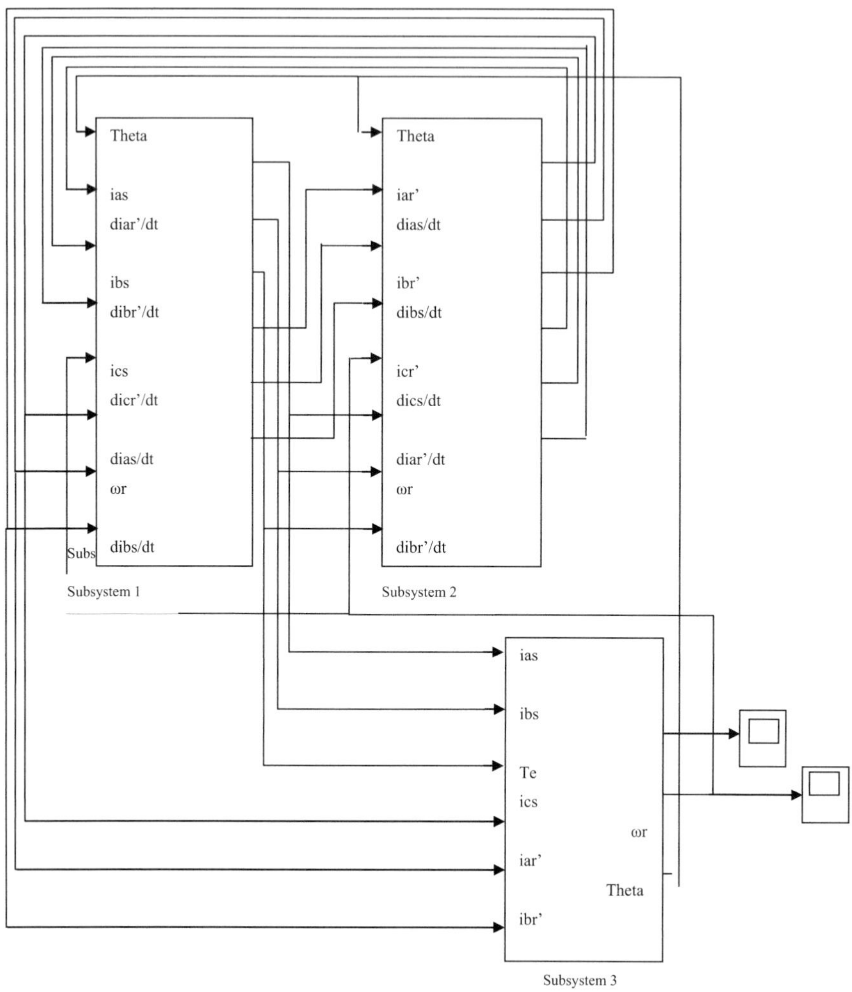

Fig. 3 Simulink model of three phase induction motor

$$T_{em} = [i_{abcs}]^T \left[\frac{dL_{sr}}{d\theta_r}\right][i_{abcr}] \tag{10}$$

The rotor mechanical equation [21] has been given in Eq. 11.

$$T_{em} = J\frac{d\omega_r}{dt} + B\omega_r + T_L \tag{11}$$

where, J is the rotor inertia, ω_r is rotor angular speed in rad/sec, B is the friction coefficient and T_L is the load torque.

Using Eqs. (6)–(11), the dynamic model of a healthy 3-Φ squirrel cage induction motor are developed in MATLAB/SIMULINK, which is shown in Fig. 3. Here, three different subsystems are created viz. subsystem 1 is for stator current calculation, subsystem 2 is for rotor current calculation and subsystem 3 is for torque and rotor speed calculation.

In this model, stator inter-turn fault is externally created by dividing the phase winding associated with inter turn fault in two series part. Considering N_s is total no. of stator turns per phase, then N_s=N_{us}+N_{sh}, where N_{sh} is no of shorted turns and N_{us} is no of unshorted turns. So, the percentage of fault can be determined by $(N_{sh}/N_s) \times 100\%$.

3.2 Simulation Result

Training data of all input parameters (SCAI and SCPI) are applied to fault detector for obtaining the optimized architecture to detect stator inter-turn fault of an induction motor. Here, up to 5% of inter turn fault has been considered as less damaged, 5–15% has been considered as medium damaged and above 15% has been considered as severely damaged condition. Figure 4 shows the ANFIS output when motor condition is less damaged inter turn fault on phase A. Here, circle line represents the target value, the star one is the actual output. Similarly, Figs. 5 and 6 will represent ANFIS output for medium damaged and severe damaged inter turn fault at phase A.

Figure 7 represents B phase medium damaged condition output, Figs. 8 and 9 represent C phase medium and severe damaged condition output.

When fault occurs on any phase A then depending upon its severity the output should equal be to one and others are zero. The result clearly shows that there is very

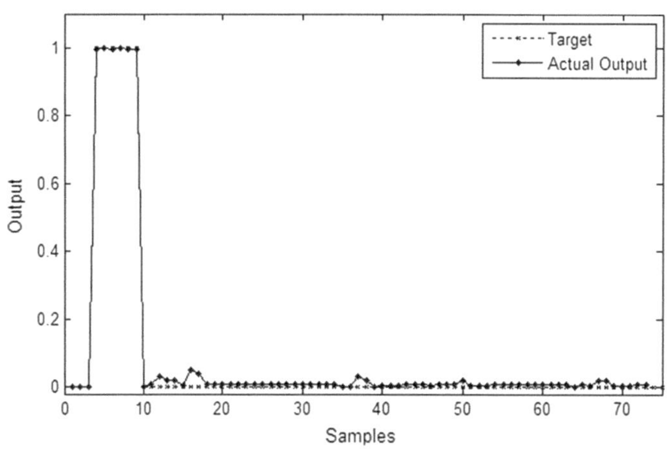

Fig. 4 ANFIS output for Stator A phase winding inter turn fault with less damaged condition

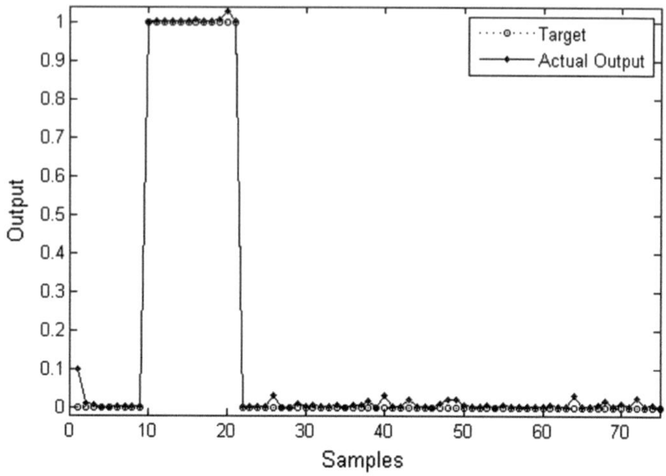

Fig. 5 ANFIS output for Stator A phase winding inter turn fault with medium damaged condition

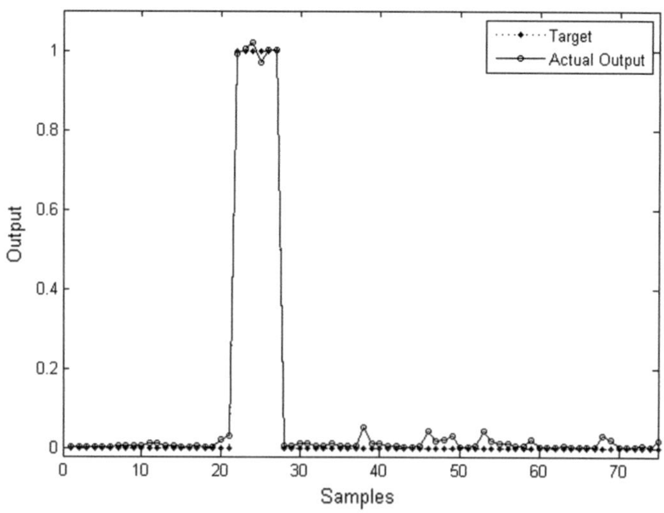

Fig. 6 ANFIS output for Stator A phase winding inter turn fault with severe damaged condition

little deviation between the target and the actual output. So the error which is the difference between the target value and the actual output is considerable.

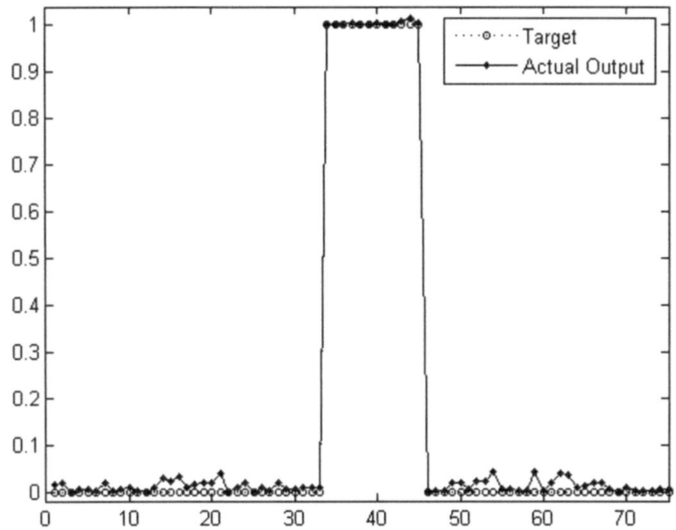

Fig. 7 ANFIS output for Stator B phase winding inter turn fault with medium damaged condition

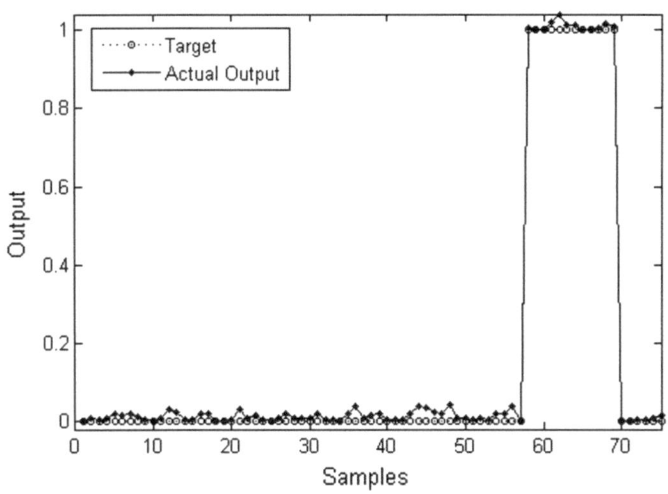

Fig. 8 ANFIS output for Stator C phase winding inter turn fault with medium damaged condition

4 Experimental Verification

To validate the proposed fault classification approach, a case-study on a 3-phase, 415 V, 0.5 HP, 50-Hz Squirrel cage induction motor was performed with no load condition. This motor was tested under healthy and Stator inter turn fault condition. 5–20% of inter turn fault data has been considered here for testing purpose. The

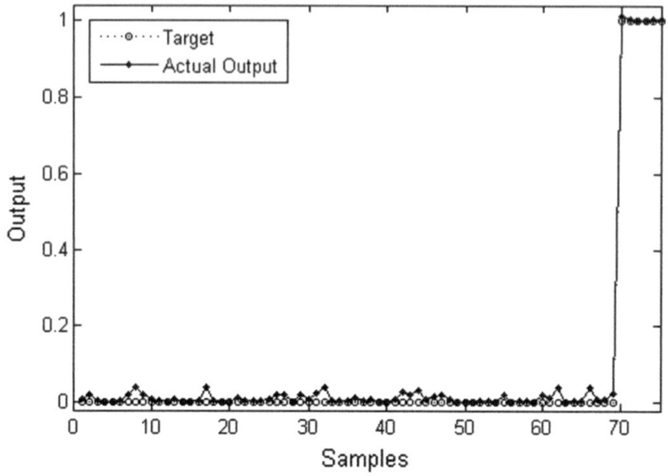

Fig. 9 ANFIS output for Stator C phase winding inter turn fault with severe damaged condition

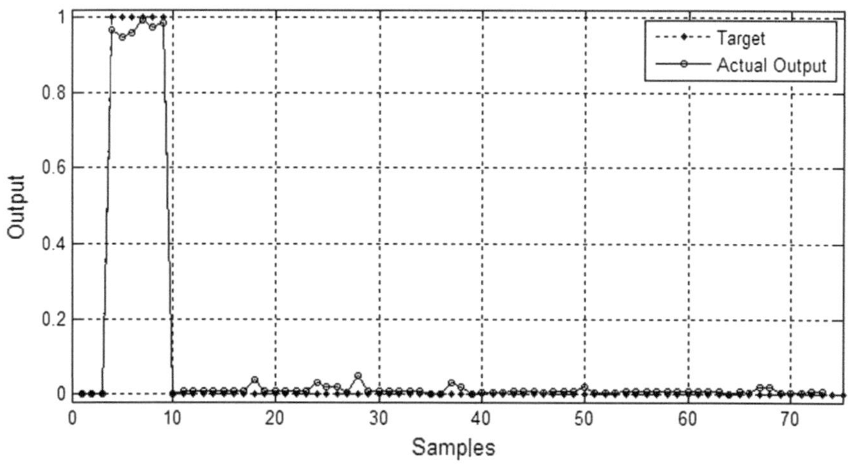

Fig. 10 ANFIS output for Stator A phase winding less damaged inter turn fault

three phase stator current samples were collected simultaneously using hall sensor and stored in PC. Then SST is applied to calculate positive and negative sequence components for every current cycle. From sequence components, SCAI and SCPI has been calculated which has been used as ANFIS input. In Figs. 10, 11 and 12, test result has been given to validate the accuracy of proposed fault detection system.

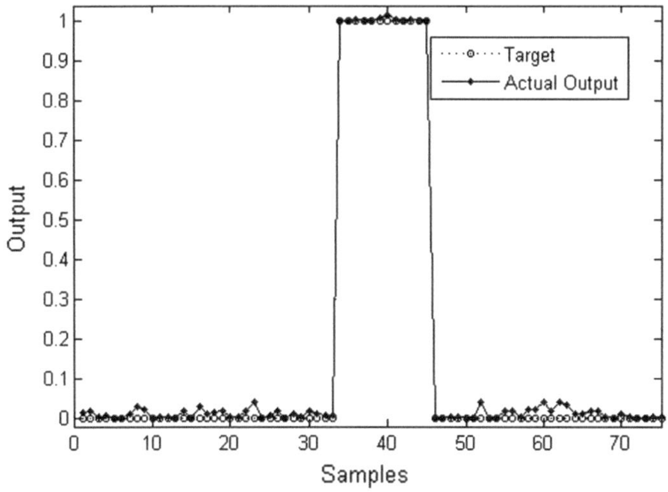

Fig. 11 ANFIS output for Stator B phase winding medium damaged inter turn fault

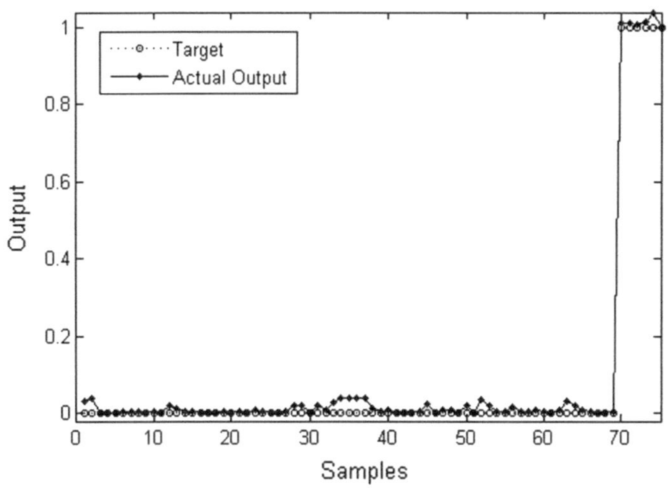

Fig. 12 ANFIS output for Stator C phase winding severe damaged inter turn fault

5 Conclusion

In this paper, ANFIS based fault diagnosis of three phase induction motor is presented. The main advantage of this work is only three phase current samples are required. Using current samples, sequence components are calculated by Sample Shifting Technique which is used for fault diagnosis. It is very simple method as sequence components are calculated by only repositioning current sample values, which can be implemented by using an ordinary microcontroller and protection cost can be reduced. Sequence current components are not so much affected by power supply quality as calculated using Sample Shifting Technique (SST). The main purpose of using the Neuro-Fuzzy approach is to automatically realize the fuzzy system by using the neural network methods. Once the system is trained for specific data over a wide range, it can be applicable to similar types of motor used in plants, and thus there is no need to train the model for each motor. So, proposed diagnosis method can be applied to any type of small and high power induction motors.

References

1. P.F. Albrecht, J.C. Appiarius, R.M. McCoy, E.L. Owen, D.K. Sharma, Assessment of the reliability of motors in utility applications—updated. IEEE Trans. Energy Convers. EC-1(1), 39–46 (1986)
2. A.H. Bonnett, G.C. Soukup, Cause and analysis of stator and rotor failures in three-phase squirrel-cage induction motors. IEEE Trans. Ind. Appl. **28**(4), 921–937 (1992)
3. A. Siddique, G.S. Yadava, A review of stator fault monitoring techniques of induction motors. IEEE Trans. Energy Convers. **20**(1), 106–114 (2005)
4. S. Rajakarunakaran, P. Venkumar, K. Devaraj, K.S.P. Rao, Artificial neural network approach for fault detection in rotary system. Appl. Soft Comput. **8**(1), 740–748 (2008)
5. P.V.J. Rodriguez, A. Arkkio, Detection of stator winding fault in induction motor using fuzzy logic. Appl. Soft Comput. **8**(2), 1112–1120 (2008)
6. V. Uraikul, C.W. Chan, P. Tontiwachwuthikul, Artificial intelligence for monitoring and supervisory control of process systems. Eng. Appl. Artif. Intell. **20**(2), 115–131 (2007)
7. M.H. Benbouzid, A simple fuzzy logic approach for induction motor stator condition monitoring, in *Electric Machines and Drives Conference, IEMDC 2001*, (IEEE International, 2001), pp. 634–639
8. X.Z. Gao, S.J. Ovaska, Soft computing methods in motor fault diagnosis. Elsevier, Appl. Soft Compu. **1**, 73–81 (2001)
9. F. Filippetti, G. Franceschini, C. Tassoni, Recent developments of induction motor drives fault diagnosis using AI techniques. IEEE Trans. Ind. Electron. **47**(5), 994–1003 (2000)
10. P.V.J. Rodriguez, A. Arkkio, Detection of stator winding fault in induction motor using fuzzy logic. Appl. Soft Comput. **8**(2), 1112–1120 (2008)
11. M.B.K. Bouzid, G. Champenois, N.M. Bellaaj, L. Signac, K. Jelassi, An effective neural approach for the automatic location of stator inter-turn faults in induction motor. IEEE Trans. Ind. Electron. **55**(12), 4277–4289 (2008)
12. W.W. Tan, H. Huo, A generic neuro-fuzzy model-based approach for detecting faults in induction motors. IEEE Trans. Ind. Electron. **52**(5), 1420–1427 (2005)
13. M.S. Ballal, Z.J. Khan, H.M. Suryawanshi, R.L. Sonolikar, Adaptive neural fuzzy inference system for the detection of inter-turn insulation and bearing wear faults in induction motor. IEEE Trans. Ind. Electron. **54**(1), 250–258 (2007)
14. J. Faiz, R. Iravani, Improved over current protection using symmetrical components. IEEE Trans. Power Deliv. **22**(2), 843–850 (2007)
15. R.E. Fehr, "A Novel Approach for Understanding Symmetrical Components and Sequence Networks of Three-Phase Power Systems", *IEEE Trans. on Power Delivery*, pp. 213– 217, 2007
16. C.F. Wagner, R.D. Evans, *Symmetrical Components* (Mc-Graw-Hill Book Company, New York, 1933)
17. C.L. Wadhwa, *Electrical Power System* (New Age International Publishers) ISBN no. 81-224-1722-1

18. R. Saha, G. Sarkar, J.N. Bera, S. Chowdhuri, A. Deb, Sample shifting technique (SST) for estimation of harmonic power in polluted environment, in *IEEE Conference Publications on Circuits, Power and Computing Technologies (ICCPCT)*, pp. 535–539 (2013)
19. M.S. Ballal, Z.J. Khan, H.M. Suryawanshi, R.L. Sonolikar, Adaptive neural fuzzy inference system for the detection of inter-turn insulation and bearing wear faults in induction motor. IEEE Trans. Ind. Electron. **54**(1), 250–258 (2007)
20. S. Altug, M.Y. Chow, H.J. Trussell, Fuzzy inference system implemented on neural architectures for motor fault detection and diagnosis. IEEE Trans. Ind. Electron. **46**(6), 1069–1079 (1999)
21. M.E. El-Hawary, *Principles of Electric Machines with Power Electronic Applications* (Prentice-Hall, Englewood Cliffs, NJ) 07632

Level Adjustment of Hydrofoil Sea-Craft Under Wave Disturbance

Sohorab Hossain, Sourish Sanyal and Amarnath Sanyal

1 Introduction

Hydrofoil sea-crafts must be capable of moving through waves of large amplitude, as such of large energy content. A hydrofoil is simply a lifting surface, or foil, that operates in water. These are similar to aerofoils used in aeroplanes. As a hydrofoil craft gains speed, the hydrofoils lift the boat's hull out of the water. It decreases drag and allows greater speeds. Hydrofoil boats incorporating the use of hydrofoil to help them propel much faster on the water was an attraction for the shipping industry for quite some time. They were created by A.G. Bell and Casey Baldwin in 1908 and were used extensively during the First World War by American troops to avoid the waters trapped by mines. The interest in hydrofoil ships flagged away after a few years as the design of other types of ships were found to be better. However, in modern times, the hydrofoil ships are again gaining momentum on account of their faster and speedier movement through the waters [1–4].

S. Hossain (✉)
EE Department, MCKV Institute of Engineering, Howrah, India
e-mail: sohorab.hossain@gmail.com

S. Sanyal
ECE Department, Cooch Behar Government Engineering College, Cooch Behar, India
e-mail: maysourish2013@gmail.com

A. Sanyal
EE Department, Calcutta Institute of Engineering and Management, Kolkata, India
e-mail: ansanyal@yahoo.co.in

S. Chattopadhyay et al. (eds.), *Modelling and Simulation in Science, Technology and Engineering Mathematics*, Advances in Intelligent Systems and Computing 749,
https://doi.org/10.1007/978-3-319-74808-5_36

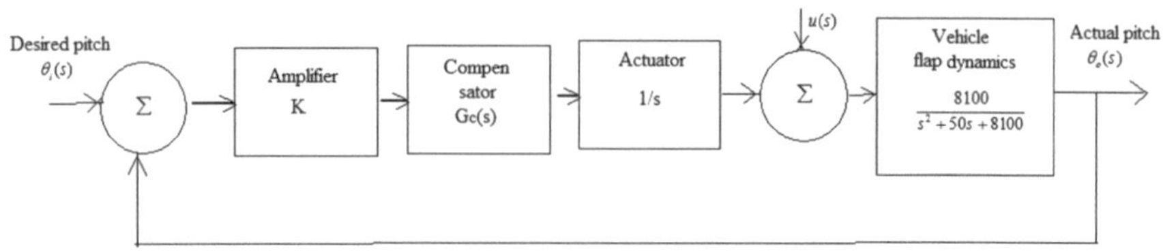

Fig. 1 Hydrofoil sea-craft control: block diagram

2 Operation of Hydrofoil Ships

A hydrofoil ship operates in a simple way: the hydrofoil based on the base allows the ship to move easily through the waters and ensures that the body of the ship (known as hull) does not come in contact with the water. As the hydrofoil works only if the ship is on the surface of the water, it prevents the ship from rising out of water. In case, the ship arises out of water, its design forces it to come back into the water and get submerged till adequate thrust is generated to lift the ship. These ships are mostly used for multiple purposes e.g. fishing in the sea etc. and for warfare. However, as the density of water is much greater than that of air, the hydrofoils are much smaller in size compared to the aerofoils and are compact [3, 4].

As of now, the use of hydrofoil ships is limited. But in foreseeable future, the hydrofoil ships may revolutionize the market for their speed advantage. The only requirement is greater efficiency and stability. The design methodology given in this paper suggests the path to make them stable with adequate margin.

H.S. Denison was the first large hydrofoil sea-craft built by United States [5]. The craft was capable of moving at a speed of 60 knots in wavy seas. The block diagram of the control system is given in Fig. 1.

3 The Control System

The control system consists of transducers to sense the craft motion and a computer to transmit commands to electro-hydraulic actuators [6]. The heave rate is fed symmetrically to the forward flaps—also the roll and the roll rate. The pitch rate is fed to the stern foil. The stabilizing circuit maintains a level flight by means of two main surface-piecing foils located ahead of the center of gravity and an all-movable submerged foil aft.

The craft is to maintain a constant level of travel in spite of wave disturbance U(s), whose energy is concentrated around 1 r/s. The pitch loop has to maintain a gain of 40 db at 1 r/s, in order to minimize the wave disturbance and a gain crossover of 160 r/s for adequate response time. The phase margin must be at least 60° and the gain margin 12 db. We are to find out the amplifier gain K and design the compensation network $G_c(s)$ and weigh the rate feedback required to fulfill the specs [7–9].

4 The Control System Design

The understanding of a control system is rather simpler than its design itself. Certain specifications, equality or inequality constraints are imposed before the designer. He has to conform to them. Very often some of the design constraints are contradictory e.g. the requirement of steady state accuracy and the stability margins—a reasonable tradeoff has to be made between the contradictory elements. Both time domain and frequency domain specifications have to be met by the design. For this reason, compensators have to be interposed in the system. The commonly used compensators are [10]:

(i) Proportional-Derivative (PD) control
(ii) Proportional-Integral (PI) control
(iii) Proportional-Derivative-Integral (PID) control
(iv) Phase-lag compensator
(v) Phase-lead compensator
(vi) Phase-lag-lead compensator etc.

or a combination of some of these elements [8]. The choice has to be judiciously made, either by mathematical or graphical procedure or by trial and error using the control system tool-box of MATLAB [9, 10].

The present problem has its own constraints and it has been solved by a combination of PD-control and phase lag using the later procedure [11].

5 Specification

The following specifications are to be fulfilled:

a. System gain is to be about 30 db at 1 r/s to combat the sea-waves. It fixes up the forward path gain: K = 31.4
b. Phase crossover is to occur at about 160 r/s and the gain margin to be more than 12 db
c. The gain crossover is to be at about 3 r/s and the phase margin is to be adequate (more than 60°)
d. The % overshoot is to be less than 5%, and the settling time less than 1 s.

6 The Mathematical Description

The open loop T.F. for the uncompensated system is given as (obtained from the manufacturer's data):

$$G_u(s) = \frac{K8100}{s\left(s^2 + 50s + 8100\right)} \tag{1}$$

For the uncompensated system, the system gain to be 30 db at 1 r/s, open loop gain must be 31.4.

The analysis is being made by using MATLAB tool [12, 13]. The t-domain response (Fig. 2) shows a peak overshoot of 12.2% which is not admissible (it has to be less than 5%).

The f-domain characteristic (Bode plot, Fig. 3) shows a negative gain margin and declares the system as closed loop unstable. The stability has to be ensured. So some sort of compensation has to be made.

To stabilize the system and to fulfill the specifications a compensator is to be cascaded, along with addition of an appropriate rate feedback. Many authors in recent times have made recourse to different compensation techniques to fulfill the requirements [13] and have become successful in implementing appropriate control.

We insert a lag compensator in the forward path and add a rate feedback. The compensator is made of static elements as shown in Fig. 4.

The transfer function of the lag compensator has been taken as:

$$G_c(s) = \frac{1 + 4s}{1 + s} \tag{2}$$

by trial and error method and that of the feedback path as:

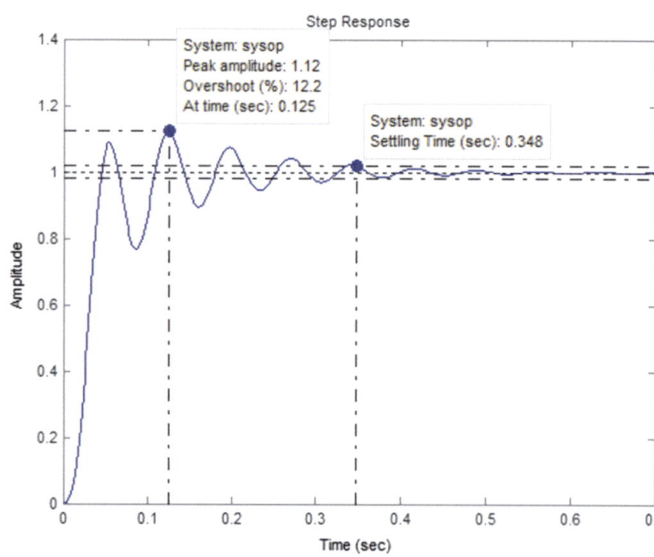

Fig. 2 t-domain response to step input

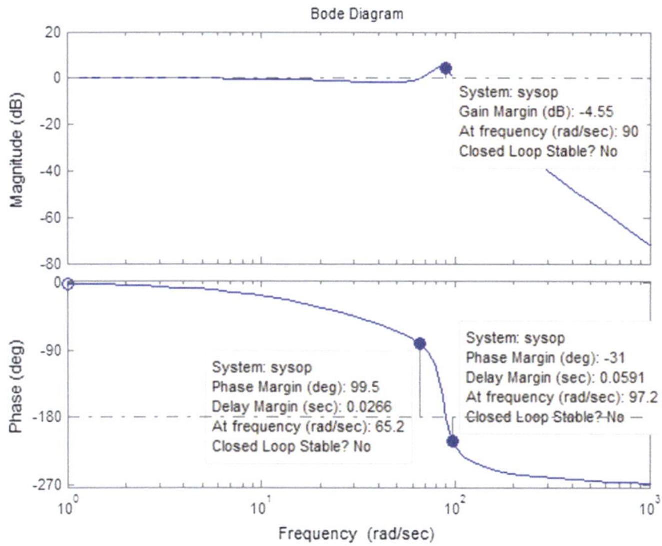

Fig. 3 Bode plot of the uncompensated system

Fig. 4 A lag compensator made of static elements

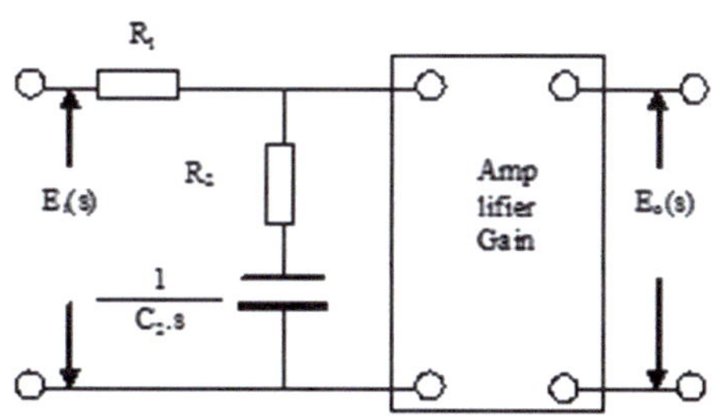

$$(1 + 0.06s) \tag{3}$$

The t-domain performance of the compensated system with PD feedback is depicted in Fig. 5.

The f-domain characteristic (Bode plot) for the compensated system is shown in Fig. 6.

Now with the lag compensator and the PD system, for the hydrofoil ship Denisson, we get the following:

System gain is 30 db at 1 r/s to combat sea-waves, which fixes up the value of open loop gain K = 31.4.
Gain margin = 12.2 db at 164 r/s; Phase margin = 166° at 2.75 r/s.;
The system is slightly underdamped—the peak overshoot is 2.11%.
The settling time is 0.553 s and rise time is 0.192 s.
So the system specifications have been fulfilled by this design.

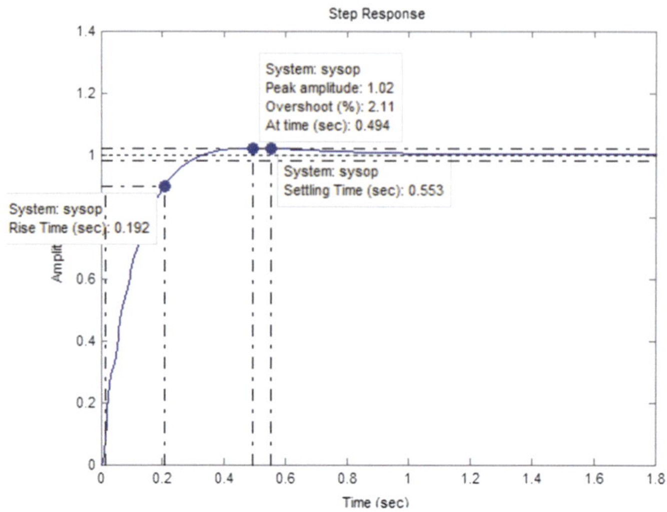

Fig. 5 t-domain response of the compensated system

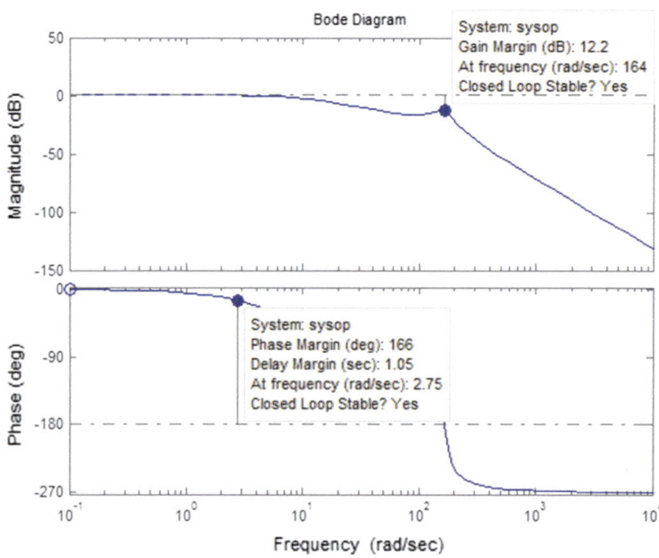

Fig. 6 Bode plot of the compensated system

7 Conclusion

The design of a control system has to be made fulfilling the requirements or the given specifications. The fulfillment has to be made optimally [14]. Often, the requirements are contradictory and a trade-off has to be made between contradictory elements. To conform to the specifications we may have to insert a compensator of a particular type or a combination of two or more of them [7, 9, 10]. Sometimes, computational intelligence is made use of to reach the optimal conditions [15]. In this paper the control system design of a large hydrofoil sea-faring aircraft "Denisson" [5] has been made. The parameters of the uncompensated system have been given—also the specifications. It is required to have a gain of 30 db at 1 r/s to combat the sea-waves. This requirement fixes up the forward path gain. With this gain, the system is found to be highly oscillatory and closed loop unstable. By using MATLAB-tools, a combination of PD-control and phase lag compensation have been found out by a trial and error procedure, which has enabled the design to match with the given specifications. Previously, the design procedure was suggested by Goswami et al. in one paper [16].

References

1. T.J. Cutler, *The Bluejacket's Manual*, 22nd edn. (Naval Institute Press, Annapolis, MD, 1999) ISBN: 1-55750-065-7
2. T.J. Cutler, *Dutton's Nautical Navigation*, 15th edn. (Naval Institute Press, Annapolis, MD, 2003) ISBN: 978-1-55750-248-3
3. G. William, *Stability and Trim for the Ship's Officer* (Cornell Maritime Press, Centreville, MD, 2005) ISBN: 978-0-87033-564-8
4. E.S. Maloney, *Chapman, "Piloting and Seamanship"*, 64th edn. (Hearst Communications Inc., New York, NY, 2003) ISBN [Spl: Book Sources/1-58816-098-0|1-58816-098-0]
5. S.M. Shinners, *Modern Control System Theory and Design* (Wiley, 1992) ISBN: 978-0-471-24906-1
6. D. Patranabis, *Sensors and Transducers*, 2nd edn. (PHI, New Delhi, 2008)
7. K. Ogata, *Modern Control Engineering* (Pearson Education) ISBN 10: 0136156738; ISBN 13: 9780136156734
8. N.S. Nise, *Control System Engineering*, 5th edn. (Wiley Student) ISBN: 978-81-265-2153-1
9. A.M. Law, W.D. Kelton, *Simulation, Modeling and Analysis*, 2nd edn. (Mcgraw-Hill, New York, 1991)
10. C.J. Herget, A.J. Laub, *Special issue on computer-aided control system design*. IEEE Control Syst. Mag. **2**(4), Dec'82
11. *Lecture Notes in Control and Information Science*, vol. 162. (Springer, 1991), pp. 106–124
12. J.J. D'Azzo, C.H. Houpis, S.N. Sheldon, *Linear Control System Analysis and Design with MATLAB*, 5th edn. (Marcel Dekker Inc. New York, BASEL)
13. A.J. Grace, N. Laub, J.N. Little, C. Thomson, *Control System Tool Box for use with MATLAB, User Guide* (Mathworks, 1990)
14. K. Deb, *Optimization for Engineering Design* (PHI, 2010)
15. D. Poole, A. Mackworth, R. Goebel, *Computational Intelligence* (Oxford University Press, 1998)
16. R. Goswami, S. Sanyal, A.N. Sanyal, Design of compensator for a hydrofoil ship. Int. J. Emerg. Technol. Adv. Eng. **2**(9), 266–271 (2012). available online

Part VI
Computation Technique

Law of Time and Mathematical Axioms

Hiran Das Mahar

1 Introduction

Mathematicians have views that in Philosophy of Time, observer has significance role to measure and use the time [1–3]. Social Sciences apply the time commercially, in regard with time management for Industrial production [4]. In Bioelectronics Dept. Manchester University, UK. make prospectus for ultra measurement of life energy in micro level i.e. bioelectric current in nano-ampere unit [5]. Deo Sanskrit University of Gayatree Peeth Haridwar applying time to increase bioelectric positive power, like physicians for exercises for meataphysics [6]. Here also bioelectric power is generated in Enthalpy (heat), therefore sweat appears. ATP, DNA and NADH are three basic molecules in life Science. All these are consist of five basic elements C, H, O, N and P of Environment, respectively representing to Earth, water, sky, air and fire on the basis of percent presence and nature (e.g. evaporation point). Time Ψ is a measure, in which events can be ordered from the past through the present into the future, and also the measure of durations of events and the intervals between them. Time is often referred to as the fourth dimension. These properties of time are also applicable for Life time.

Peter Duren et. al. [7] studied mathematical analogy between space and time to be both have nature of ever dynamic functional state. i.e., space, S = L2 {R+, (1/t)dt}. Functional space is Bergman space, applied for study. In nature there is no non-functional space, since in even vacuum, weightless particles like Cosmon, Higgs boson etc may present in functional state and cent percent vacuum is impossible [8]. Qiuhe Huang [9] introduced the logarithmic Bloch space Bqlog (Dn), a new space of analytic functions on the unit poly disc. He investigated the composition operator · C from weighted Bergman space to logarithmic Bloch space on the unit poly disc, and provides the sufficient and necessary conditions to ensure the bounding

H. D. Mahar (✉)
R.G. Govt. P.G. Autonomous College, Ambikapur, Surguja, Chhattisgarh 497001, India
e-mail: drhdmahar@gmail.com

S. Chattopadhyay et al. (eds.), *Modelling and Simulation in Science, Technology and Engineering Mathematics*, Advances in Intelligent Systems and Computing 749,
https://doi.org/10.1007/978-3-319-74808-5_37

and compactness of the composition operator · C from weighted Bergman space to logarithmic Bloch space. A.I. Rouban [10, 11] studied "Delayed time" in a sensitivity functional (SF) and formulated sensitivity coefficients (SC) of a mathematical model. He viewed that unused time is dead time in the calculation of SC and SF. Ming-Hua Wu [12] studied a pair of independent time series, in the phenomenon of spurious regression and with shifts in variance. He numerically discussed the effects of variance changes on the size of t-ratio test in model. There is a spurious regression between sequences driven by variance shifts in his regressions model and the asymptotic distribution of t-ratio test is not invariant to shifts in variance. Chen et. al. [13] studied nonparametric estimation of structural change points in volatility models for time series in econometrics. Ramans [14] formulated an additive energy of dense set of primes and monochromatic sums and he stated "When $K \geq 1$ is an integer and S is a set of prime numbers in the interval (N2N2, N] with $|S| \geq \pi^*(N)/K$, where $\pi^*(N)$ is the number of primes in this interval, we obtain an upper bound for the additive energy of S." Bas C. van Fraassen [15] viewed that the words "time" and "space" are both singular terms and John D. Barrow (2009) studied mathematics of dark energy in space [16].

2 Law of Movement

- Time is movement of Universe. The speed of time is in a proportion of the speed of universe. Time of any planet or any part of planet is depended upon the moving position of the planet or place. e.g. Earth's time is proportion to earth's movement and time of any country is depended on in position on earth. (a) Circular (b) Linear (C) Serpentine spin.

Mathematical Axiom 10. A set-theoretic point (*member* of a set).

In Vector algebra, an 'element' of a set is a representation of a coalition in intensive magnitude. A set-theoretic point (*member* of a set) is something else entirely. It is *a complete and singular composition* employed *as a unit in the process of aggregation*. It should not be called an element.

A *member* 'belongs to' a set; an *element* does not denote a rule of transformation from a 'domain set' (*A*) to another set. For example, the notation $f: A \to B$ or $f: A \to A$.

In other words, this notation is shorthand for saying that *f* defines some subset of a Cartesian product and does not *directly* denote, for instance, taking.

A immediately back into itself (as the notation $f: A \to A$ merely seems to imply).

We must not confuse the abbreviated notation for the logical train of concepts this notation is meant to convey.

3 Explanation:

3.1 Law of Opposition

Dimension of Time Ψ and Life Time Ψε In all dimensions of time is opposite to universe e.g.

- Linear (arrow of time): Universe (e.g. earth) came from past (through origin) 'is' in present and moving to future. While Time come from future is in present and went to past to be history.
- Cycle: Earth moves anti clock wise in rotational movement.
- Spin movement, Since earth and energy moves in spin, Therefore it is assumed spinning Time.

Axiom 9 (Axiom of Regularity)
Non-empty set *s* implies *x* exists such that *x* belongs to *s* and every *y* belonging to *x* implies *y* does not belong to *s*.

This axiom is meant to prohibit any set *s* from containing itself as a member. It was introduced by von Neumann in order to deal with certain highly technical issues involving unwanted circumstances that were shown to be consistent with the other axioms. One such issue was the possibility, under the system without axiom 9, of the existence of sets that did not contain some of the 'primary constituents' (such as the null set) of Zermelo's set theory. development of Euclid's axioms and Aristotle's ten 'categories,' the development of the axioms of set theory was historically an empirical and somewhat trial-and-error process. Commenting on 'system Z' (the modified Zermelo system, which is more or less like the one just described here), Fraenkel wrote.

The rather arbitrary character of the processes which are chosen in the axioms of Z as the basis of the theory is justified by the historical development of set-theory than by logical arguments. The far-reaching aim of proving the consistency of Z, which would exclude contradictions of types as yet unknown, is not likely to be attained in the present stage, and in a well-defined sense cannot be attained at all, in accordance with Gödel's incompletability theorem.

3.2 Law of Energy

Time is energy to remain irreversible movement in Continuum is mathematical Triangle (pyramid) of Energy, Time, Universe (ibe, Tinia and Uni or Vishnu, Mahesh, Brahma).

Axiom 8 (Axiom of Choice)
If '*a*' implies non-empty set '*x*' is a function defined for every *a belonging* to '*s*' then there is another function '$f(a)$' for *a* belonging to '*s*' and '$f(a)$' belongs to '*x*'.

What this axiom tries to say is we can select one member from every non-empty subset of ‘s’ and use our selections to define ‘a set x’. The axiom of choice differs from axiom 6 in that the ‘choice function’ f, is not given any specified property that defines the ‘choices’. The axiom lets us do an arbitrary ‘choosing’ even though we have no property that would define the choice function between ‘a’, ‘x’ and ‘s’.

3.3 Law of Universal

Time is continuous in continuum behaves equal for all in continuum. E.g. 7 a.m. for one individual at any place, just same for all men bethought any discriminations *in any space* Universe.

There is a “Pot theory” that universe is a oval pot. In which time energy and many continuum (spaces and galaxies) are present.

Kalashasya mukhe Wisnu, Kanthe Rudra samasita Kukhao tu sarwe sapta Deepa Wasundhara Moole tatra sthito Brahma!!! (Rom rom prati lage, koti koti brahmand.)

Above four line are for four tense depicted from Veda and Not agree with this, or rejection, or absence of pot (=Kalash) is A-tense. The wall pot is Vast Universe. And elliptical as pitcher = kalash. In bracted hindi Epic says many galaxies in each part of Universes body.

Axiom 7 (Axiom of the Power Set):
Things which are halves of the same things, are equal to one another. For any set x the set y consisting of all the subsets of x exists. The set of all subsets of x is called the ‘power set’ of x. Although y is defined by a property, namely that its members are subsets of x, this case is not covered by axiom 6 because the property is not defined as the rang of a function. For example, axiom 6b specified that the function had to be unary (a function of one member of x), whereas in axiom 7 the property ‘is a subset’ is not a unary function. Axiom 7 is another construction axiom since y will have more members in it than did x if the number of members in x is finite.

3.4 Law of 05 Tense

A great tense is a space time. It controls over first three tenses,\ simply as video recording and fore casting. Astrology and palmistry fore cast future in present. Science says, time changes in space, even, in countries and galaxies.

A-tense is defined that before birth of the space or universe, since time is space related. According to Newton’s theorem of “space time relativity”. If there was, no space, no time.

I.e. before origin of Higgs Boson, there was no time. Same, after degeneration of Universe, no space, no tense i.e. A-tense.

Metaphysically and socially may be, simplified, if officer is happy, signs in back dates. If unhappy, says, sorry! I have no time. i.e. 'A tense' But this run off from duty for a greater duty is whether for metaphysical exercises for priest or in surgical science or music. There fore a-tense is supreme time. One can criticize that this theorem of five tense is pure hypothetical. Great tense solves it. Past and future days are in this week. Present minute become past in a minute. So that, theory of three tense is depended on unit of time to be any tense Since death is definite with fixed time called "Programmed Cell Death (PCD)" Therefore, present society is future century's imp society. Since time is infinite. So that great tense is mathematically sum of past, present and future and absence of time is 'a-tense'.

Axiom 6 (Axiom of Replacement):
Things which are double of the same things, are equal to one number. This is not actually an axiom but rather a schema for an unlimited bundle of axioms. Even professional mathematicians find it difficult to re-state the formal expression in English. Roughly, the axiom says "any 'reasonable' property that can be stated in formal language can be used to define the set y of things having the stated property."

3.5 *Law of Infinite*

Time runs from infinite future to near future to present… to past to infinite past… (start of Universe, birth of Brahma).

$$\Psi_{fi} \rightarrow \Psi_{f} \rightarrow \Psi_{pr} \rightarrow \Psi_{pa} \rightarrow \Psi_{pI} \Rightarrow \Psi = \infty$$

If we assume fifth law of time, we have to add great time (e.g. dreamed time) and Atence (hypothesis of absence of time, i.e. *before and after the origin of Universe and Time*

$$\begin{aligned}
&\Psi_{\infty} = \Psi_{f} + \Psi_{pr} + \Psi_{pa} + \Psi_{g} + \Psi_{a} \\
&\therefore\ \text{Universe } U' = \infty \Leftrightarrow \text{ Time } \Psi = \infty \Leftrightarrow \text{ Energy } E = \infty \\
&\because \infty = \forall U', \psi, E\ E_{\infty} \gg \text{volume, linfetime } \Psi_{\varepsilon},\ \text{lifeenergy}
\end{aligned}$$

Universe = space + vacuum + volume + super dense black holes + inter galactic spaces + all galaxies + Milky way's + earth including Bermuda, + Jatinga, where ships dip and birds suicide).

Axiom 5 (Axiom of Infinity):
The whole is greater than the part. There is a set x that contains the empty set and that is such that if any y belongs to x then the union of y and $\{y\}$ also belongs to x. This one is a very strange axiom when expressed in its formal language. A

3.6 Law of Particle

Mathematical definition of line $L = \Sigma\infty p \rightarrow$ in a series. Line = Sigma infinite points in a series (Points == Particle).

Corpuscular Mathematical assumption in linear dimension of Time.

$\Psi\infty\ldots\ldots \rightarrow \Psi\ldots\ldots \rightarrow \Psi\infty = \Psi\prod$ (God Particle) (e.g. *Photon, thermion* Higgs Boson, Positron, Magnetron, Cosmo).

Smallest times are Femito second = ! $\times 10^{-15}$ S, and Zeta second = 1×10^{-25} S.

(While, greatest time = Era = 4 Apouches) geologists stated, Pleistocene, Eocene, Mesozoic Jurassic and Modern.

Axiom 4 (Axiom of the Sum Set or Union):
Things which are coincide with one another, are equal to one another. If x is a set of sets, the union of all its members is a set. This one might make us blink a bit. Suppose the set x is given by $\{A, B\}$. Suppose further that A is a set given as $\{a, b, c\}$ and B is a set given as $\{d, e, f\}$. The axiom says that $\{a, b, c, d, e, f\}$ is also a set. Axiom 4 is another construction axiom. Fraenkel used to describe the operation $y \cup \{y\}$ is the following. The empty set $\emptyset$ is a member of x.

Therefore $\emptyset \cup \{\emptyset\} = \{\emptyset, \{\emptyset\}\}$ is a member of x.

Therefore $\{\emptyset, \{\emptyset\}, \{\emptyset, \{\emptyset\}\}\}$ is a member of x.

3.7 Law of Uncertainty

The velocity of time is constant on any planet and similar for all living and nonliving at any place. but except for velocity, Time follows the law of uncertainty. E.g. present sun rise time may differ at other continent.

Past Time: listened, read, video, images, evidences, history, memory, tradition, rituals, etc. are records of past time. These all are based on faith and believes. Might be fluctuating with reality.

Present time: Since time is an action (move of planet around it's star) therefore present time differs place to place on one planet. Therefore there is another scales of time e.g. Greenwich, Indian Standard Time, etc. Time (action) at a place is also no certain e.g. power cut, earth quake, volcano, accidents, death, rain, gifts, prizes, etc. are uncertain.

Future time: Future time is a result of past plus present. Even though future time is uncertain sciences like Mesmerism, Astrology, Horoscopy, Metaphysics, Palmistry, etc. are developed to know the future. But yet future is uncertain.

Great time: Dream is supposed to be of Great time. It's uncertain and being truth, it is bethought evidence near to false.

A tense: Hypothesis of 'atence' (No time) is supposed to be at absence of universe, so time absents before Universe is made by big bang and after the end of Universe. So that atense (no-time) is totally hypothetical and uncertain.

Axiom 3 (Axiom of Unordered Pairs):
If equals are subtracted from equal, the remainders are equal. If x and y are sets then the set of pairs (x, y) or the set of pairs (y, x) exists. i.e. the set of all ordered pairs of members of A and members of B, $A \times B$.

3.8 Law of Life Time

$\Psi\varepsilon$ = Life time = $\Psi \rightarrow \mathrm{N} \rightarrow \mathrm{M} \rightarrow \Psi$

$\Psi\varepsilon = \Psi_{\eta} \rightarrow \Psi_{\mu} = \Psi_{\pi}$ (Here Ψ_{η} = Birth time, Ψ_{η} = Mortality time and Ψ_{π} = Life particle).

Axiom 2 (Axiom of the Null Set):
If equals are added to equal, the wholes are equal. For example, if C is the set of all cattle in Wyoming and D is the set of living dodos, we could not say $C \cup D = C^8$ if we were not permitted to use the idea of the empty set.

One theorem that comes out of the ZFS system is that the empty set (symbolize it as $\emptyset$) is *ex officio* a member of every other set. This is read as, "The union of C and D is C" or $C \cap D = \emptyset$.

3.9 Law of Quality

Morning evening midnight/\good/bad time Proportion
$\Psi\varepsilon \ldots \alpha$ Good time = +E praise/smiles = α1/Bad time = −e bad news/Hesitation
First axiom of Mathematics Differential equation, (i.e. not equal to)
Fall of the Axiom = $ABA + B < 180$.

Axiom 1 (Axiom of Extensionality):
Things which are equal to the same thing, are equal to one another. Two sets are equal if and only if they have the same members. Fraenkel remarked, The axiomatization of set-theory renounces a *definition* of the concept of set and of the relation between a set s and its elements. The latter, a dyadic relation (or predicate) is denoted by $\in$; $x \in s$ reads "x is contained in, is an element of, belongs to, the set s" … and its negation is $x \notin s$. $\in$ enters as an (undefined) *primitive* relation, *the membership relation* [BERN: 4-5].

Axiom 1 is written *formally* as

$\forall x, y\ (\forall z(z \in x \rightarrow z \in y) \rightarrow x = y)$

Fall of the Axiom = $ABA + B < 180$

4 Discussion

$$G_{\mu\nu} + \Lambda g_{\mu\nu} = \frac{8\pi G}{c^4} T_{\mu\nu}$$

Einstein in 1905

Above is the Formula of time by Einstein (1905) and followings are some models of time given below

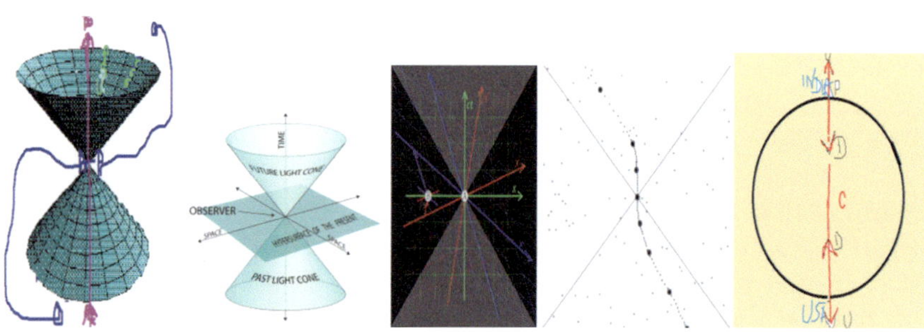

Model 1 Model 2 Model-3 Model=4 Model5

These models are described by Einstein Light cone diagram showing the world line of a moving observer in electrical experimentation [17, 18]. Y. Tian et al. [16] studied time-delay systems for a problem of static output feedback variable structure control to design of a static output feedback sliding mode surface and the variable structure controller. Although time, energy, and space are homologous and everywhere, they may be studied separately but are intermingles and unseparable called TES hypothesis [19, 20].

Acknowledgements I am gratefull to Pr. C. B. Vachon to organize MS 17 in India. I am thankfull to Pr. Surajit Chattopadhyay (Eds) for acceptance of this paper. I am also thankfull to Springer Nature for publication in this prestigious book.

References

1. C. Callender, A. Thresher, Time and the observer. University of California, San Diego, La Jolla, United States, May 3, 2018–May 4, 2018. https://philevents.org/event/show/42046/, http://ccallender7.wixsite.com/mysite/time-the-observer.
2. B. C. van Fraassen, An introduction to the philosophy of time and space. Published by Random House in 1970, Reprinted by Columbia University Press in 1985. For information contact Nousoul Digital Publishers: nousoul.dp@gmail.com. Retrived on 05.08.2018
3. V. Gijsbers, Wageningen, The philosophy of time (2017). https://www.wur.nl/upload_mm/2/c/8/b7268bad-67d9-4e4f-a984-24e798b3d773. Retrived on 05.08.2018
4. https://www.getmeashop.com/blog/importance-of-content-in-e-commerce/
5. J. Seddon, Bioelectronics study with us nano technology. Comments Apr 20, 2018. https://sites.manchester.ac.uk/bioelectronics/?s=bioelectronics
6. S. Sharma, Shahastrar aor Mooladhar: Gayatree Sadhana and Kaya me samaya pranagni ka zkheera Yug Nirman Yojana, Gayatree Tapo bhumi Mathura UP India (whole books for metaphysical excersize) 1988
7. P. Duren, E.A. Gallardo-Gutierrez, A. Montes-Rodroguez, A pale—Weiner theorem for Bergman space with application to invariant subspace. B. Lond. Math. Soc. **39**(3), 459–466 (2007). https://doi.org/10.1112/blms/bdm026
8. https://en.wikipedia.org/Istherevacuuminuniverse/. Retrived 03.09.2018.
9. Q. Huang, Composition operator from weighted Bergman space to logarithmic Bloch space on the unit polydisc. AMSE Ser. Gen. Math. Advanc. A **54**(2), 310–321 (2017). http://amsemodelling.com/publications/general_mathematics.html
10. A.I. Rouban, Sensitivity coefficients for many dimensional continuous and discontinuous dynamic systems with delay time. AMSE Ser. Gen. Math. Advanc. A **36**(2), 17–36 (1999)
11. A.I. Rouban, The construction of the sensitivity functionals in the Bolts's problem for multivariate dynamic systems described by integro–differential equations with delay time. AMSE Ser. Gen. Math. Advanc. A **51**(1), 80–99 (2014)
12. M.-H. Wu, Spurious regression of time series with shifts in variance. AMSE Ser. Gen. Math. Advanc. A **59**(1), 60—72 (2016)
13. G. Chen, Y.K. Choi, Y. Zhou, Nonparametric estimation of structural change points in volatility models for time series. J. Econom. **126**, 79–114 (2014)
14. D.S. Ramans, O. Ramans, Editive energy of dense set of primes and monochromatic sums. Israel J. Math. **199**(2), 955–974 (2014). https://link.springer.com/article/10.1007/s11856-013-0075-y
15. B.C. van Fraassen, An introduction to the philosophy of time and space. Published by Random House in 1970, reprinted by Columbia University Press in 1985, nousoul.dp@gmail.com BVF IPTS PDFAcrobat reader 142 015. Retrived on 07.09.2018
16. Y. Tian, H. Yao, F. Yuan, Static output feedback variable structure control for a class of time-delay systems. AMSE Ser. Gen. Math. Advanc. A **69**(1), 58–68 (2014)
17. Einstein's space time. https://www.einstein.stanford.edu/SPACETIME/spacetime2.html. Retrived on 07.09.2018
18. Scientists explain why time travel is possible. https://www.wired.co.uk/article/model-universeeinstein-lumpy-universe. Retrived on 07.09.2018
19. V.S. Netchitailo, 5D world-universe model space-time-energy. J. High Energy Phys. Gravit. Cosmol. **1**(1) (2015). https://doi.org/10.4236/jhepgc.2015.11003, https://www.scirp.org/journal/articles.aspx?searchCode=+Space-Time-Energy&searchField=keyword&page=1&SKID=0. Retrived on 07.09.2018

20. C.H. Tejman, Grand unified theory: Wave theory. Chapter 8 wave theory and time, space and energy (2001). http://www.grandunifiedtheory.org.il/book/time01.htm. Retrived on 07.09.2018

Analysis of Resources for the Safety and Comfort of Railway Passenger Using Analytical Hierarchy Process

Gopal Marik

1 Introduction

In India, the first railway line was laid over a stretch of 34 km from Bombay to Thane. The idea of railway transport first occurred to the then CHIEF Engineer of Bombay Government. His plans were investigated by a special Committee headed by the Chief Secretary and then approved at a meeting of the citizens of Bombay held at the Town Hall on 19th April 1845. An association was formed to execute the scheme. At the same time a company was incorporated in England by an act in 1849 under the style "The great Indian Peninsula Railways Company." The survey of this line was completed in 12 months and on 13th October 1850 the ceremony of the turning the first sod for a line connecting Bombay and Kalyan was performed at a place near Sion. A contract was awarded to an English firm for the construction of the line up to Thane. It is stated the firm employed about 100,000 workers on construction work. On 18.11.1852 the Directors of the company traveled in the train from Bombay to Thane, covering the distance in about 45 min. The formal inauguration of the railway line was later performed on 16.4.1853. when 14 railway carriages carrying about 400 guests left Bori Bonder at 3.30 pm and reached Thane at about 4.45 pm. The line was later on extended to Kallyan on 1.5.1854. The extension according to the engineering standards then existing was a difficult and a major achievement.

After the opening of the Railway up to Kalyan in 1854, further extension of the line over the Thalghats to Igatpuri was made on 30.12.1864, there by completing one of the most difficult projects in railway history and laying ground for further rapid expansion of the railway network.

In India, the origin of railways is associated with the needs of the rulers then to exploit rich agricultural raw materials like cotton and jute to the ports for transportation by sea to the factories in England. Lord Dalhouise became Governor General

G. Marik (✉)
Liluah, Howrah Department, Carriage and Wagon Workshop, Eastern Railway, Liluah, India
e-mail: gopal.marik@gmail.com

S. Chattopadhyay et al. (eds.), *Modelling and Simulation in Science, Technology and Engineering Mathematics*, Advances in Intelligent Systems and Computing 749,
https://doi.org/10.1007/978-3-319-74808-5_38

of India in 1847. Before taking up this assignment he was President of Board of Trade in England and was well conversant with the role of railways in development of industry and trade.

Indian Railways (IR) is an Indian state owned railway company headquartered in New Delhi, India. It is owned and operated by Government of India through Ministry of Railways.

In the year of India's independence, there were forty-two rail systems. In 1951 the systems were nationalized as one unit, the Indian Railways, becoming one of the largest networks in the world. IR operates both long distance and suburban rail systems on a multiple-gauge network of broad, meter and narrow gauges. It also owns locomotive and coach production facilities.

Indian Railways is a department owned and controlled by the Government of India, via the Ministry of Railways is headed by, the Union Minister for Railways, and assisted by two ministers of States for Railways. Indian Railways is administered by the Railway Board, which has a chairman, five members and a financial commissioner.

Indian Railways is divided into zones, which are further sub-divided into divisions. The number of zones increased from six to eight in 1951, nine in 1952 to sixteen in 2003 then to seventeen in 2010. Each zonal railway is made up of a certain number of divisions, each having a divisional headquarters. There are a total of sixty-eight divisions.

Each of the seventeen zones is headed by a General Manager who reports directly to the Railway Board. The zones are further divided into divisions under the control pf Divisional Railway Manager. The divisional officers of engineering, mechanical, electrical, signal & telecommunication, accounts, personnel, operating, commercial and safety branches report to the respective Divisional Manager and are in charge of operation and maintenance of assets. Further down the hierarchy tree are the Stations Masters who control individual stations and the train movement through the track territory under their stations' administration

Indian Railways is the largest railway network in Asia, and second largest in the world. The rail network consists of more than 64,000 route kilometers serviced by over 7000 block stations, 9000 locomotives, 43,500 passenger coaches, 7500 Electrical and Diesel Multiple Units and 22,000 railway wagons. With staff strength of 13.62 lakhs employees, the Indian Railways handle 8221 million passengers per year [1, 2].

Rail transport indirectly best means of transportation system for the following;

(a) The safest
(b) Least liable to cause of pollution
(c) Best suitable for mass transport and
(d) Most energy efficient mode on land.

Indian Railways has increased by about 1700% passenger kilometers but route kilometers have grown by just 23%. Railway is introducing latest technology and moving from preventive to predictive maintenance. The augmentation of physical infrastructure in terms of more railways lines, road over bridges, better signaling, safety related works etc. is impacting the punctuality of trains a little. The augmentation of passenger amenities, new train products special trains, increase in carrying capacity etc. are some of the measures to be taken to provide comfort to passengers [3].

2 Function of Operating Department

Railway operation encompasses all the activities connected with the running of a railway. However, Operating Department in particular has its role in producing a service called "Transportation". In this category, Operating department harnesses the efforts of all the departments of the Railways and optimizes usage of operational assets, viz. track, signals, fixed installations and rolling stock. Broadly the function can be categorized as under:

(a) Planning of Trans port Service
(b) Running of Trains
(c) Safety and Comfort
(d) Railway Accident and Human Elements involved
(e) Economy and Efficiency.

2.1 Planning of Transport Services

This involves both long and short term planning. A lot of spade work has to be done to run trains on day-to-day basis. Passenger trains are planned to run on a detailed time table issued once in six months. Arrangement of rolling stock and locomotives are made to meet the expected demand. Railways have to plan to meet these fluctuations in demand. Planning would involve not only the individual trains but a lot of other activities necessitating creation of capacity to form and start trains at originating stations. Though entire organization of the railways gets involved in this planning process, the basic frame work is provided by the operating department to which the plans of their departments responsible for provision and maintenance of operational assets fit in. Coaching traffic includes passengers, parcel and luggage. Passengers and their luggage are moved by the passenger trains. In the Railway parlance all Mail, Express and Passengers Trains are included in this category. Passenger trains are classified by their average speed. A faster train has fewer stops than a slower one and usually caters to long-distance travel. The passengers trains are namely as Duronto Express, Rajdhani Express, Satabdi Express, Garib Rath, Jana Shatabdi, Superfast/Mail, Express, Passenger and Fast Passenger, Suburban Trains.

Duranto Express: These are the non-stop point to point rail services introduced for the first time in 2009. These trains connect the metros and major states capitals of India and are faster than Rajdhani Express. These trains are now of the Highest Priority to the Indian Railways.

Rajdhani Express: These are all air-conditioned trains linking major cities to New Delhi. The Rajdhanis have high priority are one of the fastest trains in India, traveling at about 130 kmph. There are few stops on a Rajdhani route.

Shatabdi Express:

The Shatabdi trains are AC intercity seater-type trains for travel during day.

Garib Rath:

Fully air conditioned trains, designed for those who cannot afford to travel in the expensive Shatabdi and Rajdhani Express. Garib Rath means "Chariot of the Poor". The minimum speed is 130 kmph.

Jan Shatabdi:

These are a more affordable variety of the Shatabdi Express, which has both AC and Non-AC classes.

Superfast Mail and Express:

These are most common kind of trains in India. They have more stops in their route.

Passenger and Fast Passenger:

These are slow trains that stop at most stations along their route and they are the cheapest trains. The trains generally have unreserved seating accommodation but some night trains have cheaper sleeper class compartment.

Suburban Trains:

These trains operate in urban areas, usually stop at all stations and have unreserved seating accommodation.

Launched new train products: Humsafar, a completely air conditioned luxury service at affordable price with improved aesthetics, amenities and Antyodaya, a completely unreserved long distance service with improved coaches.

2.2 *Running of Trains*

Running of trains would involve ordering of trains booking of crew, watching that the conditions are safe for it to run and arranging various requirements on the run. Efficiency of operation of trains depends on the quality of planning whether short term or long term one. Transportation staffs both at originating and terminating as well as at the road side stations are busy round the clock to ensure timely and safe running of trains. Passengers' trains run to predetermined schedule and fluctuations in the traffic also influence their running.

2.3 Safety and Comfort

Safe transportation of passengers is the key objective of any transportation system. Railways are recognized as the safest mode of mass transportation and safety has been recognized as key issue for the railways and one of its special attributes. All business strategies emanate from this theme and service to achieve Accident Free System. Safety is therefore, the key performance index which the top managements need to monitor and take preventive steps based on the trend of accidents which are the manifestations of some of the unsafe practices on the system. Therefore, safety is the paramount importance to the Indian Railways. Highest priority is accorded to safety and the rail mode in India continues to be the safest means of transportation for public. No compromise is tolerated in Safety of Rail users and all levels of management keep reviewing the Safety performance from to time. Safety and Reliability of assets are closely linked with each other. Deterioration in the safety performance is preceded by the increase in the number of failures. Overlooking these warning signals can be disastrous as each of these is an accident waiting to happen. Although no technology is fail-proof, an error rate, however small, being inherent in any man-machine system, reliability of the equipment is the most important factor in the efficiency and safety of a transport system. If accident has to be zero level or minimized, it is imperative that the equipment in use is always kept in fine fettle.

2.4 Railway Accident and Human Elements Involved

A Rail accident is termed as any occurrence which does or may affect the safety the Railways, its engine rolling stock, permanent way, works, passengers or servants which either does or may cause delays to trains or loss to the railways. Railway accidents can be divided in the following categories: (i) Death or injury to persons, (ii) Damage to the property, (iii) Detention of traffic. In the last 10 years on Indian Railways, 62% of the accidents have been caused due to failure of Railway staff, 22% have been caused due to failure of other than Railways staff, failure of equipment has contributed 8%, sabotage has contributed 3% and balance 5% have been contributed by miscellaneous reasons. From the break-up, it is more than evident that human error from Railway staffs the major factor in causing the accidents on Indian Railways. Higher incidence of Human failures surface as technical safeguards and backups do not always replace the human effort. Though an accident occurs only when both fail but it usually gets logged as "human error" with a tendency of glossing over technical failure. Under optimum field conditions and with the best of intensions, a human being is likely to commit a mistake from time to time. This is the reason why operating rules included many redundancies in safety procedures and operating practices involve number of checks and balances. More and more automation is resorted to prevent human errors.

2.5 *Transport Economy and Efficiency*

While maintenance departments are responsible for making the assets available to the operating department in proper fettle, it is the responsibility of the latter to make the most optimum utilization thereof. Operating department is, therefore, responsible for the productivity of the system. This is measured in terms of operating indices passenger kilometers. The Indian Railways are not only a transport agency but also have a social obligation to serve the national objectives by providing necessary infrastructure for healthy economic development and rapid industrialization. In other words, railways are not only a commercial enterprise but also a public utility service catering to the needs of the society. As a result, the Railways are obliged to operate on some social services, both passenger and freight, in the interests of community. The losses accruing from such uneconomic services can be called "social burden"—as distinct from commercial deficits. As many other countries, railways in India have also certain obligations, and a definite role to play in the economic development.

3 Modernization

In the area of modernization a quantum jump in technology was achieved when RCF, association with RDSO, came out with AC Double Decker Coaches having latest features enhancing safety, comfort, passenger amenities and aesthetics all in one go besides increasing the seating capacities by more than 50%. The new Double Decker trains have become eye-candy and all geographical regions are aspiring to have one such service in their area. In the year 2011, another quantum jump towards safety features available in LHB design coaches was recognized and the High Level Safety Review Committee under the Chairmanship of Dr. Anil Kakodkar strongly recommended introduction of LHB design in all mail express trains by introducing non-AC stainless steel coaches. On the technology front, introduction of jerk free couplers on stainless steel coaches, reversible cycle air-conditioning for energy efficiency, fire and smoke detection system, passenger information systems etc. for enhanced travel experience in trains. Indian Railways face a daunting challenge ahead over its envisioned resolve to take a giant leap forward in infrastructure development and modernization of its rolling stock.

(a) Anti Collision Device
(b) Failure Indication and Brake Application Device
(c) Wheel Slide Protection.

3.1 Anti Collision Device

Anti Collision Device is a fully integrated Electronic Control system designed to minimize collisions and increase safety on Railway system. It is a non signaling system and provides additional cover of safety in train operations to prevent dangerous train collisions caused due to human errors or limitations and equipment failure. Being the non-signaling and inter locking system it does not replace any existing signaling and interlocking system and does not alter any procedures of train operations in vogue. More than 2000 Anti Collision Devices have already been installed over 2700 Route Kms of track on Indian Railways system out of which about 1900 Route Kms on North East Frontier Railway and balance are on Konkan Railway. Moreover, Hyderabad Based Company, Hyderabad Batteries Limited (HBL) developed a new anti collision safety device which is path breaking technology in ensuring safe traveling and minimize accidents. The device is based on a combination of railway signaling data with radio communications, global position, radio frequency identification devices, software and logic. During the trials, the effectiveness was demonstrated for prevention of head- on collisions, over-speeding of trains and disregard of red signal. The new anti collision device had essential features of both automatic train protection and collision prevention in one solution.

3.2 Failure Indication and Brake Application Device

There are different types of coaches which are used in Indian Railways. Some of these coaches are used for carrying passengers over long distances. In order to have smooth journey these coaches are fitted with Air Spring Bellows which absorbs the Shocks and jerks which are created due to undulation of track and many other reasons. Since Air spring bellows are filled with compressed air some mechanism is required to detect and apply the brakes in the train in case of any such failure. Most of these designs involve a large number of parts and are comparatively bulkier in size. These reasons have made it a very expensive device. In order to make it more affordable there is need of doing further research on it so that it can be made economical without compromising on its functionality and can be used on all kind of passenger trains

3.3 Wheel Slide Protection

Wheel Slide Protection equipment is generally fitted to passenger trains to manage the behavior of the wheel sets in "Low adhesion" (reduced wheel/rail friction) conditions. It is used when braking, and may be considered analogous to anti-lock braking (ABS) for cars. The system can also be used to control (or provide an input to) the traction system to control wheel spin when applying power in low adhesion conditions. Low adhesion at the rail potentially causes damage to wheels and rails. Typically, low adhesion conditions are associated with environmental causes arising from seasonal leaf fail, or industrial pollution. Occasionally the cause can be another less obvious factor such as light oxidation of the railhead or even swarms of insects. When a train is braking, the low adhesion manifests as wheel slip where the wheel set is rotating at a lower velocity (speed) that the forward speed of the train. The most extreme example of this is where the wheel stops rotating altogether (wheel slide) whilst the train is still moving and can result in a "wheel flat" caused by the softer wheel being abraded away by the harder rail steel. In traction, low adhesion may cause a wheel set to accelerate more quickly than the train (wheel spin) to the point where it can damage the train propulsion system or result in damage to the wheel and rail (rail burn).

4 Tools of Analysis

After collection of the primary and secondary data from the different sources they are to be analyzed by Analytical Hierarchy Process.

4.1 Analytical Hierarchy Process

AHP is a multi-objective decision making technique which takes care both qualitative and quantitative analysis. It was developed by American Scientist Satty [Satty, 1980]. Formation of Judgment Matrix is the most important factor in AHP. The analytic hierarchy process is a structured technique for organizing complex decisions, based on mathematics and psychology. It was developed by Thomas L. Satty in the 1980 and has been extensively used, studied and refined since then. It provides a comprehensive and rational framework for structuring a decision problem, for representing and quantifying its elements, for relating those elements to over goals and for evaluating alternative solution [4, 5]. The applications of AHP to complex decision situations have numbered in the thousands, and have produced extensive results in problem involving planning, resource allocation, priority setting, and selection among alternatives.

4.2 Steps of Analytical Hierarchy Process

Step 1 The problem and scopes are identified before analyzing relations between factors. This step signifies the key aspect of the entire process because the final result is directly related to the correctness of the logical relations of the factors.

Step 2 The layer structure and A matrix are to be established. The factors involved in each layer are made definite because the matrix will be the sticking point of this step. The difference between the factors is compared based on the criterion mentioned below:

The significance of 1–9 scales

Scale	Meanings
1	Comparing the two factors which are of same importance
3	The later is more important than the former
5	The later is important than the former
7	The later is less important than the former
9	The later is extremely less important than the former
2, 4, 6, 8	The middle value between the above results
Reciprocal	If the degree of importance between the factors "i" and "j" is "aij, then comparing "j" and "i" is "aji" which is equal to 1/aij.

The Eigen Vector or the Weight Vector is computed. The weight of each factor based on the matrix A and the vector sum of each column are computed. A new weight matrix B, is then constructed. Then normalized principal Eigen Vector will be obtained by averaging across the rows. The normalized principal Eigen Vector is also called Priority Vector. Aside from the relative weight checking of the Consistency Ratio is to be obtained. For this Principal Eigen Value (λmax) is obtained from the summation of products between each elements of Eigen Vector and the sum of the columns of the reciprocal matrix [6]. Based on the structured and Semi structured interview, views of the experts and their opinion in terms of guidelines provided in the table.

$$CI = \frac{\lambda\text{max} - n}{n - 1},\ CR = \frac{CI}{RI}$$

where, CI is the deviation degree in matrix above, which is called consistency index, CR is the consistency ratio, RI is the random ratio of consistency that is changeable with the number of factors (as per table below). When $CR < 0.1$, the result is considered to be reasonably good.

5 Objective of the Study

1. To analysis the resource available for safe and comfort of passengers during journey.
2. Improvement of quality of the service.
3. Resource planning.

First level Goal: safety and comfort of passenger						
Second level Criteria:						
Skilled man power	Availability of material	Machine	Infrastructure	Adaptation of modern technology	Socio-economic factor	Working environment
Third level Alternatives:						
1. Qualification	1. Store from outside	1. Upgragation of existing machine	1. Communication/Inform. Tech	1. Enhance safety	1. Social equity	1. Basic amenities
2. Knowledge and experience	2. In-house production	2. Purchasing of new machine	2. Better signaling	2. Reduction in-service failure	2. Reduction of fare	2. Workplace safety
3. Training	3. Logistic	3. Preventive maintenance	3. More railway lines	3. Radically reduced cost	3. Reduction of travel time	3. Pollution control
4. Salary	4. Supply chain		4. Safety related work	4. Enhance customer service	4. Reaching to common people	4. Compensation
5. Performance analysis				5. Create barrier to entry	5. Provision of extra benefit	5. Grievance

Weight analysis R-E judgment matrix

E	R1	R2	R3	R4	R5	R6	R7	W	NW	λmax	CI	RI	CR
R1	1	2	4	4	3	6	5	2.36	0.32				
R2	1/2	1	3	3	3	4	3	1.55	0.21				
R3	1/4	1/3	1	1	1/2	3	1	1.2	0.16				
R4	1/4	1/3	1	1	3	3	3	0.92	0.12	7.47	0.078	1.35	0.057
R5	1/3	1/3	2	1/3	1	5	1	0.73	0.09				
R6	1/6	1/4	1/3	1/3	1/5	1	1	0.28	0.03				
R7	1/5	1/3	1	1/3	1	1	1	0.42	0.05				

The RI changeable with the number of factors

n	1	2	3	4	5	6	7
RI	0	0	0.58	0.89	1.12	1.26	1.36

4. Modern procedure for the recruitment and selection of the personnel.
5. Financial performance of the different.
6. To study the recommendation made towards the safety and comfort passenger services.

Weight analysis R1—C1 to C5 judgment matrix

R1	C1	C2	C3	C4	C5	W	NW	λmax	CI	RI	CR
C1	1	3	1	4	3	1.72	0.34				
C2	1/3	1	2	1	3	0.925	0.18				
C3	1	1/2	1	3	7	1.08	0.22	5.107	0.027	1.12	0.024
C4	1/4	1	1/3	1	5	1.09	0.22				
C5	1/3	1/3	1/7	1/5	1	0.27	0.05				

Weight analysis R2—C6 to C9 judgment matrix

R2	C6	C7	C8	C9	W	NW	λmax	CI	RI	CR
C6	1	2	1	5	1.778	0.368				
C7	1/2	1	1	5	1.257	0.26	4.06	0.021	0.9	0.022
C8	1	1	1	5	1.495	0.309				
C9	1/5	1/5	1/5	1	0.299	0.062				

Weight analysis R3—C10 to C12 judgment matrix

R3	C10	C11	C12	W	NW	λmax	CI	RI	CR
C10	1	1/3	5	0.84	0.28				
C11	3	1	7	1.92	0.64	3.09	0.045	0.58	0.077
C12	1/5	1/7	1	0.21	0.07				

Weight analysis R4—C13 to C16 judgment matrix

R4	C13	C14	C15	C16	W	NW	λmax	CI	RI	CR
C13	1	2	1	3	1.33	0.33				
C14	1/2	1	3	2	1.09	0.27	4.15	0.05	0.9	0.05
C15	1	1/3	1	5	1.02	0.25				
C16	1/3	1/2	1/5	1	0.38	0.09				

Weight analysis R5—C17 to C21 judgment matrix

R5	C17	C18	C19	C20	C21	W	NW	λmax	CI	RI	CR
C17	1	2	3	4	5	2.39	0.55				
C18	1/2	1	2	1	2	0.92	0.17				
C19	1/3	1/2	1	2	3	0.82	0.15	5.428	0.105	1.12	0.09
C20	1/4	1	1/2	1	3	0.69	0.13				
C21	1/5	1/2	1	1/3	1	0.42	0.08				

7. To make suggestions for resources and economic planning.

6 Plan of Work

1. Identification of resources and their activity in the department attached.

Weight analysis R6—C22 to C26 judgment matrix

R6	C22	C23	C24	C25	C26	W	NW	λmax	CI	RI	CR
C22	1	2	4	3	3	1.98	0.39				
C23	1/2	1	1	2	4	1.12	0.22				
C24	1/4	1	1	3	1	0.85	0.17	5.24	0.06	1.12	0.05
C25	1/3	1/2	1/3	1	2	0.59	0.11				
C26	1/3	1/4	1	1/2	1	0.46	0.09				

Weight analysis R7—C27 to C31 judgment matrix

R7	C27	C28	C29	C30	C31	W	NW	λmax	CI	RI	CR
C27	1	3	2	4	2	1.85	0.37				
C28	1/3	1	3	2	3	1.22	0.24				
C29	1/2	1/3	1	2	3	0.92	0.18	5.42	0.105	1.12	0.09
C30	1/4	1/2	1/2	1	2	0.63	0.12				
C31	1/2	1/3	1/3	1/2	1	0.44	0.08				

2. Identification of the particular units requires modernization with the scope of work.
3. The specification of the kind of the information sought concerning with the comfort of passengers.
4. Collection of relevant data from different department and shops.
5. Delineation of the environment in which the data will be collected.
6. Analysis of the data available in order to summaries the result.
7. Interpretation of the results.
8. Drawing suggestions for implementation.

7 Conclusion

The result shows that Skilled Manpower (32%), Availability of Material (21%) and Machine (16%) are the three vital factors for the Safety and Comfort of passengers. The weight Analysis shows that infrastructure development (12%) is also a vital factor as the Railways is keenly need of proper infrastructure. Again, from expert point of view, adaptation of New and Modern technology is a crucial in the present age without which total development of Railway system is impossible. It is also experienced that implementation of new technology is very difficult. Till date the Railway people is keen to stick to previous technology and for transferring them to adopt the new technology, proper infrastructure and training is required. Moreover it is practical, due to vastness of Railway system it is not so easy job to implement the same technology to all the Workshops, Divisions and production Unit because before completion of one technology or modification another new one is came into being.

References

1. Magazine—" Indian Railways"—published by Shri Prasant Kr. Pattanaik, on behalf of Ministry of Railway (Rail Board), Rail Bhawan, New Delhi, Monthly issue—Oct. 2013, on "Integral Coach Factory", pp. 53–65 (2013)
2. Special Message of Railway Minister on the present statistics of Indian Railways on 9 Jan 2017
3. Magazine—"Indian Management"—published by All India Management Association, New Delhi—special issue on Indian Railway, pp. 18–28, 2015
4. C.L. Hwang, K.P. Yoon, *Multiple Attribute Decision Making: An Introduction* (Sage University Paper Series on Quantitative Application in Social Science, 1995), pp. 7–104
5. Dr. B. Nag, *Operation Research Applications in Railways*
6. K.D. Goepel, Implementing the analytical hierarchy process as a standard method for multi-criteria decision making in corporate enterprises, in *International Symposium on Analytic Hierarchy Process, Conference 2013*, Kuala Lumpur, Malaysia, pp. 4–8 (2013)

Electrocardiogram Signal Analysis for Diagnosis of Congestive Heart Failure

Santanu Chattopadhyay, Gautam Sarkar and Arabinda Das

1 Introduction

Congestive heart failure (CHF) has become very common type of disease observed among infant to old persons. It is a chronic progressive condition and affects the pumping power of muscles of heart. It refers to the situation when fluid builds up around the heart causing the pump inefficiently. Most common types of CHF is Left-sided CHF when left ventricle doesn't properly pump blood out to body and as a result fluid can build up in lungs, causing breathing difficult. Left-sided CHF is divided into two categories: Systolic heart failure and Diastolic failure. Causes of CHF are mainly hypertension, Coronary artery disease, Bad Valve conditions, other conditions like diabetes, thyroid disease, and obesity, etc. First stage symptoms of CHF are fatigue, swelling, weight gain, increased need for urinate. Second stage symptoms of CHF are irregular heartbeat, cough developing from lungs, wheezing, shortness of breath, etc. last stage or severe symptoms of CHF are chest pain in upper body, rapid breathing, blue skin, fainting, etc. it may be noted that Chest pain radiating through upper body may be an indication of a heart attack. For heart failure in children and infants, symptoms are poor feeding, excessive sweating, difficulty breathing, etc. CHF diagnosis is normally done by electrocardiogram, echocardiogram, MRI, Stress tests, Blood tests, Cardiac catheterization.

CHF patients require utmost care and demands early diagnosis. Research is going on to study the main cause of CHF, its effect and different methods for its diagnosis

S. Chattopadhyay (✉) · G. Sarkar · A. Das
Jadavpur University, Kolkata, India
e-mail: sansur12ct@yahoo.com

G. Sarkar
e-mail: sgautam63@gmail.com

A. Das
e-mail: adas_ee_ju@yahoo.com

S. Chattopadhyay et al. (eds.), *Modelling and Simulation in Science, Technology and Engineering Mathematics*, Advances in Intelligent Systems and Computing 749,
https://doi.org/10.1007/978-3-319-74808-5_39

[1–3]. Classification of different heart diseases has been done in [1]. Effectiveness of electrocardiogram in diagnosis of heart diseases have been found effective in [4–7]. Different monitoring scheme have been introduced for monitoring of electrocardiogram [8–12]. Time domain analysis and wavelet transform have been used for assessment of electrocardiogram in [13, 14].

In some recent work DWT based statistical parameters have been found in characterization for non-stationary signals and diagnosis of various diseases [15, 16]. However, very few works have been found which apply DWT based statistical parameter for CHF diagnosis. This has motivated authors to classify ECG signals based on DWT based statistical parameter to extract specific features for diagnosis of CHF.

2 DATA Collection

ECG signals of normal healthy person and patients suffering from congestive heart failure have been collected from well-established data bank of phisionet.com. Collected data are passed through Savizky–Goalay numerical filter for denoising. Denoised data of normal healthy person and CHF patients have been used for further assessment. De-noised ECG signal of normal healthy person and CHF patient have been shown in Fig. 1.

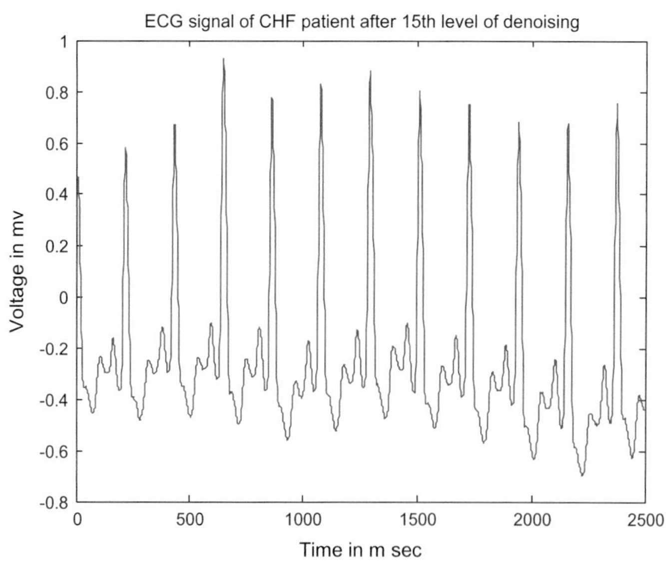

Fig. 1 De-noised ECG signal of CHF patient

3 Radar Comparison of ECG Signals

At first radars of de-noised ECG signals collected from normal person and patients suffering from congestive heart failure have been shown in following Figs. 2 and 3 respectively.

The radars of de-noised ECG signals of normal person and CHF patients as shown in Figs. 2 and 3 are different in shape and as well as in area. Thus to make visual inspection radar observation may be better than observation of ECG signals itself

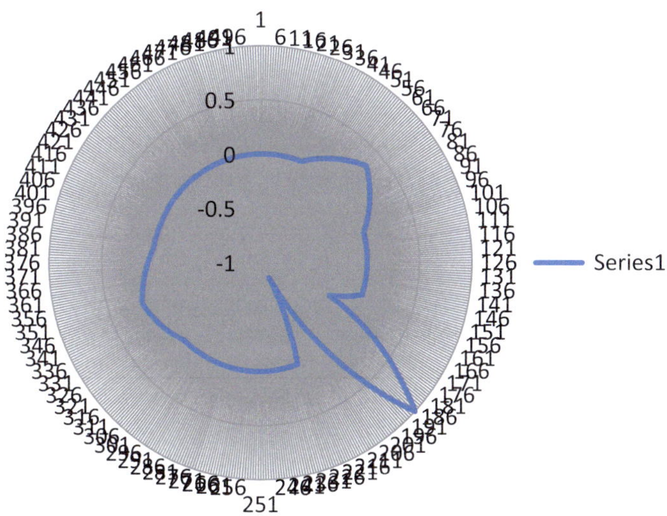

Fig. 2 Radar of ECG signal of normal healthy person

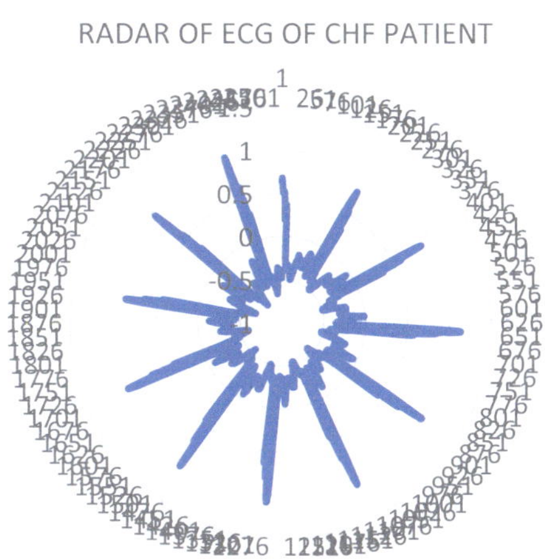

Fig. 3 Radar of ECG signal of CHF patient

which is normal practice in diagnosis as because radars (i.e. polar plot) of ECG signals are taking less area in the two dimensional frame of reference. However more information is available in the area and critical shape of the radars which are difficult to be noted in visual inspection.

4 Discrete Wavelet Transformation of ECG Signals

As ECG signals are non-stationary periodic and are collected through digital instrument, Discrete Wavelet Transform(DWT) based assessment have been performed to overcome the limitation of comparison of radars made from ECG signals as presented in previous section. Wavelet Decomposition has been performed on ECG signals for both normal person and CHF patients. In wavelet decomposition, 'db4' has been used as mother wavelet and decomposition is performed up to level 9. At each decomposition level, approximate coefficients have been calculated. To observe the variation of large number of approximate coefficient, there skewness feature have been extracted for large number of approximate coefficient single skewness value (SA) have been obtained. Thus for 9 decomposition level 9 number of SA have been obtained. Sets of 9 SA values have been obtained both for normal person and CHF patients presented in Table 1. SA versus DWT levels for normal person and CHF patients have been shown Fig. 4. The figure shows significant change of SA versus DWT level nature for congestive heart failure. SA in both cases remains constant up to level 3 and then start decreasing in irregular way. Up to DWT level 3 and at | DWT level 9, SA for CHF patient is higher than that of normal person.

Table 1 Skewness of approximate coefficients of the ECG signal at different DWT level of CHF patient

DWT level	SA of normal healthy person	SA of CHF patient
1	0.854003	2.09009
2	0.854079	2.081993
3	0.859776	2.072368
4	1.412921	1.811709
5	1.001586	1.117819
6	0.276683	0.2094
7	0.368491	−0.17752
8	−0.07648	0.158719
9	−0.7627	0.087164

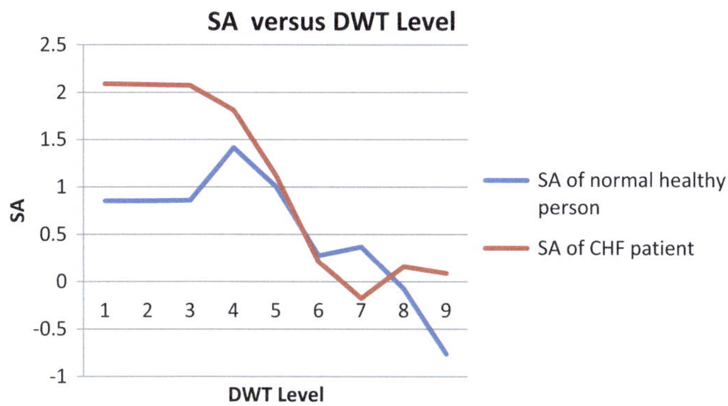

Fig. 4 SA versus DWT levels for normal person and CHF patient

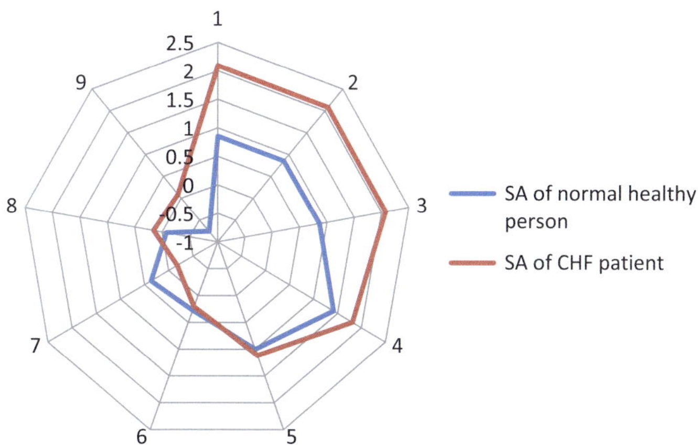

Fig. 5 Radars of SA for normal person and CHF patient

5 Comparison of Radar of SA

Radars of SA both for normal person and CHF patient have been made as shown in Fig. 5. These radars are different in shape and area. The difference was also observed in Sect. 3, but here it is easy to distinguish radar of SA of CHF patients from that of normal patients. Area of radar of SA decreases for CHF.

6 Histogram Assessment of SA

Histogram of skewness of DWT based approximate coefficients (SA) are analyzed both for normal person and CHF patient as shown in Figs. 6 and 7 respectively.

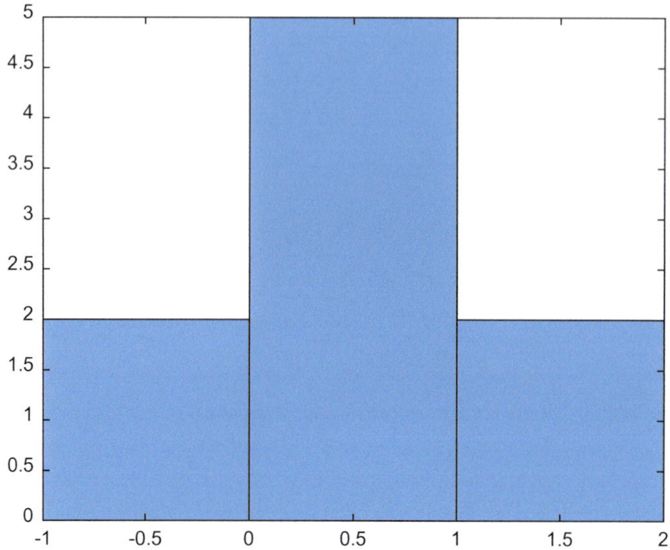

Fig. 6 Histogram of SA of normal healthy person

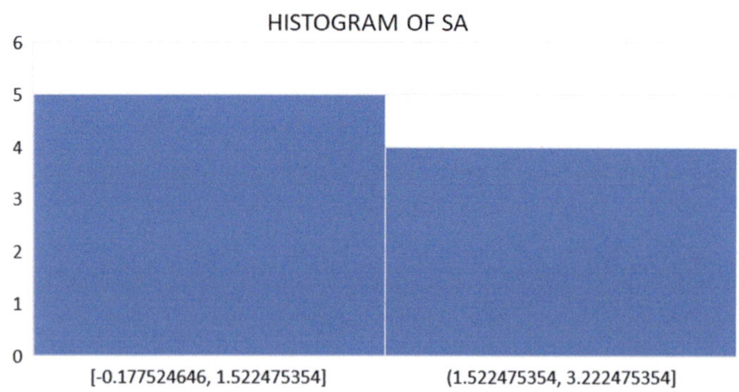

Fig. 7 Histogram of SA of normal CHF patients

Minimum value of the histogram changes due to CHF. Minimum value of histogram increases due to such disease and peak values remains almost same as that of normal person.

7 Conclusion

In this work diagnosis of congestive heart failure with the help of wavelet decomposition based approximate coefficients. These coefficients have been determined from

ECG signals both for normal person and patients suffering from congestive heart failure (CHF). Skewness property of approximate coefficients has been determined at different decomposition levels and comparative study has been carried out on SA for healthy person and CHF patient. Radars and histograms have also been made and compared. Significant difference has been observed which helps to easily distinguish ECG signals having CHF. The proposed method may be effective for diagnosis such disease.

References

1. S Mitra, M. Mitra, S Chattopadhyay, S Sengupta, An approach to a rough set decision system for classification of different heart diseases, MS-04, Lyon, France, 5–9 July, 2005, pp. 1.17–1.20
2. M. Al-Abed, M. Manry, J.R. Burk, E.A. Lucas, K. Behbehani, A method to detect obstructive sleep apnea using neural network classification of time-frequency plots of the heart rate variability, in *29th Annual International Conference of the IEEE Engineering in Medicine and Biology Society*, 2007, pp. 6101–6104. https://doi.org/10.1109/IEMBS.2007.4353741
3. O. Karadeniz, A. Yilmaz, T. Dundar, Design of portable holter recorder with MMC memory for prephase sleepapnea diagnosis, in *14th National Biomedical Engineering Meeting, 2009*. BIYOMUT, 2009, pp. 1–4. https://doi.org/10.1109/BIYOMUT.2009.5130337
4. M.O. Mendez, A.M. Bianchi, M. Matteucci, S. Cerutti, T. Penzel, Sleep apnea screening by autoregressive models from a single ECG lead. IEEE Trans. Biomed. Eng. **56**(12), pp. 2838–2850 (2009). https://doi.org/10.1109/TBME.2009.2029563
5. D. Alvarez, R. Hornero, J.V. Marcos, F. del Campo, Multivariate analysis of blood oxygen saturation recordings in obstructive sleep apnea diagnosis. IEEE Trans. Biomed. Eng. **57**(12), 2816–2824 (2010). https://doi.org/10.1109/TBME.2010.2056924
6. A. Burgos, A. Goni, A. Illarramendi, J. Bermudez, Real-time detection of apneas on a PDA. IEEE Trans. Inf. Technol. Biomed. **14**(4), 995–1002 (2010). https://doi.org/10.1109/TITB.2009.2034975
7. N.J.B.N. Mazlan, K.I. Wong, A wireless ECG sensor and a low-complexity screening algorithm for obstructive sleep apnea detection, in *2012 IEEE EMBS Conference on Biomedical Engineering and Sciences (IECBES)*, 2012, pp. 279–283. https://doi.org/10.1109/IECBES.2012.6498018
8. H. Guruler, M. Sahin, G. Ordek, A. Ferikoglu Sakarya, Sleep apnea diagnosis via single channel ECG feature selection, in *38th Annual Northeast Bioengineering Conference (NEBEC)*, 2012, pp. 159–160. https://doi.org/10.1109/NEBC.2012.6207012
9. C.-W. Wang, A. Hunter, N. Gravill, S. Matusiewicz, Unconstrained video monitoring of breathing behavior and application to diagnosis of sleep apnea. IEEE Trans. Biomed. Eng. **61**(2), 396–404 (2014). https://doi.org/10.1109/TBME.2013.2280132
10. S.H. Hwang, H.J. Lee, H.N. Yoon, D.W. Jung, Y.J.G. Lee, Y.J. Lee, D.-U. Jeong, K.S. Park, Unconstrained sleep apnea monitoring using polyvinylidene fluoride film-based sensor. IEEE Trans. Biomed. Eng. **61**(7), 2125–2134 (2014). https://doi.org/10.1109/TBME.2014.2314452
11. L. Chen, X. Zhang, C. Song, An automatic screening approach for obstructive sleep apnea diagnosis based on single-lead electrocardiogram. IEEE Trans. Autom. Sci. Eng. **12**(1), 106–115 (2015). https://doi.org/10.1109/TASE.2014.2345667
12. A. Jezzini, M. Ayache, L. Elkhansa, Z. al abidin Ibrahim, ECG classification for sleep apnea detection, in *2015 International Conference on Advances in Biomedical Engineering (ICABME)*, 2015, pp. 301–304. https://doi.org/10.1109/ICABME.2015.7323312
13. A.R. Hassan, M.A. Haque, Computer-aided sleep apnea diagnosis from single-lead electrocardiogram using dual tree complex wavelet transform and spectral features, in *International*

Conference on Electrical & Electronic Engineering (ICEEE), 2015, pp. 49–52. https://doi.org/10.1109/CEEE.2015.7428289
14. J. Jin, E. Sánchez-Sinencio, A home sleep apnea screening device with time-domain signal processing and autonomous scoring capability. IEEE Trans. Biomed. Circ. Syst. **9**(1), pp. 96–104 (2015). https://doi.org/10.1109/TBCAS.2014.2314301
15. S. Chattopadhyay, R.R. Majhi, S. Chattopadhyay, A. Ghosh, Analysis of electro-cardiogram by radar and DWT based Kurtosis comparison, in *Michael Faraday IET International Summit 2015*, p. 108 (5). https://doi.org/10.1049/cp.2015.1704, ISBN: 978-1-78561-186-5, Kolkata, India, Conference date: 12–13 Sept 2015
16. S. Chattopadhyay, R.R. Majhi, S. Chattopadhyay, A. Ghosh, Radar assessment of wavelet decomposition based skewness of ECG signals, in *Michael Faraday IET International Summit 2015*, p. 109 (5). https://doi.org/10.1049/cp.2015.1705, ISBN: 978-1-78561-186-5, Kolkata, India, Conference date: 12–13 Sept 2015

Condition Assessment of Structure Through Non Destructive Testing—A Case Study on Two Identical Buildings of Different Age

Bhaskar Chandrakar, M. K. Gupta and N. P. Dewangan

1 Introduction

1.1 Conditional Assessment

Conditional assessment of any structure is must to check its suitability for the purpose which it serves. This is done of both old and new structures for new structure it is performed to check the quality of the work carried out by the contactor. The old buildings are assessed to know the remaining life [1] or to repair the structure. Mainly the old, weird, partially damaged, abandoned structure need conditional assessment to check weather that is it still serviceable and if so then for how long or if not then can it be retrofitted and made safe for the desired purpose. Also the conditional assessment is performed when the structure has to be expanded i.e. more number of floors are to be constructed on the existing old structure. Thus it becomes very important to assess the condition [2] of the structure. Concrete is mainly tested to assess the condition of a structure since it all depends on the properties of the concrete.

B. Chandrakar (✉)
Department of Civil Engineering, Chhattisgarh Swami Vivekanand Technical University, Bhilai, Chhattisgarh, India
e-mail: chandrakar_bhaskar@yahoo.com

M. K. Gupta
Department of Civil Engineering, BIT Durg, Durg, Chhattisgarh, India

N. P. Dewangan
Department of Civil Engineering, Sarguja University, Ambikapur, Chhattisgarh, India
e-mail: npdewangan@gmail.com

S. Chattopadhyay et al. (eds.), *Modelling and Simulation in Science, Technology and Engineering Mathematics*, Advances in Intelligent Systems and Computing 749,
https://doi.org/10.1007/978-3-319-74808-5_40

1.2 Non Destructive Testing on Concrete

NDT is gaining popularity for testing concrete with a swift pace since these are mostly convenient to perform; these tests are insitu so actual concrete can be tested in its existing condition. Non-destructive testing methods [3] are being used for about four decades ago. They are now considered a powerful tool for evaluating existing concrete structures with regard to their strength and durability [4] apart from assessment and control of quality of hardened concrete. In certain cases the investigation of crack depth, micro cracks [5] and progressive deterioration are also studied by this method. NDT methods are relatively simple to perform but the analysis and interpretation of the test results is not easy.

2 Experimental Work

An extensive survey was carried out to find the suitable structure for the testing and evaluation, the task was to find out a RCC structure on which both [6] Rebound Hammer and Ultrasonic Pulse Velocity Test [7] can be carried out. Finally Residential building at Tatibandh Raipur was selected to perform the testing.

After all the necessary arrangements for carrying out the NDT such as method of testing was finalized, instruments for the tests were selected and arranged, site was also selected, the experiment had to be commenced, before that a visual inspection was needed.

2.1 Visual Inspection

Visual inspection was the first step in the conditional assessment, both the buildings were carefully observed and found that the old building whose column, beam and slab was constructed seven years ago and then left abandoned developed cracks in both micro and macro forms, reinforcement bars were also seen at some parts.

The new building is under construction on its sixth floor and no signs of physical damage appeared on any of the surfaces, it gave the same appearance of the fresh constructed building with smooth surfaces, sharp edges etc. After the thorough visual inspection outline for performing the test was planned.

2.2 *Specifications of the Instruments Used*

2.2.1 Silver Schmidt Digital Rebound Hammer

The Rebound Hammer Test was performed by the Silver Schmidt Digital Rebound Hammer manufactured by Proceq from Switzerland. The Hammerlink software compatible with the operating system of the computer gives the result in the form of bar chart when the digital Silver Schmidt Rebound Hammer is connected with the computer which has Hammerlink software installed in it (Figs. 1, 2 and 3).

2.2.2 Concrete Ultrasonic Testing Equipment

Ultrasonic Velocity Test was performed by the Concrete Ultrasonic testing equipment manufactured by Canopus Instruments.

Concrete Ultrasonic testing equipment displays the time taken by the pulse to travel from the pulse sending probe to the pulse receiving probe. The time is displayed on the screen of the equipment in micro seconds which is used to calculate the velocity of the pulse by dividing it with the distance travelled by the pulse between sending and the receiving probe.

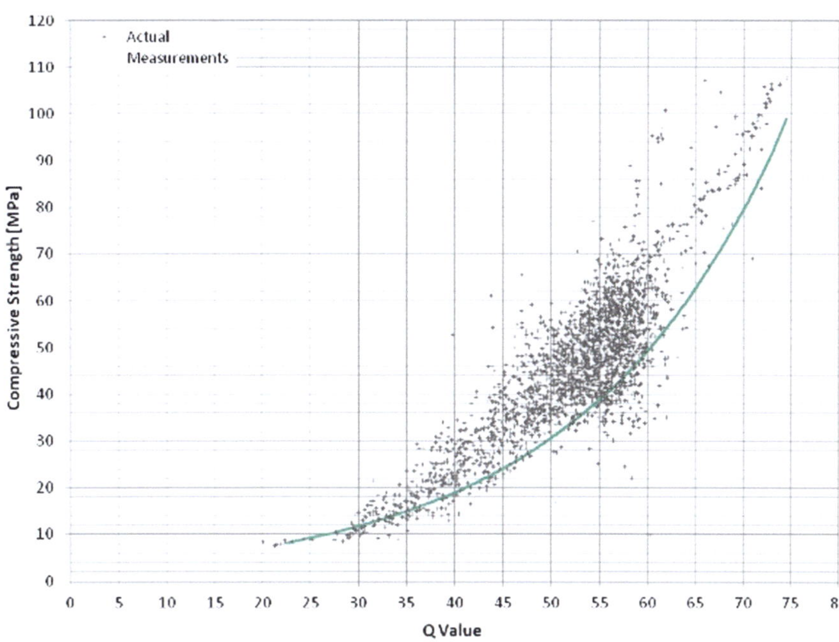

Fig. 1 This curve was used to interpret the strength from corresponding rebound coefficient

Fig. 2 Digital silver schmidt rebound hammer being calibrated on the standard Anvil

Fig. 3 Ultrasonic pulse velocity meter being calibrated before testing

2.3 Setting Out for the Tests

The bottom three floors of the building were decided to be tested; the old building had only three floors it was not possible to go higher.

48 Columns, 24 Beams and 6 Slabs to be tested were marked on both the old and new building, thus each building was to be tested for 24 column, 12 beams and 3 slabs.

After the selection of points of application for the testing instruments, it was time to test the marked parts for their strength and quality.

2.3.1 Steps Followed in Performing the Tests

(a) The gridlines were marked on the surface of columns, beams and slabs. Grids of 300 mm X 300 mm were marked on surface of column and slabs, 250 mm X 250 mm grids were marked on the surface of beams.
(b) Surface preparation was done by rubbing and smoothing the surface at the junction points of the grids.
(c) Grease was applied on the junction points of the grids.
(d) The Ultrasonic pulse velocity apparatus was connected to its probes, battery and the complete set up was made ready for the test.
(e) Before starting the test, instrument has to be calibrated this is done on a standard mild steel bar by cross probing on its ends and then calibrating the reading to 23.7 micro seconds.
(f) On successful calibration of the instrument the tests were performed on the required points that were already marked and prepared for the test.
(g) Readings were recorded by cross probing and surface probing methods depending on the accessibility of the faces of the structural members.
(h) Readings of the Ultrasonic pulse velocity test was recorded in micro seconds the value of which was displayed on the screen of the instrument.
(i) Silver Schmidt Digital Rebound hammer test was followed by the ultrasonic pulse velocity test at the same points and the readings were recorded.
(j) Readings of the Rebound Hammer test was displayed on the screen of the digital Rebound hammer which are known as rebound coefficients Q.

Performing both the tests on number of points on both the buildings was a hectic process and it took a long period for the completion of the test since around 2200 hundred points were tested in total, also weather not at all times favoured the suitability for the test conditions, other constraints also sometimes opposite, but ultimately the Non Destructive Testing of both the buildings completed (Figs. 4, 5, 6, 7, 8 and 9).

Fig. 4 Internal view of the old building to be tested

Fig. 5 Gridlines marked on the surface of column

Fig. 6 Slab surface prepared for the ultrasonic test

Fig. 7 Direct or cross probing method i.e. on the opposite faces of column

Fig. 8 Reading displayed while performing the test

Fig. 9 Silver schmidt digital rebound hammer used on the column

3 Results

The results tabulated here is the average of 2200 readings for each type of the test performed.

Comparison of cumulative result of rebound hammer test

		Old building		New building	
S. no.	Member	Actual initial design strength Mpa	Present strength Mpa	Decreased strength Mpa	Present strength Mpa
1	Column	30	28.87	**1.13**	33.88
2	Beam	30	26.82	**3.18**	33.9
3	Slab	25	23.48	**1.52**	26.68

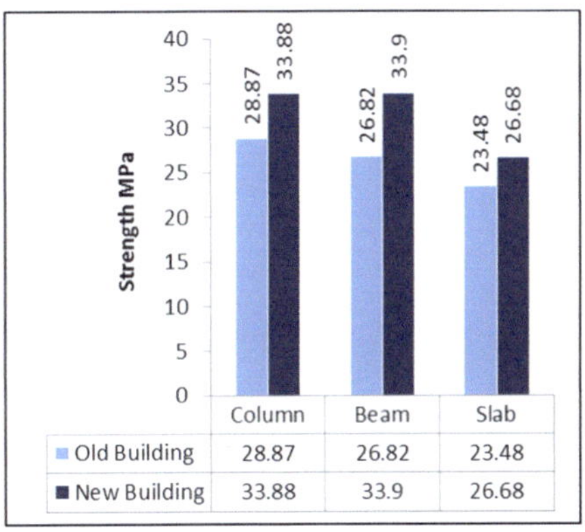

Comparison of cumulative result of ultrasonic pulse velocity test

			Old building	New building
Sl. no.	Member	Actual initial designed quality of concrete	Present quality of concrete	Present quality of concrete
1	Column	Good	Medium	Good
2	Beam	Good	Medium	Good
3	Slab	Good	Medium	Good

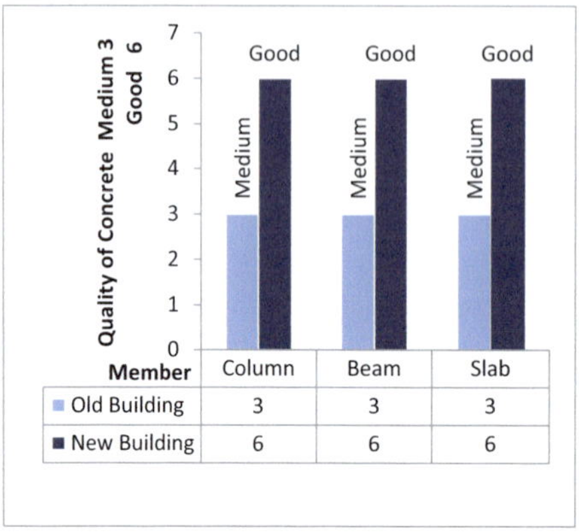

Overall comparison					
		Old building		New building	
S.no.	Member	Strength Mpa	Quality of concrete	Strength Mpa	Quality of concrete
1	Column	28.87	Medium	33.88	Good
2	Beam	26.82	Medium	33.9	Good
3	Slab	23.48	Medium	26.68	Good

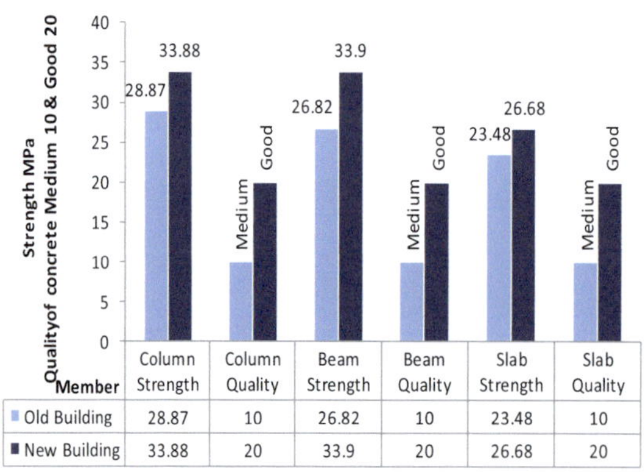

4 Discussion

The test results presented in the form of tables and charts give a clear idea of the strength [8] and quality of the concrete in the different structural parts (column, beam and slab) of both the buildings.

The result shows that there is a difference in terms of strength and quality in the old and new building and also among different structural members of the same building.

The old building shows a decrement [9] in its strength and quality while the new one has better strength and quality than its initial design grades. Since the initial design and drawing of the building was available for reference it became easier to analyze and compare the changes in the strength and quality of the buildings.

The result of the methods adopted for the test of the buildings are influenced by many factors which were taken good care off so those may not affect the test results. For example moisture content influences the readings of the Ultrasonic pulse velocity test to a large extent hence the tests were not performed during the rains and moist weather conditions.

For the most efficient results of the Ultrasonic pulse velocity test direct or cross probing method of test must be adopted but due to inaccessibility of the opposite faces in some of the cases surface probing was used and the necessary corrections has been applied.

As far as possible point of application was kept same for both the tests so that comparison of the results could be justified. The same structural members were tested on both the buildings and in the same conditions to make a fair comparison between the old and the new building.

5 Conclusion

- The result concludes a reduction of 6.81% in the strength and decrement of good to medium in the quality of concrete in the old building that was built seven years ago shows that the concrete loses its quality and strength with time.
- Carbonation due environmental effects on the bare surfaces for very long since neither plaster nor paints were applied to protect the concrete in extreme environmental conditions and internal chemical reactions in the concrete the reduction in the grade and quality of concrete although initially the strength might have increased when hydration would have been taking place.
- Since the strength is not considerably reduced but the quality of the concrete is decreased from good to medium which is a matter of concern. This also leads to conclude that the decrease in the strength is actually due to the loss of quality.
- As a result of cracking and various chemical reactions occurring internally in concrete due to its chemical composition and external environmental conditions the concrete became porous and the density reduced lead to decrease in the quality ultimately showing loss of strength.
- The building is located in vicinity of the most polluted areas and the building is affected this concludes that environmental effects are considerable during and after the construction.
- The overall strength of the new building was found 10.3% more than the initial design strength which concludes that the concrete might be still hydrating and

hence its strength is increasing since the building started construction last year and is yet under construction.

- The quality of concrete is also found to be good which concludes that the density of concrete is intact and the concrete has a good bond with all the reinforcement bars and its components.
- Finally this can be concluded that the new building is better than the old one in all respects, since the strength and quality of the old building has decreased from its initial value whereas the strength of new building is greater than its design strength.
- The comparison among different structural members of the old building shows that the flexural member. Beam lost its strength by 10.6% as compared to 3.76% loss in strength by the columns, axial stress bearers where as the slabs loss in strength is 6.08% which is intermediate to axial and flexural losses in their respective members. Hence it can be concluded that flexural members are subjected to more stresses and lose their strength faster than the other structural members which is often observed in most of the practical cases when the failure of the beams in the structure first of all shows the sign of failure.
- The owner wanted to extend more number of floors in the old building, so it is advisable that before commencing the expansion repairing through grouting or other methods must done strictly followed by other tests again to check the strength and condition after repairing.
- This concludes that the old building can be expanded of more floors but only after retrofitting the building.
- The importance and need for condition assessment of structures for evaluating their serviceability and safety have been highlighted in this piece of work.
- It can be seen that detailed visual inspection and Non Destructive Testing (NDT) plays an important role in condition assessment of existing buildings.
- An overview of the procedures and different investigations including tests involved in condition assessment and evaluation of safety is presented in a simple.

Concluding Remark

It may be emphasized here that a great deal of expertise is required for interpretation of field observations and non destructive testing results to make a proper assessment of the condition as well as for analyzing and evaluating condition of the structure. Although utmost care and every possible aspect has been considered and sincerely followed to come out with the conclusion but still the results of this work may not be considered standard or its not necessary that all structures follow the same trend.

6 Further Scope of Work

Based on the results of this work retrofitting methods for the old building could be suggested so that the building can be made suitable for extending the construction of more number of floors above the existing top floor.

After suggesting suitable methods of retrofitting when the building is repaired it can be tested again with suitable methods for its condition assessment.

Also in the present situation further investigation and testing of the building can be performed by some partial destructive methods and compare with the result of this work.

References

1. P. Shaw, A. Xu, Assessment of the deterioration of concrete in NPP- causes, effects and investigation methods. NDT.Net **3**(2) (1998)
2. N.V. Mahure, G.K. Vijh, P. Sharma, N. Sivakumar, Correlation between pulse velocity and compressive strength of concrete. Int. J. Earth Sci. Eng. **4**(6), 871–874 (2011)
3. S. Baby, T. Balasubramanian, R.J. Pardikar, M. Palaniappan, R. Subbaratnam, Time-of-flight diffraction (TOFD) technique for accurate sizingof surface-breaking cracks. Insight **45**, 426–430 (2003)
4. S.K. Verma, S.S. Bhadauria, S. Akhtar, Review of nondestructive testing methods for condition monitoring of concrete structures. J. Constr. Eng. **2013**, 1–11 (2013)
5. F. Nucera, R. Pucinotti, Destructive and non-destructive testing on reinforced concrete structure: the case study of the museum of Magna Graecia in Reggio Calabria (2009). https://www.ndt.net/article/defektoskopie2009/papers/Nucera-and-Pucinotti-8.pdf. Accessed 20 March 2014
6. S.S. Bhadauria, M. Chandra Gupta, In situ performance testing of deteriorating water tanks for durability assessment. J Perform. Constr. Fac. **21**(3), 234–239 (2007)
7. G. Pascale, A. Di Leo, V. Bonora, Nondestructive assessment of the actual compressive strength of high-strength concrete. J. Mater. Civil Eng. **15**(5), 452–459 (2003)
8. Y. Yoshida, H. Irie, in *NDT for concrete using the ultrasonic method*. Proceedings of the 12th Asia Pacific Conference of Non Destructive Testing (A-PCNDT '06), Auckland, New Zealand (2006)
9. D. Breysse, G. Klysz, X. Dérobert, C. Sirieix, J.F. Lataste, How to combine several non-destructive techniques for a better assessment of concrete structures. Cem. Concr. Res. **38**(6), 783–793 (2008)
10. M.V. Felice, A. Velichko, P.D. Wilcox, Accurate depth measurement of small surface-breaking cracks using an ultrasonic array post-processing technique. Ndt & E Int. **68**, 105–112 (2014)

A Real Time Health Monitoring and Human Tracking System Using Arduino

P. L. Lekshmy Lal, Arjun Uday, V. J. Abhijith and Parvathy R. L. Nair

1 Introduction

This project aims at health monitoring and human tracking in real time basis. Health monitoring system consists of a pulse rate and body temperature sensor. The system consists of a microcontroller based heart rate and body temperature measuring devices along with LCD output. A threshold value range will be set for heart rate and temperature. When the set threshold condition is violated, the device sends an alarm message. The threshold value is decided by the programmer. It can be amended whenever necessary. The information is transferred to the appropriate person wirelessly. A heart patient should be monitored continuously. But being at a hospital for rest of the life after a heart disease is detected is not practical option. Thus, the system can be used by the patients outside the hospital and their location can be traced in real time.

Soldiers fight in the most difficult of terrains for his country and people. By this project the system will be useful for providing health status and medical help for the needy soldiers in the battle field. The system tracks the health by focusing on soldier's heart beat and temperature [1]. If an alarming situation occurs, there will be change in the heart beat patterns and thus a message will be sent along with the location of

P. L. Lekshmy Lal (✉) · A. Uday · V. J. Abhijith · P. R. L. Nair
ECE Department, Rajadhani Instiitute of Engineering and Technology, Nagaroor, Trivandrum, Kerala, India
e-mail: lekshmylalpl@gmail.com

A. Uday
e-mail: arjunuday123@gmail.com

V. J. Abhijith
e-mail: abhijithvj@gmail.com

P. R. L. Nair
e-mail: parvathyrlnair@gmail.com

S. Chattopadhyay et al. (eds.), *Modelling and Simulation in Science, Technology and Engineering Mathematics*, Advances in Intelligent Systems and Computing 749,
https://doi.org/10.1007/978-3-319-74808-5_41

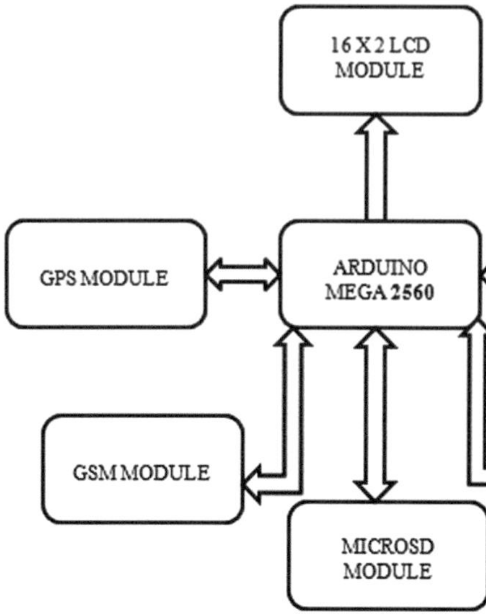

Fig. 1 Block diagram of the system

the soldier to the base camp for necessary help. The system is also equipped with a distress switch [2]. When the soldiers press this switch, a distress message will be sent to the base station or to the concerned ones [3]. ECG waveform can also be viewed from this device.

2 System Description

Block diagram of the system is shown in Fig. 1. The Arduino mega 2560 is the heart of the system through which the input output communications to the system is carried out. The 16 × 2 LCD acts as the output display device. GSM module is used for the sending and reception of messages and the GPS module is used for the 24 × 7 tracking of the user position with reference to latitudes and longitudes. The health sensors are used for the heart beat and temperature monitoring. The hex keypad is used as the input device. Micro SD module is used as the data storage system.

3 Flow Chart Description

This project has two sections, Transmitter and Receiver section.

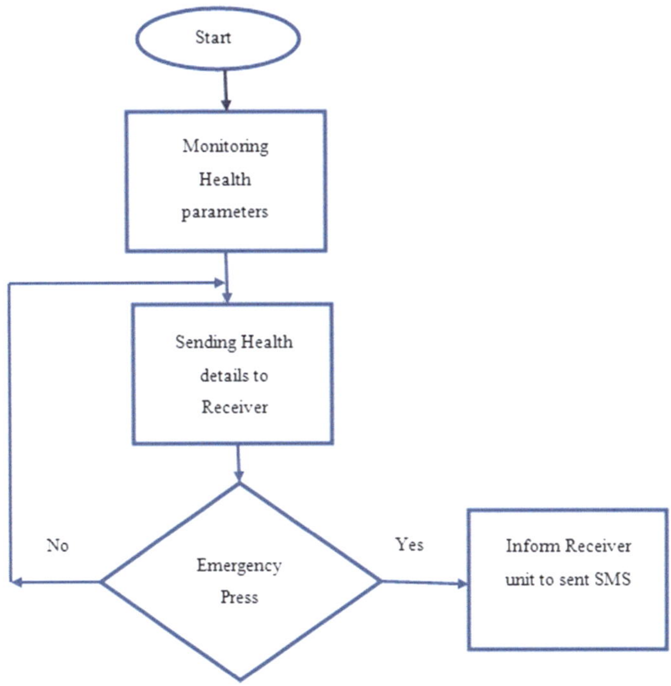

Fig. 2 Flow chart of transmitter section

3.1 Transmitter Section

Flow chart of the transmitter section is shown in Fig. 2. When the transmitter section is initialized, the system starts monitoring the health parameters. The health details will be sent to the reception unit. If an emergency press or a health problem is detected, reception unit will be informed to send reply message.

3.2 Receiver Section

The flow chart of the receiver section is shown in Fig. 3. The module gets initialized and the input details like mobile number and threshold temperature is given. Then the data will be received from the transmitter. If an emergency press at transmitter section or an SMS "BTRACK" is received or a health problem is detected, an SMS consisting of health details and location will be sent to the registered mobile number and it will be saved to an SD card module. If an SMS "STRACK" is received, the data will stop to get saved in the SD card module.

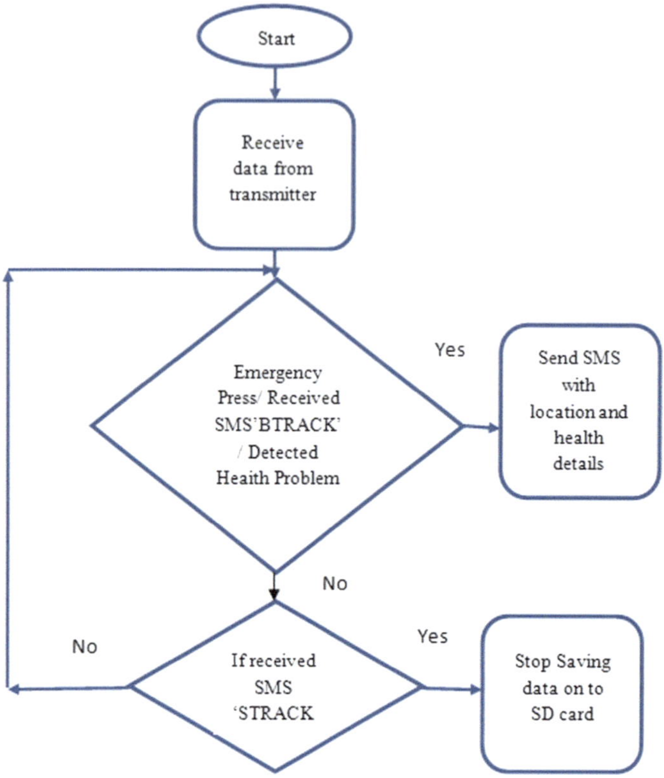

Fig. 3 Flow chart receiver section

4 Result

Figure 4 shows the hardware model of the project. Arduino Uno is the heart of the transmitter. LCD is the component that displays the output information like heart beat and temperature. The module consists of the heart beat sensor and the temperature sensor. The readings like heart beat and temperature can be transmitted at the same time. Arduino Mega 2560 is the core of receiver section. Hex keypad is provided to input the threshold values of heart beat and temperature and to register a mobile number. SD card stores the data pertaining to the health and the location. While a message "BTRACK" is sent to the receiver, the data regarding the health status and location will be sent to the registered mobile number and the server will get uploaded with the values. When a message "STRACK" is received, the data will stop getting saved to the SD card and the updation to the server will end. If an emergency press or the threshold values gets cut, then the receiver will sent a message to the registered mobile number with the health details and location automatically.

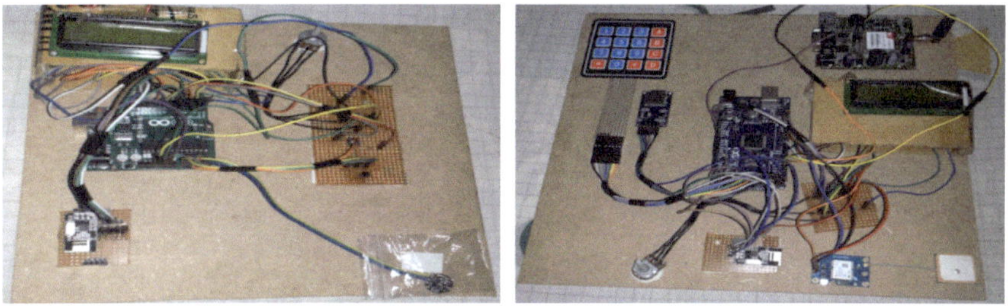

Fig. 4 Hardware model of transmitter section and receiver section

Fig. 5 Screen shot of the message exchanged by the system

Figure 5 shows the screen shot of the message exchanged by the system. When a message B TRACK is sent from the registered mobile number, first the system checks whether it is from a valid user. Once the message received is identified as that from a valid user, the system finds out the health parameters and the location and sends SMS to the valid number. When a message "STRACK" is sent from the registered user the message will not be sent. By using Google Map the location of the patient could be determined, and appeared in the SMS.

Figure 6 shows the screenshot of the health details being uploaded to the server. Server shows the last updated time and the heart beat chart and temperature chart by which the person is constantly monitored.

5 Conclusion

The health monitoring and human tracking system aims at bringing out a multi application system which is useful for the public as well as soldiers. The project is expected to provide with the functionality of locating a person and tracking him at the

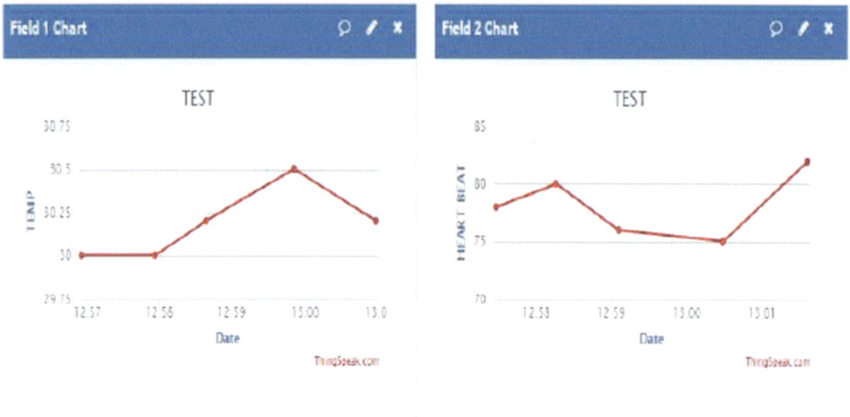

Fig. 6 Screen shot of the data exchanged by the system

same time, it gives a detailed insight on his current health condition. It is aimed that a second person related to the user will be able to monitor the user's location as well as health condition. The project aims at designing a low cost as well as low power system so that it can be used by the public. The project aims at running using a very simplified program so that only very little EPROM memory of the microcontroller is used; this in turn will increase the life and reliability of the system. The project is also expected to produce a system that can be used by aged people so that their health and location can be monitored by their near ones.

References

1. S. Atalla, K. Aziz, S. Tarapiah, S. H. Ismail, Smart real-time healthcare monitoring and tracking system using GSM/GPS technologies, in *2016 3rd MEC International Conference on Big Data and Smart City (ICBDSC)*, vol 142, No 14 (2016), pp. 19–26
2. K. Aziz, S. Tarapiah, M. Alsaedi, S.H. Is-mail, Shadi Atalla, Wireless sensor networks for road traffic monitoring. Int. J. Adv. Comput. Sci. Appl. **1**(6), 265–270 (2015)
3. C. Karthick, S. Idrissyedismail, E. Arunkailasam, S. Dhanapal, T. Devika, A Novel Based on Soldier Tracking and Health Monitoring system using Embedded Technology, in *The International Journal of Science and Technology*, vol. 3, Issue 3 (2015), pp. 212–217

Study of Arrhythmia Using Wavelet Transformation Based Statistical Parameter Computation of Electrocardiogram Signal

Santanu Chattopadhyay, Gautam Sarkar and Arabinda Das

1 Introduction

Electrocardiogram (ECG) signal has become an important medium for study and diagnosis of different types of heart diseases. By nature electrocardiogram signal is periodic but non stationary. Shapes of the ECG signals is divided into different parts whose magnitude and duration are measured in conventional method of study [1]. However with the advancement of research, different mathematical tools have been introduced for such signals. Rough set decision system based on ECG signal have been introduced for classification of different heart diseases [2]. FFT and wavelet transform based techniques have been used for analysis of electrocardiogram signals [3–5]. Dual tree complex wavelet transform have also been applied for analysis of electrocardiogram signal [3]. Application of radar is also found in classification of different types of heart diseases using electrocardiogram analysis [5]. Artificial neural network is being used for classification of heart diseases [6]. Auto regressive model has been introduced for analysis of ECG signals [7]. Compact method using unidirectional loop antenna having wideband performance having introduced for detection of heart failure [8]. Hemodynamic model using fuzzy logic has been introduced for the detection of heart failure [9]. In wavelet decomposition method large number of coefficient appear which some time makes the computation a little bit difficult to overcome this limitation use of different statistical parameters instead of using coefficients directly has become popular [10]. In this work study of arrhythmia

S. Chattopadhyay (✉) · G. Sarkar · A. Das
Jadavpur University, Kolkata, India
e-mail: sansur12ct@yahoo.com

G. Sarkar
e-mail: sgautam63@gmail.com

A. Das
e-mail: adas_ee_ju@yahoo.com

S. Chattopadhyay et al. (eds.), *Modelling and Simulation in Science, Technology and Engineering Mathematics*, Advances in Intelligent Systems and Computing 749,
https://doi.org/10.1007/978-3-319-74808-5_42

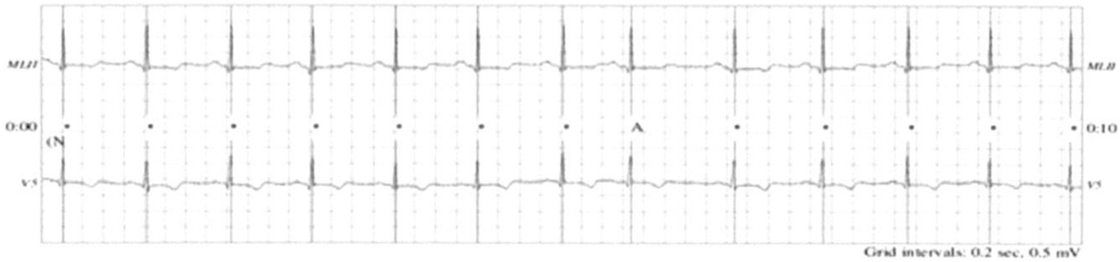

Fig. 1 ECG signal of arrhythmia patient

has been carried out into wavelet transform based statistical parameter computation of electrocardiogram signal.

2 Data Collection

From well-established data bank of phisionet.com ECG signals of normal healthy person and patients suffering from arrhythmia have been collected. Collected signals are denoised by numerical filter and the denoised signal of normal healthy person and arrhythmia patients have been used for further assessment. Sample ECG signal of arrhythmia patient has been shown in the Fig. 1.

3 Wavelet Decomposition

Denoised ECG signal are then decomposed by discrete wavelet transformation (DWT). Both for normal and arrhythmia patient, wavelet decomposition is performed up to DWT level 9. As the mother wavelet 'db 4' has been used. At each decomposition level approximate coefficient are determined and then corresponding kurtosis of approximate coefficients (KA) are calculated. KA obtained at DWT level 1–9 both for normal and arrhythmia have been presented in Table 1.

4 Observation

Kurtosis of approximate coefficient for normal and arrhythmia patient versus DWT level has been plotted has shown in Fig. 2.

It shows two different line: one for normal healthy person another for arrhythmia patient. From DWT level 1–3 KA are constant; However magnitude of KA is higher than that of normal person. Then KA decreases up to DWT level 6 and then becomes almost horizontal up to DWT level 8. After this again change of slope of KA is

Table 1 Kurtosis of approximate coefficients (KA) of ECG signal at different DWT level of normal healthy person and arrhythmia patient

DWT level	Kurtosis of approximate coefficients (KA) of normal healthy person	Kurtosis of approximate coefficients (KA) of arrhythmia patients
1	9.184569	10.14906
2	9.184136	10.15704
3	9.111071	10.14601
4	9.353461	7.792513
5	4.803909	3.812185
6	1.68599	2.577378
7	1.696927	2.876802
8	1.456951	2.954453
9	2.351525	1.687806

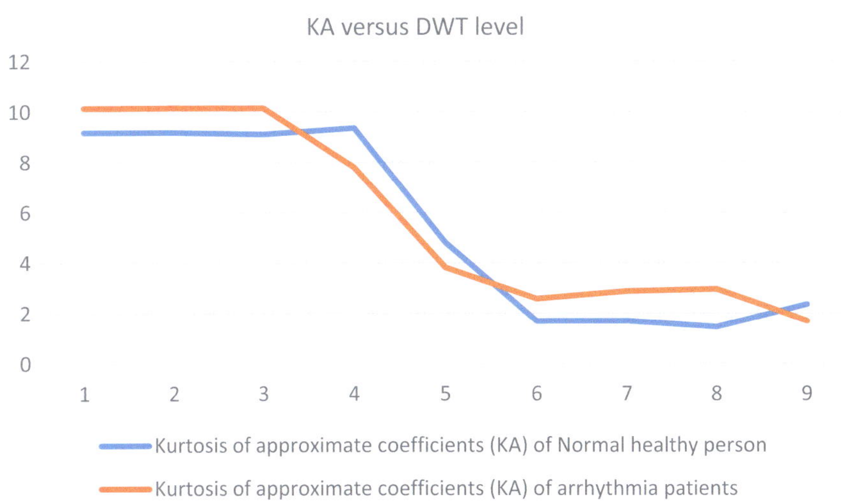

Fig. 2 KA versus DWT level of normal healthy person and arrhythmia patient

observed from DWT level 8–9. For better comparison radars of KA are formed both for normal person and arrhythmia patient in same polar plane as shown Fig. 3.

It shows two different loops of different shapes. Area of radar of KA for normal person is slightly less than that of arrhythmia patient. Thus KA analysis shows distinct difference for normal and arrhythmia patient which may be useful for the study and detection of arrhythmia.

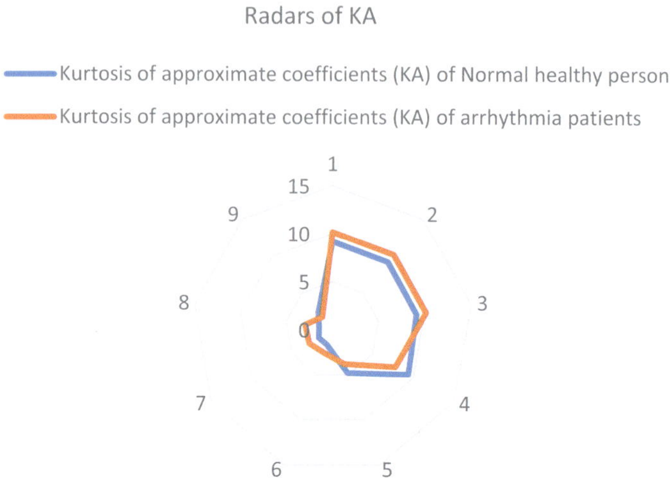

Fig. 3 Radars of KA of normal healthy person and arrhythmia patient

5 Conclusion

This work presents analysis of electrocardiogram signal based on wavelet decomposition for detection of arrhythmia. Electrocardiogram signal of normal and arrhythmia patient have been collected from normal database and denoised. After denoising wavelet decomposition has been performed up to DWT level 9. At each level kurtosis has been calculated for all approximate coefficient of that particular level. KA obtained in this way is compared. Magnitudes of KA have been found different from DWT level 1–3. Also radars of KA corresponding to normal person and arrhythmia patient are different in shape and area. These feature may be useful for detection of arrhythmia and study of electrocardiogram signal of arrhythmia patient. The work may be extended for the study of other heart diseases also.

References

1. S. Chattopadhyay, S. Chattopadhyay, A. Das, Elecctrocardiogram signal analysis for diagnosis of Apnea. AMSE J. Series: Modell. C; Chem., Geol., Environ. Bioeng., **77**(1) 28–40 (2016). ISSN: 1259-5977
2. S. Mitra, M. Mitra, S. Chattopadhyay, S. Sengupta, An approach to a rough set decision system for classification of different heart diseases, MS-04, Lyon, France, 5–9 July, 2005, pp. 1.17–1.20
3. A.R. Hassan, M.A. Haque, Computer-aided sleep apnea diagnosis from single-lead electrocardiogram using dual tree complex Wavelet Transform and spectral features. in *International Conference on Electrical & Electronic Engineering (ICEEE)*, (2015), pp. 49–52. https://doi.org/10.1109/ceee.2015.7428289

4. S. Chattopadhyay, R.R. Majhi, S. Chattopadhyay, A. Ghosh, Analysis of electro-cardiogram by radar and DWT based Kurtosis comparison. in *Michael Faraday IET International Summit 2015*, (2015), p. 108 (5 .). https://doi.org/10.1049/cp.2015.1704, ISBN: 978-1-78561-186-5, Kolkata, India, 12–13 Sept 2015
5. S. Chattopadhyay, R.R. Majhi, S. Chattopadhyay, A. Ghosh, Radar assessment of wavelet decomposition based Skewness of ECG signals. in *Michael Faraday IET International Summit 2015*, (2015), p. 109 (5 .), https://doi.org/10.1049/cp.2015.1705, ISBN: 978-1-78561-186-5, Kolkata, India, 12–13 Sept 2015
6. M. Al-Abed, M. Manry, J.R. Burk, E.A. Lucas, K. Behbehani, A Method to detect obstructive sleep apnea using neural network classification of time-frequency plots of the heart rate variability, in *29th Annual International Conference of the IEEE Engineering in Medicine and Biology Society*, (2007), pp. 6101–6104. https://doi.org/10.1109/iembs.2007.4353741
7. M.O. Mendez, A.M. Bianchi; M. Matteucci, S. Cerutti, T. Penzel, Sleep apnea screening by autoregressive models from a single ECG lead, in *IEEE Transactions on Biomedical Engineering* (2009), vol 56, Issue 12, pp. 2838–2850. https://doi.org/10.1109/tbme.2009.2029563
8. S.A. Rezaeieh, A. Zamani, K.S. Bialkowski, A.M. Abbosh, Unidirectional slot-loaded loop antenna with wideband performance and compact size for congestive heart failure detection, in *IEEE Transactions on Antennas and Propagation* (2015), vol 63, Issue 10, pp. 4557–4562. https://doi.org/10.1109/tap.2015.2457935
9. C.M. Held, R.J. Roy, Hemodynamic management of congestive heart failure by means of a multiple mode rule-based control system using fuzzy logic. IEEE Trans. Biomed. Eng. **47**(1) 115–123 (2000). https://doi.org/10.817626
10. S. Chattopadhyay, G. Sarkar, A. Das, Spider and histogram assessment of electrocardiogram for Apnea diagnosis, in *International Conference of IMBIC MSAST-2016*, vol 5 (2016), pp. 149–153. ISBN 978-81-925832-4-2

Part VII
Modelling and Simulation in General Application

Analysis of Retinal OCT Images for the Early Diagnosis of Alzheimer's Disease

C. S. Sandeep, A. Sukesh Kumar, K. Mahadevan and P. Manoj

1 Introduction

Alzheimer's disease (AD) is one of the prominent types of brain disorder related to memory that is increasing worldwide. The relevant incidence shows that the prevalence of AD increases as the age of population increases [1]. AD leads to continuous memory loss and cognitive dysfunction leading to a drop in executing routine functions and learning. The other symptoms of AD are visual abnormalities, apraxia, aphasia, and agnosia [2, 3]. The problems on vision are usually seen in AD patients that affects the routine life. The usual vision complaints seen in AD are loss of contrast responsiveness related to space, unable to identify locomotion discernment, cannot discriminate color tone and ocular loss, influencing the principal ocular cortex and other stipulated regions of the brain [4–6]. For the early diagnosis of AD, a noninvasive neuroimaging technique like Magnetic Resonance Imaging (MRI) is essential and widely used technique that provides detailed information about brain structure and cerebral imaging in AD patients. The most usual findings that are seen in MRI are degeneration of cells in the entorhinal cortex, medial temporal lobes and hippocampus.. In addition to this there is an enlargement in ventricles and depletion

C. S. Sandeep (✉) · A. Sukesh Kumar
Department of ECE, College of Engineering, Trivandrum, Kerala, India
e-mail: sandeepcs07nta@gmail.com

A. Sukesh Kumar
e-mail: drsukeshkumar@yahoo.in

K. Mahadevan
Department of Ophthalmology, SGMC&RF, Trivandrum, India
e-mail: eyemahadevan@rediffmail.com

P. Manoj
Department of Neurology, SGMC&RF, Trivandrum, India
e-mail: mnjparameswaran@gmail.com

S. Chattopadhyay et al. (eds.), *Modelling and Simulation in Science, Technology and Engineering Mathematics*, Advances in Intelligent Systems and Computing 749,
https://doi.org/10.1007/978-3-319-74808-5_43

of entire brain mass [7]. The investigations and reports made on AD were not able to explain the morphological and functional modification of brain in Alzheimer's patients. The evidence as well as researches regarding clinical and histological AD on retina suggests in such a manner that the identical neuronal disintegration affects brain through the retina. The physical changes that were observed in animals and post mortem of human retina are loss of axons and retinal ganglion cell (RGC) [8–11]. The aggregates of tau and Amyloidβeta protein build up inside the retina region and its microvasculature. By another study, the early indications of inflammation in the neurons were present in the retina [12–16]. Consequently, the studies made on number of clinical observations and microscopic structure of retina of AD shows that there is strong confirmation of retinal dysfunction in AD patients [17–19].

OCT is one of the promising non-invasive retinal imaging technique for determining the different section of eye images of neuronal disintegration. OCT scanning technique has emerged as one of the mostly used technology to detect and find out the damage in axonal region of different neuronal diseases. The damage in the axonal region can be evaluated by computing the thickness of retinal nerve fiber layer (RNFL), gives an unintended approximation of RGC layer dysfunction. The loss of neurons in the retina can be instantaneously determined by measuring thickness of macular of the eye because around 30–35% of this section mainly consists of RGCs and their fibers [20–22]. The brain and retina shares a common embryological origin eye is the outermost extension of brain. Due to this the OCT scans are becoming popular. After obtaining the OCT images of AD patient, next step is to make a database of OCT AD images. For computer based assessment of OCT images we need different processes like segmentation of the input image, after that feature extraction, feature selection and classification of OCT images should be done.

Segmentation of OCT images and dimensionality reduction is the most important process in the area of image analysis. The different Artificial Intelligence (AI) techniques used for analyzing Biomedical images are Support Vector Machine, Genetic Algorithms, Fuzzy Logic And Artificial Neural Networks (ANNs) [23, 24]. The other powerful tool used for analyzing medical images is Wavelet Networks (WNs), based on wavelets which are mathematical functions used to divide the information into its constituent frequency components. With the help of WNs, better signal to noise ratio can be obtained, the information that is outside the image can be minimized, the features of information can be extracted easily and finally it allows universal capacity calculation using Artificial Neural Network (ANN) [25–28]. Because of these out comings, WNs are used in different applications [29, 30]. In ANNs, radial basis functions (RBFs) and multilayer perceptrons (MLP) are used for building algorithm for medical image analysis. But WNs can create optimized algorithm structure than ANNs thereby reducing complexities in calculation [31]. In this research work, a new scheme for analyzing medical OCT images are developed based on WNs. Adaptive wavelet networks (AWNs) and static or Fixed-Grid Wavelet Networks (FGWNs) are the two types of WNs [32]. AWNs uses continuous wavelet transforms (CWTs), are continuous in nature whereas FGWN is a discrete wavelet transform (DWT), discrete in nature. In this research work we have used FGWNs due to the various short comings on initial value computation in AWNs [32].

The features of the proposed method based on FGWN are as follows. By using the proposed method we can find out the shift and scale parameters, as well as the number of wavelets can be calculated very easily. The parameters such as number of wavelets, scale, and shift parameters value can be determined easily. The least square algorithms are used for calculating the neuronal weights. Another feature of the proposed work is that there is no need of specifying random initial values as in AWNs. For analyzing OCT images, segmentation is the most important section. In this research work, for the segmentation of OCT images, a wavelet network with three layers is used; one input layer, one hidden layer and an output layer. In the segmentation section, first we have to select the OCT image from the database. The image is filtered by using median filter. The next step is the channel separation. After this process is over, next step is to normalization of the input image. After normalization we have introduced the Marr wavelet due to its excellent features such as simple calculation, better SNR and easier adaptation of Gaussian structures [33]. Using Marr wavelet, a static or fixed wavelet structure is formed by primary and secondary screening. Therefore we can calculate the function accurately. After this the Wavelet network is formed. Next step is the training process in which neural network training is employed. After training is completed we get the wavelet network output which is further improved using post processing. Then the segmentation of the OCT image is performed and at last the region of interest (ROI) processing can be done in this section. Next is the feature extraction of the segmented images. We have used different features like elliptical calculation, thickness ratio calculation, variance of curvature and salience, average of regional minima of OCT image, area, perimeter as well as second, third, and fourth moments. The necessary features are selected and finally the classification of OCT images has been done. The objective of this research work is to summarize the different findings on OCT images of AD patients, to consider the role of OCT in AD patients and how OCT scans involved in the retinal changes for diagnosing AD at its earlier stage. The previous related works of the author's in diagnosing AD will definitely provide a new solution with this scheme for diagnosing AD [1–5].

2 Segmentation Using Wavelet Network

Image acquisition is the first step in the analysis of OCT images. This is done with the help of OCT device as shown in Fig. 1. OCT is an eye or retina imaging technique used to find out the different eye related disorders. After image acquisition, the unwanted salt pepper noise on the obtained OCT image of the eye from the OCT device is removed by using median filter. Next step is to build a Wavelet Network (WN) using FGWN as explained in the previous section. During processing, initially a wavelet scan is performed using primary or "mother wavelet". This mother wavelet produces a multidimensional wavelet frame for providing better regularities.

In this research work for creating wavelet framework, the Marr wavelet with d-dimension is used. After creating mother wavelet, the secondary or "child wavelets"

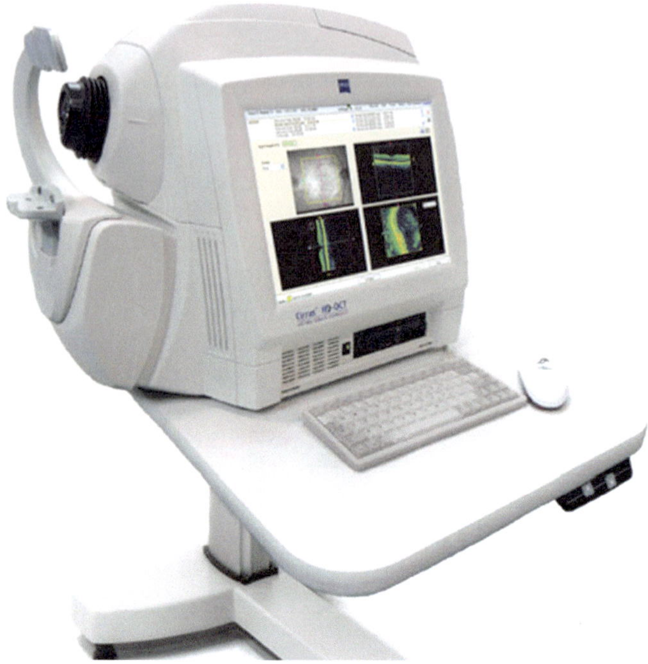

Fig. 1 A typical OCT device

are obtained through shifting and scaling of primary or mother wavelet. Wavelets are mostly helpful for reducing the size of OCT image value from a bigger one. The structure of WN is similar to ANN. The output equation of a WN with '*d*' inputs, '*q*' wavelets in the middle layer and one output as in Eq. (1).

$$\sum_{l=1}^{n} w_l \psi_{pl,ql}(X) = \sum_{l=1}^{n} w_i 2^{-p_l d/2} \psi(2^{pl} X - ql) \quad (1)$$

where weight coefficients 'w_l', ranges from $l = 1, 2, \ldots, n$, 'ψ' denotes primary wavelet function, '*pl*' denotes shift parameter '*ql*' denotes scale parameter, '$\psi pl,ql$' stands for dilated and translated versions of a mother wavelon function [31]. The WNs can construct networks with systematic algorithms easier than ANNs. [34]. After constructing the networks, the weights w_l in (1) can be find out using linear evaluation schemes that helps in constructing the proposed discrete wavelet network or FGWN. In WNs the input data changes for a broad range of values, minimize the efficiency in different investigations. Therefore in the proposed work, to limit the scattering of input data, normalization is applied as the preprocessing stage. In this stage each OCT image's red, green and blue values of RGB matrix are mapped into [0, 1] [35]. This can be done using the Eq. (2).

$$x_{n,new}^{(k)} = \frac{x_{n,old}^{(k)} - t_k}{T_k - t_k} \tag{2}$$

where $x_{n,new}^{(k)}$ is the value of each red, green and blue color matrix after normalization, t_k is the initial value and T_k is the final values of these matrices, respectively. Next step in the proposed work is to select the primary wavelet. In this work we have used multi or d-dimensional Marr wavelet is used as the primary or mother wavelet. We have chosen Marr wavelet because it gives better regularities and can produce wavelet frame with single scale multi-dimensional structure [36]. From the mother wavelet, we can produce necessary child wavelets. The Eq. (3) shows the Marr wavelet scheme employed in this research work is given as:

$$\psi(x) = \eta\|x\| = \left(d - \|x\|\hat{}2\right)e\hat{}\left(-\|x\|\hat{}2|2\right) \tag{3}$$

Next step in the proposed FGWN construction is to determine two parameters namely, scale and shift parameters. Therefore, the employment of two scale levels in this stage is required, minimum and maximum levels $[p_{\min}, p_{\max}]$ and also the shift parameter. The wavelet function with grit on the wavelon parameters space is used to find out the input vectors as in expressed in Eq. (4).

$$\psi_{pl.qj}(x) = 2^{-pld/2}\psi(2^{pl}x - q_j) \tag{4}$$

The creation of wavelon lattice structure is required in the next step. In this process we have to provide two screening stages. In the first screening stage, for every input vector, we have to create I_k set for every scale level selected. In the second screening process, the parameters, shift and scale are selected from at least two sets are calculated and set *I or* wavelet matrix is formed with input data. After the formation of set *I*, some matrix values are no longer needed. They can be eliminated in the next stage from the redundant values formed during the set *I* using an efficient algorithm as described below. In this step we have used orthogonal least squares (OLS) algorithm, a fast and reliable algorithm for creating determination models [31]. By using OLS algorithm the best subset of wavelet *W* can be selected as follows. First is to select the most important wavelets formed on the previous stage. Then the redundant wavelons are made perpendicular to the selected. As the next step in OLS, the remaining redundant wavelons are made perpendicular to the selected and so on. Thus the selected wavelets can easily isolate for the proposed WN construction [37] and the wavelelon network is defined as in Eq. (5)

$$f = \sum_{l=1}^{s} w_l \psi_l(x) \tag{5}$$

In the above equation, 's' denotes number hidden layer wavelons and '*wl*' is denotes weight of wavelons. By using wavelons, the proposed WNs hidden layer

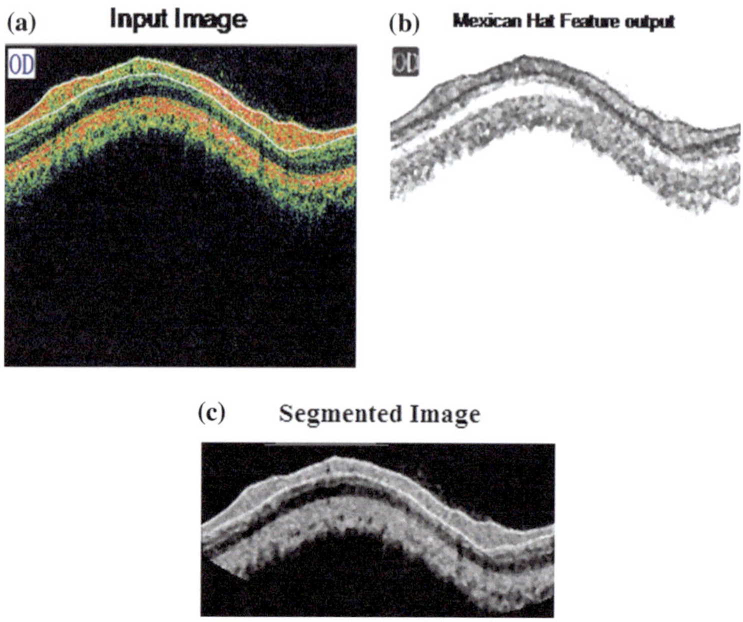

Fig. 2 **a** input image. **b** Marr Wavelet output. **c** segmented image

nodes can be constructed. Next is to find the index of the proposed wavelet network and desired error value is measured. This is calculated in Eq. (6) [8].

$$MSE = \frac{1}{P} \sum_{k=1}^{P} (\hat{f}^{(k)} - f^{(k)})^2 \tag{6}$$

Image segmentation is prominent of the different steps involved in the analysis of medical images like OCT. For the segmentation of OCT image, the proposed algorithm from the preceding stage is required. The dataset used in this work is obtained from Sree Gokulam Medical College and Research Foundation, Trivandrum, India and is not publicly available. From the OCT dataset, ten images are selected randomly for constructing the proposed wavelet network or FGWN. The network is created from the OCT images in such a way that the output is zero, if the pixel is inside the OCT image and vice versa and consequently the proposed network is constructed. The input image, Marr Wavelet output and segmented image of OCT image WN segmentation of the proposed network shown on Fig. 2.

From the above, we have segmented the OCT image with the proposed DWN called Fixed Grid Wavelet Network. In AD patients, Retinal Nerve Fiber Layer (RNFL) thinning occurs in the retina. This retinal change can be identified in an OCT image. For the automatic analysis of OCT images, this proposed method reduces complexity and very useful as it extracts the RNFL layer for further processing and

Table 1 Comparison of FGWN, NN and GVF (all values are in percentage)

Method	Accuracy	Precision	Sensitivity	Specificity
FGWN	99.65	94.77	94.32	99.82
NN	99.53	92.15	93.34	99.73
GVF	98.83	82.28	83.94	98.83

for detecting AD at its earlier stage. We have compared the segmentation process of the proposed FGWN with Neural networks (NN) and Gradient Vector Flow (GVF) for different parameters like accuracy, precision, sensitivity and specificity as shown in Table 1. The proposed approach shows better results than the other two.

3 Feature Extraction

After segmentation part is done for OCT images, the succeeding stage is to extract the features from the image. The extraction of RNFL is the most significant process after segmentation from the OCT image for diagnosing AD. In this paper, for RNFL feature extraction, at first we have removed the unwanted noise and additional parts on the background are eliminated. Thus the required RNFL is extracted using necessary morphological processes like area, perimeter, third and fourth moments, ellipticity, thickness ratio calculation, curvature variance, salience variance, and average number of regional minima [38].

4 Classification of OCT Images

After the selection of the most necessary features, the remaining process involved is the classification of images. In this research work, we have focused on Back Propagation (BP) and Radial Basis Function (RBF) of Neural Networks (NNs) for the classification.

A. *Classification using BP*

The weight parameters of input-to-hidden neurons are calculated as in Eq. (7).

$$w_{np}^{m+1} = w_{np}^{m} - \eta \frac{\partial E^{(m)}}{\partial w_{np}} \tag{7}$$

From the above, 'm' denotes the repeated sequence number; 'n, p' stands for the ratio of input neuron and hidden neuron, and 'η' denote step size. The hidden layer in the BP networks consist of more than one sigmoid neurons succeeded by linear neurons in the output layer. The BP allows more number of nonlinear neuronal layers with transfer functions for learning input and output linear-nonlinear relationships.

The output neuronal layer produces a range of values starting from negative one to positive one in the NNs [30].

B. *RBF Classification*

In RBF type classification of neural networks, supervised training method is used to accomplish function mapping as identical to multi-layer neural network. The RBF Neural network classifier consists of '*l*' values in the input layer to accept '*l*' multi-dimensional input non magnitude elements. The output layer of the network is find out by multiplying hidden layer neuron with the weighting factor *w(n, p)*. Therefore in each RBF unit '*m*', the center value is chosen as the mean that belong to class '*m*' as calculated in Eq. (8).

$$\mu_m = \frac{1}{N_m} \sum_{n=1}^{N_m} x_m^n \tag{8}$$

In the above equation, 'x_m^n' stands for the eigenvector of the '*n*th' image in class '*m*', and 'N_m' stands for trained images in class '*m*'. The input vector and vector distance is used for calculating hidden or middle layer activation function [31].

From the BP and RBF classification of neural network performed for OCT image analysis, RBF shows better results than BP in accuracy, specificity and time taken during execution.

5 Experimental Results

The dataset required for this research work of OCT images is taken from Sree Gokulam Medical College and Research Foundation (SGM & RF), Trivandrum, India and it is not publicly available. The authors are directly included in the study as per the permission received from the ethics committee. The consent from the participants has been obtained through interviewing. For making an automated computerized analysis for early diagnosis, OCT dataset is required. We cannot directly get OCT images with the support of physicians in Neurology and Ophthalmology. In this regard we have selected 50 patients for the study. The patients are initially diagnosed by the Neurologist through MRI and found that 25 are AD victims. For further investigations, all the 50 patients from the Neurologist were sent to Ophthalmologist for OCT scanning. From those 50 patients, 100 OCT images were obtained from both eyes because AD may affect in a single eye or both. The OCT images procured from the OCT machine are saved as bitmap file extension for further processing. The each OCT image obtained from OCT machine is 535,974 bytes. Consequently all the OCT images have been saved successfully for analysis. The dimension of the saved image is high and so we have reduced the images to 256 × 256 sizes for further processing. From the OCT images, unwanted noise are removed in the preprocessing stage and made free from distortion [36]. Next step is to segment the image using

Table 2 Comparison proposed method with GVF and FBSM

Method	FAR in %	FRR in %
Proposed	1	1
GVF	2	2.5
FBSM	3	3.7

the proposed wavelet network or FGWN. For the various steps included in the segmentation process that is already described in Sect. 2, out of 100 OCT images, 10 images is sufficient for the construction of wavelet lattice, calculating shift and scale parameters, and weight coefficients. The proposed method has achieved good results as compared to NN and GVF, already explained in Sect. 2. The different features like area, perimeter, third and fourth moments, ellipticity, thickness ratio calculation, curvature variance, salience variance, and average number of regional minima are extracted after the segmentation process. The classification of OCT images were done using BP and RBF, the later showed good results. In this experiment RBF classifier which we selected is compared with GVF and fuzzy based split-and-merge algorithm (FBSM) using false acceptance rate (FAR) and false rejection rate (FRR). The proposed system achieved good results compared to GVF and FBSM which is shown in Table 2.

6 Conclusion

In this research work, a new scheme is introduced using wavelet networks or FGWN for the analysis of medical OCT images for diagnosing AD at its earlier stage. Here the segmentation has been on wavelet networks and classification on Neural Networks. By comparing the RNFL between AD and control subjects we can measure the changes in retinal layer for early diagnosis. In this paper we have used Marr Wavelet function for constructing multidimensional wavelet lattice. Also the weight of wavelons, shift and scale parameters has calculated using this scheme. The optimization of the network structure for segmenting OCT images has been done using the proposed algorithm. After that different feature of OCT images are extracted. We have compared the proposed method of WN with NN and GVF, the former shows better results. In our method, the training and testing a wavelet network several times with the same data would lead to the same results improving the accuracy, specificity, precision and sensitivity. After that feature selection has done and classified the features using NN. For classification, we used Neural Network RBF classifier as it achieves good results than BP, the former needs less time for training. In this work we have compared the proposed work with GVF and FBSM using FAR and FRR, give better results than the other two. From the findings, it is clear that the proposed method on WN can be used for automatic analysis of OCT medical images for detecting AD. As for a future study an efficient algorithm could be used for WN construction.

Acknowledgements The authors in this research work are very much thankful to SGM & RF, Trivandrum, India for the support for conducting the study and providing the required dataset. The authors are also thankful to the Institutional Ethics Committee Sree Gokulam Medical College & Research Foundation standard operating Procedures (SGMC-IEC: SOPS) members, Dr.V Mohanan Nair (Chairman), Dr. Regi Jose (Member Secretary IEC), Dr. K K Manojan (Member, Institution Review Board (IRB), IEC), Dr. P.Sivasankarapillai (Chairman IRB) and Dr. Jeesha C Haran (Secretary IRB) for giving the permission for the study.

References

1. C.S. Sandeep, A.S. Kumar, A review on the early diagnosis of Alzheimer's Disease (AD) through different tests, techniques and databases. AMSE J.–2015-Series: Modell. C. **76**(1), 1–22 (2015)
2. C.S. Sandeep, A.S. Kumar, K. Mahadevan, P. Manoj, Feature extraction of MRI brain images for the early detection of Alzheimer 's disease. Bioprocess. Eng. **1**(2), 35–42 (2017). https://doi.org/10.11648/j.be.20170102.11
3. C.S. Sandeep, A.S. Kumar, K. Mahadevan, P. Manoj, Dimensionality reduction of optical coherence tomography images for the early diagnosis of Alzheimer's disease. Am. J. Electr. Electron. Eng. **5**(2), 58–63 (2017). https://doi.org/10.12691/ajeee-5-2-4
4. C.S. Sandeep, A.S. Kumar, A psychometric assessment method for the early diagnosis of Alzheimer's disease. Int. J. Sci. Eng. Res.-IJSER. **8**(3) (2017). (ISSN 2229-5518)
5. C.S. Sandeep, A.S. Kumar, "A review paper on the early diagnosis of Alzheimer's Disease (AD) through profiling of human body parameters, Scientistlink, Coimbatore, India, 2013. Int. J. Comput. Sci. Eng. Commun. (IJCSEC), **1**(1), 21–29 (2013)
6. M.P. Frosch, D.C. Anthony, U.D. Girolami, The central nervous system, in *Robbins and Cotran Pathologic Basis of Disease*, ed. by S.L. Robbins, V. Kumar, A.K. Abbas, R.S. Cotran, N. Fausto (Elsevier srl, Philadelphia, 2010), pp. 1313–1317. ISBN-10: 1416031219
7. R.A. Harvey, P.C. Champe, B.D. Fisher, *Lippincott's illustrated reviews: microbiology*, 2nd edn. (Lippincott Williams and Wilkins, 2006), p. 432. ISBN-10: 0781782155
8. J.C. Blanks, S.Y. Schmidt, Y. Torigoe, K.V. Porrello, D.R. Hinton, R.H. Blanks, Retinal pathology in Alzheimer's disease. II. Regional neuron loss and glial changes in GCL. Neurobiol. Aging **17**, 385–395 (1996)
9. J.C. Blanks, Y. Torigoe, D.R. Hinton, R.H. Blanks, Retinal pathology in Alzheimer's disease. I. Ganglion cell loss in foveal/parafoveal retina. Neurobiol. Aging **17**, 377–384 (1996)
10. D.R. Hinton, A.A. Sadun, J.C. Blanks, C.A. Miller, Optic-nerve degeneration in Alzheimer's disease. N. Engl. J. Med. **315**, 485–487 (1986)
11. A.A. Sadun, C.J. Bassi, Optic nerve damage in Alzheimer's disease. Ophthalmology **97**, 9–17 (1990)
12. R.M. Cohen, K. Rezai-Zadeh, T.M. Weitz, A. Rentsendorj, D. Gate, I. Spivak et al., A transgenic Alzheimer rat with plaques, tau pathology, behavioral impairment, oligomeric abeta, and frank neuronal loss. J. Neurosci. **33**, 6245–6256 (2013)
13. M. Koronyo-Hamaoui, Y. Koronyo, A.V. Ljubimov, C.A. Miller, M.K. Ko, K.L. Black et al., Identification of amyloid plaques in retinas from Alzheimer's patients and noninvasive in vivo optical imaging of retinal plaques in a mouse model. Neuroimage. **54**(1), S204–S217 (2011)
14. B. Liu, S. Rasool, Z. Yang, C.G. Glabe, S.S. Schreiber, J. Ge et al., Amyloid-peptide vaccinations reduce β-amyloid plaques but exacerbate vascular deposition and inflammation in the retina of Alzheimer's transgenic mice. Am. J. Pathol. **175**, 2099–2110 (2009)
15. A. Ning, J. Cui, E. To, K.H. Ashe, J. Matsubara, Amyloid-beta deposits lead to retinal degeneration in a mouse model of Alzheimer disease. Invest. Ophthalmol. Vis. Sci. **49**, 5136–5143 (2008)

16. S.E. Perez, S. Lumayag, B. Kovacs, E.J. Mufson, S. Xu, Beta-amyloid deposition and functional impairment in the retina of the APPswe/PS1DeltaE9 transgenic mouse model of Alzheimer's disease. Invest. Ophthalmol. Vis. Sci. **50**, 793–800 (2009). https://doi.org/10.1167/iovs.08-2384
17. C.A. Curcio, D.N. Drucker, Retinal ganglion cells in Alzheimer's disease and aging. Ann. Neurol. **33**, 248–257 (1993). https://doi.org/10.1002/ana.410330305
18. D.C. Davies, P. McCoubrie, B. McDonald, K.A. Jobst, Myelinated axon number in the optic nerve is unaffected by Alzheimer's disease. Br. J. Ophthalmol. **79**, 596–600 (1995)
19. V. Parisi, R. Restuccia, F. Fattapposta, C. Mina, M.G. Bucci, F. Pierelli, Morphological and functional retinal impairment in Alzheimer's disease patients. Clin. Neurophysiol. **112**, 1860–1867 (2001)
20. M.L. Monteiro, L.P. Cunha, L.V. Costa-Cunha, O.O. Maia Jr., M.K. Oyamada, Relationship between optical coherence tomography, pattern electroretinogram and automated perimetry in eyes with temporal hemianopia from chiasmal compression. Invest. Ophthalmol. Vis. Sci. **50**, 3535–3541 (2009)
21. M.L. Monteiro, D.B. Fernandes, S.L. Apostolos-Pereira, D. Callegaro, Quantification of retinal neural loss in patients with neuromyelitis optica and multiple sclerosis with or without optic neuritis using Fourier-domain optical coherence tomography. Invest. Ophthalmol. Vis. Sci. **53**, 3959–3966 (2012)
22. M.L. Monteiro, C.L. Afonso, Macular thickness measurements with frequency domain-OCT for quantification of axonal loss in chronic papilledema from pseudotumor cerebri syndrome. Eye **28**, 390–398 (2014)
23. K.-S. Cheng, J.-S. Lin, C.-W. Mao, Techniques and comparative analysis of neural network systems and fuzzy systems in medical image segmentation. Fuzzy Theor. Syst. Tech. Appl. **3**, 973–1008 (1999)
24. J. Jiang, P. Trundle, J. Ren, Medical image analysis with artificial neural networks. Comput. Med. Imag. Graph. **34**(8), 617–631 (2010)
25. R.M. Balabin, R.Z. Safieva, E.I. Lomakina, Wavelet neural network (WNN) approach for calibration model building based on gasoline near infrared (NIR) spectra. J. Chemometr. Intell. Lab. Syst. **93**(1), 58–62 (2008)
26. Q. Zhang, A. Benveniste, Wavelet networks. IEEE Trans. Neural Netw. **3**(6), 889–898 (1992)
27. Y.C. Pati, P.S. Krishnaprasad, Analysis and synthesis of feedforward neural networks using discrete affinewavelet transformations. IEEE Trans. Neur. Netw. **4**(1), 73–85 (1992)
28. H.H. Szu, B.A. Telfer, S.L. Kadambe, Neural network adaptive wavelets for signal representation and classification. Opt. Eng. **31**(9), 1907–1916 (1992)
29. H. Zhang, B. Zhang, W. Huang, Q. Tian, Gabor wavelet associative memory for face recognition. IEEE Trans. Neural Netw. **16**(1), 275–278 (2005)
30. O. Jemai, M. Zaied, C.B. Amar, M.A. Alimi, Pyramidal hybrid approach: wavelet network with OLS algorithm-based image classification. Int. J. Wavel. Multir. Inf. Process. **9**(1), 111–130 (2011)
31. R. Galvao, V.M. Becerra, M.F. Calado, Linear–wavelet networks. Int. J. Appl. Math. Comput. Sci. **14**(2), 221–232 (2004)
32. S.A. Billings, H.L. Wei, A new class of wavelet networks for nonlinear system identification. IEEE Trans. Neural Netw. **16**(4), 862–874 (2005)
33. J. Gonzalez-Nuevo, F. Argueso, M. Lopez-Caniego, L. Toffolatti, J.L. Sanz, P. Vielva, D. Herranz, The mexican hat wavelet family application to point source detection in CMB maps. Mon. Not. Roy. Astron. Soc. **369**, 1603–1610 (2006)
34. Y. Oussar, G. Dreyfus, Initialization by selection for wavelet network training. Neurocomputing **34**(1), 131–143 (2000)
35. R. Baron, B. Girau, Parameterized normalization: application to wavelet networks. Proc. IEEE Int. Conf. Neural Netw. **2**, 1433–1437 (1998)
36. Q.H. Zhang, Using wavelet network in nonparametric estimation. IEEE Trans. Neural Netw. **8**(2), 227–236 (1997)

37. M. Davanipoor, M. Zekri, F. Sheikholeslam, Fuzzy wavelet neural network with an accelerated hybrid learning algorithm. IEEE Trans. Fuzzy Syst. **20**(3), 463–470 (2012)
38. F. Mokhtarian, S. Abbasi, Shape similarity retrieval under affine transforms. Pattern Recognit. **35**(1), 31–41 (2002)

Real Time Diagnosis of Rural Cardiac Patients Through Telemedicine

R. Ramu and A. Sukesh Kumar

1 Introduction

Telemedicine in India will allow patients in rural areas to have access to specialist doctors in urban hospitals. Leading cause of death in rural areas is due to cardiac diseases. Prevalence of cardiovascular diseases in India is higher than other countries of the same region [1]. Early detection of heart disease has important significance for heart disease prevention and timely treatment. Developments in telemedicine due to the new technologies in electronics industry helps in the monitoring of patients with cardiac disorders within the home or rural hospitals. Change in the rate of heart rhythm is called arrhythmia and are difficult to obtain on an ECG tracing which are captured within few seconds. Some of the arrhythmia are dangerous like ventricular fibrillation which is the main cause of cardiac arrest or stroke. So early detection of these arrhythmias for people living in rural areas can be made possible by Telemedicine.

Telemedicine uses electronic communications to exchange medical information for improving the patient's health. There are three main categories of Telemedicine namely remote monitoring, store and forward and interactive telemedicine. In this work remote monitoring category is used for diagnosis of cardiac patients. Developments in wireless technologies leads to wireless telemedicine in which doctors can view physiological data of patients from anywhere at any time. There are different wireless technologies that are used to transmit ECG signals such as Bluetooth, Zigbee Wi-Fi and GSM. This work is an extension of the earlier works of the authors [2–4]. Wireless ECG monitoring using Bluetooth low energy (BLE) technology consist of

R. Ramu (✉)
Rajadhani Institute of Engineering and Technology, Trivandrum, Kerala, India
e-mail: mail2ramureghu@gmail.com

A. Sukesh Kumar
Rajiv Gandhi Institute of Development Studies, Trivandrum, Kerala, India
e-mail: drsukeshkumar@yahoo.in

S. Chattopadhyay et al. (eds.), *Modelling and Simulation in Science, Technology and Engineering Mathematics*, Advances in Intelligent Systems and Computing 749,
https://doi.org/10.1007/978-3-319-74808-5_44

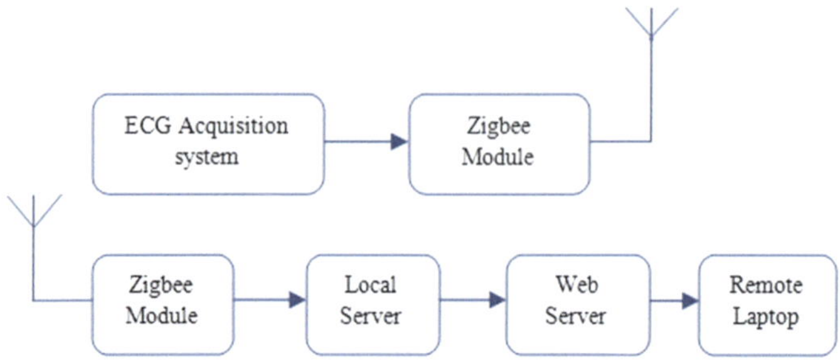

Fig. 1 Block diagram of the system

acquisition module Bluetooth module and smart phone. It capture ECG signal and transmit the ECG data via the Bluetooth wireless link and display it in a smart phone [5]. ECG transmission using Wi-Fi technology is developed which consist of a single chip ECG signal acquisition module, Wi-Fi module and a smart phone [6]. Zigbee technology is used to transmit the ECG signals in real time. Using Zigbee transmitter and receiver are used for transmission and LabVIEW is used to plot the signal in the laptop. The ECG signals transmitted to a remote laptop using zigbee are stored in the same lap top configured as server and are finally plotted in another laptop using internet [7]. In this work ECG signal from the acquisition module is transmitted to the laptop using zigbee module. ECG signal from the laptop is transmitted to the web server from the local server. The ECG signals are stored in the database of virtual private server and plotted in the browser of any laptop using internet.

2 System Description

The aim of the work is to design an inexpensive highly accurate ECG acquisition and wireless transmission system using zigbee and web server. The Fig. 1 shows the main block diagram of the system. The ECG acquisition system consists of electrodes, instrumentation amplifier, filters and microcontroller. It captures the ECG signal from the surface of the body, amplifies the signal to volt range. Filters will remove undesired signal and pass only the ECG signal.

The analog output from the filter is given to microcontroller for analog to digital conversion (ADC) and the serial data from the output of microcontroller is converted to USB standard and fed to laptop using USB cable. Wamp server is used as local server and it takes the incoming serial data and transmit it to the web server. The transmitted ECG signals are stored in the data base of the server. Virtual private server is used as server and finally, the ECG signals are then plotted in the website www.ecgtrack.com. Doctor in the specialty hospital can view the ecg signal of patient from rural area in real time and can diagnose the patient.

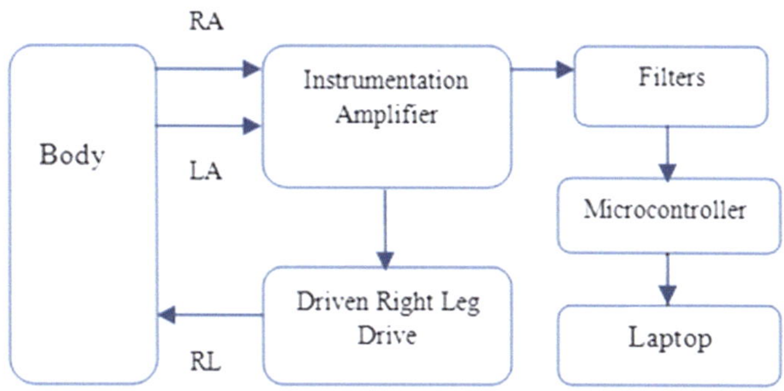

Fig. 2 Block diagram of ECG acquisition system

3 ECG Acquisition System

The signals acquired from the electrodes are amplified, filtered, digitized, and transmitted. Block diagram of ECG acquisition system is shown in Fig. 2. For three lead systems, electrodes Right arm (RA), Left arm (LA) and Left leg (LL) are used, two of the electrodes are used to form lead and the third is used as the ground.

Lead I configuration is used, it is the voltage between Left arm electrode and Right arm electrode (I = LA – RA). ECG signals vary from microvolt to the millivolt range due to this small range, the signals measured need to be amplified in order to be better interpreted [8]. Texas Instrument's instrumentation amplifier INA321 EA is used here. With an internally set gain of 5, the INA321 can be programmed for gains greater than 5 [9].

Instrumentation amplifier is a differential amplifier with additional input buffer stages. It has low offset voltage, high Common mode rejection ratio(CMRR), high input impedance and high gain. Driven Right leg drive is used to reduce the common mode interference. Filters using op-amps are used to remove the unwanted signals and line frequency noise. The last stage of the acquisition system is ATmega 328 microcontroller which is a low power CMOS 8 bit microcontroller based on AVR enhanced RISC architecture with 32 K bytes of FLASH 1 Kbytes of EEPROM and 2Kbytes of SRAM. Real time ECG signals are first digitized and converted to serial data using ATmega 328 which have an in built 10 bit ADC. The output serial data from the microcontroller is then converted to USB standard using FT232RL USB to Serial UART adapter.

The ECG acquisition system used in this work is light weight and easily portable device. This device can be placed at the patient side for capturing the real time ecg signals. For transmitting the signals to nearby PHC this acquisition system can be interfaced with wireless modules. Among the different wireless technologies available Zigbee technology is used here to transmit ecg signals to PHC.

4 Wireless Transmission

Zigbee technology is used in this work to transmit ECG signal from acquisition system to a laptop in PHC. It is a low-cost, low power, wireless mesh network standard which operates in the industrial, scientific and medical (ISM) radio bands [10]. It is less expensive and can transmit signals up to 100 m. The ZigBee network layer natively supports star, tree and mesh networks, Zigbee series 2 module is used as transmitter and receiver.

This module allow very reliable and simple communication between microcontrollers and systems. Analog ECG signals are digitized and converted to serial data by Atmega microcontroller. This serial data is fed to zigbee module and are then transmitted wirelessly using zigbee technology.

5 Reception and Display of the ECG Signal

Zigbee module and MAX232 IC are used in the receiver side. The received signal from the Zigbee module is TTL level, so it should be converted into RS232 level. For this level conversion MAX232 IC is used. The Signals from Zigbee module is connected to MAX232 IC where this TTL signals are converted to RS232. The output serial data from MAX232 are fed to PC using serial connector DB9. For connecting to laptop FT232 can be used for serial to USB conversion.

6 Ecg Display on Laptop Using Labview

Before plotting the ECG in the browser the signals are viewed in the laptop using LabVIEW. LabVIEW programs consist of two windows a front panel and block diagram. XY graph is used in this work to display the ECG in real time. The back panel, which is a block diagram, contains the graphical source code. Figure 3 shows the block diagram for receiving the serial ECG data and plotting it in real time.

7 Ecg Display on Local Server

Before loading ECG signals in the database of Virtual private server(VPS) and ploting in web page, these signals are plotted in local server for testing. ECG signals are wirelessly transmitted using Zigbee technology to a remote laptop. The Serial ECG data received by the laptop are stored in the data base of the server configured as local server. These Signals are then retrieved and plotted in the browser of the system configured as server. Web servers are computers that deliver web pages and

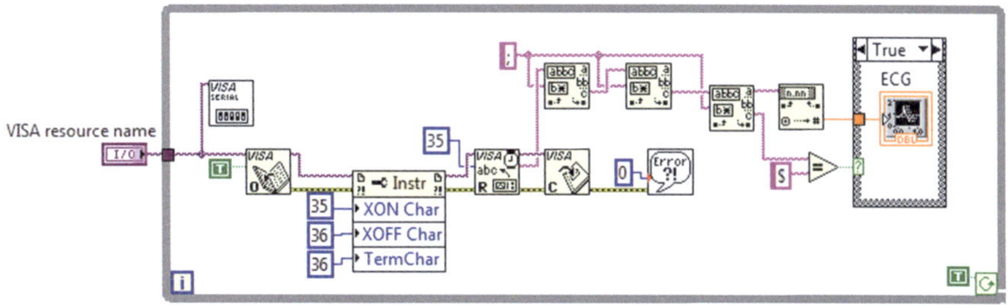

Fig. 3 Block diagram for ECG display in LabVIEW

any computer can be turned into a Web server by installing server software. In this work WAMP Server is used to make the laptop as local server. It is a Windows OS based program that installs and configures Apache web server, MySQL database and PHP scripting language [11].

8 ECG Display on Internet

ECG signals stored in the database of local server are send to the data base of web server and stored there. It is retrieved and plotted in the web browser. Domain name www.ecgtrack.com is taken and hosted on a web server. Virtual private server is used in this work for web hosting. It has many advantages over shared hosting such as complete control, can customize the appearance and settings etc. JQuery is used to plot the ECG signal in real time in the browser of the server. For viewing patients ECG in real time doctor can log on to the web site www.ecgtrack.com. First page is the login page for the doctor for authentication. Figure 4 shows the login page for doctor. After entering the required username and password it direct to display page for ECG. When the leads are properly placed and transmitting and receiving sections are ON the ECG signals will be first stored in the local server database and finally in database of virtual private server. From there it is finally plotted in this display page. JQuery is a fast JavaScript Library and Flot is a pure JavaScript plotting library for JQuery are used in this work to plot the ECG signal. It is simple to use, have attractive looks and also have interactive features [12]. ECG signal of remote cardiac patient can be transmitted from patient side to nearby PHC by Zigbee wireless technology. Signals are received with the help of zigbee receiver and are stored in the data base of laptop. configured as local sever in the PHC. On logging on to the website doctor in the specialty hospital can see the ecg of the remote cardiac patient and diagnose him.

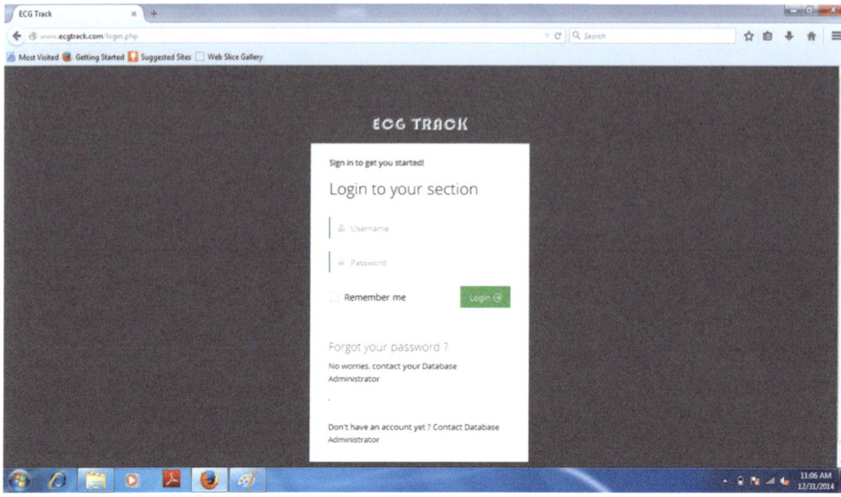

Fig. 4 Login page for doctor

9 Results

ECG signals wirelessly transmitted from patient side using Zigbee technology is plotted and displayed in the laptop using LabVIEW software. Figure 5 shows transmitted ECG signal received and plotted in PC using LabVIEW. Figure 6 shows the ECG signal plotted in the browser of the laptop configured as server. ECG signals from the data base of the local server are send to the web server and are stored in the database of web server. It is plotted in the web browser of any other laptop and can be viewed by logging on to the web site www.ecgtrack.com. Figure 7 shows the ECG signal plotted in the web browser. Five patients ecg was taken and wirelessly transmitted and displayed in the web page. It was successfully diagnosed by the doctor.

Bradycardia
Bradycardia means heart rate less than 60 beats per minute. It has regular rhythm, normal QRS duration and P wave—Visible before each QRS Complex. If heart rate computed is less than 60 beats per minute, then it is detected as bradycardia.

Trachycardia
Trachycardia is for heart beat greater than 60 beats per minute. It also has regular rhythm, normal QRS duration and P wave—Visible before each QRS Complex. Fifty patients ECG has been transmitted and diagnosed by this method.

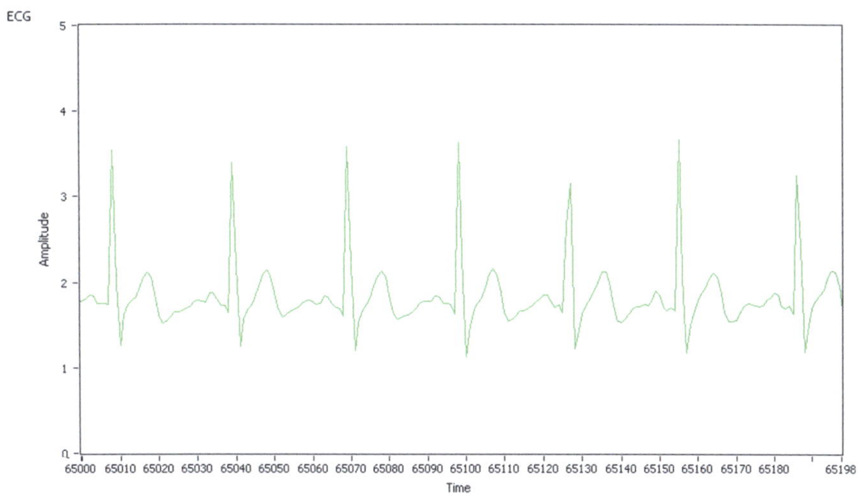

Fig. 5 Transmitted ECG signal received and plotted in PC using LabVIEW

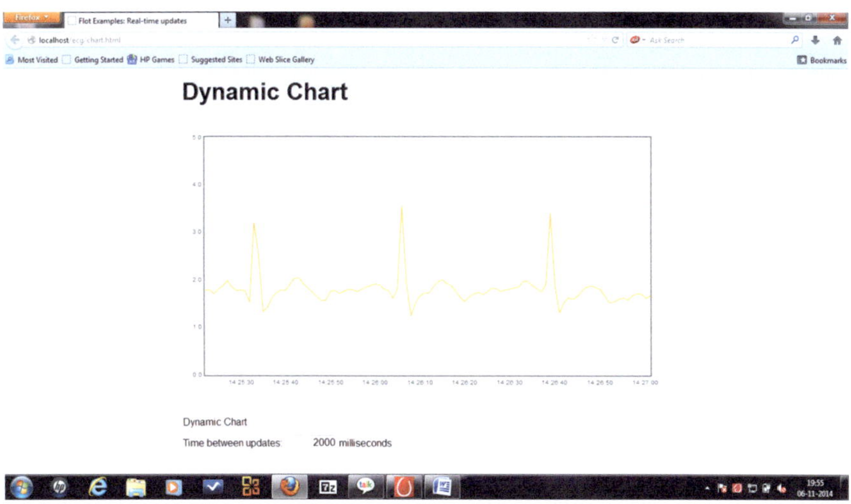

Fig. 6 ECG signal plotted in the browser of the laptop configured as server

10 Conclusion

ECG signals of patients in a rural area can be transmitted to a laptop in the nearby PHC by using Zigbee technology. Laptop in the PHC can be configured as local server and the ECG data are stored in the database of the local server. Website www.ecgtrack.com is created and hosted in a virtual private server with MySQL database. From the local server data are send to the data base of web server. Stored signals in

Fig. 7 Screen shot of the potted ECG waveform in the web browser

the MySQL data base of web server are retrieved and plotted in the web browser. So that doctor in a specialist hospital can view the ECG signal by logging on to the web site and can give medical instructions to the doctors in the PHC. Fifty patients diagnosis have been done with this system.

References

1. S. Chauhan, Dr. B.T. Aeri, Prevalence of cardiovascular disease in India and its economic impact-A review, in *International Journal of Scientific and Research Publications*, vol 3, Issue 10, October 2013
2. R. Ramu, A. Sukesh Kumar, Real-time monitoring of ECG using Zigbee Technology, in *International Journal of Engineering and Advanced Technology (IJEAT)*, vol 3, Issue 6, pp. 169–172 (2014). ISSN: 2249-8958
3. R. Ramu, A. Sukesh Kumar, Monitoring of ECG using Zigbee and wamp server, in *International Journal of Innovative Research in Technology and Science (IJIRTS)*, vol 3, Issue 4, pp. 50–54, July (2015). ISSN: 2321-1156
4. R. Ramu, A. Sukesh Kumar, Online Assessment of rural cardiac patients through Telemedicine, in *Proceedings of the 2nd National Conference on Advances in Computational Intelligence and Communication Technologies, LBSIT forWomen*, 25–26, February 2016
5. B. Yu, L. Xu, Y. Li, Bluetooth low energy based mobile electrocardiogram monitoring system, in *Proceedings of the IEEE International Conference on Information and Automation*, China 2012
6. S. Shebi Ahammed, Binu C. Pillai, *Design of Wi-Fi based mobile Electrocardiogram monitoring system on concerto platform* (Procedia Engineering, Elsevier IConDM, 2013)
7. R. Ramu, S. Kumar, Monitoring of ECG using Zigbee and Wamp Server, in *International Journal of Innovative Research in Technology and Science (IJIRTS)*, vol 3, Issue 4, pp. 50–54. ISSN 2321-1156

8. Wu Baochun, Li Min, Yang Yaning and Zhang Weiwei, ECG acquisition circuit design based on C8051F330, in *Proceedings of the IEEE-EMBS International Conference on Biomedical and Health Informatics*, China, 2012
9. INA321 Datasheet by Texas instruments
10. Zigbee/Xbee Data Sheet by Digi International
11. D. Ipswich, Setting UP a WAMP Server On Your Windows Desktop, Technology Now at Smashwords, 2011
12. S.V.O. Khanna, M. Mistry, Impact of JQuery in Web Domain, in *International Journal of Advanced Research in computer science and software engineering*, vol 1, Issue 1 (2011)

A Comparative Analysis of a Healthy Retina and Retina of a Stroke Patient

R. S. Jeena and A. Sukesh Kumar

1 Introduction

Stroke is a form of cardiovascular disease affecting the blood supply to the brain. It remains as a leading cause of disability and death for people of all races and ethnicities. Stroke [1] is a physical condition that occurs due to insufficient supply of blood to the brain cells. This damages the brain cells ultimately leading to their death. A clot in the blood vessel or a blood vessel rupture can interrupt the blood supply to brain. Stroke can be categorized into two: ischemic and hemorrhagic. Ischemic stroke accounts for nearly 85% of the cases. When a blood vessel providing blood to the cerebral framework is blocked, Ischemic stroke comes about. Hemorrhagic stroke happens when a debilitated blood vessel breaks. At the point when a hindrance happens inside a blood vessel providing blood to brain, the vessels conveying blood to eye will likewise be influenced amid the underlying stages. This is named as Retinal ischemia. People experiencing Retinal ischemia are more inclined to stroke.

The retina can be viewed and analyzed using non-invasive in vivo functional techniques. Retinal imaging permits diagnosis of different eye illnesses and also the prognosis of diabetes mellitus, blood pressure and cerebrovascular ailments like stroke. Research works show that microvasculature of retina and brain is closely linked in terms of anatomy and physiology [2]. Since retina's capacity makes it an extremely metabolically energetic tissue with two times blood supply, the retina permits direct noninvasive examination.

Cardiovascular ailment reveals itself in the retina in different ways. Blood pressure and coronary heart disease cause variations in the ratio between the diameter of reti-

R. S. Jeena (✉)
Department of ECE, College of Engineering, Trivandrum, Kerala, India
e-mail: jeena_rs@yahoo.com

A. Sukesh Kumar
Rajiv Gandhi Institute of Development Studies, Vellayambalam, Trivandrum, Kerala, India
e-mail: drsukeshkumar@yahoo.in

S. Chattopadhyay et al. (eds.), *Modelling and Simulation in Science, Technology and Engineering Mathematics*, Advances in Intelligent Systems and Computing 749,
https://doi.org/10.1007/978-3-319-74808-5_45

nal arteries and veins. Tapering of the arteries and broadening of veins is associated with an upgraded danger of stroke [2]. Arterioles of retina share practically identical anatomical, physiological and embryological features with arterioles of brain. Morphological changes in the outline of blood vessel, branching design, width, tortuosity, retinal sores, fractal measurement, branching angle and coefficient are some of the irregularities in vascular pattern of retina related with cardiovascular diseases like stroke. This work is an extension of author's earlier works for stroke prediction [3].

2 Literature Survey

The Retinex is a human-observation based computer vision design which offers steadiness of color and dynamic range compression. The possibility of Retinex was visualized by Land [4] as a model of lightness and shading view of the human vision.

Daniel J. Jobson et al. has executed multiscale retinex [5], which seals the gap between color images and the human representation of scenes. Terai et al. [6] proposed a retinex model for color image contrast improvement which lessens the computation time by handling the luminance flag. The algorithm works well for gray images.

Feng et al. [7] presented the quick Fourier calculation strategy to make the computational time speedier than that of the conventional method. The strategy functions performs well for colour pictures but not for gray scale images.

3 Methodology

Multiscale retinex has been implemented in this work for the pre processing of retinal images. Morphological operations are done for the removal of optic disc and background. For proper vessel segmentation, thresholding can be used to create binary images. To reduce all objects in the vascular map to lines, skeletonization is done. Branching points are identified and parameters like eccentricity, major axis length and orientation are computed. This has been implemented for both healthy retinas and retinas of stroke patients and the results are compared.

3.1 Preprocessing

The Retinex is an established preprocessing technique that can be applied to retinal images. It can provide sharpening, shading constancy, dynamic range compression and color rendition at the same time redressing the blurring in profound anatomical structures and heterogenity in biomedical images. It can be applied successfully for both color and gray scale images.

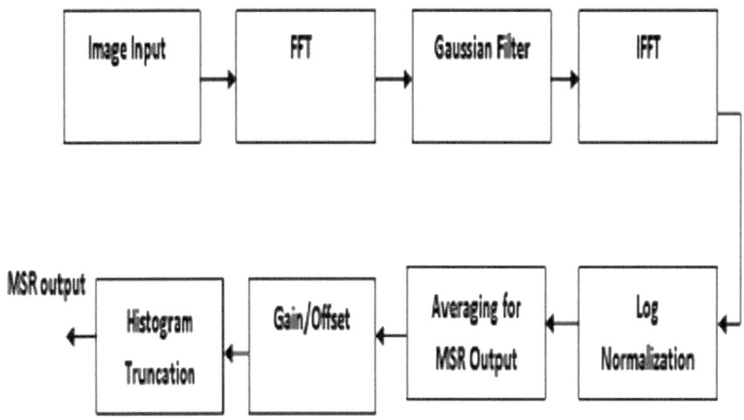

Fig. 1 Block diagram of Multiscale retinex

In Single Scale Retinex (SSR), the image is passed through the retinex filter, which is basically a Gaussian filter. Based on the filter output, the original image is scaled and then processed with a logarithmic function.

SSR is mathematically expressed as

$$S_i(x, y) = \log(F_i(x, y)) - \log(F_i(x, y) * G(x, y)) \tag{1}$$

Gaussian filter G is defined by

$$G(x, y) = k \exp\left[-\left(x^2 + y^2\right)/\sigma^2\right] \tag{2}$$

F_i Input image on the ith colour channel
S_i Retinex output image on the ith channel

Since image filtering using retinex function require various Gaussian shaped impulse response with different variance, MSR (Multiscale Retinex) technique [8] is utilized. The weighted sum of the outputs of various SSRs gives the MSR output value [6].

The MSR can be written as

$$S_{MSRi}(x, y) = \sum_k W_k.\left\{\log(F_i(x, y)) - \log\left[F_i(x, y) * G_k(x, y)\right]\right\} \tag{3}$$

N is the number of scales ($k = 1, 2 \ldots N$), W_k is the weight of each scale.

Block diagram representation of Multiscale Retinex is given in Fig. 1.

3.2 Morphological Operations

The output image obtained after retinex preprocessing is converted to grayscale and subjected to morphological operations. Structuring elements of different shapes and sizes are created using MATLAB functions. The morphological opening function with disc shaped structuring element is applied on this preprocessed image to remove the optic disk and background [9].

3.3 Skeletonization

After removing the background and optic disk, image is then converted to binary form by thresholding. The structural outline of a plane area can be reduced to a line graph called skeleton. The skeleton of the region can be obtained by a thinning algorithm. Thinning shrinks the binary image objects to a set of skeletal strokes that preserve major information about the shape of the original entity. After obtaining skeletonized output, the branching points of skeleton are detected and the lengths of branches are computed. MATLAB function computes the geodesic distance transform of the binary image. A set of features for each connected component in the binary image are computed. They include parameters [10] like eccentricity, major axis length and orientation. Parameters have been computed for both healthy retinas and retinas of stroke affected patients.

4 Result and Discussion

Figure 2 shows the input images of a healthy retinal fundus and retina of a stroke patient.

Figure 3 shows the output after applying multiscale retinex for preprocessing.

Retinex output is the converted to binary form by thresholding. Figure 4 shows the binarized output.

Figure 5 shows the skeletonized image.

After skeletonization, the branch points and end points are detected. Detection of end points and branching points are shown in Fig. 6.

Endpoints are shown in blue and branching points are shown in red.

Length of the branches are calculated and are shown in Fig. 7a, b.

Table 1 shows the values computed for both the images by analyzing its branches.

Results show that number of branching points is much higher in the case of stroke patients and branching vessels seems to be more tortuous, which can be observed visually from the processed images. This methodology has been applied to images of retinal ischemia (American retinal bank) and healthy images (DRIVE database), which substantiated the above results. Retinal ischemia proves to be an important

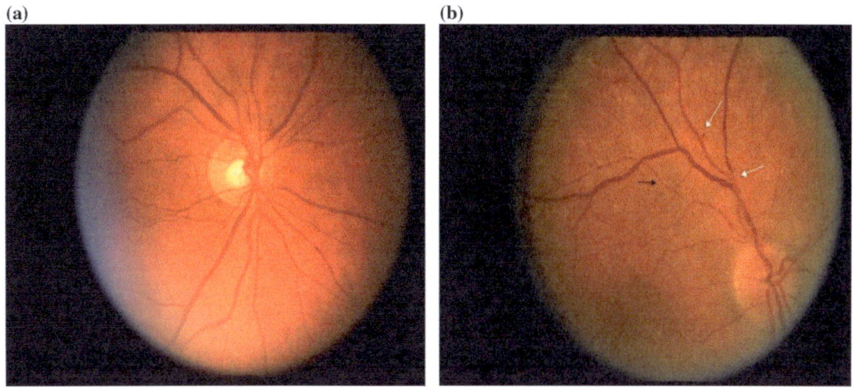

Fig. 2 **a** Healthy Retina **b** Retina of a Stroke patient. (Image Courtesy: Centre for Vision Research at Sydney University's Westmead Millennium Institute)

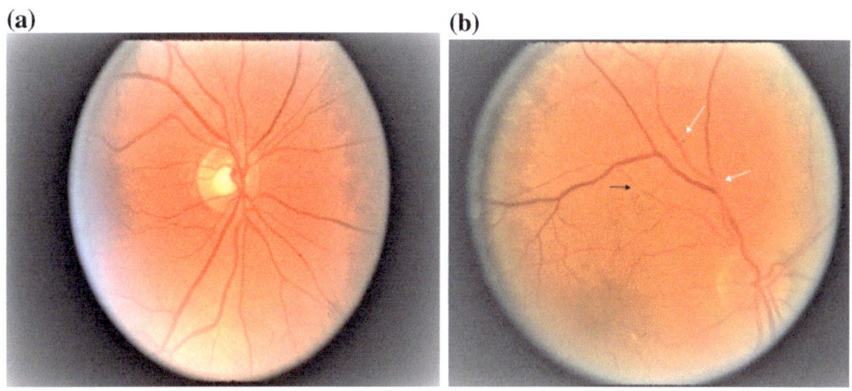

Fig. 3 **a** Retinex preprocessed healthy Retina **b** Retinex preprocessed retina of a Stroke patient

Table 1 Evaluation of parameters

Retinal fundus image	Major axis length	Eccentricity	Orientation	Number of end points	Number of branch points
Healthy person	25.78	0.722	1.16	102	115
Stroke patient	16.61	0.812	3.76	283	303

biomarker of stroke as healthy retinal fundus images seems to give a mean major axis length in the range [20–26] and fundus images of retinal ischemia gives a mean major axis length in the range [15–19].

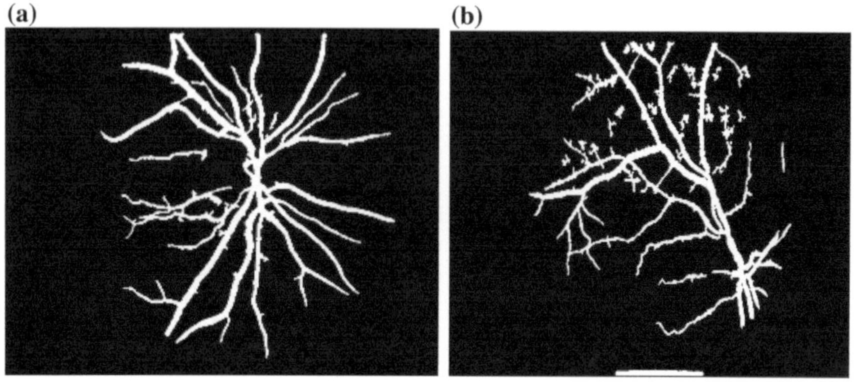

Fig. 4 **a** Binarized healthy retina **b** Binarized Retina of stroke patient

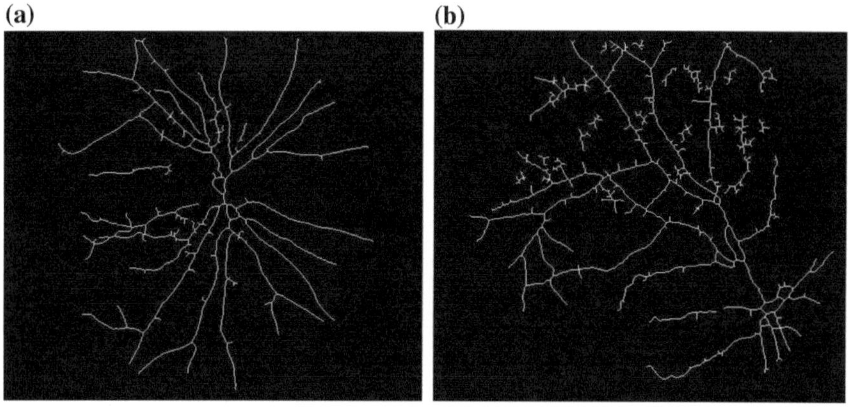

Fig. 5 **a** Skeletonized healthy retina **b** Skeletonized retina of stroke patient

5 Conclusion

Early detection of cardiovascular diseases like stroke through biomarkers derived from retinal imaging would allow patients to be treated more effectively. Retinal imaging aids in predicting the probability of stroke based on parameters evaluated from the vascular map. Performance of the system can be improved by incorporating more features like tortuosity and fractal dimension of the branching vessels and requires training from a much larger database. Interdisciplinary groups will be able to investigate the interface at the fringe between ophthalmology and neurology. Since the microvasculature of cerebral sensory system and optic framework are interlinked, it is unequivocal that retinal neurovascular variations are prognostic of microvascular variations in cerebrum.

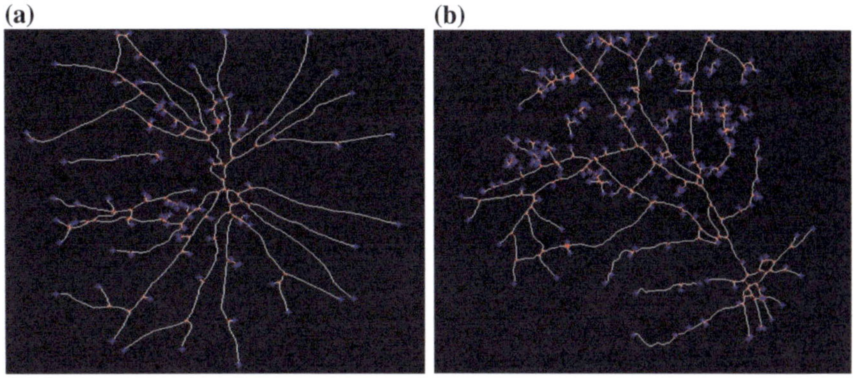

Fig. 6 **a** Detection of branching points and endpoints in Healthy Retina **b** Detection of branching points and endpoints in Retina of Stroke patient

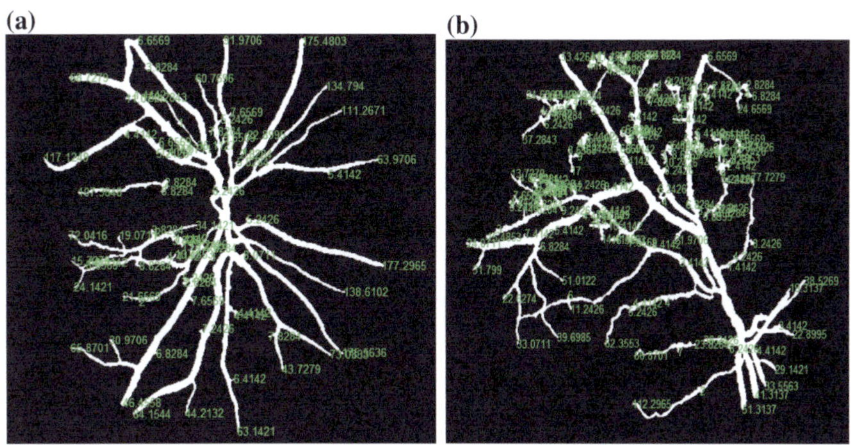

Fig. 7 **a** Length of branches computed for Healthy Retina **b** Length of branches computed for Retina of Stroke patient

Acknowledgements I would like to thank Dr. Mahadevan, Sree Gokulam Medical College and Research Foundation, Trivandrum for his valuable guidance in the smooth conduct of this work.

References

1. Barry L. Zaret, M.D., Marvin Moser, M.D., Lawrence S. Cohen, Chapter 18 *Stroke - Lawrence M. Brass*, M.D., pp 215–234
2. M.L. Baker, J.J. Wang, G. Liew, et al., Differential associations of cortical and subcortical cerebral atropy with retinal vascular signs in patients with acute stroke. Stroke **41**, 2143–50
3. R.S. Jeena, Dr. A.S. Kumar, Stroke prediction using SVM, ICCCITT 2016, Tamil Nadu

4. E. Land, An alternative technique for the computation of the designator in the retinex theory of color vision. Proc. Nat. Acad. Sci. **83**, 3078–3080 (1986)
5. D.J. Jobson, Z. Rahman, G.A. Woodell, A Multiscale retinex for bridging the gap between color images and the human observation of scenes. IEEE Trans. Image Process. **6**(7), 965–976 (1997)
6. Y. Terai, T. Goto, S. Hirano, M. Sakurai, Color image contrast enhancement by retinex model, in *Proceedings of IEEE 13th International Symposium on Consumer Electronics*, pp. 392–393, May 2009
7. Y. Feng, J. Huang, Z. Feng, M. Liu, The research and implementation of light compensation algorithm in color facial image, in *Electrical and Control Engineering Conference* (2011), pp. 2758–276
8. J.K. Priyanka, Dr. B.G. Sudarshanan, Dr. S.C. Prasanna Kumar, Dr. N. Pradhan, Development of algorithm for high resolution retinex for image enhancement. Int. J. Innovative Res. Dev. 2012 (December)
9. P. Manjiri, M. Ramesh, R. Yogesh, S. Manoj, D. Neha, Automated localization of optic disk, detection of microaneurysms and extraction of blood vessels to bypass angiography, in *Advances in Intelligent Systems and Computing*, (Springer, 2014). ISBN: 978-3-319-11933-5. https://doi.org/10.1007/978-3-319-11933-5_65
10. M.B. Patwari, R.R. Manza, Y.M. Rajput, D.D. Rathod, M. Saswade, N. Deshpande, Classification and calculation of retinal blood vessels parameters, in *IEEE's International Conferences for Convergence of Technology*, Pune, India

Square Root Quadrature Information Filters for Multiple Sensor Fusion

Aritro Dey, Smita Sadhu and Tapan Kumar Ghoshal

1 Introduction

The Information filters for nonlinear signal models is an emerging area of research and continuing interest in this form is reported in recent works [1–3]. The works of [1, 2] report square root versions of information filters based on sigma points and cubature points which have improved numerical performance over the standard error covariance form. The performance of such estimators, although performance wise less accurate compared to particle filter based approach [3], require considerably less computational effort and make real time implementation of this filters feasible. The square root approach has advantages of enhanced numerical accuracy, preservation of symmetry, availability of square root so that steps for Cholesky factorization can be avoided. Because of the above referred attributes square root approach has been recommended in previous works [1, 2, 4] over the standard error covariance form. A variety of nonlinear information filters viz., Unscented information filters [5], Central Difference information filters [6], Cubature and higher order cubature information filters [7, 8, 9] have been reported in literature which focus on multiple sensor estimation with nonlinear signal models. In these works where multi sensor estimation is carried out satisfactorily employing these estimators with information filters configuration. Information filtering algorithms are also extended for some

A. Dey (✉)
Department of Integrated Flight Control System, Aeronautical Development Agency, Bangalore 560017, Vimanapura, India
e-mail: aritro_dey@mail.ada.gov.in

S. Sadhu · T. K. Ghoshal
Department of Electrical Engineering, Jadavpur University, Raja S.C. Mallik Road, Kolkata 700032, India
e-mail: ssadhuju@gmail.com

T. K. Ghoshal
e-mail: tkghoshal@gmail.com

S. Chattopadhyay et al. (eds.), *Modelling and Simulation in Science, Technology and Engineering Mathematics*, Advances in Intelligent Systems and Computing 749,
https://doi.org/10.1007/978-3-319-74808-5_46

critical estimation problems (specifically in target tracking) where out of sequence measurements are obtained [10]. However, none of these cited algorithms include the square root framework and may terminate for some applications where accuracy of measurements are high or a linear combination of state vector components is known with better accuracy while other combinations may not be well observable [1]. However, these publications indicate continued interest in this special form of the estimation algorithm. In this paper, the authors present novel algorithms for information filters in square root framework which are based on recently reported quadrature rule for numerical approximation of Bayesian integrals. The proposed estimators are formulated using two different quadrature rules, viz., Gauss Hermite rule [11] and Cubature quadrature rule [12] so that in addition to the aspects of square root approach the accuracy of the estimators can be further improved over the existing estimators. The proposed information filters which are not yet reported in literature are evaluated using a non trivial aircraft tracking problem in sensor fusion configuration where their superiority has been demonstrated over the competing filtering available in litersture.

2 Problem Statement

We consider a nonlinear dynamic system as given below

$$\boldsymbol{x}_k = \boldsymbol{f}(\boldsymbol{x}_{k-1}) + \boldsymbol{w}_k \tag{1}$$

$$\boldsymbol{y}_k^{\zeta} = \boldsymbol{g}^{\zeta}(\boldsymbol{x}_k) + \boldsymbol{v}_k^{\zeta} \tag{2}$$

Here $\boldsymbol{x}_k \in \boldsymbol{R}^n$ is the state vector, the noise term $\boldsymbol{w}_k \in \boldsymbol{R}^n \sim (\boldsymbol{0}, \boldsymbol{Q}_k)$ indicates zero mean process noise (Gaussian). $\boldsymbol{y}_k^{\zeta} \in \boldsymbol{R}^m$ is the measurement available from the ζth sensor among M different sensors where $\zeta = 1, \ldots, M$. The measurement noise of each sensor is also considered to be white noise (Gaussian) and denoted as, $\boldsymbol{v}_k^{\zeta} \in \boldsymbol{R}^m \sim (\boldsymbol{0}, \boldsymbol{R}_k^{\zeta})$. The square root of $\boldsymbol{Q}_k$ is $\bar{\boldsymbol{S}}_k^{\boldsymbol{Q}}$ and that of $\boldsymbol{R}_k^{\zeta}$ is $\left(\bar{\boldsymbol{S}}_k^{\boldsymbol{R}}\right)^{\zeta}$.

3 Square Root Quadrature Information Filters

The proposed square root quadrature information filtering algorithm is presented in three parts. The first part presents a general approach of information filters wherein the quadrature points are to be incorporated from the second part. Sensor fusion method is presented in the third part.

3.1 General Algorithm

(i) *Initialisation:* Initialize $\hat{\boldsymbol{x}}_0, \hat{\mathbf{S}}_0, \bar{\mathbf{S}}_k^{\boldsymbol{Q}}, \left(\bar{\mathbf{S}}_k^{\boldsymbol{R}}\right)^{\zeta}$

(ii) *Time update step*:

Select the quadrature points as explained in Sect. 3.2

$$\hat{\boldsymbol{\chi}}_i = \hat{\mathbf{S}}_{k-1}\boldsymbol{q}_i + \hat{\boldsymbol{x}}_{k-1} \tag{3}$$

Compute predicted estimate as

$$\bar{\boldsymbol{x}}_k = \sum_{i=1}^{N} \boldsymbol{f}(\hat{\boldsymbol{\chi}}_i) w_i \tag{4}$$

Compute the weighted, centred (updated estimate of previous instant is subtracted off) matrix $\boldsymbol{S}_k^x$ such that ith element of $\boldsymbol{S}_k^x$ is:

$$\left(\boldsymbol{S}_k^x\right)_i = \left(\boldsymbol{f}(\hat{\boldsymbol{\chi}}_i) - \bar{\boldsymbol{x}}_k\right)\sqrt{w_i} \quad \text{for } i = 1,\ 2, \ldots, N \tag{5}$$

The estimate of the square root of predicted error covariance is obtained as

$$\bar{\mathbf{S}}_k = \text{Triangularize}\left(\left[\ \boldsymbol{S}_k^x\ \bar{\mathbf{S}}_k^{\boldsymbol{Q}}\ \right]\right) \tag{6}$$

The information vector can be obtained as

$$\bar{z}_k = \bar{\boldsymbol{S}}_k^{-T}\bar{\boldsymbol{S}}_k^{-1}\bar{\boldsymbol{x}}_k \tag{7}$$

The square root of the predicted information matrix is obtained as

$$\bar{\boldsymbol{S}}_k^{Z} = \text{Triangularize}\left(\bar{\boldsymbol{S}}_k^{-1}\right) \tag{8}$$

(iii) *Measurement update step*:

Select sigma points as

$$\bar{\boldsymbol{\chi}}_i = \bar{\mathbf{S}}_k\boldsymbol{q}_i + \bar{\boldsymbol{x}}_k \tag{9}$$

The predicted estimate of measurement becomes

$$\bar{\boldsymbol{y}}_k^{\zeta} = \sum_{i=1}^{N} \boldsymbol{g}^{\zeta}(\bar{\boldsymbol{\chi}}_i) w_i \tag{10}$$

Compute the weighted, centred (predicted estimate of measurement is subtracted off) matrix $\boldsymbol{S}_k^Y$ such that ith element is:

$$\left(\boldsymbol{S}_k^Y\right)_i = \left(\boldsymbol{g}^{\zeta}(\bar{\boldsymbol{\chi}}_i) - \bar{\boldsymbol{y}}_k^{\zeta}\right)\sqrt{w_i} \quad \text{for} \quad i = 1, 2, \ldots, N \tag{11}$$

Compute the weighted, centred (predicted estimate of state is subtracted off) matrix $\boldsymbol{S}_k^{\bar{x}}$ such that ith element of $\boldsymbol{S}_k^{\bar{x}}$ is:

$$\left(\boldsymbol{S}_k^{\bar{x}}\right)_i = \left(\bar{\boldsymbol{\chi}}_i - \bar{\boldsymbol{x}}_k\right)\sqrt{w_i} \quad \text{for} \quad i = 1, 2, \ldots, N \tag{12}$$

The cross following covariance can be computed as

$$\left(\boldsymbol{P}_k^{xz}\right)^{\zeta} = \boldsymbol{S}_k^{\bar{x}}\left(\boldsymbol{S}_k^Y\right)^T \tag{13}$$

Define the matrix

$$\boldsymbol{\Lambda}_k^{\zeta} = \bar{\boldsymbol{S}}_k^{-T}\,\bar{\boldsymbol{S}}_k^{-1}\left(\boldsymbol{P}_k^{xz}\right)^{\zeta}\left[\left(\bar{\boldsymbol{S}}_k^{R}\right)^{\zeta}\right]^{-T} \tag{14}$$

The updated estimate of information vector is obtained as

$$\hat{\boldsymbol{z}}_k = \bar{\boldsymbol{z}}_k + \boldsymbol{\Lambda}_k^{\zeta}\left[\left(\bar{\boldsymbol{S}}_k^{R}\right)^{\zeta}\right]^{-1}\left(\boldsymbol{y}_k^{\zeta} - \bar{\boldsymbol{y}}_k^{\zeta} + \left[\left(\boldsymbol{P}_k^{xz}\right)^{\zeta}\right]^T\bar{\boldsymbol{z}}_k\right) \tag{15}$$

The square root of the updated estimate of information matrix becomes

$$\hat{\boldsymbol{S}}_k^{Z} = \text{cholupdate}\left(\bar{\boldsymbol{S}}_k,\ \boldsymbol{\Lambda}_k^{\zeta}, +\right) \tag{16}$$

The square root of the corresponding error covariance matrix

$$\hat{\boldsymbol{S}}_k = \text{Triangularize}\left(\left(\hat{\boldsymbol{S}}_k^{Z}\right)^{-1}\right) \tag{17}$$

Hence the updated estimate of state becomes

$$\hat{\boldsymbol{x}}_k = \hat{\boldsymbol{S}}_k\,\hat{\boldsymbol{z}}_k \tag{18}$$

3.2 Choice of Quadrature Points

In this part two quadrature rules have been presented which generate the quadrature points and the corresponding weights.

3.2.1 Gauss Hermite Quadrature Rule [11]

Compute $\boldsymbol{J}$, a symmetric tri-diagonal, defined as $\boldsymbol{J}_{i,i} = 0$ and $\boldsymbol{J}_{i,i+1} = \sqrt{\frac{i}{2}}$ for $1 \leq i \leq N-1$ for N-quadrature points.

The quadrature points are chosen as $q_i = \sqrt{2}x_i$ where x_i are the eigen values of $\boldsymbol{J}$ matrix.

The corresponding weights (w_i) of q_i is computed as $|(v_i)_1|^2$ where $(v_i)_1$ is the first element of the ith normalized eigenvector of $\boldsymbol{J}$.

3.2.2 Cubature Quadrature Rule [12]

This rule has been recently proposed in [12]. The radial integral of spherical radial cubature rule is approximated with Gauss Laguerre Quadrature rule.

$2nn'$ number of cubature quadrature points are to be selected as

where $\boldsymbol{\xi}_i = \sqrt{2\lambda_j}\ \boldsymbol{e}_i$

λ_j is the solution of n'th order Chebyshev-Laguerre polynomial with $\alpha = n/2-1$:

$$L_{n'}^{\alpha} = \lambda^{n'} - \frac{n'}{1!}\big(n' + \alpha\big)\lambda^{n'-1} + \frac{n'(n'-1)}{2!}\big(n' + \alpha\big)\big(n' + \alpha - 1\big)\lambda^{n'-2} - \cdots = 0 \quad (19)$$

Here, $i = 1, 2, \ldots, 2nn'$, $j = 1, 2, \ldots, n'$ and $k = 1, 2, \ldots, 2n$

Corresponding weights are computed as

$$w_i = \frac{1}{2n\Gamma(n/2)} \frac{n'!\,\Gamma\big(\alpha + n' + 1\big)}{\lambda_j\ \big[\dot{L}_{n'}^{\alpha}(\lambda_j)\big]^2} \tag{20}$$

Details are provided in [12].

3.3 *Multiple Sensor Fusion*

The information contribution of ζth sensor is denoted as

$$\boldsymbol{\varphi}_k^{\zeta} = \boldsymbol{\Lambda}_k^{\zeta}\Big[\Big(\bar{\boldsymbol{S}}_k^{R}\Big)^{\zeta}\Big]^{-1}\Big(\boldsymbol{y}_k^{\zeta} - \bar{\boldsymbol{y}}_k^{\zeta} + \Big[\big(\boldsymbol{P}_k^{xz}\big)^{\zeta}\Big]^{T}\bar{\boldsymbol{z}}_k\Big) \tag{21}$$

The contributions of all the sensors starting from $\zeta = 1, \ldots, M$ are fused to obtain a more reliable estimate as

$$\hat{\boldsymbol{z}}_k = \bar{\boldsymbol{z}}_k + \sum_{\zeta=1}^{M} \boldsymbol{\varphi}_k^{\zeta} \tag{22}$$

The information matrix contribution for the ζth sensor is $\boldsymbol{\Lambda}_k^{\zeta}$. After multiple sensor fusion the square root of the a posteriori information matrix is obtained as

$$\hat{\boldsymbol{S}}_k^{Z} = \text{cholupdate}\left(\bar{\boldsymbol{S}}_k,\ \left[\boldsymbol{\Lambda}_k^1 \cdots \boldsymbol{\Lambda}_k^M\right],\ +\right) \tag{23}$$

3.4 Notes on the Algorithm

On availability of triangular matrices matrix inversion steps in the algorithm can be replaced by backward substitution symbolized by '/' as the latter is computationally economic. The proposed algorithm is for the quadrature rule with non negative weights. This has an additional advantage of unconditionally ensured positive definiteness unlike Square Root Unscented information filter [1, 8].

4 Case Study Using Aircraft Tracking Problem

For demonstration a tracking problem has been considered where an aircraft executing a maneuvering turn has to be tracked. The aircraft is tracked with bearing only measurements from two radars positioned at different locations. The system equation and the measurement equations are provided in [9].

Further investigation revealed that this tracking problem suffers from the track loss cases in the situation when the difference of bearing angles from the radars is considerably low or close to π i.e., where measurements suffer from observability problem. At this situation the lines of sight of two radars do not intersect and the filters fail to estimate satisfactorily due to non unique solution of the observation equation.

The performance of the proposed algorithms is evaluated with the help of RMSE performance and track loss count from Monte Carlo (MC) simulation with 10000 runs. Performance index is defined in [9]. The condition for track loss is defined as $\left\|\sqrt{e_x^2 + e_y^2}\right\|_{\infty} \geq 800\,m$ where e_x and e_y are the position error sequences along x axis and y axis. Note that the RMSE are calculated excluding the track loss cases. For performance comparison same seeds are used to generate the noise sequences for each candidate during MC simulation. Performance of Square root version of 2nd order Cubature Quadrature information filter (SRCQIF) & Gauss Hermite information filter (SRGHIF) is compared with recently reported Square Root Cubature information filter (SRCIF) in [2]. Figures 1, 2 and 3 show that the performances of SRCQIF and SRGHIF are superior to SRCIF as the susceptibility of track loss is less and RMSE (excluding the track loss cases) is low as well compared to SRCIF. Track loss count for SRCIF is 251 while that for SRCQIF and SRGIF are 218 (13.15% less) and 221 (11.95% less) respectively. SRGHIF and SRCQIF work equally well while the latter uses considerably less number of quadrature points (20) compared to SRGHIF (243)

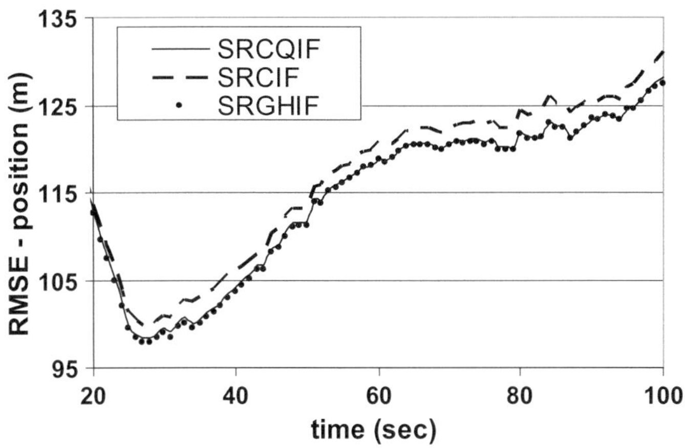

Fig. 1 RMSE of position estimation for 10,000 MC run

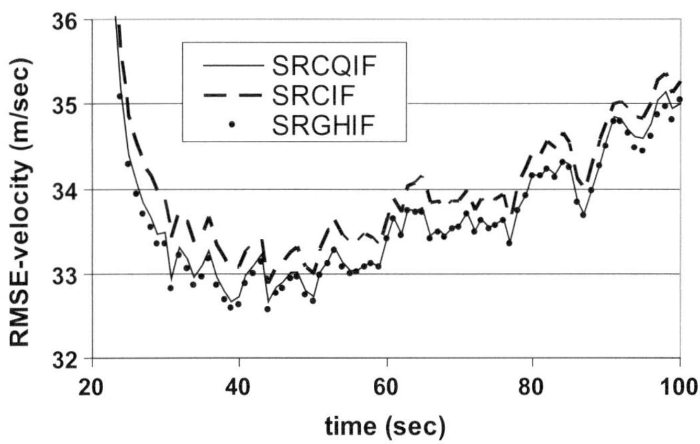

Fig. 2 RMSE of velocity estimation for 10,000 MC run

and therefore computationally economic. Average run time of SRGHIF is 0.2415 s and 8.6 times more than that for SRCQIF (0.0281 s).

5 Conclusion

Square root quadrature information filters (SRGHIF and SRCQIF) have been introduced and superiority of these newly proposed filters has been demonstrated over Square root cubature information filter with the help of a tracking problem in sensor fusion configuration. Between SRGHIF and SRCQIF the latter is advocated for

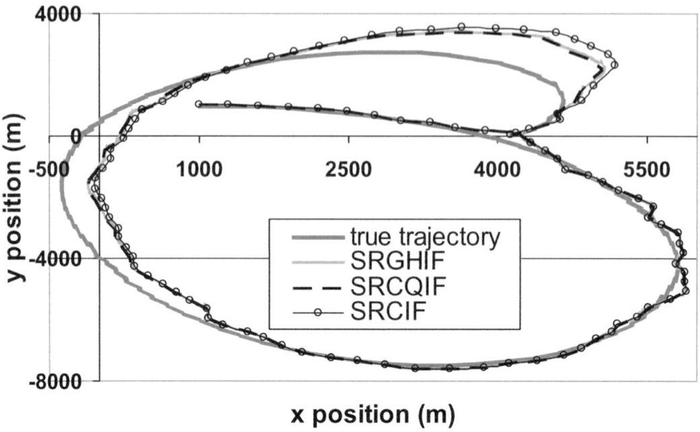

Fig. 3 True and estimated trajectory for a representative run

real time applications because of its numerical stability and substantially improved estimation accuracy at a reasonable computation effort.

Acknowledgements The first author thanks the Council of Scientific and Industrial Research (CSIR), New Delhi, India for financial support and Aeronautical Development Agency, Ministry of Defence, Bangalore, India for infrastructural support.

References

1. G. Liu, F. Worgotter, I. Markelic, Square-root sigma-point information filtering. IEEE Trans. Autom. Control **57**(11), 2945–2950 (2012)
2. K.B. Chandra, D.W. Gu, I. Postlethwaite, Square root cubature information filter. IEEE Sens. J. **13**(2), 750–758 (2013)
3. W. Zhang, J. Zuo, Q. Guo, Z. Ling, Multisensor information fusion scheme for particle filter. IET Electron. Lett. **51**(6), 486–488 (2015)
4. I. Arasaratnam, I.S. Haykin, Square-root quadrature Kalman filtering. IEEE Trans. Signal Process. **56**(6), 2589–2593 (2008)
5. G. Liu, F. Worgotter, I. Markelic, Nonlinear estimation using central difference information filter, in *Workshop on Statistical Signal Processing*, Proc. (SSP) IEEE, pp. 593–596, Nice, France, 2011
6. G. Liu, Bayes Filters with Improved Measurements for Visual Object Tracking, Doctoral dissertation, Göttingen State and University, 2012
7. K.P.B. Chandra, D.W. Gu, I. Postlethwaite, Cubature information filter and its applications, in *Proceedings of the American Control Conference (ACC)*, IEEE, San Francisco, 2011, pp. 3609–3614
8. Q. Ge, C. Wen, S. Chen, R. Sun, Y. Li, Adaptive cubature strong tracking information filter using variational Bayesian method, in *Proceedings of the 19th World Congress of the International Federation of Automatic Control*, Cape Town, South Africa, 2014, pp. 24–29
9. B. Jia, M. Xin, K. Pham, E. Blasch, G. Chen, Multiple sensor estimation using a high-degree cubature information filter, in *Sensors and systems for space applications VI*, Baltimore, Maryland, USA, April 2013, *Proc. SPIE* 8739, pp. 87390T–87390T

10. T.H. Kim, T.L. Song, H.J. Kim, Information filters with reduced data storage for out-of-sequence measurements update. IET Radar Sonar Navig. **10**(6), 1038–1045 (2016)
11. K. Ito, K. Xiong, Gaussian filters for nonlinear filtering problems. IEEE Trans. Automatic Control **45**(5), 910–927 (2000)
12. S. Bhaumik Swati, Cubature quadrature Kalman filter. IET Signal Proc. **7**(7), 533–541 (2013)

Cost Effective, Water Controlled Automated Gardening System

Piyali Mukherjee

1 Introduction

This paper introduces a cost-effective, water-controlled, automated gardening system. The objective of this paper is to design and implement a simple, efficient yet cost effective system that takes care of garden plants by constantly monitoring the moisture content of the soil and ensuring that the soil remains sufficiently hydrated without any kind of manual intervention. This intelligent system thus helps in conserving water as well as watering the plants sufficiently. The need for water conservation is well-elaborated in [1].

Decades back, watering can was used for watering the gardens at home and was later extended to fitting pipes from the tap to the garden which resulted in enormous loss of water. This led to the introduction of various automated watering systems such as sprinkler system, tube, nozzles and many others, but they are often costly to afford. Very recently, researchers have developed automated watering systems using moisture sensors, pumps and other devices but this also turned out to be costly [2–6]. Besides this, alteration of any device, if required, after they have degraded, is also difficult. The aim of this paper is thus to design a watering system that will reduce cost and can be within the reach for all Indian middle-class families, to whom the garden will no more be a luxury. Moreover the devices are easily available. It is also easy to maintain because the devices are readily available. Though design and development of automated gardening systems is in process for the last few years, not much literature is available. For instance, in [7] the authors used a water sprinkler system to build an automated plant watering system. We however, follow a different approach.

P. Mukherjee (✉)
Institute of Radiophysics and Electronics, University of Calcutta, 92, A.P.C Road, Kolkata 700009, India
e-mail: mukherjee.piyali92@gmail.com

S. Chattopadhyay et al. (eds.), *Modelling and Simulation in Science, Technology and Engineering Mathematics*, Advances in Intelligent Systems and Computing 749,
https://doi.org/10.1007/978-3-319-74808-5_47

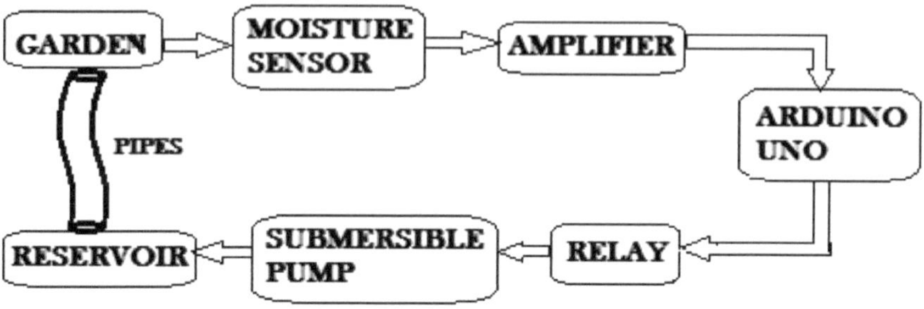

Fig. 1 System block diagram

This automated gardening system can be realized with the help of moisture sensors or moisture probes that detects the resistivity of the soil and in turn its moisture content, a microcontroller [8, 9] that will make decisions on when to water the plants, and also when to terminate the watering process after sufficient hydration, a solid state relay that acts as a switch and controls the watering mechanism and a submerged pump that is placed in a reservoir which helps in pumping water. The microcontroller used here is the Arduino Uno which is an open-source physical computing platform based on a simple microcontroller board, ATmega 328 and a development environment for writing software on the board.

2 Design Algorithm

- Moisture sensors are built using long, non-galvanised nails.
- Amplifier circuit designed using OPAMPs.
- Arduino board is then connected.
- Solid state relay is provided at the output of the Arduino.
- Submerged pump is then placed in the reservoir and proper piping is done.

3 System Block Diagram and Working

Figure 1 shows the basic block diagram of the system with its components. There are three main functional components: the moisture sensors, the solid state relay and the submerged pump.

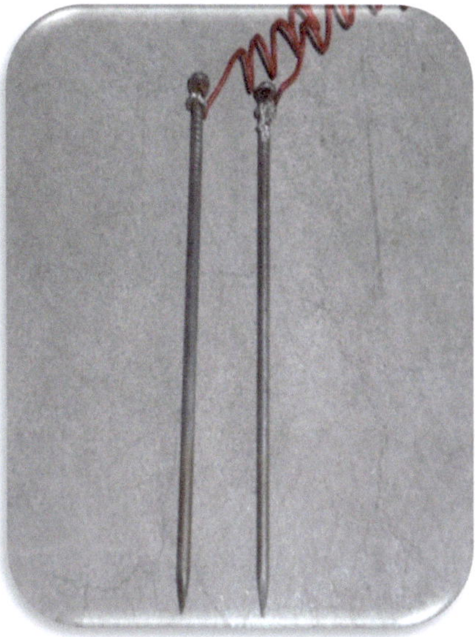

Fig. 2 Moisture sensors

3.1 *Moisture Sensors*

The moisture sensors are developed using non-galvanized nails as shown in Fig. 2. This results in reduction of cost in comparison to the available moisture sensors in the market. The moisture content of the soil controls its resistivity. This property has been used to develop moisture sensors that sense the moisture content of the soil and convert this into an equivalent voltage which is provided to a microcontroller.

3.2 *Arduino Uno*

The Arduino Uno is a microcontroller board based on the ATmega328 memory chip. It has 14 digital input/output pins (of which 6 can be used as PWM outputs), 6 analog inputs, a 16 MHz ceramic resonator, a USB connection, a power jack and a reset button. The ATmega328 is a low-power CMOS 8-bit microcontroller. The microcontroller has 32 KB memory. It also has 2 KB of SRAM (Static Random Access Memory) and 1 KB of EEPROM (which can be read and written with the EPROM library), 23 general purpose input-output lines, 32 general purpose working registers, a serial programmable USART and a 10 bit ADC [10]. The Arduino Uno can be programmed with the Arduino software. The ATmega328 on the Arduino Uno comes pre-burned with a boot loader that allows uploading new code to it without the use of an external hardware programmer.

3.3 Amplifier

The moisture sensor records voltage in the order of millivolts which is very low for the microcontroller to respond to its changes, as the board operates only on a supply of 7–20 V. So, an amplifier is required which will amplify the received voltage to a minimum voltage required for the Arduino to respond. OP-AMPs are very easily available, very cheap and due to its smaller size, they consume very little space in the circuitry. Due to these advantages it has, operational amplifiers (OP-AMPs) are used in this system.

3.4 Solid State Relay

It is an electronic switching device that switches on or off when a small external voltage is applied across its control terminals. It is used for its fast switching speed and has no physical contacts to wear out.

3.5 Submersible Pump

It has a hermetically sealed motor close-coupled to the pump body and helps in pushing water to the surface.

4 Results

The voltage equivalent of the moisture content of dry soil and moist soil (to the extent tolerable by the plants) was measured practically in our laboratory. These voltages were taken as reference voltages for program coding. It was found that the voltage level of dry soil was much lower than the voltage level of moist soil. The reason being that the resistance of dry soil is more negative compared to the resistance of moist soil.

When the garden soil is dry, the soil needs to be watered sufficiently. Accordingly, the relay is turned on by the Arduino which in turn, turns on the pump and watering is done. As watering begins, the soil turns moist increasing its resistance. When the soil becomes sufficiently hydrated, the relay is turned off by the Arduino which again turns off the pump. So watering is done only for the required period, thus saving the excess loss of water.

When the garden soil is moist, the soil does not need to be watered. Hence accordingly, the relay is not turned on by the Arduino and no watering is done.

5 Benefits that This System Provides

- Cost effective garden management: Water is applied to the garden plants on time and shut off on time. There is no need for human intervention and hence reduce manpower.
- Water saving: Watering is done at optimal times. Watering is also kept in check and excess watering is not allowed which leads to conserving water.
- The ability to be absent: One will be able to go on vacations without losing the garden. Proper care of these garden plants will be taken by this system even in the absence of manpower.
- Convenience: The watering is done at the time of necessity and not on our availability.

6 Conclusion

The system is designed in such a way that the plants are watered automatically without any need for human intervention. Moreover, the water level is also kept in check, so that the plants are watered sufficiently but not in excess and with minimum or no loss of excess water. The novelty of this system also lies is the cost-effectiveness, so that this system can be used by all middle-class Indian families. Moreover, the system is assembled in a way that any devices can be replaced easily, if required at any point of time during use. The devices are also easily available in the market at affordable rates. This developed system will not only be limited to house gardens but can also be extended to farmlands and might be of immense benefit to the farmers due to its low cost.

7 Future Scope of Work

The designed system was implemented using a reservoir that has to be filled manually. For large reservoirs, water may be filled in month's interval. But for smooth working of the system, a surface level detector may be designed that will sense the water level of the reservoir. In case the level reaches below the height of the pump, water will be pumped by a motor into the reservoir. So, no manual filling of water will be required and the process will be fully automated.

Acknowledgements The author wish to thank the Electronics and Communication Engineering Department, of Hooghly Engineering and Technology College, Hooghly, West Bengal, India, where the prototype has been developed under the guidance of Ms. Writi Mitra and to all others who have been directly or indirectly associated with this project.

References

1. W.A. Jury, H.J. Vaux, The emerging global water crisis: managing scarcity and conflict between water users. Adv. Agron. **95**, 1–76 (2007)
2. P. Archana, R. Priya, Design and implementation of automatic plant watering system. Int. J. Adv. Eng. Glob. Technol. **4**(1) (2016)
3. J. Cao, S.D. Fleming, M. Burnett, An exploration of design opportunities for 'gardening' end-user programmers' ideas, in *IEEE Symposium on Visual Languages and Human-Centric Computing* (VL/HCC), 18–22 Sept 2011
4. Birasalapati Doraswamy, Automatic irrigation system using Arduino controller. Int. J. Adv. Technol. Innovative Res. **8**(4), 635–642 (2016)
5. K. Kansara, V. Zaveri, S. Shah, S. Delwadkar, K. Jani, Sensor based automated irrigation system with IOT: a technical review. Int. J. Comput. Sci. Inf. Technol. **6**(6), 5331–5333 (2015)
6. S. Darshna, T. Sangavi, S. Mohan, A. Soundharya, S. Desikan, Smart irrigation system. IOSR J. Electron. Commun. Eng. **10**(3), 32–36 (2015) (May–Jun)
7. S.V. Devika, Sk. Khamuruddeen, Sk. Khamurunnisa, J. Thota, K. Shaik, Arduino based automatic plant watering system, Int. J. Adv. Res. Comput. Sci. Softw. Eng. **4**(10), 449–456, (2014) (Oct 10)
8. Ramesh S. Gaonkar, Microprocessor Architecture, Programming & Applications
9. B. Ram, Fundamentals of Microprocessors and Microcomputers
10. Introduction to Arduino Microcontroller, Shanghai Jiao Tong University, June 17–29, 2012

Ear Based Biometric Analysis for Human Identification

Samik Chakraborty, Anumita Mitra, Sanhita Biswas and Saurabh Pal

1 Introduction

Human biometric recognition system has been a demanding subject of research for the few decades as secure surveillance received significant attention in the present days. Biometric system based on several physiological and behavioral characteristics of human being like fingerprint, face, iris etc. are more reliable system than the conventional identification systems such as passwords and PINs [1, 2]. In spite of remarkable advancement in biometric recognition system, identify of individuals in uncontrolled and unconstrained environments remains a challenging problem like face recognition system can be affected due to illumination variance at different parts of the face or under different expressions. A change in voice due to cold and cough or other pathological conditions will make the system difficult to recognize. Recognition using finger may not be effective if the subject has dirty, deformed or cut hand. In compare to other physiological biometric feature ear has certain advantage. Contour of an ear is unique over the years [3] and also remain unchanged with change in expression [4] even person's cooperation is not necessarily required to acquire an ear image. Anatomy of human ear is shown in Fig. 1.

Burge and Burger [5] first developed ear authentication based algorithm where a graph model is constructed from for each ear from its edges and curves and introduces a graph based matching algorithm. Chang et al. [6] used PCA based ear biometric system and received 72% accuracy and also showed that due to hair or jewellary occlusion with pose variation recognition rate drops into 30%. A multimatcher system for ear recognition is proposed by Nanni and Lumini [7] where sub window is created

S. Chakraborty (✉)
AEIE Department, Heritage Institute of Technology, Kolkata, India
e-mail: samikasan0516@gmail.com

A. Mitra · S. Biswas · S. Pal
Applied Physics Department, University of Calcutta, Kolkata, India

S. Chattopadhyay et al. (eds.), *Modelling and Simulation in Science, Technology and Engineering Mathematics*, Advances in Intelligent Systems and Computing 749,
https://doi.org/10.1007/978-3-319-74808-5_48

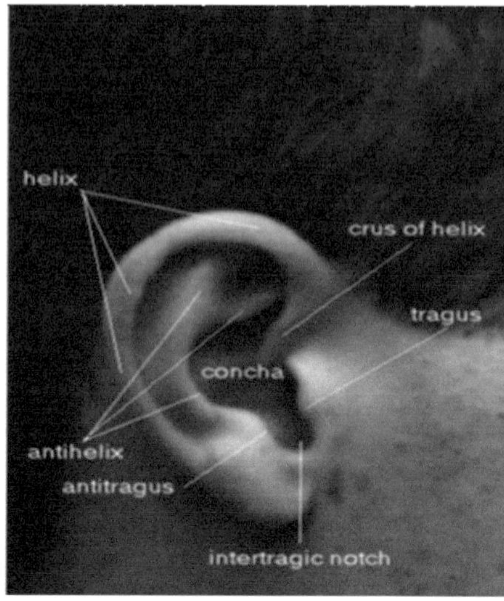

Fig. 1 Anatomy of the human ear

from each subject and features are extracted by the convolution operation with Gabor filters. This achieved 80% recognition rate. Yuan and Mu [8] approached an ear recognition system based on 2 dimensional images and achieved 94% accuracy for 0% occlusion and 85% accuracy for 35% occlusion. Gabor wavelets feature based recognition system is presented by Zhang et al. [9] to deal with partial occlusion and achieved a success rate of 96, 91 and 86% for occlusion of 15, 25 and 35% respectively. Anwar et al. [10] presented a new algorithm based on geometrical features extraction for ear recognition system and achieved 98% accuracy. Chen and Mu [11] proposed a SRC based method for improving the performance of recognition system under pose variations and occlusions and achieved an identification rate of 95.83, 91.67 and 87.5% for occlusion percentage of 15, 25 and 35 respectively. Lei et al. [12] presented an automatic localization of 3-D ear landmarks and classifying poses in profile face images where ear is represented by a novel ear tree based graph. A literature survey on ear recognition can be found in [13].

In this present work, a preliminary study of ear recognition system with 10 subjects is made where jewellary occlusion problem is counteracting by geometrical approach. The geometrical relationship between the points of the image is being established and found out when without any jewellary. There is a typical closeness among the data thus obtained. Using this data a point is predicted from that portion of the image which is occluded by jewellary. On obtaining this point the features are then found out, which are approximately similar to that of the original image. The general block diagram of ear recognition system presented in Fig. 2.

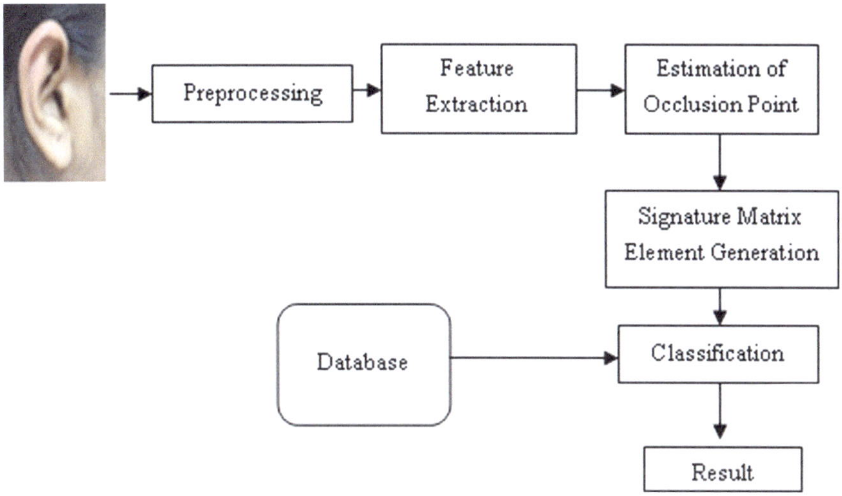

Fig. 2 Block diagram of ear based biometric system

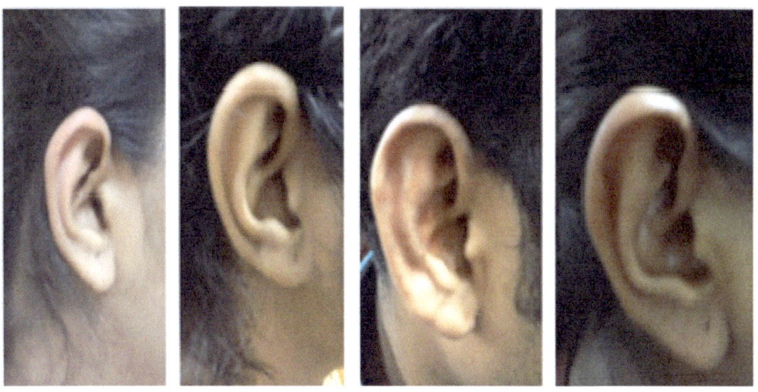

Fig. 3 Captured image

2 Methodology

The data input for the proposed ear recognition algorithm is right side ear image of a subject. In this study ear images of 10 subjects are collected through a digital camera of 5 megapixels resolution with a distance of 15–20 cm in different daylight condition without flash as shown in Fig. 3. All the data are captured with prior permission from the subjects.

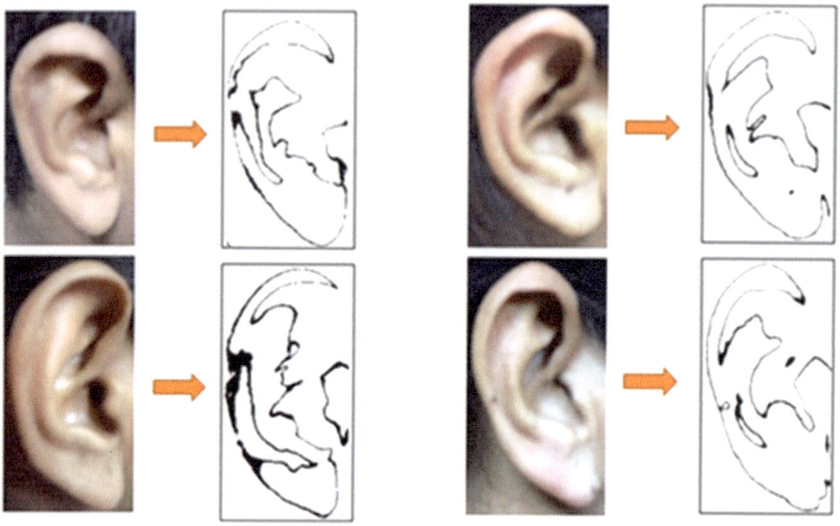

Fig. 4 Detected image boundary

2.1 Image Processing

In order to obtain only the ear image, background suppression is done by using an edge detection algorithm. In literature several well-known edge detection algorithm have been proposed [14]. Instead of using a traditional edge detector, a simple edge detection algorithm is designed where thresholding is used from 65 to 70% of the maximum intensity of an image i.e. 65% is the lower level and 70% is the upper level of thresholding as shown in Fig. 4.

2.2 Feature Extraction

After finding out the exterior contour of an ear, top most (P), bottom most (Q), left most (R) and right most (S) point of each contour are first detected and a rectangular is drawn by using these four points and cross point of two diagonals (O) of that rectangular is also detected. After detecting these five points Euclidian distance between OP, OQ, OR, OS, PR, PS, RS, RQ and SQ are determined as shown in Fig. 5. Also few other features are determined by calculating the ratios between the various distances as shown in Table 1.

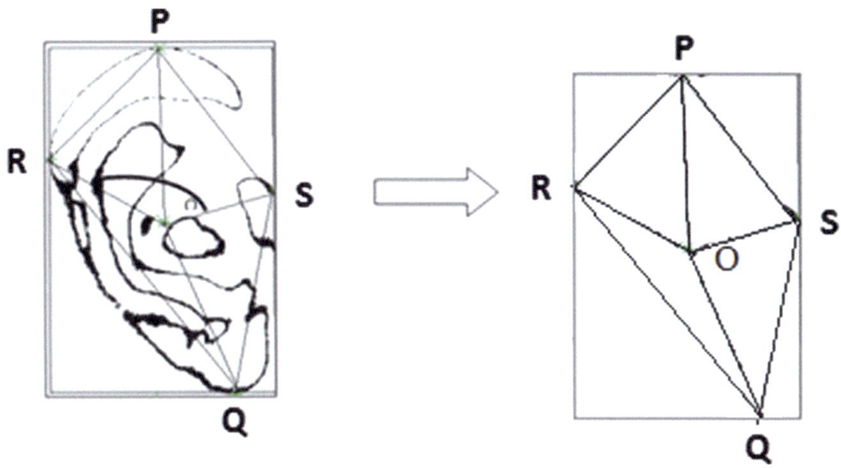

Fig. 5 Dimensionality of features extracted from ear

Table 1 Feature description

Feature symbol	Feature description
F_1	Distance between the point O and P
F_2	Distance between the point O and Q
F_3	Distance between the point O and R
F_4	Distance between the point O and S
F_5	Distance between the point R and P
F_6	Distance between the point S and P
F_7	Distance between the point R and S
F_8	Distance between the point R and Q
F_9	Distance between the point Q and S
F_{10}	Sum of the distance of the point PO and OQ
F_{11}	Ratio of distance OP to OQ
F_{12}	Ratio of distance PR to RQ
F_{13}	Ratio of distance PS to SQ
F_{14}	Ratio of distance RS to RQ
F_{15}	Ratio of distance RS to SQ
F_{16}	Ratio of distance (PO + OQ) to PS
F_{17}	Angle between RO and SO

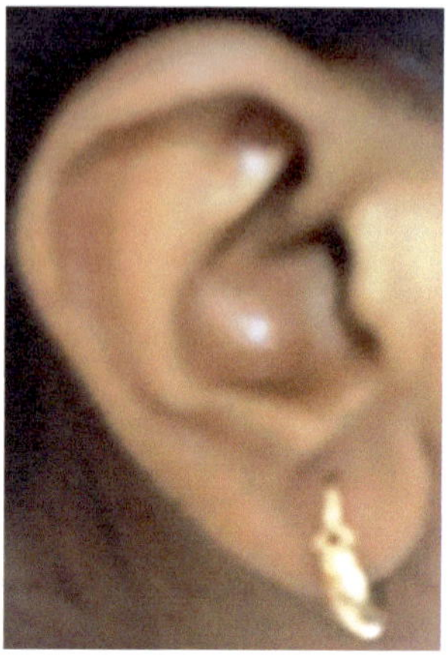

Fig. 6 Occluded ear image

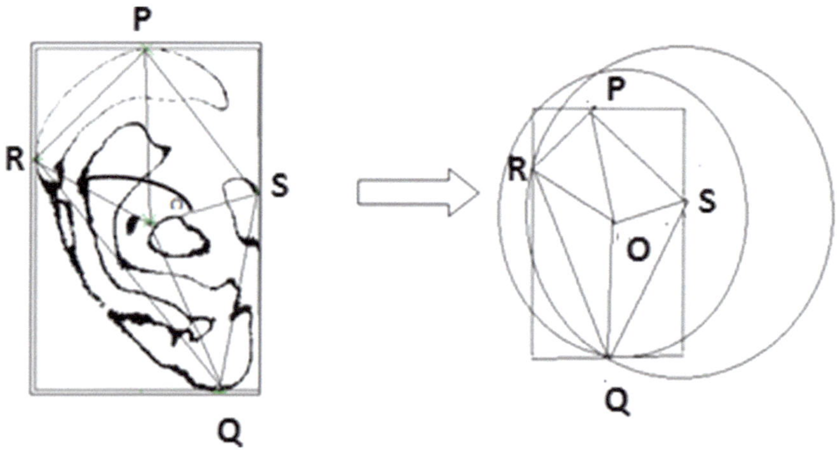

Fig. 7 Extracting features for the detection of 'Q' point

2.3 Estimation of Occlusion Point

One of the major problem arise in ear recognition system is due to occlusion of ear ring in the earlobe. As a result Q point could not be detected. An ear image occluded due to ear ring is shown in Fig. 6.

In this study a basic geometric comparison method is used to point out the approximate lower most point of the ear. The basic idea of this method is if two circles are drawn in an ear image without any occlusion, by taking OQ and SQ as radius then two intersection points of the circles will be obtained. One of them is the coordinates of Q point as shown in Fig. 7.

Table 2 Ratio of OQ/OP and SQ/RS of 7 unoccluded subjects

Subject	OQ/OP	SQ/RS
Subject 1	0.99	1.14
Subject 2	1.01	1.23
Subject 3	1.03	1.07
Subject 4	1.03	1.18
Subject 5	1.02	1.09
Subject 6	1.07	0.96
Subject 7	1.03	0.95
Average	1.03	1.09

Table 3 Comparison between actual and estimated point

Subject	Coordinate of actual Q point	Coordinate of estimated Q point	Actual OQ (distance)	Estimated OQ (distance)
Subject 1	(258,599)	(322,599)	286.7	287
Subject 2	(320,584)	(311,600)	302.7	306
Subject 3	(313,582)	(308,593)	296	303
Subject 4	(282,583)	(302,586)	300.4	292

Now to estimate Q point for an occluded image due to ear ring, averages of OQ/OP and SQ/RS of seven ear images without any occlusion is first calculated as given (Table 2).

$$OQ = 1.03 \times OP \tag{1}$$

$$SQ = 1.09 \times RS \tag{2}$$

After calculating the average values, OQ and SQ for an occluded image are calculated using these average values and the intersection point of two circles of radius OQ and SQ at center point O and S respectively is considered as coordinates of measured Q point of an occluded ear image. It is also found that this coordinate of measured Q point is in at considerable distance from the actual Q point. To minimize this error, from the measured Q point lowest distance coordinate of the earlobe is found out and considered as approximated coordinate of actual Q point as shown in Fig. 8. Coordinates of estimated Q point in case of occlusion as compared to the actual Q point for four subjects is shown in Fig. 9 and compared in Table 3.

2.4 Signature Matrix Generation

In this study, 17 different features are extracted from each ear image of 10 subjects and produce a matrix of dimension [10 × 17] and dimension reduction is done by

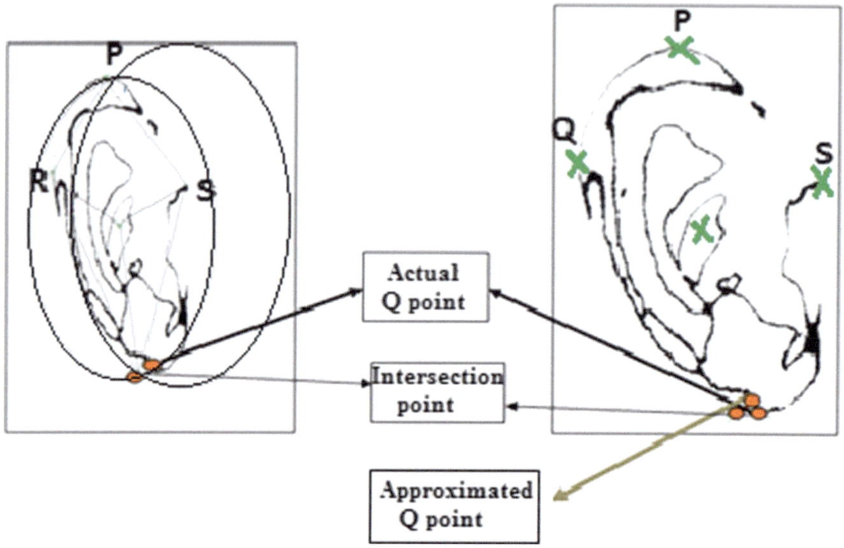

Fig. 8 Detection of a Q point

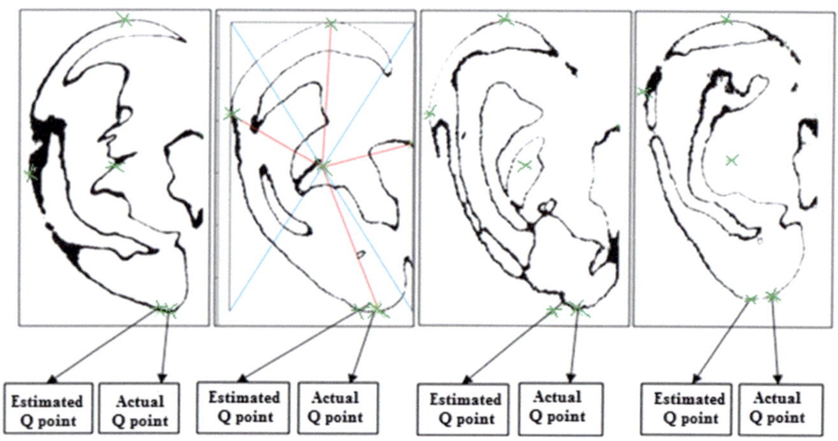

Fig. 9 Location of actual Q point and estimated Q point

using Principal Component Analysis method where using linear combination of the original feature vectors ($F_1, F_2 \ldots . F_n$) a new set of feature vectors with low dimension generates on the linear subspace.

The first principal component X_1 is given by

$$X_1 = \sum_{j=1}^{n} W_{1j} F_j \tag{3}$$

Table 4 Signature matrix element for 10 subjects

Subject	SME1	SME2	SME3
Subject 1	81.48	8.52	6.15
Subject 2	94	42	29
Subject 3	215	119	28
Subject 4	109.91	97.13	29.79
Subject 5	7199	135	24
Subject 6	1048	129	57
Subject 7	277	70	18
Subject 8	171	23	6
Subject 9	339	87	25
Subject 10	4112	209	72

where W_1 is the first weight vector and for a given constraint $W_1^T W_1 = 1$, variance of X_1 is maximum. The second principal component X_2 is found by using second weight vector W_2 so that it has maximum variance for a given constraint $W_2^T W_2 = 1$, which is uncorrelated with X_1.

The process is continued for deriving $X_3 \ldots X_n$ so that each component is uncorrelated with the previous components and the sum of variance of both original feature vectors and principal components variables are the same.

By using Principal Component Analysis method 17 feature matrix of each subject is converted into one dimensional array of 3 elements of its Eigen value and a signature matrix of 3 elements for 10 subjects are stored in the database as shown in Table 4.

3 Result and Discussion

During testing 3 elements signature matrix of a new subject is compared with each subject and the final decision is made on basis of Root Mean Square Error (RMSE) method.

$$RMSE = \sqrt{\frac{\sum_{i=1}^{n} \left(FP_{stored,i} - FP_{new,i}\right)^2}{n}} \tag{4}$$

where, $FP_{stored,i}$ is the stored feature matrix,

$FP_{new,i}$ is the feature matrix of a new entry currently under test and the number of variables of the feature matrix is considered as n.

This method is used against all 10 subjects and achieved an accuracy of 100%. The RMSE value of each testing over entire database is shown in Table 5. A comparative study of present with previously reported results is shown in Table 6.

Table 5 RMSE value between stored database and new access

	Access 1	Access 2	Access 3	Access 4	Access 5	Access 6	Access 7	Access 8	Access 9	Access 10
Subject 1	21.2	1912.3	48.9	73.5	4008.2	28.5	423.6	41.1	28.7	83.9
Subject 2	1944.3	23.7	1933.8	1981.4	2100.9	1952.4	1500.6	1964.3	1900.6	2006.9
Subject 3	45.5	1938	19.1	32.3	4037.4	37.2	450	47	54.5	64.2
Subject 4	69.4	1982.9	62	14.7	4082.5	58.5	494.9	57.1	98	33.1
Subject 5	4038.5	2125.7	4032.2	4079.6	14.4	4047.1	3598.7	4058.2	3996.6	4102.4
Subject 6	14.3	1937.9	45.1	51.5	4034.2	5.2	449.2	19	50.1	58.4
Subject 7	490.9	1441.8	481.3	528.6	3541	498.8	47.4	510.9	447	553.3
Subject 8	34.2	1963.9	62.1	44.9	4059.6	24.1	475.3	14.7	76.3	35.8
Subject 9	67.2	1874.4	52.3	97.4	3972.8	70.8	386.1	84.3	18.1	122.6
Subject 10	46.7	1975.1	65	36.2	4071.7	36.1	486.4	28.6	86.5	23.6

Table 6 Comparison table

Study	No. of ear images	Occluded image	Accuracy (%)
Chen and Mu [11]	24	Yes	95.83
Bustard and Nixon [15]	63	Yes	92
Arbab- Zavar et al. [16]	63	Yes	88
This paper	10	Yes	100

4 Conclusion

Ear biometric system is more reliable source for a unimodal biometric recognition system as the texture of external ear does not vary, in general, with age, obesity, disease etc. unlike other common biometric parameters. In this study an effort has been made to identify the subject by their ear. A major aim of this approach is to estimate the significant point for feature extraction which is within the occluded region. This work is done under a small database of 10 subjects and received a success rate of 100%. However, further study is needed using larger database having other artifacts like presence of earphone or hair and with different illumination exposure.

References

1. S. Chakraborty, M. Mitra, S. Pal, Biometric analysis using fused feature set from side face texture and electrocardiogram. IET Sci. Meas. Technol. **11**(2), 226–233 (2017)
2. S. Chakraborty, S. Pal, Photoplethysmogram signal based Biometric Recognition using Linear Discriminant, in *2nd International conference on Control, Instrumentation, Energy & Communication, CIEC'16*, Kolkata, India, pp 183–188, January 2016
3. A.V. Iannarelli, *Ear Identification (Forensic Identification Series)* (Paramont, Fremont, CA, USA, 1989)
4. J.A. Unar, W.C. Seng, A. Abbasi, A review of biometric technology along with trends and prospects. Pattern Recogn. **47**(8), 2673–2688 (2014)
5. M. Burge, W. Burger, Ear biometrics, in *Biometrics*, ed. by A.K. Jain, R. Bolle R., S. Pankanti (Springer, Boston, MA, 1996), pp. 273–285
6. K. Chang, K.W. Bowyer, S. Sarkar, B. Victor, Comparison and combination of ear and face images in appearance-based biometrics. IEEE Trans. Pattern Anal. Mach. Intell. **25**(9), 1160–1165 (2003)
7. L. Nanni, A. Lumini, A multi-matcher for ear authentication. Pattern Recogn. Lett. **28**(16), 2219–2226 (2007)
8. L. Yuan, Z. Mu, Ear recognition based on local information fusion. Pattern Recogn. Lett. **33**(2), 182–190 (2012)
9. B. Zhang, Z. Mu, C. Li, H. Zeng, Robust classification for occluded ear via Gabor scale feature-based non-negative sparse representation. Opt. Eng. **53**(6), Art. no. 061702, December 2013
10. A.S. Anwar, K.K.A. Ghany, H. Elmahdy, Human ear recognition using geometrical features extraction. Procedia Comput. Sci. **65**, 529–537 (2015)

11. L. Chen, Z. Mu, Partial data ear recognition from one sample per person. IEEE Trans. Hum. Mach. Syst. **46**(6), 799–809 (2016)
12. J. Lei, X. You, M. Abdel-Mottaleb, Automatic ear landmark localization, segmentation, and pose classification in range images. IEEE Trans. Syst. Man Cybern. Syst **46**(2), 165–176 (2016)
13. Ž. Emeršič, V. Štruc, P. Peer, Ear recognition: more than a survey. Neurocomputing **255**, 26–39 (2017)
14. J. Canny, A computational approach to edge detection. IEEE Trans. Pattern Anal. Mach. Intell. **PAMI-8**(6), 679–698 (1986)
15. J.D. Bustard, M.S. Nixon, Toward unconstrained ear recognition from two-dimensional images. IEEE Trans. Syst. Man Cybern. **40**(3), 486–494 (2010)
16. B. Arbab-Zavar, M.S. Nixon, J.N. Carter, On model-based analysis of ear biometrics, in *First IEEE International Conference on Biometrics: Theory, Applications, and Systems*, pp. 1–5, September, 2007

An Integrated Model for Early Detection and Monitoring of Diabetic Foot

K. S. Suresh and A. Sukesh Kumar

1 Introduction

Diabetes is a chronic disease that occurs either when the pancreas does not produce enough insulin or when the body cannot effectively use the insulin it produces. It is a growing global epidemic of the current century. The recent estimate of International Diabetes Federation indicates that 387 million people, have diabetes and this may rise to 592 million within the next twenty years [1]. A further 316 million with impaired glucose tolerance are at high risk from the disease, with projections indicating that over 1 billion people will be living with or at high risk of diabetes in 2035. The World Health Organisation (WHO) estimated that 9% among adults aged 18+ years are suffering from diabetes. In 2012, an estimated 1.5 million deaths were directly caused by diabetes and more than 80% of diabetes deaths occur in low and middle-income countries. WHO also projects that diabetes will be the 7th leading cause of death in 2030 [2]. According to the statistics of the International Diabetes Federation, there are nearly 65 million diabetics in India [3]. As the incidence of diabetes is on the rise, there is a proportionate rise in the complications associated with diabetes.

Diabetic Foot is one of the most threatening complications of Diabetes. This is a state that a foot exhibits any pathology that results directly from diabetes mellitus or any long-term complication of diabetes mellitus.

K. S. Suresh (✉)
Centre for Development of Imaging Technology, Trivandrum, Kerala, India
e-mail: sureshkstvm05@yahoo.com

A. Sukesh Kumar
Rajiv Gandhi Institute of Development Studies, Trivandrum, Kerala, India
e-mail: drsukeshkumar@yahoo.in

S. Chattopadhyay et al. (eds.), *Modelling and Simulation in Science, Technology and Engineering Mathematics*, Advances in Intelligent Systems and Computing 749,
https://doi.org/10.1007/978-3-319-74808-5_49

2 Diabetic Foot Infection

Different types of foot problems may develop in the diabetic cases as a result of damage to nerves and blood vessels. These problems can easily lead to infection and ulceration, which increase a person's risk of amputation. The maximum number of hospital admissions in diabetics is due to foot problems. It occurs in 15–25% of all patients with diabetes and precedes 85% of all lower leg amputations. People with diabetes are 25 times more likely to lose a leg than people without the condition.

Throughout the world, up to 70% of all leg amputations happen to people with diabetes [4]. More than one million people with diabetes lose a leg every year as a consequence of their condition. Diabetic foot problems are a common occurrence throughout the world, resulting in major economic consequences for patients, their families and society. An estimate of International Working Group on the Diabetic Foot reveals that every twenty seconds a lower limb is lost to diabetes somewhere in the world [5].

3 Key Indicators of Diabetic Foot

Peripheral neuropathies and peripheral arterial disease commonly coexist in patients with diabetes and foot ulcers [6]. Peripheral vascular diseases are circulation disorders that affect blood vessels outside of the heart and brain. These typically strikes the arteries and veins that supply the arms, legs, and organs located below the stomach. Ultrasound techniques are most widely used to identify Peripheral vascular diseases. Intermittent claudication is the most common symptom of Peripheral vascular disease.

Ankle-arm index, diastolic blood pressure, fasting plasma glucose, hemoglobin A1C, high blood pressure, medial arterial calcification, nerve conduction velocity, peripheral vascular disease, systolic blood pressure, transcutaneous oxygen tension etc. are some of the factors provide indications of diabetic foot [7].

4 Hemodynamic Model of Blood Flow

The blood is considered as a fluid and the properties depends on the suspension of particles. Blood plasma is a Newtonian fluid which consists mostly water. But the cellular elements causes significant changes in the rheological properties of blood. So the mechanical properties of whole blood is very complex and cannot be modelled as a homogeneous fluid. The size of blood vessels and the flow behaviour is also different and it is considered as a non-Newtonian fluid or Navier-Stokes fluid. Many of the characteristics can be proved with the help of Navier-Stokes equation [8].

Navier-Stokes equation for an incompressible viscous fluid.

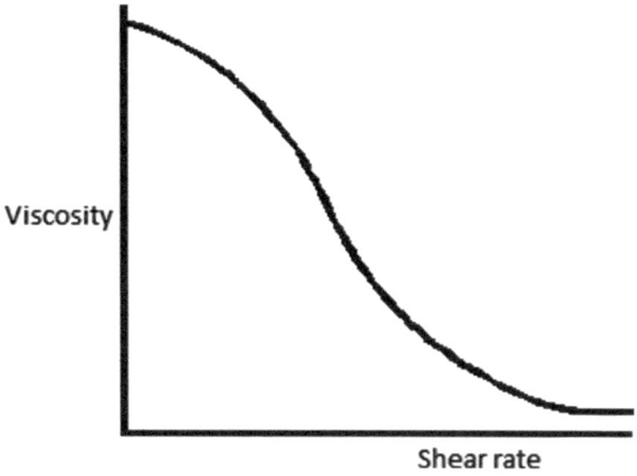

Fig. 1 Blood's non-newtonian property of shear thinning

$$\rho\frac{\partial u}{\partial t} + \rho(u.\nabla)u = -\nabla p + \nabla.t \tag{1}$$

u and ρ denotes the blood velocity and pressure.

Viscosity and shear rate (rate of deformation) of blood can also be estimated with Reiner-Rivlin equation. The factors which control the viscosity to inertia in linear flow are the velocity of the fluid, u, the diameter of the tube, d; and the kinematic viscosity of the fluid, v.

The arteries in leg are very much effected with diabetes. Atherosclerotic occlusion is most common and effecting the popliteal and femoral arteries. Diabetes effects the regular microcirculation. Foot skin blood falls while standing 80% in non diabetes and 50–70% in diabetes [9]. Resting supine foot skin blood flow 20–30% above normal in diabetes. Leukocytes also have role in changing the viscosity (Fig. 1).

A model is developed by considering the changes in blood flow and other characteristics leading to diabetic foot. This result is verified with other diagnostic methods of diabetic foot.

5 Detection Techniques

The outcome of diabetic foot is experimentally analysed with the following techniques for building the model. Preliminary results are very positive, more trials have to be accomplished for the perfect model.

5.1 Ultrasonic Measurement Techniques

Examination of the legs can give valuable information on the state of the peripheral circulation. There are a lot of indications for peripheral arterial ultrasound examination related to diabetic foot [10]. The detection of stenoses or occlusions in specified segments of the peripheral arteries in patients with suspected arterial occlusive disease. These clinical indicators, including claudication, rest pain, ischemic tissue loss, and suspected arterial embolizations may present in these patients.

One of the most popular technique used for initial diagnosis is the ankle brachial pressure index (ABPI) [11]. This is the ratio of highest pressure recorded at the ankle for that leg to the highest brachial pressure obtained for both arms. ABPI normally >1.0. ABPI <0.92 indicates arterial disease. ABPI >0.5 and <0.9 can be associated with claudication and if symptoms warrant a patient should be referred for further assessment. ABPI <0.5 indicates severe arterial disease and may be associated with gangrene, ischemic ulceration or rest pain and warrants urgent referral for a vascular opinion [12].

Figure 2 shows This patient with pain in the Left lower limb was evaluated with color and spectral Doppler ultrasound of the venous system of the left lower limb [13]. The images reveal a normal colour Doppler study of the veins, including the left femoral, popliteal, anterior and posterior tibial and peroneal veins. Often, the veins of the leg cannot be visualized in their entirety. It is often more practical to image the upper third and distal third of the anterior, posterior tibial and peroneal veins.

[Presented with permission from Dr. Joe's Ultrasound Scan, Cochin; www.ultrasound-images.com].

Figure 3 shows ultrasound and colour Doppler imaging of the right lower limb indicates early changes of diabetic arteriopathy or what is called as diabetic vasculopathy. This is characterised by the features shown in these colour Doppler ultrasound images;

1. Spectral broadening of the arterial waveform from the popliteal artery downwards
2. Mild decrease in peak systolic velocity below the popliteal artery
3. Early changes of loss of the tri-phasic spectral waveform is seen in peroneal artery

All these above-mentioned changes are typically seen in early diabetic arteriopathy suggesting mild stenosis in a diffuse fashion below the popliteal artery.

A normal pulsed wave Doppler waveform is a sharply defined tracing with a narrow Doppler Spectrum. Flow becomes turbulent at bifurcations and luminal narrowings causing spectral broadening of Doppler waveform, with filling in of the low velocity region in the spectral waveform as the blood cells move at a wide range of velocities. The normal peripheral artery waveform is triphasic [14].

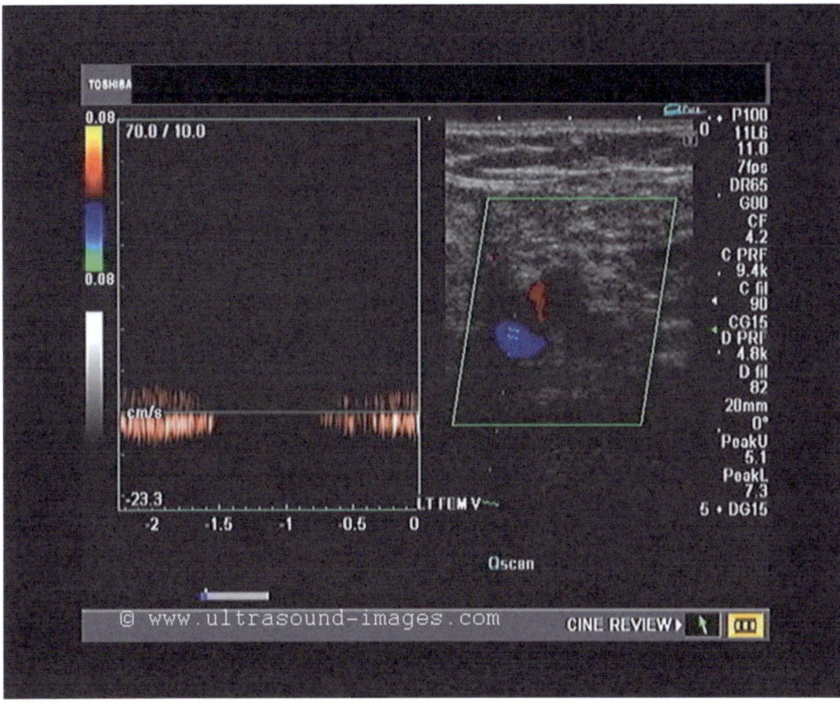

Fig. 2 Normal venous Doppler of lower limb

5.2 *Transcutaneous Oxygen Measurements*

Concentration in the blood is the so-called oxygen saturation, SO2. It indicates the rate of oxygen delivery to and consumption by the tissues [15]. The optical extinction coefficient of oxyhemoglobin differs significantly from that of deoxyhemoglobin. Thus, the spectral absorption coefficient of tissue depends on the concentration and oxygen saturation of hemoglobin within the tissue. It also established that hyperspectral tissue oximetry has the ability to identify ischemic and inflammatory complications before they are visible during a clinical examination. Retrospective analysis of hyperspectral tissue oximetry from pre-ulcerative locations showed that diabetic foot ulcer formation can be predicted with high sensitivity and specificity [16].

5.3 *Photo Plethysmography*

Photoplethysmographic method is also used to identify the characteristic of blood, which lead to diabetic foot. Hemodynamic assessment can be accomplished by this method. Photoplethysmography employs a transducer that transmits infrared light

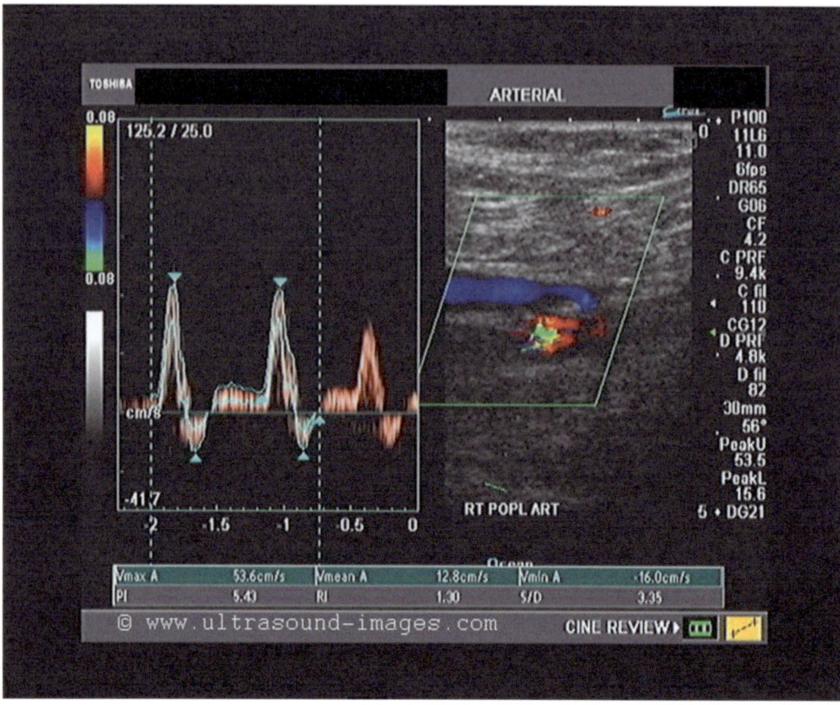

Fig. 3 Mild diabetic arteriopathy of lower limb presented with permission from Dr. Joe's Ultrasound Scan, Cochin; www.ultrasound-images.com

from an emitting diode into the tissue. Part of the transmitted light is reflected back from the blood within the cutaneous microcirculation and is received by an adjacent phototransistor. The amount of reflected light varies with the blood content of the microcirculation. This photo-transducer is taped to the end of the toe with a double-faced cellophane tape while a small digital blood pressure cuff is placed at the base of the digit. The pressure at which the waveform obliterates corresponds to the digital systolic pressure.

6 Conclusion

An integrated model is developed for early detection and monitoring of diabetic foot. The hemodynamic model is developed and verified with secondary mechanisms. Experiments have done and results are very satisfactory. More trials have to be accomplished to produce a fruitful design. Ultrasound, plethysmography and transcutaneous oxygen measurement technologies are found to be most promising.

Acknowledgements We are very much thankful to Dr. Joe's Ultrasound Scan, Cochin and Indian Institute of Diabetes, Trivandrum for providing clinical support for this research work.

References

1. International Diabetes Federation, Annual Report 2014
2. World Health Organization, Diabetes Programme, http://www.who.int/diabetes/en/, 2015
3. International Diabetes Federation, IDF Diabetes Atlas Sixth edition 2013
4. International Diabetes Federation, Diabetes and Foot Care Time to Act, 2005
5. International Working Group on the Diabetic Foot, Guidance documents, http://iwgdf.org/guidelines/ 2015
6. G.E Reiber, J.W. Lemaster, Kaufman, Epidmiology and economic impact of foot ulcers and amputations in people with diabetes, in *Levin and O'Neal's The Diabetic Foot*, 7th edn. (Mosby Elsevier, 2008), pp. 3–22
7. K.S. Suresh, Dr. A.S. Kumar, An investigation for early stage diagnosis of diabetic foot. Int. J. Enhanced Res. Sci., Technol. Eng., **5**(12), 40–43 (2016)
8. N. Bessonov, A. Sequeira, S. Simakov, Y. Vassilevskii, V. Volpert, Methods of blood flow modelling, Math. Model. Nat. Phenom. **11**(1), 1–25 (2016)
9. D.E. McMillan, Hemorheology: principles and concepts, in *Levin and O'Neal's The Diabetic Foot*, 7th edn. (Mosby Elsevier, 2008), pp. 75–88
10. AIUM practice guide line for the performance of peripheral arterial ultrasound examinations using color and spectral doppler imaging, American Institute of Ultrasound in Medicine (2010)
11. K. Vowden, P. Vowden, Hand-held doppler ultrasound: the assessment of lower limb arterial and venous disease. www.huntleigh-diagnostics.com (2002)
12. K.S. Suresh, Dr. A.S. Kumar, Clinical need of a diabetic foot infection detector, in *National Conference on Advances in Computational Intelligence & Communication Technologies*, pp. 125–128 (2016)
13. AIUM practice guide line for the performance of Peripheral venous Ultrasound examinations, American Institute of Ultrasound in Medicine, 2010
14. D. Lingegowda, S. Moorthy, K.P Sreekumar, R.R. Kannan, Imaging in diabetic ischemic foot. Int. J. Diab. Dev. Countries, **30**(4), 179:184 (2010)
15. S. Zimny, F. Dessel, M. Ehren, M. Pfohl, H. Schatz, Early detection of microcirculatory impairment in diabetic patients with foot at risk. Diab. Care **24**, 1810–1814 (2001)
16. D. Yudovsky, A. Nouvong, K. Schomacker, L. Pilon, Assessing diabetic foot ulcer development risk with hyperspectral tissue oximetry. J. Biomed. Opt. **16**(2), 026009 (2011)

Real Time Periodic Assessment of Retina of Diabetic Patients for Early Detection of Diabetic Retinopathy

P. G. Prageeth, A. Sukesh Kumar and K. Mahadevan

1 Introduction

Increase of diabetic patients in INDIA is in an alarming proportion. Uncontrolled diabetes can lead to blindness. Normally this is due to diabetic retinopathy. If it is detected earlier and required treatments are taken, blindness can be avoided. It may affect both the eyes simultaneously. Diabetic retinopathy is affected to 50% of diabetic patients. Diabetic Maculopathy and Diabetic Neuropathy are the common affected diseases in diabetic patients. The possibility of blindness for these types of patients is 25% more. Diabetic retinopathy is of two types: Non-proliferated diabetic retinopathy and Proliferated diabetic retinopathy [1, 2]. Vitreous heamorrhage and retinal detachments are the immediate complications leading to blindness, if proper treatment is not provided at the correct time. If it reaches to the complicated level and loses vision, then it is not possible to revert. Hence diabetic patients who are affected with vision may be monitored periodically to assess the real status of retina. Hence apart from a multi powered disease control approach, it is highly essential to utilize the advance technology enabling the doctors to have regular real time assessment of retinopathy in diabetic patients through telemedicine [3].

Diabetic retinopathy is a condition occurring in persons with diabetes, which causes progressive damage to the retina, the light sensitive lining at the back of the

P. G. Prageeth (✉)
Department of ECE, College of Engineering Trivandrum, University of Kerala, Thiruvananthapuram, Kerala, India
e-mail: prageethpg@cet.ac.in

A. Sukesh Kumar
Department of ECE, University of Kerala, Thiruvananthapuram, Kerala, India
e-mail: drsukeshkumara@cet.ac.in

K. Mahadevan
Department of Ophthalmology, Medical College, Thiruvananthapuram, Kerala, India
e-mail: eyemahadevan@rediffmail.com

S. Chattopadhyay et al. (eds.), *Modelling and Simulation in Science, Technology and Engineering Mathematics*, Advances in Intelligent Systems and Computing 749,
https://doi.org/10.1007/978-3-319-74808-5_50

eye. It is a serious sight-threatening complication of diabetes. Diabetes is a disease that interferes with the body's ability to use and store sugar, which can cause many health problems. Too much sugar in the blood can cause damage throughout the body, including the eyes. Over time, diabetes affects the circulatory system of the retina. Diabetic retinopathy is the result of damage to the tiny blood vessels that nourish the retina. They leak blood and other fluids that cause swelling of retinal tissue and clouding of vision. The condition usually affects both eyes. The longer a person has diabetes, the more likely they will develop diabetic retinopathy. If left untreated, diabetic retinopathy can cause blindness [4].

Symptoms of diabetic retinopathy include:

- Seeing spots or floaters in your field of vision
- Blurred vision
- Having a dark or empty spot in the center of your vision
- Difficulty seeing well at night

The ocular fundus is the only part of human body through which the vascular network can be observed directly and non-invasively. This can provide a lot of pathological information about eye diseases such as glaucoma, central serous retinopathy (CSR) and early signs of systemic diseases like diabetes, hypertension and cardiovascular diseases. All over the world, the patients having these diseases are increasing day by day. The medical imaging technology is developing in such a way that mass screening of patients is possible in ophthalmology. Nevertheless, the scarcity of medical experts in all landmarks of a country is limiting the mass screening of patients in developing countries. Hence, development of an automated system for analyzing digital fundus images and diagnosing the diseases is the greatest challenge, when mass screening of patients is needed [5]. The present work involves detection, quantification and analysis of retinal image parameters for various diseases and combined so that an intelligent system is developed for the diagnosis of glaucoma, diabetes, hypertension, stroke, heart attack and central serous retinopathy. Large numbers of retinal images of patients with necessary pathological data were taken from Regional Institute of Ophthalmology (Dept. of Ophthalmology), Medical College, Thiruvananthapuram for this work. The labelled diagram of a human eye and an image of the retina is shown in Fig. 1. Diabetes is detected from the presence of exudates and haemorrhages and changes in blood vessel parameters like arteriolar-to-venular diameter ratio (AVR). Images obtained from fundus camera are enhanced using filtering. Image segmentation is done to detect optic disc, fovea, exudates area and blood vessels. Connected component method along with concentric circle methods are used to determine the artery-vein width ratio. An algorithm is developed for the detection and quantification of the disease level from the parameters specified. The result is validated with the clinical data of the patient and achieved good results. A predictor system is developed to give the status of the patient from the analysis of the retinal image parameters using neural network techniques [6, 7].

A Fundus Camera which is used to obtain fundus images of the human retina is shown in Fig. 2. In the camera the beams propagating from two light sources, and

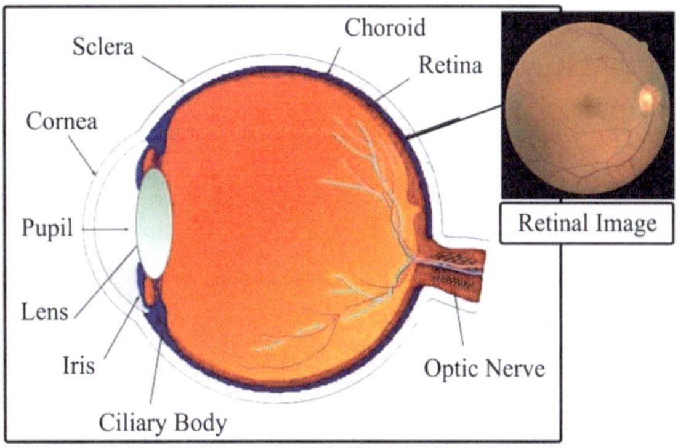

Fig. 1 Human eye cross-sectional view and a retinal image

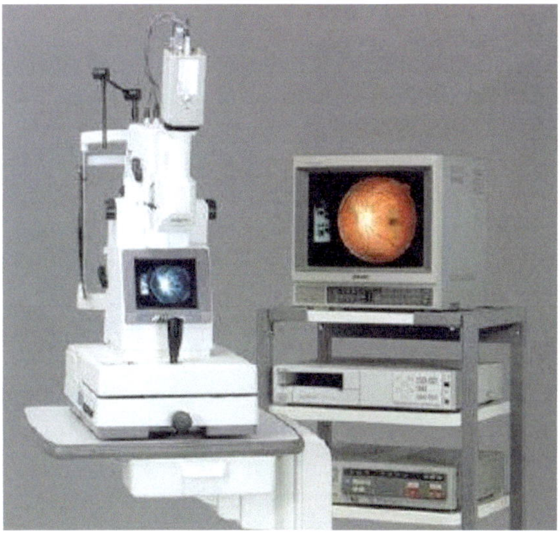

Fig. 2 A fundus camera and its accessories

incandescent lamp used for viewing the fundus and ash lamp for photography, are optically combined by a beam splitter [8].

The fundus camera images of normal and diabetic patients are shown in Figs. 3 and 4. These images are used to detect and diagnosis the disease and the method of automatic detection systems are explained.

2 Proposed Method

The Graphical User Interface (GUI) window for the detection of diabetic retinopathy is designed as shown below in Fig. 5. This disease has dark red and yellow spots in it retinal fundus images. First input the image. Then resize to required size. Find

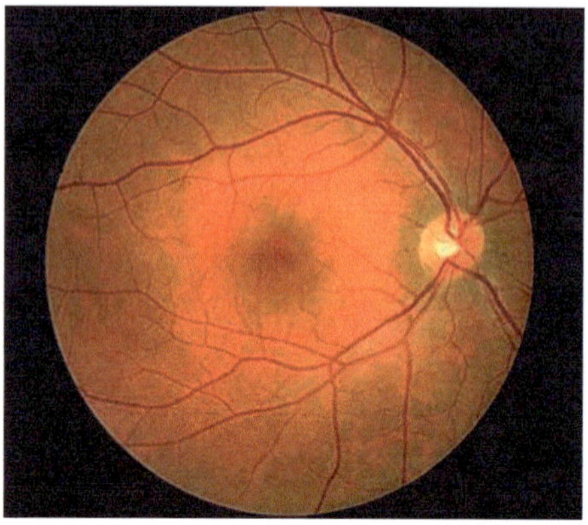

Fig. 3 Fundus image of a normal eye

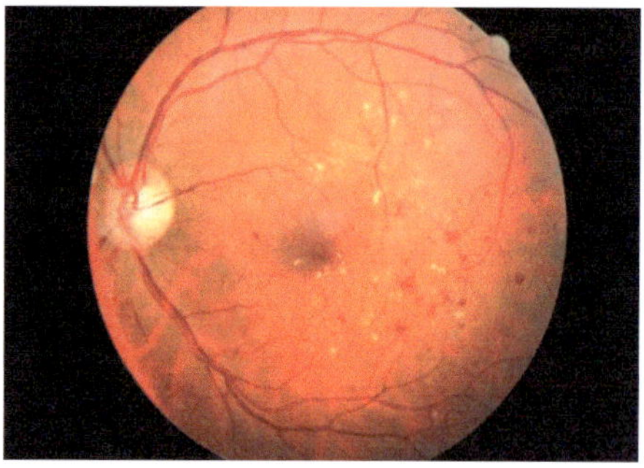

Fig. 4 Fundus image of a diabetic patient's eye

the hue, saturation and intensity of the image [9]. Find contrast, homogeneity and correlation. Find threshold value for each. Use these values for disease detection.

In machine vision, colors of the pixels are composed from Red, Green and Blue, each measured in 8 bits (2 hexadecimal bits) [10]. The complete representation of a color would be of 6 hexadecimal bits. The problem is that a family of colors (like the orange and the combinations close to it) cannot be easily known from that 6 bits representation. Moreover, the changes in brightness will cause enormous changes in the RGB representation of a certain color. Therefore, in machine vision, the most used color representation is the HSI color space, which consists of Hue angle, Color saturation and Intensity [11]. To be independent of the intensity variance, we use the HS space. This will also help in making the processing and the computing faster.

The above describes histogram equalization on a gray scale image. However it can also be used on color images by applying the same method separately to the Red, Green and Blue components of the RGB color values of the image. However, applying the same method on the Red, Green, and Blue components of an RGB image may yield dramatic changes in the image's color balance since the relative distributions of the color channels change as a result of applying the algorithm [12]. However, if the image is first converted to another color space, then the algorithm can be applied to the luminance or value channel without resulting in changes to the

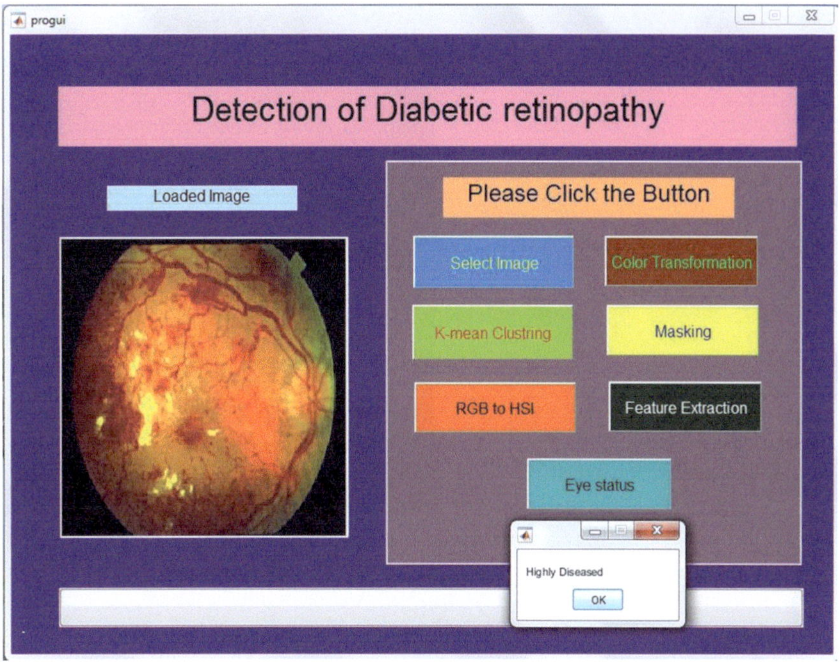

Fig. 5 Proposed GUI window for the detection of diabetic retinopathy

hue and saturation of the image. There are several histogram equalization methods in 2D and 3D space. However, it results in "whitening" where the probability of bright pixels is higher than that of dark ones [13].

With a large number of variables, K-Means may be computationally faster than hierarchical clustering (if K is small). K-Means may produce tighter clusters than hierarchical clustering, especially if the clusters are globular. Contrast is defined as the separation between the darkest and brightest areas of the image. Increase contrast and you increase the separation between dark and bright, making shadows darker and highlights brighter. Contrast enhancement plays a crucial role in image processing applications, such as digital photography, medical image analysis and scientific visualization. Image enhancement is a technique which reduces image noise, remove artifacts, and preserve details [14, 15]. Its purpose is to amplify certain image features for analysis, diagnosis and display. Contrast enhancement increases the total contrast of an image by making light colors lighter and dark colors darker at the same time. It does this by setting all color components below a specified lower bound to zero, and all color components above a specified upper bound to the maximum intensity (that is, 255). Color components between the upper and lower bounds are set to a linear ramp of values between 0 and 255. Because the upper bound must be greater than the lower bound, the lower bound must be between 0 and 254, and the upper bound must be between 1 and 255 [16, 17].

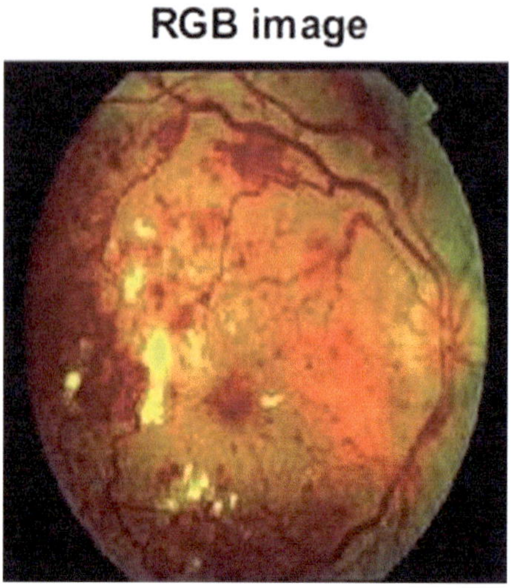

Fig. 6 Fundus image of a diabetic patient in RGB to input to the GUI window

There are several reasons for an image/video to have poor contrast:

	the poor quality of the used imaging device,
	lack of expertise of the operator and
	the adverse external conditions at the time of acquisition.

These effects result in under-utilization of the offered dynamic range. As a result, such images and videos may not reveal all the details in the captured scene, and may have a washed-out and unnatural look. In graphics and imaging, color saturation is used to describe the intensity of color in the image [18]. A saturated image has overly bright colors. Using a graphics editing program we can increase saturation on under-exposed images, or vice versa. Digital Image Correlation (DIC) is a full-field image analysis method, based on grey value digital images that can determine the contour and the displacements of an object under load in three dimensions. Due to rapid new developments in high resolution digital cameras for static as well as dynamic applications, and computer technology, the applications for this measurement method has broadened and DIC techniques have proven to be a flexible and useful tool for deformation analysis.

The image shown in Fig. 6 is that of an acute diabetic patient. Likewise we can give normal, mild, moderate, severe levels of images as the inputs to the GUI window and can get the corresponding output parameters and predictions [19, 20]. For the sample input image given above, the output images of the parameters helpful for all men to detect the disorders and the clinicians for the diagnosis, are shown in Fig. 7.

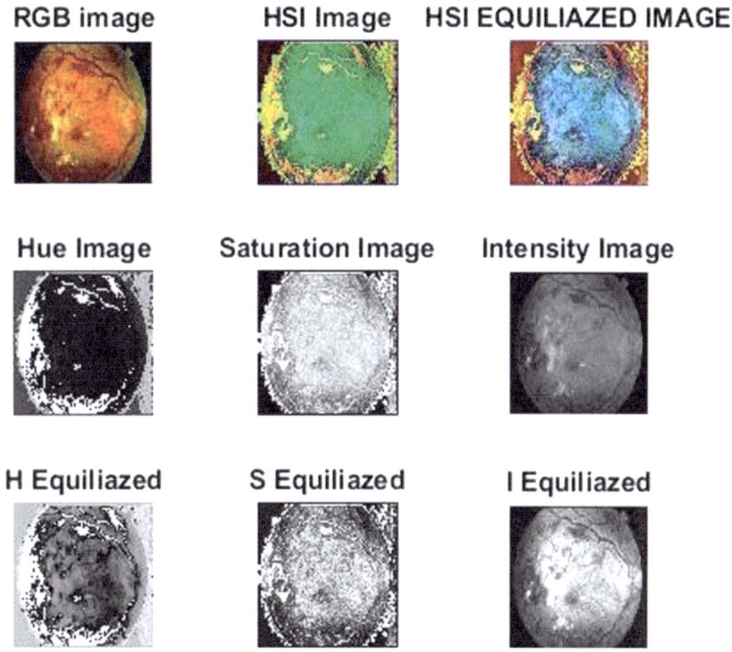

Fig. 7 Output images for evaluating the parameters of eye of a diabetic patient

3 Conclusion

According to the above images and values, we can send all the parameters to the highly qualified ophthalmologists. Final decisions and all other curing methods, preventive medicines and the consultation if needed should be done by the clinicians. The above developed system can be used by the biomedical engineers to help the doctors and the remotely resident patients. If this system is developed to work in online way, then the doctors can identify the eye conditions of the patients who are very far away from hospitals. The system can be developed to work as a *kiosk* in remote areas like ATM machines, X-ray labs, Clinical labs to test the blood and all etc. Then it will be very useful and will get a maximum coverage of the patients around the globe. The doctors need not run to all patients here and there. The patients can go to the very nearby labs and if needed only they can go the service of the doctors.

We now designed only a prototype of the system in software model only. If this system is to be developed in a hardware model, sufficient fund is needed just to make a sample. Then only we can make arrangements for the commercialization of the product.

This is going to be a great boon and boom for the common man (actual beneficiaries of this project) to get into the prevention of diabetes rather than cure as a result of periodic review using this project. Unfortunately if any person gets into the diabetic disorder, then this project will help in the periodic assessment of the various levels of the disease, thereby avoiding the progression of the disorder cent percentage.

References

1. S. Wild, G. Roglic, A. Green, R. Sicree, H. King, Global prevalence of diabetes estimates for the year 2000 and projections for 2030. Diab. Care **27**(5), 1047–1053 (2004)
2. D.A. Quillen, Common causes of vision loss in elderly patients. Am. Family Physician **60**(1), 99–108 (1999)
3. R. Klein, B.E.K. Klein, S.E. Moss, Visual impairment in diabetes. Ophthalmology **91**(1), 1–9 (1984)
4. M.D. Saleh, C. Eswaran, An automated decision support system for non-proliferative diabetic retinopathy disease based on mas and has detection. Comput. Methods Programs Biomed. **108**(1), 186–196 (2012)
5. A. Kumar, A.K. Gaur, M. Srivastava, A segment based technique for detecting exudate from retinal fundus image. Procedia Technol. **6**, 1–9 (2012)
6. M.M. Fraz, S.A. Barman, P. Remagnino, A. Hoppe, A. Basit, B. Uyyanonvara, A.R. Rudnicka, C.G. Owen, An approach to localize the retinal blood vessels using bit planes and centerline detection. Comput. Methods Programs Biomed. **108**(2), 600–616 (2012)
7. Xu Lili, Shuqian Luo, A novel method for blood vessel detection from retinal images. Biomed. Eng. Online **9**(1), 14 (2010)
8. M. Foracchia, E. Grisan, A. Ruggeri, Luminosity and contrast normalization in retinal images. Med. Image Anal. **9**(3), 179–190 (2005)
9. M. Vlachos, E. Dermatas, Multi-scale retinal vessel segmentation using line tracking. Comput. Med. Imaging Graph. **34**(3), 213–227 (2010)
10. K. Saranya, B. Ramasubramanian, S. Kaja Mohideen, A novel approach for the detection of new vessels in the retinal images for screening diabetic retinopathy, in *International Conference on Communications and Signal Processing (ICCSP)*, IEEE, 2012, pp. 57–61 (2012)
11. K. Narasimhan, V.C. Neha, K. Vijayarekha, An efficient automated system for detection of diabetic retinopathy from fundus images using support vector machine and Bayesian classifiers in Computing. in *International Conference on IEEE Electronics and Electrical Technologies (ICCEET)*, pp. 964–969 (2012)
12. K.A. Goatman, A.D. Fleming, S. Philip, G.J. Williams, J.A. Olson, P.F. Sharp, Detection of new vessels on the optic disc using retinal photographs. IEEE Trans. Med. Imaging **30**(4), 972–979 (2011)
13. B. Antal, A. Hajdu, An ensemble-based micro aneurysm detector for retinal images, in *18th IEEE International Conference on Image Processing (ICIP)*, IEEE, pp. 1621–1624 (2011)
14. M. Tamilarasi, K. Duraiswamy, Genetic based fuzzy seeded region growing segmentation for diabetic retinopathy images, in *International Conference on IEEE Computer Communication and Informatics (ICCCI)*, pp. 1–5 (2013)
15. Balint Antal, Andras Hajdu, An ensemble-based system for micro aneurysm detection and diabetic retinopathy grading. IEEE Trans. Biomed. Eng. **59**(6), 1720–1726 (2012)
16. C. Köse, U. Şevik, C. Ikibaş, H. Erdöl, Simple methods for segmentation and measurement of diabetic retinopathy lesions in retinal fundus images. Comput. Methods Programs Biomed. **107**(2), 274–293 (2012)
17. M.U. Akram, S. Khalid, A. Tariq, S.A. Khan, F. Azam, Detection and classification of retinal lesions for grading of diabetic retinopathy. Comput. Biol. Med. **45**, 161–171 (2014)
18. D. Marín, A. Aquino, M.E. Gegúndez-Arias, J.M. Bravo, A new supervised method for blood vessel segmentation in retinal images by using gray-level and moment invariants-based features, IEEE Transactions on Medical Imaging, **30**(1), 146–158 (2011)

19. A. Anzalone, F. Bizzarri, M. Parodi, M. Storace, A modular supervised algorithm for vessel segmentation in red-free retinal images. Comput. Biol. Med. **38**(8), 913–922 (2008)
20. M.U. Akram, A. Tariq, S.A. Khan, M.Y. Javed, Automated detection of exudates and macula for grading of diabetic macular edema. Comput. Methods Programs Biomed. **114**(2), 141–152 (2014)

Product Recommendation for E-Commerce Data Using Association Rule and Apriori Algorithm

Soma Bandyopadhyay, S. S. Thakur and J. K. Mandal

1 Introduction

Data mining is a multidisciplinary research field as it involves database technology, artificial intelligence (AI), machine learning (ML), statistical modeling and object oriented approach [1] etc. Data mining is used to extract useful information and discover knowledge from hidden patterns in available data [2]. With the application of mining algorithm—Apriori, purchasing habits of customers can be analyzed by finding associations among different items that customers add to their shopping cart. From the user's transaction dataset rules can be generated for customers' buying tendency. Based on product purchased by customers and their profile, recommendation about products can be suggested using association rule mining technique. The discovery of such association rules can assist the customers to fulfill their needs, as well as retailers to develop marketing strategies and increase sales. In the modern world where the numbers of choices are vast, the recommender systems assist consumers to find items of their interest. They link users with items to purchase, view, listen to, etc. by correlating the content of suggested items. Buying behavior of customers for future can be predicted by the analysis using past buying behavior. The various form of recommendation available are providing personalized product information,

S. Bandyopadhyay (✉) · S. S. Thakur
Computer Science and Engineering, MCKV Institute of Engineering,
243, G.T. Road, Liluah, Howrah 711204, India
e-mail: somabanmuk@yahoo.co.in

S. S. Thakur
e-mail: subroto_thakur@yahoo.com

J. K. Mandal
Computer Science and Engineering, University of Kalyani, Kalyani,
Nadia 741235, India
e-mail: jkm.cse@gmail.com

S. Chattopadhyay et al. (eds.), *Modelling and Simulation in Science, Technology and Engineering Mathematics*, Advances in Intelligent Systems and Computing 749,
https://doi.org/10.1007/978-3-319-74808-5_51

summarizing community opinion etc. These recommendation systems are integral part of personalization on a website as they help the site to familiarize with each customer [3].

Data mining techniques are incorporated in recommender system to make recommendations using knowledge gathered from different attributes about products and actions taken by user. The architecture of data mining consists of main components [4, 5] like data warehouse, database or other repositories of information. A server fetches the appropriate data from repositories based on user's request where knowledge base is used as guide to search according to defined constraint. Characterization, classification, clustering, association, regression and analysis evolution are the essential modules of data mining engines. To accurately achieve interested patterns, the pattern evaluation module interacts with data mining module. The user can communicate with the data mining system through graphical user interfaces.

This paper outlines the implementation of Apriori algorithm [6] by applying association rules to increase product sales. This work proposes a recommender system based on Apriori algorithm.

2 Recommender System

By correctly suggesting products to customers a relationship is created between the customers and the website, which increases consumer loyalty. These customers are expected to return to the same site where correct choices about products were recommended earlier. Recommender system, discovered from association rule is a very popular technique used, which helps to identify item-to-item correlation. More powerful systems match an entire set of items, such as those in a customer's shopping cart, should be able to identify appropriate items to recommend. These rules can also help a retailer to arrange products e.g. a consumer purchasing a child's handheld video game seeks batteries nearby.

Recommender system enhance E-commerce sales in three ways: Converting Browsers into Buyers, Increasing Cross-sell and Building Loyalty as are discussed below.

Converting Browsers into Buyers: In general, a customer or visitor browse a website first and look over a site without purchasing anything. Recommender systems can help them to find products, which they want to purchase as per their requirements. Thus, the browsers may be converted into buyers.

Increasing Cross-sell: Recommender systems [7, 8] can improve cross-sell by providing additional products for the customer, which they want to purchase in near future. If the recommendations are satisfactory, the average order may increase.

Building Loyalty: In modern world where there is a large number of competitors, gaining consumer loyalty is an important strategy as business is concern. By creating a value-added relationship between the site and the customer, recommender system can improve loyalty. Web sites invest lot of funds in learning about the customers' behavior and finally the recommender systems provide an interface that match consumer needs with products placement.

3 Proposed Approach

Product prediction and recommendation of E-commerce website emphasizes some key areas. Work flow diagram of our proposed recommender system has been shown in Fig. 1 in Sect. 3.1 for prediction and recommendation of E-commerce website. Association Rule and Apriori algorithm that is used in this work is described in Sect. 4 and 5 respectively. Experimental evaluation is discussed in Sect. 6, followed by Results and Conclusion.

3.1 Work Flow Diagram of Proposed Recommender System

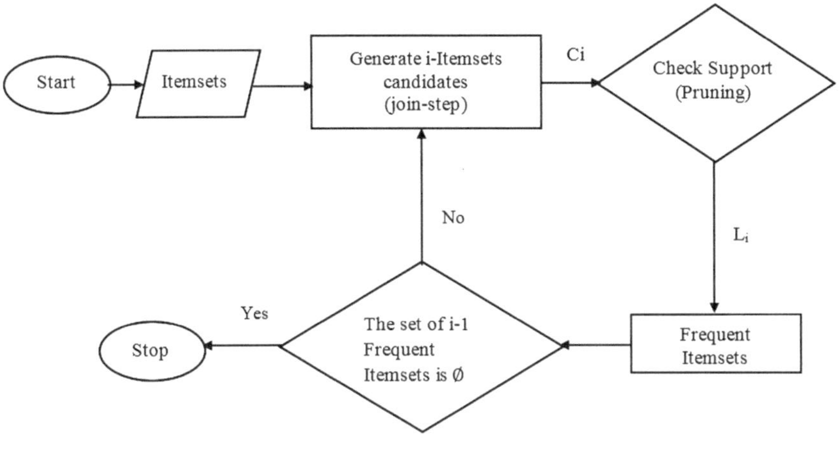

Fig. 1 Work flow diagram of proposed recommender system

4 Association Rule

Learning item-to-item correlations or 'association rule' in data mining helps recommender system to identify items frequently found in "association" with items in which a user has shown interest. Association can be based on co-purchase data, preference given by users, or any other measures. Powerful recommender systems match an entire set of items, which may be in a customer's shopping basket. For e.g., data mining may suggest that a customer who buys a video game today is likely to buy a pair of earplugs, in near future.

4.1 Meaning and Uses

Association rules have been used in retailing helps to analyze items [9] of preference of product and further suggest products to consumers based on other products they have purchased earlier. The relationship with which one product is often chosen along with other products is expressed by association rules. There is an exponential relationship between the numbers of possible association rules with the number of products in a rule. However effective search space is reduced due to constraints on confidence and support, combined with algorithms that build association rules with itemsets of n items from the rules with (n-1) items of itemsets. If shelf layout of a retail store is organized by maintaining association rule, it will definitely improve the efficiency and performance of the store. To recommend items to a customer, association rule is used as a powerful tool as domain knowledge management is concerned. So the recommendation system using association rule can attempt to predict which product can be most useful to a user.

4.2 Mining Association Rule

Association rules are required to satisfy a user-specified minimum support and minimum confidence at the same time [10, 11].

Association rule generation is usually split up into two separate steps:

1. Frequent itemset is generated by applying minimum support in a data base.
2. Using these frequent item sets' and minimum confidence constraint rules are formed.

To find all frequent itemsets in a database, it is required to search all possible itemsets (item combinations) from a database. The set of possible itemsets is the power set over I and has size 2n-1 (excluding the empty set which is not a valid itemset). Here I—mean average size of the maximum potentially large items and n means number of items. The size of the powerset grows exponentially in the number of items n in I. Efficient search is possible using the downward-closure property of support which guarantees that for a frequent itemset, all its subsets are also frequent. For an infrequent itemset, all its supersets must also be infrequent. Hence, by applying efficient algorithms, Apriori can find all frequent itemsets, and other association rules like Eclat and FP-Growth may be used.

5 Apriori Algorithm for Product Recommendation

The first pass of the algorithm counts item occurrences to determine the large 1-itemsets [12, 13]. A subsequent pass, say pass i, consists of two phases. First, the large itemsets L_{i-1} found in the (i-1)th pass are used to generate the candidate itemsets C_i, using the apriori candidate generation function. After that, the database is scanned and the support of candidates in C_i is counted. For fast counting, the candidates in C_i that are contained in a given transaction T is to be determined. A skeleton of Apriori Algorithm [14, 15] which has been used in this work is described below.

1. Collect the data from a retail shop.
2. In first pass, generate the candidate itemsets in C_1 and save the frequent itemsets in L_1.
3. From the frequent itemsets in L_{i-1} generate the candidate itemsets in C_i in ith pass.

   (a) Join L_{i-1} p with L_{i-1} q as follows:
   insert into Ci
   select p. $item_1$, p. $item_2$, ..., p. $item_{i-1}$, q. $item_{i-1}$ from L_{i-1} p, L_{i-1} q
   [here p. $item_1$ = q. $item_1$, ..., p. $item_{i-2}$ = q. $item_{i-2}$, p. $item_{k-1}$ < q. $item_{k-1}$]

(b) From the candidate itemsets in Ci generate all (i-1) subsets.
(c) From Ci prune all candidate itemsets where (i-1) subset of the candidate itemsets is not in the frequent itemsets L_{i-1}.

6 Experimental Evaluation

The dataset which was used in this work contains 4500 data samples and each record contains 9 attributes. Each transaction consists of items purchased by a customer in one interaction with developed site. Out of 4500 transactions the dataset of 15 transactions is shown in Table 1. The itemsets which was bought by the customer at least in 3 transactions i.e. 20% of the transactions have been taken for this work.

Using transaction dataset, the candidate1-itemset is shown in Table 2. Table 3 shows the details of each transaction.

From Table 2 it can be observed that most popular transaction was A5.

Table 3 shows the details of transaction and minimum support is considered as 20%. Table 4 shows the Frequent 2-itemset. Here all the itemsets having less than 20% support have been pruned.

Finally, in Table 5 the Frequent 3-itemset is shown. It is evident that this occurrence will happen most frequently that if a person purchases item A3 (Jeans) he/she will also purchase A5 (T-Shirt) and A7 (Kurti) item as shown in Table 5.

From the sample dataset, the itemset {Jeans, T-Shirt, Kurti} has a support of 3/15 = 0.20 since it occurs in 20% of all transactions. Here 3 is the number of transactions, from the dataset which contains the itemset {Jeans, T-Shirt, Kurti} while 15

Table 1 Transaction dataset

Transaction	Itemset								
	A1 Shirt	A2 Pant	A3 Jeans	A4 Salwar Kamij	A5 T-Shirt	A6 Saree	A7 Kurti	A8 Shoe	A9 Socks
1	1	0	0	0	1	1	0	1	0
2	0	1	0	1	0	0	0	1	0
3	0	0	1	1	1	0	0	0	0
4	0	1	0	0	0	0	1	0	0
5	0	0	1	0	1	1	0	0	0
6	0	1	0	1	0	0	1	0	0
7	0	1	1	0	0	1	0	0	0
8	0	0	0	0	1	0	0	0	1
9	0	0	0	0	0	0	0	1	0
10	0	0	1	0	1	0	1	0	0
11	0	0	1	0	1	0	1	0	0
12	0	0	0	0	1	1	0	1	0
13	0	1	1	1	0	1	0	0	0
14	1	0	1	0	1	0	1	0	0
15	0	1	0	0	0	0	1	0	1

Table 2 Candidate-1 itemset

Itemsets X	Transaction count	
A1	2	Deleted
A2	6	
A3	7	
A4	4	
A5	8	
A6	5	
A7	6	
A8	4	
A9	2	Deleted

Table 3 Details of transaction

Transaction ID	Items bought
T1	{A1, A5, A6, A8}
T2	{A2, A4, A8}
T3	{A3, A4, A5}
T4	{A2, A7}
T5	{A3, A5, A6}
T6	{A2, A4, A7}
T7	{A2, A3, A6}
T8	{A5, A9}
T9	{A8}
T10	{A3, A5, A7}
T11	{A3, A5, A7}
T12	{A5, A6, A8}
T13	{A2, A3, A4, A6}
T14	{A1, A3, A5, A7}
T15	{A2, A7, A9}

Table 4 Frequent 2-itemset

Itemsets X	Transaction count	Support X
A2, A4	3	0.20
A2, A7	3	0.20
A3, A5	5	0.33
A3, A6	3	0.20
A3, A7	3	0.20
A5, A6	3	0.20
A5, A7	3	0.20

Table 5 Frequent 3-itemset

Itemsets X	Transaction count	Support X	Confidence (XUY)/X
A3, A5, A7	3	0.20	0.60

represents the total number of transactions. For the rule {Jeans, T-shirt} => {Kurti} the confidence is mentioned below:

support ({Jeans, T-Shirt, Kurti})/support ({Jeans, T-Shirt}) = 0.20/0.33 = 0.60

This means the rule is correct for 60% of the transactions containing Jeans and T-Shirt.

7 Result and Conclusion

As computational complexity of the Apriori algorithm is concerned, the result shows that for 60% cases the rule is correct. In our future work by using the same dataset modified Apriori algorithm will be applied and thereafter the comparison between the performances of the two can be measured.

This system has been implemented using JSP. The presentation layer implementation is done by using CSS and HTML, the validation layer is implemented by JavaScript. The insertion and retrieval of information have been done using SQL Queries and interaction with database, implemented by MySQL, is done by Java.

Thus, taking the above facts in consideration, a recommender system has been designed which can help the customer to fulfill their needs as well as retailer to develop marketing strategies and increase sales promotion. In future, more itemsets will be added to our database and we will optimize the program code which may reduce time complexity of the developed algorithm.

Acknowledgements The authors are thankful to Director, MCKVIE and Principal, MCKVIE, for providing the Computer laboratories and other infrastructure to do the proposed work. The authors are also thankful to Santosh Sagar, student of CSE department of MCKVIE for collecting the required data for the proposed work.

References

1. Y. Li, in *The Application of Apriori Algorithm in the Area of Association Rules*. Fifth International Conference on Digital Image Processing. International Society for Optics and Photonics, 2013
2. Y. Liu, in *Study on Application of Apriori Algorithm in Data Mining*. Computer Modeling and Simulation, 2010. ICCMS'10. Second International Conference on, Vol. 3. IEEE, 2010
3. X.W. Zhao, et al., in *We Know What you Want to Buy: A Demographic-Based System for Product Recommendation on Microblogs*. Proceedings of the 20th ACM SIGKDD International Conference on Knowledge Discovery and Data Mining. ACM, 2014
4. M. Al-Maolegi, B. Arkok, An improved apriori algorithm for association rules. Int. J. Nat. Lang. Comput. (IJNLC), **3**(1), February (2014)
5. J. Han, M. Kamber, *Data Mining: Concepts and Techniques*. Morgan Kaufmann Publishers, Book (2000)
6. A. Bhandari, A. Gupta, D. Das, Improvised apriori algorithm using frequent pattern tree for real time applications in data mining. Procedia Comput. Sci. **46**, 644–651 (2015)

7. J. Bobadilla, F. Ortega, A. Hernando, A. Gutiérrez, Recommender systems survey. Knowl.-Based Syst. **46**, 109–132 (2013)
8. P. Adamopoulos, A. Tuzhilin, On Unexpectedness in Recommender Systems: Or How to Better Expect the Unexpected. Working Paper: CBA-13-03, New York University, 2013. http://hdl.handle.net/2451/31801
9. A.K. Shrivastava, R.N. Panda, Implementation of apriori algorithm using WEKA. KIET Int. J. Intell. Comput. Inform., **1**(1), 12–15, January (2014)
10. X. Fang, in *An Improved Apriori Algorithm on the Frequent Itemset*. International Conference on Education Technology and Information System (ICETIS 2013), pp. 845–848
11. P. Mandave, M. Mane, Prof. S. Patil, in *Data Mining Using Association Rule Based on Apriori Algorithm and Improved Approach with Illustration*. Int. J. Latest Trends Eng. Technol. (IJLTET), **3**(2), 107–113, November (2013)
12. R. Agrawal, R. Srikant, in *Fast Algorithms for Mining Association Rules*. Proc. the 20th VLDB[C] (1994), pp. 487–499
13. R. Agrawal, T. Imielinski, A.N. Swami, in *Mining Association Rules between Sets of Items in Large Databases*. Proc. ACM SIGMOD Int'l Conf. Management of Data, (May 1993), pp. 207–216
14. J. Yabing, Research of an improved apriori algorithm in data mining association rules. Int. J. Comput. Commun. Eng., **2**(1), January (2013)
15. X. Wu, V. Kumar, J. Ross Quinlan, J. Ghosh, Q. Yang, H. Motoda, G.J. McLachlan, A. Ng, B. Liu, P.S. Yu, Z.-H. Zhou, M. Steinbach, D.J. Hand, D. Steinberg, Top 10 algorithms in data mining. Knowl. Inf. Syst. **14**(1), 1–37 (2007)

A Comparative Analysis Between EDR and Respiration Signal: A Pilot Study with Normal Subjects

Surita Sarkar, Saurabh Pal and Parthasarathi Bhattacharyya

1 Introduction

Respiration is one of the most important physiological parameters that should be monitored and diagnosed regularly for different conditions specially in chronic diseases [1, 2]. Abnormal and changing respiration is one of the earliest indicators of major physiological instabilities [3]. Even regular follow-up of respiration can predict the risk of serious adverse events [4]. On the other hand, obstructive lung disease becomes a major problem globally with increasing rate of morbidity and mortality due to this [5–7]. Continuous exposure to smoking, pollution, dust particles and various allergens are the main factors behind this [8, 9]. However, it is unfortunate that respiration signal is often neglected in regular follow-up due to lack of awareness and unavailability of unobtrusive and accurate respiratory monitor [10, 11]. As a result most of the time the patient is unaware of the disease and does not visit the doctor until it becomes severe [6]. Equipments like spirometry, inductive plethysmography, impedance pneumography, respiratory belts, lung imaging techniques (X-ray, CT scan etc.) are commonly used [3, 12, 13, 14] and some new techniques like microelectromechanical system based acceleration sensor and low power ambulatory wearable are developed [15, 16] for measurement of respiration and lung obstruction. But most of these methods often exhibit several problems

S. Sarkar (✉) · S. Pal
Department of Applied Physics, University of Calcutta,
92 A.P.C. Road, Kolkata 700009, India
e-mail: sarkar.surita@gmail.com

S. Pal
e-mail: spal76@gmail.com

P. Bhattacharyya
Institute of Pulmocare & Research, DG-8, Action Area-1, New Town,
Kolkata 700156, India
e-mail: ipcr_india@yahoo.com

S. Chattopadhyay et al. (eds.), *Modelling and Simulation in Science, Technology and Engineering Mathematics*, Advances in Intelligent Systems and Computing 749,
https://doi.org/10.1007/978-3-319-74808-5_52

because commonly used equipments highly interfere with the patient and are often prone to noise. In this situation, various researchers found that respiration signal derived from ECG could be a useful alternative [17–19]. ECG is a common and popular test, often prescribe by the medical practitioners in regular follow-up. Moreover, the influence of respiration on the cardiac activity has already established in previous literatures [20, 21]. During respiration, cardiac axis changes due to expansion and contraction of chest [22]. Heart rate variability (HRV), on the other hand, is also effected by Respiratory Sinus Arrhythmia (RSA) due to which R-R interval decreases during inspiration and increases during expiration [21]. Using these effects, different algorithms are developed to extract ECG derived respiration signal [22, 23].

In this study, two normal subject groups- one without any present symptoms and one group with symptoms like cold, cough, etc. were taken for data collection. Features extracted from the respiration signal acquired from the respiration belt are compared with the same features extracted from the derived respiration signal from ECG (EDR). Also the statistical significance of the features was analysed.

2 Method

The whole study was done based on real time data collection of subjects followed by pre-processing of ECG and respiration signal, EDR extraction and feature extraction. Each section is described elaborately.

2.1 Data Collection

In order to evaluate EDR signals, real-time datasets were used. Simultaneous recording of ECG and respiration signal were collected using a data acquisition system called MP-45, designed by Biopac Systems Inc. [24]. ECG signal was collected using lead-II single-lead ECG, whereas, respiration signal was acquired using the respiratory effort transducer (SS5LB) by measuring the volume change in thoracic cavity. The respiration signal collected using the chest belt was used as the original respiration signal in analysis. The study was done based on a protocol approved by the Institutional Ethics Committee of Institute of Pulmocare and Research, Kolkata. Data were collected from the subjects only after they agreed to give written informed consent. Medical history of each subject was taken prior to a physical examination done by a medical expert before enrolment. All the subjects were evaluated by spirometry as per rule of the ATS guideline [25]. In this study, a total of twelve subjects with normal ECG and spirometry were included. Among them, six normal subjects are completely healthy and without symptoms. Other six subjects, though exhibited normal spirometry; they had mild symptoms like sneeze, cold, cough, etc. present at the time of data collection. Recording of each subject was done at a sampling rate of 1000 Hz and for 300 s time-duration while the patient was resting at supine condition.

The real-time data were collected from the Institute of Pulmocare and Research and further analysis was carried out in the Biomedical Instrumentation Laboratory of Department of Applied Physics, University of Calcutta. Subjects having any kind of lung and cardiac diseases were excluded from this study. The demographic parameters of the two subject groups are shown in Table 1.

2.2 *Pre-processing of ECG and Respiration Signal*

ECG and respiration signals were collected simultaneously for each subject. The design of this study is shown in Fig. 1.

Table 1 Demographic parameters of normal subjects without and with symptoms

Subject	Total population	Age (Mean ± SD)	Male:Female ratio	BMI (Mean ± SD)
Normal without symptom (N)	6	27.33 ± 7.06	1:1	22.97 ± 3.7
Normal with symptoms (NS)	6	37.83 ± 11.47	2:1	22.91 ± 2.24

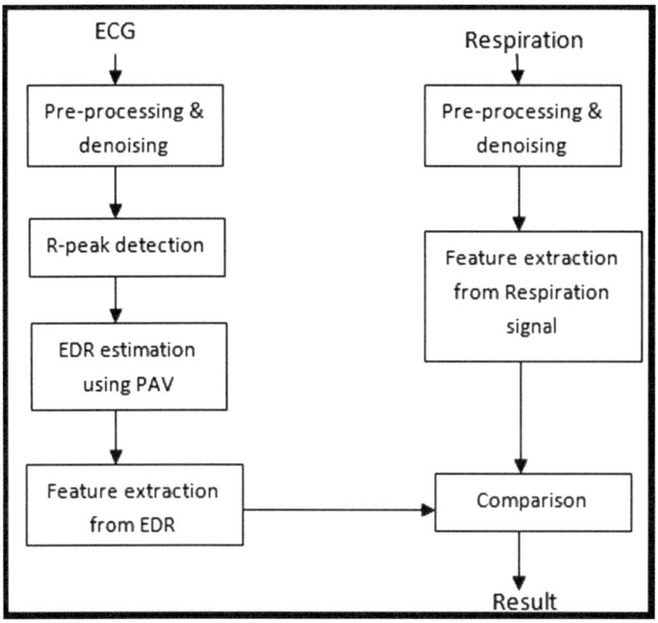

Fig. 1 Block diagram shows a comparative study between features extracted from both EDR and respiration signal

From both ECG and respiration signal, 80,000 samples were taken for data analysis purpose. Each signal was de-noised using filter. For ECG signal, a 2nd order Bandpass Butterworth filter was used for removal of noises. R-peaks were detected from the filtered ECG signal using sliding window technique for a window length of 2 s. The ECG waveform was then normalized to zero mean and unit variance. After applying a threshold to identify the potential R-peaks, a 500 ms sliding window was applied again. The peak with maximum value within the span was found out to be recognized as the actual R-peak. The whole process was done for the entire length of ECG samples taken. After the R-peaks were detected, their amplitudes were calculated against the baseline and plotted against their corresponding locations. Respiration signals were also filtered and normalized before feature extraction.

2.3 ECG Derived Respiration

To extract a surrogate respiratory signal from ECG signal, the respiration mechanism that affects the ECG signal, should be observed. The respiration activity influences the cardiac activity in different ways- one of those affects beat morphology [26, 27]. The expansion and contraction of lungs during respiration cause change in apex of heart, therefore, affecting the electrical axis direction. As a result thoracic impedance which is closely related with the change in lung volume, also changes [26]. At the time of inhalation the electrical impedance across lungs increases.

This respiratory induced change in thoracic impedance affects the QRS amplitude of ECG signal, where the QRS amplitude decreases during inspiration and the amplitude increases at the time of expiration.

There are several algorithms developed for EDR extraction. In this study, the modulation change in QRS amplitude is used for EDR extraction.

After detection, R-peaks were arranged against the time axis and cubic spline interpolation was used to create an EDR signal resembles to the original respiration signal [28]. If the original signal be, $k_i = f(g_i)$, where $i = 0, 1, 2, \ldots, m - 1$, then the output of the interpolation within a specific range of $[t_i, t_{i+1}]$, $t_0 \leq t \leq t_{m-1}$, can be written as,

$$k = uk_i + vk_i + 1 + wk_i'' + xk_i'' + 1 \tag{1}$$

With coefficients be like,

$$u = \frac{t_{i+1} - t}{t_{i+1} - t_i}, v = 1 - u, w = \frac{1}{6}(u^3 - u)(t_{i+1} - t_i)^2, x = \frac{1}{6}(v^3 - v)(t_{i+1} - t_i)^2$$

2.4 Feature Extraction

Respiration is one crucial physiological parameter. Breathing pattern changes upon different physiological changes in lungs [29]. Even common disease like cough and cold can change this pattern. Based on this, three features were extracted from both original respiration signal and extracted EDR signal.

2.4.1 Area Ratio

The locations of the starting (S), peak (P) and ending (E) points (as shown in Fig. 2) were detected using peak detection algorithm for one respiratory cycle. The location of the peak against the baseline was taken as point R. Using the points S, P and R, inspiration area was calculated and expiration area was calculated using the locations of P, R and E. Area ratio (A) was calculated for five randomly chosen cycles and the average was taken as the area ratio of that subject. The same procedure was done for EDR signal. Area ratio was calculated using the following formula (Eq. 2),

$$A = \frac{Expiration\ area}{Inspiration\ area} \tag{2}$$

2.4.2 Time Ratio

Time ratio was calculated by taking the ratio of expiration time to inspiration time. To compute inspiration and expiration time, starting (S), peak (P) and ending (E) point locations were detected for one respiratory cycle. The time ratio (T) can be written as,

$$T = \frac{\text{Time interval between points P and E}}{\text{Time interval between points S and P}} \tag{3}$$

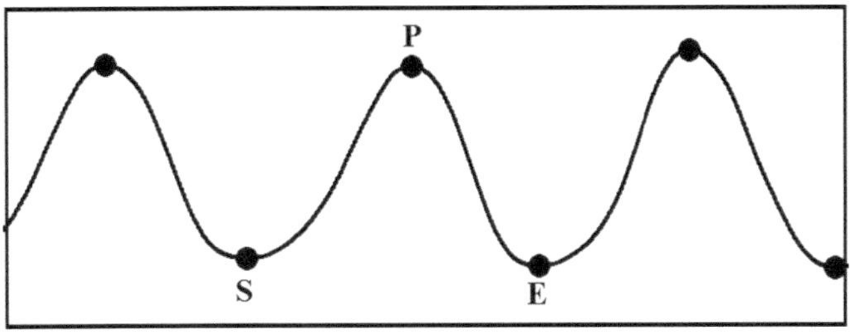

Fig. 2 'S' denotes the starting point, 'P' denotes the peak and 'E' denotes the ending point of a respiratory cycle

The whole procedure was done for five consecutive times from randomly chosen cycles. The same method was followed in case of EDR signal.

2.4.3 Respiration Rate

Respiration rate was calculated by counting the total number of one complete respiration cycle (one inspiration and one expiration) present within time interval of 60 s. The time duration between two consecutive maxima was calculated and the respiratory rate (RR) was estimated using the following formula (Eq. 4).

$$RR = \frac{60}{Time\ duration\ \text{for occurence of }[(\text{kth} - (k-1)\text{th}]\ respiratory\ \text{peaks}} \quad (4)$$

Average respiration rate was calculated by taking the average of five consecutive respiratory rates and the same method was applied for EDR signal also.

The statistical analysis was done using student t-test to see if the extracted features are significant or not.

3 Result

A total of 12 subjects were included in this study for analysis. None of them were smoker or under any kind of medication. The EDR signal derived for each subject was compared with the original respiration signal acquired of that person. Traces of EDR signal followed by the original respiration signal for both the subject groups (N and NS) are shown in Fig. 3.

From the above figures it can be seen that the respiratory pattern of subject without any symptom (Fig. 2a) differs from the respiratory pattern of the subject with symptom (Fig. 2b), whereas, no significant change can be noticed in between the EDR patterns of those subjects. The three features discussed in method section were extracted for each of the twelve subjects. The features and their corresponding p-values are shown in Table 2.

Table 2 shows that the p-values of respiratory rate calculated from both the signals are not statistically significant in case of both subject groups. But for other two features (i.e. area ratio and time ratio) derived from the EDR and the original respiration signal, the p-value significantly changes in case of subjects with symptoms according to Table 2. The p-values estimated for area ratio and time ratio from both EDR and original respiration signal turn out to be statistically insignificant for those without any symptoms.

Table 2 Features extracted from EDR and original respiration and their corresponding p-values

Name	State (N/NS)	EDR area ratio (mean)	Respiratory area ratio (mean)	P value	EDR time ratio (mean)	Respiratory time ratio (mean)	P value	EDR RR* (mean)	Original RR (mean)	P value
Subject 1	N	0.60	0.84	0.6334	0.57	0.89	0.6604	16	16	1
Subject 2	N	1.05	1.06	0.886	1.00	1.01	0.1782	20	20	1
Subject 3	N	1.01	0.92	0.4594	0.93	0.73	0.1527	15	15	0.3972
Subject 4	N	0.93	0.91	0.8231	0.74	0.65	0.5034	15	15	0.6811
Subject 5	N	1.03	0.95	0.4438	0.86	0.93	0.4416	18	18	1
Subject 6	N	1.05	0.94	0.3874	0.93	0.99	0.4654	22	22	0.6666
Subject 7	NS	1.01	1.22	0.0009	0.87	1.33	0.0001	21	21	1
Subject 8	NS	0.93	1.25	0.0002	0.95	1.25	0.0059	18	18	0.6666
Subject 9	NS	0.78	1.68	0.0329	0.79	1.89	0.0257	10	9	0.195
Subject 10	NS	0.76	1.12	0.0035	0.72	1.47	0.0021	17	17	1
Subject 11	NS	0.83	1.38	0.0428	0.84	1.68	0.0021	21	21	1
Subject 12	NS	0.83	1.34	0.0195	0.79	1.76	0.0043	15	15	0.8548

*RR denotes Respiratory Rate

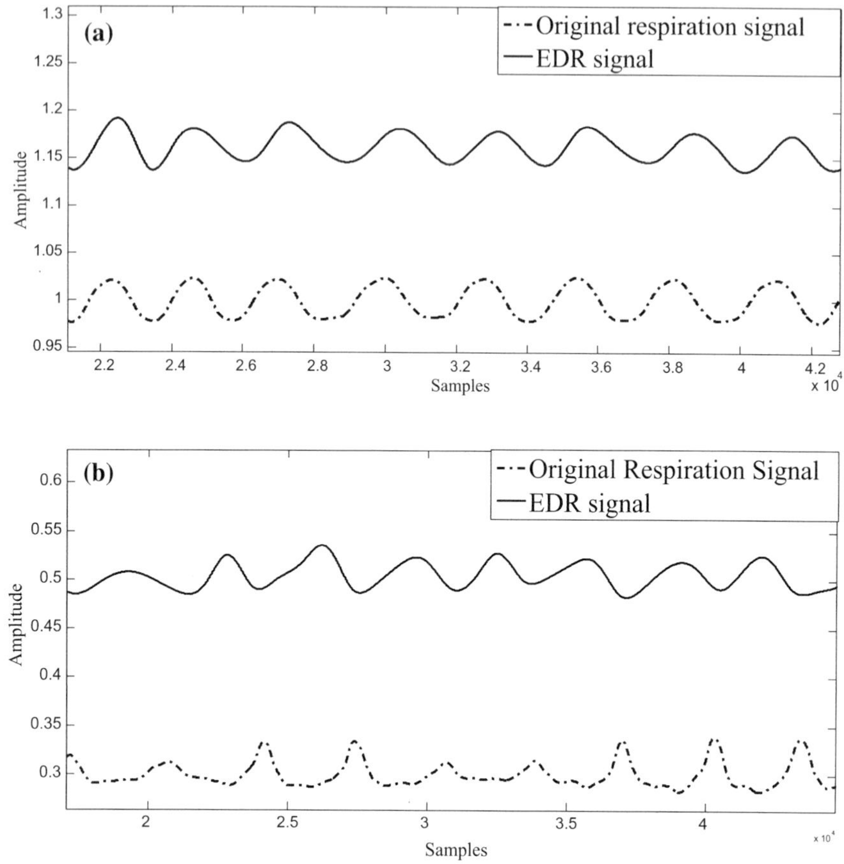

Fig. 3 Figure shows **a** The EDR and original respiration signal of a subject without symptom and **b** the EDR and respiration signal of a subject with symptom

4 Discussion

In medical diagnosis, both ECG and respiration should be monitored regularly for detecting any abnormality in cardio-pulmonary system. In countries like India, though ECG is done in routine check-up, respiration monitoring remains neglected most of the time. On the other hand, respiration has well-known influence on heart. Therefore, derivation of respiratory information from ECG will be an alternative tool to monitor both ECG and respiration. In this study, EDR signal is extracted from the ECG signal based on the R peak amplitude variation. It is interesting to note that, though all the twelve patients have shown normal spirometry, the respiration pattern for those with symptoms showed difference from that of those without symptoms. The result also reflects the same. Features extracted from the original respiration signal vary significantly with features extracted from the EDR signal in case of subjects with symptoms only.

5 Conclusion

The study shows an interesting result of applicability of EDR which can be established clinically by further analysis on bigger dataset. Despite of small dataset it can be concluded that the EDR is a better alternative for respiration monitoring compare to conventional method where respiration belt is used.

Acknowledgements The first author acknowledges the financial support obtained in the form of CSIR-SRF fellowship provided by CSIR-HRDG, Government of India.

References

1. M.A. Cretikos, R. Bellomo, K. Hillman, J. Chen, S. Finfer, A. Flabouris, Respiratory rate: the neglected vital sign. MJA **188**(11), 657–659 (2008)
2. S. Sarkar, S. Bhattacherjee, S. Pal, *Extraction of Respiration Signal from ECG for Respiratory Rate Estimation*. Proc. of Michael Faraday IET International Summit: MFIIS-2015, vol. 2 (Kolkata, India, 2015), pp. 336–40
3. A.R. Fekr, K. Radecka, Z. Zilic, Design and evolution of an intelligent remote tidal volume variability monitoring system in e-health application. IEEE J. Biomed. Health Inform. **19**(5), 1532–1548 (2015)
4. A. Garde, L. Sornmo, P. Laguna, R. Jane, S. Benito, A. Bayes-Genis, B.F. Giraldo, Assessment of respiratory flow cycle morphology in patients with chronic heart failure. Med. Biol. Eng. Comput. **55**, 245–255 (2017)
5. B. Lazovic, M.S. Zlatkovic, S. Mazic, Z. Stajic, M. Delic, Analysis of electrocardiogram in chronic obstructive pulmonary disease patients. Med. Pregl., **LXVI**(3–4), 126–129 (2013)
6. O.C.P. van Schayck, A. D'Urzo, G. Invernizzi, M. Roman, B. Stallberg, C. Urbina, Early detection of chronic obstructive pulmonary disease (COPD): the role of spirometry as a diagnostic tool in primary care. Prim. Care Respir J. **12**(3), 90–93 (2003)
7. P.J. Barnes, New concepts in chronic obstructive pulmonary disease. Annu. Rev. Med. **54**, 113–129 (2003)
8. S.A. Antoniu, Descriptors of dyspnea in obstructive lung diseases. Multi. Respir. Med. **5**(3), 216–219 (2003)
9. A.S. Buist, Similarities and differences between asthma and chronic obstructive pulmonary diseases: treatment and early outcomes. Eur. Respir. J. **39**, 30–35 (2003)
10. J.J. Liu, M.C. Huang, W. Xu, X. Zhang, L. Stevens, N. Alshurafa, M. Sarrafzadeh, Breathsens: a continuous on-bed respiratory monitoring system with torso localization using an unobtrusive pressure sensing array. IEEE J. Biomed. Health Inform. **19**(5), 1682–1688 (2015)
11. N.M. Rachaiah, J.M. Rachaiah, R.B. Krishnaswamy, A correlative study of spirometric and ECG changes in patients with chronic obstructive pulmonary disease. Int. J. Biol. Med. Res. **3**(1), 1322–1326 (2012)
12. S. Milne, G.G. King, Advanced imaging in COPD: insights into pulmonary pathophysiology. J. Thorac. Dis. **6**(11), 1570–1585 (2014)
13. G.M.F. Wallace, J.H. Winter, J.E. Winter, A. Taylor, T.W. Taylor, R.C. Cameron, Chest X-rays in COPD screening: are they worthwhile? Respir. Med. **103**, 1862–1865 (2009)

14. Global Initiative for Chronic Obstructive Lung Disease. Global strategy for the diagnosis management, and prevention of chronic obstructive pulmonary disease (2006). http://www.goldcopd.org
15. H. Sumbul, A.H. Yuzer, Development of diagnostic device for COPD: a MEMS based approach. Int. J. Comput. Sci. Netw. Secur. **17**(7), 196–203 (2017)
16. J. Acharya, A. Basu, W. Ser, Feature extraction techniques for low-power ambulatory wheeze detection wearables. *39th Annual International Conference of the IEEE Engineering in Medicine and Biology Society (EMBC 2017)*, 2017
17. C. Varon, S.V. Huffel, ECG-derived respiration for ambulatory monitoring. Comput. Cardiol. **42**, 169–172 (2015)
18. G.B. Moody, R.G. Mark, M.A. Bump, J.S. Weinstein, A.D. Berman, J.E. Mietus, A.L. Goldberger, Clinical validation of the ECG-derived respiration (EDR) technique. Comput. in Cardiol. **13**, 507–510 (1986)
19. J. Lazaro, A. Alcaine, D. Romero, E. Gill, P. Laguna, L. Sornmo, R. Bailon, Electrocardiogram derived respiration from QRS slopes: evaluation with stress testing recordings. Comput. Cardiol. **40**, 655–658 (2013)
20. L. Zhao, S. Reisman, T. Findley, Derivation of respiration from electrocardiogram during heart rate variability studies. Comput. Cardiol., 53–56 (1994)
21. S.B. Park, Y.S. Noh, S.J. Park, H.R. Yoon, An improved algorithm for respiration signal extraction from electrocardiogram measured by conductive textile electrodes using instantaneous frequency estimation. Med. Bio. Eng. Comput. **46**, 147–158 (2008)
22. A. Sobron, I. Romero, T. Lopetegi, Evaluation of methods for estimation of respiratory frequency from the ECG. Computing in Cardiology **37**, 513–516 (2010)
23. D. Labate, F.L. Foresta, G. Occhiuto, F.C. Morabito, A.L. Ekuakille, P. Vergallo, Empirical mode decomposition for the extraction of respiratory signal from single-channel ECG: a comparison. IEEE Sens. J., **13**(7), 2666–2674 (2013)
24. R. Pflanzer, W. McMullen, Electrocardiography (ECG) II Procedure. Biopac Student Lab Manual, BIOPAC Systems Inc. pp. P1–P10, 2014
25. B.R. Celli, W. MacNee, Standards for the diagnosis and treatment of patients with COPD: a summary of the ATS/ERS position paper. Eur. Respir. J. **23**(6), 932–946 (2004)
26. L. Mirmohamadsadeghi, J.M. Vesin, Respiratory rate estimation from the ECG using an instantaneous frequency tracking algorithm. Biomed. Sign. Process. Control **14**, 66–72 (2014)
27. A.M. Chan, N. Ferdosi, R. Narasimhan, Ambulatory Respiratory Rate Detection using ECG and a Triaxial Accelerometer, *35 Annual International Conference of the IEEE EMBS*, pp. 4058–4061, 2013
28. S. Sarkar, S. Pal, S. Bhattacherjee, P. Bhattacharyya, in *Characterization of Obstructive Lung Disease from Respiration Signal*. Proc. of Int. Conf. on Advancement of Computer Communication & Electrical Technology, pp. 143–147, 2016
29. S. Sarkar, S. Bhattacherjee, P. Bhattacharyya, S. Pal, Differentiation of COPD from normal population using ECG derived respiration: a pilot observation. The Pulmo-Face **XVI**(1), 12–18 (2016)

Uncertainty in Fission Product Transient Release Under Accident Condition

Subrata Bera, U. K. Paul, D. Datta and A. J. Gaikwad

1 Introduction

During power irradiation of nuclear fuel pin in Indian pressurized heavy water reactor, fission product radionuclides get accumulated in the fuel matrix. Followed by a postulated initiating event for a design extension condition without core melt [1] (e.g., loss of coolant accident with failure of emergency core cooling system), temperatures of both fuel and clad increase even after reactor shutdown caused by protection system. This fuel heat up is due to the decay heat generated in the fuel matrix by the radionuclides produced in the nuclear fission reaction and subsequent radioactive decay chain. Due to the heating up the fuel pin, the radionuclides get released to the fuel clad region by thermal diffusion process. Once fuel pin fails on exceeding the limiting clad temperature, radionuclides get released to the primary heat transport system [2]. Based on the accident sequence and its progression, radionuclides get transported to containment building through coolant vapour transport route. The

S. Bera (✉) · U. K. Paul · A. J. Gaikwad
Nuclear Safety Analysis Division, Atomic Energy Regulatory Board,
Anushaktinagar, Mumbai 400094, India
e-mail: sbera@aerb.gov.in

U. K. Paul
e-mail: ukp@aerb.gov.in

A. J. Gaikwad
e-mail: avinashg@aerb.gov.in

S. Bera
Homi Bhabha National Institute, Mumbai 400094, India

D. Datta
Radiological Physics and Advisory Division,
Bhabha Atomic Research Centre, Trombay, Mumbai 400085, India
e-mail: ddatta@barc.gov.in

S. Chattopadhyay et al. (eds.), *Modelling and Simulation in Science, Technology and Engineering Mathematics*, Advances in Intelligent Systems and Computing 749,
https://doi.org/10.1007/978-3-319-74808-5_53

pressure build up occurs in the containment due to the accumulation of steam and the associated radionuclides releases from primary coolant system. The incremental raise of pressure may lead to the release of radionuclides from containment to the environment if no mitigating accident management action is assumed to be successful. The consequent amount of radionuclides released to the environment is used for estimation of the radiological impact of a postulated accident scenario [3]. In the whole process of source term [4–6] evaluation, the time dependent release of radionuclides from fuel to the fuel-clad gap under elevated temperature condition [7, 8] is governed by thermal diffusion process. In modelling this phenomenon, first order release kinetics is considered. The rate of release is proportional to the present inventory of the radionuclides in the fuel. The proportional constant depends on the fuel temperature and its dependence is represented by Arrhenius equation with experimentally evaluated activation energy and pre-exponential factor. Thus the fractional release rate becomes an exponential of an exponential function of temperature. As this is a non-linear equation, the variability of the fractional release depends on model sensitivity parameter and variability of fuel temperature. There are many methods to assess the aleatory uncertainty such as error propagation through direct analytical expression, probabilistic method based on sampling from uncertain domain [9], emulation approach if the mathematical model is not precisely known, stochastic response surface formulation [10]. In this paper, it is attempted to quantify the variability of fractional release due to known variation of temperature.

2 Theoretical Model for Uncertainty Quantification

2.1 *Analytical Uncertainty Estimation Methodology*

The quantification of error propagation through a general function which can be linear or non-linear can be done using Taylor series expansion method. For a single input variable and single output variable, the system response function determines the sensitivity of input parameter. Based on the Taylor series expansion of a continuous function with availability of infinite number of derivatives at $\boldsymbol{x} = \boldsymbol{x}_0$, the expansion of f(x) at location $\boldsymbol{x}_0 + \boldsymbol{\sigma}_x$ can be written as given in Eq. (1). Where, $\boldsymbol{x}_0$ is the mean value of input variable x; $\boldsymbol{\sigma}_x, \boldsymbol{\sigma}_y$ are standard deviations of input variable x and output variable y respectively.

$$f(x_0 + \sigma_x) = f(x_0) + \frac{\sigma_x}{1!} f^{(1)}(x_0) + \frac{\sigma_x^2}{2!} f^{(2)}(x_0) + \cdots + \frac{\sigma_x^n}{n!} f^{(n)}(x_0) + \cdots \quad (1)$$

In the concise notation, it can be represented as given in Eq. (2).

$$f(x + \sigma_x) = \sum_{i=0}^{\infty} \frac{\sigma_x^i}{i!} f^{(i)}(x_0) \quad (2)$$

where, $\boldsymbol{f}^{(i)}(\boldsymbol{x}_0)$ represents the $\boldsymbol{i}$th derivatives of the function $\boldsymbol{f}(\boldsymbol{x})$ at $\boldsymbol{x} = \boldsymbol{x}_0$.

The uncertainty (i.e., σ_f) associated with the system response due to the variation of x can be derived from Eq. (1) as given in Eq. (3).

$$\sigma_f = f(x_0 + \sigma_x) - f(x_0) = \frac{\sigma_x}{1!} f^{(1)}(x_0) + \frac{\sigma_x^2}{2!} f^{(2)}(x_0) + \cdots + \frac{\sigma_x^n}{n!} f^{(n)}(x_0) + \cdots \tag{3}$$

In the simplified notation the standard deviation of function, the Eq. (3) can be written as Eq. (4).

$$\sigma_f = \sum_{i=1}^{\infty} \frac{\sigma_x^i}{i!} f^{(i)}(x_0) \tag{4}$$

This expansion of infinite series will be truncated depending upon the degree of nonlinearity of the system response. It can be inferred that total uncertainty depends on two things and these are how uncertain x is (i.e., σ_x) and how sensitive f is to x (e.g., $f^{(1)}(x_0),\ f^{(2)}(x_0),\ \ldots$).

2.2 *Transient Fission Product Release Model*

Fission products get generated inside the UO_2 fuel pellets, and their release to the atmosphere outside the containment during an unmitigated accident may present a health hazard. Significant release of most of the hazardous fission products from fuel occur only at very high temperatures in an unmitigated accident condition.

Fission product release, like radioactive decay, is said to be of first order when it depends on the first order of the amount present. Although not mechanistic, a model based on this principle, agrees adequately with measured releases and hence the first order model has been widely used.

In fractional release models, the transient release of radio-nuclides is proportional to the radio-nuclides inventory (M) present at any instant t, which is given in Eq. (5).

$$\frac{dM}{dt} = -kM \tag{5}$$

where, k is the release rate coefficient.

The solution of this first order release rate equation on rearrangement can be written as Eq. (6).

$$M = M_0 e^{-kt} \tag{6}$$

where, M_0 denotes the initial inventory of the radionuclide. Fractional release (f) during the time interval Δt can be represented by Eq. (7).

$$f = \frac{M_0 - M}{M_0} = \left(1 - e^{-k\Delta t}\right) \tag{7}$$

Release rate coefficients follow Arrhenius equation as given in Eq. (8):

$$k = k_0 e^{-\frac{Q}{RT}} \tag{8}$$

where, T is the fuel temperature given in K; R is the universal gas constant; k_0 is pre-multiplier; and Q is the activation energy. Release rate coefficients k_0*, and* Q depend on the radionuclide under consideration. As Iodine is one of the important radionuclides that may release from fuel during an accident, it has been considered for this study. Release rate coefficients k_0, *and* Q for Iodine is 2.0×10^5 per minute and 63.8 kcl/mol respectively. Substituting Eq. (8) in Eq. (7), the fractional release rate for radionuclide can be written as given in Eq. (9).

$$f = \frac{M_0 - M}{M_0} = \left(1 - e^{-k_0 \Delta t e^{\left(-\frac{Q}{RT}\right)}}\right) \tag{9}$$

The Eq. (9) is a non-linear equation containing an exponential of an exponential function of temperature. Therefore, fractional release is very sensitive to the temperature.

2.3 Error Propagation in Transient Fission Product Release Model

In this work, two steps process has been adapted for uncertainty estimation. In the first step, the variability of rate coefficient with respect to the variability of temperature is to be estimated. In the second step, the variability in fractional release is to be estimated using the variability of rate coefficient, which is already estimated in the first step. For the first step, it is required to determine the differential coefficient of various degrees so that Eq. (4) can be used to estimate the standard deviation of rate coefficient. It is found sensitivity analysis that up to five order of differential coefficient is sufficient to capture the variability due to the uncertain input parameter. The equations for the derivatives of rate coefficient up to order of five are derived and given in Eqs. (10)–(14).

$$\frac{dk(T)}{dT} = -\frac{kQ}{RT^2} \tag{10}$$

$$\frac{d^2k(T)}{dT^2} = k\left(\frac{2Q}{RT^3} + \frac{Q^2}{R^2T^4}\right) \tag{11}$$

$$\frac{d^3k(T)}{dT^3} = -k\left(\frac{Q^3}{R^3T^6} + \frac{6Q^2}{R^2T^5} + \frac{6Q}{RT^4}\right) \tag{12}$$

$$\frac{d^4k(T)}{dT^4} = k\left(\frac{Q^4}{R^4T^8} + \frac{12Q^3}{R^3T^7} + \frac{36Q^2}{R^2T^6} + \frac{24Q}{RT^5}\right) \tag{13}$$

$$\frac{d^5k(T)}{dT^5} = -k\left(\frac{Q^5}{R^5T^{10}} + \frac{20Q^4}{R^4T^9} + \frac{120Q^3}{R^3T^8} + \frac{240Q^2}{R^2T^7} + \frac{120Q}{RT^6}\right) \tag{14}$$

Now standard deviation of rate coefficient can be evaluated using Eq. (4) and its derivatives. The resultant equation is given in Eq. (15).

$$\sigma_k = \frac{dk(T)}{dT}\sigma_T + \frac{1}{2}\frac{d^2k(T)}{dT^2}\sigma_T^2 + \frac{1}{6}\frac{d^3k(T)}{dT^3}\sigma_T^3 + \frac{1}{24}\frac{d^4k(T)}{dT^4}\sigma_T^4 + \frac{1}{120}\frac{d^5k(T)}{dT^5}\sigma_T^5 \tag{15}$$

Similarly the equations for the derivatives up to order of five for fractional release are derived and given in Eqs. (16)–(20).

$$\frac{df(k)}{dk} = -(f-1)\Delta t \tag{16}$$

$$\frac{d^2f(k)}{dk^2} = (f-1)(\Delta t)^2 \tag{17}$$

$$\frac{d^3f(k)}{dk^3} = -(f-1)(\Delta t)^3 \tag{18}$$

$$\frac{d^4f(k)}{dk^4} = (f-1)(\Delta t)^4 \tag{19}$$

$$\frac{d^5f(k)}{dk^5} = -(f-1)(\Delta t)^5 \tag{20}$$

Now standard deviation of fractional release can be evaluated using Eq. (4) and its derivatives. The resultant equation is given in Eq. (21).

$$\sigma_f = \frac{df(k)}{dk}\sigma_k + \frac{1}{2}\frac{d^2f(k)}{dk^2}\sigma_k^2 + \frac{1}{6}\frac{d^3f(k)}{dk^3}\sigma_k^3 + \frac{1}{24}\frac{d^4f(k)}{dk^4}\sigma_k^4 + \frac{1}{120}\frac{d^5f(k)}{dk^5}\sigma_k^5 \tag{21}$$

3 Results and Discussion

Design extension condition without core melt scenario in Indian pressurised heavy water reactor may result from a postulated initiating event of loss of coolant accident with failure of emergency core cooling system with moderator cooling available. In this scenario, the fuel temperature can reach as high as 1900 K. The system dynamics modelling involves uncertainty, to captured the phenomena associated with the accident progression involves variation of fuel temperature. For the simulation purpose, it is assumed that the standard deviation associated with fuel temperature is about 50 K. The distribution function of fuel temperature is assumed to be a

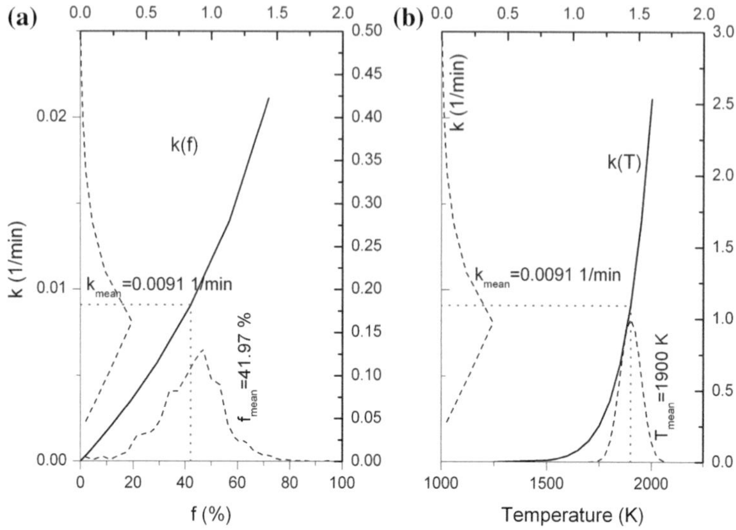

Fig. 1 Error propagation in release model

normal distribution with mean 1900 K and standard deviation of 50 K. The plot of fuel temperature distribution (i.e. pdf_T) is shown in the Fig. 1(b). In the figure, the variation of release rate coefficient (i.e. k(T)) with respect to the temperature has been shown with solid line. It is noted that the change of k(T) is very small in magnitude up to fuel temperature of 1250 K. Beyond this temperature k(T) increases very rapidly. The error due to the fuel temperature is propagated through the release rate coefficient equation (i.e. Eq. (8)) and the variability of 'k' has been estimated and found to be a normal distribution with mean 0.00907 (1/min) and standard deviation is 0.00303 (1/min). The variation of 'k' has been shown as pdf_K in the Fig. 1(a) and (b). The impact of the variation of 'k' on the fractional release has also been studied. The variation of release rate coefficient 'k' with 'f' (i.e. inverse of f(k)) has been shown as a solid line in the in Fig. 1(a). Both Fig. 1(a) and (b) have same variable ('k') and scale on the ordinate axis. In the graphical representation the variability of 'k' (i.e. dashed curve for pdf_K) is same in both left and right side graphs in Fig. 1. The error in the variable 'k' is propagated through fractional release rate equation (i.e. Eq. (7)). The estimated asymmetric distribution of 'f' due to 'k' has been shown as dashed line (i.e. pdf_F) in Fig. 1(a). The asymmetricity of the distribution is originated from non-linear transformation of Gaussian noise. The non-zero higher moments of the distribution such as skewness and kurtosis come from the existing higher order derivatives of non-linear function. The dotted lines in those graphs indicate the mean value of the corresponding distribution.

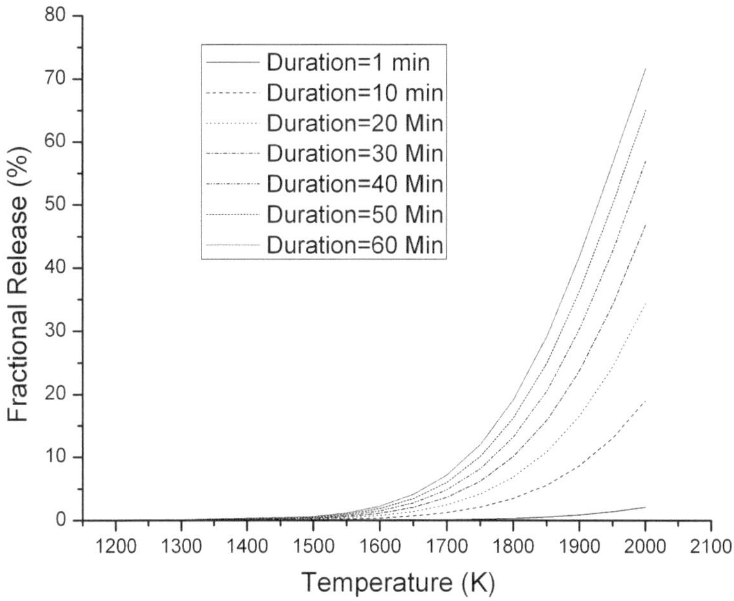

Fig. 2 Variation of fractional release with temperature for different durations

The variation of fractional release with fuel temperature is shown in Fig. 2. It is found that the fractional release of radionuclide is very small if the fuel temperature is below 1500 K. Hence, thermal diffusion of radionuclide below this temperature does not play a significant role. However, for fuel temperature beyond 1500 K, the fractional release increases very rapidly. The fractional release also depends on the time duration for which fuel temperature persisted. The variations of fractional releases with fuel temperatures for different durations are also shown in Fig. 2. Form this figure, it is clear that longer the duration, more is the amount of radionuclide release into the fuel pellet-clad gap. Similarly, the variations of fractional releases with duration for different fuel temperatures have been shown in Fig. 3. Based on Figs. 2 and 3, the safety importance of the fuel temperature and its persistence duration can be drawn with respect to the radionuclide release aspect. For this purpose, a contour plot of fuel temperature and duration has been made for various release fraction (given in %) and shown in Fig. 4. In this evaluation, low fractional release is considered to be safe and high fractional release will be considered as unsafe. The contour with lower fractional release represents lower radiological impact. However, contour with higher fractional release implies higher radiological impact.

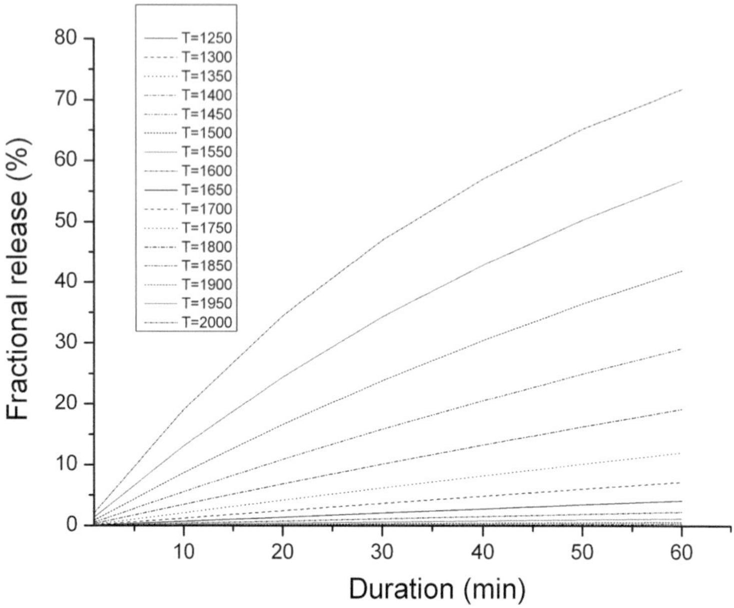

Fig. 3 Variation of fractional release with duration for different temperatures

Fig. 4 Safety importance of temperature and duration with respect to fractional release

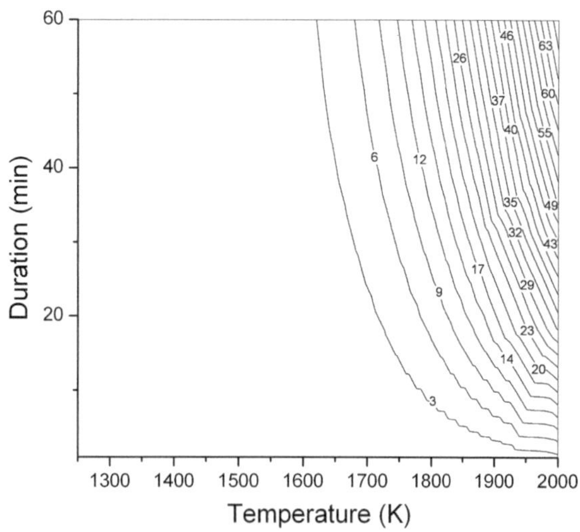

4 Conclusion

The fractional release of radionuclides from fuel matrix to the pellet-clad gap base on thermal diffusion phenomena is an exponential of an exponential function of fuel temperature. The fractional release is found to be very small if fuel temperature is below the 1500 K. Fuel temperature persistent duration increases fractional release. Analytical uncertainty estimation methodology for an exponential of an exponential function has been developed and demonstrated for radionuclide release phenomena. The safety importance of the fuel temperature and duration from the radionuclide release point of view is also highlighted.

References

1. AERB Safety Code, *Design of light water reactor based nuclear power plants*. AERB/NPP-LWR/SC/D (2015)
2. AERB Safety Guide, *Fuel design for pressurized heavy water reactor*. AERB/SG/D-6 (2003)
3. AERB Safety Guide, *Design basis events for PHWR*. AERB/SG/D-5 (2000)
4. U.K. Viswanathan et al., Measurement of fission gas release, internal pressure and cladding creep rate in the fuel pins of PHWR bundle of normal discharge burnup. J. Nucl. Mater. **392**, 545–551 (2009)
5. D.N. Sah et al., Post irradiation examination of thermal reactor fuels. J. Nucl. Mater. **383**, 45–53 (2008)
6. D.N. Sah et al., Blind prediction exercise on modelling of PHWR fuel at extended burnup. J. Nucl. Mater. **383**, 144–149 (2008)
7. B. Gera et al., Estimation of source term and related consequences for some postulated severe accident scenarios of Indian PHWR. Nucl. Eng. Des. **240**, 3529–3538 (2010)
8. S. Bera et al., *Estimation of radionuclide release from PHT to containment towards RIA guideline formulation: case study for PHWR-700 DBA*. AERB Report, AERB/NSAD/TR/2016/18 (2016)
9. S. Bera, D.B. Nagrale, A.J. Gaikwad, R. Kumar, D. Datta, *Safety margin assessment in handling fissile material using probabilistic approach*, in SRESA National Conference on reliability and Safety Engineering, Anna University, BIT Campus, Tiruchirappalli, Tamilnadu, 13–15 Feb 2014
10. D.B. Nagrale, S. Bera, M. Prasad, R.S. Rao, D. Datta, A.J. Gaikwad, *Safety margin estimation and its comparison using various uncertainty techniques for small break loss of coolant accident sequences*, in International Conference on Facets of Uncertainty and Application (ICFUA-2013), Ramkrishna Mission Institute of Culture, Kolkata, India, 05–07 Dec 2013

Statistical Aggregation of Extreme Value Analysis Models

Subrata Bera, Dhanesh B. Nagrale, U. K. Paul, D. Datta and A. J. Gaikwad

1 Introduction

Probability and statistics are the greatest inventions made in the eighteenth century [1, 2]. The probability and statistics are in extensive use in the design of engineering structures and safety assessment. If an experiment is repeated for large number of times then based on central limit theory the mean of these experimental values will follow the normal distribution [3]. However, for some specific cases, those means are not important for design consideration. Designer may want to know the extreme value such as minimum or maximum value of the hazard parameters to justify the design basis of the engineering structure of nuclear facility. The design basis flood level, tsunami height, and wind load to the tall structure such as chimney and industrial stack are few examples where extreme value is important in design [4]. These extreme

S. Bera (✉) · D. B. Nagrale · U. K. Paul · A. J. Gaikwad
Nuclear Safety Analysis Division (NSAD), Atomic Energy Regulatory Board (AERB), Anushaktinagar, Mumbai 400094, India
e-mail: sbera@aerb.gov.in

D. B. Nagrale
e-mail: dbnagrale@aerb.gov.in

U. K. Paul
e-mail: ukp@aerb.gov.in

A. J. Gaikwad
e-mail: avinashg@aerb.gov.in

D. Datta
Radiological Physics and Advisory Division, Bhabha Atomic Research Centre, Trombay, Mumbai 400085, India
e-mail: ddatta@barc.gov.in

S. Bera
Homi Bhabha National Institute, Mumbai 400094, India

S. Chattopadhyay et al. (eds.), *Modelling and Simulation in Science, Technology and Engineering Mathematics*, Advances in Intelligent Systems and Computing 749,
https://doi.org/10.1007/978-3-319-74808-5_54

values do not follow the central limit theory. The distribution of extreme value does not follow the normal distribution [5]. When the minimum value is of interest, the distribution is skewed toward lower values. Similarly, the distribution skewed towards higher value in case when the maximum value is the desired quantity. There are two approaches to estimate the extreme value: (1) block maxima approach [6] and (2) peak over threshold approach. There are many extreme value distribution functions such as Gumbel distribution, Weibull distribution and Generalised Extreme Value (GEV) distribution function, etc. [7, 8] used in block maxima approach. However, generalised Pareto distribution is used in peak over threshold approach [9, 10]. These distributions are utilised for estimation of extreme values. Most of the extreme value analyses are carried out using the generalised extreme value distribution function [11]. In the present analytical study, wind speed is considered to be an extreme value variable. For the extreme value analysis, wind speed data for several years are required for the analysis. For such analysis, the data based on measurement in the nearest meteorological stations are used. In the study reported here the year wise maximum wind speed data are fitted with generalised extreme value distribution function. With known fitted parameters, the extreme value analysis model is generated to extrapolate the extreme value for the desired return period. If data from multiple stations are used and those stations are significantly away from each other, then the wind profile will be significantly different. Extreme value analysis model can vary from station to station and result in variation of fitted parameters based on station data. Thus individual model may either under estimate or overestimate the extreme value parameters. For conservative design, the overestimating model for the extreme value can be used. However, for a realistic prediction it requires statistical aggregation of extreme values obtained from models based on various station data. A demonstration case study of statistical aggregation of extreme value analysis models has been carried out while preserving the statistical properties.

In this case study, data generated by four stations are considered as for wind speed data collected over several years and the data related to maximum value of wind speed in a year is fitted with the generalised extreme value distribution. The model uncertainty of each model has also been analysed and a statistical aggregation methodology has been developed to obtain an average model from the models generated for data obtained from the four stations.

2 Theoretical Methodology

2.1 Generalized Extreme Value Analysis Model

Genaralised extreme value (GEV) distribution is a family of continuous distribution to combine Gumbel, Frechet and Weibull distributions. The three parameter distribution function of the standard GEV is given in Eq. (1).

$$F_{k,m,s}(x) = \begin{cases} exp\left(-\left(1 + k\frac{x-m}{s}\right)^{-\frac{1}{k}}\right) & if\ k \neq 0 \\ exp\left(-\frac{x-m}{s}\right) & if\ k = 0 \end{cases} \quad (1)$$

where, 'k', 'm', 's' are known as shape parameter, location parameter, scale parameter respectively. Shape parameter determines the nature of the tail distribution. The estimation of these three parameters are carried out using various methods such as probability plots, moment methods, least square fitting method, maximum likelihood methods, etc. [9]. The extreme value distribution in Eq. (1) is generalised in the sense that parametric form subsumes three types of distributions which are known by other names according to the value of 'k'. If the data related to the year wise maximum value of wind speed is obtained for 'n' number of years, then distribution function F(x) will be calculated based on the empirical density estimation methodology. This is an approximation of a population density function that is derived from a sample and has no unknown parameters is the empirical density. This is basically an order statistics [12, 13]. In order statistics, year wise wind speed data ($i.e.\ X = \{x_1, x_2, \ldots, x_n\}$) are random in nature. These data arranged in the ascending order. After shorting, the sequence of wind speed can be represented as given in Eq. (2).

$$x_{(1)} \leq x_{(2)} \leq \cdots \leq x_{(n)} \quad (2)$$

In particular, the equation of minimum and maximum value of the order is given in Eqs. (3) and (4) respectively.

$$x_{(1)} = Min\{x_1, x_2, \ldots, x_n\} \quad (3)$$

$$x_{(n)} = Max\{x_1, x_2, \ldots, x_n\} \quad (4)$$

In the order statistics, all the data is considered equally probable. The corresponding density function is called empirical density f(x) and given in Eq. (5).

$$f(x) = \begin{cases} \frac{1}{n}\ for\ X = \{x_1, x_2, \ldots, x_n\} \\ 0 \qquad\qquad eleswhere \end{cases} \quad (5)$$

The empirical distribution function F(x) is given in Eq. (6).

$$F(x) = \begin{cases} 0\ \ for\ x < x_{(1)} \\ \frac{i}{n}\ \ for\ x = x_{(i)} \\ 1\ \ for\ x > x_{(n)} \end{cases} \quad (6)$$

From the station specific measured data, empirical distribution is calculated using Eq. (6). This distribution is used to develop the GEV model by fitting the Eq. (1). Each GEV model will have specific set of {k, m, s} values.

2.2 Probabilistic Methodology for Estimate the Model Uncertainty

The GEV model is three parameter (i.e., k, m, s) continuous distribution function. While fitting the Eq. (1) with measured data, it is obtained the mean value of three parameters along with their standard deviation. Model uncertainty is the estimation error due to the variation of the model parameters (i.e., k, m, s) not for the input parameter such as wind speed. Probabilistic method has been used to estimate the model uncertainty. In this method, random sample has been taken uncertainty domain of three parameters with normal distribution. The model uncertainty analysis methodology has been shown in the Fig. 1.

2.3 Statistical Aggregation Methodology

Four stations generate the four GEV models with different set of three parameters (i.e. k, m, s). Average model from these three models has been generated weighted average of quantile data of each model. Quantile data is mathematically represented as inverse of the Eq. (1). Mathematically the quantile information is represented in Eq. (7).

$$x = \left(m - \frac{s}{k}\right) + \frac{s}{k}(-\ln F)^{-k} \tag{7}$$

For a given 'F', four values of 'x' can be generated for four models using Eq. (7). If the weight given for each model is 'w', then the average value of 'x' will be as given in Eq. (8).

$$\bar{x} = \frac{w_1 x_1 + w_2 x_2 + w_3 x_3 + w_4 x_4}{w_1 + w_2 + w_3 + w_4} \tag{8}$$

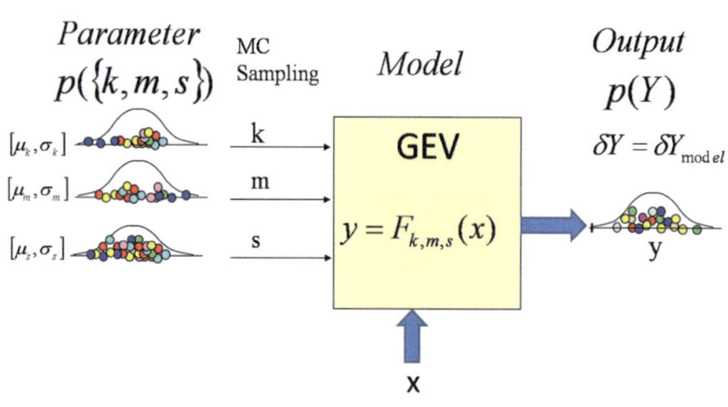

Fig. 1 Model uncertainty analysis methodology

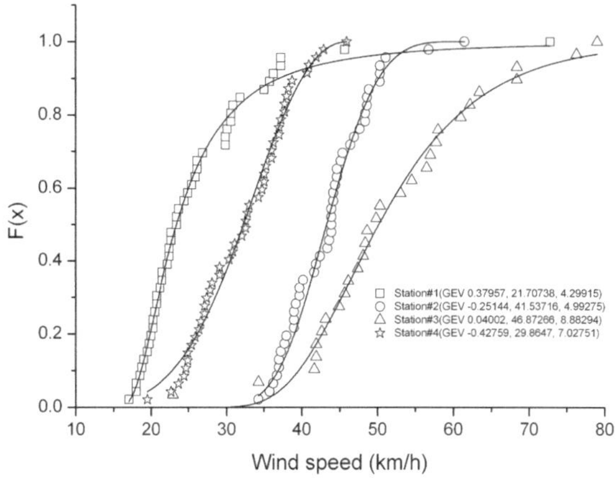

Fig. 2 GEV models for four stations

Weights can be decided based on various attributes such as the distance of the measuring station from the site, no of data points available and reliability of the data, etc. Weights can also be generated based on expert elicitation method/process. The average value of 'x' will be generated for different value of 'F'. These data can be used to regenerate the average model of four models. The average GEV model will have different set of three parameter data compared to the four station wise GEV models. The generated average model preserved the statistical property of the GEV.

3 Results and Discussions

Year wise maximum wind speed data given in km/h unit for four stations have been plotted in the Fig. 2. The distribution functions for four stations are different due to different value of shape parameter, location parameter and scale parameter in the probability density function for wind speed. It is noted that the lowest 50th percentile value is found in the measured data in station#1. Highest 50th percentile value is found in station#3 data. Station#1 and station#3 have enveloping distribution function for the range of wind speed from 10 km/h to 80 km/h. Station#2 and station#4 follow the distribution in between the enveloping distribution of station#1 and station#3. Each data have been fitted with GEV distribution function as given in Eq. (1) in Sect. 2.1 to obtain the shape parameter, location parameter and scale parameter.

The estimated mean value and standard deviation of these three parameters are given in Table 1. The goodness of fit has (i.e. R^2) also been included in the Table 1. The best fit is obtained for the station#1 with $R^2 = 0.99441$. Highest shape parameter is found in station#1 GEV model equal to 0.37957. Lowest shape parameter is found to in station#4 GEV model. However, highest scale parameter is found in station#3 GEV model. Lowest scale parameter is found in station#1 GEV model.

The variation of three parameters in GEV model may change the model prediction. The model uncertainty due to the standard deviation of three model parameter as given in Table 1 has been assessed using probabilistic methodology. Samples of three parameter data set (k, m, s) have been generated from their uncertain domain considering normal distribution represented as $\mathcal{N}(\mu, \sigma)$. Model uncertainties for four GEV models due to the three parameters have been shown in the Fig. 3. Solid line for each model corresponds to the mean value of three parameters of each model. Scatter data represent the variation of distribution function due to the model uncertainty. It is noted that the model uncertainty is found to be higher in the fourth quantiles. These uncertainties are strongly dependent on the variation of scale parameter. Highest model uncertainty is found for station#3 due to its highest standard deviation of scale parameter (i.e. 4.85% of its mean value). Lowest model uncertainty is found to be for station#1 due to the lowest standard deviation of scale parameter (i.e. 2.45% of its mean value).

For the statistical aggregation needs weights of individual GEV models. In this statistical aggregation study, equal weights are given for the each GEV model. The estimated average GEV model is shown with circular symbol in Fig. 4. It is noted that the average GEV distribution follows in between the four individual GEV models. Again this average GEV model data has been fitted with GEV distribution function to obtain the three parameters. The estimated three parameters i.e. k, m, s are −0.04216, 34.9559, 6.34007 respectively.

Table 1 GEV fitting coefficients for all stations data

Station	$k \pm \sigma_k$	$m \pm \sigma_m$	$s \pm \sigma_s$	R^2
Station#1	0.37957 ± 0.04058	21.70738 ± 0.05952	4.29915 ± 0.10526 (2.45%)	0.99441
Station#2	−0.25144 ± 0.06549	41.53716 ± 0.12877	4.99275 ± 0.1987 (3.98%)	0.98143
Station#3	0.04002 ± 0.07333	46.87266 ± 0.20452	8.88294 ± 0.43103 (4.85%)	0.98973
Station#4	−0.42759 ± 0.05479	29.8647 ± 0.13544	7.02751 ± 0.23419 (3.33%)	0.98832

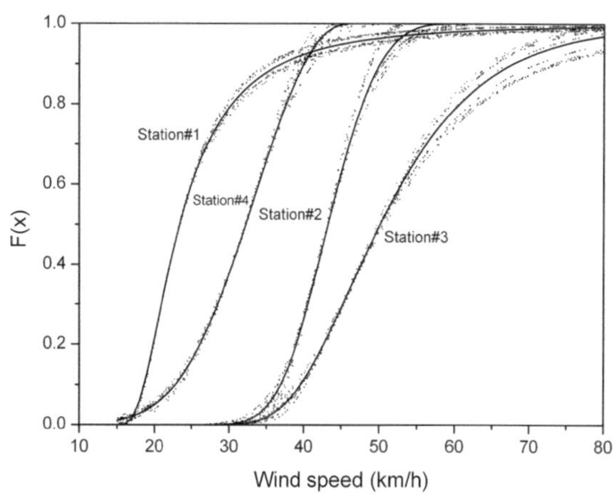

Fig. 3 Model uncertainty due to variation of model parameters

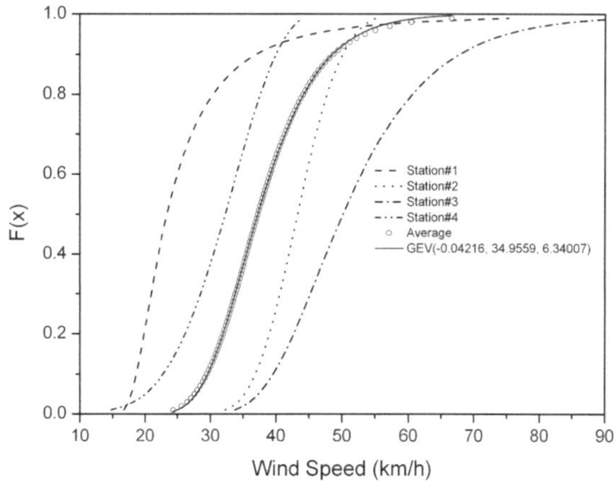

Fig. 4 Average GEV model with statistical averaging

4 Conclusion

Wind data collected from four measuring stations are used to carry out extreme value analysis. Year wised extreme data are fitted with generalised extreme value distribution function that results model. The lowest 50th percentile value is found in the measured data in station#1. Highest 50th percentile value is found in station#3 data. Model uncertainty due to the variation of model parameters is also estimated. The model uncertainty is found to be higher in the fourth quantiles. These uncertainties are strongly dependent on the variation of scale parameter. Highest model uncertainty is found for station#3 due to its highest standard deviation of scale parameter (i.e., 4.85% of its mean value). Lowest model uncertainty is found to be for station#1 due to the lowest standard deviation of scale parameter (i.e. 2.45% of its mean value). Multiple models are developed for each measuring stations. A methodology for statistical aggregation of multiple models is developed with preserving the statistical properties and demonstrated with considering four measuring stations. The average GEV model is developed with model parameters i.e., 'k', 'm', 's' equal to -0.04216, 34.9559, 6.34007 respectively.

References

1. J. Galambos, S. Kotz, *Characterizations of Probability Distributions* (Springer, Berlin, 1978)
2. D.R. Cox, D.V. Hinkley, *Theoretical Statistics* (Chapman and Hall, London, 1974)
3. L. de Hann, A. Ferreira, *Extreme Value Theory an Introduction* (Springer, 2006)
4. AERB Safety Guide, "Extreme value of mateorological parameters", AERB/NF/SG/S-3 (2008)
5. R.D. Reiss, M. Thomas, *Statistical Analysis of Extreme Values with Applications to Insurance, Finance, Hydrology and Other Fields* (Birkhauser Verlag, 2007)
6. R.A. Fisher, L.H.C. Tippett, Limiting forms of the frequency distribution of the largest or smallest member of a sample. Proc. Camb. Philos. Soc. **24**, 180–290 (1928)

7. S. Kotz, S. Nadarajah, *Extreme Value Distributions Theory and Applications* (Imperial College Press, 2000)
8. S. Coles, *An Introduction to Statistical Modeling of Extreme Values* (Springer, 2001)
9. E. Castillo, A.S. Hadi, Fitting the generalized Pareto distribution to data. J. Am. Stat. Assoc. **92**, 1609–1620 (1997)
10. A.C. Davison, R.L. Smith, Models for exceedances over high thresholds (with discussion). J. Roy. Stat. Soc. B **52**, 393–442 (1990)
11. J.R.M. Hosking, J.R. Wallis, E.F. Wood, Estimation of the generalized extreme value distribution by the method of probability-weighted moments. Technometrics **27**, 251–261 (1985)
12. H. Finner, M. Roters, On the limit behavior of the joint distribution function of order-statistics. Ann. Inst. Stat. Math. **46**, 343–349 (1994)
13. J. Galambos, A statistical test for extreme value distributions, in *Nonparametric Statistical Inference*, ed. by B.V. Gnedenko, M.L. Puri, I. Vincze (Amsterdam: North-Holland, 2000), pp. 221–230

Electroosmotic Effects on Rough Wall Micro-channel Flow

Nisat Nowroz Anika and L. Djenidi

1 Introduction

The improvement of mixing in laminar flow is the major challenge to deal with microfluidic. However, the astonishing characteristic of microfluidic is smallness-having dimension ranging millimeter to microns μ. Such miniature technology allows the small volume of order of micro to nano-liter. The growing availability of the devices allows miniature technology that has reliable capability to deal with Micro-electro-mechanical system (MEMS), Lab-on-chip devises to analyze the biological and chemical applications outcome. In chemical engineering, some reactant/solvent is inadequate in the nature. To carry out such investigations, of small volume of those samples require large interfacial area to bring two species together as well as to generate instability in a micro-device. The generated instability could help to perform mixing process inside the micro-channel in which flow can be insensitive to accept turbulence. In laboratory, the surface of the micro-devices may not be smooth, rather rough. Therefore, requires strong driving force to transport the sample species through the rough micro-channel. Because roughness along can induce pressure perturbation in bulk flow to conserve the rate of mass transfer. Electrokinetic transport phenomenon- one of the fast growing ubiquitous community- has been receiving immense attraction dealing with electronics micro-fabrication.

The micro-electro mechanical process is critical when mixing plays a major part. Inside the microchannels, the flow remains in a laminar regime where mixing only occurs at molecular level (molecular diffusion) at somewhat reduced rate. In this present study, our aim is to generate and enhance turbulence incorporating the both

N. N. Anika (✉) · L. Djenidi
Discipline of Mechanical Engineering, School of Engineering,
University of Newcastle, Callaghan NSW-2308, Australia
e-mail: NisatNowroz.Anika@uon.edu.au

L. Djenidi
e-mail: Lyazid.Djenidi@newcastle.edu.au

S. Chattopadhyay et al. (eds.), *Modelling and Simulation in Science, Technology and Engineering Mathematics*, Advances in Intelligent Systems and Computing 749,
https://doi.org/10.1007/978-3-319-74808-5_55

active and passive methods which can help to increase the exchange area as the best strategy to prolong mixing at Reynolds number much lower than its critical value ($Re < 10.$). Relatively recent numerical work of [1] observed that the conservation of energy in the process of effective mixing occurred in the combination of active and passive mixer. The strategy was to develop turbulence in the rough wall with an active triggering of jets at low Reynolds number $Re = 700$. The flow was driven by the pressure difference and the parabolic profile of velocity eventually affected by the interaction of local turbulence.

Our present simulation study is motivated by the aforesaid work of [1] to accelerate mixing process at low Reynolds number $Re \leq 7$ where the active method is replaced by an external electric force. Also, the computational domain has been fixed with parallel plates separated by a distance h, h being the height of the channel. The surface of both the channel's walls is roughened by 2D square bars. The height of the roughness is denoted by k. The ratio between heights is numerically fixed at $2k/h = 0.1$. The separation p between roughness elements is homogeneous for both walls. Leonardi et al. [2] investigated fully developed turbulent flow with transverse square roughness only at the bottom wall. They performed three dimensional numerical parametric studies on p/k ranging from $0.33 \leq p/k \leq 19$ to aid optimizing the value ($p/k = 7$) at which viscous drag can be minimum. Form drags helped to inject turbulence having nearly zero viscous drag. The ratio of roughness to channel height was fixed at $2k/h = 0.1$ to reduce the blockage effects in the main channel flow. In our study, the separation is 4 times of the height of the roughness elements at $Re = 7$ (based on $h/2$; half height of the channel). As the increasing of p/k, the viscous drag reduces. And $p/k = 4$ is the transitional value to get viscous and form drag due to the presence of induced pressure around each roughness elements. Hu et al. [3] carried out finite-volume method to investigate electro-kinetic transport phenomena on rough channel at micro-scale. Symmetric and asymmetric arrangements of roughness elements have been considered where the $2k/h$ ratio varies between $0.2 \leq 2k/h = 0.4$. From their study, it can be concluded that the induced pressure around the 3D roughness elements reduces the electoosmotic flow rate and the effects were immense with further increase of roughness height. For symmetric arrangement, the author had less consideration reducing the blockage effect of the roughness elements. At our best knowledge, there was no study performed regarding electroosmotic microchannel flow to enhance mixing considering $2k/h \leq 0.2$.

Motivated by the aforesaid knowledge gap, a direct numerical simulation is performed to investigate the electrically forced two-dimensional channel flow driven by pressure including both active and passive methods unitedly. The passive method involves the roughness elements where the channel height is 10 times bigger than the height of roughness element to neglect the blockage effect on the electric flow field. Numerical studies have been carried out on both smooth and rough channels to see the effect of external force on roughness elements. Also, the calculations have been analyzed for p/k ranging from 3 to 7 with electric field strength.

2 Numerical Procedure

A direct numerical Simulation is performed based on novel numerical scheme Lattice Boltzmann method (LBM). The evolution equation of Lattice Boltzmann model results a macroscopic velocity filed which solves Naiver-Stokes equation of second order accuracy with sufficient amount of lattice symmetry. Boltzmann equation discretized on a lattice to construct a kinetic model that describes the underlying physics of mesoscopic averaged properties. In this study, the two-dimensional square lattice model has been discretized on continuous Boltzmann equation. The lattice model is composed of $D2Q9$ four particles ($i = 1, 2, 3, 4$) directed along vertical and horizontal directions, another four ($i = 5, 6, 7, 8$) are directed diagonally and one particle at rest ($i = 0$) residing at the center of that square lattice. Therefore for $D2Q9$ lattice model, there are three different velocities and abscissas acting to describe the corresponding weighing factors ω_i.

	e_0	e_1	e_2	e_3	e_4	e_5	e_6	e_7	e_8
e_x	0	1	0	−1	0	1	−1	−1	1
e_y	0	0	1	0	−1	1	1	−1	−1

and

$$\omega_i = \left.\begin{matrix} \frac{4}{9} & \text{for } i = 0 \\ \frac{1}{9} & \text{for } i = 1, 2, 3, 4 \\ \frac{1}{36} & \text{for } i = 5, 6, 7, 8 \end{matrix}\right\}$$

For the evolution equation of lattice Boltzmann model with external force term of local collision operator, we always present the general form. The equation reads-

$$f_i(\vec{x} + \vec{e}_i \nabla t, t + \nabla t) = f_i(\vec{x}, t) - \frac{1}{\tau}\left[f_i(\vec{x}, t) - f_i^{eql}(\vec{x}, t)\right] + \vec{F}_i \tag{1}$$

In the above equation Eq. [1], f_i denotes the microscopic distribution function of ith discrete velocity nodes of the set of microscopic velocities ($\vec{e}_i$), position ($\vec{x}$) and time (t). The gradient, ∇ is the representation of position and velocity spacing and ϖ denotes the frequency of collision, $\varpi = \frac{1}{\tau}$ where, $\tau = 3\nu + \frac{1}{2}$ represents relaxation time due to the collision. ν is the kinematic viscosity.

The right-hand side represents the Bhatnagar-Gross-krook collision approximation. The equilibrium distribution function f_i^{eql} for 2D nine velocities can be written as-

$$f_i^{eql} = \omega_i \rho \left[1 + \frac{3\vec{e}_i.\vec{u}}{c^2} + 9\frac{(\vec{e}_i.\vec{u})^2}{2c^4} - \frac{3u^2}{2c^2}\right] \tag{2}$$

where c the lattice speed denoted by $c = \frac{\Delta t}{\Delta x}$. $\vec{F}_i$: an electrical force term can be expressed as-

$$\vec{F}_i = \omega_i \rho_e \left[\frac{\vec{e}_i - \vec{u}}{\rho c_s^2} \right] \vec{E} f_i^{eql} \tag{3}$$

c_s is the sound speed, for $D2Q9$, $c_s = 1$, ρ is characteristic fluid density $\left(\rho = \sum f_i\right)$, $\vec{u}$ is the macroscopic velocity $\left(\rho u = \sum f_i\right)$. ρ_e defines the net charge density. For single ionic bond, the ρ_e reads-

$$\rho_e = (-1)ze2n_0 \sinh\left(\frac{ze\psi}{k_b T}\right) \tag{4}$$

The all physical parameters have their usual meaning. Noticed that all qualities have been expressed in the lattice units [4]. Here, $n_0 = c_0 N_0$ is the bulk ion concentration with $c_0 = 1.0 \times 10^{-7}$ mole/volume, Avogadro number, $N_0 = 6.023 \times 10^{23}$.

This study solves thin electric double layer to develop a plugged-like (zero at the centerline and maximum at the wall) electrical potential ψ. As soon as we received ψ, the external electrical field ($\vec{E}$ volt/length) applied along the stream wise direction. At the inlet and outlet, we put $\nabla\psi$ to force the EOS flow. For hydrodynamics boundary condition, periodic condition implemented along stream wise direction and no-slip at the wall. The Reynolds number is calculated based on half-height of channel $h/2$, mean centerline velocity $\left(\bar{U}_c\right)$ and viscosity and the value at about Re $= 7$. The velocity boundary condition based on non-equilibrium distribution of bounce-back [5] has been invoked. For calculating the net charge density, the zeta potential at the wall region of electric double layer (EDL) is about $\psi_w = -0.027$ Volt. The external electric field strength is about 2.7 V per length. The computational domain is two dimensional having mesh size $h\pi \times h$. 2D mesh increments are $\Delta x = \Delta t = 1$. The macroscopic velocities have been calculated with the presence of electric potential everywhere in between both smooth and rough walls. The smooth wall simulation runs for longer time for single species fluid density with external electric force and results have been taken as a reference which helps understanding the effects of electro-osmotic flow on rough micro-channel.

3 Results and Discussions

In present LBM model, all the statistical analysis has been performed with ensemble averaged calculations. The time averaged kinetic energy has been calculated against time variation first.

It is clearly evident from the Fig. 1 that the thin electric double layer and the distribution is maximum at the wall and zero at the middle of the channel's height for a given small electro-static zeta potential, ψ_w. The ψ_w is homogeneous and equal for both the walls and so for the roughness elements.

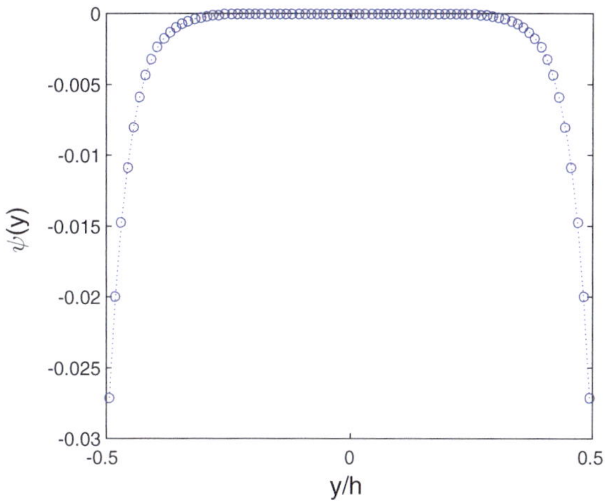

Fig. 1 Distribution of electric potential across the channel

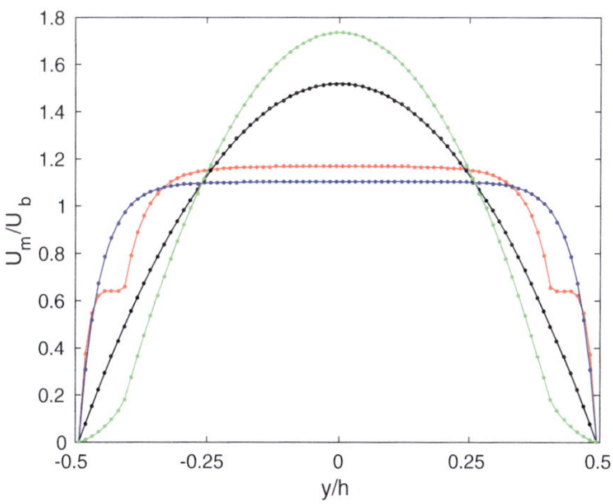

Fig. 2 Mean electroosmotic flow (EOF) profile, black: smooth wall Poiseuille flow; blue: smooth wall with potential; red: rough wall EOF flow and green: rough wall without EOF

In Fig. 2, the linear axis represents the mean velocity normalized by U_b-the bulk velocity. The profile of mean velocities reveals the better understanding at the near wall region where the entire mechanism relies on. It is clearly depicted that the mean velocity for smooth channel flow without any external force is the Poiseuille distribution with a peak value at the center of the channel. The peak is transformed by a flatness at the middle of the channel for the EOF on both smooth and rough wall. The situation is more pronounce on the smooth wall (symbol with blue symbol line in Fig. 2) as compared to the rough wall (red line symbol). This transformation is due to the presence of electric field on the near wall EDL region [6].

The effect of $\frac{2k}{h}$ ratio has been examined first on rough wall for $p/k = 3.7$ with EOF across the channel. Figure 3 presents the visualization of velocity flow direction and their corresponding vectors for $\frac{2k}{h}$ varies form 0.2 to 0.1, where the

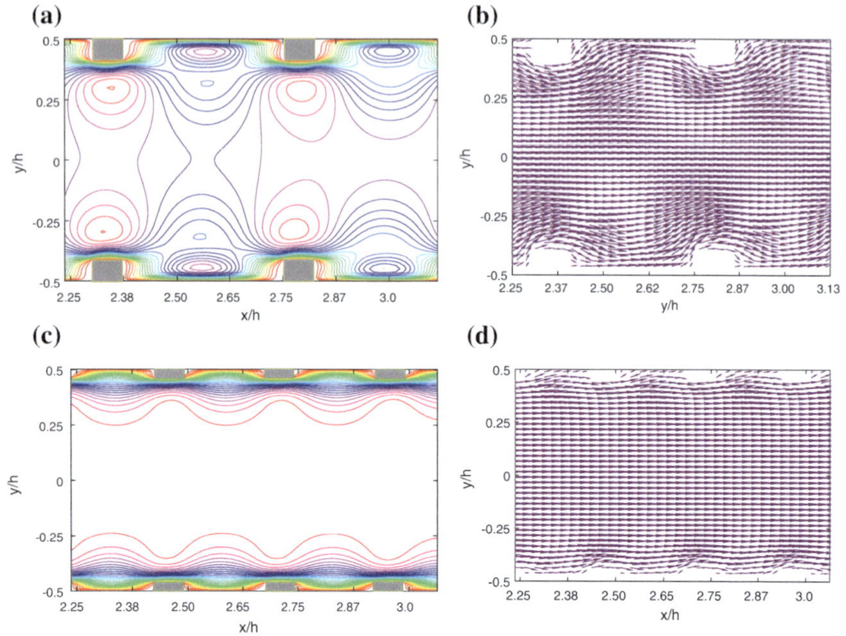

Fig. 3 Instantaneous velocity-contour and vector repetition of u and v on rough wall EOF channel flow. **a**, **b** $\frac{2k}{h} = 0.2$ and **c**, **d** $\frac{2k}{h} = 0.1$. And $p/k = 3.7$

other parameters are kept exactly same. To stress out better the results have been magnified in between $2.25 \leq 2x/h \leq 3.05$ along the stream-wise direction. The strong ejection on the crest of elements and suction at the cavity occurred in the flow filed at the vicinity of top and bottom walls. In between two consecutive elements, the flow is strongly disturbed by the electric field. Far from the wall, the flow behavior is most likely laminar but contains small oscillation due to the external electric field strength. The effects are also present for $\frac{2k}{h} = 0.1$ as we see in Fig. 3c, d. The EOF is disturbed only at the wall region in a small amount compared to the rough wall (Fig. 3a; $\frac{2k}{h} = 0.2$) EO flow but large compared to the rough wall laminar flow (not shown here). For better understanding, it would worth to calculate rms for these cases.

Figure 4 showed the root-mean-square of the velocity fluctuations. It is clearly evident that the mean centerline fluctuation is less in magnitude to that of wall fluctuations. As we see that at $y/h = 0$ (bottom wall) and $y/h = 2$ (top wall), the $u'_1(u_{rms})$ is maximum. The fluctuation is less evident on the rough wall $\left(\frac{2k}{h} = 0.1\right)$. It can be well understood that for the rough wall at $\frac{2k}{h} = 0.2$, there is a possibility to

Fig. 4 Root mean square of the (with electro-osmotic on smooth and rough wall) velocity fluctuation $(u_1' = u_{rms})$. Black symbol: $\frac{2k}{h} = 0.2$ and $p/k = 3.7$; green symbol: $\frac{2k}{h} = 0.1$ and $p/k = 3.7$; blue symbol: smooth wall EOF flow

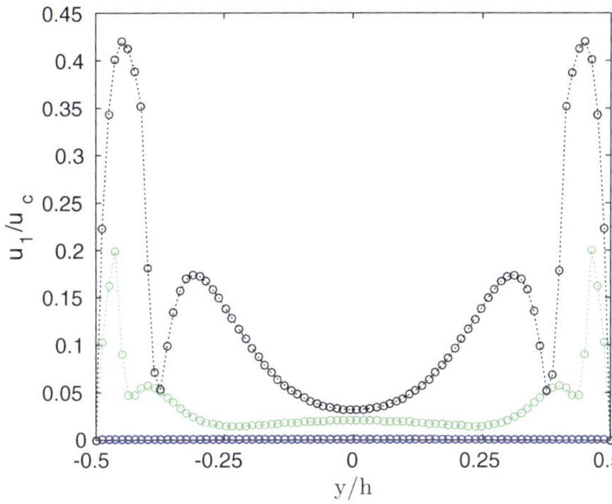

achieve turbulence-like motion where there is nearly zero instability for rough wall Poiseuille flow (not shown here).

With the increase of separation distance between two roughness elements $(p/k = 6.7)$ there was more interfacial are to exchange energy but comparatively low in magnitude. It is obvious that the EOF feels the surrounding pressure from the dipole roughness elements better for the case when the ratio lies between $3 \leq p/k \leq 4$ for $\frac{2k}{h} = 0.2$.

4 Conclusion

A direct numerical study based on novel technique Lattice-Boltzmann model have been carried out to simulate the electric potential distribution on rough wall channel flow at Reynolds number $\text{Re} = 7$, based on half height, $h/2$ and mean centerline velocity. The simulation not only suggested that the effect of EOF changes with separation length to height ratio of roughness elements but also the mean velocity affected by shear layer for $\frac{2k}{h}$ ratio. There lies a great possibility that the mixing can be achieved at laminar micro-channel by generating turbulent-like motion actuated by the combination of passive and active method of mixing. The EOF based on thin electric double layer that has been observed by analyzing the root-mean-square fluctuation velocities and flow visualizations. The parameters $3 \leq p/k \leq 4$ and $\frac{2k}{h} = 0.2$ showed us better effect of EOF on rough microchannel that could help minimize the blockage effects on microfluidic based application outcomes. In future, it would be interesting to investigate the electroosmostic effects on laminar flow in the presence of wall jet following the recent paper of Anika et al. [7].

References

1. N. Anika, L. Djenidi, S. Tardu, Pulsed jets in laminar smooth and rough wall channel flow, ETMM11, 2016
2. S. Leonardi, P. Orlandi, R. Smalley, L. Djenidi, R. Antonia, Direct numerical simulations of turbulent channel flow with transverse square bars on one wall. J. Fluid Mech. **491**, 229–238 (2003)
3. Y. Hu, C. Werner, D. Li, Electrokinetic transport through rough microchannels. Anal. Chem. **75**(21), 5747 (2003)
4. L. Djenidi, Lattice-Boltzmann simulation for grid generated turbulence. J. Fluid Mech. **552**, 13–35 (2006)
5. Q. Zou, X. He, On pressure and velocity boundary conditions for the lattice boltzmann BGK model. Phys. Fluids **9**(6), 1591–1598 (1997)
6. G.H. Tang, X.F. Li, W.Q.Tao He, Electroosmotic flow of non-Newtonian fluid in microchannels. J. Nonnewton. Fluid Mech. **157**, 133–137 (2009)
7. N.N. Anika, L. Djenidi, S. Tardu, Bypass transition mechanism in a rough wall channel flow. Phys. Rev. Fluids **3**(8) (2018)

Comparative Study on Fuzzy Based Linearization Technique Between MATLAB and LABVIEW Platform

Joyanta Kumar Roy and Bansari Deb Majumder

1 Introduction

In the process industry like the power plant, Boiler drum level is one of the critical parameters to be measured and controlled. The boiler is a process where water is converted into steam, and the steam is used to turn a steam turbine and eventually electricity is generated. Water level control is an essential parameter for operation of boiler efficiency. Different techniques like conductivity probe type, sight gauges, magnetic type level transmitters etc. are the sensors preferred in boiler drum level measurement [1–3]. Unfortunately boiler drum level control is complicated by changes in electrical load requirements or variation in the fuel and air supply. All the available methods of measuring the boiler drum level is discrete in nature.

Admittance type level transmitter is a continuous method which can be implemented in boiler drum level measurement. It can be both single electrode type and double electrode type method of level measurement [4, 5]. But because of the presence of its significant cross sensitivity of liquid temperature and liquid property, the analysis is not accurate [6, 7]. Hence a suitable tool needs to be developed to eliminate this cross-sensitivity effect. Linearization is the method of error removal from the measured value of the process parameter. Fuzzy based linearization using MATLAB and fuzzy based linearization using LABVIEW [8, 9] in the earlier work which found to be satisfactory. In this work, a comparative study has been made between the two advanced linearization methods. From the competitive data, the optimum way of linearization has been developed.

J. K. Roy (✉)
MCKV Institute of Engineering, Kolkata, India
e-mail: jkroy.cal51@gmail.com

B. D. Majumder
Department of Electronics and Instrumentation Engineering,
Narula Institute of Technology, Agarpara, India
e-mail: bansari.debmajumder@nit.ac.in

S. Chattopadhyay et al. (eds.), *Modelling and Simulation in Science, Technology and Engineering Mathematics*, Advances in Intelligent Systems and Computing 749,
https://doi.org/10.1007/978-3-319-74808-5_56

Table 1 Input and output variables for fuzzy based linearizer in MATLAB platform

Input variable	Output variable
Admittance without error	Corrected admittance value
Admittance with the effect of temperature	
Admittance with the effect of ionic concentration	

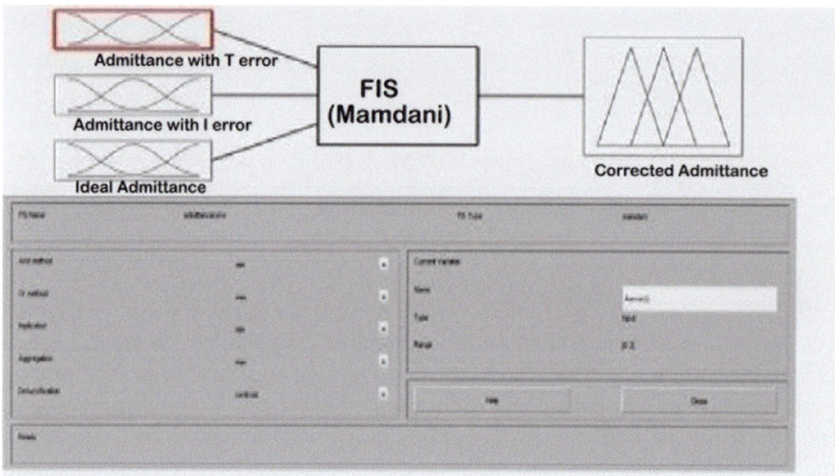

Fig. 1 Fuzzy-based linearizer in Matlab window [8]

2 Fuzzy Based Linearization in MATLAB

Fuzzy logic deals with knowledge-based computation technique [10, 11] can be suitable for linearization, error analysis and elimination [8]. Fuzzy linearizer includes input variables, rule base or inference engine and output variables. There exist knowledge-based systems which related the nature of input and output variables in different condition. In this linearizer, three inputs are given from the experimental setup. And the fuzzy-based linearizer is generating one output. So, in this fuzzy model, there are three input variables and one output variable shown in Table 1.

The fuzzy based linearizer is designed in MATLAB window as shown in Fig. 1. In the window, the input variables are fed to the fuzzy inference engine. The FIS consists of the rule base, and the rules are set according to Mamdani rules of the fuzzy system. The fuzzy data generated by the FIS is defuzzified to create crisp value. The method of de-fuzzification used is centroid method of defuzzification. After defuzzification, the crisp data is generated and recorded finally.

The membership functions define each of the input variables and these membership functions are framed according to the relation of the variables. Now the rule base is set following the Mamdali rule. Finally, the corrected admittance value is recorded

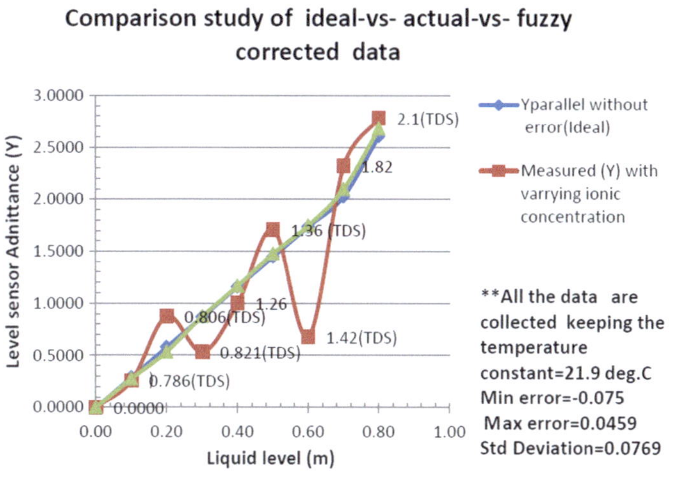

Fig. 2 Comparison between fuzzy corrected versus ideal admittance versus measured admittance value keeping temperature constant [3]

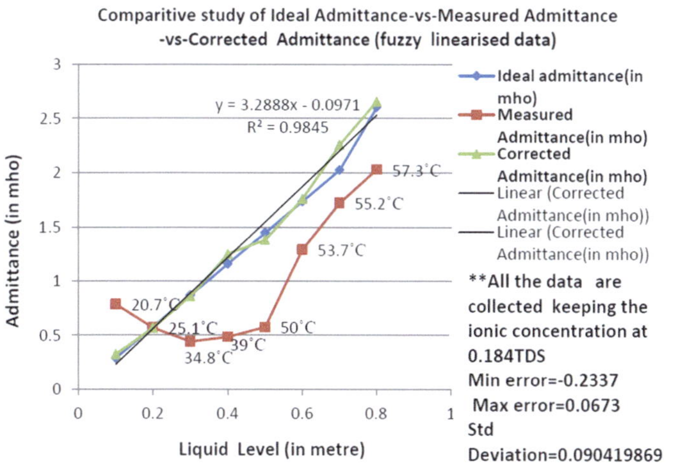

Fig. 3 Comparison between fuzzy corrected versus ideal admittance versus measured admittance value keeping ionic concentration constant [8]

from the output window. The recorded data of corrected admittance value is plotted against the error admittance keeping either temperature or ionic concentration constant.

Figure 2 shows the graph of ideal, actual and corrected fuzzy admittance data at the constant temperature of 21.9 °C. Similarly, Fig. 3 shows a graph of ideal, actual and corrected fuzzy admittance data at constant ionic concentration of 0.184TDS.

The statistical analysis was made on data available from Matlab base simulation and is shown in Table 2.

Table 2 Comparative Statistical analysis of Level measurement error

MATLAB based analysis		
	Corrected, ideal and measured admittance (Temperature constant)	Corrected, ideal and measured admittance (Ionic concentration constant)
Min. Error	0.075	−0.2337
Max. Error	0.0459	0.0673
Standard deviation	0.0769	0.0904

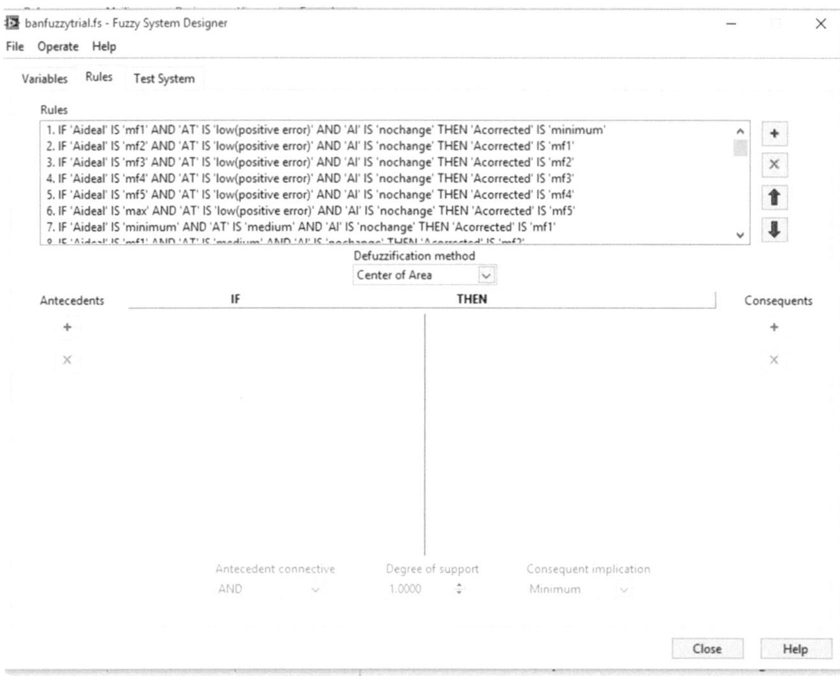

Fig. 4 Rule viewer page of fuzzy system designer [9]

3 Fuzzy Based Linearization in LABVIEW

The real-time data from the admittance lever transmitter is acquired using NI-DAQ and send to the personal computer. In the personal computer, the LABVIEW software is installed. Labview is a software platform of National instruments (NI) for analysis of data and plays a very vital role in virtual instrumentation. In the NI Labview 2013 version, the fuzzy system designer is selected from control design and simulation toolbox. In the fuzzy system designer, the input variables and output variables are chosen as per Table 1. In the next step, the rule base between the fuzzified data is set as shown in Fig. 4.

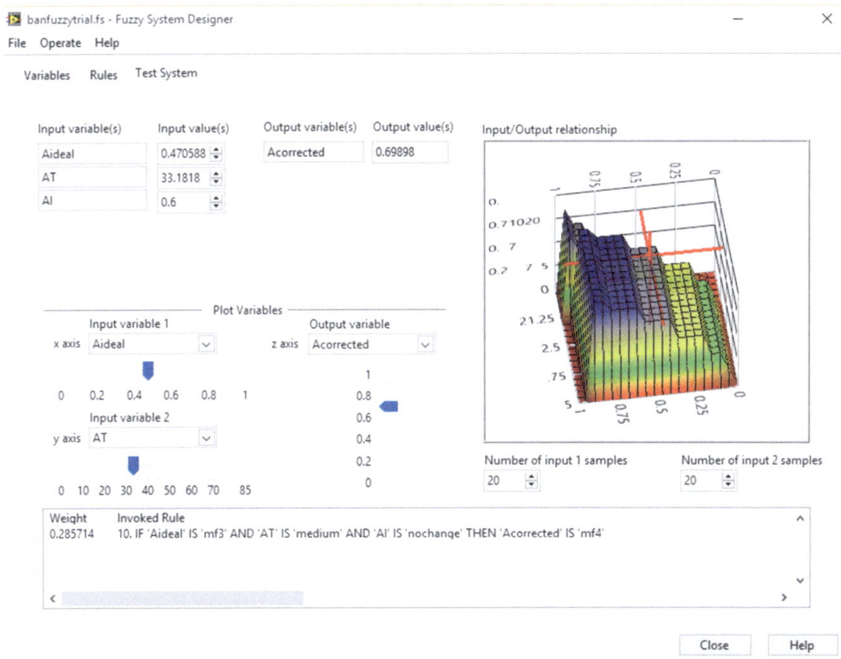

Fig. 5 Test system window [9]

After simulating the test system, the output values can be recorded and stored for further analysis. Figure 5 shows the test system of the fuzzy designer.

NI Labview has two programming window. One is block diagram window where the program is to be written, and other is front panel window where the result is to be analyzed. The block diagram window of the fuzzy-based linearizer is shown in Fig. 6.

In the block diagram window, all the input variables are given to the Fuzzy based linearizer. The input values are acquired through NI DAQ 6211 card. The output of the fuzzy-based linearizer is corrected admittance value. The data generated can be recorded in an excel file using the write to measurement block. Figure 7 shows the front panel window of indicators of input and output values.

From the front panel window, the corrected value of admittance can be noted. Now the comparison chart has been prepared for the fuzzy corrected value and the ideal value of admittance data generated from fuzzy based linearizer in Labview. The data are recorded by considering the parameters of temperature and ionic concentration which is shown in Figs. 8 and 9.

In Fig. 8, the temperature is kept constant at 24.3 degree Celsius. The ionic concentration is varied, and corresponding corrected admittance value is recorded. The same method is repeated maintaining the ionic concentration constant at 0.203TDS, and the temperature is varied. Eventually, the corrected admittance data is recorded.

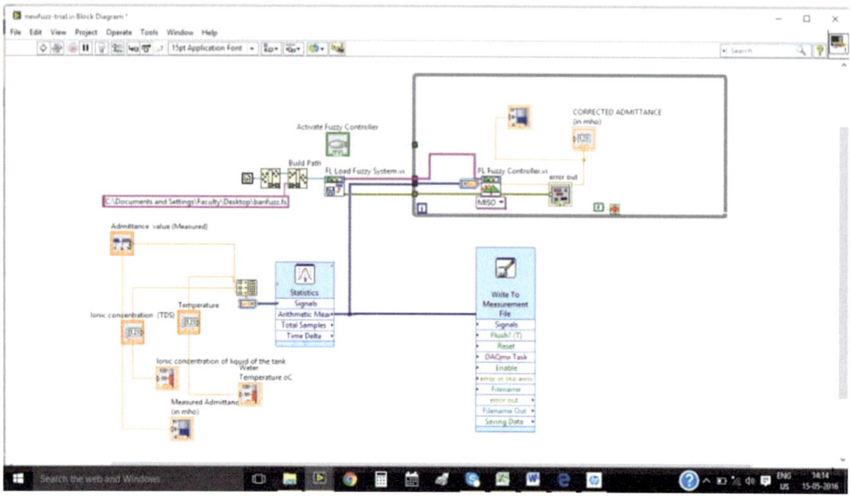

Fig. 6 Block diagram window [9]

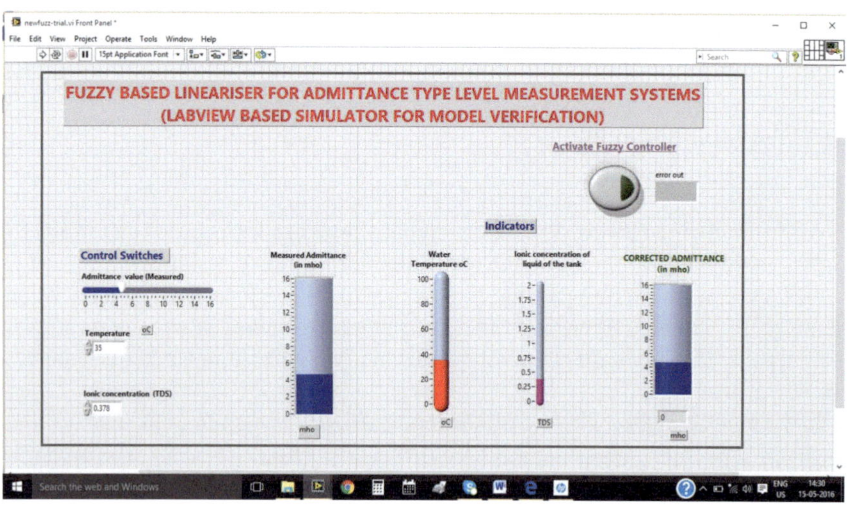

Fig. 7 Front panel window [9]

The statistical comparison is listed in Table 3.

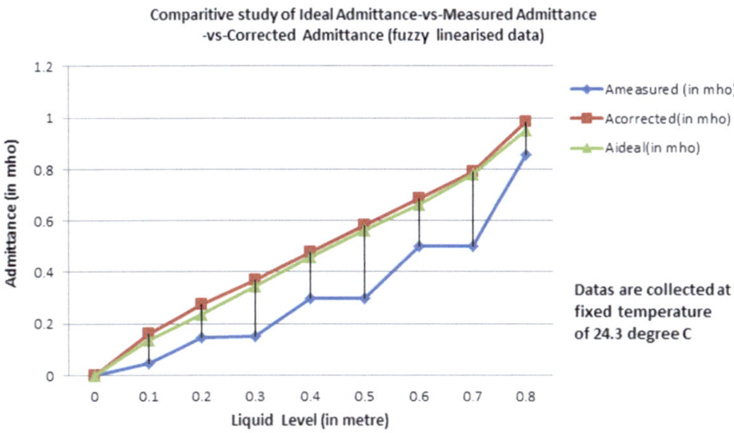

Fig. 8 Comparison chart at fixed temperature [9]

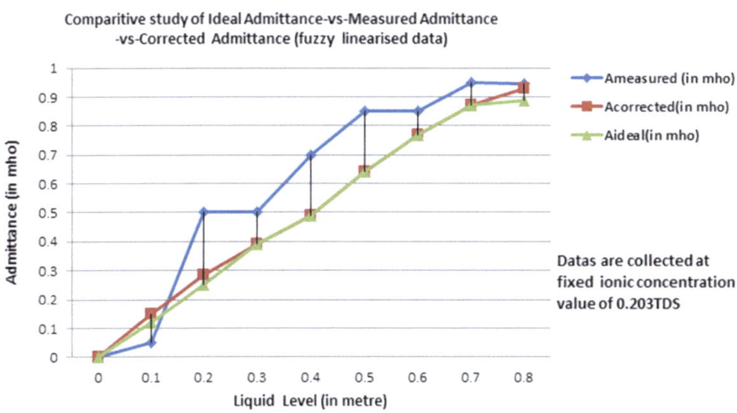

Fig. 9 Comparison chart at fixed ionic concentration [9]

Table 3 Statistical analysis chart

LABVIEW based analysis		
	Comparison of corrected, ideal and measured admittance (Temperature constant)	Comparison of corrected, ideal and measured admittance (Ionic concentration constant)
Min. Error	0.063	−0.0327
Max. Error	0.0329	0.0345
Standard deviation	0.0543	0.0652

4 Conclusion

The statistical analysis has compared the two methods of fuzzy based linearizer. From the data of Tables 2 and 3 it can be concluded that MATLAB is much better for computation than LabVIEW. The classical program code is much more suitable for calculations than block diagrams. On the contrary, the most significant advantage of LabVIEW is fast and simple construction of the graphical user interface that facilitates the updating of parameters (no need to interfere with the code) and elegant presentation of the results. Comparing the LabVIEW, the Creation of a comparable user interface in MATLAB may be more painful and limited. The another advantage of LabVIEW is that most MATLAB functions, that are accessible from LabVIEW via the MathScript Node, which can pass data to m code, execute it and get results back. Hence LabVIEW based linearizer has more advantage. From the comparative study, it has been found that in both cases (offline in/Matlab and online in Labview) accuracies of the measurement is very close to each other. Therefore the NI Labview in real-time analysis, the accuracy will be good enough in developing physical level measuring instrument using admittance method. It can be incorporated into the measurement system by real-time basis. With the help of NIDAQ 6009, the real-time data can be acquired. It has the advantage of signal analysis and processing simultaneously. In the later stage cross sensitivity effect of the sensor can be used to measure three different parameters simultaneously. This phenomenon is called multi-function sensing. In the future work, the primary focus will be on the development of multifunction sensor. Some of the researchers [12] already started developing a multifunction sensor for level measurement in industrial applications. Hence this is the significant area of research in the recent days.

References

1. W. Skierucha, Time Domain Reflectometry: Temperature-dependent Measurements of Soil Dielectric Permittivity, in *Electromagnetic waves* (Institute of Agrophysics, Polish Academy of Sciences, Poland, 2011), Chapter 17, pp. 374–379
2. L. Guirong, Z. Xianshan, A new proposal for monitoring oil-temperature and oil level, ISBN:978-1-4244-8158-3, pp. 350–353, ICEMI-2011
3. S.C. Bera, J.K. Ray, S. Chattopadhyay, A low-cost noncontact capacitance-type level transducer for a conducting liquid. IEEE Transac. Instrum. Measur., **55**(3), 778–786 (2006)
4. J. Kumar Roy, Low-Cost Sensing Techniques of Industrial Process Variables, ISBN: 978-3-659-11192-1LAMBERT Academic Publishing, TP-395, 2012
5. S.C. Bera, J.K. Roy, Study of an admittance type single electrode transducer for continuous monitoring of liquid level in a metallic storage tank. J Instn. of Eng. (I), **83**, 56–60, Jan (2003)
6. J. Kumar Roy, B. Deb, *Investigation of Cross-sensitivity of a Single and Double Electrode of Admittance Type Level Measurement*. Sixth International Conference on Sensing Technology, Kolkata, India, Dec 2012. Proceedings published in IEEE Digital Xplore, ISBN: 978-1-4673-22454, pp. 234–237 (2012)
7. J. Kumar Roy, B. Deb Majumder, *Cross-sensitivity of Ionic Concentration on Admittance Type Level Measurement*. Eighth International Conference on Sensing Technology, Liverpool,

UK September 2014. Proceedings published in International Journal on Smart and Intelligent Systems, ISSN: 1178–5608, pp. 41–45 (2013)
8. J.K. Roy, B. Deb Majumder, Elimination of Cross-sensitivity in Admittance Type Level Measurement Using Fuzzy Based Linearizer, on Smart Sensing and Intelligent Systems. Scopus Indexed Journal **7**(4), ISSN: 11785608 H Index:8, December (2014)
9. J. Kumar Roy, B. Deb Majumder, *Real-Time Measurement of Water Level Using Admittance Method and Fuzzy Based Linearizer*. Tenth International Conference on Sensing Technology, Nanjing, China. Proceedings published in IEEE Digital Xplore, ISBN: 978-1-5090-0796-7, Nov (2016)
10. L.A. Zedeh, Knowledge representation in fuzzy logic. IEEE Trans. Knowl. Data Eng. **1**(1), 89–100 (1989)
11. C. Carlsson, Fuzzy logic and hyper knowledge: a new, effective paradigm for active DSS. IEEE environmental management, vol. 5, pp. 324–333, Print ISBN: 0-81867743, 1997
12. G. Lu, in *A new proposal of multi-functional level meter*, ISBN:0-7803-7987-X. IEEE International Conference on Multisensory Fusion and Integration for Intelligent Systems, pp. 209–212, 2003

Automated Identification of Myocardial Infarction Using a Single Vectorcardiographic Feature

Deboleena Sadhukhan, Jayita Datta, Saurabh Pal and Madhuchhanda Mitra

1 Introduction

According to recent health reports [1], Myocardial Infarction (MI), more commonly known as heart attack, continues to be the predominant cause of death all over the world. Early stage detection and medication can largely reduce the risk of mortality. MI is caused by occlusion of the coronary arteries which causes insufficient blood flow to the heart muscle cells or myocardium in different regions leading to their damage (ischemia) or complete death (necrosis) [2]. The dysfunctional tissues cause a disruption of the heart's electrical activity which in turn affects the synchronous contraction and relaxation of the different heart chambers (atria and ventricles). Electro-cardiogram (ECG), the recording of the heart's electrical activity, is the most predominant tool used for cardiac diagnosis. The presence of MI is manifested as changes in the morphological features of the ECG such as the shape of T-wave, Q-wave and ST-segment [3]. Changes in these morphological and temporal wave features from a standard 12 lead ECG system are normally used to diagnose MI development.

D. Sadhukhan (✉) · S. Pal · M. Mitra
Department of Applied Physics, University of Calcutta,
92 A.P.C. Road, Kolkata 700009, India
e-mail: dsaphy_rs@caluniv.ac.in

S. Pal
e-mail: spaphy@caluniv.ac.in

M. Mitra
e-mail: mmaphy@caluniv.ac.in

J. Datta
Department of Electronics & Instrumentation Engineering,
Guru Nanak Institute of Technology, 157/F, Nilgunj Road,
Panihati, Kolkata 700114, India
e-mail: jayita.datta@gnit.ac.in

S. Chattopadhyay et al. (eds.), *Modelling and Simulation in Science, Technology and Engineering Mathematics*, Advances in Intelligent Systems and Computing 749,
https://doi.org/10.1007/978-3-319-74808-5_57

The need for early and reliable MI identification has led to immense research to automate the process of cardiac analysis. Automated cardiac analysis is based on extraction of clinically significant features from the cardiac data and employs machine intelligence techniques for classifying these features. Most of the reported automated MI identification tools use ECG features. Techniques [4, 5] based on the explicit time plane features (the wave amplitudes and time durations) and ST segment analysis relies on accurate detection of the ECG wave segments which is difficult to achieve due to the large morphological variation of the ECG waveform and presence of noise. Use of advanced signal processing tools have also been applied to extract time-frequency ECG features based on Discrete wavelet transform [6, 7] and Cross wavelet transform [8] to achieve high classification performance. But these techniques are not only computationally complex but also suffer from the "curse of dimensionality" due to the use large number of features. Although the standard 12-lead ECG is sufficient to represent the spatiotemporal activity of the heart in different perspectives, but there is a loss of spatial information in each temporal ECG tracing. Moreover, analysis of all 12 ECG leads adds to the computation burden of the automated software and much of the information within it is still redundant.

In recent years the vector-cardiogram or VCG has drawn significant attention in cardiac analysis. VCG enables spatio-temporal visualization by monitoring heart's electrical activity in three spatial planes (horizontal, frontal and sagittal) generating different loops (shown in Fig. 1) for the P, QRS and T waves representing the activation and relaxation of the different heart chambers. It not only gives a clearer spatial orientation of the cardiac activity but also uses a reduced lead set (V_X, V_Y and V_Z). Computerized VCG analysis hence proves to be more advantageous and also provides higher specificity, sensitivity and accuracy as compared with conventional ECG for the diagnosis of different cardiac pathologies [9].

Significant amount of literatures [10–20] are available on the topic of MI detection using the VCG features. Most of the approaches [10–15] are based on the use of different morphological features of the VCG loop which includes areas, perimeters and angles of the QRS and T loops [11–13], vector magnitudes of the Q-wave, R-wave, T-wave [12, 14], ST change vector magnitude [10], vector magnitude differences between the loops [14], angle between R and T vector [11], azimuth angle of the vectors [11] and also octant based features [15]. Extraction of such features needs segmentation of the different waves including the isoelectric point for the angle measurements. Moreover, almost all of them consider separate features for each wave, which significantly increases the feature dimension. Also the computation of the angles and the 2-d area needs complex measurements achieved by projecting 3-d VCG loops in different planes. Use of the more advanced signal processing tools like principal component analysis (PCA) and independent component analysis (ICA) [16], recurrence quantification analysis [17], random walk network [18], wavelet coherence analysis [19], self-organizing visualization and pattern matching [20] to extract different VCG features involves intensive mathematical operations which makes them difficult and time consuming to implement.

In this paper we propose a new a VCG feature which combines both the QRS and ST-T loops morphological changes into one single feature, thus reducing the

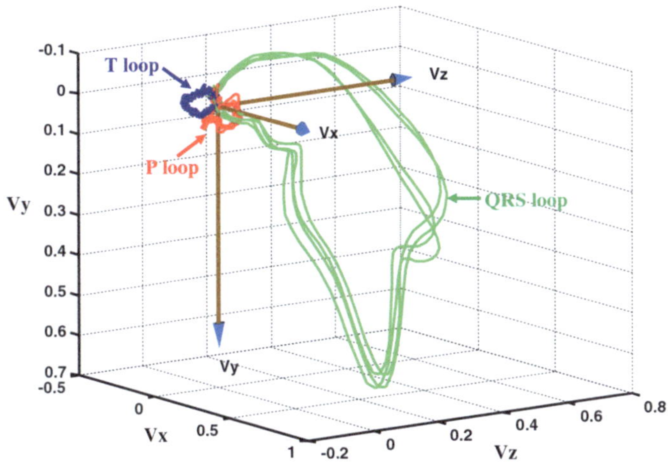

Fig. 1 Representation of Vector cardiogram (VCG). VCG vector loops contain 3D recurring, near-periodic P, QRS, and T wave activities representing each heart cycle

feature dimensionality problem. Instead of representing the morphological changes in the loops by their angles (which requires identification of iso-electric point) or areas (which needs projections in different planes), we simply consider the QRS and the ST-T loop volumes evaluated by representing the loops using 3-d convex-hull technique. Statistical analysis shows that the ratio of the QRS and ST-T loop volumes is significantly different for healthy and the MI subjects as tested with the PTB diagnostic ECG database [21]. Hence this single parameter has sufficient discriminative power to identify MI.

The detailed processing steps of the algorithm are described in Sect. 2. In Sect. 3 the performance of the algorithm is validated with the PTB diagnostic database. Finally Sect. 4 concludes the paper by analyzing the relative advantages and disadvantages of the proposed technique.

2 Methodology

Figure 2 shows the detailed block diagram of the proposed MI identification technique using the VCG signal. The key concept is to extract cardiac beats from the 3 VCG leads (V_X, V_Y and V_Z) and segment it into the QRS and the ST-T to represent the corresponding loops. The volumes of the loops are computed by representing them with 3-d convex-hull. The volume ratio of the two loops is used to classify the healthy and the MI records.

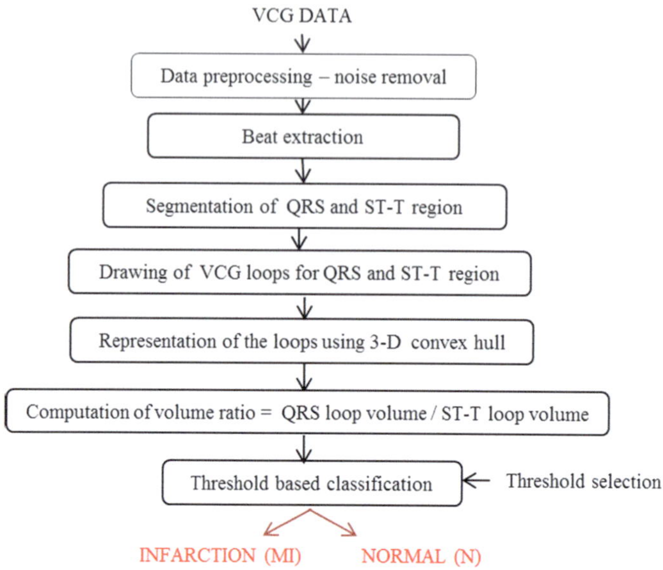

Fig. 2 Block diagram representing the processing steps of the proposed technique

2.1 Data Pre-processing

The pre-processing step involves the elimination of noise from the 3 orthogonal VCG lead data. The very low frequency baseline drifts (below 0.2 Hz), the high frequency noises (above 80 Hz) and the power-line interference (of 50 Hz) are eliminated by means of the Fourier co-efficient suppression technique proposed by us in [22].

2.2 Beat Extraction

One single cardiac cycle comprising of the P, QRS and the T waves is sufficient to provide information of the presence of MI abnormalities. The R peak is the most distinctly identifiable feature of the cardiac cycle and hence location is used for beat extraction. The R peaks of lead V_X are detected using the method proposed in [23] based on double differencing. To account for the heart rate variation, a single beat is extracted based on the previous and the consequent RR intervals instead of using fixed window length. The beat start point is selected to be 1/3rd of the previous RR interval before the current peak so as to include the P peak. The selected end point is 2/3rd of the consequent RR interval after the current peak position so as to include the T peak. The same start and end points are used to extract time aligned beats from the three leads.

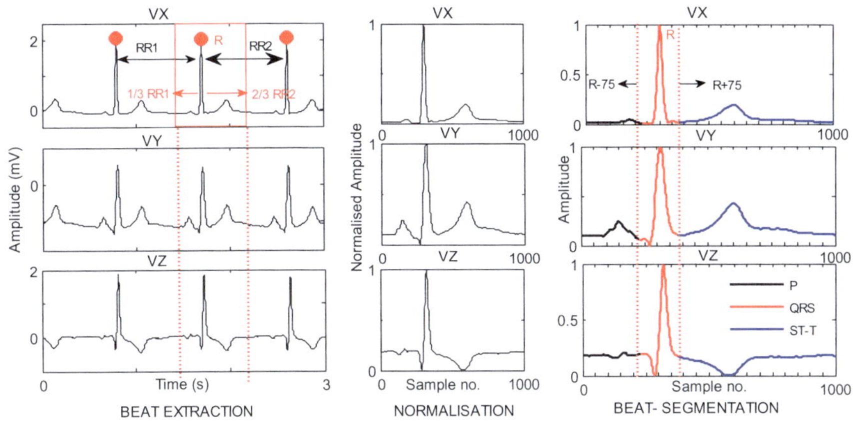

Fig. 3 Beat extraction and segmentation process. Red dots denote the detected R peaks. One beat is extracted from 1/3rd of the previous RR interval (RR1) to 2/3rd of the consequent interval (RR2). The beats are then normalized both in time and amplitude domain. The QRS segment is identified within the interval of ±75 data points around the R position. The remaining beat succeeding the QRS region is considered to be ST-T segment

2.3 Beat Segmentation

Presence of infarction significantly alters the morphological pattern of the QRS region, ST segment and T wave [3]. So, instead of the whole ECG beats the QRS regions and the ST-T regions are considered separately to detect the identifiable changes. To ensure parity of the loop amplitudes, the beat amplitudes are normalized in the range of 0–1. For extraction of the QRS and ST-T regions from each beat, it is required to identify the start and the end points of each wave by accurate delineation algorithms which can significantly increase the computational burden. To avoid this, all the beats are first time normalized to contain 1000 data points each using the FFT interpolation technique. This helps in alignment of all the beats in time axis. Next the QRS regions are identified to be within ±75 data points around the R peak location. The remaining beat from the QRS end point is considered to be the ST-T region.

Figure 3 explains the beat extraction and segmentation process.

2.4 Drawing of VCG Loops

The QRS and the ST-T loops are obtained by drawing simultaneously on 3-D plot the instantaneous amplitudes of the orthogonal leads for every sample of the temporal interval corresponding to each detected QRS-complex and ST-T segments respectively. Figure 4 illustrates QRS and the ST-T loops obtained from a healthy record

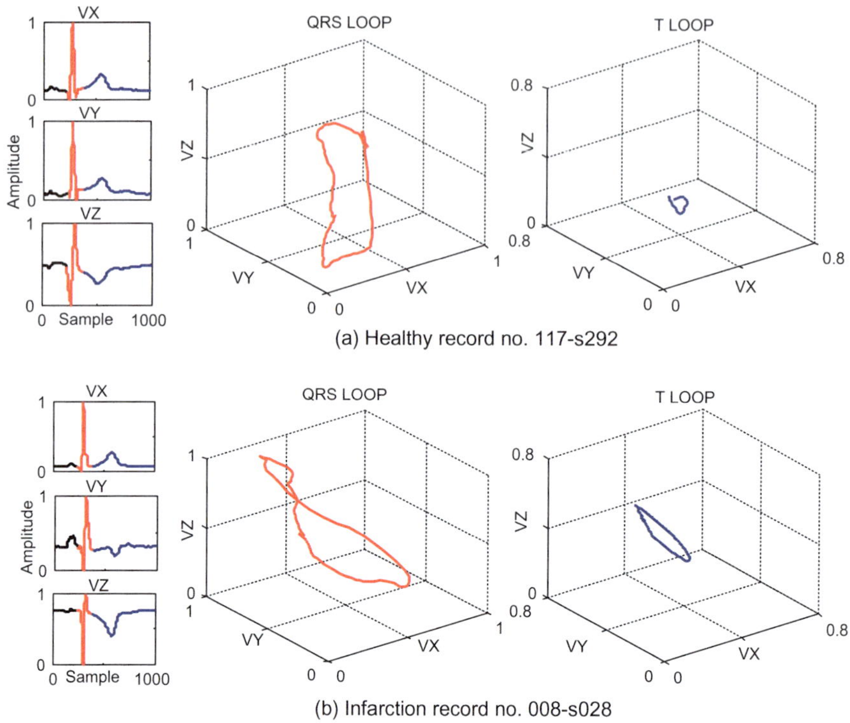

Fig. 4 QRS and ST-T loops. The corresponding beats are displayed in the left hand side. **a** QRS and ST-T loops for a healthy record. **b** QRS and ST-T loops for an infarction record. The QRS loop volume of the healthy record is higher than that for the infarction case whereas the ST-T volume is significantly lower

and an infarction record. It can be seen that both loop morphologies are significantly dissimilar for the 2 cases.

2.5 Computation of Volume Ratio

As seen from Fig. 4, there is a wide variation of the 3-d QRS and ST-T loops for healthy and MI data. Instead of using multiple features including the angles and loop areas in different projected planes, we quantify these variations using the loop volumes.

To accurately estimate the volume of the 3-D loops, they are first represented by the set of points that produce the minimum convex volume containing all the points of the loop using the "Convex Hull" algorithm [24]. This representation is illustrated in Fig. 5. Then, the volume of the polyhedron is evaluated for each loop. To combine the QRS and ST-T changes into a single parameter the volume ratio of the two loops

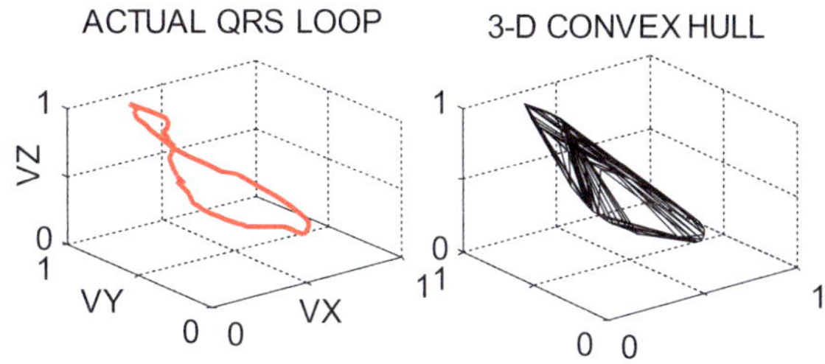

Fig. 5 3-D convex hull representation of a QRS loop. The loop is represented using a polyhedron having the minimum convex volume and containing all the points of the loop

(QRS volume/ST-T volume) is considered as the final feature to discriminate between the normal and MI data.

3 Experimental Results

3.1 Used Data

The development and validation of the proposed technique has been done with VCG data extracted from the PTB Diagnostic ECG database available under Physionet [21] as it contains well classified records both from healthy (H) and infarction patients (MI). Each recording contains 15 simultaneous recorded signals from the conventional 12-lead ECG and the 3-lead VCG. Data is sampled at 1 kHz with 16 bit resolution. For our work we have used VCG recordings from 70 healthy data and 150 MI data (including inferior and anterior infarction). From these, beats extracted from 50 records from each group is used for the identification of the feature threshold value. Then randomly extracted beats from all the H and MI records are used validation of the proposed technique.

3.2 Performance Evaluation Parameters

The overall discrimination ability of the proposed VCG parameter is evaluated using the most commonly used performance metric- Accuracy (Acc), Sensitivity (Se) and Specificity (Sp) defined as follows:

$$Acc = \frac{TP + TN}{TP + TN + FP + FN} \tag{1}$$

$$Se = \frac{TP}{TP + FN} \tag{2}$$

$$Sp = \frac{TN}{FP + TN} \tag{3}$$

where *TP* denotes the number of MI beats correctly detected, *TN* denotes the number of H beats correctly classified, *FP* denotes the number of H beats misclassified as MI, and *FN* denotes the number of MI beats misclassified as H.

3.3 Performance Evaluation

As seen from Fig. 4, the QRS loop volume for the healthy records is higher than that of the infarction cases, whereas ST-T loop volume is significantly lower. This corresponds to higher loop volume ratios for the healthy records. The statistical parameters (the mean and the standard deviation values) of the volume ratio for the 500 H and 500 MI beats extracted from the training data set are shown in Table 1. To visualize the within class variation of the proposed VCG feature the box plots of the feature value for both the classes are shown in Fig. 6. The red line in the middle of boxplot represents the median, the blue box shows the lower quartile and upper quartile of data distributions, and the black dash lines represent the most extreme values within 1.5 times the interquartile range. Outliers are shown as red crosses in the box plot. Both Table 1 and Fig. 6 strongly indicates the distinct difference between the mean values of the feature for the H and MI cases.

From the obtained feature values for the training data set, a volume ratio value of 220 is selected as the threshold value for the classification of the H and the MI data, giving 99% coverage to the MI data in the training set. QRS and ST-T volume ratio below this threshold value indicates the presence of MI.

This threshold value is then used to classify 2000 beats randomly extracted from each class in the entire dataset used for the work (eliminating the beats used in the training phase). The classification results are displayed in Table 2.

For medical diagnosis applications misdetections of positive cases (FN) can be more fatal. To reduce misdetections of MI data, the feature threshold value is selected

Table 1 Volume ratio values for healthy and MI data

Feature	Healthy		Infarction	
	Mean	*SD*	*Mean*	*SD*
QRS volume/ST-T volume	489.41	351.11	150.13	111.80

SD denotes standard deviation

Table 2 Overall classification performance

True class	Predicted Class		Acc (%)	Se (%)	Sp (%)
	MI	Healthy			
MI	*TP* = 1976	*FN* = 24	97.5	98.8	96.2
Healthy	*FP* = 76	*TN* = 1924			

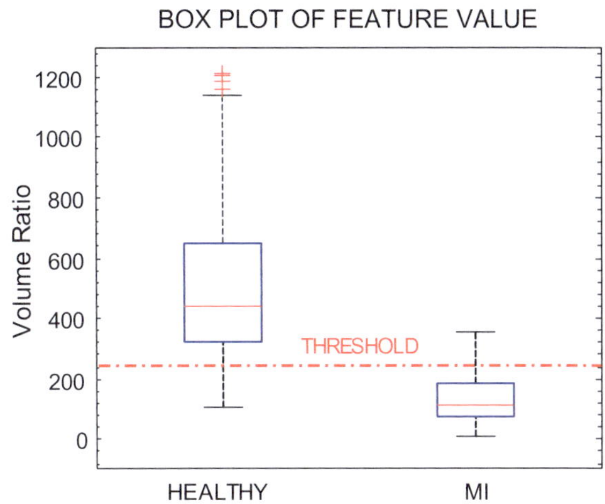

Fig. 6 Box plot showing the variation of volume ratio value for healthy and MI data. The red dotted line shows the threshold feature value selected for MI classification

giving maximum coverage to the MI data. Hence, the proposed technique achieves a fairly high detection sensitivity of 98.8%, whereas the specificity falls to 96.2%.

4 Conclusion and Discussion

The vector-cardiogram (VCG) proves to be a more informative and low dimensional alternative of the 12 lead Electrocardiogram (ECG) for MI diagnosis. The automated VCG analysis tools, reported till date, utilize a large number of features based on the sizes, area and orientation of the QRS and the T loops. This paper proposes a novel VCG feature—the volume ratio of the 3-d QRS and the ST-T loop. This feature can appropriately incorporate all the morphological changes of the VCG into a single parameter. The obtained classification performance reveals that this feature alone can provide significantly high detection sensitivity. Moreover, the use of this single feature for cardiac analysis can significantly reduce the computational burden and the detection time, thus enabling more reliable and faster identification of MI. These results are thus strongly indicative of the potential of QRS ST-T volume ratio to be used as a MI detection parameter in the automated cardiac analysis tools.

However, in this paper we have restricted our analysis to two types of MI only (inferior and anterior). But for actual clinical application the study needs to be extended to incorporate all other types of MI as well as other cardiac diseases. Moreover, MI localization (identification of the zone of infarction) has not been considered for the present study.

Acknowledgements The first author acknowledges the financial support obtained in the form of DST INSPIRE Fellowship provided by the Department of Science & Technology, Government of India.

References

1. WHO Fact Sheet, The top ten causes of death, Fact sheet N°310, 2012 [Online]. Available: http://www.who.int/mediacentre/factsheets/fs310/en/
2. K. Thygesen et al., Third universal definition of myocardial infarction. Circulation **126**(16), 2020–2035 (2012)
3. A.L. Goldberger, *Clinical Electrocardiography: A Simplified Approach* (Elsevier Health Sciences, Amsterdam, 2012)
4. S. Mitra, M. Mitra, B.B. Chaudhuri, A rough-set-based inference engine for ECG classification. IEEE Trans. Instrum. Meas. **55**(6), 2198–2206 (2006)
5. Jocelyne Fayn, A classification tree approach for cardiac ischemia detection using spatiotemporal information from three standard ECG leads. IEEE Trans. Biomed. Eng. **58**(1), 95–102 (2011)
6. L. Sharma, R. Tripathy, S. Dandapat, Multiscale energy and eigen space approach to detection and localization of myocardial infarction. IEEE Trans. Biomed. Eng. **62**(7), 1827–1837 (2015)
7. S. Padhy, S. Dandapat, Third-order tensor based analysis of multilead ECG for classification of myocardial infarction. Biomed. Signal Process. Control **31**, 71–78 (2017)
8. S. Banerjee, M. Mitra, Application of cross wavelet transform for ecg pattern analysis and classification. IEEE Trans. Instrum. Meas. **63**(2), 326–333 (2014)
9. A.R.P. Riera, A.H. Uchida, C.F. Filho, A. Meneghini, C. Ferrerira, E. Schapacknik et al., Significance of VCG in the cardiological diagnosis of the 21st century. Clin. Cardiol. **30**, 319–323 (2007)
10. M. Dellborg, H. Emanuelsson, M. Riha, K. Swedberg, Dynamic QRS-complex and ST-segment monitoring by continuous vectorcardiography during coronary angioplasty. Coron. Artery Dis. **2**(1), 43–53 (1991)
11. G. Bortolan, I. Christov, Myocardial infarction and ischemia characterization from t-loop morphology in VCG, Computers in Cardiology, pp. 633–636, 2001
12. Raúl Correa et al., Novel set of vectorcardiographic parameters for the identification of ischemic patients. Med. Eng. Phys. **35**(1), 16–22 (2013)
13. Correa Raúl et al., Acute myocardial ischemia monitoring before and during angioplasty by a novel vectorcardiographic parameter set. J. Electrocardiol. **46**(6), 635–643 (2013)
14. R. Correa, P.D. Arini, L.S. Correa, M. Valentinuzzi, E. Laciar, Novel technique for st-t interval characterization in patients with acute myocardial ischemia. Comp. Biol. Med. **50**, 49–55 (2014)
15. H. Yang, S.T. Bukkapatnam, T. Le, R. Komanduri, Identification of myocardial infarction using spatio-temporal heart dynamics. Med. Eng. Phy. **34**(4), 485–497 (2012)
16. A.R.M. Dehnavi, I. Farahabadi, H. Rabbani, A. Farahabadi, M.P. Mahjoob, N.R. Dehnavi, Detection and classification of cardiac ischemia using vectorcardiogram signal via neural network. J. Res. Med. Sci. **16**(2), 136–142 (2011)
17. H. Yang, Multiscale recurrence quantification analysis of spatial cardiac vectorcardiogram signals. IEEE Trans. BME **58**(2), 339–347 (2011)
18. T.Q. Le, S.T.S. Bukkapatnam, B.A. Benjamin, B.A. Wilkins, R. Komanduri, Topology and random-walk network representation of cardiac dynamics for localization of myocardial infarction. IEEE Trans. BME **60**(8), 2325–2331 (2013)
19. S.M. Dima et al., On the detection of myocadial scar based on ECG/VCG analysis. IEEE Trans. BME **60**(12), 3399–3409 (2013)
20. Hui Yang, Fabio Leonelli, Self-organizing visualization and pattern matching of vectorcardiographic QRS waveforms. Comp. Biol. Med. **79**, 1–9 (2016)
21. PTB Diagnostic ECG Database Directory, Physiobank Archive Index, PTB Diagnostic ECG Database [Online]. Available: http://physionet.org/physiobank/database
22. D. Sadhukhan, M. Mitra, in *ECG Noise Reduction Using Fourier Coefficient Suppression*. International Conference on Control, Instrumentation, Energy and Communication (CIEC) 2014 (Kolkata, India, 2014), pp. 142–146

23. D. Sadhukhan, M. Mitra, in *R-Peak Detection Algorithm for ECG Using Double Difference and RR- Interval Processing*. 2nd International Conference on Computer Science, Communication and Control Engineering, Academy of Technology, vol. 4 (Procedia Technology, Kolkata, India, 2012), pp. 873–877
24. B. Chazelle, An optimal convex hull algorithm in any fixed dimension. Discrete Comp. Geom. **10**(1), 377–409 (1993)

Content Extraction Studies for Multilingual Unstructured Web Documents

Kolla Bhanu Prakash and M. A. Dorai Rangaswamy

1 Introduction

Recent developments in communication and internet have brought in significant changes in scientific, engineering and societal context and wide range of user-oriented mobile applications like whatsapp, twitter etc. have added new dimension to modern living and thought process. Simultaneously, the reach of these developments is still a long way to go as long as the gap between human communication and computer-based communication is not bridged fully. There are many barriers to overcome like language, dialect, tradition, way of living etc.

This is where; conventional data mining approaches need to be elevated to media-mining or content extraction approaches. Content extraction is the process of identifying main content of a web page which may consist of different forms of data in an unstructured and non-homogeneous manner [1–3]. Added to this is the ability of including region and language based information, thanks to the exponential growth in use of cellular communication.

Text based information has reached different levels with different languages forming the text either as a computer-generated data or acquired data through images forming most of the pages. All these aspects bring in a necessity of using a more general approach to extraction of information and it has become very important to consider different kinds of web pages. A typical web page in present day context is shown in Fig. 1. This web page has text-based information in two different lan-

K. B. Prakash (✉)
Department of Computer Science and Engineering, Koneru Lakshmaiah Education Foundation, Vaddeswaram, Guntur, India
e-mail: drkbp@kluniversity.in

M. A. Dorai Rangaswamy
St.Peters University, Avadi, Chennai, India
e-mail: drdorairs@yahoo.co.in

S. Chattopadhyay et al. (eds.), *Modelling and Simulation in Science, Technology and Engineering Mathematics*, Advances in Intelligent Systems and Computing 749,
https://doi.org/10.1007/978-3-319-74808-5_58

Fig. 1 Variations in form, text and language levels in web page

guages—content may or may not be just translated one- and also different kinds of images which may be a photo or computer-generated drawings.

This web page conveys information in the form of content even to a person who does not know any of the languages as the images convey more than the text. So, content extraction for such web pages can be considered as a pre-processing step for text mining and Web information retrieval. Furthermore, such main contents are very valuable as an input for many devices that have limited presentation capacity, such as mobile phones, speech readers, etc. [4, 5]. The focus of the present study is to develop a generic content extraction approach which is based on the unstructured, non-homogeneous and text and/or non-text based data, as that of the web page shown in Fig. 1. A typical block diagram for content extraction is shown in Fig. 2.

This is a major difference to be looked into when one considers Asian web pages, which contain language and information, which are older than those used in European web pages and this aspect gets much more complex in Indian context, where dialect and text differ widely even in small regions. The present study is an attempt to develop

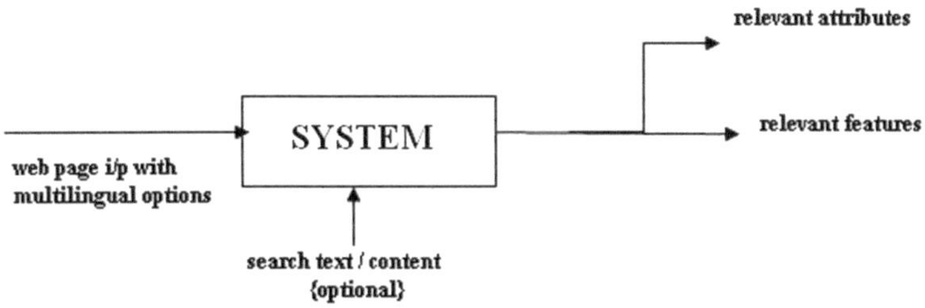

Fig. 2 Typical block diagram for content extraction

pixel-based approach-which gives flexibility in dealing with any language or media-and start from generic text level to a hybrid unstructured level.

2 Content Extraction Techniques

As content extraction is different from text or data mining, where a set of keywords form the basis, most of the previous approaches use HTML tags to separate the main content from the extraneous items. This implies the need to employ a parser for the entire Web page. Consequently, the computation costs of these main content extraction approaches include the overhead for the parser.

In the early stage of the Internet, the contents of Web pages were written only in English language. Now, especially in the last decade, a large part of information is being published in regional or native languages, like for example Spanish, German, French, Tamizh, Arabic, Urdu, Hindi with native tongue and usage reflected in the text. Except for the non-English languages mentioned here, there are also many languages using non-ASCII codes for their characters.

Typically, a modern web page for commercial intent looks like the one shown in Fig. 3 and this is taken from a web magazine. A collage of data in the form of text in different languages and sizes, numerals, images and blocks, forms the web page with the intent that content is reached to the web-surfer, who may be from different country with different languages and dialect and culture. But, the content in terms of shoes or dresses reaches him so that he/she can follow and get more details. This is typically an unstructured, heterogeneous and hyper media web page. Extracting content requires language, text and image processing. Extracting main content from web page is pre-processing of web information system. The content extraction approach based on wrapper is limited to one specific information source, and greatly depends on web page structure. It is seldom employed in practice. So, a generic model employing basic features of data is needed and the proposed model is from basic pixel level making it applicable to any kind of data or text or image or even media to assess the content in a short period of time.

Fig. 3 Multilingual and multi-tasking modern web magazine

3 Nature and Features in Web Documents

As seen earlier web pages are unstructured-not conforming to any document form-, non-homogeneous with information and data presented in different forms from text to images to video, and multi-lingual depending on the audience and their location. This gets more complex and involved when Asian or Indian regional web pages display information. Indian languages are very much different from European -or other Asian languages—like Japanese or Persian—in that regional customs and practices bring in certain commonalities like the scripts of Tamizh or Telugu or Kannada have similarities of different kinds as compared to the northern Hindi or Punjabi scripts. But English being the link language both in oral and written communication and forms the basis in higher education, some complexities in migrating from English to regional language or vice versa exist like the ones shown in Fig. 4. Figure 4 shows a typical web page displaying news on the same day and here web pages in Malayalam, English and Hindi are shown and one can see that even the news content varies as the region and language change.

Fig. 4 Web documents in regional India—different languages, different contents

For example, if one compares Malayalam with Hindi, there is variation in news content and one finds regional news dominating over national in both the cases. So, if one wants to continue surfing and later interact with, the content of the web is the only way to go about and if figures or images are not there, content needs to be given in a short period of time. It may be seen clearly that regional web documents pose different problems in terms of comprehension, understanding and interaction in other language regions. Even if one looks at script or character level, or even word level, complexities are many-fold, as the web pages try to present information in easily understandable form using words freely from different languages. As an example, a word 'Computer' in English translated in other languages like Hindi, Arabic, Tamizh and Telugu is shown in Fig. 5a. But many times, popular words in one language are used as they are like word 'computer' in English is written in local scripts as in Fig. 5b. Also one can see clearly the variations in structure of text in different forms and these do not form part of the local language dialect.

Hence, it is preferable to assess the content even before looking at the document fully. Images and figures do help, but, many times texts and sketches with words pose problems as they reflect local dialect and flavour. So, it is necessary to assess the content irrespective of the language and the way text is produced.

Hence, the objective is to develop a generic model and later apply for complexities to check whether it is possible to assess the content in a short period of time.

(a) Word 'computer' translated in other languages

(b) Word 'computer' written in other languages

Fig. 5 Complexities in Indian and Foreign languages with English

4 Text and Character Issues

One of the basic steps in any content extraction or mining approach is in processing the data as it is, and as the data may be in different types like pictures, texts or in different forms like media or audio or in different formats like .bmp or .jpg in either full form or compressed form. So, the pixel map of any data can form the basis for any form or format of data as computer processes at this level. But in unstructured and non-homogeneous documents, complexities begin at character level and later extend to word or document or web page form. Even in texts or documents, which are well structured with words and sentences, language brings in variations and this is true in education where text books written by authors in regional languages are digitized and used in web learning. These are becoming a major source of on-line education in different levels. To cite an example, Fig. 6 shows a Physics web page in two different languages English and Arabic used in schools.

Here one can see free mixing of words in Arab and English in both forms of documents. Keeping in view of all the above mentioned issues, it is preferable to consider extracting the content of the document rather than translation or data-mining.

The present study aims at developing a generic tool based on pixel map data, to extract content in a web page and later, using reduced attributes and features of pixel maps, a pattern matching approach is used to assess the content.

5 Development of Pixel Map Attributes

A web document may contain texts, images, audio/video files; and in some regional documents, scanned copies of hand-written texts or images are found. So, it is necessary to look at the generic level of data which is used by computer for processing. Any pixel map can be seen as a matrix of columns and rows with each element giving the color scheme for the pixel. So, the characteristic and attribute of any pixel map can

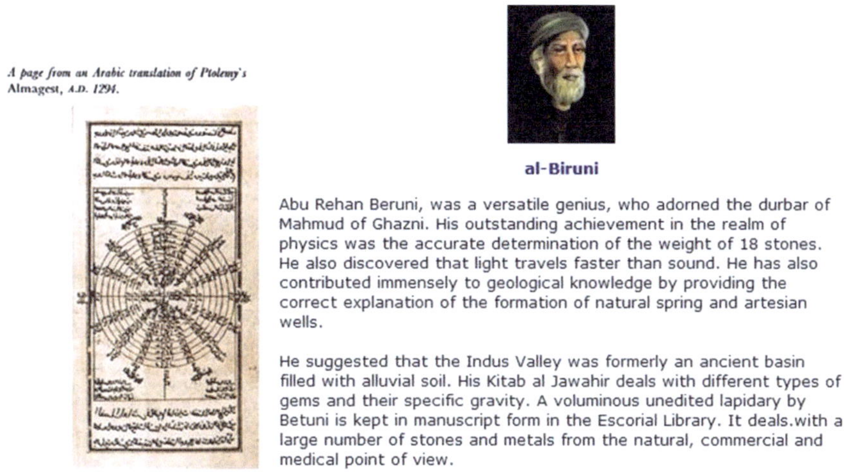

A page from an Arabic translation of Ptolemy's Almagest, *A.D. 1294.*

al-Biruni

Abu Rehan Beruni, was a versatile genius, who adorned the durbar of Mahmud of Ghazni. His outstanding achievement in the realm of physics was the accurate determination of the weight of 18 stones. He also discovered that light travels faster than sound. He has also contributed immensely to geological knowledge by providing the correct explanation of the formation of natural spring and artesian wells.

He suggested that the Indus Valley was formerly an ancient basin filled with alluvial soil. His Kitab al Jawahir deals with different types of gems and their specific gravity. A voluminous unedited lapidary by Betuni is kept in manuscript form in the Escorial Library. It deals.with a large number of stones and metals from the natural, commercial and medical point of view.

Fig. 6 Text book page in two languages—Arabic and English

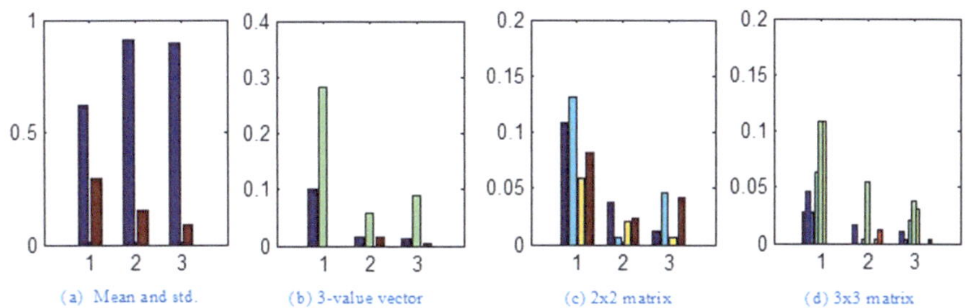

Fig. 7 Variation of pixel map attributes for letter 'a'

be deduced from these three values and most of image processing and data mining techniques depend on this basic matrix. The matrix size being large, it is preferable to reduce it by converting into grey-scale or binary form giving 0–7 or 0–1 values in the matrix. Typically a letter 'a' in English has [$10 \times 11 \times 3$] matrix and this is reduced to [10×11] with 0 and 1 value and even then there are 110 values to reflect the matrix fully. Table 1 gives pixel map attributes for letter 'a' in three languages, English, Arabic and Urdu. Here, only three sets are given like Mean and Standard deviation (std), 3-row vector attributes and 2×2 matrix attributes. Similarly, 3×3 matrix attributes can also be generated.

Figure 7 gives a comparison of features of pixel map attributes for letter 'a' in English, Arabic and Urdu, all normalized with area of pixel map to get consistency.

The bar chart shows variation of values for the three pixel maps. Similarly, words or images can also be used to generate pixel map attributes as shown in Fig. 7 and typically, for a word like 'computer' translated in three languages—English, Hindi and Arabic are shown in Fig. 8.

Table 1 Pixel map attributes for letter 'a' in three languages

Input	Mean	Std.	Vector Attributes				2 × 2 matrix attributes		
Eng. letter 'a'	0.3818	0.2960	0.1000	0.2818	0	0.1091	0.1318	0.0591	0.0818
Arab letter 'a'	0.0889	0.1496	0.0159	0.0571	0.0159	0.0381	0.0063	0.0206	0.0238
Urdu letter 'a'	0.1414	0.0994	0.0074	0.0565	0.0774	0.0179	0.0179	0.0097	0.0960

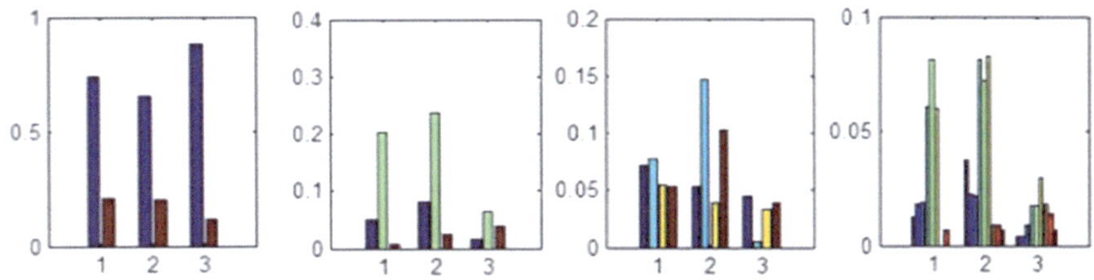

(a) Mean and std. (b) 3-value vector (c) 2x2 matrix (d) 3x3matrix

Fig. 8 Pixel map variations for word 'computer' translated in three languages

But, contents of the matrices are different and if processed in terms of either non-zero values—which gives the pattern, or vector matrix values with content being same. This gives a clear idea of feature extraction.

Since, Asian language letters have characters surrounding the main body; the pixel map may be divided into three segments like 25% top, 50% middle and 25% bottom. Letters 'g' and 'y' in English have bottom 25% for example. And in the case of Arabic fonts, most of them have occupancy in top and bottom halves also.

Even though a letter 'a' has these values in different scripts, its usage also differs as in English 'a' can be a letter or a word. So, processing of text and documents ultimately has to be considered as a problem related to the content and context and natural language understanding.

6 Content Extraction—Results and Discussion

The method described earlier is used with pattern recognition to compare whether any new input in the form of letter or word or image can relate to the content of base patterns. The proposed technique is purely data driven and does not make use of domain dependent background information, nor does it rely on predefined document categories or a given list of topics. Character 'a' which is unique in content, similar in many languages—Arabic, Hindi, Telugu, Tamizh and English. Uniqueness of letter 'a' is that, it has same meaning or content in all the above mentioned languages. But, this trend completely changes when one has a character like 'e' which is a vowel by itself like 'a' but it is not unique in any language. As an individual character 'e' doesn't give any meaning, unlike 'a' which gives some meaning in English and regional languages. The attribute variations of 'e' in comparison with 'a' are given in Fig. 9.

The purpose of choosing 'x' in English is that it is not unique in any language by any context, 'x' is not a vowel and as an individual character 'x' doesn't give any meaning, unlike 'a' which gives some meaning in English and regional languages. 'x' to be written in other languages it requires more than one character, which is

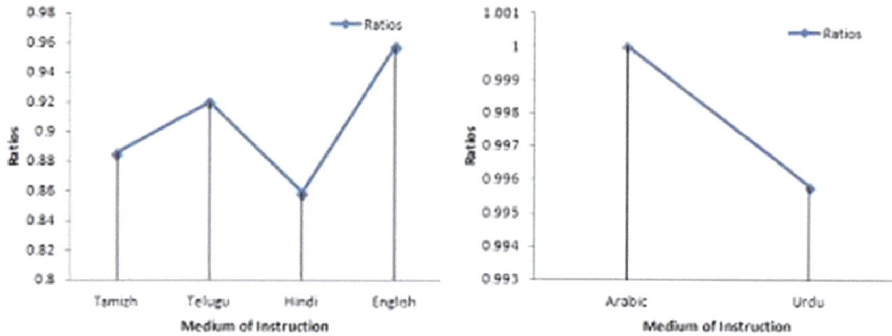

(a) Indian languages (b) Arabic and Urdu languages

Fig. 9 Content extraction for ‘a’ and ‘e’

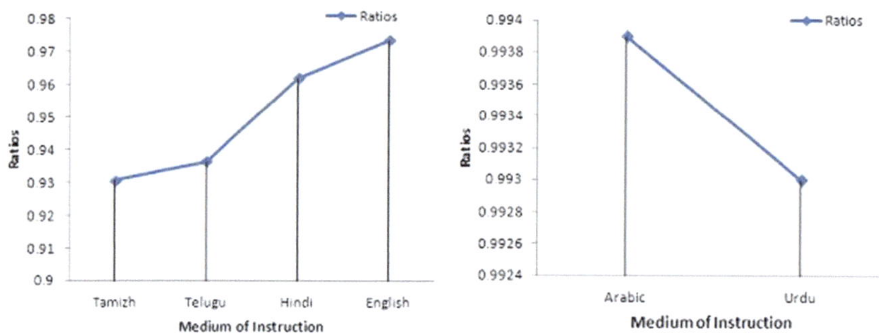

Fig. 10 Content extraction for ‘a’ and ‘x’

another interesting feature in comparison with ‘a’ and ‘e’. The attribute variations of ‘x’ in comparison with ‘a’ are given in Fig. 10.

Extending this basis to words, a typical data set of words in English relating to the same content ‘magnetism’ are chosen, which is considered as data set-2 and using pixel map attributes as basis, comparisons with a new data ‘magnet’ in Arabic, related to the content and ‘flower’ in Arabic not related to content are shown in Table 2 and Fig. 11.

One can see clearly that even though pixel map variations are significant, matching patterns can help in identifying the content.

Table 2 Pixel map attribute variations for data set-2

S. No	Pixel-map	Attribute value
1	Diamagnet	0.2344
2	Dipole	0.2326
3	Ferri	0.2249
4	Filings	0.2333
5	Moment	0.2316
6	Monopole	0.232
7	Magnet-arabic	0.2181
8	Flower-arabic	0.2154

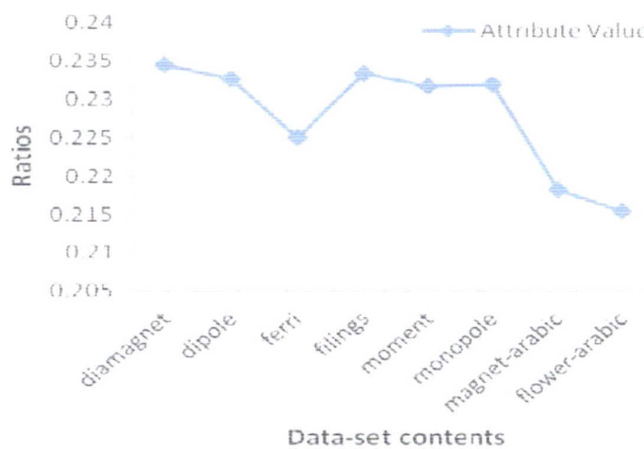

Fig. 11 Pixel map attribute variations for data set-2

7 Conclusion

A generic model for Content Extraction for regional web documents is developed based on the basic data system in computers, namely pixel maps. Beginning with complexities in letters, different methods of generating attributes are presented which form the basis for pattern matching and later for neural modeling. Some preliminary test results are given for pattern matching of features, for letter and word level relating to the same content. This preliminary study is focused to bring out the complexities in regional web documents and how a generic tool based on pixel maps—which do not have language or form of data as inputs—can be used for either text mining or content extraction. Further enhancements and techniques are to be suitably generated to account for the vagaries, so that, web content is extractable in any region.

References

1. T. Gottron, Content code blurring: A new approach to content extraction, DEXA '08: 19th International Workshop on Database and Expert Systems Applications. IEEE Computer Society, pp. 29–33 (2008)

2. S. Gupta, G. Kaiser, D. Neistadt, G. Grimm, in *DOM Based Content Extraction of HTML Documents*. WWW '03: Proceedings of the 12th International Conference on World Wide Web (ACM Press, New York, NY, USA, 2003), pp. 207–214
3. J. Moreno, K. Deschacht, M. Moens, in *Language Independent Content Extraction from Web Pages*. Proceeding of the 9th Dutch-Belgian Information Retrieval Workshop, pp. 50–55, 2009
4. D. Pinto, M. Branstein, R. Coleman, W.B. Croft, M. King, W. Li, X. Wei, in *QuASM: A System for Question Answering Using Semi-structured Data*. JCDL '02: Proceedings of the 2nd ACM/IEEE-CS Joint Conference on Digital libraries (ACM Press, New York, NY, USA, 2002), pp. 46–55
5. C. Mantratzis, M. Orgun, S. Cassidy, in *Separating XHTML Content from Navigation Clutter Using DOM-structure Block Analysis*. HYPERTEXT '05: Proceedings of the Sixteenth ACM Conference on Hypertext and Hypermedia (ACM Press, New York, NY, USA, 2005), pp. 145–147

Potentiality of Retina for Disease Diagnosis Through Retinal Image Processing Techniques

P. G. Prageeth, A. Sukesh Kumar, C. S. Sandeep and R. S. Jeena

1 Introduction

Human eye is one of the most important organs in the body. It is estimated that in every 5 seconds, one person goes blind somewhere in the world. There are several diseases of the eye which when properly diagnosed could save the sight of the patient [1]. According to the estimates of the World Health Organisation, about 80% of human blindness is avoidable. In spite of highly effective treatment, cure rates are unsatisfactorily low in most developing and developed countries. Hence there is great need for the effective implementation of modern technology and investigation into the field of eye care for a social cause.

Age-related Macular Degeneration (AMD) and Diabetic Retinopathy (DR) are the leading causes for preventable vision loss in the country [2]. In our earlier works, we have developed a neural network based tool from retinal images for the early detection of AMD and diabetic retinopathy. Also we developed expert system to detect these eye abnormalities earlier. We have done an extensive work for the early detection of the above eye diseases through retinal image analysis and processing [3]. It was convinced that retina is the vital part of the eye, is a potential one to provide vital information on the eye diseases and the extraction of this information through retinal images will definitely help to prevent the abnormalities of the eye [4, 5]. Based on the above works, currently research works are being pursued in the

P. G. Prageeth
Department of Electronics & Communication Engineering, College of Engineering Trivandrum, University of Kerala, Thiruvananthapuram, Kerala, India
e-mail: prageethpg@cet.ac.in

A. Sukesh Kumar (✉) · C. S. Sandeep · R. S. Jeena
Electronics & Communication Engineering, Faculty of Engineering & Technology, University of Kerala, Thiruvananthapuram, Kerala, India
e-mail: drsukeshkumara@cet.ac.in

S. Chattopadhyay et al. (eds.), *Modelling and Simulation in Science, Technology and Engineering Mathematics*, Advances in Intelligent Systems and Computing 749,
https://doi.org/10.1007/978-3-319-74808-5_59

area of retinal image analysis for the detection of most of the eye diseases which can be detected from retinal image processing.

2 Retina Based Eye Diseases and Its Detection

Currently we are concentrating on the early detection of the following eye diseases through retina image processing.

(1) Cataract

This section focuses on fundus image analysis and fully automatic cataract classification. Its goal is to reduce the burden of scarce resources and improve the effectiveness and efficiency of fundus image review, through which to enable active and enhanced healthcare services. Studies on fundus image analysis have been made for years. Segmentation and location of retinal structures, such as retinal lesions, vessels, optic disc and fovea have been widely studied. Based on these techniques, researchers are also trying to develop diagnose systems for specific retina-related diseases including micro aneurysms, diabetic retinopathy, age-related macular degeneration, glaucoma, cardiovascular diseases [6]. It has made an effort to classify and diagnose specific cataract automatically by split image and retro-illumination image, including nuclear cataract, cortical cataract and posterior sub-capsular cataract. However, there is little work reported on cataract classification and grading by using fundus images. Figure 1 shows the fundus images of non-cataract and cataract persons in different grading. In the image (a) without cataract, the blood vessels can be shown very clearly, even the capillary ones. The more severe cataract the patients have, the more cloud will be in the lens, resulting in that less vessels can be observed from the fundus image. There are less vessels details in mild cataract patients' eye fundus image, while only the trunk vessel and little details in the moderate cataract ones'.

Furthermore, there is hardly anything in the severe cataract ones.

(2) Glaucoma

Glaucoma is a chronic disease often called "silent thief of sight" as it has no symptoms and if not detected at an early stage it may cause permanent blindness. Glaucoma

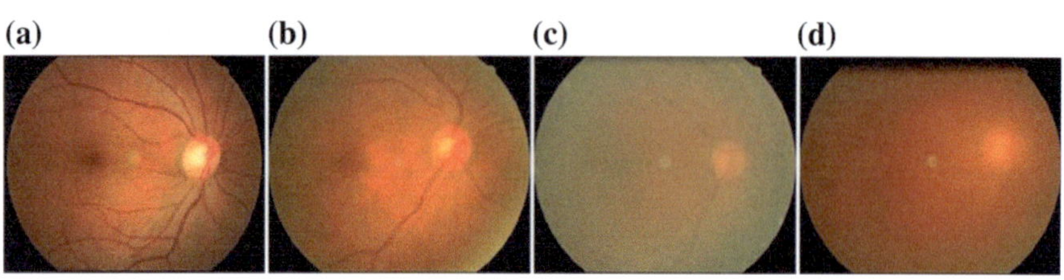

Fig. 1 Fundus images of non-cataract and cataract in different grading. **a** Non-cataract **b** mild **c** moderate and **d** severe

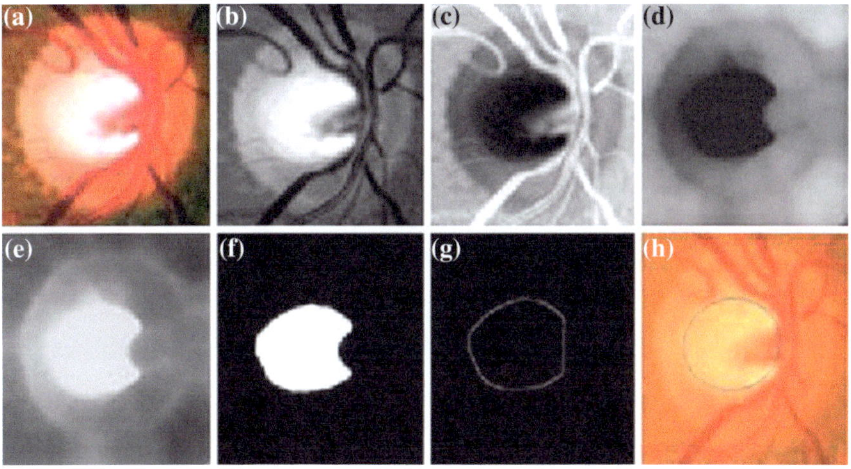

Fig. 2 **a** Contrast enhanced RGB colour space image **b** extracted green plane **c** negative transform **d** vessel removal using opening **e** negative transform **f** region growing **g** edge extraction of convex hull from **f** and **h** detected cup after ellipse fitting

progression precedes some structural changes in the retina which aid ophthalmologists to detect glaucoma at an early stage and stop its progression. Fundoscopy is among one of the biomedical imaging techniques to analyze the internal structure of retina. Our proposed technique provides a novel algorithm to detect glaucoma from digital fundus image using a hybrid feature set. This section proposes a novel combination of structural (cup to disc ratio) and non-structural (texture and intensity) features to improve the accuracy of automated diagnosis of glaucoma. The proposed method introduces a suspect class in automated diagnosis in case of any conflict in decision from structural and non-structural features [7]. Figure 2 shows the images of qualitative and quantitative evaluations of different parameters which help in the early detection of glaucoma. The evaluation of proposed algorithm is performed using a local database containing fundus images from 100 patients. This system is designed to refer glaucoma cases from rural areas to specialists and the motivation behind introducing suspect class is to ensure high sensitivity of proposed system. The average sensitivity and specificity of proposed system are 100 and 87% respectively.

(3) Diabetic Retinopathy

Diabetes is detected from the presence of exudates and haemorrhages and changes in blood vessel parameters like arteriolar-to-venular diameter ratio (AVR). Images obtained from fundus camera are enhanced using filtering. Image segmentation is done to detect optic disc, fovea, exudates area and blood vessels. Connected component method along with concentric circle methods are used to determine the artery-vein width ratio [8]. An algorithm is developed for the detection and quantification of the disease level from the parameters specified. The result is validated with the clinical data of the patient and achieved good results. A predictor system is developed to

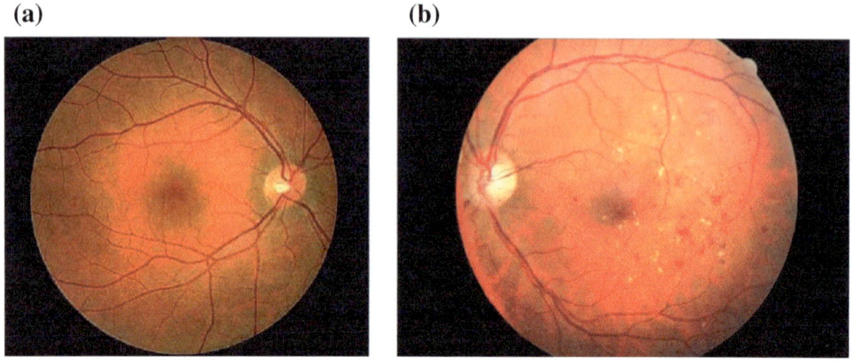

Fig. 3 **a** Fundus image of a normal eye. **b** Fundus image of a diabetic patient

give the status of the patient from the analysis of the retinal image parameters using neural network techniques [9]. Figure 3 shows the clear cut differences between the images of a normal person and a diabetic patient.

(4) Retinitis Pigmentosa

The paper has referenced retinitis pigmentosa for the analysis purpose. Automated approach for detection of micro aneurysms in digital colour retinal fundus photographs helps ophthalmologist to detect the emergence of its initial symptoms and determine the next immediate action step for the patient. A similar mechanism for automated early disease detection method is proposed featuring identification of dark pigments like minute features, exudate and micro aneurysm detection and these features extracted can prove to a greater extent as primary instances for defectiveness of eye [10]. A good number of images along with the response from the ophthalmologist has proved to be a great help towards the observation as derived from this mechanism and discussed in the paper. The proposed mechanism can be extended up to the limit of supervised learning so as to automate the practical responses as obtained from the ophthalmologist in real time scenario. Figure 4 shows the fundus image of a retinitis pigmentosa patient.

(5) Retinopathy of Prematurity

Retinal imaging with remote interpretation could decrease the number of diagnostic eye examinations that premature infants need for the detection of retinopathy of prematurity and thus decrease the time demand on the relatively small pool of ophthalmologists who perform retinopathy of prematurity examinations. Our goal was to review systematically the evidence regarding the reliability, validity, safety, costs and benefits of retinal imaging to screen infants who are at risk for retinopathy of prematurity. We searched Medline, the Cochrane library, CINAHL, and the bibliographies of all relevant articles [11]. All English-language studies regardless of design with primary data about our study questions were included. We excluded (1) studies that only included subjects with retinopathy of prematurity, (2) hypothetical models

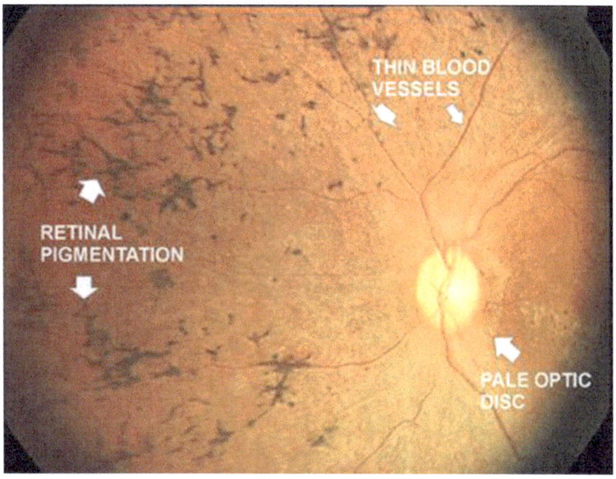

Fig. 4 Retinitis pigmentosa fundus image

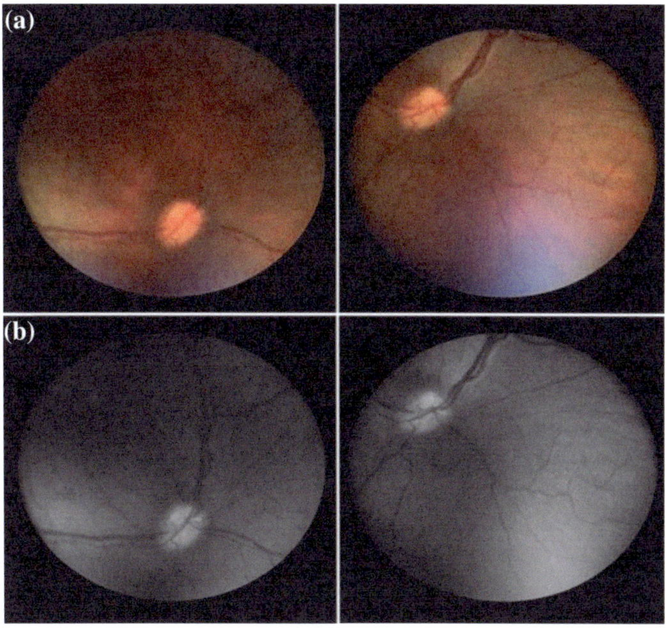

Fig. 5 Fundus images of evaluation of retinopathy of prematurity

other than cost-effectiveness studies and (3) validity studies without sufficient data to determine prevalence, sensitivity and specificity or that only evaluated subjects for 1 component of retinopathy of prematurity (eg: plus disease only). Figure 5 depicts the different evaluation windows of retinopathy of prematurity.

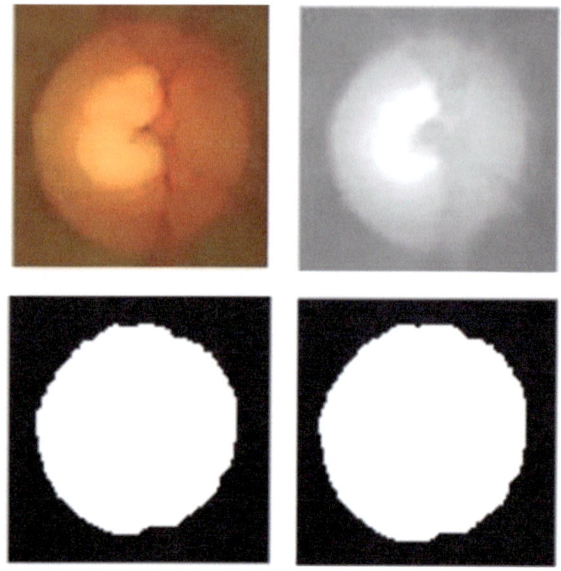

Fig. 6 Fundus images of retinal detachment patient

(6) Retinal Detachment

Arterial macro aneurysms are dilated places in the arteries of the retina, the lining of the back of the eye. Macro aneurysms are weak spots, which can leak clear fluid into the retina, causing gradually developing blurred vision. They also can pop, with bleeding inside the eye and sudden visual loss. There is no pain associated with macro aneurysms. Pictures of a normal retina and of a retina with a macro aneurysm are shown below. What Causes Arterial Macro aneurysms? The causes of macro aneurysms are unknown, but we know of certain associated risk factors [12]. Macro aneurysms tend to occur more commonly in women than men (3:1 ratio), occur late in life, and often occur in patients with high blood pressure and other forms of vascular disease, such as heart attacks and strokes. From these clues, we think that hormones, wear and tear over time, and extra stress from high blood pressure may contribute to macro aneurysms. How Are Macro aneurysms Discovered? Sometimes they can be found by your ophthalmologist simply looking inside the eye. Other times they may be covered by blood, and dye pictures may be taken to help in finding them. Occasionally, after cases of rupture, they become evident only after blood has reabsorbed. Figure 6 depicts the qualitative and quantitative detachments of retina in a patient.

(7) Stargardt's disease

Eye diseases are the burning diseases nowadays. Eye diseases detection is one of the imperative problems in computer vision. It has much relevance such as face live detection and driver fatigue analysis. In this paper first, the captured images are collected from different patients and are processed for enhancement. Figure 7 shows

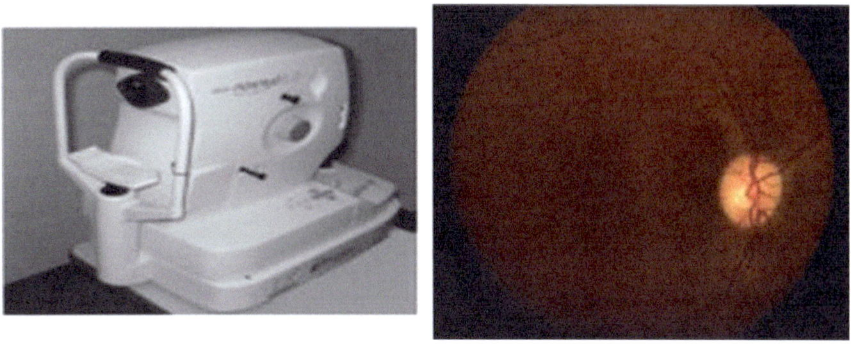

Fig. 7 Fundus camera and fundus image of a retina detached patient

a fundus camera and fundus image of a retina detached patient [13]. Then image segmentation is carried out to get target regions (disease spots). Finally, analysis of the target regions (disease spots) based on covariance approach to finding the phase of the disease and then the treatment consultative module can easily be prepared on the lookout for human being. The captured infected eye images are collected from different patients and are processed for enhancement. Using the covariance approach and scoring scale technique to exact intensity pattern to anterior disease accordingly it is then possible to analyze the different Eye diseases. Then image segmentation is carried out to get target regions (disease spots). Finally, analysis of the target regions (disease spots) based on covariance approach to finding the phase of the disease and then the treatment consultative module can easily be prepared on the lookout for human being [14]. The result from the preliminary study indicated that the proposed strategy is effective to assess disease.

(8) Cone Dystrophy

The results for each performance of the sampling bright lesion detection method is good even for lesion based evaluation, as the proposed hybrid microaneurysm detection method resulted in a very high sensitivity with reasonable specificity, an ophthalmologist can take its assistance in detecting Microaneurysms, exudates and cotton wool spot in the mass screening of diabetic retinopathy [15]. Figure 8 shows the images of feature extractions of cone dystrophy. It achieves a sensitivity of 94% and a specificity of 94.87% and accuracy of 95.38%. The performance of the microaneurysm detection method can be enhanced further by augmenting the amount of training data for the microaneurysm candidate object classification.

(9) Cancer in relation to the Retina

Choroidal Melanoma (CM) is the most common primary malignancy of the eye. The overall incidence is approximately 5–7 cases per million per year. In this paper the new technique for tumours tissue structure evaluation using ultrasound spectral analysis is presented. Based on the obtained results, it can be said that radio frequency (RF) ultrasound signals parameters at the healthy tissue area and the area with the

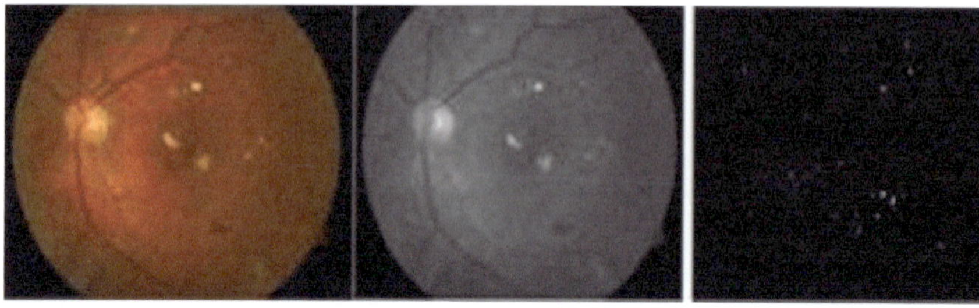

Fig. 8 Fundus images of feature extractions of cone dystrophy

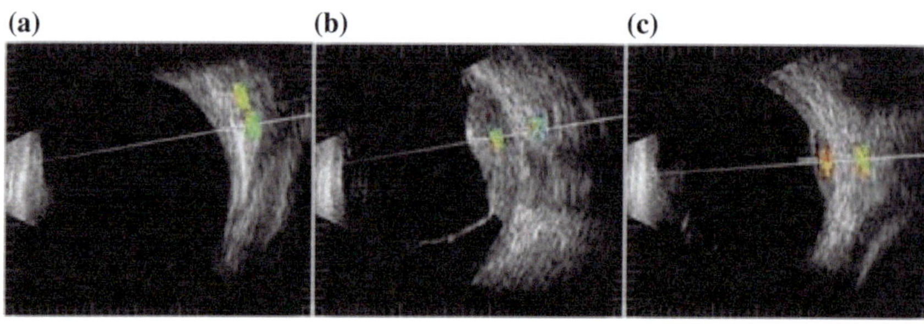

Fig. 9 ROI selection in one direction of B scan images **a** in healthy (control) eye **b** in melanoma eye before treatment **c** in melanoma after brachytherapy treatment

intraocular tumour—melanoma before and after treatment statistically are significantly different [16]. Application of spectral analysis using non-invasive ultrasound expert system, provides the new opportunities in early diagnosis, differentiation of tumours, evaluation of the treatment effectiveness.

This study has shown that the lower amplitude, lower spectral intercept, high spectral slope and high momentary bandwidth are typical for choroidal melanoma if compared with healthy tissues and the lower momentary bandwidth are typical for choroidal melanoma after treatment if compared with melanoma before treatment [17]. Figure 9 shows the ROI selection in one direction of B scan images (a) in healthy (control) eye (b) in melanoma eye before treatment (c) in melanoma after brachytherapy treatment.

3 Early Detection of Stroke Through Retinal Image Analysis

Stroke is a form of cardiovascular disease affecting the blood supply to the brain. It remains as a leading cause of disability and death for people of all races and ethnicities [18]. Stroke can be subdivided into two types: ischemic and haemorrhagic. Ischemic stroke accounts for almost 85% of the cases. The retina can be viewed and analysed using non-invasive, in-vivo functional techniques. The retina is a layered tissue lining the inner part of the eye that enables the conversion of incoming light into a neural signal that is appropriate for further processing in the brain. It is therefore an extension of human brain. Research works show that microvasculature of retina and brain is closely linked in terms of anatomy and physiology [19]. Morphological changes in blood vessel shape, branching pattern, width, tortuosity, appearance of retinal lesions, branching angle, branching coefficient and fractal dimension are some of the abnormalities in vascular pattern of retina associated with cardiovascular diseases like stroke.

The current research work focuses on the prediction of retinal ischemia from retinal fundus images and thereby predicting the occurrence of stroke. Pre-processing of retinal images is done by retinex processing [20] and morphological operations are done to remove noisy background. Branching points are detected and various features like major axis length, mean diameter, orientation, eccentricity, fractal dimension and tortuosity are computed. This has been compared with a set of healthy retinal images for the prediction of the possibility of retinal ischemia.

Figure 10 shows the various output stages of a healthy retinal fundus image. Retinal imaging aids in predicting the probability of stroke based parameters evaluated from the vascular map of retinal ischemia. Early detection of cardiovascular diseases like stoke through biomarkers derived from retinal imaging would allow patients to be treated more effectively. This work is an extension of author's other works for the prediction of stroke [21].

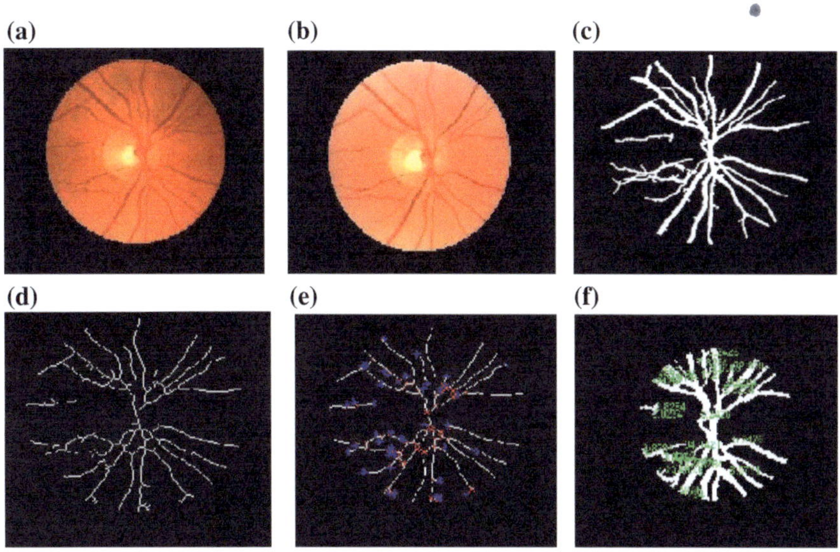

Fig. 10 Various output stages of a healthy retinal fundus image

4 Early Detection of Alzheimer's Disease Through Retinal Image Processing

Alzheimer's disease is an irreversible progressive neurodegenerative disorder, which means once it is affected cannot be cured. It is a memory and behavioural disturbance which leads to intense and eternal loss of cognitive impairment. The two most important causes of AD are plaques and tangles that on the neurons which blocks the signals from brain to neuron and vice versa. AD is the most common cause of dementia and its incidence is increasing worldwide associated with population ageing [22]. AD is characterised by progressive cognitive impairment such as memory deficit, decline in learning and executive functioning, aphasia, apraxia, agnosia and visual abnormalities [23–25]. There are lot of tests and imaging modalities to be performed for an effective diagnosis of the disease. Conventional clinical decision making systems are more manual in nature and ultimate conclusion in terms of exact diagnosis is remote. The American Academy of Neurology recommend that the clinical tests which are in connection with AD includes total blood count, electrolytes, calcium, glucose, liver function tests, thyroid function tests, sedimentation rate, urine analysis and imaging modalities. But today profiling of human body parameters (clinical test results) using computers can be utilized for the earlier prognosis of AD. But it is clinically established that all the changes taking place in brain neurons will be available from retina of the eye of the patients also [26]. Currently in our works we are including the retina image results also as a biomarker. Retina image analysis for the early detection of AD is currently going on. More details on this will be presented in the conference.

5 Conclusions

With the help of retinal image processing, eye diseases can be diagnosed well in advance. An expert system for the early detection of the above eye diseases has already been developed and it is in clinical use in local ophthalmic hospitals. Retinal image can be utilised for the prediction of stroke also. Currently, developed a system for the prediction of retinal ischemia from retinal fundus images for the prediction of ischemic stroke from the global databases. Clinical trials are now going on. The profiling of human body parameters (clinical test results) using computers for the early detection of AD is currently done. Now the retinal image parameters are also included in the profile for further studies.

Acknowledgements We are very thankful to Prof. (Dr). Mahadevan, a leading ophthalmology surgeon, ophthalmology teacher and a serious and sincere researcher, who provide all the helps to carry out the above research works successfully. Also extend our heartfelt thanks to (1) Ahalia Hospital (2) Precise Eye Hospital and (3) Gokulam Medical College, Thiruvananthapuram for being given all the facilities to carry out the above research work under the supervision of Prof. (Dr). Mahadevan. K.

References

1. S. Wild, G. Roglic, A. Green, R. Sicree, H. King, Global prevalence of diabetes estimates for the year 2000 and projections for 2030. Diabetes Care **27**(5), 1047–1053 (2004)
2. D.A. Quillen, *Common causes of vision loss in elderly patients*. Am. Fam. Physician **60**(1), 99–108 (1999)
3. R. Klein, B.E.K. Klein, S.E. Moss, Visual impairment in diabetes. Ophthalmology, **91**(1), 1–9 (1984)
4. O. Sheeba, *Retinal Image Analysis and Development of Neural Model for the Diagnosis of Eye Diseases*, PhD Thesis, University of Kerala, Thiruvananthapuram, Kerala, INDIA, Guided by Prof. (Dr). Sukesh Kumar. A, 2010
5. J. David, *Medical Expert System using Retinal Image Parameter Analysis*, PhD Thesis, University of Kerala, Thiruvananthapuram, Kerala, INDIA, Guided by Prof. (Dr). Sukesh Kumar. A, 20
6. M.D. Saleh, C. Eswaran, An automated decision support system for non-proliferative diabetic retinopathy disease based on mas and has detection. Comput. Methods Programs Biomed., **108**(1), 186–196 (2012)
7. A. Kumar, A.K. Gaur, M. Srivastava, A segment based technique for detecting exudate from retinal fundus image. Procedia Techno. **6**, 1–9 (2012)
8. M.M. Fraz, S.A. Barman, P. Remagnino, A. Hoppe, A. Basit, B. Uyyanonvara, A.R. Rudnicka, C.G. Owen, An approach to localize the retinal blood vessels using bit planes and centre line detection. Comput. Methods Programs Biomed. **108**(2), 600–616 (2012)
9. L. Xu, S. Luo, A novel method for blood vessel detection from retinal images. Biomed. Eng. Online **9**(1), 14 (2010)
10. M. Foracchia, E. Grisan, A. Ruggeri, Luminosity and contrast normalization in retinal images. Med. Image Anal. **9**(3), 179–190 (2005)
11. M. Vlachos, E. Dermatas, Multi-scale retinal vessel segmentation using line tracking. Comput. Med. Imaging Graph. **34**(3), 213–227 (2010)
12. K. Narasimhan, V.C. Neha, K. Vijayarekha, in *An Efficient Automated System for Detection Of Diabetic Retinopathy from Fundus Images Using Support Vector Machine and Bayesian Classifiers in Computing*. Electronics and Electrical Technologies (ICCEET), International Conference on IEEE, 964–969 (2012)
13. K.A. Goatman, A.D. Fleming, S. Philip, G.J. Williams, J.A. Olson, P.F. Sharp, Detection of new vessels on the optic disc using retinal photographs. IEEE Transac. Med. Imaging **30**(4) 972–979 (2011)
14. M. Tamilarasi, K. Duraiswamy, in *Genetic Based Fuzzy Seeded Region Growing Segmentation for Diabetic Retinopathy Images*. Computer Communication and Informatics (ICCCI), International Conference on IEEE, pp. 1–5, 2013
15. B. Antal, A. Hajdu, An ensemble-based system for micro aneurysm detection and diabetic retinopathy grading. IEEE Transac. Biomed. Eng. **59**(6), 1720–1726 (2012)
16. D. Marín, A. Aquino, M.E. Gegúndez-Arias, J.M. Bravo, A new supervised method for blood vessel segmentation in retinal images by using gray-level and moment invariants-based features. Med. Imaging, IEEE Transac. **30**(1), 146–158 (2011)
17. A. Anzalone, F. Bizzarri, M. Parodi, M. Storace, A modular supervised algorithm for vessel segmentation in red-free retinal images. Comput. Biol. Med. **38**(8), 913–922 (2008)
18. Barry. L. Zaret MD, Marvin Moser MD., Lawrence. S. Cohen, Chapter 18, "*Stroke - Lawrence M. Brass, MD*", pp. 215–234
19. M.L. Baker, J.J. Wang, G. Liew, et al., Differential associations of cortical and subcortical cerebral atrophy with retinal vascular signs in patients with acute stroke. Stroke **41**(10), 2143–2150 (2010)
20. D.J. Jobson, Z. Rahman, G.A. Woodell, A multiscale retinex for bridging the gap between color images and the human observation of scenes. IEEE Transac. Image Process. **6**(7), 965–976, July (1997)

21. R.S. Jeena, A. Sukesh Kumar, in *Artificial Neural Networks in Stroke Prediction*. International Conference on Innovative Systems, Bangalore, December 2016
22. K. Ohno-Matsu, Parallel findings in age-related macular degeneration and Alzheimer's disease. Prog. Retin Eye Res. **30**, 217–238 (2011)
23. C.S. Sandeep, A. Sukesh Kumar, A review on the early diagnosis of Alzheimer's Disease (AD) through different tests, techniques and databases. AMSE JOURNALS 2015 Series: Modelling C **76**(1), 1–22, February (2015)
24. C.S. Sandeep, A. Sukesh Kumar, M.J. Susanth, The online datasets used to classify the different stages for the early diagnosis of Alzheimer's Disease (AD). Int. J. Eng. Adv. Technol., ISSN: 2249–8958, **6**(4), 38–45, April (2017)
25. C.S. Sandeep, A. Sukesh Kumar, M.J. Susanth, The early diagnosis of Alzheimer's Disease (AD) using CAMD, TREAD and NAAC databases. Int. J. Sci. Adv. Res. Technol. (IJSART), ISSN Online 2359–1052, **3**(3), 366–371, March (2017)
26. D.A. Valenti, in *Anterior Visual System and circadian function with reference to Alzheimer's disease*, ed. by A. Cronin-Golomb, P.R. Hof. Vision in Alzheimer's Disease—Interdisciplinary Topics in Gerontology, Vol. 34, pp. 1–29, 2004

Generalized LFT Modeling of an Uncertain MIMO System

Tamal Roy, Ranjit Kumar Barai and Rajeeb Dey

1 Introduction

In the past few years, a growing interest has been devoted to formulate control oriented modeling of real physical system from the inherent need for the modeling quality improvement and truly integrates control objectives into the system identification process from the experimental input-output test data set [1] from the standpoint of control system design [2], where the role of the system identification is to condense the plant uncertainty such that the design and implementation of a robust controller achieves the performance specifications even in the face of the plant uncertainty and disturbances. Over the last two decades, there has been a widespread interest to design robust controller where system model is considered to be consisting of a nominal model and a model uncertainty part [3]. The classical H_{∞}-control based robust control design technique has become a challenging task and very effective design tools guaranteeing to meet the specifications provided the model of the system under consideration leads to a system model in the form of a linear fractional transformation (LFT) modeling. The performance of the robust controller depends on the appropriate representation of the model uncertainty. This motivates the robust-control-oriented system identification to explicitly consider the robust control performance require-

T. Roy (✉)
Electrical Engineering Department, MCKV Institute of Engineering,
Liluah, Howrah 711204, India
e-mail: tamalroy77@gmail.com

R. K. Barai
Electrical Engineering Department, Jadavpur University,
Kolkata 700032, India
e-mail: ranjit.k.barai@gmail.com

R. Dey
Electrical Engineering Department, NIT, Silchar 788 010, Assam, India
e-mail: rajeeb.iitkgp@gmail.com

S. Chattopadhyay et al. (eds.), *Modelling and Simulation in Science, Technology and Engineering Mathematics*, Advances in Intelligent Systems and Computing 749,
https://doi.org/10.1007/978-3-319-74808-5_60

ments during the system identification step. The model of nonlinear systems may vary due to changes in system configuration and operating conditions. This system variation can be characterized as model uncertainties and can be represented in the linearized model by expressing the system state-matrices as matrix polynomials in the uncertain parameters in the form of a linear fractional transform (LFT) [3, 4]. This LFT based model uncertainty representation of the nonlinear system is essential for the application of modern robust control technique like μ-analysis and synthesis [5] in addition to H_∞-control and H_∞-Loop Shaping [4, 6–8]. Linearization of uncertain nonlinear systems as LFT model relates each of the uncertainty with a physically meaningful parameter of the actual system [9]. Linear Fractional Transform (LFT) technique offers a unified framework for parameter identification problems in [10]. In the LFT framework, a wide variety of identification problems concerning structured nonlinear systems, linear parameter varying (LPV) systems, and also the various parametric linear system model structures can be accommodated due to its general nature. This paper presents an uncertainty modeling algorithm of a generic linear multi-input multi-output system with coupled dynamics in LFT framework for implementing the classical H_∞-control law. During the formulation of the modeling algorithm, the effect of model uncertainty has been explicitly described by a possible mismatch between the mathematical model and the real physical system, the presence of disturbance signal and the possible model order reduction.

The essential contribution for the derivation of LFT model of a coupled dynamic system has been developing a comprehensive model consisting of the nominal system model and an unknown transfer function matrix consisting various uncertainties introduced due to unmolded dynamics, system parameters changes due to environmental variation, the presence of disturbance signal, model order reduction etc. Different uncertainty modeling technique in LFT framework has been reported in the literature. A generalized descripted type LFT-based modeling approach consisting rationally dependent parametric matrices in terms of multi-variable functions has been discussed in [11]. An uncertainty modeling formulation of nonlinear systems whose models parameters vary due to change in the system configuration and operating conditions have been represented in LFT framework in [4]. A symbolic LFT modeling techniques for nonlinear systems has been presented by combining symbolic modeling and LFT technique in [8]. The best LFT uncertainty model has been proposed by minimizing the H_∞ norm of the uncertainty set with respect to a nominal model known as input- output data [12]. However, LFT modeling technique is only applicable for those nonlinear systems where linearization of the mathematical modeling is possible. In this paper, a novel methodology has been developed to formulate generalized LFT modeling of a multi-variable dynamic system with equal number of input output consisting a comprehensive nominal model and model uncertainties expressed by an unknown transfer function matrix, accumulating usual dynamics of the system represents in a form that required in μ-synthesis-based H_∞ controller design technique. To the best of the knowledge of the authors, such compact and effective uncertainty modeling approach in generic LFT modeling framework, compatible with H_∞ controller design for the linear multi-dimensional system has never been addressed in the literature.

The effectiveness of proposed generalized control oriented LFT modeling algorithm for the linear multivariable system has been verified on a Two-DOF mass-spring-dashpot dynamic system and the recommended LFT structure has been validated in the frequency domain in the context of H_∞ based robust control design. The LFT modeling structure of the Two-DOF mass-spring-dashpot dynamic system achieves satisfactory performance criterion in μ-synthesis based frequency domain validation.

2 Generalized LFT Modeling of Linear MIMO System

Most of the modern real physical systems result from the synergetic integration of different subsystems. The overall system is very complicated due to the cross coupling of the various subsystems and any attempt to derive a generalized mathematical model for such a highly coupled system results into a very big and complex and the system models vary due to change in the system configuration and operating conditions. The generalized LFT modeling of such linear coupled multi-variable system results in a compact and manageable modeling algorithm is suitable for its implementation in robust control theory.

2.1 *Problem Formulation*

A linear coupled dynamic multi-variable system consisting of an equal number of input and output is considered for formulating the control oriented modeling in LFT framework.

The generalized model of the coupled dynamic MIMO system is considered as

$$D\ddot{\mathbf{y}} + C\dot{\mathbf{y}} + E\mathbf{y} + \mathbf{z} = K\mathbf{u} \tag{1}$$

where, input, disturbance and output vectors $\mathbf{u} = \begin{bmatrix} u_1 \\ u_2 \\ \vdots \\ u_m \end{bmatrix}_{m\times 1}$, $\mathbf{z} = \begin{bmatrix} z_1 \\ z_2 \\ \vdots \\ z_p \end{bmatrix}_{p\times 1}$ and $\mathbf{y} = \begin{bmatrix} y_1 \\ y_2 \\ \vdots \\ y_p \end{bmatrix}_{p\times 1}$ respectively with $m = p$ and the associated system parameters are

$$D = \begin{bmatrix} d_{11} & d_{12} & \cdots & d_{1p} \\ d_{21} & d_{22} & \cdots & d_{2p} \\ \vdots & \vdots & \ddots & \vdots \\ d_{p1} & d_{p2} & \cdots & d_{pp} \end{bmatrix}_{p\times p}, \quad C = \begin{bmatrix} c_{11} & c_{12} & \cdots & c_{1p} \\ c_{21} & c_{22} & \cdots & c_{2p} \\ \vdots & \vdots & \ddots & \vdots \\ c_{p1} & c_{p2} & \cdots & c_{pp} \end{bmatrix}_{p\times p},$$

$$E = \begin{bmatrix} e_{11} & e_{12} & \cdots & e_{1p} \\ e_{21} & e_{22} & \cdots & e_{2p} \\ \vdots & \vdots & \ddots & \vdots \\ e_{p1} & e_{p2} & \cdots & e_{pp} \end{bmatrix}_{p\times p}, \quad K = \begin{bmatrix} k_{11} & k_{12} & \cdots & k_{1m} \\ k_{21} & k_{22} & \cdots & k_{2m} \\ \vdots & \vdots & \ddots & \vdots \\ k_{m1} & k_{m2} & \cdots & k_{mm} \end{bmatrix}_{m\times m}$$

In real situations, system parameter values of the above matrices are not known exactly and it is assumed to be varying within certain known intervals.

2.2 LFT Modeling Algorithm for Multiplicative Uncertainty Structure

This section represents the derivation of the proposed LFT modeling algorithm for a linear coupled dynamic MIMO system in a systematic and generalized manner. The linear dynamic MIMO system expressed in (1) is considered for representing uncertainty modeling in LFT framework. The parametric uncertainties of the system are expressed in multiplicative uncertainty structure representation, which is further decomposed appropriately to get an LFT structure. The generalized framework for obtaining LFT structure of system represented in (1) for the case where $m = p$. The details modeling algorithm is described as following:

Step 1: The linear dynamic system express by Eq. (1) can be written as

$$\ddot{\mathbf{y}} = -D^{-1}C\dot{\mathbf{y}} - D^{-1}E\mathbf{y} - D^{-1}\mathbf{z} + D^{-1}K\mathbf{u} \tag{2}$$

The Eq. (2) survives provided D^{-1} exists, the block diagram representation of the system described by Eq. (2) is shown in Fig. 1.

Step 2: The uncertain matrices D, C and E characterizes the model uncertainties leads to the variation in the system parameters. Let us assumed for generic system the uncertain elements of the D matrix placed diagonally can be expressed as

$$d_{ii} = \bar{d}_{ii}(1 + s_{d_{ii}}\delta_{d_{ii}}) \quad \text{where } i = 1, 2, \ldots, p \tag{3}$$

where $\bar{d}_{ii}$ is the nominal values of the system parameter, $s_{d_{ii}}$ is the corresponding maximum relative parameter uncertainty and lies within a bound $-1 \le \delta_{d_{ii}} \le 1$. Matrix D can be decomposed and the terms containing nominal and uncertain parts given as

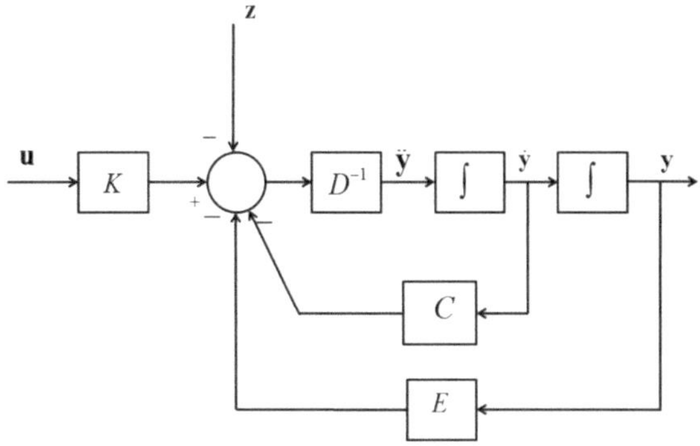

Fig. 1 Block diagram representation of the linear multidimensional system

$$D = \overline{D} + D_r \Delta_d \tag{4}$$

where

$$\overline{D} = \begin{bmatrix} \bar{d}_{11} & \bar{d}_{12} & \cdots & \bar{d}_{1p} \\ \bar{d}_{21} & \bar{d}_{22} & \cdots & \bar{d}_{2p} \\ \vdots & \vdots & \ddots & \vdots \\ \bar{d}_{p1} & \bar{d}_{p2} & \cdots & \bar{d}_{pp} \end{bmatrix}_{p\times p}, \quad D_r = \begin{bmatrix} \bar{d}_{11}s_{d_{11}} & 0 & \cdots & 0 \\ 0 & \bar{d}_{22}s_{d_{22}} & \cdots & 0 \\ \vdots & \vdots & \ddots & \vdots \\ 0 & 0 & \cdots & \bar{d}_{pp}s_{d_{pp}} \end{bmatrix}_{p\times p}$$

$$\text{and } \Delta_d = \begin{bmatrix} \delta_{d_{11}} & 0 & \cdots & 0 \\ 0 & \delta_{d_{22}} & \cdots & 0 \\ \vdots & \vdots & \ddots & \vdots \\ 0 & 0 & \cdots & \delta_{d_{pp}} \end{bmatrix}_{p\times p}$$

Now the term D^{-1} can be written by using matrix inversion lemma

$$D^{-1} = (D_r^{-1}\overline{D} + \Delta_d)^{-1} D_r^{-1} \tag{5}$$

An upper LFT representation of D^{-1} can be written as

$$D^{-1} = F_U(Q_d, \Delta_d) = Q_{d_{22}} + Q_{d_{21}}\Delta_d(I_p - Q_{d_{11}}\Delta_d)^{-1}Q_{d_{12}} \tag{6}$$

where, $Q_{d_{11}} = -\overline{D}^{-1}D_r$, $Q_{d_{12}} = \overline{D}^{-1}$, $Q_{d_{21}} = -\overline{D}^{-1}D_r$ and $Q_{d_{22}} = \overline{D}^{-1}$

The block partition matrix Q_d is express as

$$Q_d = \left[\begin{array}{c|c} Q_{d_{11}} & Q_{d_{12}} \\ \hline Q_{d_{21}} & Q_{d_{22}} \end{array}\right] = \left[\begin{array}{c|c} -\overline{D}^{-1}D_r & \overline{D}^{-1} \\ \hline -\overline{D}^{-1}D_r & \overline{D}^{-1} \end{array}\right]_{2p\times 2p}$$

Now treating respective uncertain elements of matrix 'C' in (2) is express in terms of the uncertain parametric representation can be described by

$$c_{ii} = \bar{c}_{ii}(1 + s_{c_{ii}}\delta_{c_{ii}}) \quad \text{where } i = 1, 2, \ldots, p \tag{7}$$

where $\bar{c}_{ii}$ is the nominal values of the system parameter, $s_{c_{ii}}$ is the corresponding maximum relative parameter uncertainty and lies within a bound $-1 \leq \delta_{c_{ii}} \leq 1$.

Similarly, the matrix C can be decomposed as

$$C = \overline{C} + \Delta C \tag{8}$$

The uncertainty matrix ΔC is decomposed as

$$\Delta C = C_f \Delta_c C_g \tag{9}$$

where Δ_c is a diagonal uncertainty matrix. The elements of the matrix C_g (may be identity matrix or unit matrix) depends on the position of uncertain parameters in the matrix ΔC. In generalized approach, it is assumed that the uncertain parameters in ΔC is located diagonally than the matrices C_f and C_g can be expressed as

$$C_f = \begin{bmatrix} \bar{c}_{11}s_{c_{11}} & 0 & \cdots & 0 \\ 0 & \bar{c}_{22}s_{c_{22}} & \cdots & 0 \\ \vdots & \vdots & \ddots & \vdots \\ 0 & 0 & \cdots & \bar{c}_{pp}s_{c_{pp}} \end{bmatrix}_{p\times p}, \quad \Delta_c = \begin{bmatrix} \delta_{c_{11}} & 0 & \cdots & 0 \\ 0 & \delta_{c_{22}} & \cdots & 0 \\ \vdots & \vdots & \ddots & \vdots \\ 0 & 0 & \cdots & \delta_{c_{pp}} \end{bmatrix}_{p\times p}$$

and $C_g = \begin{bmatrix} 1 & 0 & \cdots & 0 \\ 0 & 1 & \cdots & 0 \\ \vdots & \vdots & \ddots & \vdots \\ 0 & 0 & \cdots & 1 \end{bmatrix}_{p\times p}$

and the nominal part $\overline{C}$ is given as

$$\overline{C} = \begin{bmatrix} \bar{c}_{11} & \bar{c}_{12} & \cdots & \bar{c}_{1p} \\ \bar{c}_{21} & \bar{c}_{22} & \cdots & \bar{c}_{2p} \\ \vdots & \vdots & \ddots & \vdots \\ \bar{c}_{p1} & \bar{c}_{p2} & \cdots & \bar{c}_{pp} \end{bmatrix}_{p\times p}$$

Upper LFT representation of matrix C can be express as

$$C = F_U(Q_c, \Delta_c) = Q_{c_{22}} + Q_{c_{21}}\Delta_c(I_p - Q_{c_{11}}\Delta_c)^{-1}Q_{c_{12}} \tag{10}$$

Fig. 2 Block diagram representation of the system with LFT structure

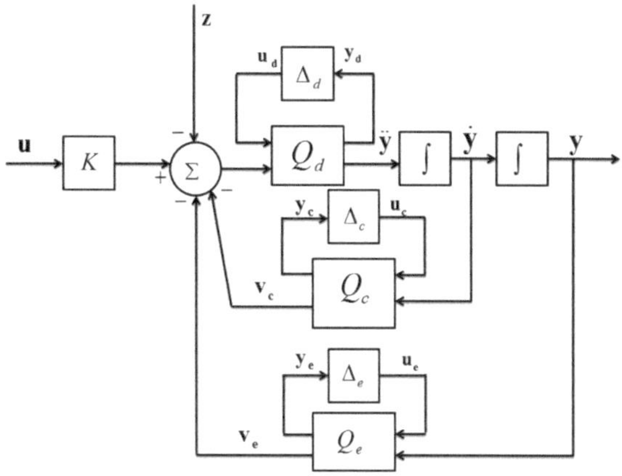

where, $Q_c = \left[\begin{array}{c:c} Q_{c_{11}} & Q_{c_{12}} \\ \hdashline Q_{c_{21}} & Q_{c_{22}} \end{array}\right] = \left[\begin{array}{c:c} 0_{p\times p} & C_g \\ \hdashline C_f & \bar{C} \end{array}\right]_{2p\times 2p}$

The similar problem of uncertainty representation as elements of the E can express as

$$e_{ii} = \bar{e}_{ii}(1 + s_{e_{ii}}\delta_{e_{ii}}) \quad \text{where } i = 1, 2, \ldots, p \tag{11}$$

The upper LFT representation of matrix can be express as

$$E = F_U(Q_e, \Delta_e) = Q_{e_{22}} + Q_{e_{21}}\Delta_e(I_p - Q_{e_{11}}\Delta_e)^{-1}Q_{e_{12}} \tag{12}$$

where $Q_e = \left[\begin{array}{c:c} Q_{e_{11}} & Q_{e_{12}} \\ \hdashline Q_{e_{21}} & Q_{e_{22}} \end{array}\right] = \left[\begin{array}{c:c} 0_{p\times p} & E_g \\ \hdashline E_f & \bar{E} \end{array}\right]_{2p\times 2p}$

Step 3: The block diagram representation of the linear multi-dimensional system is redrawn in Fig. 2 treating u_d, u_c and u_e to be the outputs of uncertainty blocks Δ_d, Δ_c and Δ_e are fed to the nominal blocks Q_d, Q_c and Q_e respectively. Similarly y_d, y_c and y_e, outputs of the nominal block Q_d, Q_c and Q_e are fed back to the uncertainty blocks Δ_d, Δ_c and Δ_e respectively.

Now consider the state vector of the system as

$$\begin{bmatrix} x_1 \\ x_2 \\ \vdots \\ x_p \end{bmatrix} \triangleq \begin{bmatrix} \mathbf{y_1} \\ \mathbf{y_2} \\ \vdots \\ \mathbf{y_p} \end{bmatrix} \text{ and } \begin{bmatrix} x_{p+1} \\ x_{p+2} \\ \vdots \\ x_{p+p} \end{bmatrix} = \begin{bmatrix} \dot{\mathbf{y}}_1 \\ \dot{\mathbf{y}}_2 \\ \vdots \\ \dot{\mathbf{y}}_p \end{bmatrix} \tag{13}$$

The generalized upper LFT representation of the linear dynamic MIMO system considering input output of all block partition matrices Q_d, Q_c and Q_e can be is represented as

$$\begin{bmatrix} \dot{\mathbf{x}}_1 \\ \dot{\mathbf{x}}_2 \\ \vdots \\ \dot{\mathbf{x}}_p \\ \hdashline \dot{\mathbf{x}}_{p+1} \\ \dot{\mathbf{x}}_{p+2} \\ \vdots \\ \dot{\mathbf{x}}_{p+p} \\ \hdashline \mathbf{y}_d \\ \mathbf{y}_c \\ \mathbf{y}_e \\ \hdashline \mathbf{y} \\ \mathbf{y}_n \end{bmatrix} = \Pi \begin{bmatrix} \mathbf{x}_1 \\ \mathbf{x}_2 \\ \vdots \\ \mathbf{x}_p \\ \hdashline \mathbf{x}_{p+1} \\ \mathbf{x}_{p+2} \\ \vdots \\ \mathbf{x}_{p+p} \\ \hdashline \mathbf{u}_d \\ \mathbf{u}_c \\ \mathbf{u}_e \\ \hdashline \mathbf{z} \\ \mathbf{u} \end{bmatrix} \tag{14}$$

where, $\Pi = \begin{bmatrix} 0_{p\times p} & I_{p\times p} & 0_{p\times p} & 0_{p\times p} & 0_{p\times p} & 0_{p\times p} & 0_{p\times p} \\ -\overline{D}^{-1}\overline{E} & \overline{D}^{-1}\overline{C} & \overline{D}^{-1}D_r & \overline{D}^{-1}C_f & -\overline{D}^{-1}E_f & -\overline{D}^{-1} & \overline{D}^{-1}K \\ \hline -\overline{D}^{-1}\overline{E} & \overline{D}^{-1}\overline{C} & \overline{D}^{-1}D_r & \overline{D}^{-1}C_f & -\overline{D}^{-1}E_f & -\overline{D}^{-1} & \overline{D}^{-1}K \\ 0_{p\times p} & C_g & 0_{p\times p} & 0_{p\times p} & 0_{p\times p} & 0_{p\times p} & 0_{p\times p} \\ E_g & 0_{p\times p} & 0_{p\times p} & 0_{p\times p} & 0_{p\times p} & 0_{p\times p} & 0_{p\times p} \\ \hline I_{p\times p} & 0_{p\times p} & 0_{p\times p} & 0_{p\times p} & 0_{p\times p} & 0_{p\times p} & 0_{p\times p} \\ \Gamma_n & 0_{p\times p} & 0_{p\times p} & 0_{p\times p} & 0_{p\times p} & 0_{p\times p} & 0_{p\times p} \end{bmatrix}_{7p\times 7p}$

The input output representation of the uncertainty matrix can be expressed as

$$\begin{bmatrix} \mathbf{u_d} \\ \mathbf{u_c} \\ \mathbf{u_e} \end{bmatrix} = \Delta_{sys} \begin{bmatrix} \mathbf{y_d} \\ \mathbf{y_c} \\ \mathbf{y_e} \end{bmatrix} \tag{15}$$

where, $\Delta_{sys} = \begin{bmatrix} \Delta_d & 0 & 0 \\ 0 & \Delta_c & 0 \\ 0 & 0 & \Delta_e \end{bmatrix}_{3p\times 3p}$

The state space representation of the linear multivariable system is expressed as

$$\begin{bmatrix} \dot{\mathbf{x}} \\ \mathbf{y_d} \\ \mathbf{y_c} \\ \mathbf{y_e} \\ \mathbf{y} \\ \mathbf{y_n} \end{bmatrix} = G_{sys} \begin{bmatrix} \mathbf{x} \\ \mathbf{u_d} \\ \mathbf{u_c} \\ \mathbf{u_e} \\ \mathbf{z} \\ \mathbf{u} \end{bmatrix} \tag{16}$$

where, $G_{sys} = \left[\begin{array}{c:cc} A & B_1 & B_2 \\ \hdashline C_1 & D_{11} & D_{12} \\ C_2 & D_{21} & D_{22} \end{array}\right]_{7p\times 7p}$

Now,

$$A = \begin{bmatrix} 0_{p\times p} & I_{p\times p} \\ -\overline{D}^{-1}E & -\overline{D}^{-1}\overline{C} \end{bmatrix}_{2p\times 2p}, \quad B_1 = \begin{bmatrix} 0_{p\times p} & 0_{p\times p} & 0_{p\times p} \\ -\overline{D}^{-1}D_r & -\overline{D}^{-1}C_f & -\overline{D}^{-1}E_f \end{bmatrix}_{2p\times 3p},$$

$$B_2 = \begin{bmatrix} 0_{p\times p} & 0_{p\times p} \\ \overline{D}^{-1}T & -\overline{D}^{-1}K \end{bmatrix}_{2p\times 2p}, \quad C_1 = \begin{bmatrix} -\overline{D}^{-1}E & -\overline{D}^{-1}\overline{C} \\ 0_{p\times p} & C_g \\ E_g & 0_{p\times p} \end{bmatrix}_{3p\times 2p},$$

$$D_{11} = \begin{bmatrix} -\overline{D}^{-1}D_r & -\overline{D}^{-1}C_f & -\overline{D}^{-1}E_f \\ 0_{p\times p} & 0_{p\times p} & 0_{p\times p} \\ 0_{p\times p} & 0_{p\times p} & 0_{p\times p} \end{bmatrix}_{3p\times 3p}, \quad D_{12} = \begin{bmatrix} \overline{D}^{-1}T & -\overline{D}^{-1}K \\ 0_{p\times p} & 0_{p\times p} \\ 0_{p\times p} & 0_{p\times p} \end{bmatrix}_{3p\times 2p},$$

$$C_2 = \begin{bmatrix} \Gamma_n & 0_{p\times p} \\ I_{p\times p} & 0_{p\times p} \end{bmatrix}_{2p\times 2p} \quad D_{21} = \begin{bmatrix} 0_{p\times p} & 0_{p\times p} \\ 0_{p\times p} & 0_{p\times p} \end{bmatrix}_{2p\times 2p},$$

$$D_{22} = \begin{bmatrix} 0_{p\times p} & 0_{p\times p} \\ 0_{p\times p} & 0_{p\times p} \end{bmatrix}_{2p\times 2p}$$

The generalized perturbed linear dynamic MIMO system in Upper LFT (Fig. 3) framework can be described by

$$\begin{bmatrix} \mathbf{y} \\ \mathbf{y_n} \end{bmatrix} = F_U(G_{sys}, \Delta_{sys}) \begin{bmatrix} \mathbf{z} \\ \mathbf{u} \end{bmatrix} \tag{17}$$

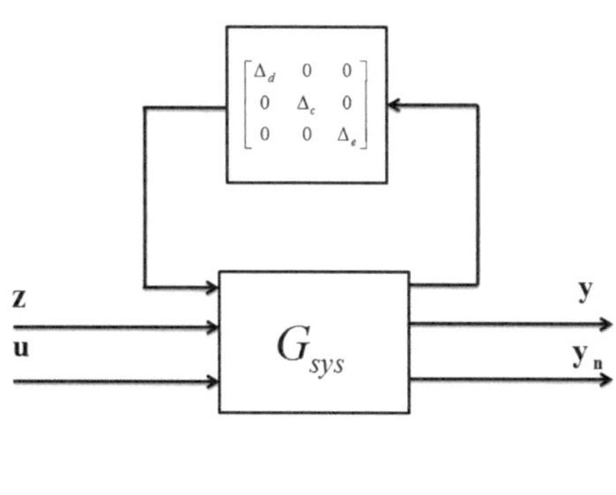

Fig. 3 Upper LFT representation of the linear dynamic system

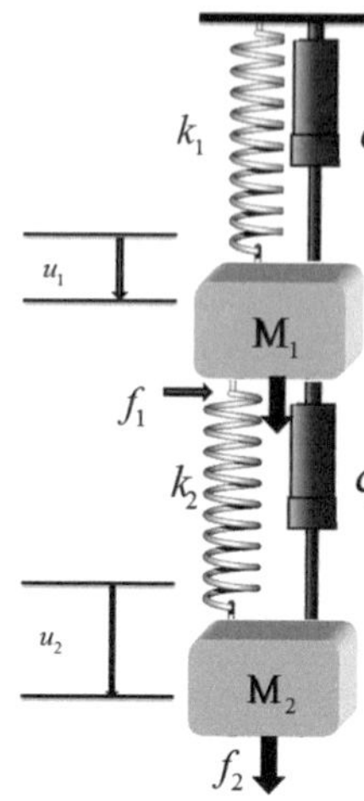

Fig. 4 Two-DOF MSD dynamic system

with uncertain block $\Delta_{sys} = \begin{bmatrix} \Delta_d & 0 & 0 \\ 0 & \Delta_c & 0 \\ 0 & 0 & \Delta_e \end{bmatrix}_{3p \times 3p}$.

3 LFT Modeling of Two-DOF Mass-Spring-Dashpot Dynamic System

An interesting linear two-DOF mass-spring-dashpot (MSD) dynamic system, pose challenges for many linear classical control techniques, has been considered as a candidate system to validate the generalized algorithm of LFT modeling for uncertain MIMO system [13]. A point masses m_2 are connected by a spring-dashpot pair contains spring stiffness constants k_2 and damping coefficient c_2 respectively. Another point mass m_1 is linked to the ground by another spring-dashpot pair contains spring stiffness constants k_1 and damping coefficient c_1 respectively. Two known dynamic forces f_1 and f_2 are applied on the two point masses to create displacement u_1 and u_2 from their equilibrium positions respectively (Fig. 4).

The equation of motion (EOM) of the two-DOF MSD dynamics system is represented by

$$\begin{bmatrix} m_1 & 0 \\ 0 & m_2 \end{bmatrix} \begin{bmatrix} \ddot{u}_1 \\ \ddot{u}_2 \end{bmatrix} + \begin{bmatrix} c_1 + c_2 & -c_2 \\ -c_2 & c_2 \end{bmatrix} \begin{bmatrix} \dot{u}_1 \\ \dot{u}_2 \end{bmatrix} + \begin{bmatrix} k_1 + k_2 & -k_2 \\ -k_2 & k_2 \end{bmatrix} \begin{bmatrix} u_1 \\ u_2 \end{bmatrix} = \begin{bmatrix} f_1 \\ f_2 \end{bmatrix} \quad (18)$$

The compact matrix notation of EOM of two DOF mass-springs-dashpot systems is represented by

Fig. 5 Block diagram of Two-DOF MSD dynamic system

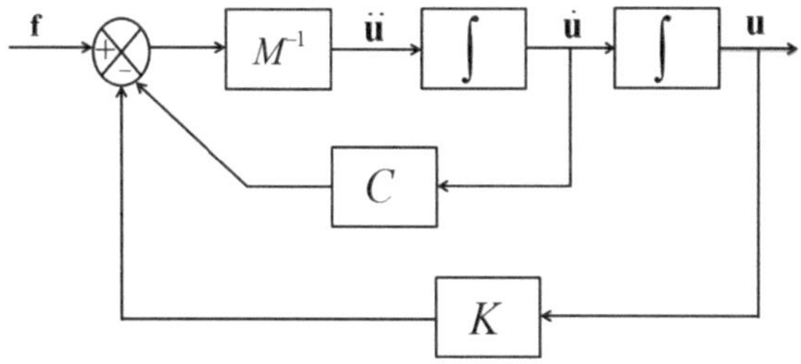

$$M\ddot{\mathbf{u}} + C\dot{\mathbf{u}} + K\mathbf{u} = \mathbf{f} \tag{19}$$

where, M, C, and K denote the mass, damping and stiffness matrices respectively and f, u, $\dot{u}$ and $\ddot{u}$ are the force, displacement, velocity and acceleration vectors respectively. Now, the block diagram representation of the system described in (19) is shown in Fig. 5, provided M^{-1} exists.

In a realistic system, variation of the physical parameters mass (m_1 and m_2), damping coefficients (c_1 and c_2) and spring stiffness constants (k_1 and k_2) are considered as uncertain parameters in the candidate system. It is assumed that the mass leads to 40% variation, spring stiffness constants represents up to 30% variation and damping coefficients varying 20% around the nominal value.

The actual mass of the system with all possible uncertainties are represented as

$$m_i = \bar{m}_i(1 + p_{m_i}\delta_{m_i}),\ i = 1, 2 \tag{20}$$

where $\bar{m}_i$ are the nominal value of the corresponding mass, $p_{m_i} = 0.4$ is the maximum relative uncertainties in each of the mass and $-1 \leq \delta_{m_i} \leq 1$.

As mass m_i contained in the matrix M, then matrix M decomposed as

$$M = \overline{M} + M_p\Delta_m \tag{21}$$

where,

$$\overline{M} = \begin{bmatrix} \bar{m}_1 & 0 \\ 0 & \bar{m}_2 \end{bmatrix}_{2\times 2},\quad M_p = \begin{bmatrix} \bar{m}_1 p_{m_1} & 0 \\ 0 & \bar{m}_2 p_{m_2} \end{bmatrix} \text{ and } \Delta_m = \begin{bmatrix} \delta_{m_1} & 0 \\ 0 & \delta_{m_2} \end{bmatrix}$$

The block partition matrix Q_m is represented as

$$Q_m = \left[\begin{array}{c:c} Q_{m_{11}} & Q_{m_{12}} \\ \hdashline Q_{m_{21}} & Q_{m_{22}} \end{array}\right] = \left[\begin{array}{c:c} -\overline{M}^{-1}M_p & \overline{M}^{-1} \\ \hdashline -\overline{M}^{-1}M_p & M^{-1} \end{array}\right]_{4\times 4} \tag{22}$$

The damping coefficients of the system with all possible uncertainties are represented as

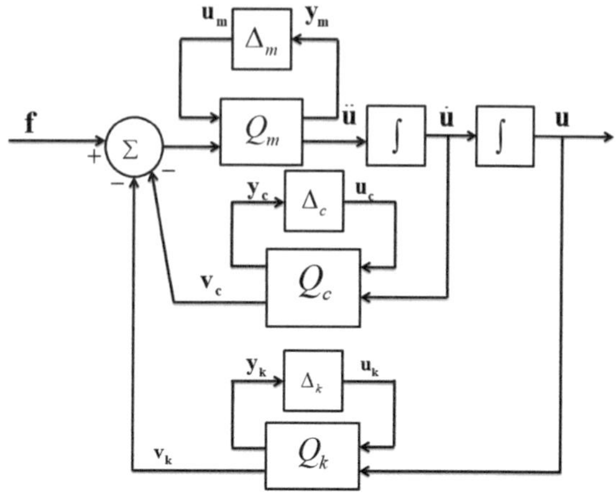

Fig. 6 Block diagram representation of Two-DOF MSD dynamic system

$$c_i = \bar{c}_i(1 + p_{c_i}\delta_{c_i}), \quad i = 1, 2 \tag{23}$$

where $\bar{c}_i$ are the nominal value of the corresponding mass, $p_{c_i} = 0.3$ is the maximum relative uncertainties in each of the mass and $-1 \leq \delta_{c_i} \leq 1$.

The block partition matrix Q_c is represented as

$$Q_c = \left[\begin{array}{c:c} Q_{c_{11}} & Q_{c_{12}} \\ \hdashline Q_{c_{21}} & Q_{c_{22}} \end{array}\right] = \left[\begin{array}{c:c} 0_{2\times 2} & C_g \\ \hdashline C_f & \bar{C} \end{array}\right]_{4\times 4} \tag{24}$$

Similarly, the actual damping coefficients of the system with all possible uncertainties are represented as

$$k_i = \overline{k_i}(1 + p_{k_i}\delta_{k_i}), i = 1, 2 \tag{25}$$

where $\overline{k_i}$ is the nominal value of the corresponding spring stiffness constant, $p_{k_i} = 0.2$ is the maximum relative uncertainty in each of this coefficient and $-1 \leq \delta_{k_i} \leq 1$.

The block partition matrix Q_k is represented as

$$Q_k = \left[\begin{array}{c:c} Q_{k_{11}} & Q_{k_{12}} \\ \hdashline Q_{k_{21}} & Q_{k_{22}} \end{array}\right] = \left[\begin{array}{c:c} 0_{2\times 2} & K_g \\ \hdashline K_f & \bar{K} \end{array}\right]_{4\times 4} \tag{26}$$

The block diagram representation of a two-DOF mass-spring-dashpot dynamic system in Fig. 6 with uncertain parameters treating $\mathbf{u_m}$, $\mathbf{u_c}$ and $\mathbf{u_k}$ to be the output of uncertain Δ_m, Δ_c and Δ_k block respectively that are fed as input to the nominal blocks Q_m, Q_c and Q_k respectively. Similarly, $\mathbf{y_m}$, $\mathbf{y_c}$ and $\mathbf{y_k}$ are outputs of Q_m, Q_c and Q_k are fed as inputs to the Δ_m, Δ_c and Δ_k respectively.

Now, the state vector for the two-DOF mass-spring-dashpot dynamic system can be defined as

$$\mathbf{X} \triangleq \begin{bmatrix} x_1 & x_2 & x_3 & x_4 \end{bmatrix}^T \tag{27}$$

where, $x_1 \triangleq \mathbf{u_1}$, $x_2 \triangleq \mathbf{u_2}$, $x_3 \triangleq \dot{\mathbf{u}}_\mathbf{1}$, $x_4 \triangleq \dot{\mathbf{u}}_\mathbf{2}$

The output vector defined in terms of state variable as

$$\mathbf{y} \triangleq \begin{bmatrix} x_1 & x_2 \end{bmatrix}^T = \begin{bmatrix} y_1 & y_2 \end{bmatrix}^T = \begin{bmatrix} \mathbf{u_1} & \mathbf{u_2} \end{bmatrix}^T \tag{28}$$

The LFT representation of two-DOF mass-spring-dashpot dynamic system considering input output of all block partition matrixes Q_m, Q_c and Q_k can be represented as

$$\begin{bmatrix} \dot{\mathbf{x}}_1 \\ \dot{\mathbf{x}}_2 \\ \dot{\mathbf{x}}_3 \\ \dot{\mathbf{x}}_4 \\ \hline \mathbf{y_m} \\ \mathbf{y_c} \\ \mathbf{y_k} \\ \hline \mathbf{y} \end{bmatrix} = \Pi \begin{bmatrix} \mathbf{x}_1 \\ \mathbf{x}_2 \\ \mathbf{x}_3 \\ \mathbf{x}_4 \\ \hline \mathbf{u_m} \\ \mathbf{u_c} \\ \mathbf{u_k} \\ \hline \mathbf{f} \end{bmatrix} \tag{29}$$

where, $\Pi = \left[\begin{array}{cc:ccc:c} 0_{2\times2} & I_{2\times2} & 0_{2\times2} & 0_{2\times2} & 0_{2\times2} & 0_{2\times2} \\ -\bar{M}^{-1}\bar{K} & -\bar{M}^{-1}\bar{C} & -\bar{M}^{-1}M_p & -\bar{M}^{-1}C_f & -\bar{M}^{-1}K_f & \bar{M}^{-1} \\ \hdashline -\bar{M}^{-1}\bar{K} & -\bar{M}^{-1}\bar{C} & -\bar{M}^{-1}M_p & -\bar{M}^{-1}C_f & -\bar{M}^{-1}K_f & \bar{M}^{-1} \\ 0_{2\times2} & C_g & 0_{2\times2} & 0_{2\times2} & 0_{2\times2} & 0_{2\times2} \\ K_g & 0_{2\times2} & 0_{2\times2} & 0_{2\times2} & 0_{2\times2} & 0_{2\times2} \\ \hdashline I_{2\times2} & 0_{2\times2} & 0_{2\times2} & 0_{2\times2} & 0_{2\times2} & 0_{2\times2} \end{array}\right]_{12\times12}$

The input output representation of the uncertainty matrix can be expressed as

$$\begin{bmatrix} \mathbf{u_m} \\ \mathbf{u_c} \\ \mathbf{u_k} \end{bmatrix} = \Delta_{smd} \begin{bmatrix} \mathbf{y_m} \\ \mathbf{y_c} \\ \mathbf{y_k} \end{bmatrix} \tag{30}$$

where, $\Delta_{smd} = \begin{bmatrix} \Delta_m & 0 & 0 \\ 0 & \Delta_c & 0 \\ 0 & 0 & \Delta_k \end{bmatrix}_{6\times6}$

The state space representation of two-DOF mass-spring-dashpot system is expressed as

$$\begin{bmatrix} \dot{x} \\ \mathbf{y_m} \\ \mathbf{y_c} \\ \mathbf{y_k} \\ \mathbf{y} \end{bmatrix} = G_{smd} \begin{bmatrix} x \\ \mathbf{u_m} \\ \mathbf{u_c} \\ \mathbf{u_k} \\ \mathbf{f} \end{bmatrix} \tag{31}$$

where, $G_{smd} = \left[\begin{array}{c|cc} A & B_1 & B_2 \\ \hline C_1 & D_{11} & D_{12} \\ C_2 & D_{21} & D_{22} \end{array}\right]_{12\times 12}$

Now,

$$A = \begin{bmatrix} 0_{2\times 2} & I_{2\times 2} \\ -\overline{M}^{-1}\overline{K} & -\overline{M}^{-1}\overline{C} \end{bmatrix}_{4\times 4}, \quad B_1 = \begin{bmatrix} 0_{2\times 2} & 0_{2\times 2} & 0_{2\times 2} \\ -\overline{M}^{-1}M_p & -\overline{M}^{-1}C_f & -\overline{M}^{-1}K_f \end{bmatrix},$$

$$B_2 = \begin{bmatrix} 0_{2\times 2} \\ \overline{M}^{-1} \end{bmatrix}_{4\times 2}, \quad C_1 = \begin{bmatrix} -\overline{M}^{-1}\overline{K} & -\overline{M}^{-1}\overline{C} \\ 0_{2\times 2} & C_g \\ K_g & 0_{2\times 2} \end{bmatrix},$$

$$D_{11} = \begin{bmatrix} -\overline{M}^{-1}M_p & -\overline{M}^{-1}C_f & -\overline{M}^{-1}K_f \\ 0_{2\times 2} & 0_{2\times 2} & 0_{2\times 2} \\ 0_{2\times 2} & 0_{2\times 2} & 0_{2\times 2} \end{bmatrix}_{6\times 6}, \quad D_{12} = \begin{bmatrix} -\overline{M}^{-1} \\ 0_{2\times 2} \\ 0_{2\times 2} \end{bmatrix}_{6\times 2}$$

$$C_2 = \begin{bmatrix} I_{2\times 2} & 0_{2\times 2} \end{bmatrix}_{2\times 4}, \quad D_{21} = \begin{bmatrix} 0_{2\times 2} & 0_{2\times 2} & 0_{2\times 2} \end{bmatrix}_{2\times 3}, \quad D_{22} = \begin{bmatrix} 0_{2\times 2} \end{bmatrix}_{2\times 2}$$

The generalized perturbed two-DOF mass-spring-dashpot dynamic can be described by Upper

LFT framework (Fig. 7).

$$\mathbf{y} = F_U(G_{smd}, \Delta_{smd})\mathbf{f} \tag{32}$$

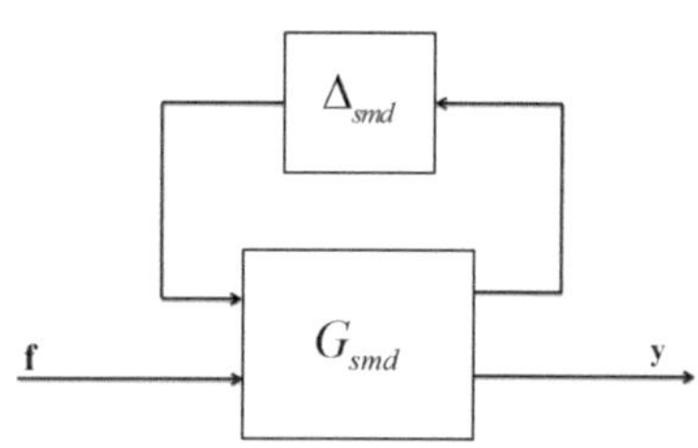

Fig. 7 Upper LFT representation of the Two-DOF MSD dynamic system

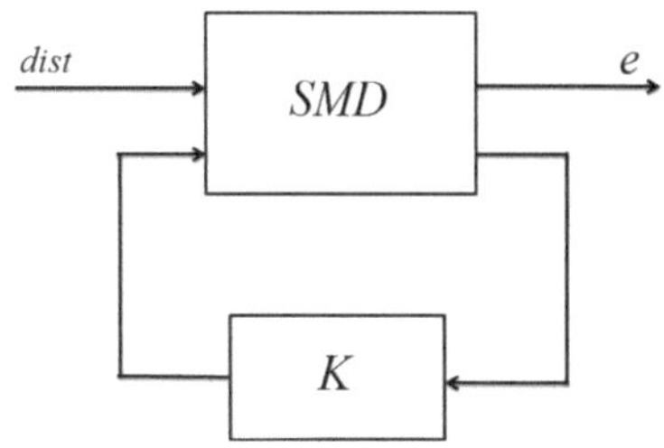

Fig. 8 Closed loop LFT in H_∞ design

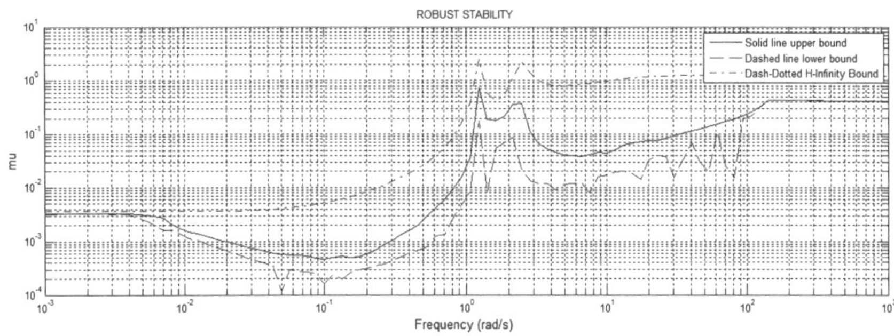

Fig. 9 Robust stability test of MSD with H_∞ controller

4 H_∞ Control Based Frequency Domain Validation

The frequency domain validation of uncertainty modeling in LFT framework has been investigated in the context of robust control theory [14]. The objective is to design an H_∞ controller that achieves certain performance specification and remain stable in the presence of all possible uncertainties.

4.1 Simulation Results

An H_∞ sub- optimal control law has been implemented for closed loop interconnected system shown in the Fig. 8. The H_∞ controller K minimizes the $\| . \|_\infty$ norm of the nominal transfer function matrix $F_L(P, K)$ from the disturbance $(dist)$ to the weighted output e.

The interval of γ iteration for H_∞ control law has been chosen in between 0 to 10 with a tolerance of 0.001. The H_∞ controller of the closed loop system achieves the $\| . \|_\infty$ norm equal to 1.0005. All stable poles of the designed H_∞ controller make the closed loop system more acceptable in practice. The closed loop system with H_∞ controller achieves the robust stability and the maximum value of μ is 0.73995 shown in Fig. 9 and also achieve the robust performance with a maximum value of the μ is 0.97076 shown in Fig. 10.

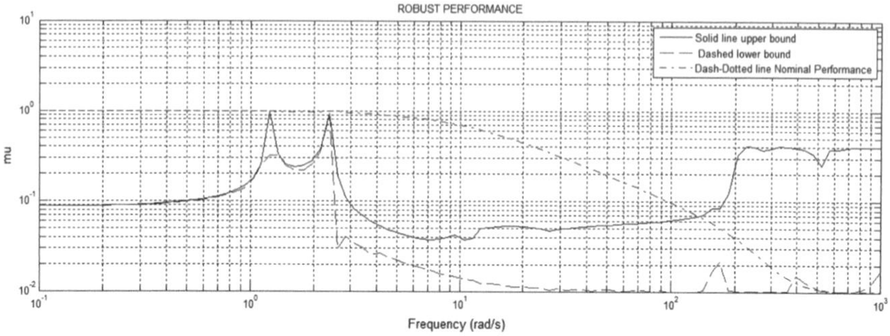

Fig. 10 Robust performance of MSD with H_∞ controller

Table 1 Numerical values of Two-DOF MSD dynamics system

Symbol	Value	Unit	Symbol	Value	Unit
m_1	2	Kg	k_1	6	N/m
m_2	1	Kg	k_2	3	N/m
c_1	0.1	N/m/s	f_1	$\sin t$	**N**
c_2	0.3	N/m/s	f_2	$\sin t$	N

Numerical values of the two-DOF- MSD dynamics system are given in Table 1.

5 Conclusions

This paper presents the mathematical generalization of control oriented LFT modeling of an uncertain coupled multi-input multi-output system with an equal number of input outputs. The uncertain physical parameters are not exactly known and it can be assumed that the parameters values are known within certain intervals and the uncertain parameters express in terms of possible relative error. LFT modeling of a given linear multi input multi output system is not necessarily minimal. For the minimal realization of any multidimensional system refers to as a smallest possible representation of the uncertainty matrix Δ. It is entirely dependent on the field of realization. The proposed generalized LFT modeling structure for linear multi input multi output dynamic system is applicable only to a system having an equal number of inputs and outputs. Proposed generalized control oriented LFT modeling algorithm has been implemented in a two-DOF linear mass-spring-dashpot dynamic system and the formulated model in LFT framework has been validated in the context of H_∞ Control law. H_∞ control based frequency domain validation process shows satisfactory results for validating uncertainty model of the two-DOF linear mass-spring-dashpot dynamic system in LFT framework.

References

1. C. Q. Zhan, K. Tsakalis, *A new robust-control-oriented system identification method*, in Proceedings of the 17th World Congress The International Federation of Automatic Control Seoul, Korea, pp. 12371–12376, 6–11 July 2008
2. A.J. Helmicki, C.A. Jacobson, C.N. Nett, Control oriented system identification: a worst-case/deterministic approach in H_∞. IEEE Trans. Autom. Control **36**(10), 1163–1176 (1991)
3. A. Marcos, D.G. Bates, I. Postlethwaite, *Exact nonlinear modeling using symbolic linear fractional transformation*, in Proceedings of the 16th IFAC World Congress, vol. 16, no. 1, pp. 31–36, 4–8 July 2005
4. G.E. Boukarim, J.H. Chow, *Modeling of nonlinear system uncertainties using a linear fractional transformation approach*. in Proceedings of the American Control Conference Philadelphia, Pennsylvania, pp. 2973–2979, 21–26 June 1998
5. Z. Xingfeng, Z. Zhiqiang, *Uncertainty modeling and robustness analysis of flight control using μ-analysis techniques*, in Proceedings of Innovation Computing Information and Control (ICICIC 2008), pp. 490–497, 18–20 June 2008
6. R. Swaminathan, K. Ramkumar, S.R. Kumar, *Design of linear fractional transformation based robust control for interacting pressure tank process*, in Proceedings of IEEE International Conference on Advances in Engineering, Science and Management (ICAESM-2012), pp. 136–140, 30–31 Mar 2012
7. A.J. Helmicki, C.A. Jacobson, C.N. Nett, Control oriented system identification: a worst-case/deterministic approach in H_∞. IEEE Trans. Autom. Control **36**(10), 1163–1176 (1991)
8. S. Hecker, A. Varga, Symbolic manipulation techniques for low order LFT-based parametric uncertainty modeling. Int. J. Control **79**(11), 1485–1494 (2006)
9. Y.C. Paw, G.J. Balas, *Parametric uncertainty modeling for LFT model realization*, in Proceedings of IEEE International Conference on Computer-Aided Control System Design San Antonio, pp. 834–839, 3–5 Sept 2008
10. G. Wolodkin, S. Rangan, K. Poolla, *An LFT approach to parameter estimation*, in Proceedings of the American Control Conference, vol. 3, pp. 2088–2092 (1997)
11. S. Hecker, A. Varga, Generalized LFT—based representation of parametric uncertain models. Eur. J. Control **10**(4), 326–337 (2004)
12. K.E. Haggbiom, *Finding an LFT uncertainty model with minimal uncertainty*, in European Control Conference (ECC), Zurich, Switzerland 2013, pp. 1107–1113 (2013)
13. ASEN 3112-Structures, *A two-DOF mass-spring-dashpot dynamics system*. Lecture 1 -Slide 1–24
14. G. Gu, Remarks on validation of uncertainty models using frequency response data. IEEE Trans. Autom. Control **47**(3), 486–490 (2002)

MIX
Papier aus verantwortungsvollen Quellen
Paper from responsible sources
FSC® C105338

If you have any concerns about our products,
you can contact us on
ProductSafety@springernature.com

In case Publisher is established outside the EU,
the EU authorized representative is:
Springer Nature Customer Service Center GmbH
Europaplatz 3, 69115 Heidelberg, Germany

Printed by Libri Plureos GmbH
in Hamburg, Germany